Integration formulas

$$\int \frac{dx}{a^2 + x^2} = \frac{1}{a} \tan^{-1} \frac{x}{a} + C$$

$$\int \frac{dx}{a^2 - x^2} = \frac{1}{2a} \ln \left| \frac{x + a}{x - a} \right| + C$$

$$\int \frac{dx}{\sqrt{a^2 + x^2}} = \ln |x + \sqrt{a^2 + x^2}| + C$$

$$\int \frac{dx}{\sqrt{a^2 - x^2}} = \sin^{-1} \frac{x}{a} + C$$

$$\int \sqrt{a^2 - x^2}\, dx = \frac{x}{2} \sqrt{a^2 - x^2} + \frac{a^2}{2} \sin^{-1} \frac{x}{a} + C$$

$$\int \frac{dx}{\sqrt{x^2 - a^2}} = \ln |x + \sqrt{x^2 - a^2}| + C$$

$$\int \sin ax\, dx = -\frac{1}{a} \cos ax + C$$

$$\int \cos ax\, dx = \frac{1}{a} \sin ax + C$$

$$\int \sin^2 ax\, dx = \frac{x}{2} - \frac{\sin 2ax}{4a} + C$$

$$\int \cos^2 ax\, dx = \frac{x}{2} + \frac{\sin 2ax}{4a} + C$$

$$\int \tan ax\, dx = -\frac{1}{a} \ln |\cos ax| + C$$

$$\int \cot ax\, dx = \frac{1}{a} \ln |\sin ax| + C$$

$$\int \sec ax\, dx = \frac{1}{a} \ln |\sec ax + \tan ax| + C$$

$$\int \csc ax\, dx = -\frac{1}{a} \ln |\csc ax + \cot ax| + C$$

$$\int e^{ax}\, dx = \frac{1}{a} e^{ax} + C$$

$$\int b^{ax}\, dx = \frac{1}{a} \frac{b^{ax}}{\ln b} + C, \quad b > 0, b \neq 1$$

$$\int xe^{ax}\, dx = \frac{e^{ax}}{a^2} (ax - 1) + C$$

$$\int \ln ax\, dx = x \ln ax - x + C$$

CONTENTS

Preface xi

Chapter 1 REVIEW OF BASIC FACTS 1

1.1 Exponents 2
1.2 Inequalities 6
1.3 Intervals and Inequalities 10
1.4 Absolute Values 14
1.5 Coordinates in the Plane 17
1.6 Distance Formula and Circles 20
    Quiz 1 23
    Quiz 2 24

Chapter 2 FUNCTIONS AND LIMITS 25

2.1 What Are Functions? 25
2.2 Graphs 29
2.3 Linear Functions 33
2.4 Tangents to $y = x^2$ 39
2.5 Tangents to $y = f(x)$ 42
2.6 Limits 47
    Quiz 1 54
    Quiz 2 55

Chapter 3 DERIVATIVES 56

3.1 Introduction 56
3.2 Derivatives of Polynomials 61
3.3 Derivatives of Products and Quotients 64
3.4 Derivatives of Powers 70
3.5 Higher Derivatives, Velocity, and Acceleration 76
3.6 The Chain Rule and Implicit Functions 80
    Quiz 1 87
    Quiz 2 87

Chapter 4 GRAPHS AND CONTINUITY 89

4.1 The First Derivative and the Shape of a Curve 89
4.2 The Second Derivative and the Shape of a Curve 99
4.3 Rational Functions and Asymptotes 105
4.4 One-Sided Limits 110
4.5 Continuity 113
4.6 The Mean Value Theorem 119
    Quiz 1 125
    Quiz 2 126

Chapter 5 APPLICATIONS 127

5.1 Marginal Cost and Marginal Revenue 127
5.2 Maximum and Minimum Problems 132
5.3 Rate of Change 138
5.4 Newton's Method 143
5.5 Linear Approximations 146
5.6 Differentials 150
    Quiz 1 152
    Quiz 2 152

Chapter 6 INTEGRALS 154

6.1 Introduction 154
6.2 Existence of the Definite Integral 161
6.3 Properties of the Integral 164
6.4 The Evaluation of Definite Integrals 173
6.5 The Fundamental Theorem of Calculus 178
6.6 Indefinite Integrals 183
6.7 Integration by Substitution 187
6.8 Piecewise Continuous Functions and Absolute Values 194
    Appendix 200
    Quiz 1 201
    Quiz 2 202

Chapter 7 NATURAL LOGARITHMS AND EXPONENTIALS 203

7.1 Natural Logarithms 203
7.2 The Derivative of ln x 210
7.3 The Exponential Function 213
7.4 General Exponents and Logarithms 218
7.5 Derivatives and Integrals of Exponentials 224
7.6 Integration by Parts 229
    Quiz 1 236
    Quiz 2 236

Chapter 8 APPLICATIONS OF INTEGRATION 238

8.1 Integration of Rates 238
8.2 Growth and Decay 246
8.3 Volumes of Solids of Revolution 251
8.4 Volume by Slicing 260
8.5 Improper Integrals over Infinite Intervals 266
8.6 Improper Integrals with Unbounded Integrands 271
8.7 Numerical Integration 274
    Quiz 1 283
    Quiz 2 284

Chapter 9 TRIGONOMETRIC FUNCTIONS 286

9.1 Sine and Cosine 286
9.2 The Derivatives of Sine and Cosine 295
9.3 Further Trigonometric Functions 303
9.4 Inverse Functions 306
9.5 The Inverse Trigonometric Functions 311
9.6 Applications 318
9.7 Integration of Trigonometric Functions 326
9.8 Techniques of Integration 331
    Quiz 1 337
    Quiz 2 337

Chapter 10 TECHNIQUES OF INTEGRATION 339

10.1 Substitution with Differentials 339
10.2 Some Trigonometric Integrals 342
10.3 Trigonometric Substitutions 348
10.4 Partial Fractions 352
10.5 Rationalizing Substitutions 359
    Quiz 1 364
    Quiz 2 364

Chapter 11 TAYLOR APPROXIMATIONS AND INFINITE SERIES 366

- 11.1 Sequences 366
- 11.2 Properties of Sequences 374
- 11.3 The Error in Linear Approximations 380
- 11.4 Taylor Approximations 385
- 11.5 The Error in Taylor Approximations 390
- 11.6 Infinite Series 399
- 11.7 Properties of Infinite Series 407
- 11.8 Convergence Tests 414
- 11.9 Power Series 423
- 11.10 Indeterminate Forms 431
  - Quiz 1 438
  - Quiz 2 439

Chapter 12 PLANE GEOMETRY 441

- 12.1 Curves in Parametric Form 441
- 12.2 Polar Coordinates 450
- 12.3 Symmetries and Graphing in Polar Coordinates 456
- 12.4 Area and Length in Polar Coordinates 461
- 12.5 Parabolas 465
- 12.6 Ellipses 472
- 12.7 Hyperbolas 481
- 12.8 Rotation of Coordinate Axes 489
  - Quiz 1 494
  - Quiz 2 495

Chapter 13 FUNCTIONS OF SEVERAL VARIABLES 497

- 13.1 Coordinates in Three Dimensions 497
- 13.2 Functions of Two Variables 502
- 13.3 Cross Sections and Surfaces 508
- 13.4 Partial Derivatives 513
- 13.5 Limits and Continuity 517
- 13.6 Functions of Three or More Variables 522
- 13.7 Linear Approximations 528
- 13.8 The Chain Rule 533
- 13.9 Higher-Order Partial Derivatives 540
- 13.10 Maxima and Minima 546
  - Quiz 1 553
  - Quiz 2 553

Chapter 14 MULTIPLE INTEGRALS 555

- 14.1 Introduction to Double Integrals 555

14.2 Double Integrals as Limits   561
14.3 Integration over General Regions   566
14.4 Mass, Density, and Center of Gravity   572
14.5 Double Integrals in Polar Coordinates   576
14.6 Triple Integrals   582
14.7 Mass, Density, and Center of Gravity of Solids   588
14.8 Cylindrical and Spherical Coordinates   592
    Quiz 1   599
    Quiz 2   599

Chapter 15 VECTORS   600

15.1 Introduction to Vectors   600
15.2 Geometric Properties of Vectors   606
15.3 Direction Cosines and the Scalar Product   612
15.4 Vector Functions and Their Derivatives   618
15.5 Tangents and Normals to Curves   626
15.6 Curvature, Velocity, and Acceleration   631
15.7 Planes   636
15.8 Tangent Planes and the Gradient   639
15.9 Directional Derivatives   646
15.10 The Vector Product   650
15.11 An Application of Vector Products to Curvature   655
    Quiz 1   658
    Quiz 2   658

Table 1 Powers and Roots   661
Table 2 Natural Logarithms   662
Table 3 Common Logarithms   664
Table 4 Exponential Functions   666
Table 5 Trigonometric Functions of
    Certain Angles   668
Table 6 Trigonometric Functions   669
Table 7 Differentiation Formulas   670
Table 8 Table of Integrals   671

Answers to Odd-Numbered Exercises   679

Index   731

PREFACE

This text contains the material usually covered in a basic three-semester or five-quarter calculus course. Chapters 2–12 discuss calculus of one variable, and Chapters 13–15 discuss two and more variables. All the necessary introductory and review material is contained in Chapter 1, so that the text is completely self-contained.

It has been our purpose to present the essential concepts and techniques of calculus and their applicability to various fields of endeavor in an interesting way. Even without the multitude of important applications, calculus holds great interest as a body of knowledge, and it is a challenging area of introductory mathematics. We hope that we succeeded in communicating some of its appeal and our enthusiasm for the subject.

In setting a goal for this text, we kept in mind that the student will be expected to learn the basic concepts of calculus and their use, while gaining proficiency in certain techniques. For this reason, the discussion of new concepts, definitions, and theorems is interwoven with numerous illustrative worked-out examples. There are over 600 of these, and they are also used to illustrate and teach techniques for doing the exercises. There is a worked-out example for virtually every type of exercise that the student is expected to do. This integrated treatment of theory and problem solving involves the student in the act of "doing mathematics." It develops his appreciation of mathematical reasoning and methods of proof, while building up his capabilities for dealing with more and more diversified problems.

There are over 2500 exercises in this text, ranging from routine drill exercises to more challenging ones. The exercises, in general, are on the level on which the majority of students is expected to perform. There is sufficient variety, however, to permit a selection for different levels of performance.

Applications were taken from such fields as biology, business, ecology, economics, physics, sociology, and other parts of mathematics. These applications are an integral part of the exposition. They are used to illustrate various concepts and techniques, and thus are interspersed throughout the text. They are discussed at a nontechnical level for the purpose of acquainting the student with the language and type of problems that can be solved.

Whenever possible, geometric intuition was used as a pedagogical device to further enhance understanding. Over 700 figures have been included to accomplish this. The idea of many theorems and proofs can be understood from the figures, and this makes the technical details more transparent and easier to follow. Proofs that do not add to the understanding of the material have been postponed, as a rule, to the ends of sections, and these can be omitted in many cases.

The student who can pass Test 1 on page 1 meets the prerequisites for this course. This test could be used to indicate if Chapter 1, in whole or in part, should be omitted. The quizzes at the end of each chapter can be used for self-testing, for determining the degree of retention and proficiency, and for discovering weak areas.

Applications may be covered selectively. The chapters and sections that deal exclusively with applications are Chapters 5, 8, Sections 4.1, 4.2, 9.6, 12.4, 13.10, 14.4, and 14.7. Chapter 10, dealing with more specialized techniques of integration, is optional.

Many people contributed their talent and effort during the writing and production of this book. Barbara Federman prepared excellent sketches for the artwork, contributed, along with Dr. James Holmes, most of the answers to the exercises, and otherwise contributed her considerable skills and devotion. The typing was done by a team which accepts only perfection as the standard; it consisted of Delores Brannon, Louise Kraus, and Sonia Ospina. We are also grateful to Professor Philip Ostrand for some invaluable suggestions. Finally, we owe special gratitude to George Telecki, Mathematics Editor, Lois Wernick, Project Editor, and the other members of the excellent staff of Harper & Row.

P. F.
D. A. S.

CALCULUS
AND ANALYTIC GEOMETRY

CHAPTER 1
REVIEW OF BASIC FACTS

This text presupposes a basic knowledge of high school algebra. A good grasp of this material is essential because it is the foundation of the many theoretical arguments and related techniques that comprise the calculus. As a preparation to learning calculus, many students find it necessary to review the basic concepts and facts concerning numbers, and to brush-up on manipulative techniques. The relevant review material is contained in this chapter in abbreviated form. To determine the sections on which review should be concentrated, complete Test 1 that follows. Class performance on this test can be used to set the beginning level of the course.

TEST 1

The following problems should be completed in 50 min.

1. Insert one of the symbols $=$ or $\neq$ into each box to produce a true statement.

   (a) $\dfrac{1}{0.05}$ ☐ 2000 (b) 0.05×0.05 ☐ 0.00005×5

   (c) $\sqrt{4+16}$ ☐ $\sqrt{4}+\sqrt{16}$ (d) $\dfrac{a}{b}$ ☐ $\dfrac{1}{b/a}$

2 REVIEW OF BASIC FACTS

(e) $\sqrt{\frac{2}{3}}$ ☐ $\frac{\sqrt{6}}{3}$ (f) 5^0 ☐ 0

(g) $\frac{3 \times 1000}{5 \times 0.01}$ ☐ 6×10^6 (h) a^{2^3} ☐ $(a^2)^3$

(i) $\frac{a}{b} + \frac{c}{d}$ ☐ $\frac{ad+bc}{bd}$ (j) $\frac{b-1}{b+1}$ ☐ $1 - \frac{2}{b+1}$

(k) $\frac{1}{a^2-b^2}$ ☐ $\frac{1}{2b}\left(\frac{1}{a-b} + \frac{1}{a+b}\right)$

(l) $(a^{1/3})$ ☐ $\frac{1}{a^3}$

(m) $\sqrt{a} - \sqrt{b}$ ☐ $\frac{a+b}{\sqrt{a}+\sqrt{b}}$

2. Simplify the numbers below as much as possible.

   (a) $\left(\frac{2}{3}\right)^{-1}\left(\frac{3}{2}\right)^{-2}$

   (b) $\frac{(a+h)^2 - a^2}{h}$

   (c) $\frac{(a+h)^3 - a^3}{h}$

3. Give the intervals where the following inequalities are true.

   (a) $2x + 1 > 0$ (b) $x^2 - 2 < 0$

   (c) $|x - 1| < \frac{1}{2}$ (d) $|2x + 3| > 1$

4. A rectangle with sides parallel to the coordinate axes has vertices $(-1, -1)$ and $(3, 4)$ (see Figure 1.1, top). Find the remaining two vertices.

5. Find the equation of the circle passing through the point $(8, 5)$ and having center $(4, 2)$ (see Figure 1.1, bottom).

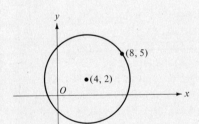

Figure 1.1

1.1 EXPONENTS

Numbers are well-known objects to us. In this book we shall regard *real numbers* as decimals

$$a.a_1a_2a_3a_4\cdots$$

where a is any integer (positive, negative, or zero), and each a_k is one of the digits $0, 1, 2, 3, \ldots, 9$. The association between the real numbers and the points on a straight line is explained at the end of this section.

Manipulating long decimals can become very awkward, and the exponential notation is thus very useful.

DEFINITION 1
If m is a positive integer, then

$$a^m = \underbrace{a \cdot a \cdot a \cdot \cdots \cdot a}_{m \text{ times}}$$

If $a \neq 0$, then

$$a^{-m} = \frac{1}{a^m}$$

and

$$a^0 = 1$$

Exponents (powers) play a very important part in mathematics. General exponents are discussed later in this section. First we want to show the advantage of using integral exponents in manipulating decimals. *Remember that 0^0 is undefined.*

Example 1

Consider the product

$$0.00000125 \times 0.005$$

Writing

$$0.00000125 = 1.25 \times 10^{-6}$$
$$0.005 = 5 \times 10^{-3}$$

gives

$$0.00000125 \times 0.005 = (1.25 \times 10^{-6}) \times (5 \times 10^{-3})$$
$$= (1.25 \times 5) \times (10^{-6} \times 10^{-3})$$
$$= 6.25 \times 10^{-9}$$

The answer is usually left in this form. This notation is common to all scientific writing, and is often called scientific notation.

SCIENTIFIC NOTATION
A number B is said to be written in *scientific notation* if it is written in the form

$$B = \pm A \times 10^n$$

where

$$1 \leq A < 10$$

and n is any integer.

Example 2

Evaluate

$$\frac{12000}{0.000016}$$

Solution
Using scientific notation,

$$\frac{12000}{0.000016} = \frac{1.2 \times 10^4}{1.6 \times 10^{-5}} = \frac{1.2}{1.6} \times \frac{10^4}{10^{-5}} = 0.75 \times (10^4 \times 10^5) = 0.75 \times 10^9$$

$$= 7.5 \times 10^8$$

We now augment Definition 1 to include fractional exponents.

DEFINITION 2
If $a > 0$ and n is a positive integer, then $a^{1/n}$, the *nth root* of a, is the positive number b with the property $b^n = a$. The notation $\sqrt[n]{a} = a^{1/n}$ is also used. If m is a positive integer, then

EXAMPLES
$8^{2/3} = (8^2)^{1/3} = (64)^{1/3} = 4$

$8^{-2/3} = \dfrac{1}{8^{2/3}} = \dfrac{1}{4}$

$$a^{m/n} = (a^m)^{1/n}$$

$$a^{-m/n} = \frac{1}{a^{m/n}}$$

We said nothing here about the meaning of a^r when r is an irrational number, such as $\sqrt{2}$, $\sqrt{3}$, $\sqrt[3]{2}$, or π. This will have to wait until we introduce logarithms (see Chapter 7). Without proof, we list here the basic properties of exponents.

PROPERTIES OF EXPONENTS
The following properties hold whenever the expressions are meaningful:

EXAMPLES
$2^{1/4} \cdot 2^{3/4} = 2^{1/4+3/4} = 2$

$\dfrac{2^{1/4}}{2^{3/4}} = 2^{1/4-3/4} = 2^{-2/4} = 2^{-1/2}$

$(2^{1/4})^{3/4} = 2^{1/4 \cdot 3/4} = 2^{3/16}$

$2^{1/5} \cdot 3^{1/5} = (2 \cdot 3)^{1/5} = 6^{1/5}$

$\sqrt{\tfrac{1}{3}} < \sqrt{\tfrac{1}{2}}$ since $0 < \tfrac{1}{3} < \tfrac{1}{2}$

1. $a^r \cdot a^s = a^{r+s}$
2. $\dfrac{a^r}{a^s} = a^{r-s}$
3. $(a^r)^s = a^{r \cdot s}$
4. $a^r \cdot b^r = (a \cdot b)^r$
5. $a^r < b^r$ if $0 < a < b$ and $r > 0$

It is customary to write a^{rs} instead of $a^{(rs)}$.

THE NUMBER LINE
Decimals are associated with points on a straight line as follows: An arbitrary point is selected as the origin and marked 0. A second

point is chosen to its right and marked 1. The line segment $\overline{01}$ is called *unit of length*. By means of this unit of length, we mark off equally spaced points to the right and to the left of 0. The point lying n units to the right of 0 is labeled n, and the point lying n units to the left of 0 is labeled $-n$. Dividing each segment so created into ten equal parts gives the points $a.a_1$, where a is any integer and a_1 is one of the digits $0, 1, 2, \ldots, 9$. Successive subdivisions of each segment into ten equal parts give the points labeled $a.a_1a_2$, $a.a_1a_2a_3$, $a.a_1a_2a_3a_4$, and so on. In this way we achieve an association between all points of the line and the decimals $a.a_1a_2a_3a_4\ldots$. When the points of a line are so identified with numbers, we call the line a *number line* or the *real line*.

EXERCISES

1. Use the following exercises to test your understanding of exponents. Insert into each box one of the symbols $=$ or $\neq$ to produce a true statement.

   (a) $(-a)^5 \;\square\; -a^5$ (b) $(-a)^6 \;\square\; a^6$

   (c) $(a^r)^2 \;\square\; a^{r^2}$ (d) $(a^3)^{1/2} \;\square\; (a^{1/2})^3$

   (e) $(-1)^n \;\square\; -1$ (f) $(-1)^{2n} \;\square\; 1$

   (g) $\left(\dfrac{1}{a}\right)^r \;\square\; a^{-r}$ (h) $(\sqrt{3}-\sqrt{2})(\sqrt{3}+\sqrt{2}) \;\square\; 1$

   (i) $a^{-1/r} \;\square\; a^r$ (j) $a^r \cdot b^s \;\square\; (a \cdot b)^{r \cdot s}$

   (k) $a^r \cdot b^s \;\square\; (a \cdot b)^r \cdot b^{s-r}$ (l) $a^r + a^s \;\square\; a^{r+s}$

   (m) $a^r + a^{-r} \;\square\; 1$ (n) $(-a)^{-r} \;\square\; a^r$

   (o) $\sqrt{11} - \sqrt{10} \;\square\; \dfrac{1}{\sqrt{11}+\sqrt{10}}$

2. Use the following exercises to test your manipulative skills. Verify the given identities.

   (a) $\dfrac{8^5}{2^{12}} = 2^3$ (b) $(2^2 \cdot 16)^7 = 2^{42}$

   (c) $2^{42} = 4^{21}$ (d) $\dfrac{10^{-3}}{10^{-5}} = 10^2$

   (e) $\left(\tfrac{1}{2}\right)^{-5} = 2^5$ (f) $(a^{-1})^{-1} \cdot a^{-1} = 1$

   (g) $\dfrac{(a+h)^3 - a^3}{h} = 3a^2 + (3a+h)h$

   (h) $\dfrac{\sqrt{b} - \sqrt{b-h}}{h} = \dfrac{1}{\sqrt{b}+\sqrt{b-h}}$

6 REVIEW OF BASIC FACTS

(i) $a^4 - b^4 = (a-b)(a^3 + a^2b + ab^2 + a^3)$

(j) $\dfrac{\sqrt{b} - \sqrt{a}}{b - a} = \dfrac{1}{\sqrt{b} + \sqrt{a}}$

3. Write the following numbers in scientific notation.

(a) 0.1 (b) 0.01
(c) 0.0909 (d) 0.09090
(e) 0.0000012 (f) 1275
(g) −1275.25 (h) 100000000
(i) −0.500001 (j) 10.000001
(k) $(3 \times 10^2) \times (4 \times 10^{-5})$ (l) $10^4 + 10^5$
(m) $1257 \times 10^4 \times 10^4$ (n) 0.5×0.5
(o) 0.04×0.001 (p) $\dfrac{3 \times 10^2}{4 \times 10^{-5}}$
(q) $\dfrac{10^{10}}{125}$

4. Write the following numbers in decimal form.

(a) 1.5×10^{-5} (b) 1.5×10^5 (c) -1250×10^{-4} (d) 10×10^{-1}
(e) -1×10^{-7} (f) 0.03×10^{-3} (g) 0.03×10^3

1.2 INEQUALITIES   Of fundamental importance in calculus are the concepts "greater than" and "less than" as applied to real numbers. This ordering is expressed with the familiar symbols $<$ (less than) and $>$ (greater than). Given numbers a and b, then of the relations

$a < b \qquad a = b \qquad a > b$

exactly one is true and two are false. The symbols $\leq$ and $\geq$ are used in the following sense:

$a \leq b$ stands for "$a < b$ or $a = b$,"
$a \geq b$ stands for "$a > b$ or $a = b$."

Example 1

The statement

"$5 < 5 \quad$ or $\quad 5 = 5$"

is true because $5 = 5$. Hence we have $5 \leq 5$. The statement

"$-5 < 0$ or $-5 = 0$"

is true because $-5 < 0$. Hence $-5 \leq 0$.

TERMINOLOGY

$a > b$ is read "a is greater than b,"
$a \geq b$ is read "a is greater than or equal to b,"
$a < b$ is read "a is less than b,"
$a \leq b$ is read "a is less than or equal to b."

Certain properties of inequalities are often called axioms, an *axiom* being a statement whose truth is unquestioned. A list of axioms is chosen in such a way that all other properties of inequalities can be derived therefrom. A typical list is given below.

AXIOMS FOR INEQUALITIES

1. Any two numbers a and b are related by exactly one of the relations $a < b$, $a = b$, or $a > b$.
2. If $a < b$ and $b < c$, then $a < c$.
3. If $a < b$, then $a + c < b + c$ no matter what c is.
4. If $a < b$ and $c > 0$, then $ac < bc$.

Example 2

Verify that

$$0 < b - a \quad \text{when} \quad a < b$$

and conversely

$$a < b \quad \text{when} \quad 0 < b - a$$

Solution

If $a < b$, then adding $-a$ to both sides of the inequality gives, by axiom 3, $0 < b - a$. Conversely, if $0 < b - a$, then the same axiom enables us to add a to both sides of this inequality, thereby giving $a < b$.

Example 3

Show that

If $a < b$ and $c < 0$, then $bc < ac$.

Solution
We begin with the observation

If $c < 0$, then $-c > 0$.

This follows from axiom 3 by adding $-c$ to both sides of the inequality $c < 0$.

From this observation and Example 2, we see that

$$0 < b - a \quad \text{and} \quad 0 < -c$$

if $a < b$ and $c < 0$. By axiom 4,

$$0 < (b - a) \cdot (-c) = -bc + ac$$

and adding bc to both sides of this inequality (axiom 3) shows that

$$bc < ac$$

REMARK

The statement "$a < b$ and $b < c$" is usually abbreviated and expressed in the form $a < b < c$. Similarly, we shall encounter inequalities such as $a > b > c$, $a > b > c > d$, and so on.

EXERCISES

1. Test your knowledge of inequalities with the following set of true-false questions. If a statement is false, you should be able to give a counterexample. Thus, "if $t > 0$ then $t > 1$" is false. A counterexample is $t = \frac{1}{2}$, because $\frac{1}{2} > 0$ is true, but $\frac{1}{2} > 1$ is false. It takes only *one* counterexample to make a statement false!

   (a) $-1 \leq -1$

   (b) $a \leq a + 1$

   (c) $-a - 1 \leq -a$

   (d) $-2 < 2 \leq 3$

   (e) $-a < -2a$

   (f) if $a < 0$, then $a^2 > 0$

   (g) if $a^2 \neq 0$, then $a^2 > 0$

   (h) if $a = -a$, then $a = 0$

   (i) if $a < -b$, then $a < b$

   (j) if $a + b < a - b$, then $a < b$

   (k) if $-t > 0$, then $t < 0$

   (l) if $t = 16$, then $4/t \leq 2$

   (m) if $t < -1$, then $t < 1$

   (n) if $a^2 = 0$, then $a = 0$

(o) if $a < b$, then $5/a < 5/b$
(p) if $a \leq b$, then $a < b$
(q) if $a < b$, then $a \leq b$
(r) if $a < -a$, then $7 < -7$
(s) if $t < 5$, then $t \neq 4$
(t) if $a + c < b + c$, then $a < b$
(u) if $t = 7$, then $1/t \leq 1/7$
(v) if $a > b$, then $-1/a > -1/b$

2. Which of the following statements say the same as "$a < b$": $b > a$, $b - a < 0$, $a - b < 0$, $b - a > 0$, $a - b \leq 0$, $b - a$ is positive.

3. Insert one of the symbols $\leq$, $=$, or $\geq$ into each box to produce a true statement.

(a) $a^2 \ \square \ 0$

(b) $\sqrt{a^2 + 1} \ \square \ \dfrac{1}{\sqrt{a^2 + 1}}$

(c) $a^2 + b^2 \ \square \ 2ab$

(d) $\dfrac{x^2}{x^2 + 1} \ \square \ 1 - \dfrac{1}{x^2 + 1}$

(e) $\dfrac{a^3 - 1}{a - 1} \ \square \ a^2 + a + 1$

(f) $\dfrac{a^4 - 1}{a - 1} \ \square \ a^3 + a^2 + a + 1$

(g) $\dfrac{1}{1 - x} \ \square \ x^2 + x + 1 \quad (0 < x < 1)$

(h) $\dfrac{1}{1 - x} \ \square \ x^3 + x^2 + x + 1 \quad (0 < x < 1)$

(i) $\dfrac{1}{1 - x} \ \square \ x^4 + x^3 + x^2 + x + 1 \quad (x > 1)$

(j) $\dfrac{\sqrt{a} - \sqrt{b}}{a - b} \ \square \ \dfrac{1}{\sqrt{a} + \sqrt{b}}$

(k) $\dfrac{1}{a^2 + b^2} \ \square \ \dfrac{1}{a^2} + \dfrac{1}{b^2}$

(l) $2 \cdot 3 \cdot 4 \cdots n \ \square \ \underbrace{n \cdot n \cdot n \cdots n}_{n - 1 \text{ times}}$

1.3 INTERVALS AND INEQUALITIES

Inequalities can be described graphically by means of line segments. Finite or infinite line segments (such as a ray or the whole line) will be called *intervals*.

Example 1

Consider the pair of inequalities

$$0 < x \quad \text{and} \quad x < 1$$

The inequality $x > 0$ specifies the positive real line (the point 0 being excluded). The inequality $x < 1$ specifies the ray originating at 1 and lying to its left (see Figure 1.2). The inequalities are both true for values on the line segment from 0 to 1.

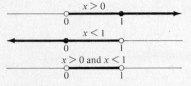

Figure 1.2
The inequalities $x > 0$ and $x < 1$. An arrow indicates that the interval goes on and on in the given direction.

The statement "$0 < x$ and $x < 1$" is expressed in the compact form "$0 < x < 1$." The interval determined by the inequalities $0 < x < 1$ is called an *open interval*, and the symbol representing it is $(0, 1)$.

In general, there are four types of finite intervals. These are given in Figure 1.3.

Inequalities	Interval	Symbol	Name
$a < x < b$	○────○	(a, b)	open interval
$a \leq x \leq b$	●────●	$[a, b]$	closed interval
$a < x \leq b$	○────●	$(a, b]$	half-open interval or
$a \leq x < b$	●────○	$[a, b)$	half-closed interval

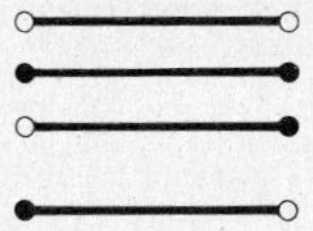

Figure 1.3
The finite intervals. An open circle indicates an *excluded* point. A solid circle indicates an *included* point.

Example 2

Graph the inequality $x^2 \leq 2$.

Solution
By graphing an inequality we mean plotting all points for which it is true.

According to Example 2 of Section 1.2, $x^2 \leq 2$ is the same as $2 - x^2 \geq 0$, but this can be written as

$$(\sqrt{2} + x)(\sqrt{2} - x) \geq 0$$

This inequality holds when either one of the following statements is true:

1. $\sqrt{2} + x \geq 0$ and $\sqrt{2} - x \geq 0$
2. $\sqrt{2} + x \leq 0$ and $\sqrt{2} - x \leq 0$

The inequalities in (1) can be expressed as

$$-\sqrt{2} \leq x \quad \text{and} \quad x \leq \sqrt{2}$$

and these two inequalities combine into

$$-\sqrt{2} \leq x \leq \sqrt{2}$$

The two inequalities in (2), which can be expressed as $x \leq -\sqrt{2}$ and $x \geq \sqrt{2}$, cannot hold simultaneously, since they require x to be positive and negative at the same time. Hence,

$$x^2 \leq 2 \quad \text{is the same as} \quad -\sqrt{2} \leq x \leq \sqrt{2}$$

and the inequality holds for x in the closed interval $[-\sqrt{2}, \sqrt{2}]$ (see Figure 1.4).

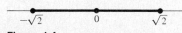

Figure 1.4
The graph of $x^2 \leq 2$.

Let us now consider infinite intervals. By these we mean line segments extending indefinitely in at least one direction. To describe such intervals symbolically we introduce two special symbols, ∞ and $-\infty$, read *infinity* and *minus infinity*, respectively. The use of these symbols is indicated in Figure 1.5.

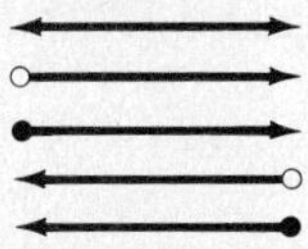

Inequalities	Alternative inequalities	Interval	Symbol	Name
$-\infty < x < \infty$	none	←——————→	$(-\infty, \infty)$	real line R
$a < x < \infty$	$x > a$	○——————→	(a, ∞)	open half-line
$a \leq x < \infty$	$x \geq a$	●——————→	$[a, \infty)$	closed half-line
$-\infty < x < a$	$x < a$	←——————○	$(-\infty, a)$	open half-line
$-\infty < x \leq a$	$x \leq a$	←——————●	$(-\infty, a]$	closed half-line

Figure 1.5
The infinite intervals. An arrow indicates that the interval continues indefinitely in the given direction.

WARNING
The symbols ∞ and $-\infty$ do not represent numbers.

Example 3

Graph the inequality $x^2 > 2$.

Solution
We recall that for any real number a, exactly one of the following relations holds: Either $a \leq 2$ or $a > 2$. Likewise, either $x^2 \leq 2$ or $x^2 > 2$. Since $x^2 \leq 2$ is true only for x in the closed interval $[-\sqrt{2}, \sqrt{2}]$ (see Example 2), it follows that $x^2 > 2$ is true outside of this interval. This consists of the intervals $(-\infty, -\sqrt{2})$ and $(\sqrt{2}, \infty)$.

Example 4

Graph the inequality $x(x-1)(x-2)(x-3) < 0$.

Solution

The above expression is 0 at the points $x = 0, 1, 2$, and 3. On the intervals determined by these points we determine the sign as shown below.

	x	$x-1$	$x-2$	$x-3$	$x(x-1)(x-2)(x-3)$
$x < 0$	−	−	−	−	+
$0 < x < 1$	+	−	−	−	−
$1 < x < 2$	+	+	−	−	+
$2 < x < 3$	+	+	+	−	−
$x > 3$	+	+	+	+	+

According to this table the inequality is true for x in the open intervals $(0, 1)$ and $(2, 3)$. Our findings are presented schematically in Figure 1.6.

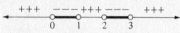

Figure 1.6
The graph of $x(x-1)(x-2)(x-3) < 0$.

In describing the various intervals, we used the terms "open" and "closed." We shall explain now a very important difference between intervals of the two categories.

Consider a closed interval $[a, b]$. The point a is the *first* point of this interval, since

1. a is in the interval,
2. $a < x$ for any point $x \neq a$ in the interval.

Similarly, b is the *last* point of the interval, since

3. b is in the interval,
4. $x < b$ for any point $x \neq b$ in the interval.

By contrast, we have the following fact: *An open interval (a, b) does not have a first or a last point.* This statement is clear from Figure 1.7. Namely, for any point A in (a, b), the point $\frac{1}{2}(a+A)$ is in (a, A). Hence, no point A can be the first point. Similarly, $\frac{1}{2}(A+b)$ is in (A, b), and so no point A can be the last point of (a, b).

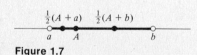

Figure 1.7

EXERCISES

Graph the inequalities in Exercises 1 and 2.

Ans.

1. (a) $2x + 1 > 0$ (b) $2x - 1 > 3x + 1$

(c) $x^2 - 2 < 0$
(d) $x^2 - 2 > 0$
(e) $(x - 2)^2 \geq 0$
(f) $1/x \leq 1$
(g) $(x - 1)(x + 1) \leq 0$
(h) $x < -3$ or $x > 3$

2. (a) $x^2 > x$
(b) $x > -5$ and $x < 1$
(c) $(2x - 1)^2 \geq 5$
(d) $x(x + 1) \geq 0$

Hint: Use Example 4.

(e) $x(x - 1)(x - 2)(x - 3) \geq 0$

3. Express symbolically each of the intervals below.
   (a) the interval of all negative numbers
   (b) the interval of all nonnegative numbers
   (c) the interval of all numbers greater than -20.

Warning: $-\infty$ is not a number.

4. Explain why the open half-line $(-\infty, a)$ has no first (smallest) point.

Warning: Remember that ∞ is not a number.

5. Explain why the open half-line (a, ∞) has no last (largest) point.

6. Explain why there is no smallest positive number.

7. What interval contains all points of $(-1, 1]$ and of $[0, \frac{1}{2})$?

8. Which of the following statements are true?
   (a) if $-1 < x < 1$, then $-1 \leq x \leq 1$
   (b) if $-2 < x < 2$, then $x^2 < 4$
   (c) if $x \leq -1$, then $x \leq 0$
   (d) if $x \leq 3$, then $x > 0$
   (e) if $-x < 3$, then $x < 3$
   (f) if $0 < t < 1$, then $-1 < -t < 0$
   (g) if $1 \leq t \leq 8$, then $2 \leq t \leq 8$
   (h) if $a \leq t \leq a$, then $t = a$
   (i) if $-4 \leq t \leq 4$, then $0 \leq t^2 \leq 16$
   (j) if $-a < t < a$, then $|a| < t$

9. Which of the following statements are true?
   (a) if $x < 7$, then $-x < 7$
   (b) if $1 < x < 5$, then $x > 1$
   (c) if $2 < x < 3$, then $-3 < -x < -2$
   (d) if $2 < -x < 3$, then $-3 < x < -2$

14 REVIEW OF BASIC FACTS

(e) if $1 < t < 8$, then $1 \leq t \leq 8$

(f) if $1 < t < 4$, and $1 < u < 4$, then $1 < t + u < 4$

1.4 ABSOLUTE VALUES

The number that gives the distance of a point a on the number line from the origin 0 is the absolute value of a.

DEFINITION 1
The *absolute value* of a, written $|a|$, is such that

$$|a| = \begin{cases} a & \text{when } a \geq 0 \\ -a & \text{when } a < 0 \end{cases}$$

Thus, absolute value (and hence distance) is positive or zero.

Example 1

$|4| = 4$ because $4 \geq 0$
$|-4| = -(-4) = 4$ because $-4 < 0$
$|0| = 0$ because $0 \geq 0$

The basic properties of absolute values are given in the following theorem.

THEOREM 1
For any numbers a and b, we have

1. $\left|\dfrac{1}{a}\right| = \dfrac{1}{|a|}$ when $a \neq 0$

2. $|a + b| \leq |a| + |b|$

3. $|a \cdot b| \leq |a| \cdot |b|$

The proof of this theorem is straightforward but lengthy. It is given at the end of this section.

From the definition, we see that if $b > 0$, then the equality $|x| = b$ is satisfied by the two points $x = b$ and $x = -b$. If we now consider any point u, such that $0 < u < b$, then $|x| = u$ is satisfied by the points $x = u$ and $x = -u$. These points lie in the open interval $(-b, b)$ (see Figure 1.8). We thus conclude that the inequality $|x| < b$ is satisfied by precisely the points in the open interval $(-b, b)$ (see Figure 1.9). Hence:

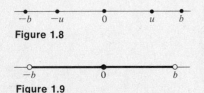

Figure 1.8

Figure 1.9
The graph of $|x| < b$.

$$|x| < b \text{ is the same as } -b < x < b. \tag{1}$$

Accordingly, the inequality $|x| > b$ is satisfied by all points lying outside the closed interval $[-b, b]$. These points constitute the two intervals $(-\infty, -b)$ and (b, ∞) (see Figure 1.10). Hence:

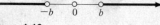
Figure 1.10
The graph of $|x| > b$.

$$|x| > b \text{ is the same as } x < -b \text{ or } x > b. \tag{2}$$

Now consider the inequality

$$|x - a| < b$$

From (1) this is the same as

$$-b < x - a < b$$

and, adding a throughout, these inequalities become

$$a - b < x < a + b$$

Hence:

Figure 1.11
The graph of $|x - a| < b$.

$$|x - a| < b \text{ is the same as } a - b < x < a + b. \tag{3}$$

The graph of the inequality $|x - a| < b$ is an interval of length $2b$ which is symmetric about a (see Figure 1.11).

Example 2

Graph the inequality $|x + 1| > 1$.

Solution
Using (3) with $a = -1$ and $b = 1$ shows that $|x + 1| < 1$ is the same as $-2 < x < 0$. We conclude from this that the inequality $|x + 1| > 1$ is satisfied by the points lying outside the closed interval $[-2, 0]$, that is, by points satisfying the inequality $x < -2$ or $x > 0$ (see Figure 1.12).

Figure 1.12
The graph of $|x + 1| > 1$.

In general, we see (Figure 1.13) that:

Figure 1.13
The graph of $|x - a| > b$.

$$|x - a| > b \text{ is the same as } x < a - b \text{ or } x > a + b.$$

Proof of Theorem 1

1. The proof is broken up into the cases $a > 0$ and $a < 0$.

   Case 1
   If $a > 0$, then $1/a > 0$ and $a = |a|$. Hence $|1/a| = 1/a$ and $1/a = 1/|a|$, so that $|1/a| = 1/|a|$ follows.

Case 2

If $a < 0$, then $1/a < 0$ and $-a = |a|$. Hence $|1/a| = -1/a$ and $-1/a = 1/-a = 1/|a|$, so that $|1/a| = 1/|a|$ follows once more.

2. We know that the relation $|a| \leq |a|$ is always true. Substituting $x = a$ and $b = |a|$ in (1) shows, therefore, that

$$-|a| \leq a \leq |a| \tag{4}$$

Likewise, we see that

$$-|b| \leq b \leq |b| \tag{5}$$

Adding (4) and (5) gives

$$-(|a| + |b|) \leq a + b \leq (|a| + |b|)$$

but this is the same as

$$|a + b| \leq |a| + |b|$$

3. The proof of this part can be carried out by the student by considering the four cases $a > 0$ and $b > 0$, $a < 0$ and $b > 0$, $a > 0$ and $b < 0$, and $a < 0$ and $b < 0$.

EXERCISES

Graph the inequalities in Exercises 1 and 2.

1. (a) $|x - 1| < \frac{1}{2}$ (b) $|2x| \geq \frac{1}{2}$
   (c) $|2x - 1| \leq 3$ (d) $x < |x|$
   (e) $\left|\dfrac{1}{x - 2}\right| < 1$ (f) $|x - 2| + \frac{1}{2} \leq 2|x - 2|$

2. (a) $|4x + 8| \leq |x + 2| + 3$ (b) $|x||x + 2| \leq 2$
   (c) $|x||x + 2| > 2$ (d) $|-x + 1| \leq 7$
   (e) $|x^2 - 4| \geq 1$ (f) $x^2 < |2x + 1|$

3. Write symbolically each of the following intervals:

   (a) the interval of all negative numbers

   (b) the interval of all numbers

   (c) the interval of all numbers not greater than 5

   (d) the interval of all nonnegative numbers

   (e) the interval of all numbers greater than -5

Armed with your knowledge of absolute values, see how you stand with the true-false questions in Exercises 4 and 5.

4. (a) $|-1| = 1$
   (b) $|-5| > 0$
   (c) $|-3| \leq 0$
   (d) if $a < b$, then $|a| < |b|$
   (e) if $a + b > 0$, then $|a + b| > 0$
   (f) $\dfrac{t + |t|}{2} \geq t$
   (g) if $2 < t < 5$, then $|t - 2| < 3$
   (h) if $|t - 2| < 3$, then $2 < t < 5$
   (i) if $|t - 1| > 1$, then $t - 1 < 1$

5. (a) if $t - 1 < 1$, then $|t - 1| > 1$
   (b) if $|t - 5| < 0.01$, then $|t^2 - 5| < 0.001$
   (c) $|2.02 \times 3.03 - 6| < \frac{1}{5}$
   (d) if $|t - 7| < a$, then $|t - 7| \leq a$
   (e) if $|t - 7| \leq a$, then $|t - 7| < a$
   (f) if $|t - 3| > 1$, then $t - 3 > 1$ or $t - 3 < -1$
   (g) if $|t - 3| > 1$, then $|t - 5| > 0$
   (h) if $|t - 3| < 1$, then $|t| \leq 4$
   (i) if $|t - 3| \leq 1$, then $|t| < 4$

6. Using Theorem 1, verify the following facts.
   (a) $\left|\dfrac{1}{a \cdot b}\right| = \dfrac{1}{|a| \cdot |b|}$
   (b) $|a \cdot b \cdot c| = |a| \cdot |b| \cdot |c|$
   (c) $|a + b + c| \leq |a| + |b| + |c|$

1.5 COORDINATES IN THE PLANE

The association of points on a straight line with numbers is now used to associate points in the plane with pairs of numbers. This is done by constructing a two-dimensional coordinate system, as follows.

We select two perpendicular number lines in the plane intersecting in the origin, O, of each. One of these lines is customarily drawn horizontal and labeled the *x axis*; the other line, which is vertical, is labeled the *y axis*. The two axes are positioned as shown in Figure 1.14.

18 REVIEW OF BASIC FACTS

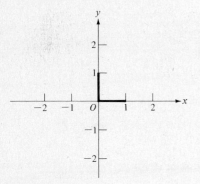

Figure 1.14
A coordinate system in the plane.

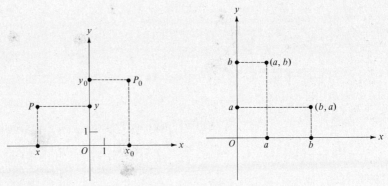

Figure 1.15
Plotting points in the plane.

Figure 1.16

> A system of two number lines intersecting in the point O is called a *coordinate system*.

The terms *rectangular coordinate system* and *cartesian coordinate system* are also used. The plane containing an *xy* coordinate system is called the *xy plane*.

To explain the association of points with pairs of numbers, pick any point P_0 in the *xy* plane and draw through P_0 perpendicular lines to the *x* and *y* axes (see Figure 1.15). These lines intersect the respective axes in points x_0 and y_0. Conversely, the perpendicular lines through x_0 and y_0 intersect in the point P_0. It follows that P_0 can be identified by writing $P_0 = (x_0, y_0)$. We deduce that every point *P* in the plane is uniquely determined by a pair of numbers (x, y), and conversely. That is, if (x, y) and (x', y') represent points in the plane, then:

> $(x, y) = (x', y')$ when and only when $x = x'$ and $y = y'$.

A pair (x, y) is called an *ordered pair*. The numbers *x* and *y* are referred to as the *coordinates* of the point $P = (x, y)$, *x* being called the *first* or *x coordinate*, *y* the *second* or *y coordinate*. The term ordered pair is justified in view of the fact that if $a \neq b$, then $(a, b) \neq (b, a)$; that is, (a, b) and (b, a) specify distinct points (see Figure 1.16). Thus, when a coordinate system is specified in the plane, each point can be identified with an ordered pair of real numbers.

Example 1

A rectangle with sides parallel to the coordinate axes has two vertices, $(-2, -1)$ and $(2, 1)$. Find the other two vertices.

1.5 COORDINATES IN THE PLANE

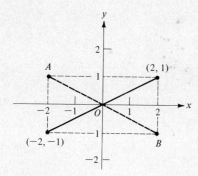

Figure 1.17

Solution
Consult Figure 1.17. Since the rectangle has sides parallel to the coordinate axes, it follows that vertex A must lie 2 units to the left of the y axis [like the vertex $(-2, -1)$] and 1 unit above the x axis [like the vertex $(2, 1)$]. We thus have $A = (-2, 1)$. Similarly, it is found that $B = (2, -1)$.

Example 2

Find four points whose distance from $(2, 0)$ is 3 units.

Solution
Since the point $(2, 0)$ lies on the x axis, it is easiest to locate two of the desired points on the same axis. Moving 3 units to the left and to the right of $(2, 0)$ gives the points $(2 - 3, 0) = (-1, 0)$ and $(2 + 3, 0) = (5, 0)$. Two more points can be located on the perpendicular line through $(2, 0)$. The desired points are $(2, 0 + 3) = (2, 3)$, and $(2, 0 - 3) = (2, -3)$ (see Figure 1.18).

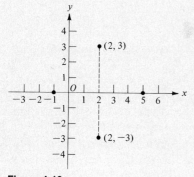

Figure 1.18

Our coordinate system divides the plane into four *quadrants*. The sign of the coordinates of a given point reveal its location relative to these quadrants and the coordinate axes. This information is summarized in Figure 1.19.

We have to still account for points with one or two zero coordinates.

Example 3

What is the equation of the x axis?

Solution
Let $P = (x, y)$ be any point on the x axis. Then necessarily $y = 0$. Conversely, if $y = 0$, then $P = (x, y) = (x, 0)$ lies on the x axis. We deduce that:

> The equation of the x axis is $y = 0$.

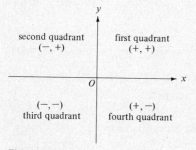

Figure 1.19
The location of points in the plane.

	location of $P = (x, y)$
$x > 0 \quad y > 0$	first quadrant
$x < 0 \quad y > 0$	second quadrant
$x < 0 \quad y < 0$	third quadrant
$x > 0 \quad y < 0$	fourth quadrant

EXERCISES

1. Give the location in terms of quadrants or coordinate axes of each of the following points: $A = (-1, -1)$; $B = (1, 0)$; $C = (0, 1)$; $D = (1, -1)$; $E = (0, 0)$.

2. Give the equations of the following lines:

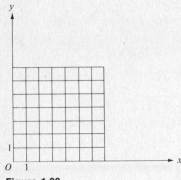

Figure 1.20

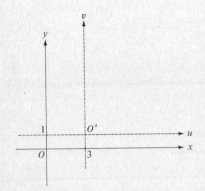

Figure 1.21

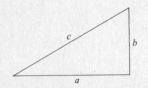

Figure 1.22

(a) the y axis

(b) the horizontal line passing through $(0, -2)$

(c) the horizontal line passing through $(5, -2)$

(d) the vertical line passing through $(1, 1)$

3. Give the remaining two vertices of the square of sides of length 1 when two vertices are $(1, 3)$ and $(2, 4)$. (See Figure 1.20.)

4. Consider a right triangle with two sides parallel to the coordinate axes and of lengths 3 and 4 units, respectively. Find the vertices of all such right triangles if one vertex is $(-2, 2)$, and the triangle lies entirely in the second quadrant.

5. Find the coordinates of a third point on the line segment joining the points $(0, 1)$ and $(5, 6)$.

6. Draw all squares with one vertex at $(1, 1)$ and sides of length 5 that are parallel to the coordinate axes. Find the remaining vertices of each square.

7. Consider the point $P = (-3, -4)$ and the line segment $\overline{PQ}$ joining it to the point $Q = (x, y)$. What can you say about the coordinates of Q when

  (a) $\overline{PQ}$ is parallel to the line $x = 1$

  (b) $\overline{PQ}$ is parallel to the y axis

  (c) $\overline{PQ}$ is parallel to the line passing through the points $(1, 1)$ and $(2, 0)$

8. Consult Figure 1.21. You are given an xy coordinate system and a uv coordinate system in the plane, with the x axis parallel to the u axis, and the y axis parallel to the v axis. The origin O' of the uv coordinate system has coordinates $(3, 1)$ relative to the xy coordinates. The table below lists points in one coordinate system. Find the coordinates in the other coordinate system.

(x, y)	$(0, 0)$	$(-2, 3)$	$(1, 1)$	$(0, 1)$				
(u, v)					$(0, 0)$	$(-2, 3)$	$(1, 1)$	$(-7, 7)$

1.6 DISTANCE FORMULA AND CIRCLES

In deriving a formula for the distance between two points, we make use of the *theorem of Pythagoras,* which states that if a right triangle (Figure 1.22) has sides of lengths a and b, and hypotenuse of length c, then

$$a^2 + b^2 = c^2$$

1.6 DISTANCE FORMULA AND CIRCLES

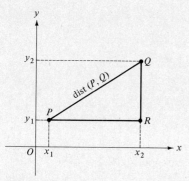

Figure 1.23

Consider points $P = (x_1, y_1)$ and $Q = (x_2, y_2)$. The distance between their x coordinates is $|x_2 - x_1|$, the distance between their y coordinates is $|y_2 - y_1|$. With the point $R = (x_2, y_1)$, we have a right triangle with hypotenuse $\overline{PQ}$ whose length corresponds to the distance between P and Q (see Figure 1.23). If we use the notation

$$\text{dist}(P, Q) = \text{distance between } P \text{ and } Q$$

then, by the theorem of Pythagoras,

$$\text{dist}(P, Q) = \sqrt{|x_2 - x_1|^2 + |y_2 - y_1|^2}$$

The absolute value signs under the square root can be omitted, since $|x_2 - x_1|^2 = (x_2 - x_1)^2$ and $|y_2 - y_1|^2 = (y_2 - y_1)^2$. Summarizing, we have the following formula:

DISTANCE FORMULA
If $P = (x_1, y_1)$ and $Q = (x_2, y_2)$ are any two points in the plane, then the distance between them is

$$\text{dist}(P, Q) = \sqrt{(x_2 - x_1)^2 + (y_2 - y_1)^2} \tag{1}$$

Example 1

Find the distance between the points $P = (1, 1)$ and $Q = (5, 4)$.

Solution
Put $(x_1, y_1) = (1, 1)$ and $(x_2, y_2) = (5, 4)$. Then the distance formula becomes

$$\text{dist}(P, Q) = \sqrt{(5 - 1)^2 + (4 - 1)^2} = \sqrt{4^2 + 3^2}$$
$$= \sqrt{25} = 5$$

CIRCLES
From elementary geometry we recall that:

A *circle* is the locus of all points in the plane whose distance from a fixed point is constant.

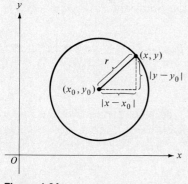

Figure 1.24

This description is easily transformed into a simple formula by means of the distance formula. Namely, consider the fixed point $C = (x_0, y_0)$ and an arbitrary point $P = (x, y)$ at a distance r from it (see Figure 1.24). Then we have the two formulas

$$\text{dist}(C, P) = r$$
$$\text{dist}(C, P) = \sqrt{(x-x_0)^2 + (y-y_0)^2}$$

and hence

$$\sqrt{(x-x_0)^2 + (y-y_0)^2} = r$$

Squaring both sides of this equation gives:

> The equation of a circle with center at (x_0, y_0) and radius r is
> $$(x-x_0)^2 + (y-y_0)^2 = r^2$$

Example 2

Find the equation of the circle with center at the origin and radius 1.

Solution
Here $(x_0, y_0) = (0, 0)$ and $r = 1$. The equation of the circle in question is, therefore,

$$x^2 + y^2 = 1$$

This circle is given in Figure 1.25 and is called the *unit circle*.

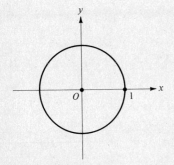

Figure 1.25

Example 3

Find the equation of the circle that passes through the point $A = (-5, 0)$ and has center $C = (-2, 4)$.

Solution
The radius of the circle in question is, by the distance formula,

$$r = \text{dist}(A, C) = \sqrt{[-2-(-5)]^2 + (4-0)^2} = \sqrt{3^2 + 4^2} = \sqrt{25} = 5$$

Using $(x_0, y_0) = (-2, 4)$ gives for the equation of the circle

$$[x-(-2)]^2 + (y-4)^2 = 5^2$$

or

$$(x+2)^2 + (y-4)^2 = 5^2$$

This circle is given in Figure 1.26.

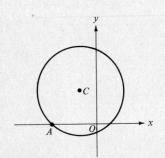

Figure 1.26

EXERCISES

1. Find the distance between the following pairs of points.

    (a) $(3, 3)$ and $(0, 0)$

    (b) $(3, 3)$ and $(-3, -3)$

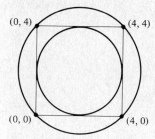

Figure 1.27

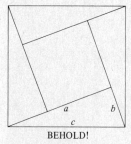

BEHOLD!

Figure 1.28

The Pythagorean theorem states that for a right triangle with sides a and b, and hypotenuse c, the relation $a^2 + b^2 = c^2$ always holds.

  (c) $(1, 0)$ and $(0, 1)$

  (d) (a, b) and $(a + 1, b + 2)$

  (e) $(\sqrt{2}, 0)$ and $(0, \sqrt{3})$

  (f) $(1 - a, 1 + a)$ and $(1 + b, 1 - b)$

  (g) (x_0, y_0) and $(5x_0, 5y_0)$

2. What is the equation of the circle passing through the points $(0, 0)$, $(4, 0)$, $(4, 4)$, and $(0, 4)$?

3. Find the equation of the circle that passes through the point A and has center C when

  (a) $A = (-3, 4)$ and $C = (4, -3)$

  (b) $A = (1, -1)$ and $C = (-1, 1)$

  (c) $A = (0, 0)$ and $C = (0, 9)$

4. You are given the square with vertices $(0, 0)$, $(4, 0)$, $(4, 4)$, and $(0, 4)$ (see Figure 1.27).

  (a) What is the equation of the *largest* circle that can be inscribed in the square?

  (b) What is the equation of the *smallest* circle that can be circumscribed about the square?

5. The Indian mathematician Bhaskara (1114–1158) published a one-word proof of the Pythagorean theorem. This proof was already known to the ancient Chinese. Can you show that the "proof" reproduced in Figure 1.28 is, indeed, a proof?

QUIZ 1

Insert into each box one of the symbols $=$ or $\neq$ to produce a true statement.

1. $\dfrac{\sqrt{a + h} - \sqrt{a}}{h}\ \square\ \dfrac{1}{\sqrt{a + h} + \sqrt{a}}$

2. $\dfrac{1}{\sqrt{a} + \sqrt{b}}\ \square\ \dfrac{1}{\sqrt{a}} + \dfrac{1}{\sqrt{b}}$

3. $(a + b)^r\ \square\ a^r + b^r$

4. $a^{-m/n}\ \square\ a^{n/m}$

5. $(a^{1/q})^p\ \square\ (a^p)^{1/q}$

6. $0.000901\ \square\ 9.01 \times 10^{-3}$

7. $10^5 + 10^6$ ☐ 11×10^5

8. $|ab|$ ☐ $|a| \cdot |b|$

9. Graph the inequality $(2x + 1)^2 > 5$.

10. Graph the inequality $|-2x + 1| \leq 1$.

11. Give the equation of the circle passing through the point $(-1, -1)$ and having center at $(2, 3)$.

12. Find the remaining vertices of all squares having the line joining $(3, 0)$ and $(0, 3)$ as one side.

QUIZ 2

Insert into each box one of the symbols $\leq$, $=$, or $\geq$ to produce a true statement.

1. $\dfrac{a + |a|}{2}$ ☐ a

2. $\sqrt{a^2 + 1} - \sqrt{a^2}$ ☐ $\dfrac{1}{2\sqrt{a^2}}$

3. $\dfrac{a^4 - 1}{a - 1}$ ☐ $a^3 + a^2 + a + 1$

4. 1.000001 ☐ $(1.000001)^5$

5. 0.99999 ☐ $(0.99999)^4$

6. Write $(7 \times 10^4)/(210 \times 10^{-6})$ in scientific notation.

7. Graph the inequality $x(x - 1)(x + 1) > 0$.

8. Graph the inequality $x^2 < x$.

9. Find all circles passing through the points $P = (-2, 0)$ and $Q = (0, 2)$ and having radius 2.

10. Find the coordinates of a third point on the line segment joining the points $(0, 5)$ and $(5, 0)$.

Hint: Use the distance formula.

11. Do the following three points lie on a straight line: $A = (0, 0)$, $B = (3, 1)$, $C = (25, 8)$?

CHAPTER 2

FUNCTIONS AND LIMITS

2.1 WHAT ARE FUNCTIONS?

The term "function" refers to a special kind of relation between variable quantities. To the economist, a function may be a relation between supply and demand; to the biologist, a function may be a relation between the size of a bacteria population and time; to the psychologist, a function may be a relation between intelligence and achievement.

Intuitively, a function can be thought of as a relation that converts one variable quantity into another, like an electric heater converting current into heat. Before making the concept of function precise, we clarify it with two examples.

Example 1

Let us obtain a formula for converting temperatures in the Fahrenheit scale to temperatures in the Celsius scale. The situation is described in Figure 2.1. Since the interval between the freezing and boiling temperatures is divided into 100 equal intervals in the Celsius scale, and into $212 - 32 = 180$ equal intervals in the Fahrenheit scale, it follows that the conversion factor from Fahrenheit to Celsius is $\frac{100}{180} = \frac{5}{9}$. In other words,

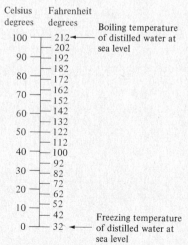

Figure 2.1

26 FUNCTIONS AND LIMITS

a change of t degrees Fahrenheit corresponds to a change of $\frac{5}{9}t$ degrees Celsius.

This and the fact that

$$0°C = 32°F$$

suggests that we put

$$y = \tfrac{5}{9}(x - 32) \tag{1}$$

If x is given in the Fahrenheit scale, then y gives the corresponding reading in the Celsius scale. This relation is a function that converts Fahrenheit degrees into Celsius degrees. When all points (x, y) for which x and y are related by equation (1) are plotted in the xy plane, we obtain the graph of a straight line (Figure 2.2).

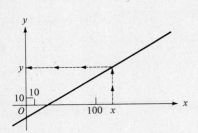

Figure 2.2
The graph of $y = \tfrac{5}{9}(x - 32)$.

Example 2

In figuring the cost of producing x tons of fertilizer, a manufacturer has to take into account the following expenses:

cost of x tons raw material: $10x - \sqrt{x}$ dollars
labor cost of processing x tons: $45x$ dollars
plant cost of processing x tons: $2\sqrt{x} + 1000$ dollars

The total cost, y, of producing x tons of fertilizer is determined by the relation

$$y = 55x + \sqrt{x} + 1000 \tag{2}$$

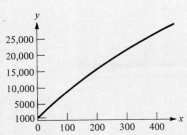

Figure 2.3
The graph of $y = 55x + \sqrt{x} + 1000$.

Thus, cost y is a function of x that converts tons into dollars. The graph of this function consists of all points (x, y) satisfying relation (2); it is given in Figure 2.3.

In the two examples just considered, we arrived at relations that converted values x from one set into unique values y from another set. This is described graphically in Figure 2.4, where the function is described as a machine. The function concept is made precise in the following definition.

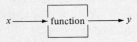

Figure 2.4
A function as a machine: x goes into the machine, which converts it into y.

DEFINITION 1
A *function* is a relation that associates with each value x in a set X exactly one value y in a set Y.

This definition is explained pictorially in Figure 2.5.

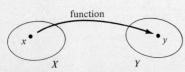

Figure 2.5
A function associating values in X with values in Y.

TERMINOLOGY AND NOTATION
In equations (1) and (2), the letter x stands for an unspecified number; it is a *variable*. y is also a variable, but it depends on x. Each

value of x determines a single value which is denoted y. For this reason y is called a *dependent* variable and x an *independent* variable. The formula

$$y = f(x)$$

read "y equals f of x," simply tells us that y is a function of x. Formulas such as $y = g(x)$ and $y = h(x)$ are also used for this purpose.

Applying the functional notation $y = f(x)$ to the function in Example 1, we write

$$f(x) = \tfrac{5}{9}(x - 32) \tag{3}$$

and say that this formula defines the function f. For simplicity we shall speak of "the function $f(x) = \tfrac{5}{9}(x-32)$" or "the function $f(x)$" or "the function $\tfrac{5}{9}(x - 32)$," when we mean the function f defined by equation (3).

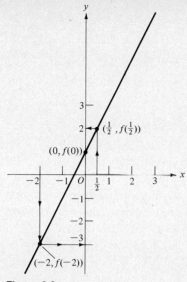

Figure 2.6
The function $f(x) = 2x + 1$.

Example 3

Let us illustrate the use of the functional notation $y = f(x)$ with the function

$$f(x) = 2x + 1$$

When this function is evaluated at particular points, it is found that

$$f(-2) = 2 \cdot (-2) + 1 = -3$$
$$f(0) = 2 \cdot 0 + 1 = 1$$
$$f(\tfrac{1}{2}) = 2 \cdot \tfrac{1}{2} + 1 = 2$$

and so on (see Figure 2.6).

A relation that converts x into y is a function only when the conversion is unique, that is, only when each x gives only one y. Let us consider a relation between x and y where this is not so.

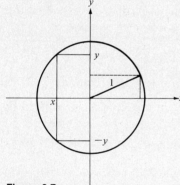

Figure 2.7
The circle $x^2 + y^2 = 1$.

Example 4

We have seen in Section 1.6 that the equation $x^2 + y^2 = 1$ describes a circle with center at the origin and radius 1. This equation can be written as $y^2 = 1 - x^2$ (see Figure 2.7). It *does not* specify y as a function of x, because two values of y may correspond to one value of x. The two solutions for y are

$$y = \sqrt{1 - x^2} \quad \text{and} \quad y = -\sqrt{1 - x^2}$$

Each of these equations defines a function. The first gives the upper half of a circle, the second its lower half (see Figure 2.8).

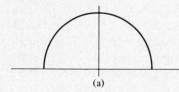

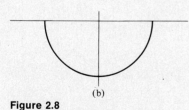

Figure 2.8
(a) The graph of $y = \sqrt{1 - x^2}$.
(b) The graph of $y = -\sqrt{1 - x^2}$.

THE DOMAIN AND RANGE OF A FUNCTION

The set of points x on which a function f is defined and the corresponding set of values $f(x)$ have special names.

DEFINITION 2

The *domain* of the function f is the set of values x for which $f(x)$ is defined. The *range* of f is the set of values $f(x)$ taken on by f as x varies over its domain.

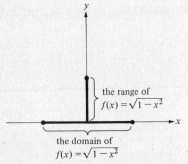

Figure 2.9

Example 5

Let us examine the function
$$f(x) = \sqrt{1 - x^2}$$
Since a square root is defined for nonnegative numbers only, it follows that by necessity $0 \leq 1 - x^2$, which is the same as $-1 \leq x \leq 1$. The interval $[-1, 1]$ is called the *domain* of the function $f(x)$. When x is restricted by the inequalities $-1 \leq x \leq 1$, $f(x)$ is restricted by the inequalities $0 \leq f(x) \leq 1$. The interval $[0, 1]$ is called the *range* of $f(x)$ (see Figure 2.9).

Example 6

Find the domain and range of $f(x) = x^2/(x^2 + 1)$.

Solution
This function is defined for all real numbers x, and hence its domain consists of the entire real line. To find its range we note that
$$0 \leq \frac{x^2}{x^2 + 1} < 1$$
Since $f(0) = 0$ and $f(x) < 1$ for any value of x, we see that the range of f is the interval $[0, 1)$.

EXERCISES

In Exercises 1–10, find the indicated values of the given functions and describe their domains and ranges.

1. $f(x) = 1/x$. Find $f(\tfrac{1}{2})$; $f(-\tfrac{1}{2})$; $f(2)$; $f(-2)$
2. $f(x) = x - |x|$. Find $f(0)$; $f(1)$; $f(-1)$; $f(-\tfrac{1}{2})$
3. $g(t) = t - |t|$. Find $g(0)$; $g(1)$; $g(-1)$; $g(-\tfrac{1}{2})$
4. $h(u) = u/|u|$. Find $h(-2)$; $h(-\tfrac{1}{2})$; $h(\tfrac{1}{2})$; $h(2)$
5. $f(x) = |x - 1|$. Find $f(a)$; $f(a + 1)$; $f(1)$; $f(-1)$

6. $k(x) = x^2$. Find $k(2) \cdot k(\frac{1}{2})$; $k(2) \cdot k(-2)$; $k(2) - k(\sqrt{2})$
7. $m(x) = x^3$. Find $m(2) - m(-2)$; $m(2) + m(-2)$; $m(3^{1/2})$
8. $p(x) = 1$. Find $p(0)$; $p(1)$; $p(a)$
9. $f(x) = x^3 - x$. Find $f(1)$; $f(-1)$; $f(\sqrt{a})$
10. $f(x) = x^{1/2}$. Find $f(1)$; $f(4)$; $f(a^{1/2})$; $f(b^{2/3})$

11. A cube has sides of length x (see Figure 2.10). Express the volume v of the cube as a function of x: $v = f(x)$. Find $f(1)$ and $f(2)$.

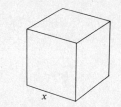
Figure 2.10

12. A cube has volume v. Express the length x of its sides as a function of v: $x = g(v)$. Find $g(1)$ and $g(8)$.

13. A closed rectangular metal box has dimensions t by t by $2t$ (see Figure 2.11). Give the surface area s of the box as a function of t: $s = f(t)$. Find $f(1)$ and $f(\frac{1}{2})$.

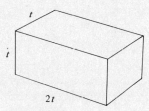
Figure 2.11

14. A closed cylindrical can is to have height $1/h$ and base of radius h (see Figure 2.12). Give the volume v as a function of h: $v = f(h)$. Find $f(\pi)$.

15. Is there any difference between the functions

$$f(x) = \frac{x^2 - 1}{x - 1}$$

and

$$g(x) = \begin{cases} \frac{x^2 - 1}{x - 1} & \text{for } x \neq 1 \\ 2 & \text{for } x = 1 \end{cases}$$

Explain your answer.

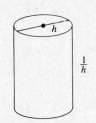
Figure 2.12

16. Referring to Example 1, at what temperature is the reading the same in the Fahrenheit and Celsius scales?

17. Referring to Example 1, at what temperature is the reading in Celsius degrees twice as large as the reading in Fahrenheit degrees?

2.2 GRAPHS

The *graph* of a function $y = f(x)$ is the collection of points $(x, y) = (x, f(x))$, where x varies over the domain of the function. In Chapter 4 we shall study properties of graphs in great detail and develop methods for obtaining accurate graphs. For the time being we construct graphs from tables of values for x and $y = f(x)$.

Example 1

Plot the graph of $f(x) = (x - 1)^2$.

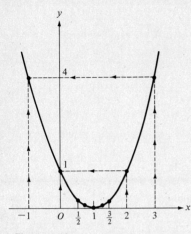

Figure 2.13
The graph of $f(x) = (x - 1)^2$.

x	$f(x)$
-1	4
0	1
$\frac{1}{2}$	$\frac{1}{4}$
$\frac{3}{4}$	$\frac{1}{16}$
1	0
$\frac{5}{4}$	$\frac{1}{16}$
$\frac{3}{2}$	$\frac{1}{4}$
2	1
3	4

Solution
Since $(x - 1)^2 \geq 0$ for all values of x, we see that points from the x axis are associated with points on the nonnegative y axis. The graph is obtained in Figure 2.13 from the accompanying table of values. The correspondence between some values x and $f(x)$ is indicated with broken lines.

Example 2

Find the domain and range of the function

$$f(x) = \frac{1}{x^2}$$

and plot its graph.

Solution
This function is defined for all values $x \neq 0$. Its *domain* consists therefore of the *two* intervals $(-\infty, 0)$ and $(0, \infty)$. The range of $f(x)$ is easily determined here from the graph, which shows it to be $(0, \infty)$. The graph is obtained with the help of the table of values in Figure 2.14. Observe that this graph consists of two separate curves. If the point (x, y) lies on the graph, then so does the point $(-x, y)$. This graph is *symmetric* with respect to the y axis.

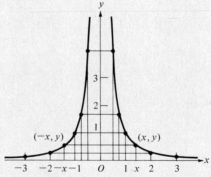

x	$f(x)$
± 3	$\frac{1}{9}$
± 2	$\frac{1}{4}$
± 1	1
$\pm \frac{3}{4}$	$\frac{16}{9}$
$\pm \frac{1}{2}$	4

Figure 2.14
The graph of $f(x) = 1/x^2$.

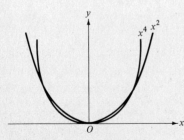

Figure 2.15
The graphs of $y = x^2$ and $y = x^4$.

EVEN AND ODD FUNCTIONS
Consider the functions $f_n(x) = x^n$ for $n = 2, 4, 6, \ldots$ (see Figure 2.15). The graphs, which all look alike, are *symmetric about* the y axis, which is to say that the portion lying to the left of the y axis is the mirror image of the portion lying to the right. Expressed more precisely, if (x, y) lies on the graph, so does $(-x, y)$. Any function with this property is said to be even (see Figure 2.16).

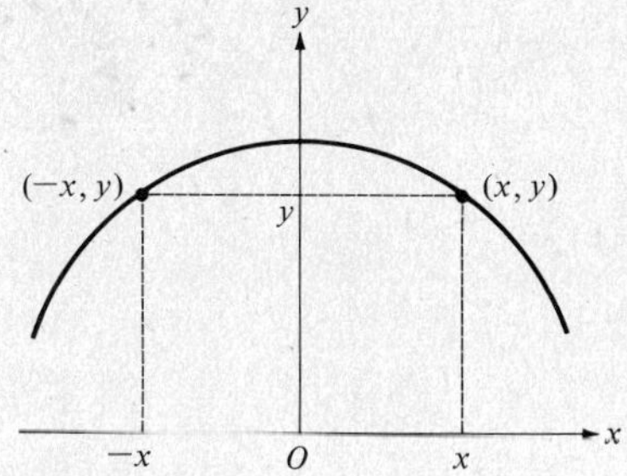

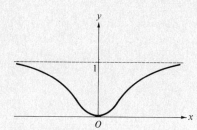

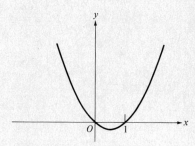

Figure 2.16
The graph of an even function.

Figure 2.17
The graph of $f(x) = x^2/(x^2 + 1)$.

Figure 2.18
The graph of $g(x) = x^2 - x$.

DEFINITION 1
The function f is *even* when

$$f(x) = f(-x)$$

for all values x for which f is defined.

Example 3
Which of the following functions is even?

$$f(x) = \frac{x^2}{x^2 + 1}$$

$$g(x) = x^2 - x$$

Solution
Since

$$f(-x) = \frac{(-x)^2}{(-x)^2 + 1} = \frac{x^2}{x^2 + 1} = f(x)$$

for all values of x, we find that f is even (see Figure 2.17). Regarding the function g, we see that

$$g(-x) = (-x)^2 - (-x) = x^2 + x$$

Hence

$$g(-x) \neq g(x)$$

and so g is not even. Indeed, the graph of $y = g(x)$ is not symmetric about the y axis (see Figure 2.18).

Now consider the functions $f_n(x) = x^n$ for $n = 1, 3, 5, 7, \ldots$ (see Figure 2.19). As in the case of even powers (Figure 2.15), these graphs look alike. They are said to be *symmetric about the origin* because if (x, y) lies on the graph, so does $(-x, -y)$ (see Figure 2.20). Functions with this property are called odd.

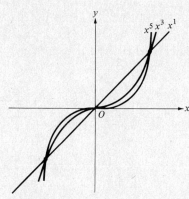

Figure 2.19
The graphs of $y = x^n$ for $n = 1, 3, 5$.

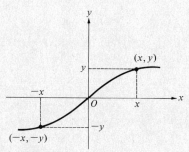

Figure 2.20
The graph of an odd function.

DEFINITION 2
The function f is *odd* when
$$f(-x) = -f(x)$$
for all values of x for which f is defined.

Example 4
Which of the following functions is odd?
$$f(x) = 1 - \frac{1}{x}$$
$$g(x) = x + \frac{1}{x}$$

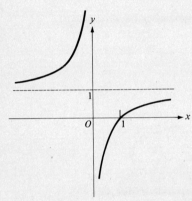

Figure 2.21
The graph of $f(x) = 1 - 1/x$.

Solution
Since, for all values of x, $x \neq 0$,
$$f(-x) = 1 - \frac{1}{(-x)} = 1 + \frac{1}{x}$$
and
$$-f(x) = -\left(1 - \frac{1}{x}\right) = -1 + \frac{1}{x}$$
we find that $f(-x) \neq -f(x)$ and hence this function is not odd (see Figure 2.21).

For the function g we have
$$g(-x) = -x + \frac{1}{-x} = -x - \frac{1}{x} = -\left(x + \frac{1}{x}\right) = -g(x)$$
and hence this function is odd (see Figure 2.22).

Observe that neither function in this example is defined for $x = 0$, and that their graphs consist of two separate branches.

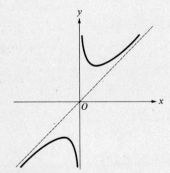

Figure 2.22
The graph of $g(x) = x + 1/x$.

EXERCISES

Sketch the graph of the functions in Exercises 1–11.

1. $y = x + x^2$
2. $y = x - x^2$
3. $y = 1 + |x|$
4. $y = x - |x|$
5. $y = |x - 1|$
6. $y = (x - 1)(x + 1)$
7. $y = x \cdot |x|$
8. $y = \begin{cases} x^2 & \text{for } 0 \leq x \leq 1 \\ 0 & \text{for } -2 \leq x \leq 0 \end{cases}$

9. $y = \begin{cases} -1 & \text{for } -1 \leq x \leq 0 \\ 1 & \text{for } 0 < x \leq 1 \end{cases}$ 10. $y = \sqrt{x^2 - 1}$

11. $y = \sqrt{4 - x^2}$

In Exercises 12–21, decide which functions are even, odd, or neither even nor odd.

12. $f(x) = \sqrt{x}$
13. $f(x) = 1 + |x| + |x|^2$
14. $f(x) = 1 + x^2 + x^4$
15. $f(x) = 1 + x + x^3$
16. $f(x) = |x|$
17. $f(x) = \dfrac{x}{|x|}$
18. $f(x) = \dfrac{x-1}{x+1}$
19. $f(x) = \sqrt{\dfrac{x^2+1}{x^2-1}}$
20. $f(x) = x + x^2 + x^3 + x^4$
21. $f(x) = \dfrac{1}{x} + \dfrac{1}{x^3}$

22. For what values of m and b will the function $L(x) = mx + b$ be odd?

23. Is there an even function $L(x) = mx + b$?

24. A *curve* is *symmetric about the x axis* if the point $(x, -y)$ lies on the curve when (x, y) lies on it. Is there a function whose graph is symmetric about the x axis? If so, give an example; if not, explain why.

25. For what positive integers n are the curves $y = x^n$ cut by *every* straight line in not more than two points?

26. Is the product of odd functions necessarily odd? Even? Neither odd nor even? Explain your answer.

27. If f is even, is $5 \cdot f$ even? Why?

28. Suppose that g is odd. Is the function $1/g$ odd?

29. If h is odd, can you conclude that $|h|$ is odd? Even?

30. For what values of a, b, and c is the function $f(x) = ax^2 + bx + c$ even?

31. For what values of a, b, c, and d is the function $f(x) = ax^3 + bx^2 + cx + d$ odd?

2.3 LINEAR FUNCTIONS

A *linear function* is a relation $f(x) = mx + b$, where m and b are given constants. The graph $y = mx + b$ of such a function is a straight line. The number m is called the *slope* of the line; it has an important interpretation:

Let (x_1, y_1) and (x_2, y_2) be any two distinct points on the line $y = mx + b$. Then

$$y_1 = mx_1 + b \quad \text{and} \quad y_2 = mx_2 + b$$

and hence

$$\frac{y_2 - y_1}{x_2 - x_1} = \frac{(mx_2 + b) - (mx_1 + b)}{x_2 - x_1} = \frac{mx_2 - mx_1}{x_2 - x_1} = m$$

We observe that

$$\text{slope} = \frac{\text{change along vertical axis}}{\text{change along horizontal axis}}$$

In particular, we see that the slope of a line is determined by any two points on it.

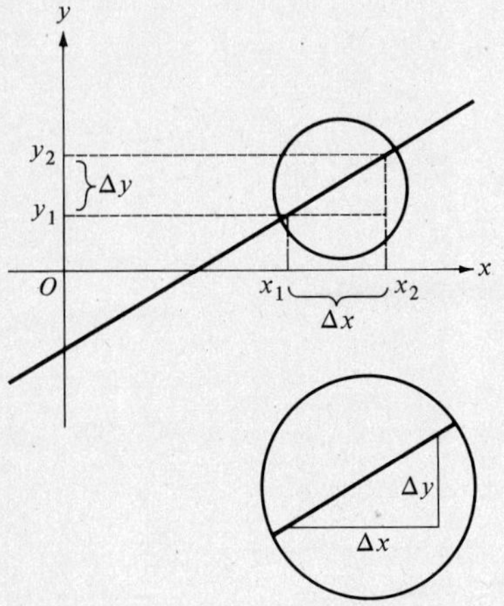

INCREMENTS

The differences $x_2 - x_1$ and $y_2 - y_1$ are called *increments* and are usually designated

$\Delta x = x_2 - x_1 \quad$ Δx is read "delta x"
$\Delta y = y_2 - y_1 \quad$ Δy is read "delta y"

(see Figure 2.23).

Figure 2.23
The increment Δx represents change along the x axis; the increment Δy represents change along the y axis.

Example 1

Let $y = 5x - 7$.
  If $x_1 = 2$ and $x_2 = 3$, then

$\Delta x = 3 - 2 = 1$
$\Delta y = (5 \times 3 - 7) - (5 \times 2 - 7) = 5$

If $x_1 = 2$ and $x_2 = 0$, then

$\Delta x = 0 - 2 = -2$
$\Delta y = (5 \times 0 - 7) - (5 \times 2 - 7) = -10$

If $x_1 = x_2$, then $\Delta x = \Delta y = 0$.

We thus see:

Increments can be positive, negative, or zero.

Example 2

What is the slope of the line passing through the points $(-1, -1)$ and $(2, 3)$?

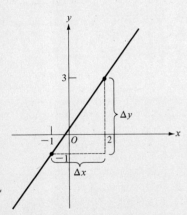

Figure 2.24
The line passing through the points $(-1, -1)$ and $(2, 3)$.

Solution
We have $\Delta y = 3 - (-1) = 4$ and $\Delta x = 2 - (-1) = 3$ (see Figure 2.24).

Hence, the slope of the line is

$$m = \frac{\Delta y}{\Delta x} = \frac{4}{3}$$

SLOPE-INTERCEPT FORMULA
In the equation $y = mx + b$, we have $y = b$ when $x = 0$. Hence, this line intersects the y axis in the point $(0, b)$, called its y *intercept*. Thus:

> $y = mx + b$ is the equation of a line with slope m and y intercept $(0, b)$.

Example 3

Find the equation of the line whose slope is -4 and whose y intercept is $(0, -3)$. Draw its graph.

Solution
Apply the slope-intercept formula with $m = -4$ and $b = -3$. This gives

$$y = -4x + (-3) = -4x - 3$$

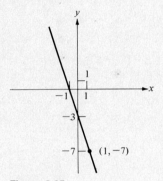

Figure 2.25
The line $y = -4x - 3$.

To draw the graph we need two points on the line. We are given that $(0, -3)$ is one point on the line. Another point is obtained by taking, say, $x = 1$: We get $y = -4 \times 1 - 3 = -7$, and thus the point $(1, -7)$ is a second point on the line (see Figure 2.25).

POINT-SLOPE FORMULA
If (x_0, y_0) and (x, y) are distinct points on a line with slope m, then

$$\frac{y - y_0}{x - x_0} = m$$

Multiplying both sides by $x - x_0$ gives the equation

$$y - y_0 = m(x - x_0)$$

This equation holds for all points (x, y) on the line. Thus:

> $y - y_0 = m(x - x_0)$ is the equation of a line having slope m and passing through the point (x_0, y_0).

Example 4

Find an equation of the line going through the point $(1, -3)$ and having slope $m = -2$.

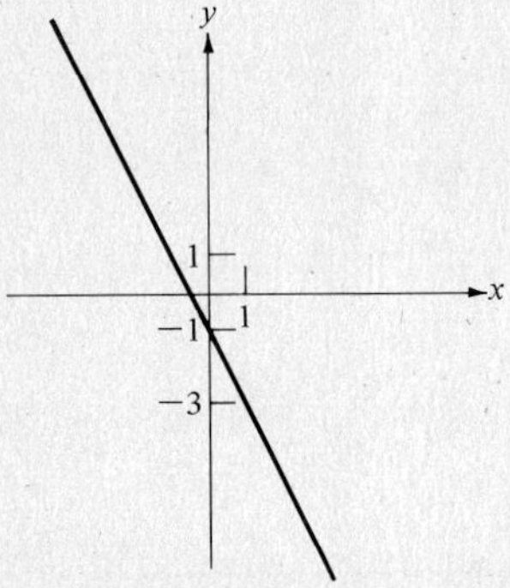

Solution
We use the point-slope formula with $x_0 = 1$, $y_0 = -3$, and $m = -2$. This gives $y - (-3) = -2(x - 1)$, and this simplifies to

$$y + 3 = -2(x - 1)$$

(see Figure 2.26).

Figure 2.26
The line $y + 3 = -2(x - 1)$.

TWO-POINT FORMULA
If (x_0, y_0), (x_1, y_1), and (x, y) are distinct points on a line with slope m, then

$$m = \frac{y - y_0}{x - x_0} = \frac{y_0 - y_1}{x_0 - x_1}$$

This formula holds for all points (x, y) on the line such that $x \neq x_0$. Thus:

> $\dfrac{y - y_0}{x - x_0} = \dfrac{y_0 - y_1}{x_0 - x_1}$ is the equation of a line passing through the points (x_0, y_0) and (x_1, y_1).

Example 5

Find an equation of the line passing through the points $(0, 1)$ and $(4, 2)$. Draw its graph.

Solution
We first find the slope as in Example 2.

$$\Delta y = 2 - 1 = 1$$
$$\Delta x = 4 - 0 = 4$$

and hence the slope is

$$m = \frac{\Delta y}{\Delta x} = \frac{1}{4}$$

We now use the point-slope formula with $x_0 = 0$, $y_0 = 1$, and $m = \frac{1}{4}$. This gives

$$y - 1 = \tfrac{1}{4}(x - 0)$$

or

$$y - 1 = \tfrac{1}{4} x$$

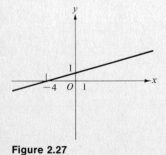

Figure 2.27
The line $y - 1 = \tfrac{1}{4}x$.

The graph is given in Figure 2.27. You should verify that we would get this same equation if we used $x_0 = 4$ and $y_0 = 2$ instead.

2.3 LINEAR FUNCTIONS

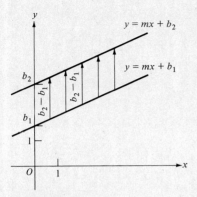

Figure 2.28
Parallel lines ($b_2 > b_1$).

PARALLEL LINES
Consider two lines

$$y = mx + b_1 \quad \text{and} \quad y = mx + b_2$$

with the same slope m (see Figure 2.28). Since

$$(mx + b_2) = (mx + b_1) + (b_2 - b_1)$$

we see that for any number x_0, the point $(x_0, mx_0 + b_2)$ on the line $y = mx + b_2$ is $(b_2 - b_1)$ units above or below the point $(x_0, mx_0 + b_1)$ on the line $y = mx + b_1$. We thus conclude that the line $y = mx + b_2$ is obtained from the line $y = mx + b_1$ by pushing it $b_2 - b_1$ units in the y direction. Thus:

> Lines with the same slope are parallel.

EXERCISES

1. The slope of a line $y = mx + c$ was defined as the ratio $m = \Delta y / \Delta x$. For a vertical line, all increments Δx are zero. Since we may not divide by zero,

> The slope of a vertical line is undefined.

and such a line cannot be written in the form $y = mx + c$. For each of the pairs of points listed below, find the slope m of the line determined by them; when a line is vertical, write "m is undefined."

(a) $P(1, 2)$ $Q(2, 1)$

(b) $P(1, 0)$ $Q(0, 1)$

(c) $P(-1, 0)$ $Q(0, 1)$

(d) $P(1, 1)$ $Q(1, 2)$

(e) $P(1, 1)$ $Q(2, 1)$

(f) $P(a, 0)$ $Q(b, b)$ $(b \neq 0)$

(g) $P(x, -1)$ $Q(x, 1)$

(h) $P(-4, -4)$ $Q(-2, -5)$

(i) $P(-1, -6)$ $Q(-2, -5)$

(j) $P(0, 0)$ $Q(1, m)$

(k) $P(\pi, \log_{10} 2)$ $Q(\pi + 1, \log_{10} 2 + m)$

(l) $P(0, 0)$ $Q(1/m, 1)$

2. If P, Q, and R are three distinct points, then any pair of them determines a line. If the slope of the line determined by P and Q equals the slope of the line determined by Q and R, then the three points all lie on the same line (the points are said to be collinear). Using this information, determine which of the following groups of points lie on a straight line.

   (a) $P(1, 1)$ $Q(2, 2)$ $R(-1, -1)$
   (b) $P(1, 2)$ $Q(2, 2)$ $R(0, 0)$
   (c) $P(0, -2)$ $Q(1, 1)$ $R(-1, -4)$
   (d) $P(0, -1)$ $Q(1, -4)$ $R(-1, 2)$
   (e) $P(1, -1)$ $Q(1, 0)$ $R(1, 1)$
   (f) $P(-4, -3)$ $Q(2, -1)$ $R(-10, -7)$

3. In the table that follows, write in the given forms the equation of the line passing through the indicated point (x_0, y_0) and having prescribed slope m.

(x_0, y_0)	m	$y - y_0 = m(x - x_0)$	$y = mx + b$	$Ax + By + C = 0$, A and B integers
$(-1, 0)$	$\frac{1}{2}$			
$(0, 0)$	1			
$(0, 4)$	10			
$(5, 0)$	$-\frac{1}{2}$			

4. Find the equation of the line that passes through the point $(-1, 1)$ and is parallel to the line

   (a) $y + 1 = 3(x - 1)$ (b) $y = -2x + 3$
   (c) $y = 7$ (d) $2y - 5x = 9$

5. Consider lines

   $$y = m_1 x + c_1$$
   $$y = m_2 x + c_2$$

   (see Figure 2.29). When we put

   $$m_1 x + c_1 = m_2 x + c_2$$

   we find that this equation can be solved for x, provided that $m_1 \neq m_2$. The solution is, of course,

   $$x = \frac{c_2 - c_1}{m_1 - m_2}$$

   This tells us that the above lines intersect in the point (x_0, y_0),

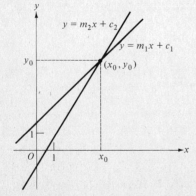

Figure 2.29

where $x_0 = (c_2 - c_1)/(m_1 - m_2)$, and $y_0 = m_1 x_0 + c_1$ (notice that $y_0 = m_2 x_0 + c_2$ is also true). Find the point of intersection for each of the following pairs of lines.

(a) $y = 3x + 2$; $y = -x - 2$

(b) $y = 1$; $y = x$

(c) $y = -x - 1$; $y = x - 1$

(d) $2y + 3x + 1 = 0$; $x + y - 1 = 0$

(e) $y - 4x - 1 = 0$; $y - 3 = \frac{1}{2}x$

(f) $y - 1 = \frac{1}{10}(x - 1)$; $y - 1 = -10(x - 1)$

6. Referring to Exercise 5, plot the pairs of lines (a), (e), and (f).

7. The lines L_1 and L_2 are *perpendicular* when there is a right angle between them. Given the line $y = 2x$, find the line L that passes through the origin, O, and is perpendicular to it.

Procedure

In this exercise you are asked to complete the steps. To determine the line L, it suffices to find its slope m, since any line through O can be written as $y = mx$. To begin, pick the point $P(1, 2)$ on the line $y = 2x$ and the point $Q(1, m)$ on the line $y = mx$ (see Figure 2.30).

1. Use the distance formula to find

$[\text{dist}(O, P)]^2 =$
$[\text{dist}(O, Q)]^2 =$
$[\text{dist}(P, Q)]^2 =$

2. The angle POQ is a right angle when

$[\text{dist}(P, Q)]^2 = [\text{dist}(O, P)]^2 + [\text{dist}(O, Q)]^2;$

substitute into this equation the expressions computed in step (1) and solve for m.

3. If you made no errors, your answer should be $m = -\frac{1}{2}$.
4. Repeat the above procedure to find a line through O that is perpendicular to the line $y = Mx$ (rather than to $y = 2x$).
5. State your conclusion regarding the relation between the slopes of two perpendicular lines.

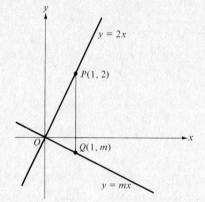

Figure 2.30

2.4 TANGENTS TO $y = x^2$

We now come to the major innovation in calculus — the limit concept. Special limits that are of particular importance in mathematics and its applications are derivatives and integrals. These are studied in great detail in the coming chapters. In this and the next section

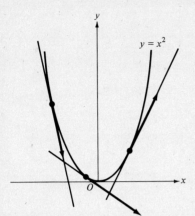

Figure 2.31
The curve $y = x^2$ as traced by a moving point.

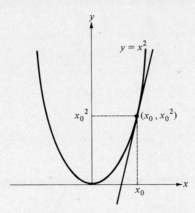

Figure 2.32
The tangent to $y = x^2$ at (x_0, x_0^2).

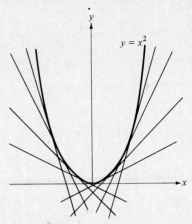

Figure 2.33
Tangents to $y = x^2$.

we motivate the limit concept by developing a procedure for finding tangents to the curve $y = x^2$. If this curve is imagined as being traced out by a moving point, then the tangents give the direction of motion of this point (see Figure 2.31).

For the curve $y = x^2$, it is easy to give an intuitive description of the notion of tangent: Consider any point x_0 on the x axis. Then the point (x_0, x_0^2) lies on the curve $y = x^2$ (see Figure 2.32). The tangent line to $y = x^2$ at (x_0, x_0^2) is the unique straight line that meets the curve at this point without cutting it (that is, the curve lies to one side of the tangent line, as indicated in Figure 2.33). For more general curves, however, such a description is not possible.

In Section 2.3 we learned that any straight line through the point (x_0, x_0^2) can be represented by an equation

$$y - x_0^2 = m(x - x_0)$$

where m is the slope of the line. Since x_0 is a given number, the only unknown quantity in this equation is the slope m. *The problem of finding the tangent is therefore the problem of finding the slope m.* We shall see later that the slope of a tangent line is one of the many interpretations of derivatives.

We propose the following procedure for finding the slope m. For a small increment $\Delta x \neq 0$, we look at the line through the points $P(x_0, y_0)$ and $Q(x_0 + \Delta x, y_0 + \Delta y)$ on the curve $y = x^2$ (see Figure 2.34). This line, which we designate $\overline{PQ}$, has slope $\Delta y/\Delta x$. For different increments Δx that approach zero, the line $\overline{PQ}$ approaches the tangent line and the slope $\Delta y/\Delta x$ approaches the slope m of the tangent line (see Figure 2.35). To find this limiting value of $\Delta y/\Delta x$ we perform some calculations.

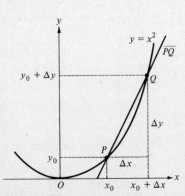

Figure 2.34
The line $\overline{PQ}$ with slope $\Delta y/\Delta x$.

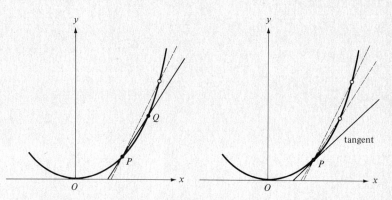

Figure 2.35
Finding the tangent to $y = x^2$ at P.

Table 2.1
The Slopes $\Delta y/\Delta x$ for Different Values of Δx.

$\Delta x > 0$	$\Delta y/\Delta x$
1.0	$2x_0 + 1$
0.1	$2x_0 + 0.1$
0.01	$2x_0 + 0.01$
0.001	$2x_0 + 0.001$
0.0001	$2x_0 + 0.0001$
0.00001	$2x_0 + 0.00001$

$\Delta x < 0$	$\Delta y/\Delta x$
-1.0	$2x_0 - 1$
-0.1	$2x_0 - 0.1$
-0.01	$2x_0 - 0.01$
-0.001	$2x_0 - 0.001$
-0.0001	$2x_0 - 0.0001$
-0.00001	$2x_0 - 0.00001$

THE SLOPES OF TANGENTS TO $y = x^2$
Consult Figure 2.34 again. Since $y = x^2$, we have

$$y_0 = x_0^2 \quad \text{and} \quad y_0 + \Delta y = (x_0 + \Delta x)^2$$

From this we find that

$$\Delta y = (x_0 + \Delta x)^2 - x_0^2 = x_0^2 + 2x_0 \cdot \Delta x + (\Delta x)^2 - x_0^2$$
$$= 2x_0 \cdot \Delta x + (\Delta x)^2 = (2x_0 + \Delta x) \cdot \Delta x$$

and consequently

$$\frac{\Delta y}{\Delta x} = \frac{(x_0 + \Delta x)^2 - x_0^2}{\Delta x} = \frac{(2x_0 + \Delta x) \cdot \Delta x}{\Delta x} = 2x_0 + \Delta x \tag{1}$$

It is clear from Table 2.1 that $\Delta y/\Delta x$ approaches $2x_0$ as Δx approaches 0. Symbolically, this is expressed as

$$\frac{\Delta y}{\Delta x} \to 2x_0 \quad \text{as} \quad \Delta x \to 0 \tag{2}$$

where "$\to$" is read "approaches."

DEFINITION 1
The number $2x_0$ is called the *limit* (limiting value) of the slopes $\Delta y/\Delta x$ as Δx approaches 0. This is expressed symbolically by the statement

$$\lim_{\Delta x \to 0} \frac{\Delta y}{\Delta x} = 2x_0$$

read "the limit of $\Delta y/\Delta x$ as Δx approaches 0 is $2x_0$." The *derivative* of x^2 at x_0 is $2x_0$.

The number $2x_0$ is the *slope* of the tangent to the curve $y = x^2$ at (x_0, x_0^2).

Example 1

The slope of the tangent to $y = x^2$ at $(x_0, x_0^2) = (3, 9)$ is $2 \cdot 3 = 6$. The tangent at this point is, therefore, given by $y - 9 = 6(x - 3)$.

We can now state precisely what we mean by a tangent to the curve $y = x^2$.

> **DEFINITION 2**
> The line
> $$y - x_0^2 = 2x_0(x - x_0)$$
> is the *tangent* to the curve $y = x^2$ at $P(x_0, x_0^2)$.

EXERCISES

The exercises below pertain to the curve $y = x^2$.

1. Find the tangent at $P(-\frac{1}{2}, \frac{1}{4})$.
2. Find the point on the curve at which the tangent has slope 5.
3. Find the tangent that is parallel to the line $y = x$.
4. Find the tangent that is parallel to the line $y = 2x - 1$.
5. Does the curve have two points at which the tangents are parallel?
6. Find that point on the curve at which the tangent has slope 0.
7. Find the two tangents to the curve that pass through $P(2, -5)$ (observe that this point does not lie on the curve).
8. Find that tangent to the curve that is perpendicular to the tangent $y - 1 = 2(x - 1)$ (the lines $y = m_1 x + b_1$ and $y = m_2 x + b_2$ are perpendicular when $m_1 m_2 = -1$).
9. Does the curve have a tangent that is perpendicular to the x axis?

2.5 TANGENTS TO $y = f(x)$

To find the tangent to a curve $y = f(x)$ at $P(x_0, f(x_0))$ (see Figure 2.36), we follow the procedure to find tangents to $y = x^2$. Namely, the tangent line can be represented by an equation

$$y - f(x_0) = m(x - x_0)$$

where the slope m is to be determined. Choosing an increment $\Delta x \neq 0$, we put

$$\Delta y = f(x_0 + \Delta x) - f(x_0)$$

2.5 TANGENTS TO $y = f(x)$

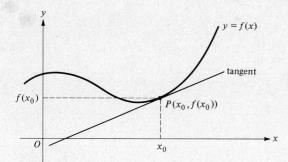

Figure 2.36
The tangent to $y = f(x)$ at $P(x_0, f(x_0))$.

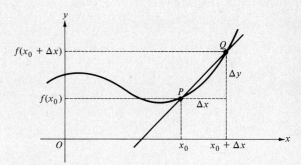

Figure 2.37
The line $\overline{PQ}$ with slope $\Delta y/\Delta x$.

(see Figure 2.37) and examine the line $\overline{PQ}$ with slope

$$\frac{\Delta y}{\Delta x} = \frac{f(x_0 + \Delta x) - f(x_0)}{\Delta x} \tag{1}$$

If $\Delta y/\Delta x$ approaches a number m as we use different increments Δx that approach zero, then we call m the *slope* of the tangent to $y = f(x)$ at $(x_0, f(x_0))$. We write

$$\frac{\Delta y}{\Delta x} \to m \text{ as } \Delta x \to 0$$

or, equivalently,

$$\lim_{\Delta x \to 0} \frac{\Delta y}{\Delta x} = m$$

provided, of course, that a limit (limiting value) m exists.

DEFINITION 1
The *tangent* to $y = f(x)$ at $(x_0, f(x_0))$ is the line

$$y - f(x_0) = m(x - x_0)$$

where the slope m is given by

$$m = \lim_{\Delta x \to 0} \frac{f(x_0 + \Delta x) - f(x_0)}{\Delta x}$$

provided that the limit exists.

The number m is also called the *derivative* of f at x_0 (see Chapter 3); it will be denoted $m = f'(x_0)$ later on.

Example 1

Find the tangent to $y = x^3$ at $(2, 8)$.

44 FUNCTIONS AND LIMITS

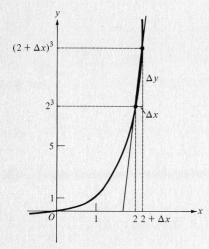

Figure 2.38
Finding the tangent to $y = x^3$ at $(2, 8)$.

Table 2.2
The Slopes $\Delta y/\Delta x$ for Different Values of Δx

$\Delta x > 0$	$\Delta y/\Delta x$
0.1	12.61
0.01	12.0601
0.001	12.006001
0.0001	12.00060001

$\Delta x < 0$	$\Delta y/\Delta x$
-0.1	11.41
-0.01	11.9401
-0.001	11.994001
-0.0001	11.99940001

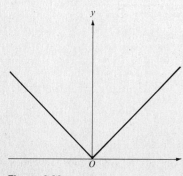

Figure 2.39
The graph of $y = |x|$.

Solution
Taking $f(x) = x^3$ and $x_0 = 2$, then $f(x_0) = f(2) = 2^3 = 8$, and we have for the equation of the tangent

$$y - 8 = m(x - 2)$$

To compute the slope m, choose an increment $\Delta x \neq 0$ (see Figure 2.38). Then

$$\Delta y = f(2 + \Delta x) - f(2) = (2 + \Delta x)^3 - 2^3$$
$$= 2^3 + 3 \cdot 2^2 (\Delta x) + 3 \cdot 2(\Delta x)^2 + (\Delta x)^3 - 2^3$$
$$= [3 \cdot 2^2 + 3 \cdot 2(\Delta x) + (\Delta x)^2] \Delta x = [12 + 6(\Delta x) + (\Delta x)^2] \Delta x$$

and, using equation (1),

$$\frac{\Delta y}{\Delta x} = \frac{f(2 + \Delta x) - f(2)}{\Delta x} = \frac{[12 + 6(\Delta x) + (\Delta x)^2] \Delta x}{\Delta x}$$

$$= 12 + (6 + \Delta x) \Delta x$$

Table 2.2 makes it clear that

$$\frac{\Delta y}{\Delta x} \to 12 \quad \text{as} \quad \Delta x \to 0$$

or

$$\lim_{\Delta x \to 0} \frac{(2 + \Delta x)^3 - 2^3}{\Delta x} = 12$$

Hence, the slope of the tangent to $y = x^3$ at $(2, 8)$ is

$$m = 12$$

and the equation of the tangent is

$$y - 8 = 12(x - 2)$$

It is implicit in our definition of tangent to a curve $y = f(x)$ at $(x_0, f(x_0))$ that the slopes

$$\frac{\Delta y}{\Delta x} = \frac{f(x_0 + \Delta x) - f(x_0)}{\Delta x}$$

approach the same number (which we shall call the *derivative* of f at x_0) regardless of the way in which the numbers Δx approach 0. Indeed, this is not always true, and some functions do not possess a tangent at certain points. Two such cases are considered next.

Example 2

The curve $y = |x|$ fails to have a tangent at the point $(0, 0)$ (see Figure 2.39). To see why this is so, we must show that the slopes

$$\frac{\Delta y}{\Delta x} = \frac{f(0 + \Delta x) - f(0)}{\Delta x}$$

do not have a limit as Δx approaches zero. Now,

$$f(0) = 0$$
$$f(0 + \Delta x) = |\Delta x|$$
$$\Delta y = f(0 + \Delta x) - f(0) = |\Delta x|$$

and hence

$$\frac{\Delta y}{\Delta x} = \frac{|\Delta x|}{\Delta x}$$

where, of course, $\Delta x \neq 0$.

Since $|\Delta x|$ is always positive, we conclude that

$$\frac{|\Delta x|}{\Delta x} = \begin{cases} 1 & \text{for all } \Delta x > 0 \\ -1 & \text{for all } \Delta x < 0 \end{cases}$$

This tells us that no matter how small Δx is taken to be, $\Delta y/\Delta x$ does not approach any single number (see Figure 2.40). Instead, if we let Δx assume both positive and negative values, then $\Delta y/\Delta x$ oscillates between the two numbers $+1$ and -1.

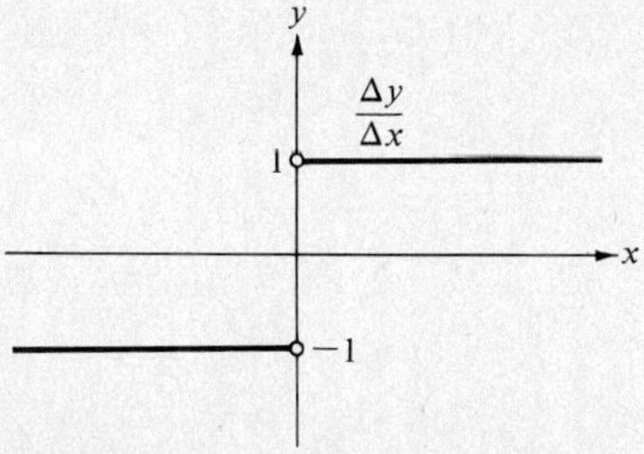

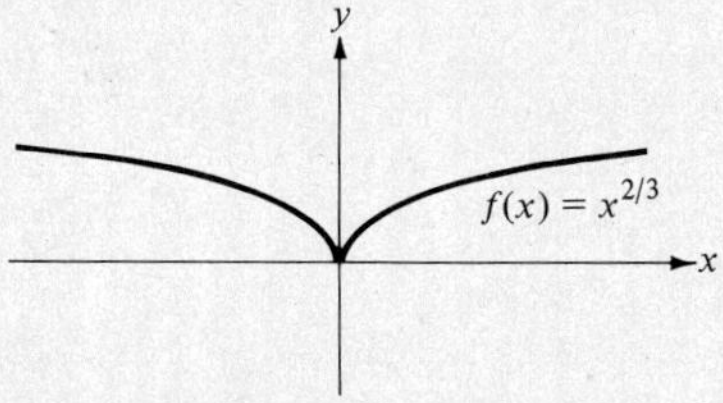

Figure 2.40 The graph of $\Delta y/\Delta x$.

Example 3

Another curve that fails to have a tangent at $(0, 0)$ is $y = x^{2/3}$ (see Figure 2.41). This is seen from the following facts:

$$f(0) = 0$$
$$f(0 + \Delta x) = (\Delta x)^{2/3}$$
$$\Delta y = f(0 + \Delta x) - f(0) = (\Delta x)^{2/3}$$

so that

$$\frac{\Delta y}{\Delta x} = \frac{(\Delta x)^{2/3}}{\Delta x} = \frac{1}{(\Delta x)^{1/3}}$$

Inspecting Table 2.3, we see that $\Delta y/\Delta x$ increases beyond all bounds in the positive direction when $\Delta x \to 0$ through positive values, in the negative direction when $\Delta x \to 0$ through negative values. Hence the slopes $\Delta y/\Delta x$ do not approach a number as $\Delta x \to 0$.

Figure 2.41 The graph of $y = x^{2/3}$. Note that this curve is not smooth at the origin, but has a cusp.

Table 2.3

Δx	$\Delta y/\Delta x$
± 1	± 1
$\pm \frac{1}{10^3}$	± 10
$\pm \frac{1}{10^6}$	± 100
$\pm \frac{1}{10^9}$	$\pm 1{,}000$
$\pm \frac{1}{10^{12}}$	$\pm 10{,}000$
$\pm \frac{1}{10^{15}}$	$\pm 100{,}000$

Example 4

Find the tangent to $y = 1/x$ at $P(1, 1)$.

Solution

Taking $f(x) = 1/x$ and $x_0 = 1$, we have for the equation of the tangent

46 FUNCTIONS AND LIMITS

Table 2.4

$\Delta x > 0$	$\Delta y/\Delta x$
1.0	-0.5
0.1	-0.90909090
0.01	$-0.99009900\ldots$
0.001	-0.999000999000
0.0001	$-0.99990000\ldots$

$\Delta x < 0$	$\Delta y/\Delta x$
-0.1	$-1.111\ldots$
-0.01	$-1.0101\ldots$
-0.001	$-1.001001\ldots$
-0.0001	$-1.00010001\ldots$

$$y - 1 = m(x - 1)$$

To compute m, we observe that, for an increment $\Delta x \neq 0$,

$$\Delta y = f(1 + \Delta x) - f(1) = \frac{1}{1 + \Delta x} - 1 = -\frac{\Delta x}{1 + \Delta x}$$

and hence

$$\frac{\Delta y}{\Delta x} = -\frac{1}{1 + \Delta x}$$

It is evident from Table 2.4 that

$$m = \lim_{\Delta x \to 0} \frac{\Delta y}{\Delta x} = -1$$

EXERCISES

Exercises 1–6 pertain to the curve $y = x^3$.

1. Verify that the tangent to $y = x^3$ at (x_0, x_0^3) has slope $m = 3x_0^2$.

2. Find the tangent to the curve at $(-1, -1)$.

3. Find the tangent to the curve at $(0, 0)$. Is there anything unusual about this tangent?

4. Find the two tangents that are parallel to the line $y = x$.

5. Does this curve have a tangent that is perpendicular to the x axis?

6. Find the two points on the curve at which the tangent has slope $m = 12$.

Exercises 7–12 pertain to the curve $y = 1/x$.

7. Verify that the tangent to $y = 1/x$ at $(x_0, 1/x_0)$ has slope $m = -1/x_0^2$.

8. Find the tangent to the curve at $(-\frac{1}{2}, -2)$.

9. Find the two points on the curve at which the tangents have slope $-\frac{1}{2}$.

10. Find the two tangents passing through the point $(\frac{3}{2}, \frac{1}{2})$.

11. Is there a point (x, y) that lies on more than two tangents to the curve?

12. Find one tangent to the curve that is parallel to the line $y = -3x$.

In Exercises 13–15, use numerical tables to show that the given curves have no tangent at the indicated points.

13. $y = |x - 1|$ at $(1, 0)$

14. $y = |x^2 - 1|$ at $(-1, 0)$

15. $y = \begin{cases} x & \text{for } x \geq 0 \\ 0 & \text{for } x \leq 0 \end{cases}$ at $(0, 0)$

16. When $f(x) = 5x^2 + 3$, simplify the quotient

$$\frac{\Delta y}{\Delta x} = \frac{f(x_0 + \Delta x) - f(x_0)}{\Delta x}$$

as much as possible, then evaluate

$$\lim_{\Delta x \to 0} \frac{\Delta y}{\Delta x} =$$

17. When $g(x) = 5x + 3$, simplify the quotient

$$\frac{\Delta y}{\Delta x} = \frac{g(x_0 + \Delta x) - g(x_0)}{\Delta x}$$

as much as possible, then evaluate

$$\lim_{\Delta x \to 0} \frac{\Delta y}{\Delta x} =$$

What is the tangent to $y = 5x + 3$ at $(2, 13)$?

2.6 LIMITS

In defining the derivative (the slope of a tangent to curve), we introduced the limit notation in the following form:

$$\lim_{\Delta x \to 0} \frac{f(x_0 + \Delta x) - f(x_0)}{\Delta x} = m$$

Here we took a limit of the function

$$F(\Delta x) = \frac{f(x_0 + \Delta x) - f(x_0)}{\Delta x} \qquad (\Delta x \neq 0)$$

of the variable Δx. The concept of limit has many other profound and important applications in calculus and elsewhere in mathematics and its applications. In the next chapter we shall develop properties of derivatives and formulas for their computation. For this we shall use general properties of limits that will subsequently be important also in other parts of our subject. In the statement of these properties, it is customary to use functions of the variable x. As we shall point out, however, the limit does not depend on the variable used.

> **DEFINITION 1**
> If the function f approaches a single number L as x approaches x_0, we write
> $$\lim_{x \to x_0} f(x) = L$$
> and say that *f has the limit L at x_0*.

It is implicit in the formulation that f be defined in some interval $(x_0 - \delta, x_0 + \delta)$, except possibly at x_0.

This definition is quite intuitive, but it is at this level that we work most of the time (a more formal definition, which is used in proofs, is given in Definition 2, below). It is implicitly understood that f approaches the number L no matter how x approaches x_0.

Example 1

$$\lim_{x \to \pi} x^2 = \pi^2$$

$$\lim_{x \to 2} x^3 = 2^3 = 8$$

$$\lim_{x \to 3} (x^2 - 2x + 5) = 3^2 - 2 \cdot 3 + 5 = 8$$

Finding these limits was simple, because each function was defined at the limit point x_0. In calculating arbitrary limits, we make use of some general properties of limits. Before stating these, however, we have to explain how functions can be combined to form new functions.

From given functions f and g, we obtain the new functions $f + g$, $f - g$, $f \cdot g$, and f/g, through the following definition:

$$(f + g)(x) = f(x) + g(x)$$
$$(f - g)(x) = f(x) - g(x)$$
$$(f \cdot g)(x) = f(x) \cdot g(x)$$
$$\left(\frac{f}{g}\right)(x) = \frac{f(x)}{g(x)} \quad \text{wherever } g(x) \neq 0$$

The domain of the sum, difference, and product functions is the common domain of f and g; the domain of the quotient function excludes all points where g vanishes. These rules are, of course, used implicitly throughout this chapter.

Example 2

Let $f(x) = x$, $g(x) = \sqrt{1 - x^2}$. Then

$$(f+g)(x) = x + \sqrt{1-x^2}$$
$$(f-g)(x) = x - \sqrt{1-x^2}$$
$$(f \cdot g)(x) = x\sqrt{1-x^2}$$
$$\left(\frac{f}{g}\right)(x) = \frac{x}{\sqrt{1-x^2}}$$

Observe that the domain of f consists of the entire real line whereas the domain of g is the closed interval $[-1, 1]$. The domain of the first three functions is $[-1, 1]$, and the domain of f/g is the open interval $(-1, 1)$ (why?).

We can now list the following general properties of limits:

THEOREM 1: PROPERTIES OF LIMITS

1. If c is any constant, then $\lim_{x \to x_0} c = c$.

2. $\lim_{x \to x_0} x^r = x_0^r$ for all numbers $r \neq 0$.

   If $\lim_{x \to x_0} f(x) = A$, $\lim_{x \to x_0} g(x) = B$, then

3. $\lim_{x \to x_0} d \cdot f(x) = d \cdot A$ for any constant d.

4. $\lim_{x \to x_0} [f(x) + g(x)] = A + B$.

5. $\lim_{x \to x_0} [f(x) - g(x)] = A - B$.

6. $\lim_{x \to x_0} [f(x) \cdot g(x)] = A \cdot B$.

7. $\lim_{x \to x_0} \frac{f(x)}{g(x)} = \frac{A}{B}$, provided that $B \neq 0$.

The proof, which involves Definition 2 below, is omitted here.

Example 3

Find $\lim_{x \to \frac{1}{2}} (2x - 1)$.

Solution

$$\lim_{x \to \frac{1}{2}} (2x - 1) = \lim_{x \to \frac{1}{2}} 2x - \lim_{x \to \frac{1}{2}} 1 \quad \text{(Property 5)}$$

$$= 2 \lim_{x \to \frac{1}{2}} x - \lim_{x \to \frac{1}{2}} 1 \quad \text{(Property 3)}$$

$$= 2 \cdot \tfrac{1}{2} - 1 \quad \text{(Property (2) with } r = 1 \text{ and}$$
$$= 1 - 1 = 0 \quad \quad \text{Property (1) with } c = 1)$$

Example 4

Find $\lim_{x \to \frac{1}{2}} \sqrt{1 - x^2}$ and $\lim_{x \to \frac{1}{2}} \frac{x}{\sqrt{1 - x^2}}$.

Solution
We use the following steps:

$$\lim_{x \to \frac{1}{2}} x^2 = \left(\frac{1}{2}\right)^2 = \frac{1}{4} \quad \text{(Property 2, with } r = 2\text{)}$$

$$\lim_{x \to \frac{1}{2}} 1 - x^2 = 1 - \frac{1}{4} = \frac{3}{4} \quad \text{(Properties 5 and 1)}$$

$$\lim_{x \to \frac{1}{2}} \sqrt{1 - x^2} = \sqrt{\frac{3}{4}} = \frac{\sqrt{3}}{2} \quad \left(\text{Property 2 with } r = \frac{1}{2}\right)$$

$$\lim_{x \to \frac{1}{2}} \frac{x}{\sqrt{1 - x^2}} = \frac{1/2}{\sqrt{3/2}} = \frac{1}{\sqrt{3}} \quad \text{(Property 7)}$$

Example 5

Find $\lim_{t \to 4} \frac{\sqrt{t} - 2}{t - 4}$

Solution
Since $\lim_{t \to 4} (t - 4) = 0$, we cannot apply property 7 to this problem and we must use a trick. It is this: As long as $t \neq 4$, we have

$$\frac{\sqrt{t} - 2}{t - 4} = \frac{(\sqrt{t} - 2)(\sqrt{t} + 2)}{(t - 4)(\sqrt{t} + 2)} = \frac{(\sqrt{t})^2 - 2\sqrt{t} + 2\sqrt{t} - 2^2}{(t - 4)(\sqrt{t} + 2)}$$

$$= \left(\frac{t - 4}{t - 4}\right)\left(\frac{1}{\sqrt{t} + 2}\right) = \frac{1}{\sqrt{t} + 2}$$

But by Property 2, with $r = \frac{1}{2}$, $\lim_{t \to 4} \sqrt{t} = 2$; by Properties 5 and 7, therefore,

$$\lim_{t \to 4} \frac{\sqrt{t} - 2}{t - 4} = \lim_{t \to 4} \frac{1}{\sqrt{t} + 2} = \frac{1}{2 + 2} = \frac{1}{4}$$

As in the case of derivatives, when the variable approaches 0, we often use Δx.

Example 6

Find $\lim_{\Delta x \to 0} \left[2 + \Delta x \left(\Delta x + \frac{1}{2}\right)\right]^{1/2}$.

Solution
Since $\lim_{\Delta x \to 0} \Delta x \left(\Delta x + \frac{1}{2}\right) = 0 \cdot \frac{1}{2} = 0$ (Properties 2 and 6)

we have

$$\lim_{\Delta x \to 0} \left[2 + \Delta x \left(\Delta x + \frac{1}{2} \right) \right]^{1/2} = (2 + 0)^{1/2} = \sqrt{2}$$

The following formal definition of limit will give you further insight into this concept; it will be useful in later sections.

DEFINITION 2
The function f has the limit L at x_0 if, for each error $\epsilon > 0$, no matter how small, there is a $\delta > 0$, such that

$$|f(x) - L| < \epsilon \quad \text{whenever} \quad 0 < |x - x_0| < \delta$$

Geometrically, the situation is like that described in Figure 2.42, with $f(x_0)$ replaced by L. The latter is necessary because $f(x_0)$ may not be defined.

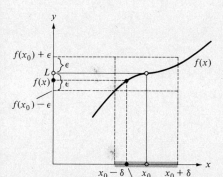

Figure 2.42

LIMITS AT INFINITY

Let us examine the function $f(x) = 1/x$ for large values of x. Looking at Figure 2.43 and the accompanying table, we see that $1/x$ approaches 0 as x becomes larger and larger. We write

$$\frac{1}{x} \to 0 \quad \text{as} \quad x \to \infty$$

where the symbol $x \to \infty$ is read "as x tends to infinity" or "as x increases beyond all bounds." This observation motivates us to extend Definition 1 as follows:

DEFINITION 3
If f approaches a single number L as x approaches ∞, we write

$$\lim_{x \to \infty} f(x) = L$$

and say that f has *limit L at infinity*. If f approaches a single number L as x approaches $-\infty$, we write

$$\lim_{x \to -\infty} f(x) = L$$

and say that f has *limit L at minus* infinity.

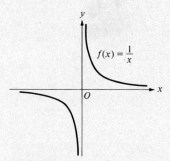

Figure 2.43
The graph of $y = 1/x$.

x	$1/x$
1	1
10	0.1
100	0.01
1000	0.001
10000	0.0001

The basic properties of limits, with the exception of Property 2, continue to hold when x_0 is replaced by ∞ or $-\infty$.

Example 7

Find the limit of $f(x) = \dfrac{x-1}{x+1}$ as $x \to \infty$ and as $x \to -\infty$.

Solution

When x is large, both $x - 1$ and $x + 1$ are large, but looking at the quotient of two large numbers does not tell us much. We take a hint from the fact that $1/x \to 0$ as $x \to \infty$: If we multiply both numerator and denominator of $f(x)$ by $1/x$, we get

$$f(x) = \frac{1 - (1/x)}{1 + (1/x)}$$

This time we get a meaningful answer, since $1 - 1/x \to 1$ and $1 + 1/x \to 1$ as $x \to \infty$. Hence

$$\lim_{x \to \infty} \frac{x - 1}{x + 1} = 1$$

Since $1/x \to 0$ as $x \to -\infty$, we see also that

$$\lim_{x \to -\infty} \frac{x - 1}{x + 1} = 1$$

Example 8

Find the limit of $g(x) = (2x^2 + 5)/(x^3 + 1)$ as $x \to \infty$ and as $x \to -\infty$.

Solution

We begin with the general observation that

$$\lim_{x \to \infty} \frac{1}{x^n} = 0 \quad \text{and} \quad \lim_{x \to -\infty} \frac{1}{x^n} = 0 \quad \text{for all integers } n > 0$$

Trying the procedure of the preceding problem, we write

$$g(x) = \frac{(1/x)(2x^2 + 5)}{(1/x)(x^3 + 1)} = \frac{2x + (5/x)}{x^2 + (1/x)}$$

This leads to no conclusion, since we still have the quotient of two large numbers when x is large. We realize that we should have used $1/x^3$ instead of $1/x$:

$$g(x) = \frac{(1/x^3)(2x^2 + 5)}{(1/x^3)(x^3 + 1)} = \frac{(2/x) + (5/x^3)}{1 + (1/x^3)}$$

Since $(2/x) + (5/x^3) \to 0$ as $x \to \infty$ whereas $1 + 1/x^3 \to 1$ as $x \to \infty$, we conclude that

$$\lim_{x \to \infty} \frac{2x^2 + 5}{x^3 + 1} = 0$$

The same argument can also be used to show that

$$\lim_{x \to -\infty} \frac{2x^2 + 5}{x^3 + 1} = 0$$

To give you some idea of what a rigorous treatment of limits at infinity entails, we conclude this section with the following formal definition:

DEFINITION 4
We write

$$\lim_{x \to \infty} f(x) = L$$

if, for any error $\epsilon > 0$, there is a number M such that

$$|f(x) - L| < \epsilon \quad \text{whenever } x > M$$

The definition for the case $x \to -\infty$ is similar.

EXERCISES

Evaluate the following limits.

1. $\lim_{x \to 7} (x^2 - 7x + 1) =$
2. $\lim_{y \to -2} \dfrac{-y - 1}{-y + 2} =$
3. $\lim_{x \to -3} 5 =$
4. $\lim_{x \to x_0} (x^3 - x^2 + 1) =$
5. $\lim_{t \to 0} 3t =$
6. $\lim_{x \to -2} (x - 1)(-x - 1) =$
7. $\lim_{t \to -8} \sqrt[3]{t} =$
8. $\lim_{t \to t_0^2} \sqrt{t} =$
9. $\lim_{x \to \frac{1}{2}} \dfrac{x^2 - (1/2^2)}{x - \frac{1}{2}} =$
10. $\lim_{\Delta x \to 0} \dfrac{(\frac{1}{2} + \Delta x)^2 - (1/2^2)}{\Delta x} =$
11. $\lim_{\Delta x \to 0} \sqrt{\Delta x + 4} =$
12. $\lim_{h \to 0} \dfrac{2h}{h} =$
13. $\lim_{x \to 3} \dfrac{x - 3}{x^2 - 3^2} =$
14. $\lim_{x \to 1} \dfrac{x^2 - 2x + 1}{x - 1} =$
15. $\lim_{u \to 0} \dfrac{4u^3 - 2u^2}{4u^2 + 2u^3} =$
16. $\lim_{x \to 0} \dfrac{x}{(1/x)} =$
17. $\lim_{x \to 0} \dfrac{(1/x^2) - (3/x) + 5}{(1/x^3) - 7} =$
18. $\lim_{x \to -2} (x + 1)^{19} =$
19. $\lim_{x \to 1} \dfrac{x}{|x|} =$
20. $\lim_{x \to -1} \dfrac{x}{|x|} =$
21. $\lim_{\Delta x \to 0} \dfrac{2 + \Delta x}{2 - \Delta x} =$

Evaluate each of the following limits in which x increases beyond all bounds.

22. $\lim_{x\to\infty} \dfrac{1}{-x} =$

23. $\lim_{x\to-\infty} \dfrac{1}{x} =$

24. $\lim_{x\to\infty} \sqrt{x}\cdot \dfrac{x}{x^2+1} =$

25. $\lim_{x\to\infty} \dfrac{\sqrt{x+1}}{\sqrt{x-1}} =$

26. $\lim_{x\to\infty} \dfrac{x+(1/x)}{-x+(1/x)} =$

27. $\lim_{x\to-\infty} \dfrac{x+(1/x)}{-x+(1/x)} =$

28. $\lim_{x\to\infty} \left(1+\dfrac{1}{x}\right)\left(1-\dfrac{1}{x}\right) =$

29. $\lim_{x\to\infty} \dfrac{x^2+x+1}{x^3+x^2+x+1} =$

30. $\lim_{x\to\infty} \dfrac{(1/x^2)}{(1/x)+1} =$

31. $\lim_{x\to\infty} \dfrac{[(1/2)+(1/x)]^2 - (1/2^2)}{(1/x)} =$

32. $\lim_{x\to\infty} \dfrac{x}{|x|} =$

33. $\lim_{x\to-\infty} \dfrac{x}{|x|} =$

QUIZ 1

1. Find the domain and range of the function $f(x) = 1/(x^2-1)$.

2. Let $f(x) = 1/x$. Sketch the graphs of $y = f(x)$ and $y = f(x-1)$.

3. Decide which of the following functions is even, odd, or neither even nor odd:

    (a) $f(x) = \dfrac{x^2-1}{x^2+1}$

    (b) $g(x) = x\cdot |x|$

    (c) $h(x) = \dfrac{x-1}{x+1}$

4. Find the point of intersection of the two lines
$$y = 2x+2$$
$$y = -\tfrac{1}{4}x + \tfrac{5}{4}$$

5. Find the line that passes through the point $P(1, 1)$ and is parallel to the line $y = 5x + 3$.

6. Find the tangent to the curve $y = x^2$ at $P(-2, 4)$.

7. Evaluate the following limits:

    (a) $\lim_{x\to 2} \dfrac{x^4-16}{x-2}$

    (b) $\lim_{\Delta x\to 0} \dfrac{\sqrt{9+\Delta x}-\sqrt{9}}{\Delta x}$

    (c) $\lim_{t\to\infty} \dfrac{2t+1}{3t-1}$

QUIZ 2

1. Find the domain and range of the function
$$f(x) = \sqrt{\frac{x^2 - 1}{x^2 + 1}}$$

2. Let $f(x) = x(x+2)$. Sketch the graphs of $y = f(x)$ and $y = f(x-2)$.

3. Decide which of the following functions is even, odd, or neither even nor odd:

   (a) $f(x) = x$

   (b) $g(x) = x^2 - x^4$

   (c) $h(x) = \dfrac{x^3}{x^2 + 1}$

4. Find the line that passes through the point $P(5, 0)$ and is perpendicular to the line $2x + 3y = 5$.

5. Find the tangent to the curve $y = x^3$ at $P(-1, -1)$.

6. Evaluate the following limits:

$$\lim_{x \to 2} \frac{x^4 - 16}{x - 2}$$

$$\lim_{h \to 0} \frac{(1-h)^4 - 1}{h}$$

$$\lim_{x \to -\infty} \frac{1 + x + 2x^2}{1 - x - 3x^2}$$

CHAPTER 3

DERIVATIVES

3.1 INTRODUCTION In discussing the tangent to a curve $y = f(x)$ (see Section 2.5), we found that its slope m at a point $(x_0, f(x_0))$ could be obtained as a limit

$$m = \lim_{\Delta x \to 0} \frac{f(x_0 + \Delta x) - f(x_0)}{\Delta x} \tag{1}$$

This limit appears in many parts of calculus, and it has many important interpretations of which the slope is but one. The limit is called the derivative of f at x_0.

> **DEFINITION 1**
> The function f has a *derivative* at x_0 if the limit
> $$\lim_{\Delta x \to 0} \frac{f(x_0 + \Delta x) - f(x_0)}{\Delta x}$$
> exists. A general symbol for the derivative is $f'(x_0)$, read "f prime at x_0." If $f'(x_0)$ exists, then f is said to be *differentiable* at x_0.

In this chapter we shall study the properties of derivatives and general techniques for calculating them. In Chapter 4 the concept of derivative is applied to graphing (curve plotting) and velocity. Many other applications will be discussed in Chapter 5.

Example 1

Find the derivative of $f(x) = x^2$ at x_0. What is the derivative at $-\frac{7}{2}$?

Solution
We know the answer from Section 2.4; when $f(x) = x^2$, we have

$$\frac{f(x_0 + \Delta x) - f(x_0)}{\Delta x} = \frac{(x_0 + \Delta x)^2 - x_0^2}{\Delta x} = 2x_0 + \Delta x$$

and hence

$$\lim_{\Delta x \to 0} \frac{f(x_0 + \Delta x) - f(x_0)}{\Delta x} = \lim_{\Delta x \to 0} (2x_0 + \Delta x) = 2x_0$$

Thus,

$$f'(x_0) = 2x_0 \tag{2}$$

and when $x_0 = -\frac{7}{2}$, we have

$$f'(-\tfrac{7}{2}) = -7$$

Our first general result is given in the following theorem.

THEOREM 1
If $f(x) = x^r$, then $f'(x_0) = rx_0^{r-1}$ for any number x_0.

In Section 3.4 we shall be able to give a very simple proof of this theorem for the case when r is a rational number. After defining general exponents (see Chapter 7), we shall be able to prove the theorem for any real number r. We observe that for $r = 2$ we arrive at equation (2) above.

Example 2

Find the derivative of $f(x) = x^r$ at x_0 when $r = \frac{1}{2}, -\frac{1}{2}, -1$.

Solution
The derivatives are easily written down.

$$r = \frac{1}{2}: \quad f'(x_0) = \frac{1}{2}x_0^{(1/2)-1} = \frac{1}{2}x_0^{-1/2} = \frac{1}{2\sqrt{x_0}}$$

$$r = -\frac{1}{2}: \quad f'(x_0) = -\frac{1}{2}x_0^{(-1/2)-1} = -\frac{1}{2}x_0^{-3/2} = -\frac{1}{2\sqrt{x_0^3}}$$

$$r = -1: \quad f'(x_0) = -1 x_0^{-1-1} = -x_0^{-2} = -\frac{1}{x_0^2}$$

Note that the formula for $r = -1$ is consistent with Exercise 7 of Section 2.5.

The derivatives of $f(x) = x^r$ are of particular importance when $r = 0$ and $r = 1$. For $r = 1$, Theorem 1 gives the formula

$$f'(x_0) = 1 \cdot x_0^{1-1} = 1 \cdot x_0^0 = 1 \cdot 1 = 1$$

and thus:

> If $f(x) = x$, then $f'(x_0) = 1$ for all values of x_0. (3)

When $r = 0$, then $f(x) = x^0 = 1$, but this is only a special function of the form $f(x) = c$, where c is any constant. Such a function is called a *constant function*. We prove the following theorem.

THEOREM 2
If $f(x) = c$ for any constant c, then $f'(x_0) = 0$ for all values of x_0.

Proof
If $f(x) = c$ for all values of x, then $f(x_0) = c$ and $f(x_0 + \Delta x) = c$. Hence, for $\Delta x \neq 0$,

$$\frac{f(x_0 + \Delta x) - f(x_0)}{\Delta x} = \frac{c - c}{\Delta x} = \frac{0}{\Delta x} = 0$$

and consequently,

$$\lim_{\Delta x \to 0} \frac{f(x_0 + \Delta x) - f(x_0)}{\Delta x} = \lim_{\Delta x \to 0} 0 = 0$$

Thus, $f'(x_0) = 0$, as was to be shown.

AN IMPORTANT REMARK AND NEW NOTATION
For the case where $f(x) = x^r$, we have seen that $f'(x_0) = r x_0^{r-1}$ no matter what x_0 is. This being the case, there is no need to refer to the derivative at a particular point x_0, and we write instead

$$f'(x) = r x^{r-1}$$

We now see that the derivative of f is simply a new function, f', whose value at x_0 is the derivative of f at x_0. This is also the case with arbitrary functions f, and consequently we shall adopt the notation $f'(x)$ without further comment.

Various symbols are used to designate derivatives. As the course progresses, you will see that no single symbol is best. One of the most commonly used symbols for the derivative of $y = f(x)$ is dy/dx.

To explain this notation, recall that when
$$y = f(x)$$
then
$$\Delta y = f(x + \Delta x) - f(x)$$
and
$$f'(x) = \lim_{\Delta x \to 0} \frac{\Delta y}{\Delta x}$$

We thus define

$$\frac{dy}{dx} = \lim_{\Delta x \to 0} \frac{\Delta y}{\Delta x}$$

NOTATION FOR THE DERIVATIVE

$y' = f'(x)$ y' is read "y prime"

$\dfrac{dy}{dx} = f'(x)$ $\dfrac{dy}{dx}$ is read "d–y d–x"

$\dfrac{df}{dx} = f'(x)$ $f'(x)$ is read "f prime at x"

$\dfrac{d}{dx}[f(x)] = f'(x)$

In particular, we have

$$\frac{d}{dx}(x^r) = rx^{r-1}$$

$$\frac{d}{dx}(x) = 1$$

$$\frac{d}{dx}(c) = 0$$

Example 3

Find dy/dx when $y = \sqrt{x}$.

Solution
Since $\sqrt{x} = x^{1/2}$, we have

$$\frac{dy}{dx} = \frac{1}{2}x^{-1/2} \quad \text{(see Example 2)}$$

In the next two examples, we illustrate finding derivatives when variables other than x and y are used.

Example 4

If $s = t^{7/4}$, find ds/dt.

Solution

$$\frac{ds}{dt} = \frac{7}{4}t^{(7/4)-1} = \frac{7}{4}t^{3/4}$$

Example 5

Find $\frac{d}{dy}(y^2)$ and $\frac{d}{du}(u^2)$.

Solution

$$\frac{d}{dy}(y^2) = 2y \quad \text{and} \quad \frac{d}{du}(u^2) = 2u$$

EXERCISES

Find the derivatives in Exercises 1–10.

1. $\dfrac{d}{dx}(x^{10})$ 2. $\dfrac{d}{dx}\left(\dfrac{1}{x}\right)$

3. $\dfrac{d}{dx}(\sqrt[3]{x})$ 4. $\dfrac{d}{dt}\left(\dfrac{1}{t^3}\right)$

5. $\dfrac{d}{dz}(z^{-5})$ 6. $\dfrac{d}{dx}(1)$

7. $\dfrac{d}{du}(u^{1.1})$ 8. $\dfrac{d}{dx}(x^{100})$

9. $\dfrac{d}{dx}(x^{-100})$ 10. $\dfrac{d}{dx}(4^2)$

Evaluate the following derivatives at the given values.

11. $f(x) = 25$ $f'(5) =$ $f'(-5) =$

12. $f(t) = \dfrac{1}{t}$ $f'(2) =$ $f'(\tfrac{1}{2}) =$

13. $g(x) = \sqrt[4]{x}$ $g'(1) =$ $g'(5) =$

14. $f(x) = x^{-5}$ $f'(\tfrac{1}{2}) =$ $f'(-\tfrac{1}{2}) =$

15. $f(x) = x^{11}$ $f'(0) =$ $f'(-\tfrac{1}{2}) =$

16. What is the geometric interpretation of formula (3)?

17. Can you give a geometric interpretation for Theorem 2?

18. Find the tangent to the curve $y = x^3$ at $(2, 8)$.

19. Find the tangent to the curve $y = \sqrt{x}$ at $(2, \sqrt{2})$.

20. Find the point (x_0, y_0) on the curve $y = x^{-2}$ at which the line $y = -2x + 3$ is tangent to the curve.

3.2 DERIVATIVES OF POLYNOMIALS

In the preceding section we learned how to find the derivatives of $u = x^4$ and $v = x^2$; but what about the derivative of $w = u + v = x^4 + x^2$? The answer to this is contained in the following theorem.

THEOREM 1
If $u = f(x)$ and $v = g(x)$ are differentiable functions, then

$$\frac{d}{dx}(u + v) = \frac{du}{dx} + \frac{dv}{dx} \tag{1}$$

$$\frac{d}{dx}(u - v) = \frac{du}{dx} - \frac{dv}{dx} \tag{2}$$

Loosely speaking, this theorem states that the derivative of a sum is the sum of the derivatives, and the derivative of a difference is the difference of the derivatives.

Proof
To prove formula (1), where we are finding the derivative of $u + v = f(x) + g(x)$, put

$$h(x) = f(x) + g(x)$$

Then

$$h(x + \Delta x) - h(x) = [f(x + \Delta x) + g(x + \Delta x)] - [f(x) + g(x)]$$
$$= [f(x + \Delta x) - f(x)] + [g(x + \Delta x) - g(x)]$$

and simple arithmetic gives

$$\frac{h(x + \Delta x) - h(x)}{\Delta x} = \frac{f(x + \Delta x) - f(x)}{\Delta x} + \frac{g(x + \Delta x) - g(x)}{\Delta x}$$

Since the functions f and g are differentiable, we have

$$\lim_{\Delta x \to 0} \frac{f(x + \Delta x) - f(x)}{\Delta x} = f'(x) = \frac{du}{dx}$$

$$\lim_{\Delta x \to 0} \frac{g(x + \Delta x) - g(x)}{\Delta x} = g'(x) = \frac{dv}{dx}$$

and hence, by Theorem 1 of Section 2.6,

$$\lim_{\Delta x \to 0} \frac{h(x + \Delta x) - h(x)}{\Delta x} = \frac{du}{dx} + \frac{dv}{dx} \qquad (3)$$

Since the limit in (3) also equals $h'(x) = (d/dx)(u + v)$, we get formula (1). Formula (2) is established in the same way, using the function $k(x) = f(x) - g(x)$ instead of $h(x)$.

Example 1

If $u = x^4$ and $v = x^2$, then

$$\frac{d}{dx}(u + v) = 4x^3 + 2x$$

$$\frac{d}{dx}(u - v) = 4x^3 - 2x$$

Formulas similar to (1) and (2) can be established for any number of functions. Thus, if

$$u_1 = f_1(x), u_2 = f_2(x), \ldots, u_n = f_n(x)$$

are differentiable functions, then

$$\frac{d}{dx}(u_1 + u_2 + \cdots + u_n) = \frac{du_1}{dx} + \frac{du_2}{dx} + \cdots + \frac{du_n}{dx}$$

Example 2

Let $u_1 = 1, u_2 = x, u_3 = x^2, \ldots, u_{n+1} = x^n$. Then

$$\frac{d}{dx}(u_1 + u_2 + \cdots + u_{n+1}) = 0 + 1 + 2x + \cdots + nx^{n-1}$$

THEOREM 2

If $u = f(x)$ is a differentiable function and c is any constant, then

$$\frac{d}{dx}(cu) = c\frac{du}{dx}$$

Proof

This property is a consequence of the fact that

$$\Delta(cu) = cu(x_0 + \Delta x) - cu(x_0) = c[u(x_0 + \Delta x) - u(x_0)] = c\,\Delta u$$

because it shows that

$$\frac{\Delta(cu)}{\Delta x} = c\,\frac{\Delta u}{\Delta x}$$

Now,

$$\lim_{\Delta x \to 0} \frac{\Delta(cu)}{\Delta x} = \frac{d}{dx}(cu) \quad \text{and} \quad \lim_{\Delta x \to 0} c\,\frac{\Delta u}{\Delta x} = c \lim_{\Delta x \to 0} \frac{\Delta u}{\Delta x} = c\,\frac{du}{dx}$$

and the desired relation follows.

Example 3

$$\frac{d}{dx}(x - 2x^2 + 3x^3) = \frac{d}{dx}(x) - \frac{d}{dx}(2x^2) + \frac{d}{dx}(3x^3) \qquad \text{(by Theorem 1)}$$

$$= 1 - 4x + 9x^2 \qquad \text{(by Theorem 2)}$$

Combining the results of this and the last section, we shall derive a general formula for differentiating polynomials. We recall from high-school algebra that a *polynomial* in x is a function of the form

$$y = a_0 + a_1 x + a_2 x^2 + \cdots + a_n x^n$$

where $a_0, a_1, a_2, \ldots, a_n$ are given numbers, positive, negative, or zero; these numbers are called *coefficients*. The number n is a non-negative integer, and it will be assumed that $a_n \neq 0$. A polynomial with $a_n \neq 0$ is said to be of *degree n*. For example,

$$y = 1 - x^3 + \pi x^4$$

is a polynomial of degree 4; it can be written in the form

$$y = a_0 + a_1 x + a_2 x^2 + a_3 x^3 + a_4 x^4$$

with $a_0 = 1$, $a_1 = a_2 = 0$, $a_3 = -1$, and $a_4 = \pi$.

DERIVATIVES OF POLYNOMIALS

If

$$y = a_0 + a_1 x + a_2 x^2 + \cdots + a_n x^n$$

then

$$\frac{dy}{dx} = a_1 + 2a_2 x + \cdots + n a_n x^{n-1}$$

Thus, the derivative of the polynomial $y = 1 - x^3 + \pi x^4$ is

$$\frac{dy}{dx} = -3x^2 + 4\pi x^3$$

EXERCISES

Find the derivatives of the functions in Exercises 1–15.

64 DERIVATIVES

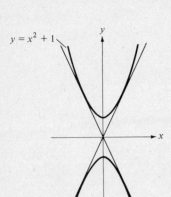

Figure 3.1

1. $y = 2$
2. $y = 2x^2$
3. $y = x^4 + 2x - 6$
4. $y = -x^5 + x^3$
5. $y = \dfrac{x}{a+b} - \dfrac{x^2}{a-b}$
6. $y = \dfrac{x}{c} + \dfrac{c}{x}$
7. $y = ax^2 + \dfrac{1}{a}x$
8. $y = \dfrac{x^2 + 2}{x}$
9. $s = \sqrt{2t}$
10. $u = (5x)^{10}$
11. $s = u + \dfrac{1}{u}$
12. $y = x^2 - x^2$
13. $y = ax^3 + bx^2 + cx + d$
14. $y = x^4 - x^3 + x^2 - x + 1$
15. $y = \dfrac{x^4 - x^3 + x^2 - x + 1}{x^4}$

16. Find the points on the curve $y = \tfrac{1}{3}x^3 - x - 1$ at which the tangent is perpendicular to the line $y = -x$.

17. Consider the curves $y = x^2 + 1$ and $y = -x^2 - 1$ (Figure 3.1). There are two lines that are tangent to both curves. Find them.

18. Consider the polynomials

$$p(x) = a_0 + a_1 x + a_2 x^2 + a_3 x^3 + a_4 x^4$$
$$p'(x) = -2 + 3x - 4x^2 + 12x^3$$

For what coefficients $a_0, a_1, \ldots, a_4$ is $p'(x)$ the derivative of $p(x)$?

19. Find the two points on the curve $y = \tfrac{1}{3}x^3 - x + 5$ at which it has a horizontal tangent.

3.3 DERIVATIVES OF PRODUCTS AND QUOTIENTS

In this section we shall learn how to find the derivatives of the following types of functions, where $u = f(x)$ and $v = g(x)$ are differentiable functions.

products: $uv = f(x) \cdot g(x)$

reciprocals: $\dfrac{1}{v} = \dfrac{1}{g(x)}$

quotients: $\dfrac{u}{v} = \dfrac{f(x)}{g(x)}$

A summary of differentiation formulas is given at the end of Section 3.4.

3.3 DERIVATIVES OF PRODUCTS AND QUOTIENTS

> **THEOREM 1. PRODUCT RULE**
> If $u = f(x)$ and $v = g(x)$ are differentiable functions, then
> $$\frac{d}{dx}(uv) = u\frac{dv}{dx} + v\frac{du}{dx} \tag{1}$$

We precede the proof with an example.

Example 1

Find
$$\frac{d}{dx}(2x-1)(3x+2)$$

$(2x-1)(3) + (3x+2)(2)$

Solution
Putting
$$u = 2x - 1 \quad \text{and} \quad v = 3x + 2$$
we have
$$\frac{du}{dx} = 2 \quad \text{and} \quad \frac{dv}{dx} = 3$$

Using formula (1) gives
$$\frac{d}{dx}(2x-1)(3x+2) = (2x-1)\frac{d}{dx}(3x+2) + (3x+2)\frac{d}{dx}(2x-1)$$
$$= (2x-1)3 + (3x+2)2 = (6x-3) + (6x+4)$$
$$= 12x + 1$$

One can check this result by multiplying out $(2x - 1) \cdot (3x + 2)$ and differentiating it.

Proof of Theorem 1
In proving the product rule we use the identity
$$AB - ab = a(B - b) + b(A - a) + (A - a)(B - b)$$
which can be interpreted in terms of areas of rectangles (see Figure 3.2). The validity of this equation is seen by expanding and simplifying the right side.

Let $F(x) = f(x)g(x)$ and put
$$\Delta u = f(x_0 + \Delta x) - f(x_0)$$
$$\Delta v = g(x_0 + \Delta x) - g(x_0)$$

To use the above identity, let
$$A = f(x_0 + \Delta x) \qquad a = f(x_0)$$
$$B = g(x_0 + \Delta x) \qquad b = g(x_0)$$

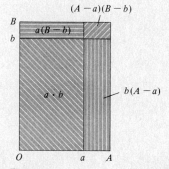

Figure 3.2

We note that
$$A - a = \Delta u \quad \text{and} \quad B - b = \Delta v$$
and hence
$$F(x_0 + \Delta x) - F(x_0) = f(x_0 + \Delta x)g(x_0 + \Delta x) - f(x_0)g(x_0)$$
$$= f(x_0)\Delta v + g(x_0)\Delta u + \Delta u \Delta v$$

Dividing by Δx and writing $\Delta v = (\Delta v / \Delta x)\Delta x$ gives
$$\frac{F(x_0 + \Delta x) - F(x_0)}{\Delta x} = f(x_0)\frac{\Delta v}{\Delta x} + g(x_0)\frac{\Delta u}{\Delta x} + \frac{\Delta u}{\Delta x}\frac{\Delta v}{\Delta x}\Delta x$$

But
$$\lim_{\Delta x \to 0} \frac{\Delta v}{\Delta x} = g'(x_0) = \frac{dv}{dx}$$
and
$$\lim_{\Delta x \to 0} \frac{\Delta u}{\Delta x} = f'(x_0) = \frac{du}{dx}$$

According to Theorem 1(6) in Section 2.6, therefore,
$$\lim_{\Delta x \to 0} \frac{\Delta u}{\Delta x}\frac{\Delta v}{\Delta x}\Delta x = f'(x_0)g'(x_0)0 = 0$$
and from the same theorem it now follows that
$$\lim_{\Delta x \to 0} \frac{F(x_0 + \Delta x) - F(x_0)}{\Delta x} = f(x_0)\frac{dv}{dx} + g(x_0)\frac{du}{dx} = u\frac{dv}{dx} + v\frac{du}{dx}$$

The left side gives, by definition, $d(uv)/dx$ and this proves formula (1).

THEOREM 2. RECIPROCAL RULE
If $v = g(x)$ is differentiable, then
$$\frac{d}{dx}\left(\frac{1}{v}\right) = -\frac{1}{v^2} \cdot \frac{dv}{dx} \tag{2}$$

Again, we illustrate this theorem with an example before proving it.

Example 2
Find
$$\frac{d}{dx}\left(\frac{1}{x^2 - 3x + 1}\right) = -\frac{1}{(x^2 - 3x + 1)^2} \cdot (2x - 3)$$

Solution
Putting
$$v = x^2 - 3x + 1$$
gives
$$\frac{dv}{dx} = 2x - 3$$
and hence, by formula (2),
$$\frac{d}{dx}\left(\frac{1}{x^2 - 3x + 1}\right) = -\frac{1}{(x^2 - 3x + 1)^2} \frac{d}{dx}(x^2 - 3x + 1)$$
$$= -\frac{1}{(x^2 - 3x + 1)^2}(2x - 3) = -\frac{2x - 3}{(x^2 - 3x + 1)^2}$$

Proof of Theorem 2
Put
$$y = \frac{1}{g(x)} \qquad [g(x) \neq 0]$$
$$y + \Delta y = \frac{1}{g(x + \Delta x)} \qquad [g(x + \Delta x) \neq 0]$$
Then
$$\Delta y = \frac{1}{g(x + \Delta x)} - \frac{1}{g(x)} = \frac{g(x) - g(x + \Delta x)}{g(x + \Delta x) \cdot g(x)}$$
$$= -[g(x + \Delta x) - g(x)] \frac{1}{g(x + \Delta x) \cdot g(x)}$$
Hence
$$\frac{\Delta y}{\Delta x} = -\frac{g(x + \Delta x) - g(x)}{\Delta x} \cdot \frac{1}{g(x + \Delta x) \cdot g(x)}$$
Since
$$\lim_{\Delta x \to 0}\left[-\frac{g(x + \Delta x) - g(x)}{\Delta x}\right] = -\lim_{\Delta x \to 0} \frac{g(x + \Delta x) - g(x)}{\Delta x} = -g'(x)$$
$$= -\frac{dv}{dx}$$
and
$$\lim_{\Delta x \to 0} [g(x + \Delta x) \cdot g(x)] = g(x) \lim_{\Delta x \to 0} g(x + \Delta x) = g(x) \cdot g(x)$$
$$= [g(x)]^2 = v^2$$
we conclude from Section 2.6 that
$$\frac{dy}{dx} = \lim_{\Delta x \to 0} \frac{\Delta y}{\Delta x} = -\frac{dv}{dx} \cdot \frac{1}{v^2}$$

The fact that

$$y = \frac{1}{g(x)} = \frac{1}{v}$$

tells us that

$$\frac{dy}{dx} = \frac{d}{dx}\left(\frac{1}{v}\right)$$

and this now gives us the desired formula (2).

THEOREM 3. QUOTIENT RULE
If $u = f(x)$ and $v = g(x)$ are differentiable functions, then

$$\frac{d}{dx}\left(\frac{u}{v}\right) = \frac{v\dfrac{du}{dx} - u\dfrac{dv}{dx}}{v^2} \qquad (3)$$

Observe that when $u = f(x) = 1$, then $du/dx = f'(x) = 0$ and formula (3) becomes formula (2).

Example 3
Find

$$\frac{d}{dx}\left(\frac{2x^2 - x + 5}{3x^2 + 5x - 7}\right)$$

[handwritten: $\dfrac{(3x^2+5x-7)(4x-1) - (2x^2-x+5)}{(3x^2+5x-7)^2}$]

Solution
Put

$$u = 2x^2 - x + 5 \quad \text{and} \quad v = 3x^2 + 5x - 7$$

Then

$$\frac{du}{dx} = 4x - 1 \quad \text{and} \quad \frac{dv}{dx} = 6x + 5$$

Using formula (3), we get

$$\frac{d}{dx}\left(\frac{2x^2 - x + 5}{3x^2 + 5x - 7}\right) = \frac{(3x^2 + 5x - 7)(4x - 1) - (2x^2 - x + 5)(6x + 5)}{(3x^2 + 5x - 7)^2}$$

$$= \frac{13x^2 - 58x - 18}{(3x^2 + 5x - 7)^2}$$

Proof of Theorem 3
Using the product rule for derivatives gives

$$\frac{d}{dx}\left(\frac{u}{v}\right) = \frac{d}{dx}\left(u\,\frac{1}{v}\right) = \frac{1}{v}\frac{du}{dx} + u\frac{d}{dx}\left(\frac{1}{v}\right)$$

By the reciprocal rule (2),

$$\frac{d}{dx}\left(\frac{1}{v}\right) = -\frac{1}{v^2}\frac{dv}{dx}$$

and hence

$$\frac{d}{dx}\left(\frac{u}{v}\right) = \frac{1}{v}\frac{du}{dx} + u\left(-\frac{1}{v^2}\frac{dv}{dx}\right) = \frac{1}{v^2}\left(v\frac{du}{dx}\right) - \frac{1}{v^2}\left(u\frac{dv}{dx}\right)$$

$$= \frac{v(du/dx) - u(dv/dx)}{v^2}$$

EXERCISES

Find the derivatives of the functions in Exercises 1–15.

1. $y = \dfrac{x^2 + 2}{x}$
2. $y = \dfrac{x^2 - 1}{x^2 + 1}$
3. $y = (x^2 - x - 1)(x^2 + x + 1)$
4. $s = t^2(t^3 - t^2 + t - 1)$
5. $y = x(x - 1)(x - 2)$
6. $y = x(1 - x^2)(1 - x^3)$
7. $w = \dfrac{az + b}{cz + d}$
8. $y = \dfrac{1}{x^2 - 1}$
9. $y = \dfrac{1 - x}{1 + x}$
10. $y = \left(1 - \dfrac{1}{x}\right)\left(1 + \dfrac{1}{1 - x}\right)$
11. $u = x(x + 1)$
12. $y = \dfrac{x - 1}{x + 1}$
13. $y = (x^{1/3} + x^{-1/3})(x^3 - 1)$
14. $y = \dfrac{u^{1/2} + u^{-1/2}}{u^{1/2} - u^{-1/2}}$
15. $y = \dfrac{1}{1 + x^{-1/3}}$

16. Find a formula for dy/dx when $y = xf(x)$.

17. Find a formula for dy/dx when $y = x/g(x)$.

18. Use the product rule for derivatives to obtain the formula

$$\frac{d}{dx}(uvw) = vw\frac{du}{dx} + uw\frac{dv}{dx} + uv\frac{dw}{dx}$$

19. Use Exercise 18 to find a formula for $d(uvwz)/dx$.

20. Find the points on the curve $y = (x - 1)(x - 2)(x - 3)$ at which it has a horizontal tangent.

21. Find the tangent to the curve $y = (x - 1)/(x + 1)$ at the point $(2, \tfrac{1}{3})$.

3.4 DERIVATIVES OF POWERS

The last formula for derivatives to be discussed in this chapter is given in the following theorem.

> **THEOREM 1. POWER RULE**
> If $v = g(x)$ is a differentiable function, then
> $$\frac{d}{dx}(v^r) = rv^{r-1}\frac{dv}{dx} \qquad (1)$$

Following several applications of the power rule, we shall give a proof for the case when r is a rational number. A general proof that includes all numbers r must await our treatment of the exponential function in Section 7.5.

Example 1

Find

$$\frac{d}{dx}(\sqrt{x^2 - 1})$$

Solution
Put

$$v = x^2 - 1 \quad \text{and} \quad r = \tfrac{1}{2}$$

Then

$$\frac{dv}{dx} = 2x$$

and formula (1) gives

$$\frac{d}{dx}(\sqrt{x^2 - 1}) = \frac{d}{dx}[(x^2 - 1)^{1/2}] = \tfrac{1}{2}(x^2 - 1)^{-1/2} 2x$$

$$= \frac{x}{\sqrt{x^2 - 1}}$$

Example 2

Find

$$\frac{d}{dx}(2x^2 + 5x - 3)^{100}$$

Solution
We use $v = 2x^2 + 5x - 3$ and $r = 100$. Since

$$\frac{dv}{dx} = 4x + 5$$

Theorem 1 gives

$$\frac{d}{dx}(2x^2 + 5x - 3)^{100} = 100\,(2x^2 + 5x - 3)^{99}(4x + 5)$$

Example 3

Find dy/dx when

$$y = \left(\frac{x-1}{x+1}\right)^{3/2}$$

Solution
Put $v = (x-1)/(x+1)$. Then $y = v^{3/2}$, and using Theorem 1 gives

$$\frac{dy}{dx} = \frac{3}{2}v^{1/2}\frac{dv}{dx} = \frac{3}{2}\left(\frac{x-1}{x+1}\right)^{1/2}\frac{dv}{dx}$$

Using the quotient rule, which is contained in our list of differentiation formulas at the end of this section,

$$\frac{d}{dx}\left(\frac{x-1}{x+1}\right) = \frac{(x+1)[(d/dx)(x-1) - (x-1)(d/dx)(x+1)]}{(x+1)^2}$$

$$= \frac{(x+1) - (x-1)}{(x+1)^2} = \frac{2}{(x+1)^2}$$

and hence

$$\frac{dy}{dx} = \frac{3}{2}\left(\frac{x-1}{x+1}\right)^{1/2} \cdot \frac{2}{(x+1)^2} = \frac{3(x-1)^{1/2}}{(x+1)^{5/2}}$$

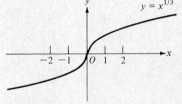

Figure 3.3
At 0 the tangent to $y = x^{1/3}$ coincides with the y axis. A vertical line, however, has no slope.

Example 4

Consider the function $f(x) = x^{1/3}$ (see Figure 3.3). This function is defined for all x, and we shall attempt to compute the derivative of $f(x)$ at 0. To do this, we proceed via the definition:

$$f(0) = 0^{1/3} = 0$$

$$f(0 + \Delta x) = (0 + \Delta x)^{1/3} = (\Delta x)^{1/3}$$

$$\frac{f(0 + \Delta x) - f(0)}{\Delta x} = \frac{(\Delta x)^{1/3}}{\Delta x} = (\Delta x)^{-2/3}$$

But $(\Delta x)^{-2/3} \to \infty$ as $\Delta x \to 0$ and hence the derivative does not exist at 0. In this connection see also Example 3 in Section 2.5.

Example 5

Find the tangent to $f(x) = x^{1/3}$ at $P(8, 2)$.

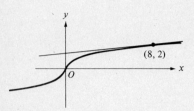

Figure 3.4
The tangent to $y = x^{1/3}$ at $P(8, 2)$.

Solution
Recall that any line passing through $P(8, 2)$ can be written as

$$y - 2 = m(x - 8)$$

The slope m at $x = 8$ is

$$m = f'(8) = \frac{1}{3} 8^{-2/3} = \frac{1}{12}$$

Hence, the tangent is

$$y - 2 = \frac{1}{12}(x - 8)$$

(see Figure 3.4).

Proof of Theorem 1
The proof of this theorem is divided into three steps.

Step 1: r is a positive integer
We want to establish formula (1) for $r = 1, 2, 3, \ldots$. This is done with the help of *mathematical induction*. This is a general method of proof that is used when a formula is to be established for all positive integers r. The method consists of two stages:

First stage: The formula is proved for $r = 1$.
Second stage: The formula is assumed to be true for an arbitrary integer n and, with this assumption, it is proved for $n + 1$.

First stage carried out: When $r = 1$, formula (1) becomes

$$\frac{dv}{dx} = 1 \cdot v^{1-1} \frac{dv}{dx} = v^0 \frac{dv}{dx} = \frac{dv}{dx}$$

Second stage carried out: Assume that

$$\frac{d}{dx}(v^n) = nv^{n-1} \frac{dv}{dx}$$

is true. Using the product rule (Section 3.3) with $u = v^n$ gives

$$\frac{d}{dx}(v^{n+1}) = \frac{d}{dx}(v^n \cdot v) = v^n \cdot \frac{dv}{dx} + v \frac{d}{dx}(v^n)$$

$$= v^n \frac{dv}{dx} + v \cdot nv^{n-1} \frac{dv}{dx} = v^n \frac{dv}{dx} + nv^n \frac{dv}{dx} = (n+1)v^n \frac{dv}{dx}$$

Thus, with the assumption that formula (1) holds for $r = n$, we have established the formula

$$\frac{d}{dx}(v^{n+1}) = (n+1)v^n \frac{dv}{dx}$$

which is the original formula with r replaced by $n + 1$.

Step 2: r is a negative integer
We want to establish now the formula

$$\frac{d}{dx}(v^{-n}) = -nv^{-n-1}\frac{dv}{dx} \quad \text{for} \quad n = 1, 2, 3, \ldots.$$

Since $v^{-n} = 1/v^n$, we can use the reciprocal rule (Section 3.3), which gives in this case

$$\frac{d}{dx}\left(\frac{1}{v^n}\right) = -\frac{1}{v^{2n}} \cdot \frac{d}{dx}(v^n)$$

According to Step 1,

$$\frac{d}{dx}(v^n) = nv^{n-1}\frac{dv}{dx}$$

and hence

$$\frac{d}{dx}\left(\frac{1}{v^n}\right) = -\frac{1}{v^{2n}}nv^{n-1}\frac{dv}{dx} = -nv^{n-1-2n}\frac{dv}{dx} = -n \cdot v^{-n-1}\frac{dv}{dx}$$

Step 3: r is a rational number
Let $r = m/n$, where m and n are integers and $n \neq 0$. We want now to establish the formula

$$\frac{d}{dx}(v^{m/n}) = \frac{m}{n}v^{(m/n)-1}\frac{dv}{dx} \tag{2}$$

If we let

$$u = v^{m/n} \tag{3}$$

then raising each side to the nth power gives

$$u^n = v^m \tag{4}$$

Differentiating both sides of (4) with respect to x and using the results of Steps 1 and 2 gives

$$nu^{n-1}\frac{du}{dx} = mv^{m-1}\frac{dv}{dx}$$

Thus,

$$\frac{du}{dx} = \frac{m}{n}\frac{v^{m-1}}{u^{n-1}}\frac{dv}{dx}$$

To complete the proof, it remains to show that

$$\frac{v^{m-1}}{u^{n-1}} = v^{(m/n)-1}$$

This, however, follows easily from (3).

We conclude this section with a list of the differentiation formulas that have been developed thus far.

DIFFERENTIATION FORMULAS

1. $\dfrac{dc}{dx} = 0$ for any constant c.

2. $\dfrac{dx}{dx} = 1$.

If $u = f(x)$ and $v = g(x)$, then

3. $\dfrac{d}{dx}(cu) = c\dfrac{du}{dx}$ for any constant c.

4. $\dfrac{d}{dx}(u + v) = \dfrac{du}{dx} + \dfrac{dv}{dx}$

5. $\dfrac{d}{dx}(u - v) = \dfrac{du}{dx} - \dfrac{dv}{dx}$

6. $\dfrac{d}{dx}(uv) = u\dfrac{dv}{dx} + v\dfrac{du}{dx}$

7. $\dfrac{d}{dx}\left(\dfrac{1}{v}\right) = -\dfrac{1}{v^2}\dfrac{dv}{dx}$

8. $\dfrac{d}{dx}\left(\dfrac{u}{v}\right) = \dfrac{v\,du/dx - u\,dv/dx}{v^2}$

9. $\dfrac{d}{dx}(v^r) = rv^{r-1}\dfrac{dv}{dx}$

EXERCISES

Find dy/dx in Exercises 1–51.

1. $y = \sqrt{\dfrac{1}{x}}$
2. $y = \sqrt{\dfrac{x-1}{x+1}}$
3. $y = (x^2 - 1)^3(x^2 + 1)^5$
4. $y = (x - 1)^3(x + 1)^6$
5. $y = [x(x - 1)(x - 2)]^3$
6. $y = [(x + 4)^2 + 1]^2$
7. $y = (x^2 - 5)^7 x^5$
8. $y = \sqrt{1 + \sqrt{x}}$
9. $y = (x + 1/x)^{1/3}$
10. $y = \dfrac{1}{(x^2 + 1)^3}$
11. $y = (x^4 - x + 1)^{-1}$
12. $y = (x^4 - x + 1)^{-2}$
13. $y = \left(\dfrac{1}{x+1}\right)^{-3}$
14. $y = [f(x) + 2g(x) - h(x)]^5$
15. $y = [f(x) \cdot g(x)]^3$
16. $y = \sqrt{f(x)}$

17. $y = \sqrt{\dfrac{f(x)}{g(x)}}$

18. $y = (x+1)^{\sqrt{2}}$

19. $y = (2x+1)^{\sqrt{2}}$

20. $y = (3x^2+1)^{\sqrt{2}}$

21. $y = (x+1)^{-\sqrt{2}}$

22. $y = (3x^{-2}+1)^{-\sqrt{2}}$

23. $y = \dfrac{1}{x^2 - 2x + 3}$

24. $y = \left(\dfrac{1}{x^2 - 2x + 3}\right)^{100}$

25. $y = \dfrac{a-x}{a+x}$

26. $y = ax^2 + ax^{-2}$

27. $y = \dfrac{x}{x^2+1}$

28. $y = a_0 + a_1 x + a_2 x^2 + a_3 x^3$

29. $y = \dfrac{ax^2}{b + c/x^2}$

30. $y = x(x-1)^2(x+2)^3$

31. $y = \dfrac{x^4}{a^2 - x^2}$

32. $y = \dfrac{x^m}{1+x^n}$

33. $y = x^{1/3} + x^{-1/3}$

34. $y = \dfrac{1}{\sqrt{x}}$

35. $y = x^2 + x^{1/2} + x^{-1/2}$

36. $y = (a+x)\sqrt{a-x}$

37. $y = \dfrac{\sqrt{x}}{\sqrt{x}+1}$

38. $y = x\sqrt{x}$

39. $y = x^{-1/2}(1+x)^{1/2}$

40. $y = \sqrt[3]{x^5}$

41. $y = (x^2 + x + 1)^{7/8}$

42. $y = \sqrt[5]{x} - 3\sqrt[3]{x} + 1$

43. $y = (x + x^{-1})^{-1}$

44. $y = x^{-1/2}(x+1)(x-1)^{1/2}$

45. $y = \sqrt{x-1}\sqrt{x+1}$

46. $y = \dfrac{1}{1+x^{-2}}$

47. $y = \sqrt{x + \sqrt{x + \sqrt{x}}}$

48. $y = \left(1 + \dfrac{x}{1+\sqrt{x}}\right)^{1/2}$

49. $y = \left[\dfrac{x^2 - 2x + 1}{(x-1)(x-2)}\right]^2$

50. $y = x^2\sqrt{x^2 - a^2}$

51. $y = \dfrac{x}{\sqrt{1-x^2}}$

52. Find the tangent to the function $y = \sqrt{x-1}$ at $P(2, 1)$.

53. Find the point $P(x, y)$ on $f(x) = x^{1/2} + x^{-1/2}$ at which $f(x)$ has a horizontal tangent.

54. Find the point $P(x, y)$ on $f(x) = x^{3/2} - x$ at which the tangent is perpendicular to the line $y = -\tfrac{1}{2}x$.

3.5 HIGHER DERIVATIVES, VELOCITY, AND ACCELERATION

In concluding Section 3.1 we observed that the derivative of a function can itself be looked upon as a function. Thus, for example, when

$$y = x^3 - x + 2$$

then

$$\frac{dy}{dx} = 3x^2 - 1$$

Putting

$$z = 3x^2 - 1$$

we can again form a derivative,

$$\frac{dz}{dx} = \frac{d}{dx}\left(\frac{dy}{dx}\right) = 6x$$

DEFINITION 1

The derivative of $\frac{dy}{dx} = f'(x)$ is called the *second derivative* of $y = f(x)$, designated $\frac{d^2y}{dx^2} = f''(x)$.

The symbol $\frac{d^2y}{dx^2}$ is read "*d* two-*y d x*-square"; the symbol $f''(x)$ is read "*f* double prime of *x*."

Example 1

If

$$y = 2x^6 + 6x^2$$

then

$$\frac{dy}{dx} = 12x^5 + 12x$$

and

$$\frac{d^2y}{dx^2} = 60x^4 + 12$$

It seems clear that in some instances the differentiation process can be carried on even more than twice.

DEFINITION 2

The *n*th *derivative* of $f(x)$ is the result of *n* successive differentiations of *f* and is designated with one of the symbols $\frac{d^n y}{dx^n}$, $f^{(n)}(x)$, or $y^{(n)}$.

The symbol $\frac{d^n y}{dx^n}$ is read "*d n-y dx* to the *n*"; $f^{(n)}(x)$ is called "*f-n* of *x*"; $y^{(n)}$ is called "*y-n*."

3.5 HIGHER DERIVATIVES, VELOCITY, AND ACCELERATION

Example 2

Find the fourth derivative of $y = x^4 - x^3 + x^2 - x + 1$.

Solution

We perform four successive differentiations:

$$\frac{dy}{dx} = 4x^3 - 3x^2 + 2x - 1$$

$$\frac{d^2y}{dx^2} = 12x^2 - 6x + 2$$

$$\frac{d^3y}{dx^3} = 24x - 6$$

$$\frac{d^4y}{dx^4} = 24$$

Example 3

Find a formula for the nth derivative of $f(x) = 1/x$.

Solution

Writing $f(x) = x^{-1}$ and using the power formula (Section 3.3) gives

$$f'(x) = -1 \cdot x^{-2}$$
$$f''(x) = (-1) \cdot (-2) \cdot x^{-3} = 2x^{-3}$$
$$f^{(3)}(x) = 2 \cdot (-3)x^{-4} = -2 \cdot 3 \cdot x^{-4}$$
$$f^{(4)}(x) = -2 \cdot 3 \cdot (-4) \cdot x^{-5} = 2 \cdot 3 \cdot 4 \cdot x^{-5}$$

In general, we have

$$f^{(n)}(x) = (-1)^n \cdot 2 \cdot 3 \cdot 4 \cdot \cdots \cdot n \cdot x^{-n-1}$$

VELOCITY

Let a particle move along a straight line, starting from a point 0 at time $t = 0$; suppose that at time $t > 0$ its distance from 0 is $s = f(t)$. The *average velocity* over a time interval is defined as

$$\text{average velocity} = \frac{\text{displacement}}{\text{travel time}}$$

Since displacement (distance covered) over the time interval from t to $t + \Delta t$ is $s = f(t + \Delta t) - f(t)$, we have

$$\text{average velocity} = \frac{\Delta s}{\Delta t} = \frac{f(t + \Delta t) - f(t)}{\Delta t}$$

DEFINITION 3

The *velocity* of the particle, also called *instantaneous velocity*, is

$$v = \lim_{\Delta t \to 0} \frac{\Delta s}{\Delta t} = \frac{ds}{dt}$$

Thus, velocity is given by a derivative, and it measures the rate of change of distance.

Example 4

Suppose that a particle travels along a straight line. It starts from a point 0 at time $t = 0$, and at time $t > 0$ it is at a distance $s = \frac{1}{2}t^2$ from 0. Graphically, the situation is described in Figure 3.5. The (instantaneous) velocity of the particle at time $t > 0$ is

$$v = \frac{ds}{dt} = \frac{1}{2} \cdot 2t = t$$

If distance is measured in feet and time is measured in seconds, then we shall say that the velocity at time t is v feet per second, written v ft/sec.

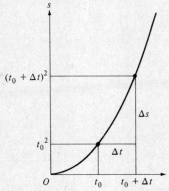

Figure 3.5

ACCELERATION

Velocity measures the rate of change of distance, and acceleration measures the rate of change of velocity. Specifically,

$$\text{average acceleration} = \frac{\text{change in velocity}}{\text{travel time}}$$

but change in velocity over the interval from t to $t + \Delta t$ is $\Delta v = v(t + \Delta t) - v(t)$, and accordingly,

$$\text{average acceleration} = \frac{v(t + \Delta t) - v(t)}{\Delta t}$$

Acceleration is defined as

$$a = \lim_{\Delta t \to 0} \frac{\Delta v}{\Delta t} = \frac{dv}{dt}$$

Thus,

$$\text{velocity} = v = \frac{ds}{dt}$$

$$\text{acceleration} = a = \frac{dv}{dt} = \frac{d^2s}{dt^2}$$

Example 5

A free-falling body starts from rest at time $t = 0$. If the distance covered in time t is given by the formula

3.5 HIGHER DERIVATIVES, VELOCITY, AND ACCELERATION

$$s = 16t^2$$

find

(a) the velocity v at time $t > 0$;
(b) the acceleration a at time $t > 0$.

Solution
According to the above,

$$v = \frac{ds}{dt} = 32t$$

and

$$a = \frac{dv}{dt} = 32$$

We note that here velocity increases with time whereas acceleration is constant. The functions $s = 16t^2$, $v = 32t$, and $a = 32$ are graphed in Figure 3.6.

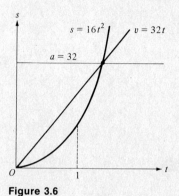

Figure 3.6

EXERCISES

Find y' and y'' in Exercises 1–11.

1. $y = (x - 1)(x + 1)$
2. $y = (2x + 3)^5$
3. $y = x^5(x + 1)^6$
4. $y = x(x - 1)(x + 1)$
5. $y = x^{10} - x^7 + x^3$
6. $y = 5$
7. $y = 5x$
8. $y = x + 1/x$
9. $y = [f(x)]^2$
10. $y = f(x)g(x)$
11. $y = xf(x)$

In Exercises 12–15, s represents the position at time t of a moving body. Find the velocity and acceleration in each case.

12. $s = \frac{1}{2}gt^2$, g being a constant
13. $s = t^2 - 2t$
14. $s = mt + c$
15. $s = gt^2 + bt + c$
16. The *Leibnitz formula* for derivatives states that

$$(fg)'' = f''g + 2f'g' + fg''$$
$$(fg)''' = f'''g + 3f''g' + 3f'g'' + fg'''$$

Verify these formulas by successive differentiations of the product $fg = f(x) \cdot g(x)$.

80 DERIVATIVES

17. A particle starting from 0 at time $t = 0$ covers the distance $s = t^3 - t^2 + 1$ in t sec.

   (a) Find the velocity and acceleration of the particle.

   (b) At what time is the acceleration 0?

   (c) During the first 10 sec, when is the velocity minimum and maximum? When is acceleration maximum?

3.6 THE CHAIN RULE AND IMPLICIT FUNCTIONS

COMPOSITION OF FUNCTIONS

Two functions are often combined to produce a new function. Consider, for example, the functions

$$y = \frac{u}{1 + \sqrt{u}}$$

$$u = 1 - x^2$$

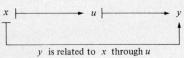

y is related to *x* through *u*

Figure 3.7

(see Figure 3.7). With a simple substitution,

$$y = \frac{1 - x^2}{1 + \sqrt{1 - x^2}}$$

Using functional notation, if

$$f(u) = \frac{u}{1 + \sqrt{u}} \qquad g(x) = 1 - x^2 \qquad h(x) = \frac{1 - x^2}{1 + \sqrt{1 - x^2}}$$

then the process of substitution is symbolized

$$h(x) = f[g(x)]$$

(see Figure 3.8). The function $h(x)$ is called the *composition* of f with g. It is important to observe that while $g(x)$ is defined for all x, $f[g(x)]$ is defined only for the values x for which $1 - x^2 \geq 0$.

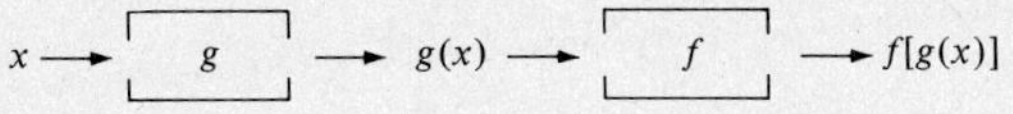

$x \longrightarrow \boxed{g} \longrightarrow g(x) \longrightarrow \boxed{f} \longrightarrow f[g(x)]$

Figure 3.8

Example 1

What are the functions $f[g(x)]$ and $g[f(x)]$ when $f(x) = 1 + \sqrt{x}$ and $g(x) = x^2$? Find the domain of definition of each of them.

Solution

To form $f[g(x)]$, we replace x in $f(x)$ by $g(x)$:

$$f[g(x)] = 1 + \sqrt{g(x)} = 1 + \sqrt{x^2} = 1 + |x|$$

To form $g[f(x)]$, we replace x in $g(x)$ by $f(x)$:
$$g[f(x)] = [f(x)]^2 = (1 + \sqrt{x})^2 = 1 + x + 2\sqrt{x}$$
The following information is easily checked out:

Function	Domain of definition
$f(x) = 1 + \sqrt{x}$	$x \geq 0$
$g(x) = x^2$	all x
$f[g(x)] = 1 + |x|$	all x
$g[f(x)] = 1 + x + 2\sqrt{x}$	$x \geq 0$

The derivatives of composite functions obey the so-called *chain rule:*

THEOREM 1: THE CHAIN RULE
If $y = f(u)$, $u = g(x)$, and $y = f[g(x)]$ is defined, then
$$\frac{dy}{dx} = \frac{dy}{du} \cdot \frac{du}{dx} \quad \text{whenever} \quad \frac{dy}{du} \quad \text{and} \quad \frac{du}{dx} \quad \text{exist.}$$

This rule is often expressed with a different notation: If we put $h(x) = f[g(x)]$, then
$$\frac{dy}{dx} = h'(x), \quad \frac{dy}{du} = f'(u) = f'[g(x)], \quad \text{and} \quad \frac{du}{dx} = g'(x)$$

This leads us to the formula

$$h'(x) = f'[g(x)] \cdot g'(x)$$

which is an alternative form of the chain rule. Before getting involved with the proof, we consider some applications.

Example 2

Find dy/dx when
$$y = \frac{1 - x^2}{1 + \sqrt{1 - x^2}}$$

Solution
This is the example considered at the beginning of this section, where we used the notation $y = u/(1 + \sqrt{u})$ and $u = 1 - x^2$. Now,

$$\frac{dy}{du} = \frac{(1+\sqrt{u}) - u \cdot 1/(2\sqrt{u})}{(1+\sqrt{u})^2} = \frac{1+\frac{1}{2}\sqrt{u}}{(1+\sqrt{u})^2}$$

$$\frac{du}{dx} = -2x$$

and the chain rule gives

$$\frac{dy}{dx} = \frac{dy}{du}\frac{du}{dx} = \frac{1+\frac{1}{2}\sqrt{u}}{(1+\sqrt{u})^2}(-2x) = -2x\frac{1+\frac{1}{2}\sqrt{1-x^2}}{(1+\sqrt{1-x^2})^2}$$

Example 3

Find dy/dx when $y = u^{10}$ and $u = x^2 + x + 1$.

Solution
Expressing y as a function of x, we get $y = (x^2 + x + 1)^{10}$. Since $dy/du = 10u^9$ and $du/dx = 2x + 1$, we find by the chain rule that

$$\frac{dy}{dx} = \frac{dy}{du}\frac{du}{dx} = 10u^9(2x+1) = 10(x^2+x+1)^9(2x+1)$$

Observe that this problem could also be solved with the formula

$$\frac{d}{dx}(u^n) = nu^{n-1}\frac{du}{dx}$$

which is a special case of the chain rule.

Proof of Theorem 1
To differentiate $y = f(u)$, we put

$$y + \Delta y = f(u + \Delta u)$$

and consider the quotient

$$\frac{\Delta y}{\Delta u} = \frac{f(u+\Delta u) - f(u)}{\Delta u}$$

To differentiate $u = g(x)$, we put

$$u + \Delta u = g(x + \Delta x) \tag{1}$$

and consider the quotient

$$\frac{\Delta u}{\Delta x} = \frac{g(x+\Delta x) - g(x)}{\Delta x}$$

For $y = f[g(x)]$, we thus have

$$y + \Delta y = f[g(x + \Delta x)]$$

and this yields the quotient

$$\frac{\Delta y}{\Delta x} = \frac{f[g(x+\Delta x)] - f[g(x)]}{\Delta x}$$

By (1) we can write

$$\frac{\Delta y}{\Delta x} = \frac{f(u+\Delta u) - f(u)}{\Delta x} = \frac{f(u+\Delta u) - f(u)}{\Delta u} \cdot \frac{\Delta u}{\Delta x}$$

$$= \frac{\Delta y}{\Delta u} \cdot \frac{\Delta u}{\Delta x}$$

if $\Delta x \neq 0$ and $\Delta u \neq 0$. According to (1), however, Δu depends on Δx, since $\Delta u = g(x+\Delta x) - g(x)$, and hence we cannot guarantee that $\Delta u \neq 0$. To avoid this difficulty we proceed as follows: Since

$$\frac{dy}{du} = \lim_{\Delta u \to 0} \frac{\Delta y}{\Delta u}$$

we can put

$$\frac{\Delta y}{\Delta u} = \frac{dy}{du} + \epsilon$$

where $\epsilon \to 0$ as $\Delta u \to 0$. Hence, instead of

$$\frac{\Delta y}{\Delta x} = \frac{\Delta y}{\Delta u} \frac{\Delta u}{\Delta x}$$

we write

$$\frac{\Delta y}{\Delta x} = \left(\frac{dy}{du} + \epsilon\right) \frac{\Delta u}{\Delta x}$$

and so

$$\frac{dy}{dx} = \lim_{\Delta x \to 0} \frac{\Delta y}{\Delta x} = \lim_{\Delta x \to 0} \left[\left(\frac{dy}{du} + \epsilon\right)\frac{\Delta u}{\Delta x}\right] = \frac{dy}{du} \cdot \frac{du}{dx}$$

In this demonstration we assumed that $\Delta u = g(x+\Delta x) - g(x) \to 0$ as $\Delta x \to 0$. While this may be intuitively true, it does require justification. This will actually be a simple consequence of the concept of continuity discussed in Section 4.5 (see, in particular, Theorem 1 there).

FUNCTIONS DEFINED IMPLICITLY
Consider the equation

$$xy + 2 = y \tag{2}$$

It can be solved for y in terms of x:

$$xy - y = -2$$
$$y(x - 1) = -2$$
$$y = \frac{-2}{x - 1} = \frac{2}{1 - x} \tag{3}$$

Equation (3) is said to define y as a function of x *explicitly*, equation (2) *implicitly*.

Consider the equation $x^2 + y^2 = 1$ (see Figure 3.9). Solving it for y yields the two equations

$$y = \sqrt{1 - x^2} \quad \text{or} \quad y = -\sqrt{1 - x^2}$$

Each of these equations represents a function, the first describing the upper half-circle, the second the lower half-circle.

As another example of an implicit relation, examine the equation $x^2 + y^2 = -1$. Since the square of a real number is positive or zero, it follows that this equation has no solution. It therefore does *not* define a function.

The equation

$$y^5 + y = x^5 + x^2$$

defines y in terms of x implicitly, but this time we are unable to solve for y. That is, this function cannot be written in the form $y = f(x)$. Despite this, we can find dy/dx. The technique for achieving this is called *implicit differentiation*.

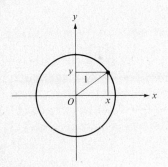

Figure 3.9
The circle $x^2 + y^2 = 1$.

Example 4

Find dy/dx at the point $(x, y) = (1, 1)$ when $y^5 + y = x^5 + x^2$.

Solution
Differentiate both sides of the equation with respect to x. Since y is assumed to be a function of x, we have by the chain rule

$$\frac{d}{dx}(y^5 + y) = \frac{d}{dx}(y^5) + \frac{dy}{dx} = 5y^4 \frac{dy}{dx} + \frac{dy}{dx} = (5y^4 + 1)\frac{dy}{dx}$$

Also,

$$\frac{d}{dx}(x^5 + x^2) = 5x^4 + 2x$$

Hence,

$$(5y^4 + 1)\frac{dy}{dx} = 5x^4 + 2x$$

or

$$\frac{dy}{dx} = \frac{5x^4 + 2x}{5y^4 + 1} \qquad (4)$$

At $(x, y) = (1, 1)$, we get

$$\frac{dy}{dx} = \frac{5 \cdot 1^4 + 2 \cdot 1}{5 \cdot 1^4 + 1} = \frac{7}{6} \qquad (5)$$

REMARK
Up to this point, dy/dx has always been expressed in terms of x alone, but this is not the case in (4). No substitution for y is possible, because the equation $y^5 + y = x^5 + x^2$ cannot be solved for y in terms of x explicitly. For this reason (4) is accepted as the final answer. When, as in (5), the derivative is evaluated at a point (x_0, y_0), substitution is restricted to points that satisfy the original equation.

Example 5

Find dy/dx at $(x, y) = \left(2, \frac{\sqrt{21}}{2}\right)$ when $(y^2 + 1)^{1/2} = x + 1/x$.

Solution
Differentiate both sides with respect to x:

$$\frac{1}{2}(y^2 + 1)^{-1/2} 2y \frac{dy}{dx} = 1 - \frac{1}{x^2}$$

Upon simplification this becomes

$$\frac{dy}{dx} = \frac{(x^2 - 1)\sqrt{y^2 + 1}}{x^2 y}$$

At $(2, \sqrt{21}/2)$ this gives

$$\frac{dy}{dx} = \frac{15}{4\sqrt{21}} = \frac{5\sqrt{21}}{28}$$

Example 6

Find dy/dx at $(1, \sqrt{3})$ and $(1, -\sqrt{3})$ when $x^2 + y^2 = 4$.

Solution
Differentiating both sides of the equation with respect to x results in $2x + 2y\, dy/dx = 0$, and hence

$$\frac{dy}{dx} = -\frac{x}{y}$$

86 DERIVATIVES

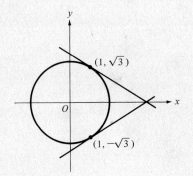

Figure 3.10
Tangents to $x^2 + y^2 = 4$.

At $(1, \sqrt{3})$ we have $dy/dx = -\sqrt{3}/3$; at $(1, -\sqrt{3})$ we have $dy/dx = \sqrt{3}/3$ (see Figure 3.10).

EXERCISES

In Exercises 1–10, find dy/dx.

1. $y = \sqrt{\dfrac{x+1}{x-1}}$

2. $y = t + \dfrac{1}{t},\ t = x + \dfrac{1}{x}$

3. $y = \dfrac{1}{1 + 1/u},\ u = 1 - x^2$

4. $y = \left(\dfrac{x^2 - 1}{x^2 + 1}\right)^{6/5}$

5. $y = (x^2 + 1)^{-1/3} + (x^2 + 1)^{1/3}$

6. $y = \sqrt{x + \sqrt{x}}$

7. $y = \dfrac{\sqrt{u}}{1 + u},\ u = \dfrac{x+1}{x-1}$

8. $y = \dfrac{1}{(1 + x^3)\sqrt{1 + x^3}}$

9. $y = (1 + x^{1/3})^{1/3}$

10. $y = (x^{10} + x)^{10} + 1$

In Exercises 11–23, find dy/dx at the indicated points.

11. $y^2 = 4px$ at $(1, 2\sqrt{p})$

12. $8x^2 + 4y^2 = 1$ at $(0, \tfrac{1}{2})$

13. $\sqrt{x} + \sqrt{y} = \sqrt{a}$ at $(a/4, a/4)$

14. $y + xy + x = 5$ at (a, b)

15. $x^3 + 3x^2y + 3xy^2 + y^3 = 0$ at (x_0, y_0)

16. $x^{10} + y^{10} = 2^{10}$ at $(0, 2)$

17. $\dfrac{x^2}{a^2} + \dfrac{y^2}{b^2} = \dfrac{1}{c^2}$ at (x_0, y_0)

18. $\dfrac{x+y}{x-y} = x$ at (x_0, y_0)

19. $\dfrac{x+y}{x-y} = y$ at (x_0, y_0)

20. $\dfrac{xy+1}{xy-1} = xy$ at (x_0, y_0)

21. $y + \dfrac{1}{y} = \sqrt{x + \dfrac{1}{x}}$ at (x_0, y_0)

22. $x^2 = y^2$ at $(-2, 2)$

23. $y = \sqrt{1 + \sqrt{x + y}}$ at (x_0, y_0)

24. Find d^2y/dx^2 at $(1, \sqrt{3})$ when $x^2 + y^2 = 2^2$.

25. Find d^2y/dx^2 at (x_0, y_0) when $(x - y) \cdot (dy/dx) = 1$.

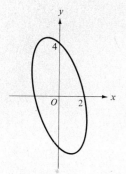

Figure 3.11

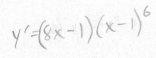

26. Generalize the chain rule to the case in which $y = f(u)$, $u = g(t)$, $t = h(x)$, and $y = f[g(h(x))]$ is defined. Find dy/dx when $f(u) = u^2 + u + 1$, $g(t) = t^2 + t + 1$, and $h(x) = x^2 + x + 1$.

27. The equation $4x^2 + 2xy + y^2 = 16$ describes an *ellipse* (see Figure 3.11). Find the points on the curve where $dy/dx = 0$ and the points where dy/dx is undefined.

28. If $x^2 y = 10$, find d^3y/dx^3 at $(2, \tfrac{5}{2})$.

QUIZ 1

1. Find the derivative of each function below.

   (a) $y = x(x-1)^7$ (b) $y = (x^4 + x^{-4})^{-1}$

   (c) $y = \dfrac{\sqrt{x}-1}{\sqrt{x}+1}$ (d) $y = \sqrt{x + \sqrt{x}}$

   (e) $y = (x-1)^5(x+5)^8$ (f) $y = \sqrt{\dfrac{x-1}{x+1}}$

   (g) $y = x^{2/3}(3x^2+1)^{1/3}$

2. Find the tangent to $y = (x-1)/(x+1)$ at the point $(1, 0)$.

3. Find $f'(4)$ and $f'''(4)$ if $f(x) = x - 1/x$.

4. Find dy/dx using implicit differentiation.

   (a) $y^3 + xy + x^4 = 5$

   (b) $\sqrt{(x+y)} - \sqrt{x} = y$

5. Find $f[g(x)]$ and $g[f(x)]$ when $f(x) = x^2 + 4$ and $g(x) = (x-2)/(x+2)$.

QUIZ 2

1. Find the derivative of each function below.

   (a) $y = x\sqrt{x^2+4}$ (b) $y = x^{-3}(x+6)^3$

   (c) $y = (x^{-2} + x^2)^{-3}$ (d) $y = \sqrt{\dfrac{x-1}{x+1}}$

   (e) $y = \dfrac{x^2}{\sqrt{1-x^2}}$ (f) $y = \dfrac{1}{x^3 + x^{-3}}$

2. Using implicit differentiation, find dy/dx when $x + xy + y^4 = 0$.

3. Find the tangent to $y = x(x+2)^8$ at the point $(-2, 0)$.

4. Find $f^{(8)}(1)$ and $f^{(10)}(1)$ if $f(x) = x^{10} - 90x^8 + 20x^6 - x$.

5. Find $f[g(x)]$ and $g[f(x)]$ when $f(x) = \sqrt{x}/(x^2 + x + 1)$ and $g(x) = x + 1/x$.

6. When $y = u + 1/(u^2 + 1)$ and $u = [1/(t-1)] - [1/(t+1)]$ find dy/dt.

CHAPTER 4

GRAPHS AND CONTINUITY

4.1 THE FIRST DERIVATIVE AND THE SHAPE OF A CURVE

Basically this chapter deals with the application of derivatives and limits to the study of curves $y = f(x)$. By way of motivation, consider the polynomial

$$y = \tfrac{1}{3}x^3 - 2x^2 + 3x + 2$$

whose graph is given in Figure 4.1, and observe the following facts:

The graph has a maximum (peak) at A, a minimum (dip) at B;
The curve increases (rises) as x increases to 1; it decreases (falls) as x increases from 1 to 3, then increases again with further increasing x.

Our objective is to discover methods for finding peaks and dips, and where the graph increases or decreases.

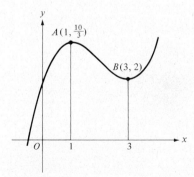

Figure 4.1
The polynomial $y = \tfrac{1}{3}x^3 - 2x^2 + 3x + 2$.

INCREASING AND DECREASING FUNCTIONS

Let us begin by giving precise meaning to the intuitive concepts of rising and falling (see Figure 4.2).

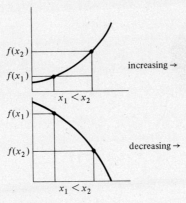

Figure 4.2

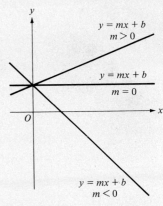

Figure 4.3

> **DEFINITION 1**
> Let $f(x)$ be defined on an interval I, finite or infinite. If $f(x_1) < f(x_2)$ for all points $x_1 < x_2$ in I, then $f(x)$ is *increasing*; if $f(x_1) > f(x_2)$ for all points $x_1 < x_2$ in I, then $f(x)$ is *decreasing*.

Note that we speak of increasing and decreasing functions for x moving from left to right.

Example 1

If $f(x) = mx + b$ then, on the interval $I = (-\infty, \infty)$,

$f(x)$ is increasing when $m > 0$,
$f(x)$ is decreasing when $m < 0$,
$f(x)$ is neither increasing nor decreasing when $m = 0$ (see Figure 4.3).

To justify these claims, pick any two numbers $x_1 < x_2$.

When $m > 0$, then $mx_1 < mx_2$ and hence

$$f(x_1) = mx_1 + b < mx_2 + b = f(x_2)$$

When $m < 0$, then $mx_1 > mx_2$ and hence

$$f(x_1) = mx_1 + b > mx_2 + b = f(x_2)$$

When $m = 0$, then $mx_1 = mx_2 = 0$ and hence

$$f(x_1) = b = f(x_2)$$

Example 2

Consider the function $f(x) = x^2$ (see Figure 4.4).

4.1 THE FIRST DERIVATIVE AND THE SHAPE OF A CURVE

$f(x)$ is decreasing on the interval $(-\infty, 0]$,
$f(x)$ is increasing on the interval $[0, \infty)$.

This follows from the observation that

$x_1^2 > x_2^2$ when $x_1 < x_2 \leq 0$, and
$x_1^2 < x_2^2$ when $0 \leq x_1 < x_2$.

On any interval $(-a, a)$ containing the origin, the function is not decreasing and is not increasing.

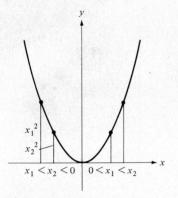

Figure 4.4

We now state a simple condition that ensures that $f(x)$ is increasing or decreasing on an interval. Note, however, that this condition requires that $f(x)$ be differentiable.

CRITERION FOR $f(x)$ TO BE INCREASING
If $f'(x) > 0$ for all x in the open interval (a, b), then $f(x)$ is increasing on the closed interval $[a, b]$.

CRITERION FOR $f(x)$ TO BE DECREASING
If $f'(x) < 0$ for all x in the open interval (a, b), then $f(x)$ is decreasing on the closed interval $[a, b]$.

The proof of these statements is given in Section 4.6 because we do not yet have the necessary tools to carry it through.

Example 3

Determine the intervals where

$$f(x) = \tfrac{1}{3}x^3 - 2x^2 + 3x + 2$$

is increasing, and where it is decreasing.

Solution
Compute the derivative of $f(x)$:

$$f'(x) = x^2 - 4x + 3 = (x - 1)(x - 3)$$

Now look at the following table:

$x < 1$	$x - 1 < 0$	$x - 3 < 0$	$f'(x) > 0$
$1 < x < 3$	$x - 1 > 0$	$x - 3 < 0$	$f'(x) < 0$
$x > 3$	$x - 1 > 0$	$x - 3 > 0$	$f'(x) > 0$

According to the criteria above, therefore,

$f(x)$ is increasing on the intervals $(-\infty, 1]$ and $[3, \infty)$, and

$f(x)$ is decreasing on the interval $[1, 3]$ (see Figure 4.5).

The sign of the derivative $f'(x)$ is easily determined from the graph in Figure 4.6.

MAXIMA AND MINIMA

For the function $f(x) = \frac{1}{3}x^3 - 2x^2 + 3x + 2$, we found in Example 3 that $f'(x) = (x-1)(x-3)$ (see Figure 4.6). It follows that $f'(1) = 0$ and $f'(3) = 0$. The values $x = 1$ and $x = 3$ correspond to the points $A(1, \frac{10}{3})$ and $B(3, 2)$ on the graph of $y = f(x)$. At these points the curve turns and the tangents are horizontal (see Figure 4.7).

> **BASIC OBSERVATION**
> If $f'(x_0) = 0$, then the tangent to $y = f(x)$ at $P(x_0, f(x_0))$ has slope 0.

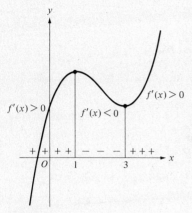

Figure 4.5

In the above example, $f'(x) = 0$ at those points where the curve $y = f(x)$ has a peak or a dip.

> **DEFINITION 2**
> Let the function $f(x)$ be defined on an interval I, finite or infinite.
> If $f(x) < f(x_0)$ for all $x \neq x_0$ in some interval $[x_0 - \delta, x_0 + \delta]$, then $f(x)$ has a *local maximum* at x_0.
> If $f(x) > f(x_0)$ for all $x \neq x_0$ in some interval $[x_0 - \delta, x_0 + \delta]$, then $f(x)$ has a *local minimum* at x_0.

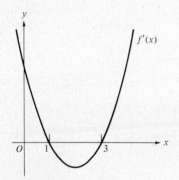

Figure 4.6
The graph of $y = f'(x)$.

Example 4

Consider once more the function $f(x) = \frac{1}{3}x^3 - 2x^2 + 3x + 2$. It has a local maximum at $x = 1$, since $f(x) < f(1)$ for all $x \neq 1$ in the interval $[0, 2]$; $f(x)$ has a local minimum at $x = 3$, since $f(x) > f(3)$ for all $x \neq 3$ in the interval $[2, 4]$.

Example 5

Investigate the function
$$f(x) = -x^2 + 5x - 6$$
and give a rough sketch of its graph.

Solution
The derivative of $f(x)$ is
$$f'(x) = -2x + 5$$

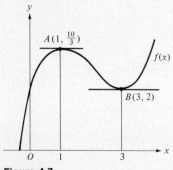

Figure 4.7

4.1 THE FIRST DERIVATIVE AND THE SHAPE OF A CURVE

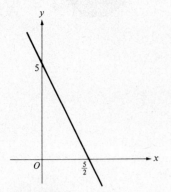

Figure 4.8
The graph of $f'(x)$.

(see Figure 4.8). A simple computation gives

$f'(x) > 0$ when $x < \frac{5}{2}$
$f'(x) = 0$ when $x = \frac{5}{2}$
$f'(x) < 0$ when $x > \frac{5}{2}$

This tells us that

$f(x)$ is increasing on $(-\infty, \frac{5}{2}]$,
$f(x)$ is decreasing on $[\frac{5}{2}, \infty)$,

and hence $f(x)$ has a local maximum at $x = \frac{5}{2}$.

To get a rough picture of $f(x)$ (Figure 4.9), we note that

$$f(x) = -(x^2 - 5x + 6) = -(x-2)(x-3)$$

This helps in compiling a table of values for $f(x)$.

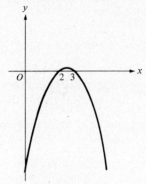

x	$f(x)$
0	-6
1	-2
2	0
$\frac{5}{2}$	$\frac{1}{4}$
3	0
4	-2

Figure 4.9
The graph of $f(x) = -x^2 + 5x - 6$.

Example 6

Investigate the function

$$g(x) = x^2 + 2ax + b$$

where a and b are any numbers.

Solution
Proceeding as in Example 5, we compute $g'(x)$:

$$g'(x) = 2x + 2a = 2(x + a)$$

We find that (see Figure 4.10)

$g'(x) < 0$ for $x < -a$ and hence $g(x)$ is decreasing on $(-\infty, -a]$;
$g'(x) = 0$ at $x = -a$;
$g'(x) > 0$ for $x > -a$ and hence $g(x)$ is increasing on $[-a, \infty)$.

A rough graph of $g(x)$ is given in Figure 4.11.

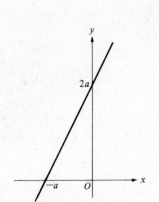

Figure 4.10
The graph of $g'(x)$.

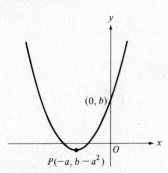

Figure 4.11
The graph of $g(x) = x^2 + 2ax + b$.

94 GRAPHS AND CONTINUITY

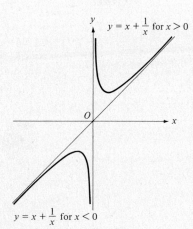

Figure 4.12

Figure 4.13
The graph of $y = x + (1/x)$.

x	$x + (1/x)$
$\pm \frac{1}{8}$	$\pm(8 + \frac{1}{8})$
$\pm \frac{1}{4}$	$\pm(4 + \frac{1}{4})$
$\pm \frac{1}{2}$	$\pm(2 + \frac{1}{2})$
± 1	± 2
± 2	$\pm(2 + \frac{1}{2})$
± 4	$\pm(4 + \frac{1}{4})$
± 8	$\pm(8 + \frac{1}{8})$

Example 7

Give a rough graph of the function

$$h(x) = x + \frac{1}{x}$$

Solution

We make some preliminary observations:

$h(x) > x$ when $x > 0$,
$h(x) < x$ when $x < 0$.

In addition, $h(0)$ is undefined, and all this tells us that

the graph consists of two pieces
lying in the shaded area in Figure 4.12. (1)

Now we compute $h'(x)$:

$$h'(x) = 1 - \frac{1}{x^2} = \frac{x^2 - 1}{x^2} \quad (x \neq 0)$$

Since $x^2 > 0$ when $x \neq 0$, it follows that $h'(x) > 0$ when $x^2 - 1 > 0$. But $x^2 - 1 < 0$ when $-1 < x < 0$ or $0 < x < 1$. Hence:

$h'(x) > 0$ and $h(x)$ is increasing on $(-\infty, -1]$ and $[1, \infty)$,
$h'(x) < 0$ and $h(x)$ is decreasing on $[-1, 0)$ and $(0, 1]$. (2)

From this and the fact that $h'(-1) = h'(1) = 0$, we conclude that

$h(x)$ has a local maximum at $x = -1$,
a local minimum at $x = 1$. (3)

Remember that $x + (1/x) > x$ when $x > 0$, whereas $x + (1/x) < x$ when $x < 0$. The vertical distance between the curves $y = x + (1/x)$ and $y = x$ is $|x + (1/x) - x| = 1/|x|$. This distance approaches 0 as $x \to \pm\infty$; that is, the curve $y = x + (1/x)$ comes closer and closer to the line $y = x$ as x increases beyond all bounds in the positive or the negative direction. The line $y = x$ is said to be an *asymptote* to the curve $y = x + (1/x)$. Also, the y axis is an asymptote to the curve, since $x + (1/x)$ approaches the line $x = 0$ as $x \to 0$.

With the information in (1), (2), and (3), we can now plot the graph (Figure 4.13).

Example 8

There are situations other than maxima or minima at which $f'(x)$ may vanish. When

$$f(x) = x^3$$

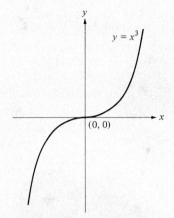

Figure 4.14
The graph of $y = x^3$.

then
$$f'(0) = 3 \cdot 0^2 = 0$$

Hence, the tangent to the curve at $(0, 0)$ is the line $y = 0 \cdot x + 0 = 0$, that is, the tangent is the x axis. The point $(0, 0)$ does not correspond to a peak or a dip in this case, and we see that the tangent crosses the curve (see Figure 4.14).

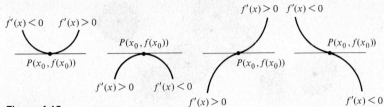

Figure 4.15
Situations in which $f'(x) = 0$.

Thus, four cases for the vanishing of a derivative present themselves, as depicted in Figure 4.15.

ABSOLUTE MAXIMA AND MINIMA

The function $f(x)$ was said to have a local maximum at x_0 when $f(x_0) > f(x)$ for all $x \neq x_0$ sufficiently close to x_0. In Example 5, with $f(x) = -x^2 + 5x - 6$, we found that $f(\frac{5}{2}) > f(x)$ for *all* $x \neq \frac{5}{2}$; in this case, $f(x)$ is said to have an absolute maximum at $\frac{5}{2}$ (see Figure 4.16a). Similarly, the function $g(x)$ in Example 6 has an absolute minimum at $-a$, since $g(-a) < g(x)$ for all $x \neq a$ (see Figure 4.16b). Formally, we lay down this definition:

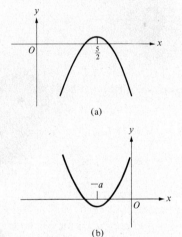

Figure 4.16

> **DEFINITION 3**
> Let $f(x)$ be defined on an interval I, finite or infinite.
>
> If $f(x) \leq f(x_0)$ for all values $x \neq x_0$ in I, then $f(x)$ has an *absolute maximum* at x_0.
> If $f(x) \geq f(x_0)$ for all values $x \neq x_0$ in I, then $f(x)$ has an *absolute minimum* at x_0.

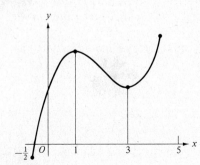

Figure 4.17
The maxima and minima of $f(x) = \frac{1}{3}x^3 - 2x^2 + 3x + 2$ for $-\frac{1}{2} \leq x \leq 5$.

Example 9

Restricted to the interval $[-\frac{1}{2}, 5]$, the function
$$f(x) = \frac{1}{3}x^3 - 2x^2 + 3x + 2$$
has the following maxima and minima (see Figure 4.17):

absolute minimum at $x = -\frac{1}{2}$,
absolute maximum at $x = 5$,
local minimum at $x = 3$,
local maximum at $x = 1$.

Do all functions have maxima and minima? An answer to this inquiry is contained in Theorem 1 below. This theorem applies to so-called *continuous functions,* which are intuitively those functions whose graphs have no breaks. Polynomials are examples of continuous functions; the function $f(x) = x + (1/x)$ (see Example 7) is an example of a *discontinuous* function, since its graph has a break at $x = 0$.

The subject of continuity will be discussed in some detail in Sections 4.5 and 4.6. For the moment we simply mention the following definition.

DEFINITION 4

A function f is *continuous on an interval I* if

1. f is defined at each point of I;
2. if x_0 belongs to I and $x \to x_0$, then $f(x) \to f(x_0)$.

For intervals that are closed and bounded (finite), we can state the following theorem.

THEOREM 1

If $f(x)$ is continuous on the closed interval I, then there are two points x_{min} and x_{max} in I such that

$$f(x_{min}) \leq f(x) \leq f(x_{max})$$

for all points x in I.

This theorem will be repeated in Section 4.5, where it is used in the discussion of continuity.

CRITICAL POINTS

A procedure for locating maximum and minimum points of a curve $y = f(x)$ is this: We find those points

1. where $f'(x) = 0$, or
2. where $f'(x)$ does not exist, or
3. that are end points of the interval on which $f(x)$ is defined.

These candidates are called *critical points*. Maximum and minimum points are always found among the critical points.

Example 10

Locate all maxima and minima of

$$f(x) = |x^2 - 1| \quad -2 \leq x \leq 2$$

4.1 THE FIRST DERIVATIVE AND THE SHAPE OF A CURVE

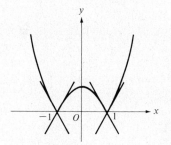

Figure 4.18
The graph of $f(x) = |x^2 - 1|$. At the points $x = 1$ and $x = -1$, the curve has different tangents on the left and the right side of these points. The derivative is not defined at these points.

Solution
Writing $f(x)$ in the form

$$f(x) = \begin{cases} x^2 - 1 & \text{for } -2 \leq x \leq -1 \text{ and } 1 \leq x \leq 2 \\ 1 - x^2 & \text{for } -1 \leq x \leq 1 \end{cases}$$

makes it easy to write down the following table (see Figure 4.18).

Interval	$f'(x)$	Behavior of $f(x)$
$[-2, -1)$	$2x < 0$	decreasing
$(-1, 0)$	$-2x > 0$	increasing
$(0, 1)$	$-2x < 0$	decreasing
$(1, 2]$	$2x > 0$	increasing

Observe that nothing was said about $f'(x)$ at $x = -1$ and $x = 1$. This is because $f'(x)$ fails to exist at these points (see Figure 4.18). To see this, recall the definition

$$f'(x) = \lim_{\Delta x \to 0} \frac{f(x + \Delta x) - f(x)}{\Delta x}$$

where Δx *may be positive or negative*. Take $x = 1$. Then, for $\Delta x > 0$, $1 + \Delta x > 1$, so that $f(1 + \Delta x) = (1 + \Delta x)^2 - 1 = 2\Delta x + (\Delta x)^2$; since $f(1) = 0$.

$$\frac{f(1 + \Delta x) - f(1)}{\Delta x} = \frac{2\Delta x + (\Delta x)^2}{\Delta x} = 2 + \Delta x \to 2 \quad \text{as} \quad \Delta x \to 0$$

For $\Delta x < 0$, however, $1 + \Delta x < 1$, so that $f(1 + \Delta x) = 1 - (1 + \Delta x)^2 = -2\Delta x - (\Delta x)^2$. Hence

$$\frac{f(1 + \Delta x) - f(1)}{\Delta x} = \frac{-2\Delta x - (\Delta x)^2}{\Delta x} = -2 - \Delta x \to -2 \quad \text{as} \quad \Delta x \to 0$$

The derivative at $x = 1$ is not defined, because the answer depends on the manner in which Δx approaches 0. The nonexistence of the derivative at $x = -1$ is demonstrated in the same way. We are now able to list the critical points of $f(x)$:

Critical points	Reason
$x = -2$ and $x = 2$	end points of the domain of $f(x)$
$x = 0$	$f'(0) = 0$
$x = -1$ and $x = 1$	$f'(x)$ does not exist at these points

It is easily seen that $f(x)$ has an absolute maximum at $x = -2$ and $x = 2$, an absolute minimum at $x = -1$ and $x = 1$, and a local maximum at $x = 0$.

EXERCISES

In Exercises 1–12, find $f'(x)$ (where it exists) and determine the following:

(a) the intervals where $f(x)$ is increasing;

(b) the intervals where $f(x)$ is decreasing;

(c) maximum and minimum points of $f(x)$;

(d) a rough sketch of the curve $y = f(x)$.

	$f(x)$	domain
1.	$x^2 - x + 1$	$(-\infty, \infty)$
2.	$-x^2 + x + 1$	$[0, \infty)$
3.	$x^3 - 1$	$(-\infty, \infty)$
4.	$x^4 - x^2$	$(-2, 2)$
5.	$\frac{1}{3}x^3 + \frac{1}{2}x^2 - 6x$	$(-\infty, \infty)$
6.	$1/x$	$(-\infty, 0)$ and $(0, \infty)$
7.	$x - (1/x)$	$(-\infty, 0)$ and $(0, \infty)$
8.	$|x|$	$[-1, 2]$
9.	$x - |x|$	$[-2, 3]$
10.	$x|x|$	$[-1, 1]$
11.	$|x - (1/x)|$	$(-\infty, 0)$ and $(0, \infty)$
12.	$|x|^3$	$(-\infty, \infty)$

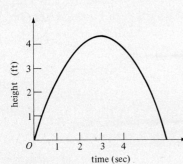

Figure 4.19

13. The trajectory of a projectile is described by $h = -\frac{1}{2}t^2 + 3t$, where h represents height at time t (Figure 4.19). Find the projectile's velocity and acceleration, and the maximum height it reaches.

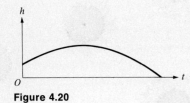

Figure 4.20

14. If the trajectory of a projectile is $h = -16t^2 + 64t + 1$ (Figure 4.20), what is the maximum height the projectile reaches?

15. Give an example of a continuous function $f(x)$ such that
$$f'(x) = \begin{cases} 0 & \text{for } -\infty < x < 0 \\ \frac{1}{2} & \text{for } 0 < x < \infty \end{cases}$$

16. Consider the function
$$f(x) = (x-1)(x-2)(x-3)(x-4)(x-5)$$

Obtain a rough sketch of $f(x)$ from the following information:

(a) List the points x where $f(x) = 0$.

(b) Enter in the table the signs of the factors $x - k$ in each interval and with this obtain the sign of $f(x)$.

Factor Interval	$x-1$	$x-2$	$x-3$	$x-4$	$x-5$	$f(x)$
$(-\infty, 1)$	$-$					
$(1, 2)$	$+$					
$(2, 3)$						
$(3, 4)$						
$(4, 5)$						
$(5, \infty)$						

4.2 THE SECOND DERIVATIVE AND THE SHAPE OF A CURVE

It was observed in the preceding section that a positive derivative on an interval belongs to an increasing function, a negative derivative to a decreasing one. Let $f(x)$ have a second derivative on (a, b). If we apply the same observation to the function $f'(x)$, then we arrive at the following basic principle (see Figure 4.21).

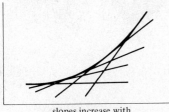

slopes increase with increasing x

slopes decrease with increasing x

Figure 4.21

> If $f''(x) > 0$ for all x in (a, b), then $f'(x)$ is *increasing* on (a, b).
> If $f''(x) < 0$ for all x in (a, b), then $f'(x)$ is *decreasing* on (a, b).

TESTS FOR LOCAL MAXIMA AND MINIMA

We observe that if $f''(x) > 0$ in an interval, then the curve $y = f(x)$ lies above its tangents in the interval. In particular, if $f'(x_0) = 0$, then the tangent at $(x_0, f(x_0))$ is the horizontal line $y = f(x_0)$ (see Figure 4.22); for $y = f(x)$ to lie above this line means that $f(x)$ has a local minimum at x_0. This discussion leads to the following test.

> TEST FOR LOCAL MINIMUM
> If $f'(x_0) = 0$ and $f''(x_0) > 0$, then $f(x)$ has a local minimum at x_0.
>
> TEST FOR LOCAL MAXIMUM
> If $f'(x_0) = 0$ and $f''(x_0) < 0$, then $f(x)$ has a local maximum at x_0.

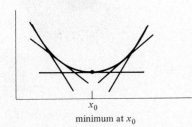

minimum at x_0

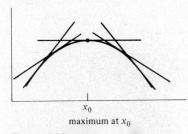

maximum at x_0

Figure 4.22

The proof of these tests will be found in Section 11.3.

Example 1

Consider once more the function

$$f(x) = \tfrac{1}{3}x^3 - 2x^2 + 3x + 2$$

Here

$$f'(x) = x^2 - 4x + 3 = (x-1)(x-3)$$

and

$$f''(x) = 2x - 4$$

Since $f'(1) = 0$ and $f''(1) = -2 < 0$, it follows that $f(x)$ has a local maximum at $x = 1$; since $f'(3) = 0$ and $f''(3) = 2 > 0$, it follows that $f(x)$ has a local minimum at $x = 3$.

Example 2

Find all maxima and minima of
$$g(x) = \tfrac{1}{4}x^4 - \tfrac{1}{2}x^2$$

Solution
From the derivatives
$$g'(x) = x^3 - x = x(x-1)(x+1)$$
$$g''(x) = 3x^2 - 1$$
we obtain the following information:

x	$g'(x)$	$g''(x)$	$g(x)$
-1	0	$2 > 0$	local minimum
0	0	$-1 < 0$	local maximum
1	0	$2 > 0$	local minimum

The graph is given in Figure 4.23.

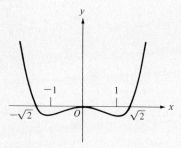

Figure 4.23
The graph of $y = \tfrac{1}{4}x^4 - \tfrac{1}{2}x^2$.

$f(x) = x^2$ lies *above* every tangent to the curve.

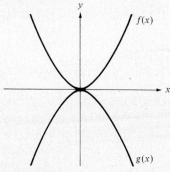

Figure 4.24
$g(x) = -x^2$ lies *below* every tangent to the curve.

CONCAVE UP AND CONCAVE DOWN
Consult Figure 4.24. The curve $f(x) = x^2$ lies above all its tangents: For this function, $f''(x) = 2 > 0$; the curve $g(x) = -x^2$ lies below all its tangents, and for this function $g''(x) = -2 < 0$. The following principle is, in fact, valid:

> If $f''(x_0) > 0$, then the curve $y = f(x)$ lies *above* the tangent at $(x_0, f(x_0))$ for all values of x in some interval $[x_0 - \delta, x_0 + \delta]$.
> If $f''(x_0) < 0$, then the curve $y = f(x)$ lies *below* the tangent at $(x_0, f(x_0))$ for all values of x in some interval $[x_0 - \delta, x_0 + \delta]$.

When a curve lies above its tangents, it is oriented like a bowl with its opening facing upward; when a curve lies below its tangents, it is oriented like an overturned bowl with its opening facing downward (see Figure 4.25). For this reason it is customary to say that:

> **DEFINITION 1**
> $f(x)$ is *concave up* on $[x_0 - \delta, x_0 + \delta]$ when $f''(x) > 0$ on that interval.
> $f(x)$ is *concave down* on $[x_0 - \delta, x_0 + \delta]$ when $f''(x) < 0$ on that interval.

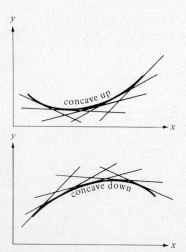

Figure 4.25

4.2 THE SECOND DERIVATIVE AND THE SHAPE OF A CURVE

Example 3

Find the intervals where $f(x) = \frac{1}{3}x^3 - 2x^2 + 3x + 2$ is concave up and where it is concave down.

Solution
By Example 1,

$$f''(x) = 2x - 4$$

Hence,

$f''(x) > 0$ when $2x - 4 > 0$ or $x > 2$
$f''(x) < 0$ when $2x - 4 < 0$ or $x < 2$

Accordingly, $f(x)$ is concave up on $[2, \infty)$, concave down on $(-\infty, 2]$ (see Figure 4.26).

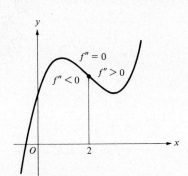

Figure 4.26
The graph of $y = \frac{1}{3}x^3 - 2x^2 + 3x + 2$.

INFLECTION POINTS

From Example 3 we note that

> for $x < 2$ the curve lies *below* its tangents;
> for $x > 2$ the curve lies *above* its tangents;
> at $x = 2$ the curve *crosses* the tangent (see Figure 4.27).

The point $(2, \frac{8}{3})$ is called an inflection point of the curve. More generally:

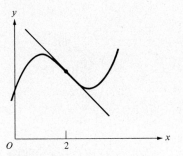

Figure 4.27
The tangent at $(2, \frac{8}{3})$.

DEFINITION 2
The curve $y = f(x)$ has an *inflection point* at $(x_0, f(x_0))$ if it changes its direction of concavity at this point from concave up to concave down, or conversely.

In Example 3, $f''(x) = 0$ at the inflection point, but this is not always true. We shall see later that a function may have an inflection point at which the derivative does not exist.

TEST FOR INFLECTION POINTS
$f'(x)$ has an inflection point at x_0 if, for some $\delta > 0$,

$f''(x) < 0$ for $x_0 - \delta < x < x_0$
$f''(x_0) = 0$
$f''(x) > 0$ for $x_0 < x < x_0 + \delta$

or

$f''(x) > 0$ for $x_0 - \delta < x < x_0$
$f''(x_0) = 0$
$f''(x) < 0$ for $x_0 < x < x_0 + \delta$

This test can be verbalized as follows: $f(x)$ has an inflection point at x_0 if $f''(x)$ changes sign as x changes from below x_0 to above x_0.

Example 4

Locate all maxima, minima, and inflection points of
$$f(x) = \tfrac{1}{20}x^5 - \tfrac{1}{6}x^3$$

Solution
As in the other examples in this section, we begin by finding the first two derivatives of $f(x)$:

$$f'(x) = \tfrac{1}{4}x^4 - \tfrac{1}{2}x^2 = \tfrac{1}{2}x^2(\tfrac{1}{2}x^2 - 1)$$
$$f''(x) = x^3 - x = x(x-1)(x+1)$$

The following information is now easily compiled:

	$f'(x)$	$f''(x)$	$f(x)$
$x = -\sqrt{2}$	0	$-\sqrt{2}(-\sqrt{2}-1)(1-\sqrt{2}) < 0$	relative maximum
$x = \sqrt{2}$	0	$\sqrt{2}(\sqrt{2}-1)(\sqrt{2}+1) > 0$	relative minimum
$-\infty < x < -1$		< 0	
$-1 < x < 0$		> 0	
$0 < x < 1$		< 0	
$1 < x < \infty$		> 0	
$x = -1, 0, 1$		0	inflection points

The graph is given in Figure 4.28.

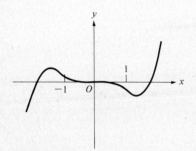

Figure 4.28
The graph of $y = \tfrac{1}{20}x^5 - \tfrac{1}{6}x^3$.

A SECOND TEST FOR LOCAL MAXIMA AND MINIMA
Consider the functions $f_1(x) = x^3$, $f_2(x) = x^4$, and $f_3(x) = -x^4$. Then $f_k'(0) = f_k''(0) = 0$ for $k = 1, 2, 3$, yet each function behaves differently at the origin: $f_1(x)$ has an inflection point, $f_2(x)$ has a minimum, and $f_3(x)$ has a maximum at $x = 0$. To account for such situations we have the following generalization of the earlier test.

TEST FOR LOCAL MAXIMA AND MINIMA
Suppose that $f'(x_0) = f''(x_0) = \cdots = f^{(n-1)}(x_0) = 0$ and $f^{(n)}(x_0) \neq 0$. If n is even, then

$f^{(n)}(x_0) > 0$ gives a local minimum at x_0,
$f^{(n)}(x_0) < 0$ gives a local maximum at x_0;

if n is odd, then $f(x)$ has neither a local minimum nor a local maximum at x_0.

Example 5

Consider the function

$$f(x) = (2x - 1)^4$$

Here

$$f^{(1)}(x) = 2 \cdot 4(2x - 1)^3$$
$$f^{(2)}(x) = 2^2 \cdot 3 \cdot 4(2x - 1)^2$$
$$f^{(3)}(x) = 2^3 \cdot 2 \cdot 3 \cdot 4(2x - 1)$$
$$f^{(4)}(x) = 2^4 \cdot 2 \cdot 3 \cdot 4 > 0$$

Since $f^{(1)}(\frac{1}{2}) = f^{(2)}(\frac{1}{2}) = f^{(3)}(\frac{1}{2}) = 0$ and $f^{(4)}(\frac{1}{2}) > 0$, it follows that $f(x)$ has a local minimum at $x = \frac{1}{2}$.

PLOTTING OF POLYNOMIALS

The following rules for plotting polynomials summarize the work of the last two sections on the first and second derivatives. To plot a polynomial $y = f(x)$, follow these steps:

Step 1
Calculate $f'(x)$ and $f''(x)$.

Step 2
Find those points $a_1, a_2, \ldots, a_m$ where $f''(a_k) = 0, k = 1, 2, \ldots, m$.

Step 3
Determine the sign of $f''(x)$ on each interval (a_k, a_{k+1}). You now know the inflection points and the concavity of $f(x)$.

Step 4
Find those points $b_1, b_2, \ldots, b_n$ where $f'(b_j) = 0, j = 1, 2, \ldots, n$. With step 3 this enables you to determine the local maxima and minima of $f(x)$. If necessary, resort to the second test for local maxima and minima.

Step 5
Determine the sign of $f'(x)$ on each interval (b_j, b_{j+1}). This tells you where $f(x)$ is increasing and where it is decreasing.

Step 6
Compile a table of values of $f(x)$, including the values $f(a_k)$ and $f(b_j)$, and as many other values as are necessary to get a good idea of the shape of the graph.

Step 7
Choosing an appropriate scale on each coordinate axis, plot the

points $(x, f(x))$ computed in step 5 and sketch the graph with the help of the information in steps 2-5.

Example 6

Plot the polynomial
$$f(x) = x^3 - 9x^2 + 24x - 7$$

Solution

Write down the first and second derivatives of $f(x)$:
$$f'(x) = 3x^2 - 18x + 24 = 3(x-2)(x-4)$$
$$f''(x) = 6x - 18 = 6(x-3)$$

Following the procedure in steps 2-5, you obtain the following facts about $f(x)$:

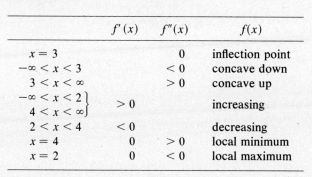

	$f'(x)$	$f''(x)$	$f(x)$
$x = 3$		0	inflection point
$-\infty < x < 3$		< 0	concave down
$3 < x < \infty$		> 0	concave up
$-\infty < x < 2$ $4 < x < \infty$	> 0		increasing
$2 < x < 4$	< 0		decreasing
$x = 4$	0	> 0	local minimum
$x = 2$	0	< 0	local maximum

The graph is given in Figure 4.29.

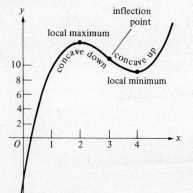

Figure 4.29
The graph of
$f(x) = x^3 - 9x^2 + 24x - 7$.

x	$f(x)$
0	-7
$\frac{1}{2}$	$2\frac{7}{8}$
1	9
2	13
3	11
4	9
5	13

EXERCISES

Using the procedure outlined in the section above on plotting polynomials, plot the polynomials in Exercises 1-10.

1. $y = x^3 - x$
2. $y = x^3 - x^2$
3. $y = x^3 - 6x^2 + 9x + 6$
4. $y = 1 - x^2$
5. $y = 1 - x^3$
6. $y = (x-1)^3(x+1)^2$
7. $y = \frac{1}{56}x^8 - \frac{1}{20}x^5$
8. $y = (x+1)^5$
9. $y = x^4 - 2x^2 + 2$
10. $y = \frac{1}{10}x^5 - \frac{1}{3}x^4 + \frac{1}{3}x^3$

Use the results of the last two sections to plot the functions in Exercises 11-15.

11. $y = |x^3 - x|$
12. $y = |(x-1)(x-2)(x-3)|$

13. $y = |x^4 - 2x^2 + 2|$ 14. $y = |x^2 - 1| - |x^2 + 1|$ 15. $y = |x^3| - 1$

In Exercises 16–20, you are given the derivatives $f'(x)$ of $f(x)$. Determine the local maxima and minima of $f(x)$ in each case.

16. $f'(x) = (x - 1)^3(x + 1)$
17. $f'(x) = (x - 1)^2(x + 1)^2$
18. $f'(x) = x^4(2x + 1)$
19. $f'(x) = x^5(2x + 1)$
20. $f'(x) = (x + 1)(x + 2)(x - 1)(x - 2)$

21. For which of the coefficients a, b, and c is the graph of $y = ax^2 + bx + c$ concave upward?

22. When does $y = ax^3 + bx^2 + cx + d$ have a local maximum and a local minimum?

In Exercises 23–26, find the numbers m for which

   (a) $f(x)$ has two tangents with slope m;

   (b) $f(x)$ has exactly one tangent with slope m;

   (c) $f(x)$ has no tangents with slope m.

23. $f(x) = x^3$
24. $f(x) = x(x - 1)(x + 1)$
25. $f(x) = ax^3 - b$
26. $f(x) = ax^3 + bx^2 + cx + d$

4.3 RATIONAL FUNCTIONS AND ASYMPTOTES

The sum, difference, or product of two polynomials is again a polynomial, but the quotient of two polynomials need not be a polynomial. Functions such as $1/x$ or $(2x^2 - 1)/(x^3 - x + 5)$ are examples of rational functions.

DEFINITION 1
A *rational function* $r(x)$ is the quotient of two polynomials,

$$r(x) = \frac{p(x)}{q(x)}$$

The polynomial $q(x)$ may be a nonzero constant (a polynomial of degree zero). In this particular case $p(x)/q(x)$ is again a polynomial. Thus, polynomials are also rational functions; but whereas polynomials are defined for all values of x, the rational function $p(x)/q(x)$ is *not* defined at those points at which $q(x) = 0$.

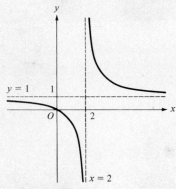

Figure 4.30
The graph of $y = x/(x-2)$. Observe that $x/(x-2) = 1 + 2/(x-2)$.

In Section 3.3 we learned how to find the derivatives of rational functions. We have the formula

$$r'(x) = \frac{q(x)p'(x) - p(x)q'(x)}{[q(x)]^2}$$

Since p and q are polynomials, $qp' - pq'$ is also a polynomial; also, polynomials are differentiable. We thus conclude the following:

1. Rational functions are differentiable.
2. The derivative of a rational function is a rational function.

ASYMPTOTES

Looking at the graph of $y = x/(x-2)$ in Figure 4.30, you will see that it approaches the horizontal line $y = 1$ as $x \to \infty$. The line $y = 1$ is said to be a *horizontal asymptote*. We know that the curve approaches the line $y = 1$ because

$$\lim_{x \to \infty} \frac{x}{x-2} = \lim_{x \to \infty} \left(\frac{1}{1 - 2/x}\right) = 1$$

and the same is true when $x \to -\infty$.

The following is a general definition.

DEFINITION 2

The line $y = b$ is a *horizontal asymptote* of the curve $y = f(x)$ if

$$\lim_{x \to \infty} f(x) = b \quad \text{or} \quad \lim_{x \to -\infty} f(x) = b$$

When x approaches 2 from the right in the above example, then $x/(x-2) \to \infty$ [since $x/(x-2) > 0$ for $x > 2$], and the curve approaches the vertical line $x = 2$. Symbolically, this is expressed as

$$\lim_{x \to 2+} \frac{x}{x-2} = \infty$$

When x approaches 2 from the left, then $x/(x-2) \to -\infty$ [since $x/(x-2) < 0$ for $0 < x < 2$], and again the curve approaches the line $x = 2$. This is expressed as

$$\lim_{x \to 2-} \frac{x}{x-2} = -\infty$$

We give the following general definition.

DEFINITION 3

The line $x = a$ is a *vertical asymptote* of the curve $y = f(x)$ if any one of the following conditions is true:

$$\lim_{x \to a+} f(x) = \infty \qquad \lim_{x \to a-} f(x) = \infty$$

$$\lim_{x \to a+} f(x) = -\infty \qquad \lim_{x \to a-} f(x) = -\infty$$

4.3 RATIONAL FUNCTIONS AND ASYMPTOTES

Limits as $x \to a^+$ or $x \to a^-$ are fully discussed in Section 4.4.

Vertical asymptotes are easy to describe for some quotients of functions (such as rational functions):

> The line $x = a$ is a vertical asymptote of the curve $y = f(x)/g(x)$ if $g(a) = 0$ and $f(a) \neq 0$.

Example 1

Find the vertical and horizontal asymptotes of

$$y = \frac{x^2 + 1}{x^2 - 2x - 3}$$

Solution
Put $p(x) = x^2 + 1$ and $q(x) = x^2 - 2x - 3 = (x+1)(x-3)$. Then $q(-1) = 0$ and $q(3) = 0$, while $p(-1) = 2 \neq 0$ and $p(3) = 10 \neq 0$. Hence, the lines $x = -1$ and $x = 3$ are vertical asymptotes of the curve. Writing

$$f(x) = \frac{x^2 + 1}{x^2 - 2x - 3} = \frac{1 + 1/x^2}{1 - (2/x) - (3/x^2)}$$

we see that

$$\lim_{x \to \infty} f(x) = 1 \quad \text{and} \quad \lim_{x \to -\infty} f(x) = 1$$

Hence, the line $y = 1$ is a horizontal asymptote of the curve (see Figure 4.31).

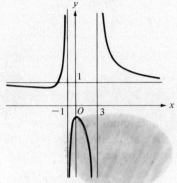

Figure 4.31
The graph and asymptotes of $y = \dfrac{x^2 + 1}{x^2 - 2x - 3}$

Example 2

Find the asymptotes of

$$y = \frac{x - 3}{x^2 - 3x}$$

Solution
Put $p(x) = x - 3$ and $q(x) = x^2 - 3x$. Then $q(0) = 0$ and $q(3) = 0$. Since $p(0) = -3 \neq 0$, the line $x = 0$ is an asymptote, but $p(3) = 0$ and so the line $x = 3$ is *not* an asymptote. This is explained as follows: When $x \neq 0$ and $x \neq 3$,

$$y = \frac{x - 3}{x^2 - 3x} = \frac{x - 3}{x - 3} \frac{1}{x} = \frac{1}{x}$$

This means that the curve $y = (x - 3)/(x^2 - 3x)$ is the *same* as the curve $y = 1/x$, except that it is not defined at $x = 3$ (see Figure 4.32). The x axis is seen to be a horizontal asymptote.

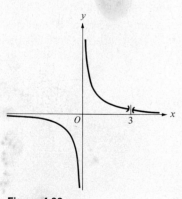

Figure 4.32
The graph of $y = \dfrac{x - 3}{x^2 - 3x}$

Example 3

Find the asymptotes of

$$y = \frac{x^2}{x^2 + 1}$$

Solution
Since $x^2 + 1 \neq 0$ for all x, we conclude that no vertical asymptotes exist.

To find horizontal asymptotes, we write

$$\frac{x^2}{x^2 + 1} = \frac{1}{1 + 1/x^2}$$

and see that

$$\lim_{x \to \infty} \frac{x^2}{x^2 + 1} = 1$$

Hence, the line $y = 1$ is a horizontal asymptote (see Figure 4.33).

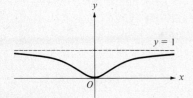

Figure 4.33
The graph of $y = x^2/(x^2 + 1)$.

PLOTTING OF RATIONAL FUNCTIONS
The scheme of the last section for plotting polynomials can be used for plotting rational functions with a single modification: We first locate the asymptotes of the function, then apply the scheme to the individual portions of the graph as separated by vertical asymptotes.

Example 4

Plot the function

$$f(x) = \frac{1}{x^2 - 1}$$

Solution
It is convenient to put our findings in table form. We first list the asymptotes:

Behavior of x	Behavior of $f(x) = 1/(x^2 - 1)$	Asymptotes
$\to \infty$ $\to -\infty$	$\to 0$	$y = 0$
$\to (-1)-$ $\to (-1)+$	$\to \infty$ $\to -\infty$	$x = -1$
$\to 1-$ $\to 1+$	$\to -\infty$ $\to \infty$	$x = 1$

We now compute $f'(x)$ and $f''(x)$:

$$f'(x) = \frac{-2x}{(x^2-1)^2}$$

$$f''(x) = \frac{6x^2+2}{(x^2-1)^3}$$

	$f'(x)$	$f''(x)$	$f(x)$
$x < -1$		> 0	concave up
	> 0		increasing
$-1 < x < 1$		< 0	concave down
$-1 < x < 0$	> 0		increasing
$0 < x < 1$	< 0		decreasing
$x = 0$	$= 0$	< 0	local maximum
$x > 1$		> 0	concave up
	< 0		decreasing

The graph is given in Figure 4.34.

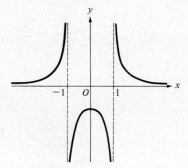

Figure 4.34
The graph of $f(x) = 1/(x^2 - 1)$.

x	$f(x)$
-2	$\frac{1}{3}$
$-\frac{3}{2}$	$\frac{4}{5}$
$-\frac{5}{4}$	$\frac{16}{9}$
$-\frac{3}{4}$	$-\frac{16}{7}$
$-\frac{1}{2}$	$-\frac{4}{3}$
0	-1
$\frac{1}{2}$	$-\frac{4}{3}$
$\frac{3}{4}$	$-\frac{16}{7}$
$\frac{5}{4}$	$\frac{16}{9}$
$\frac{3}{2}$	$\frac{4}{5}$
2	$\frac{1}{3}$

EXERCISES

Find dy/dx in Exercises 1–6.

1. $y = \dfrac{a - x}{a + x}$

2. $y = \dfrac{x^m}{1 + x^n}$

3. $y = \dfrac{x^2 + x + 1}{1 + x + x^2}$

4. $y = \dfrac{x^2 - 4}{x^2 + 4}$

5. $y = \dfrac{x^4}{c^2 - x^2}$

6. $y = \dfrac{x^4 - x}{x + 1}$

In Exercises 7–10, u and v are functions of x. Find the indicated derivatives.

7. $\dfrac{d}{dx}\left(\dfrac{v-1}{v+1}\right) =$

8. $\dfrac{d}{dx}\left(\dfrac{u}{v}\right)^2 =$

9. $\dfrac{d}{dx}\left(\dfrac{v-u}{v+u}\right) =$

10. $\dfrac{d}{dx}\left(\dfrac{u^2-v^2}{u^2+v^2}\right) =$

Find the vertical and horizontal asymptotes in Exercises 11–16.

11. $y = \dfrac{x^2 + 1}{x^2 - x}$

12. $y = \dfrac{x^4}{2^2 - x^2}$

13. $y = \dfrac{4x^2 - 1}{x^3 - 4}$

14. $y = \dfrac{1}{x^2 + 1}$

15. $y = \begin{cases} \dfrac{x^2}{x^2+1} & \text{for } x \geq 0 \\ x^2 & \text{for } x < 0 \end{cases}$

16. $y = \begin{cases} \dfrac{1}{x^2-1} & \text{for } -1 < x < 1 \\ \dfrac{1}{1-x} & \text{for } x < -1 \\ \dfrac{x}{x+1} & \text{for } x > 1 \end{cases}$

17. Let $q(x) = x(x-3)(x-5)$. Find the vertical and horizontal asymptotes of the following functions:

   (a) $f_0(x) = 1/q(x)$ (b) $f_1(x) = x/q(x)$
   (c) $f_2(x) = x^2/q(x)$ (d) $f_3(x) = x^3/q(x)$
   (e) $f_4(x) = x^4/q(x)$

Plot the graphs of the functions in Exercises 18–25, indicating their asymptotes.

18. $y = \dfrac{1}{x^2+1}$ 19. $y = x^2 + \dfrac{1}{x^2}$

20. $y = \dfrac{1}{x^2-4}$ 21. $y = \dfrac{x}{x+2}$

22. $y = \dfrac{1-x}{1+x}$ 23. $y = \left|\dfrac{1-x}{1+x}\right|$

24. $y = \left|\dfrac{x}{x+2}\right|$ 25. $y = \left|\dfrac{1+x^2}{1-x^2}\right|$

4.4 ONE-SIDED LIMITS

As a prelude to discussing continuity on closed intervals, we must refine the concept of limit. In discussing limits we used the notation $x \to x_0$ to convey the idea that x approaches x_0 (see Figure 4.35); no restrictions were imposed on the manner of approach. Thus, when we consider a function such as $f(x) = \sqrt{x}$, we cannot speak of $\lim_{x \to 0} f(x)$ since f is undefined for values $x < 0$. To accommodate such situations, we introduce *one-sided limits,* which are limits when x approaches a given number from one side only, either from the left or from the right.

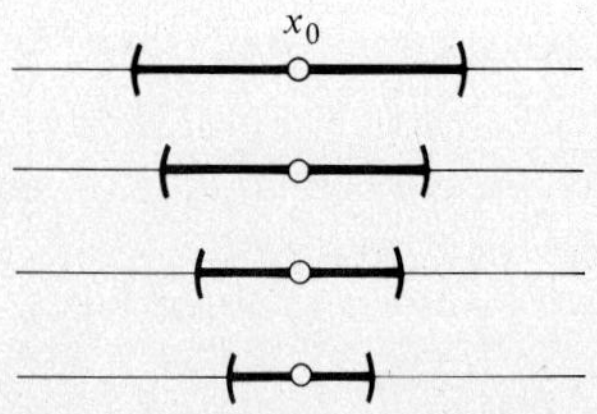

Figure 4.35
$x \to x_0$. x is restricted to smaller and smaller intervals centered at x_0.

NOTATION

$x \to x_0+$ is read "x approaches x_0 from the right";
$x \to x_0-$ is read "x approaches x_0 from the left."

4.4 ONE-SIDED LIMITS 111

This notation is explained graphically in Figures 4.36 and 4.37. Before defining one-sided limits, we consider an example.

Example 1

Consider the function

$$f(x) = \frac{|x|}{x}(x^2 + 1)$$

which is defined for all values $x \neq 0$. The graph of $y = f(x)$ is easily obtained when we recall that

$$\frac{|x|}{x} = \begin{cases} 1 & \text{when } x > 0 \\ -1 & \text{when } x < 0 \end{cases}$$

and, consequently,

$$f(x) = \begin{cases} x^2 + 1 & \text{when } x > 0 \\ -(x^2 + 1) & \text{when } x < 0 \end{cases}$$

The graph of this function is given in Figure 4.38. Examining this graph, we discover the following facts:

$f(x)$ approaches 1 as x approaches 0 from the right;
$f(x)$ approaches -1 as x approaches 0 from the left.

These facts are expressed symbolically as

$$\lim_{x \to 0+} f(x) = 1 \quad \text{or} \quad f(x) \to 1 \text{ as } x \to 0+$$

$$\lim_{x \to 0-} f(x) = -1 \quad \text{or} \quad f(x) \to -1 \text{ as } x \to 0-$$

Thus, we have a function that has no limit at $x = 0$, but it does have "one-sided" limits. Formally we lay down this definition:

DEFINITION 1

If f approaches a single number L as x approaches x_0 from the right, then we write

$$\lim_{x \to x_0+} f(x) = L$$

and say that f has a *right-hand limit* at x_0.

If f approaches a single number L as x approaches x_0 from the left, then we write

$$\lim_{x \to x_0-} f(x) = L$$

and say that f has a *left-hand limit* at x_0.

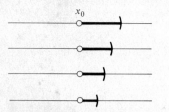

Figure 4.36
$x \to x_0+$. x is restricted to smaller and smaller intervals lying to the right of x_0.

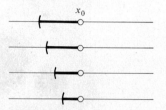

Figure 4.37
$x \to x_0-$. x is restricted to smaller and smaller intervals lying to the left of x_0.

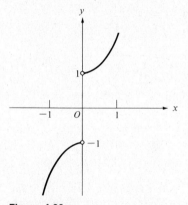

Figure 4.38
The graph of $f(x) = (|x|/x)(x^2+1)$. Observe that $f(0)$ is undefined; in fact, there is a jump of two units as x goes through 0.

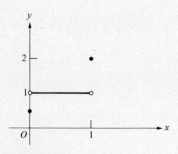

Figure 4.39
The graph of $y = g(x)$.

Example 2

Let $g(x)$ be defined on the closed interval $[0, 1]$ as follows:

$$g(x) = \begin{cases} \frac{1}{2} & \text{for } x = 0 \\ 1 & \text{for } 0 < x < 1 \\ 2 & \text{for } x = 1 \end{cases}$$

(see Figure 4.39). Then $g(x) = 1$ whenever $0 < x < 1$, so that

$$\lim_{x \to 0+} g(x) = 1 \quad \text{and} \quad \lim_{x \to 1-} g(x) = 1$$

It should be observed that

1. $\lim_{x \to 0+} g(x) \neq g(0) = \frac{1}{2}$ and $\lim_{x \to 1-} g(x) \neq g(1) = 2$.
2. $g(x)$ has no left-hand limit at 0 and no right-hand limit at 1, since $g(x)$ is not defined outside the interval $[0, 1]$.

Example 3

Find the right-hand and left-hand limits of $f(x)$ at $x = 0$ and $x = 1$ when $f(x) = x$ for $0 < x \leq 1$.

Solution
Here $f(0)$ is undefined, but

$$\lim_{x \to 0+} f(x) = \lim_{x \to 0+} x = 0$$

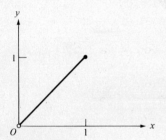

Figure 4.40
The graph of $y = f(x)$.

(see Figure 4.40). Also,

$$\lim_{x \to 1-} f(x) = \lim_{x \to 1-} x = 1$$

Since $f(1) = 1$, we actually have

$$\lim_{x \to 1-} f(x) = f(1)$$

The following theorem relates limits and one-sided limits.

THEOREM 1
Let f be defined on an interval $(x_0 - \delta, x_0 + \delta)$ except possibly at x_0. Then

$$\lim_{x \to x_0} f(x) = L$$

if and only if

$$\lim_{x \to x_0+} f(x) = L \quad \text{and} \quad \lim_{x \to x_0-} f(x) = L$$

EXERCISES

In Exercises 1–7, find the right-hand and left-hand limits at the indicated points. When no such limit exists, explain why.

1. $f(x) = x - \dfrac{x}{|x|}$ $\lim\limits_{x \to 0+} f(x) =$ $\lim\limits_{x \to 0-} f(x) =$

2. $f(x) = x - \dfrac{x}{|x|}$ $\lim\limits_{x \to 2+} f(x) =$ $\lim\limits_{x \to 2-} f(x) =$

3. $g(x) = \begin{cases} 0 & \text{for } x \leq 0 \\ 1/x & \text{for } x > 0 \end{cases}$ $\lim\limits_{x \to 0+} g(x) =$ $\lim\limits_{x \to 0-} g(x) =$

4. $h(x) = \begin{cases} 1 & \text{for } x \neq 1 \\ 0 & \text{for } x = 1 \end{cases}$ $\lim\limits_{x \to 1+} h(x) =$ $\lim\limits_{x \to 1-} h(x) =$

5. $k(x) = \begin{cases} 1 & \text{for } x = 1/n, \\ & n = 1, 2, 3, \ldots \\ 0 & \text{otherwise} \end{cases}$ $\lim\limits_{x \to 0+} k(x) =$ $\lim\limits_{x \to 0-} k(x) =$

6. $u(x) = \dfrac{x}{x-2}$ $\lim\limits_{x \to 2+} u(x) =$ $\lim\limits_{x \to 2-} u(x) =$

7. $v(x) = x\sqrt{1 - x^2}$ $\lim\limits_{x \to 0+} v(x) =$ $\lim\limits_{x \to 0-} v(x) =$

In Exercises 8–11, find the points at which f has no limit. Explain your answer.

8. $f(x) = \dfrac{x}{|x|}$

9. $f(x) = \begin{cases} x/|x| & \text{for } x \neq 0 \\ 1 & \text{for } x = 0 \end{cases}$

10. $f(x) = \dfrac{x^2 - 1}{x - 1}$

11. $f(x) = 4 - x^2$

4.5 CONTINUITY

The concept of continuity was mentioned briefly in Section 4.1, where it was stated that such functions possess a maximum and a minimum on a closed interval. In this section we look at some properties of continuous functions that are of fundamental importance. Definition 4 of Section 4.1 will be restated below, but first we backtrack to define continuity at a point and on an open interval.

DEFINITION 1

A function $f(x)$ is *continuous at* x_0 if it is defined on an interval $(x_0 - \delta, x_0 + \delta)$ and

$$\lim_{x \to x_0} f(x) = f(x_0)$$

Our first theorem shows that differentiability implies continuity (the converse, however, is false).

THEOREM 1
If $f(x)$ has a derivative at x_0, then $f(x)$ is continuous at x_0.

Proof
For $x \neq x_0$, we have

$$f(x) - f(x_0) = (x - x_0) \frac{f(x) - f(x_0)}{x - x_0}$$

Since

$$\lim_{x \to x_0} (x - x_0) = 0$$

and

$$\lim_{x \to x_0} \frac{f(x) - f(x_0)}{x - x_0} = f'(x_0)$$

we have

$$\lim_{x \to x_0} [f(x) - f(x_0)] = \lim_{x \to x_0} \left[(x - x_0) \frac{f(x) - f(x_0)}{x - x_0} \right]$$

$$= \lim_{x \to x_0} (x - x_0) \cdot \lim_{x \to x_0} \frac{f(x) - f(x_0)}{x - x_0}$$

$$= 0 \cdot f'(x_0) = 0$$

Hence

$$\lim_{x \to x_0} f(x) = f(x_0)$$

Example 1

Polynomials have derivatives at all points x_0. Theorem 1 proves, therefore, that polynomials are continuous at all points x_0.

Example 2

The function

$$f(x) = x^{2/3}$$

is continuous at 0, since $f(0) = 0$ and

$$\lim_{x \to 0} x^{2/3} = 0^{2/3} = 0$$

It is not differentiable at 0, however, as we saw in Example 3, Section 2.5. This example shows that a function may be continuous at a point without having a derivative there.

The properties of limits in Theorem 1 of Section 2.6 yield the following properties of continuity.

THEOREM 2
Let $f(x)$ and $g(x)$ be continuous at x_0. Then $f(x)+g(x)$, $f(x)-g(x)$, and $f(x)g(x)$ are continuous at x_0, and $f(x)/g(x)$ is continuous at x_0 if $g(x_0) \neq 0$.

We now extend the concept of continuity to open intervals.

DEFINITION 2
A function f is *continuous on an open interval I* if f is continuous at each point of I.

Equivalently, this definition says that f is continuous on the open interval I if $\lim_{x \to x_0} f(x) = f(x_0)$ for each x_0 in I. We mention that I may be of one of the forms (a, b), $(-\infty, b)$, (a, ∞), or $(-\infty, \infty)$.

Example 3

Let $f(x) = 1/x$. For each point x_0 in the interval $(0, \infty)$, we have

$$\lim_{x \to x_0} f(x) = \lim_{x \to x_0} \frac{1}{x} = \frac{1}{x_0} = f(x_0)$$

Hence, f is continuous on the interval $(0, \infty)$.

Definition 4 of Section 4.1 is now restated as follows:

DEFINITION 3
A function f is continuous on a closed interval $I = [a, b]$ if

1. f is continuous on the open interval (a, b).
2. $\lim_{x \to a+} f(x) = f(a)$ and $\lim_{x \to b-} f(x) = f(b)$.

Example 4

Polynomials are continuous on the entire line $(-\infty, \infty)$. Hence they are continuous on any closed interval $[a, b]$.

Example 5

Let

$$g(x) = \begin{cases} 1/x & \text{for } 0 < x \leq 1 \\ 0 & \text{for } x = 0 \end{cases}$$

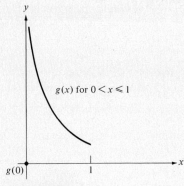

Figure 4.41
The graph of $g(x)$.

(see Figure 4.41). Then $g(x)$ is defined on the closed interval $[0, 1]$, but it is not continuous there since

$$\lim_{x \to 0+} g(x) \neq 0$$

In fact, no matter how $g(0)$ is defined, the resulting function will not be continuous since $\lim_{x \to 0+} g(x)$ does not exist.

BOUNDS

In considering functions $f(x)$ defined on an interval I, it is often important to know whether the values of $f(x)$ as x varies in I are bounded.

DEFINITION 4

The function $f(x)$ is *bounded* on the interval I if there are constants m and M such that

$$m \leq f(x) \leq M \text{ for all } x \text{ in } I$$

The number m is called a *lower bound*, M an *upper bound*.

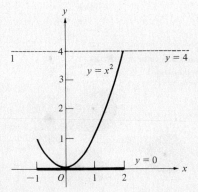

Figure 4.42
The graph of $f(x) = x^2$ on $[-1, 2]$.

Example 6

The function $f(x) = x^2$ on the interval $[-1, 2]$ is bounded. As x varies from -1 to 2, $f(x) = x^2$ decreases from 1 to 0, and then increases to 4 (see Figure 4.42). Thus,

$$0 \leq f(x) \leq 4$$

so that $m = 0$ is a lower bound, and $M = 4$ is an upper bound. Observe that any number less than 0 is likewise a lower bound and any number greater than 4 is an upper bound.

It is worthwhile noting that, geometrically, Definition 4 means that the graph $y = f(x)$ lies between the horizontal lines $y = m$ and $y = M$.

Example 7

Is -5 a lower bound for $f(x) = x^2$ on $[-1, 2]$?

Solution
Yes, since $-5 \leq f(x)$ for all x in the given interval.

Example 8

Find bounds for the function

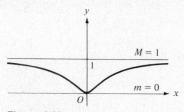

Figure 4.43
The graph of $f(x) = \dfrac{x^2}{x^2+1}$.

$$f(x) = \frac{x^2}{x^2+1}$$

defined for all values of x.

Solution
Consult Figure 4.43. Since

$$0 \le x^2 < x^2 + 1$$

we find, upon division by $x^2 + 1$, that

$$0 \le \frac{x^2}{x^2+1} < 1$$

Hence, 0 and 1 are lower and upper bounds, respectively.

Example 9

Consider the function $f(x)$ defined on $[0, 1]$ by

$$f(x) = \begin{cases} 0 & \text{for } x = 0 \\ 1/x^2 & \text{for } 0 < x \le 1 \end{cases}$$

Is this function bounded on the interval $[0, 1]$?

Solution
It is apparent from the graph and numerical table in Figure 4.44 that $f(x)$ is *unbounded* on $[0, 1]$. This is also easy to prove, because if $M > 1$ is any number, then $0 < 1/M < 1$. Hence, taking $x = 1/M$, we have

$$\frac{1}{x^2} = \frac{1}{(1/M)^2} = M^2 > M$$

This tells us that for no number M is $f(x) \le M$ for all x in $[0, 1]$.

Returning to continuous functions, we have the following fact, which was stated in Section 4.1.

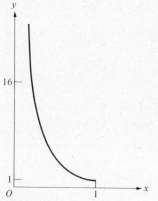

Figure 4.44
The graph of $y = f(x)$.

x	$f(x)$
1	1
$\frac{1}{10}$	100
$\frac{1}{100}$	10000
$\frac{1}{1000}$	1000000

> **THEOREM 3**
> If $f(x)$ is continuous on the closed interval $[a, b]$, then there are two points $x_{\min}$ and $x_{\max}$ in $[a, b]$ such that
> $$f(x_{\min}) \le f(x) \le f(x_{\max}) \text{ for all points } x \text{ in } [a, b]$$

This theorem tells us that a continuous function on an interval $[a, b]$ is bounded and that it actually takes on its bounds.

We conclude this section with a basic fact that will be used several times in the next section.

GRAPHS AND CONTINUITY

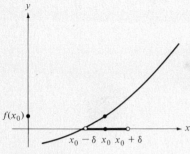

Figure 4.45

THEOREM 4
If $f(x)$ is continuous at x_0 and $f(x_0) > 0$, then there is an open interval $I = (x_0 - \delta, x_0 + \delta)$ such that $f(x) > 0$ for all points x in I.

Proof
Consult Figure 4.45. The statement

$$\lim_{x \to x_0} f(x) = f(x_0)$$

means that for any error $\epsilon > 0$ there is a number $\delta > 0$ such that

$$|f(x) - f(x_0)| < \epsilon \quad \text{whenever} \quad |x - x_0| < \delta$$

If we choose the particular error $\epsilon = \frac{1}{2} f(x_0)$, then this reads

$$|f(x) - f(x_0)| < \tfrac{1}{2} f(x_0) \quad \text{whenever} \quad |x - x_0| < \delta$$

The last statement is equivalent to

$$\tfrac{1}{2} f(x_0) < f(x) < \tfrac{3}{2} f(x_0) \quad \text{whenever} \quad |x - x_0| < \delta$$

In particular, since $f(x_0) > 0$, it follows that

$$f(x) > 0 \quad \text{whenever} \quad |x - x_0| < \delta$$

and the theorem is thus established.

EXERCISES

A function $f(x)$ may *fail* to be continuous at a point x_0 for any *one* of the following three reasons:

(a) $f(x_0)$ is not defined.

(b) $f(x_0)$ is defined, but $\lim_{x \to x_0} f(x)$ does not exist.

(c) $f(x_0)$ is defined, but $\lim_{x \to x_0} f(x) \neq f(x_0)$.

Of the following functions, find which are discontinuous, where, and give a reason why.

1. $f(x) = \dfrac{1}{x - 1}$

2. $g(x) = x^2 + 2x + 1$

3. $h(x) = \dfrac{x^2 - 1}{x - 1}$

4. $k(x) = \dfrac{1}{(x - 1)(x^2 + 2x + 1)}$

5. $u(x) = \dfrac{x}{|x|}$

6. $v(x) = \dfrac{x}{1 + x^2}$

7. $f(x) = \dfrac{1}{x^2 + 1}$

8. $g(x) = x^2 - 1$

9. $h(x) = \dfrac{x^2 - 1}{x^2 + 1}$ 　　　　10. $k(x) = \dfrac{x^2 + 1}{x^2 - 1}$

11. $u(x) = 1$ 　　　　12. $v(x) = \dfrac{1}{x(x-2)}$

In Exercises 13–17, find a bound M such that $|f(x)| \leq M$ for the indicated range of $f(x)$.

13. $f(x) = \sqrt{x}$ 　　$0 \leq x \leq 4$ 　　14. $f(x) = |x + 1|$ 　　$-2 < x < 2$

15. $f(x) = x^2 - 9$ 　　$-4 \leq x \leq 4$ 　　16. $f(x) = 9 - x^2$ 　　$-4 \leq x \leq 4$

17. $f(x) = \dfrac{x - |x|}{1 + |x|}$

Find lower and upper bounds for the functions in Exercises 18–27. When one bound fails to exist, say that the function is unbounded.

18. $f(x) = \dfrac{x - 1}{x + 1}$ 　　$0 \leq x \leq 1$ 　　19. $g(x) = 2$

20. $h(x) = \sqrt{x}$ 　　　　21. $k(x) = |x|$

22. $f(x) = \dfrac{x}{|x|}$ 　　$x \neq 0$ 　　23. $g(x) = x - |x|$

24. $h(x) = \dfrac{x + |x|}{x}$ 　　$x > 0$ 　　25. $k(x) = -x^2$ 　　$-4 \leq x \leq 4$

26. $m(x) = |x - 2|$ 　　$-2 < x < 2$ 　　27. $p(x) = \begin{cases} 1/x & x \neq 0 \\ 0 & x = 0 \end{cases}$

4.6 THE MEAN VALUE THEOREM

This section centers around an intuitively simple result known as *Rolle's theorem* (see Fig. 4.46). Roughly stated, it says that a differentiable function that meets the x axis in two points must have a horizontal tangent at an intermediate point. This theorem is the first step in a major advance in our knowledge. It will enable us to prove several assertions made earlier in the text, as well as to solve many new kinds of problems.

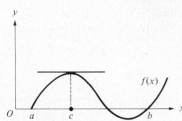

Figure 4.46
An illustration of Rolle's theorem. $f(x)$ meets the x axis at $x = a$ and $x = b$, and has a horizontal tangent at the point $x = c$ lying between a and b.

ROLLE'S THEOREM
Let the function $f(x)$, defined on the closed interval $[a, b]$, satisfy the following conditions:

(i) $f(x)$ is continuous on the closed interval $[a, b]$.
(ii) $f(x)$ is differentiable on the open interval (a, b).
(iii) $f(a) = f(b) = 0$.

120 GRAPHS AND CONTINUITY

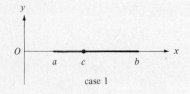
case 1

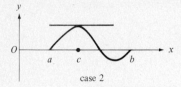
case 2

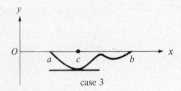
case 3

Figure 4.47
The three cases in the proof of Rolle's theorem.

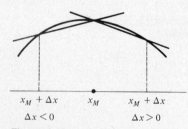

Figure 4.48
$\Delta x \to 0+$ means that Δx approaches 0 through positive values; $\Delta x \to 0-$ means that Δx approaches 0 through negative values (see Section 4.4).

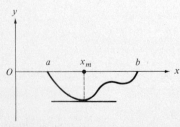

Figure 4.49
The graph of $y = f(x)$.

Then there is a point c, $a < c < b$, such that
$$f'(c) = 0$$

Proof
Three cases are distinguished in the proof (see Figure 4.47).

Case 1
If $f(x)$ is the constant function $f(x) \equiv 0$, then $f'(x) = 0$ for each point x in (a, b), and c can be chosen to be any point in (a, b).

Case 2
Suppose that $f(x) > 0$ for some x in (a, b). Then $f(x)$ takes on its maximum value at a point x_M in (a, b); that is,

$$f(x) \leq f(x_M) \tag{1}$$

for all x in $[a, b]$. We now show that $f'(x_M) = 0$ and this enables us to take $c = x_M$. By definition,

$$f'(x_M) = \lim_{\Delta x \to 0} \frac{f(x_M + \Delta x) - f(x_M)}{\Delta x} \tag{2}$$

and owing to (1), $f(x_M + \Delta x) \leq f(x_M)$, which is the same as $f(x_M + \Delta x) - f(x_M) \leq 0$. Hence (see Figure 4.48),

$$\frac{f(x_M + \Delta x) - f(x_M)}{\Delta x} \begin{cases} \leq 0 & \text{when } \Delta x > 0 \\ \geq 0 & \text{when } \Delta x < 0 \end{cases}$$

and accordingly,

$$\lim_{\Delta x \to 0+} \frac{f(x_M + \Delta x) - f(x_M)}{\Delta x} \leq 0$$

whereas

$$\lim_{\Delta x \to 0-} \frac{f(x_M + \Delta x) - f(x_M)}{\Delta x} \geq 0$$

Since, according to (2), each of these limits equals $f'(x_M)$, it follows that
$$f'(x_M) \leq 0 \quad \text{and} \quad f'(x_M) \geq 0$$
and hence $f'(x_M) = 0$.

Case 3
When neither case 1 nor case 2 holds, then $f(x) < 0$ at some point x_0 in (a, b) (see Figure 4.49). If we put $g(x) = -f(x)$, then $g(x)$ satisfies the hypotheses of the theorem and, moreover, $g(x_0) > 0$. We can therefore repeat the argument in case 2, with $f(x)$ replaced

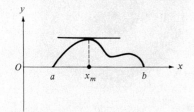

Figure 4.50
The graph of $y = -f(x)$.

by $g(x)$, leading to the conclusion that there is a point x_m in (a, b) such that $g'(x_m) = 0$ (see Figure 4.50). Since, however, $g'(x) = -f'(x)$ for all x in (a, b), we find that $f'(x_m) = 0$.

Let us rephrase Rolle's theorem. The theorem states that under the stipulated conditions there is a tangent to $y = f(x)$ at $x = c$ that is parallel to the line joining $(a, f(a))$ and $(b, f(b))$ (see Figure 4.51). It seems that this statement should also be true when the condition $f(a) = f(b) = 0$ is violated. This is the subject of the next theorem (see Figure 4.52).

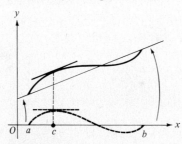

Figure 4.51
Motivating the mean value theorem.

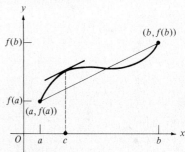

Figure 4.52
Illustrating the mean value theorem.

MEAN VALUE THEOREM
Let the function $f(x)$, defined on the closed interval $[a, b]$, satisfy the following conditions:

(i) $f(x)$ is continuous on the closed interval $[a, b]$.
(ii) $f(x)$ is differentiable on the open interval (a, b).

Then there is a point c, $a < c < b$, such that

$$f'(c) = \frac{f(b) - f(a)}{b - a} \tag{3}$$

Proof
We observe that conditions (i) and (ii) of Rolle's theorem are satisfied here, and this makes it natural to try to reduce this theorem to the former. Observing that the line joining the points $(a, f(a))$ and $(b, f(b))$ can be expressed as

$$L(x) = \frac{f(b) - f(a)}{b - a}(x - a) + f(a)$$

(see Figure 4.53), we consider the function

$$g(x) = f(x) - L(x)$$

Since $f(x)$ and $L(x)$ are continuous on $[a, b]$ and differentiable on (a, b), so is their difference $g(x)$. Furthermore,

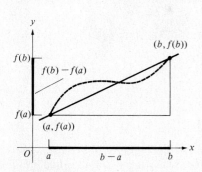

Figure 4.53
The line $y = L(x)$.

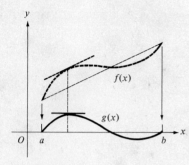

Figure 4.54
The graph of $y = f(x)$ and $y = g(x)$.

$$g(a) = f(a) - L(a) = f(a) - \left[\frac{f(b) - f(a)}{b - a}(a - a) + f(a)\right]$$
$$= f(a) - f(a) = 0$$

and

$$g(b) = f(b) - L(b) = f(b) - \left[\frac{f(b) - f(a)}{b - a}(b - a) + f(a)\right]$$
$$= f(b) - [f(b) - f(a) + f(a)] = 0$$

and we see that $g(x)$ satisfies the three conditions of Rolle's theorem (see Figure 4.54). Accordingly, there is a point c, $a < c < b$, such that $g'(c) = 0$. But

$$g'(x) = f'(x) - L'(x) = f'(x) - \frac{f(b) - f(a)}{b - a}$$

so that, in particular,

$$g'(c) = f'(c) - \frac{f(b) - f(a)}{b - a} = 0$$

This, however, is the conclusion we were after.

Example 1

The mean value theorem has an interesting interpretation in terms of velocity. Recall that if $s = f(t)$ is the equation of motion of an object, then $f'(t_0)$ is its *velocity* at time t_0, and $[f(b) - f(a)]/(b - a)$ is its *average velocity* over the time interval $a \leq t \leq b$. In this framework the mean value theorem states that the average velocity is actually taken on at some time t_0. If

$$f(t) = t^2 - 6t + 8 \qquad [a, b] = [2, 6]$$

(see Figure 4.55), find a time t_0 at which the average velocity is assumed.

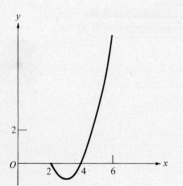

Figure 4.55
The graph of $f(t) = t^2 - 6t + 8$.

Solution
The average velocity over the time interval $2 \leq t \leq 6$ is

$$\frac{f(6) - f(2)}{6 - 2} = \frac{8 - 0}{4} = 2$$

If we put

$$f'(t_0) = 2t_0 - 6 = 2$$

we find that $t_0 = 4$. Hence, the average velocity is taken on at time $t_0 = 4$.

Example 2

Does the mean value theorem apply to the function

$$f(x) = (x-1)^{2/3}$$

when $0 \leq x \leq 2$?

Solution

Consult Figure 4.56. Writing $f(x)$ in the form $f(x) = [(x-1)^2]^{1/3}$ shows that it is continuous at each x; in particular, this is so for $0 \leq x \leq 2$. We find, however, that

$$f'(x) = \tfrac{2}{3}(x-1)^{-1/3}$$

is not defined at $x = 1$. Indeed,

$$\frac{f(1+\Delta x) - f(1)}{\Delta x} = \frac{(\Delta x)^{2/3} - 0^{2/3}}{\Delta x} = \frac{1}{(\Delta x)^{1/3}}$$

The fact that $1/(\Delta x)^{1/3} \to \infty$ as $\Delta x \to 0$ tells us that

$$\lim_{\Delta x \to 0} \frac{f(1+\Delta x) - f(1)}{\Delta x}$$

does not exist, but this says that $f'(1)$ is undefined. Hence, the hypothesis of the mean value theorem that $f(x)$ be differentiable on $(a, b) = (0, 2)$ is not satisfied, and the theorem does not apply to the function at hand.

We note, incidentally, that

$$\frac{f(2) - f(0)}{2 - 0} = \frac{1 - 1}{2} = 0$$

and clearly there is no point c for which $f'(c) = 0$.

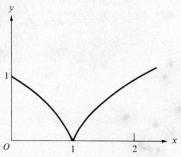

Figure 4.56
The graph of $f(x) = (x-1)^{2/3}$.

Several important conclusions can be obtained from the mean value theorem. In deriving these, we use the following observation: If x is any point in the interval $(a, b]$, then the theorem can be applied to the interval $[a, x]$. Formula (3) then becomes

$$f'(c) = \frac{f(x) - f(a)}{x - a}$$

where, this time, $a < c < x$; equivalently,

$$\boxed{f(x) = f(a) + (x-a)f'(c) \qquad a < c < x} \qquad (4)$$

THEOREM 1

Let $f(x)$ be continuous on the closed interval $[a, b]$. If $f'(x) = 0$ on the open interval (a, b), then $f(x) \equiv$ constant.

Proof

If x is any point in $(a, b]$, then $f(x)$ is continuous on $[a, x]$ and differentiable on (a, x). Using the mean value theorem, we obtain formula (4). Since $f'(x) = 0$ for all x in (a, b) it follows, in particular, that $f'(c) = 0$. Hence, $f(x) = f(a)$. Since this is true for *each* point x in $(a, b]$, it follows that $f(x) \equiv f(a)$; that is, $f(x) =$ constant.

The following result, used earlier in this chapter, is now easy to prove.

THEOREM 2

Let $f(x)$ be continuous on the closed interval $[a, b]$.

(i) If $f'(x) > 0$ on the open interval (a, b), then $f(x)$ is increasing on $[a, b]$.
(ii) If $f'(x) < 0$ on the open interval (a, b), then $f(x)$ is decreasing on $[a, b]$.

We recall that $f(x)$ is *increasing* on $[a, b]$ if $f(x_1) < f(x_2)$ for all points $x_1 < x_2$ in $[a, b]$; $f(x)$ is *decreasing* on $[a, b]$ if $f(x_1) > f(x_2)$ for all points $x_1 < x_2$ in $[a, b]$.

Proof

If $x_1 < x_2$ are arbitrary points in $[a, b]$, then the mean value theorem can be applied to the interval $[x_1, x_2]$ to yield the formula

$$f(x_2) = f(x_1) + (x_2 - x_1)f'(c)$$

for some point c such that $x_1 < c < x_2$. If (i) is the case, then $f'(c) > 0$. Since $x_1 < x_2$, we have $x_2 - x_1 > 0$, and it follows that $f(x_2) > f(x_1)$. If (ii) is the case, then $f'(c) < 0$. Since $x_2 - x_1 > 0$, it follows that $f(x_2) < f(x_1)$.

Further applications of the mean value theorem will be found in Chapters 5 and 11.

EXERCISES

In Exercises 1–6, determine when Rolle's theorem applies. When it applies, find a point c where $f'(c) = 0$; when it does not apply, state the conditions of the theorem that are not satisfied.

1. $f(x) = |x| - \frac{1}{2}$ $\qquad$ $[a, b] = [-\frac{1}{2}, \frac{1}{2}]$
2. $f(x) = x(x-1)(x+1)$ $\qquad$ $[a, b] = [-1, 0]$
3. $f(x) = \begin{cases} x & \text{for } 0 \le x < 1 \\ 0 & \text{for } x = 1 \end{cases}$ $\qquad$ $[a, b] = [0, 1]$

4. $f(x) = (x - \frac{1}{2})^2 - 1$ $[a, b] = [-1, 1]$
5. $f(x) = x - |x|$ $[a, b] = [1, 2]$
6. $f(x) = (x^2 - 1)(x^2 - \frac{1}{2})$ $[a, b] = [-1, 1]$

In Exercises 7–11, determine when the mean value theorem applies. When it applies, find an appropriate point c; when it does not apply, state the conditions of the theorem that are not satisfied.

7. $f(x) = x^{1/3}$ $[a, b] = [-1, 1]$
8. $f(x) = x^{1/3}$ $[a, b] = [0, 1]$
9. $f(x) = (x - 2)(x - 3)$ $[a, b] = [1, 3]$
10. $f(t) = (t^2 - 2t + 1)$ $[a, b] = [-1, 1]$
11. $f(x) = \begin{cases} x^3 & -2 \le x < 1 \\ -10 & x = 1 \end{cases}$ $[a, b] = [-2, 1]$

Hint: Write $h(x) = g(x) - f(x)$.

12. Show that if $f(x)$ and $g(x)$ are defined on an open interval (a, b) and $f'(x) = g'(x)$ on (a, b), then there is a constant c such that $g(x) \equiv f(x) + c$.

QUIZ 1

1. Consider the polynomial $y = x^3 - \frac{9}{2}x^2 + \frac{9}{2}x$

   (a) Determine the local maximum and local minimum points.

   (b) Determine the inflection points.

   (c) Find where the function is concave up, where it is concave down.

   (d) Plot the graph.

2. Consider the polynomial $y = \frac{1}{3}x^3 - x^2 + \frac{10}{3}$.

   (a) Determine its absolute and local maxima and minima.

   (b) Find where the function is concave up, where it is concave down.

   (c) Plot the graph.

3. (a) Find the asymptotes of $y = x/(x + 1)$ and (b) plot the graph.

4. If $f(x) = (|x| - 1)/(x - 1)$, find $\lim_{x \to 1+} f(x)$ and $\lim_{x \to 1-} f(x)$.

5. (a) Give an example of a function that is defined on the closed interval $[-1, 1]$ and discontinuous exactly at the points -1 and 1.

(b) Graph your function.

6. Find the points x where $g(x) = (x-2)/(x^3-2)$ is discontinuous.
7. Find lower and upper bounds for the function $h(x) = x^2/(x^2+1)$.

QUIZ 2

1. Consider the polynomial $y = x^3 - 6x^2 + 9x + 6$.

    (a) Find all local maximum and local minimum points.

    (b) Find the inflection points.

    (c) Determine where the function is concave up, where it is concave down.

    (d) Plot the graph.

2. Find the asymptotes of the following functions, and plot their graphs.

    (a) $y = \left| \dfrac{x}{x-2} \right|$

    (b) $y = \dfrac{x+1}{x(x-1)(x-2)}$

3. If $f(x) = (x - |x|)/(x + |x|)$, find $\lim_{x \to 0+} f(x)$ and $\lim_{x \to 1-} f(x)$.

4. Give an example of a function that is defined on the closed interval $[0, 2]$ and discontinuous exactly at the points 0, 1, and 2. Graph the function.

5. Find the points x where $f(x) = 1/(x^3 - |x|^3)$ is discontinuous.

6. Find lower and upper bounds for the function $f(x) = x/(|x|+1)$.

CHAPTER 5

APPLICATIONS

5.1 MARGINAL COST AND MARGINAL REVENUE

Economics is one of the many fields in which the differential calculus has been used to great advantage. Economists have a special name for the application of derivatives to problems in economics — it is *marginal analysis*. Thus, whenever the term *marginal* appears in a discussion, it signals the presence of derivatives in the background.

COST

Let c be the cost in dollars for a weekly production of q tons of steel. Let the function $c = f(q)$ give the dependence of c on q. The names given $f(q)$ and its derivative are these:

$c = f(q)$ is the *cost* of production,

$\dfrac{dc}{dq} = f'(q)$ is the *marginal cost*.

In practice, the function $f(q)$ is obtained by complicated accounting procedures. The marginal cost is seen to depend on q; it is expressed in dollars per ton in the above situation. An approximate value of marginal cost is obtained as follows: Since

$$f'(q) = \lim_{\Delta q \to 0} \frac{f(q + \Delta q) - f(q)}{\Delta q}$$

it follows that

$$f'(q) \approx \frac{f(q+\Delta q) - f(q)}{\Delta q}$$

for small increments Δq; in words,

$$f'(q) \approx \frac{\text{cost of increasing production from } q \text{ to } q + \Delta q}{\text{increment } \Delta q}$$

In many practical situations we deal with large q, relative to which $\Delta q = 1$ is small. In such cases

$$f'(q) \approx f(q+1) - f(q) = \text{cost of increasing production by one unit.} \quad (1)$$

Many economists think of marginal cost in this way.

Example 1

Suppose that the cost in dollars for a weekly production of steel is given by the formula

$$c = \tfrac{1}{10} q^2 + 5q + 2000$$

(a) Find a formula for the marginal cost.

Ans. $\dfrac{dc}{dq} = \tfrac{1}{5} q + 5$

(b) Find the cost and marginal cost when $q = 1000$ tons.

Ans.

$c = \tfrac{1}{10}(1000)^2 + 5 \cdot 1000 + 2000 = 107{,}000$ dollars

$\dfrac{dc}{dq} = \tfrac{1}{5} \cdot 1000 + 5 = 205$ dollars/ton

(c) Compare the marginal cost with the approximate formula (1).

Ans.

$f'(1000) \approx f(1001) - f(1000)$

$\quad = (\tfrac{1}{10} \cdot 1001^2 + 5 \cdot 1001 + 2000)$
$\quad\quad - (\tfrac{1}{10} \cdot 1000^2 + 5 \cdot 1000 + 2000)$

$\quad = \tfrac{1}{10}(1001 + 1000)(1001 - 1000) + 5$

$\quad = 205.1$ dollars/ton

We observe that the two computations differ only by 0.1 dollars/ton, which is less than $\tfrac{1}{20}$th of 1 percent.

(d) Draw the *cost curve* $c = \tfrac{1}{10} q^2 + 5q + 2000$ and interpret dc/dq geometrically.

Ans. The curve (see Figure 5.1) is obtained with a table of values. $dc/dq = 205$ is the slope of the tangent to the cost curve at $(1000, 107{,}000)$.

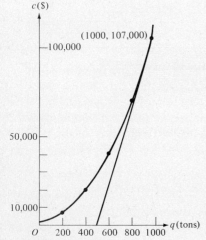

Figure 5.1
The cost curve
$c = \tfrac{1}{10} q^2 + 5q + 2000$.

q	c
0	2000
200	7000
400	20,000
600	41,000
800	70,000
1000	107,000

REVENUE

The money received by our hypothetical steel manufacturer when he sells his product is called *revenue*. Specifically, let r be the *total revenue* in dollars received from selling q tons of steel in one week. Let the function $r = m(q)$ give the dependence of r on q. Then

> $r = m(q)$ is the *total revenue*,
>
> $\dfrac{dr}{dq} = m'(q)$ is the *marginal revenue*.

Marginal revenue, like marginal cost, is measured in dollars per ton. An approximate value for $m'(q)$ is obtained by noting again that

$$m'(q) \approx \frac{m(q + \Delta q) - m(q)}{\Delta q}$$

for small increments Δq. When q is large, and $\Delta q = 1$ is small by comparison, we get

> $m'(q) \approx m(q + 1) - m(q)$ = revenue resulting from sale of one more unit. (2)

This is the way in which many economists think of marginal revenue.

Example 2

Suppose that the total revenue for a weekly sale of steel is given by the formula

$$r = q^2 + 5q + 2000$$

(a) Find the total and marginal revenue when $q = 1000$ tons.

Ans. Since $dr/dq = 2q + 5$, we have

$r = 1000^2 + 5 \cdot 1000 + 2000 = 1{,}007{,}000$ dollars

$\dfrac{dr}{dq} = 2 \cdot 1000 + 5 = 2005$ dollars/ton

(b) Compare the marginal revenue with the approximate formula (2).

Ans. Here $m(q) = q^2 + 5q + 2000$, and we find upon simplification that

$$m(q + 1) - m(q) = [(q + 1)^2 + 5(q + 1) + 2000] \\ - [q^2 + 5q + 2000]$$
$$= 2q + 6$$

Thus,

$$m'(1000) \approx 2006 \text{ dollars/ton}$$

and we observe that the two computations of marginal revenue differ by only 1 dollar, which is about $\frac{1}{20}$ th of 1 percent.

PRICE

The *average total revenue*, r/q, is usually called the *price*, p.

$$p = r/q = \text{average revenue} = \text{price}$$

The price p is clearly a function of q, and we observe the following relations to be true:

$$p = \frac{m(q)}{q} = \text{price}$$

$$r = pq = \text{total revenue}$$

$$\frac{dr}{dq} = p + q\frac{dp}{dq} = \text{marginal revenue}$$

Very often the price function is established from past experience, and the total and marginal revenue are then determined from the last two equations above.

Example 3

Suppose that the price p of q tons of steel is given by the formula

$$p = \frac{1200}{q + 20} - 10$$

(a) Find the formulas for the total and marginal revenue.

Ans. The total revenue is

$$r = pq = \frac{1200q}{q + 20} - 10q$$

Since

$$\frac{dp}{dq} = -\frac{1200}{(q + 20)^2}$$

we see that the marginal revenue is

$$\frac{dr}{dq} = p + q\frac{dp}{dq} = \frac{1200}{q + 20} - 10 + q\left(-\frac{1200}{(q + 20)^2}\right)$$

$$= \frac{1200}{q + 20}\left(1 - \frac{q}{q + 20}\right) - 10$$

$$= \frac{24{,}000}{(q + 20)^2} - 10$$

(b) Plot the price curve

$$p = \frac{1200}{q + 20} - 10$$

Ans. For the problem to have meaning we must have $q \geq 0$ (see Figure 5.2).

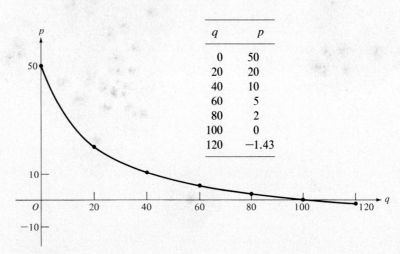

Figure 5.2
The curve $p = 1200/(q + 20) - 10$. Note in the graph that $p < 0$ when $q > 100$. This may happen because the company must pay to have the excess (unsold) steel hauled away.

EXERCISES

1. The cost of producing q gallons of a lubricant is $c = 20q^{3/2} + 300$ dollars.

    (a) Find the marginal cost when $q = 100$ and $q = 10{,}000$ gallons.

    (b) Compare the marginal cost with the approximate formula (1) for $q = 100$.

    (c) The *average cost* is defined by the formula

    $$\bar{c} = \frac{c}{q}$$

    Find the average cost for the given values of q in (a).

2. The price of q tons of Portland cement is given by the formula

    $$p = 20q^2 - 10q + 1000 \text{ dollars}$$

    (a) Find formulas for the total and marginal revenue.

    (b) Compare the marginal revenue with the approximate formula (2) when $q = 100$ tons.

3. Suppose that the cost and price for a certain product are given by the formulas

$$c = 50q + 30{,}000$$
$$p = 100 - 0.01q$$

(a) Find the formulas for total revenue, marginal revenue, and marginal cost.

(b) The definition of *profit*, P, is revenue minus cost:

$$\boxed{\text{profit} = P = r - c}$$

Find the level of production q at which profit is maximum.

5.2 MAXIMUM AND MINIMUM PROBLEMS

The solution of various problems involves finding maximum or minimum values of functions. The basic tests for locating maximum and minimum points were discussed in Sections 4.1 and 4.2; they are repeated here for convenience. The following example is a typical problem.

Example 1

Find the sides of that rectangle with perimeter p whose area is maximum.

The following general procedure for solving this type of problem is offered as a guide:

1. Draw even the simplest diagram illustrating the situation.
2. Write down the function f to be maximized or minimized, and whatever other equations there are relating the various unknowns. If possible, express f in terms of a single variable.
3. Candidates for maximum or minimum points are those points x_0

   where $f'(x_0) = 0$
   where $f'(x_0)$ does not exist
   that are end points of the interval of definition (domain) of f, if the function is not defined for all x.

Recall that

$f'(x_0) = 0$ and $f''(x_0) < 0$ implies that $f(x)$ has a local maximum at x_0,
$f'(x_0) = 0$ and $f''(x_0) > 0$ implies that $f(x)$ has a local minimum at x_0.

Figure 5.3

5.2 MAXIMUM AND MINIMUM PROBLEMS

Solution of Example 1
Let the rectangle have sides x and y, respectively (Figure 5.3). Then we have the relations

$$\text{area} = A = xy$$
$$2x + 2y = p$$

Solving the second equation for y and substituting into the first gives

$$A = x\left(\frac{p}{2} - x\right)$$

This function is defined and differentiable for all x, and we have

$$\frac{dA}{dx} = \frac{p}{2} - 2x = 0 \quad \text{when} \quad x = \frac{p}{4}$$

$$\frac{d^2 A}{dx^2} = -2 < 0 \quad \text{for all values of } x$$

We conclude that there is a maximum at $x = p/4$. To find the shape of the rectangle, we observe that

$$y = \frac{p}{2} - x = \frac{p}{2} - \frac{p}{4} = \frac{p}{4}$$

and hence the rectangle of maximum area with perimeter p is a square with sides of length $p/4$.

Example 2

Find the dimensions of the largest rectangle that can be inscribed in a circle of radius r.

Solution
With x and y as in Figure 5.4, we have the relations

$$\text{area} = A = 4xy$$
$$y = \sqrt{r^2 - x^2}$$

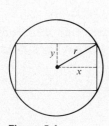

Figure 5.4

A substitution for y gives

$$A = 4x\sqrt{r^2 - x^2}$$

Since x represents length, $x \geq 0$, and hence this function is defined for $0 \leq x \leq r$. Differentiating A with respect to x gives

$$\frac{dA}{dx} = 4\sqrt{r^2 - x^2} - 4x\frac{x}{\sqrt{r^2 - x^2}} = \frac{4}{\sqrt{r^2 - x^2}}(r^2 - 2x^2) = 0$$

$$\text{when } x = \frac{r}{\sqrt{2}}$$

$$\frac{d^2A}{dx^2} = 4 \frac{d}{dx}[(r^2-x^2)^{-1/2}(r^2-2x^2)]$$

$$= \frac{4}{\sqrt{r^2-x^2}}\left[-4x + \frac{(r^2-2x^2)x}{r^2-x^2}\right] = -16 \quad \text{when } x = \frac{r}{\sqrt{2}}$$

We conclude that a maximum occurs when $x = r/\sqrt{2}$. Since

$$y = \sqrt{r^2 - x^2} = \sqrt{r^2 - \frac{r^2}{2}} = \frac{r}{\sqrt{2}}$$

we see that the rectangle with largest area is, as in Example 1, a square.

EXERCISES

1. A closed rectangular box (Figure 5.5) is to be constructed such that the total area of all six sides is 20 ft². What is the largest possible volume?

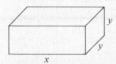

Figure 5.5

2. An open box is to be constructed from a square sheet of metal with a 10-in. side (Figure 5.6). Find the dimensions of the box of maximum volume.

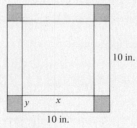

Figure 5.6

3. A closed cylindrical can (Figure 5.7) is to have a total surface area (side, top, and bottom) of 1 ft². What dimensions will yield maximum volume?

Figure 5.7

4. Find the area of the largest rectangle that can be inscribed in the right triangle with sides of length 3 and 4 in., respectively (Figure 5.8).

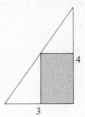

Figure 5.8

5. A book page is to have 1-in. margins (Figure 5.9). What dimensions of the page will yield the least margin area if the total page area is 120 in.²?

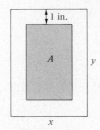

Figure 5.9

6. Find the point $P(x, y)$ on the curve $y = \sqrt{x}$ that is closest to the point $Q(1, 0)$ (Figure 5.10).

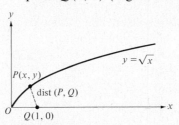

Figure 5.10

7. Find two positive numbers A and B whose sum is 10 and whose product is maximum.

8. Find the point Q on the hypotenuse of a right triangle (Figure 5.11) with sides of lengths 3 and 4 in., respectively, such that $a^2 + b^2 + c^2$ is minimum.

9. Find the sides of that rectangle of area A whose perimeter is maximum.

10. Find the sides of that rectangle of area A whose perimeter is minimum.

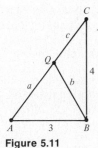

Figure 5.11

11. A length of wire is cut into two pieces, one of which is bent to form a circle, the other an equilateral triangle (see Figure 5.12). What are the lengths of the pieces if the total area enclosed by the circle and triangle is to be minimum?

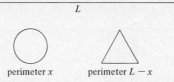

perimeter x perimeter $L - x$

Figure 5.12

12. What are the lengths of the pieces of wire in Exercise 11 if the total enclosed area is to be maximum?

13. Find the shortest diagonal of a rectangle having area 144 in.^{2} (see Figure 5.13).

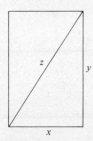

Figure 5.13

14. A semicircular stained glass window has a rectangular clear glass insert that passes twice as much light as the surrounding stained glass (Figure 5.14). Find the dimensions of the rectangular insert for which the entire window passes maximum light.

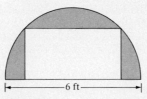

Figure 5.14

15. Let $a > 0$ and $b > 0$ be given numbers. Find the minimum of $f(x) = ax + b/x$ for $x > 0$.

16. Let $\overline{PQ}$ be the shortest line segment that can be drawn from the *fixed* point $Q(x_0, 0)$ to the curve $y = \sqrt{x}$ (Figure 5.15). Prove that this line segment is perpendicular to the tangent to $y = f(x)$

at $P(x, y)$. (See in this connection Exercise 7 in Section 2.3.)

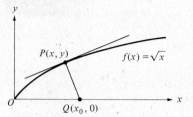

Figure 5.15

17. A cylindrical metal can is to have a capacity of 27 in.^{3}. Its top and bottom are cut from square sheets, and the residual metal is wasted (Figure 5.16). For what dimension can will the total amount of metal used (including waste) be minimum?

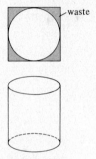

Figure 5.16

18. Suppose a radiation reading on a meter is directly proportional to the strength of the source, and inversely proportional to the distance between the meter and the source. If sources A and B are 100 ft apart and source A is three times stronger than source B, at what point between A and B will the meter register minimum radiation (Figure 5.17)?

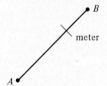

Figure 5.17

19. The cost of producing x tons of steel/week is

$$c = \tfrac{1}{20}x^2 + 700x + 500 \text{ dollars}$$

When x tons/week is sold, the manufacturer can sell the steel at a price of

$$p = 2700x - \tfrac{3}{20}x^2 \text{ dollars/ton}$$

If profit is defined by the formula

$P = \text{profit} = \text{price minus cost}$

find the production in tons that yields the biggest profit.

5.3 RATE OF CHANGE

The derivative as introduced previously has been interpreted as the slope of a tangent line to a curve and as velocity. In each interpretation it measured the rate at which the differentiated function was changing. For this reason the derivative is often called *rate of change*. Thus, when a function $y = f(x)$ changes by an amount $\Delta y = f(x + \Delta x) - f(x)$, its *average rate of change* with respect to x is $\Delta y/\Delta x$. We say that:

$$\frac{dy}{dx} = \lim_{\Delta x \to 0} \frac{\Delta y}{\Delta x} \text{ is the rate of change of } y \text{ with respect to } x.$$

This interpretation of the derivative is rich in applications. A representative variety of problems solved with derivatives is studied below.

Example 1

When the radius of a sphere is increasing, so is its volume. What is the rate of change of the volume of a sphere with respect to its radius r when $r = 3$ ft?

Solution

1. The volume of a sphere of radius r is

    $V = \frac{4}{3}\pi r^3$ ft^3

2. The rate of change of V with respect to r is

    $\frac{dV}{dr} = 4\pi r^2$ ft^2

3. When $r = 3$, then

    $\frac{dV}{dr} = 36\pi$ ft^2

Example 2

How fast is the volume of a sphere changing when its radius is increasing at the rate of 0.01 in./sec?

Solution

As in the previous problem $V = \frac{4}{3}\pi r^3$, but here both r and V are regarded as functions of time t. According to the statement of the problem, we are asked to relate dV/dt, the rate of change of V with respect to time, to dr/dt, the rate of change of r with respect to time. Now, by the chain rule,

$$\frac{dV}{dt} = \frac{d}{dt}\left(\frac{4}{3}\pi r^3\right) = \frac{4}{3}\pi \frac{d}{dt}(r^3) = \frac{4}{3}\pi\left(3r^2 \frac{dr}{dt}\right) = 4\pi r^2 \frac{dr}{dt}$$

that is, we have the relation

$$\frac{dV}{dt} = 4\pi r^2 \frac{dr}{dt}$$

According to the problem, $dr/dt = 0.01$ in./sec, and hence

$$\frac{dV}{dt} = 0.04\pi r^2 \text{ in.}^2/\text{sec}$$

Observe that the rate of change dV/dt depends on r. Observe further that we found dV/dt without ever knowing the explicit form of the dependence of V on t. This is an example of a *related rate* problem. A general procedure for handling related rate problems is illustrated next.

Example 3

You are walking away from a 200-ft tall tower (see Figure 5.18).

(a) If your distance from the base of the tower is x and your distance from its top is z, what is the relation between dx/dt, the rate at which you are leaving the base, and dz/dt, the rate at which your distance from the top is changing?

(b) Suppose that $dx/dt = 2$ ft/sec when $x = 100$ ft. What is dz/dt at this time?

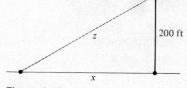

Figure 5.18

Solution

This problem is solved in three steps. Notice, however, the following:

Important warning

The data in (b) can be used only after the general problem in (a) has been solved.

Step 1

Find an equation relating x and z: From Figure 5.18 we deduce that

$$x^2 + 200^2 = z^2$$

Step 2
Regarding x and z as functions of time t, differentiate both sides of the equation just obtained with respect to t:

$$\frac{d}{dt}(x^2 + 200^2) = \frac{d}{dt}(z^2)$$

The chain rule gives

$$2x\frac{dx}{dt} = 2z\frac{dz}{dt}$$

and solving for dz/dt gives

$$\frac{dz}{dt} = \frac{x}{z}\frac{dx}{dt}$$

This answers part (a).

Step 3
Substitute the numerical data in (b) into the last equation in step 2; but first we must find z. This is done with step 1. Since when $x = 100$, then

$$z^2 = 100^2 + 200^2 = 50{,}000$$

or

$$z = \sqrt{50{,}000} = 100\sqrt{5}$$

We are now ready to answer part (b):

$$\frac{dz}{dt} = \frac{100}{100\sqrt{5}} \cdot 2 = \frac{2}{\sqrt{5}} = \frac{2\sqrt{5}}{5} \text{ ft/sec}$$

Example 4

Imagine a point $P(x, y)$ sliding down the curve $y = x^3$ (see Figure 5.19). As this point moves along the curve, its "shadows" $P(x, 0)$ and $P(0, y)$ move along the x and y axes, respectively. Find the velocity of $P(0, y)$ when $x = 2$ in. and the velocity of $P(x, 0)$ is -3 in./sec.

Solution
The velocity of $P(x, 0)$ is dx/dt, that of $P(0, y)$ is dy/dt. Differentiating both sides of the equation $y = x^3$ with respect to t, we have

$$\frac{dy}{dt} = 3x^2 \frac{dx}{dt}$$

According to the data specified in the problem, $dx/dt = -3$ in./sec

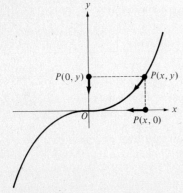

Figure 5.19
x and y are decreasing; dx/dt and dy/dt are therefore negative.

when $x = 2$ in. Hence

$$\frac{dy}{dt} = 3 \cdot 2^2(-3) = -36 \text{ in./sec}$$

EXERCISES

Hint: $V = \frac{4}{3}\pi r^3$, $S = 4\pi r^2$, and
$$\frac{dV}{dt} = \frac{dV}{dr}\frac{dr}{dt}.$$

1. A spherical weather balloon is expanding due to heating of its gas by solar radiation. As the balloon's radius r reaches 20 ft, its area S is expanding at the rate of 1 ft²/sec. At what rate is the volume V changing?

2. You are standing on top of a 10-ft ladder that is leaning against a vertical wall (Figure 5.20). The foot of the ladder is slipping away from the wall at the rate of $\frac{1}{2}$ ft/sec. At what rate is the top of the ladder coming down when it is 5 ft off the ground?

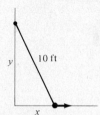

Figure 5.20

3. A 6-ft high light source is moving at a velocity of 5 ft/min toward a 3-ft wall (Figure 5.21). How fast is the shadow receding when the light is 4 ft from the wall?

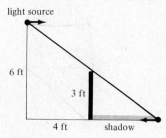

Figure 5.21

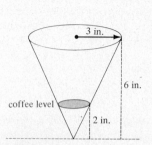

Figure 5.22
The volume of a cone of height h and radius r is $V = \frac{1}{3}\pi r^2 h$. You must compute dh/dt.

4. Consider the conical paper filter in Figure 5.22. Coffee drips out at the rate of 1 in.³/min. How fast is the coffee level decreasing when it is 2 in. deep?

5. Two airplanes leave Chicago airport at the same time, one flying due north with velocity 180 mph, the other flying due east with velocity 220 mph (see Figure 5.23). How fast is the distance R between the airplanes increasing at the elapsed flying time of $\frac{1}{2}$ hr?

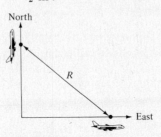

Figure 5.23

6. A 6-ft tall man is walking away from a 15-ft lamp post at the rate of 50 yards/min (Figure 5.24). How fast is the man's shadow lengthening when he is 100 yards from the post?

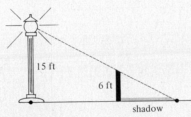

Figure 5.24

7. A test tube of radius $\frac{1}{2}$ in. has a piston inserted at its open end (Figure 5.25). If the piston is pulled out at the rate of $\frac{1}{10}$ in./sec, what is the rate of change in volume when the piston is 12 in. above the base of the tube?

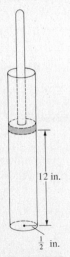

Figure 5.25

8. As the diagonal of a square changes, so does its area. Find the rate of change of the area of a square with respect to its diagonal.

9. According to the special theory of relativity, the mass of a particle moving at velocity v is

$$m = \frac{m_0}{\sqrt{1 - v^2/c^2}}$$

where m_0 is the mass of the particle when at rest and c is the velocity of light. Find the rate of change of mass with respect to velocity when $v = \frac{1}{2}c$.

10. How does the area of a closed 10-in.-high cylindrical can (Figure 5.26) change with respect to its radius?

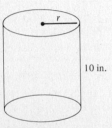

Figure 5.26

5.4 NEWTON'S METHOD

By way of motivation, consider the polynomial

$$p(x) = x^3 + x^2 + x - 1$$

Since $p(\frac{1}{2}) = -\frac{1}{8}$ and $p(1) = 2$, we know that the graph of $y = p(x)$ crosses the x axis at some point a between $-\frac{1}{8}$ and 2; that is, $p(a) = 0$. How can we locate a? The method described below dates back to Newton, and it uses tangents in a very interesting way.

Shifting our attention to a general situation, suppose that $y = f(x)$ meets the x axis at a point a. From the graph or other information, we know the approximate location of a. Let x_0 be in this location and examine the tangent to $y = f(x)$ at the point $(x_0, f(x_0))$ (see Figure 5.27). This tangent is given by

$$y - f(x_0) = f'(x_0)(x - x_0)$$

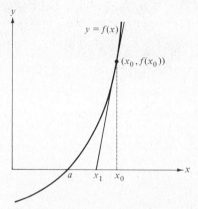

Figure 5.27

If the point of intersection of the tangent with the x axis is marked x_1, then

$$0 - f(x_0) = f'(x_0)(x_1 - x_0)$$

A solution for x_1 gives

$$x_1 = x_0 - \frac{f(x_0)}{f'(x_0)}$$

Under the right conditions, x_1 will be closer to a than x_0. If a better approximation to a is desired, we repeat this procedure by constructing the tangent to $y = f(x)$ at $(x_1, f(x_1))$ (see Figure 5.28). The intersection of this tangent with the x axis is given by

$$x_2 = x_1 - \frac{f(x_1)}{f'(x_1)}$$

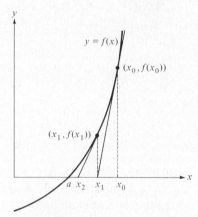

Figure 5.28

This procedure can be repeated over and over. It is an example of an *iterative method*. The step from x_0 to x_1 is called the *first iteration*, the step from x_1 to x_2 is called the *second iteration*, and so on. If x_k is the result of the kth iteration, then the $(k+1)$st iteration leads from x_k to x_{k+1} with the equation

$$\boxed{x_{k+1} = x_k - \frac{f(x_k)}{f'(x_k)}}$$

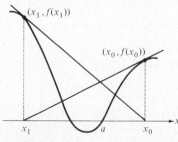

Figure 5.29
A situation in which the Newton method fails to work.

A case in which the iteration may fail to work is illustrated in Figure 5.29. In this situation, successive iterations give, in turn, the points x_0 and x_1 over and over again. Note, however, that the method could have worked if the initial guess x_0 were closer to a.

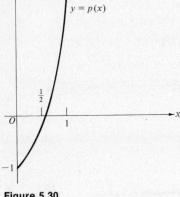

Figure 5.30
The graph of $p(x) = x^3 + x^2 + x - 1$.

Assuming that the method works, we apply it below in some typical cases.

Example 1

Let us consider once more the polynomial

$$p(x) = x^3 + x^2 + x - 1$$

Since

$$p'(x) = 3x^2 + 2x + 1 > 0 \qquad \text{for } x \geq 0$$

and

$$p''(x) = 6x + 2 > 0 \qquad \text{for } x \geq 0$$

we conclude that $p(x)$ is increasing and concave up for $x > 0$ (Figure 5.30). From this and the fact that $p(\frac{1}{2}) = -0.125$ and $p(1) = 2$, we conclude that $p(x) = 0$ has exactly one root in the interval $(\frac{1}{2}, 1)$. We perform the first four iterations, selecting as initial point $x_0 = 1$.

$$x_1 = 1 - \frac{p(1)}{p'(1)} \approx 0.666667 \qquad p(x_1) \approx 0.407407$$

$$x_2 = \frac{2}{3} - \frac{p(\frac{2}{3})}{p'(\frac{2}{3})} \approx 0.555555 \qquad p(x_2) \approx 0.035665$$

$$x_3 = \frac{5}{9} - \frac{p(\frac{5}{9})}{p'(\frac{5}{9})} \approx 0.548260 \qquad p(x_3) \approx 0.013651$$

$$x_4 \approx 0.543707 \qquad p(x_4) \approx 0.000053$$

We see that four iterations get us close to the solution of $p(x) = 0$.

FINDING SQUARE ROOTS

Let us apply Newton's method to the problem of finding the square root of a number $b > 0$; that is, we use the method to solve the equation $x^2 - b = 0$.

If we put $f(x) = x^2 - b$, then $f'(x) = 2x$. With x_0 as initial guess, the first step of the iteration gives

$$x_1 = x_0 - \frac{x_0^2 - b}{2x_0}$$

The right side simplifies as follows:

$$x_0 - \frac{x_0^2 - b}{2x_0} = x_0 - \frac{x_0^2}{2x_0} + \frac{b}{2x_0} = \frac{1}{2}x_0 + \frac{1}{2}\frac{b}{x_0} = \frac{1}{2}\left(x_0 + \frac{b}{x_0}\right)$$

and hence

$$x_1 = \frac{1}{2}\left(x_0 + \frac{b}{x_0}\right)$$

The $(k+1)$st step gives

$$x_{k+1} = \frac{1}{2}\left(x_k + \frac{b}{x_k}\right) \tag{1}$$

Example 2

Approximate $\sqrt{2}$ using three iterations.

Solution
We apply formula (1) with $b = 2$. Letting also $x_0 = 2$ gives

$$x_1 = \frac{1}{2}\left(2 + \frac{2}{2}\right) = \frac{3}{2} = 1.5$$

$$x_2 = \frac{1}{2}\left(\frac{3}{2} + \frac{2}{3/2}\right) = \frac{17}{12} \approx 1.417$$

$$x_3 = \frac{1}{2}\left(\frac{17}{12} + \frac{2}{17/12}\right) = \frac{577}{408} \approx 1.414215686$$

The number $\sqrt{2}$ correct to ten significant figures is

$$\sqrt{2} \approx 1.414213562$$

Thus, three iterations yielded the square root of 2 with an error less than $0.000003 = 3 \times 10^{-6}$.

EXERCISES

1. Obtain the iteration formula for finding cube roots. Starting with $x_0 = 2$, use three iterations to approximate $\sqrt[3]{2}$.

2. Show that the polynomial equation $x^3 + x - \frac{1}{2} = 0$ has one positive root, then approximate this root using three iterations.

3. Show that the polynomial equation $x^3 - 2x^2 + 1 = 0$ has one negative root, then approximate this root using three iterations. Use $x_0 = -\frac{1}{2}$.

4. For finding the square root of b, we had the formula

$$x_{k+1} = \frac{1}{2}\left(x_k + \frac{b}{x_k}\right)$$

Verify the following facts.

(a) $x_{k+1} - \sqrt{b} = \dfrac{(x_k - \sqrt{b})^2}{2x_k}$

(b) $x_1 > \sqrt{b}$ no matter what x_0 is, as long as $x_0 > 0$.

146 APPLICATIONS

(c) $x_k > \sqrt{b}$ for $k = 1, 2, 3, 4, \ldots$.

(d) $x_1 > x_2 > x_3 > x_4 > \ldots$.

(e) $x_{k+1} - \sqrt{b} < \frac{1}{2}(x_k - \sqrt{b})^2$ (assume that $b > 1$)

Hint: $x_2 - \sqrt{b} < \frac{1}{2}(x_1 - \sqrt{b})^2$ by (e)

(f) $x_{k+1} - \sqrt{b} < \frac{1}{2^k}(x_1 - \sqrt{b})^{2^k}$ (assume that $b > 1$)

$x_3 - \sqrt{b} < \frac{1}{2}(x_2 - \sqrt{b})^2$ by (e)

$< \frac{1}{2^2}(x_1 - \sqrt{b})^{2^2}$

5. If $x_1 - \sqrt{b} < \frac{1}{10}$, how many more iterations will give the estimate $x_k - \sqrt{b} < 5 \cdot 10^{-20}$? Use part (f) of Exercise 4.

5.5 LINEAR APPROXIMATIONS

In certain computations it is often desirable to replace a given function $f(x)$ near a point $(x_0, f(x_0))$ by its tangent at that point. For values of x close to x_0, the tangent gives a useful approximation to the function (see Figure 5.31). This idea was used in the last section.

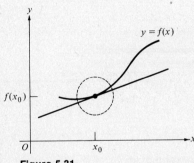

Figure 5.31

Consider a function f having a derivative at x_0. The tangent to $y = f(x)$ at $(x_0, f(x_0))$ is given by the formula

$$y = f(x_0) + f'(x_0)(x - x_0)$$

(see Section 2.5). Putting $y = f(x)$, we obtain the approximate formula

$$f(x) \approx f(x_0) + f'(x_0)(x - x_0) \qquad \text{when } x \approx x_0 \qquad (1)$$

The geometric meaning of this approximation is given in Figure 5.32. Before using (1), let us see how it is related to the definition

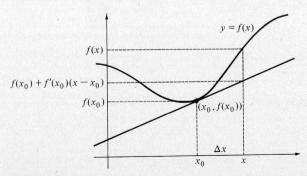

Figure 5.32

of derivative. We recall that

$$\lim_{\Delta x \to 0} \frac{f(x_0 + \Delta x) - f(x_0)}{\Delta x} = f'(x_0)$$

Using approximate equality, this relation yields the relation

$$\frac{f(x_0 + \Delta x) - f(x_0)}{\Delta x} \approx f'(x_0) \quad \text{when } \Delta x \approx 0$$

We emphasize that $\Delta x \neq 0$. Multiplying by Δx gives

$$f(x_0 + \Delta x) - f(x_0) \approx f'(x_0) \Delta x \quad \text{when } \Delta x \approx 0 \tag{2}$$

Letting $x = x_0 + \Delta x$, we have $\Delta x = x - x_0$ and (2) becomes

$$f(x) - f(x_0) \approx f'(x_0)(x - x_0) \quad \text{when } x - x_0 \approx 0$$

This, however, is equivalent to (1).

When x_0 is set equal to 0 in formula (1), we have

$$f(x) \approx f(0) + f'(0)x \quad \text{when } x \approx 0 \tag{3}$$

Example 1

Find an approximation for $f(x) = 1/(1 - x)$ when $x \approx 0$.

Solution

We use (3) with $f(x) = 1/(1 - x)$. We have

$$f'(x) = \frac{1}{(1 - x)^2}$$

so that

$$f(0) = 1 \quad \text{and} \quad f'(0) = 1$$

Hence, a substitution in (3) gives

$$\frac{1}{1 - x} \approx 1 + 1 \cdot x = 1 + x \quad \text{when } x \approx 0$$

Thus, when $x = 0.01$, this says that

$$\frac{1}{0.99} \approx 1.01$$

Actually,

$$\frac{1}{0.99} = 1.01010101 \ldots$$

and we see that the approximation is quite good.

In the examples below, we apply formula (1) to a variety of approximation problems. In Chapter 11 we shall see how to estimate the magnitude of the errors resulting from such approximations.

Example 2

Find an approximation to $(1 + x)^r$ for values of x close to 0.

Solution
If
$$f(x) = (1 + x)^r$$
then
$$f'(x) = r(1 + x)^{r-1}$$
and, in particular,
$$f(0) = 1 \quad \text{and} \quad f'(0) = r$$
Applying formula (3) to this function gives
$$(1 + x)^r \approx 1 + rx \quad \text{when } x \approx 0 \tag{4}$$
For example, taking $r = \frac{1}{2}$ gives the approximation
$$\sqrt{1 + x} \approx 1 + \tfrac{1}{2}x \quad \text{when } x \approx 0$$
For $x = 0.01$ this gives
$$\sqrt{1.01} \approx 1.005$$
To seven significant figures, the answer is
$$\sqrt{1.01} = 1.004988$$
showing that our approximation is quite good.

Example 3

Find an approximation to $\sqrt{\dfrac{1+x}{1-\frac{1}{2}x}}$ for values of x close to 1.

Solution
Put
$$f(x) = \sqrt{\frac{1+x}{1-\tfrac{1}{2}x}} = \left(\frac{1+x}{1-\tfrac{1}{2}x}\right)^{1/2} \quad \text{and} \quad x_0 = 1$$
Then
$$f'(x) = \frac{1}{2}\left(\frac{1+x}{1-\tfrac{1}{2}x}\right)^{-1/2} \frac{d}{dx}\left(\frac{1+x}{1-\tfrac{1}{2}x}\right)$$
$$= \frac{1}{2}\left(\frac{1+x}{1-\tfrac{1}{2}x}\right)^{-1/2} \frac{3}{2(1-\tfrac{1}{2}x)^2}$$
Hence
$$f(1) = \left(\frac{2}{\tfrac{1}{2}}\right)^{1/2} = 2 \quad \text{and} \quad f'(1) = \frac{1}{2} \cdot \frac{1}{2} \cdot \frac{3}{2 \cdot \tfrac{1}{4}} = \frac{3}{2}$$

and accordingly the formula $f(x) \approx f(1) + f'(1)(x-1)$ when $x \approx 1$ becomes

$$\sqrt{\frac{1+x}{1-\frac{1}{2}x}} \approx 2 + \frac{3}{2}(x-1) \qquad \text{when } x \approx 1$$

EXERCISES

1. Find an approximation to $\sqrt{(1+x)/(1-\frac{1}{2}x)}$ for values of x close to 0.

In Exercises 2–8, find an approximate value for the given expressions.

2. $\sqrt{65}$ (65 is close to $x_0 = 64$)
3. $\sqrt{0.24}$
4. $\sqrt[3]{0.120}$ $[0.5^3 = 0.125]$
5. $1.1^{2/5}$
6. $\dfrac{0.01}{1.99}$
7. $\sqrt{16.056}$
8. $8.014^{2/3}$
9. In Example 2 we derived the formula

$$\frac{1}{1-x} \approx 1 + x \qquad \text{when } x \approx 0$$

The equation

$$\frac{1}{1-x} = 1 + x + \frac{x^2}{1-x}$$

which is verified with a multiplication by $1-x$, tells us that the *error* in the approximate equality is

$$E(x) = \frac{x^2}{1-x}$$

(a) If x is an arbitrary number such that $|x| < 0.01$, how big can the error $E(x)$ become?

(b) How big an interval $-\delta \le x \le \delta$ can we take if $E(x)$ is not to exceed 0.01?

5.6 DIFFERENTIALS

The concept of differential is related to the linear approximation discussed in Section 5.5. Consider formula (2) of that section:

$$f(x_0 + \Delta x) - f(x_0) \approx f'(x_0) \Delta x$$

With the notation

$$\Delta y = f(x_0 + \Delta x) - f(x_0)$$

this formula becomes

$$\Delta y \approx f'(x_0) \Delta x$$

A very important formula used in integration (see Section 6.7) is obtained from this approximation by using the right-hand side. Assuming x_0 to be fixed, we define the differential dy for each value of Δx by

$$dy = f'(x_0) \Delta x$$

Notice that the symbol Δx is treated simply as an independent variable, dy as a dependent variable. To simplify notation, it is customary to replace the symbol Δx by dx.

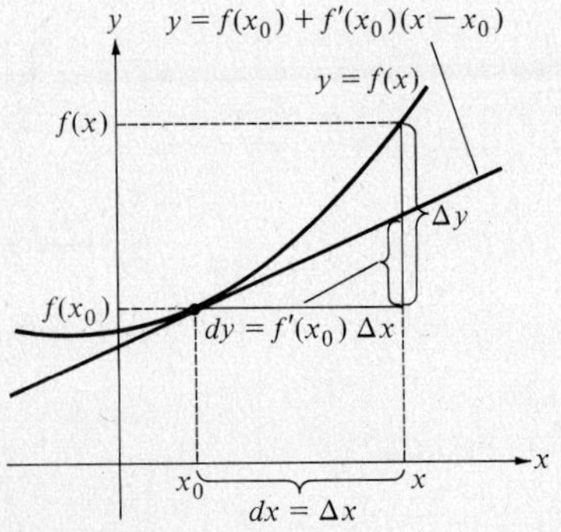

DEFINITION 1
If $y = f(x)$, then the *differential* of f at x is defined by the formula

$$dy = f'(x) \, dx \qquad (1)$$

The quantity dx is called the differential of x.

The relation between Δy and dy is described in Figure 5.33. The roles of dy and dx as dependent and independent variables, respectively, is made clear in the following examples.

Figure 5.33
The differentials dy and dx.

Example 1

Find dy when $y = x^2$.

Solution
With $f(x) = x^2$, we have $f'(x) = 2x$ and hence

$$dy = 2x \, dx$$

Example 2

Find dy when $y = (x + 1)^r$.

Solution
With $f(x) = (x + 1)^r$, we have $f'(x) = r(x + 1)^{r-1}$ and hence

$$dy = r(x + 1)^{r-1} \, dx$$

Example 3

Suppose that $z = u^3$ and $u = x^2 + 5x + 4$. Then the differentials dz and du are given by the formulas

$$dz = 3u^2\, du \tag{2}$$

$$du = (2x + 5)\, dx \tag{3}$$

Substituting $u = x^2 + 5x + 4$ and $du = (2x + 5)\, dx$ into equation (2) gives

$$dz = 3(x^2 + 5x + 4)^2 (2x + 5)\, dx \tag{4}$$

We obtain the same result by substituting $u = x^2 + 5x + 4$ into $z = u^3$. For then

$$z = (x^2 + 5x + 4)^3$$

and from this we again get (4) by using the chain rule. In general, when $z = f(u)$, then $dz = f'(u)\, du$ even when u itself is a dependent variable.

EXERCISES

Find the differential of each of the functions in Exercises 1–10.

1. $y = \dfrac{1}{x}$

2. $y = |x|$ $\quad (x > 0)$

3. $y = |x|$ $\quad (x < 0)$

4. $u = \dfrac{x - 1}{x + 1}$

5. $v = \tfrac{4}{3}\pi r^3$

6. $s = t + \dfrac{1}{t}$

7. $y = x(x + 1)^{10}$

8. $y = x\sqrt{x + 1}$

9. $p = \sqrt{z^2 + 1}$

10. $E = mc^2$, where c is a constant.

Proceed as in Example 3 to find dy in terms of x and dx.

11. $y = u^2 \quad u = x + \dfrac{1}{x}$

12. $y = t^3 \quad t = x + \dfrac{1}{x}$

13. $y = (v^2 + 1)(v^2 - 1) \quad v = x^2 + 1$

14. $y = \dfrac{1}{u} \quad u = x + \dfrac{x - 1}{x + 1}$

15. $y = |u| \quad u = x^3 + x + 1 \quad (x > 0)$

QUIZ 1

1. The radius of a sphere of volume V is $r = (\frac{3}{4}V/\pi)^{1/3}$. Find the rate of change of radius with respect to volume when $V = 10$ ft^3.

2. How fast is the radius of a sphere changing when its volume is increasing at the rate of $\frac{1}{2}$ ft^3/sec and the volume itself is 2 ft^3?

3. Find the dimensions of the rectangle of maximum area that can be inscribed inside a circle of radius 10 in.

4. Use the first two iterations in Newton's method to find approximate values for the real root of the polynomial equation $x^3 + x - 4 = 0$; use $x = 1$ as your starting point.

5. The cost c in dollars for a weekly production of q tons of steel is given by the formula

$$c = \frac{1}{10,000}q^3 + \frac{1}{5}q^2 + 2000$$

   (a) Find a formula for the marginal cost.

   (b) What is the marginal cost when 1000 tons are produced per week?

6. The equations $y = u/(1 + u^2)$ and $u = (x - 2)/(x + 2)$ define y as a function of x. Find the linear approximation for $x \approx 2$.

7. If $y = \sqrt{(x - 2)/(x + 2)}$, find dy.

QUIZ 2

1. Find two positive numbers whose product is 36 and whose sum is minimal.

2. The total cost of producing x typewriters is $c = x^2 + 10x + 1$ dollars; the selling price per typewriter is $s = 50 - x$ dollars. How many typewriters must be produced to maximize profit? Recall that profit = revenue − cost. (Fractional typewriters are an impossibility.)

3. Ship A is sailing due south at velocity 20 mph and ship B, which is 40 miles south of A, is sailing due east at velocity 15 mph. At what rate is the distance between the ships changing at the end of 1 hr?

4. Use the first two iterations in Newton's method to evaluate $\sqrt{5}$; take $x = 2$ as your starting point.

5. The total revenue r in dollars from a weekly sale of q tons of steel is given by the formula

$$r = \frac{1}{10,000}q^3 + \frac{1}{2}q^2 + 1500$$

   (a) Find a formula for the marginal revenue.

   (b) Find the marginal revenue when 1000 tons are sold per week.

6. Find a linear approximation for $\sqrt{(x-2)/(x+2)}$ for x close to 3.

7. The equations $y = u/(1+u^2)$, $u = (x-2)/(x+2)$ define y as a function of x. Find dy in terms of u, x, and dx.

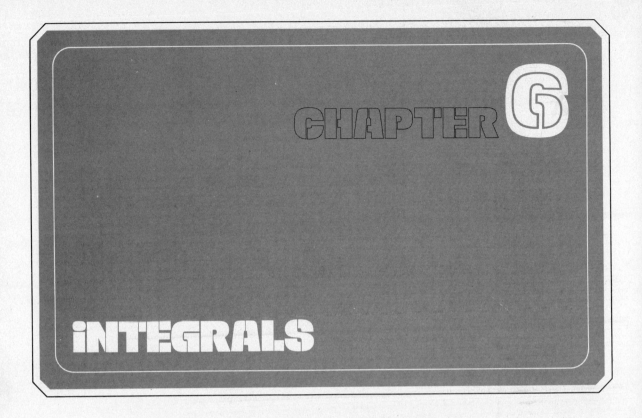

CHAPTER 6

INTEGRALS

6.1 INTRODUCTION

The integral is a twin concept to the derivative. We introduce the integral as the area under a curve. This leads to a definition of the integral as a limit of sums. Although area is the main interpretation of the integral, the concept is also used to find and define volumes, mass, probability, work, length of a curve, and electric charge, to name but a few.

As the chapter progresses, you will discover that numerical values can easily be found for many integrals by using a close connection between integrals and derivatives. It is quite amazing that a theoretical result, called the *fundamental theorem of calculus*, linking integrals and derivatives can be used to find numerical values of areas, volumes, and so on.

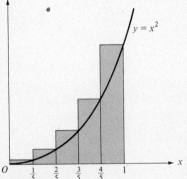

Figure 6.1
Approximating the area under $y = x^2$ from 0 to 1.

Example 1

Let us approximate the area under the curve $y = x^2$ and between the lines $x = 0$ and $x = 1$ by dividing the interval $[0, 1]$ into five subintervals and using the rectangles illustrated in Figure 6.1. All these rectangles have base $\frac{1}{5}$. The first has height $(\frac{1}{5})^2$ and area $\frac{1}{5}(\frac{1}{5})^2$; the second has height $(\frac{2}{5})^2$ and area $\frac{1}{5}(\frac{2}{5})^2$, and so on. As approximation to the area we take the sum of the areas of these five rectangles. Designating this sum by S_5, we have

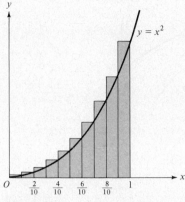

Figure 6.2
Improving the approximation to the area under $y = x^2$ from 0 to 1.

$$S_5 = \tfrac{1}{5}(\tfrac{1}{5})^2 + \tfrac{1}{5}(\tfrac{2}{5})^2 + \tfrac{1}{5}(\tfrac{3}{5})^2 + \tfrac{1}{5}(\tfrac{4}{5})^2 + \tfrac{1}{5}(\tfrac{5}{5})^2$$

$$= \frac{1}{5^3}(1^2 + 2^2 + 3^2 + 4^2 + 5^2) = \frac{55}{125} = 0.044$$

To improve this result, let us repeat the procedure with a subdivision into ten subintervals (see Figure 6.2). This gives rectangles having base $\tfrac{1}{10}$ and heights $(\tfrac{1}{10})^2$, $(\tfrac{2}{10})^2$, ..., $(\tfrac{10}{10})^2$. Designating by S_{10} the sum of these ten rectangles, we have

$$S_{10} = \tfrac{1}{10}(\tfrac{1}{10})^2 + \tfrac{1}{10}(\tfrac{2}{10})^2 + \tfrac{1}{10}(\tfrac{3}{10})^2 + \cdots + \tfrac{1}{10}(\tfrac{10}{10})^2$$

$$= \frac{1}{10^3}(1^2 + 2^2 + 3^2 + \cdots + 10^2) = \frac{385}{1000} = 0.385$$

These approximations lead to the idea of finding the desired area by letting the number of rectangles increase beyond all bounds. We develop this idea for the more general interval $[0, b]$ instead of $[0, 1]$.

Example 2

We now find nth approximating sums to the area under the curve $y = x^2$ and between the line $x = 0$ and $x = b$; this will be followed by finding the limit as n increases beyond all bounds ($n \to \infty$).

The interval $[0, b]$ is divided into n equal subintervals by means of the points

$$\frac{b}{n}, \ 2\frac{b}{n}, \ 3\frac{b}{n}, \ \ldots, \ n\frac{b}{n} = b$$

As illustrated in Figure 6.3, we now approximate the area with rectangles having base b/n and heights $[b/n]^2$, $[2(b/n)]^2$, ..., $[n(b/n)]^2$, respectively.

Letting S_n designate the sum of the areas of these n rectangles, we have

$$S_n = \frac{b}{n}\left(\frac{b}{n}\right)^2 + \frac{b}{n}\left(2\frac{b}{n}\right)^2 + \frac{b}{n}\left(3\frac{b}{n}\right)^2 + \cdots + \frac{b}{n}\left(n\frac{b}{n}\right)^2$$

$$= \frac{b^3}{n^3}(1^2 + 2^2 + 3^2 + \cdots + n^2)$$

In the appendix to this chapter we establish the formula

$$1^2 + 2^2 + 3^2 + \cdots + n^2 = \tfrac{1}{6}n(n + 1)(2n + 1)$$

This gives

$$S_n = \frac{b^3}{n^3}[\tfrac{1}{6}n(n + 1)(2n + 1)] = \tfrac{1}{6}b^3 \frac{n}{n} \frac{n+1}{n} \frac{2n+1}{n}$$

$$= \tfrac{1}{6}b^3\left(1 + \frac{1}{n}\right)\left(2 + \frac{1}{n}\right)$$

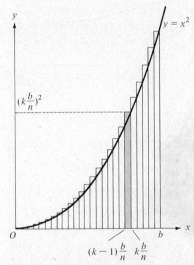

Figure 6.3
Approximating the area under $y = x^2$ from 0 to b with n rectangles.

We see at once that
$$\lim_{n \to \infty} \left(1 + \frac{1}{n}\right) = 1$$
and
$$\lim_{n \to \infty} \left(2 + \frac{1}{n}\right) = 2$$
Hence,
$$\lim_{n \to \infty} S_n = \tfrac{1}{3} b^3$$
For $b = 1$ this gives
$$\lim_{n \to \infty} S_n = \tfrac{1}{3}$$
and this is taken to be the area under the curve from $x = 0$ to $x = 1$.

A REMARK ABOUT LIMITS

General limits as $n \to \infty$ are formally introduced in Section 11.1. In this chapter we use these limits in exactly the same way that limits as $x \to \infty$ were used in Section 2.6. We mention the following properties of limits.

1. $\lim\limits_{n \to \infty}(a_n + b_n) = \lim\limits_{n \to \infty} a_n + \lim\limits_{n \to \infty} b_n$

2. $\lim\limits_{n \to \infty} c a_n = c \lim\limits_{n \to \infty} a_n$

3. $\lim\limits_{n \to \infty} a_n = c$ if $a_n = c$ for all n;

    this is often expressed by
    $$\lim_{n \to \infty} c = c$$

4. $\lim\limits_{n \to \infty} \dfrac{1}{n^\alpha} = 0$ for $\alpha > 0$

Let us point out that in taking the limit
$$\lim_{x \to \infty} f(x)$$
x varies over all real numbers, whereas in the limit
$$\lim_{n \to \infty} a_n$$
n varies only over the positive integers.

The approximation scheme described above can be used for finding the area under a curve $y = f(x)$ from $x = a$ to $x = b$ (see Figure 6.4). It is customary to use the notation

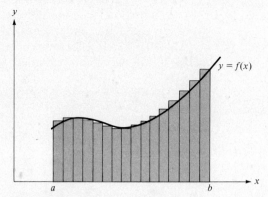

Figure 6.4
Approximating the area under $y = f(x)$ from a to b.

$$\int_a^b f(x)\, dx$$

read "the integral of $f(x)$ from a to b." In setting up the general scheme, the following notation is used: We let

$$\Delta x = \frac{b - a}{n}$$

and get the following points of subdivision of the interval $[a, b]$:

$x_1 = a + \Delta x$
$x_2 = a + 2\,\Delta x$
$x_3 = a + 3\,\Delta x$
$\vdots$
$x_n = a + n\,\Delta x = b$

The corresponding rectangles approximating the area have base Δx and heights $f(x_1), f(x_2), f(x_3), \ldots, f(x_n)$.

DEFINITION 1
The *integral* of $f(x)$ from a to b, written

$$\int_a^b f(x)\, dx$$

is the limit of the sums

$$S_n = f(x_1)\,\Delta x + f(x_2)\,\Delta x + \cdots + f(x_n)\,\Delta x$$

when the limit exists.

This definition, which was motivated with functions $f(x) \geq 0$, makes sense also for functions that are negative in part or all of the interval [a, b]. When $f(x) \geq 0$ for all points x in the interval $[a, b]$, the integral gives the area under the curve $y = f(x)$ lying between the lines $x = a$ and $x = b$. In effect, this gives a *definition* of area as a limit of sums. Using the integral notation, we have from Example 2

$$\int_0^b x^2 \, dx = \tfrac{1}{3} b^3$$

In the next example, the *integrand* $f(x)$ is negative on part of the interval of integration. An interpretation of such a situation is given following Example 4.

Example 3

Using the definition, find $\int_a^b x \, dx$.

Solution
As in the procedure outlined earlier, let

$$\Delta x = \frac{b - a}{n}$$

$$x_k = a + k\Delta x$$

For $f(x) = x$, we then have

$$f(x_k) = x_k = a + k\Delta x$$

and hence (see Figure 6.5)

$$\begin{aligned} S_n &= f(x_1)\Delta x + \cdots + f(x_n)\Delta x \\ &= [(a + \Delta x) + (a + 2\Delta x) + \cdots + (a + n\Delta x)]\,\Delta x \\ &= [na + (1 + 2 + \cdots + n)\Delta x]\Delta x \\ &= na\Delta x + (1 + 2 + \cdots + n)(\Delta x)^2 \end{aligned}$$

We now use the formula (see appendix to this chapter)

$$1 + 2 + \cdots + n = \frac{n(n+1)}{2}$$

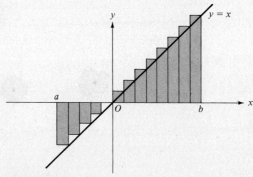

Figure 6.5
Approximation to $\int_a^b x \, dx$.

which gives

$$S_n = na\frac{b-a}{n} + \frac{n(n+1)}{2}\frac{(b-a)^2}{n^2}$$

$$= a(b-a) + \frac{n+1}{n}\frac{(b-a)^2}{2}$$

Since $\lim_{n\to\infty}\frac{n+1}{n} = 1$, we have

$$\lim_{n\to\infty} S_n = a(b-a) + \frac{(b-a)^2}{2} = \frac{b^2-a^2}{2}$$

and hence

$$\int_a^b x\,dx = \tfrac{1}{2}(b^2 - a^2) \tag{1}$$

(see Figure 6.6).

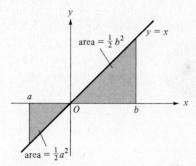

Figure 6.6
Illustrating formula (1).

Example 4

Evaluate $\int_{-1}^{1} x\,dx$.

Solution
Using formula (1) with $a = -1$ and $b = 1$ gives

$$\int_{-1}^{1} x\,dx = \tfrac{1}{2}[1^2 - (-1)^2] = 0$$

This result will now be explained.

When $f(x) < 0$ on $[a, b]$, then $f(x_k)\,\Delta x$ is a negative number. Since $-f(x_k)\,\Delta x$ is positive and hence the area of a rectangle of height $-f(x_k)$ and base Δx, it follows that $f(x_k)\,\Delta x$ can be thought of as minus the area of the rectangle (Figure 6.7). Thus, S_n approaches minus the area bounded by the x axis, the curve $y = f(x)$, and the lines $x = a$ and $x = b$. The following term is convenient: If $f(x) \geq 0$ on $[a, b]$ or $f(x) \leq 0$ on $[a, b]$, we call the integral $\int_a^b f(x)\,dx$ *signed area*. Thus, signed area is negative when it lies below the x axis (see Figure 6.8), and it coincides with area when it lies above the x axis. When $f(x)$ is sometimes positive and sometimes negative on $[a, b]$, then the contribution of $f(x_k) \cdot \Delta x$ to S_n is a signed area, positive when $f(x_k) > 0$ and negative when $f(x_k) < 0$. This explains why

$$\int_{-1}^{1} x\,dx = 0$$

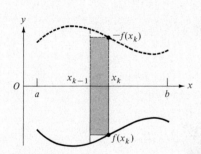

Figure 6.7

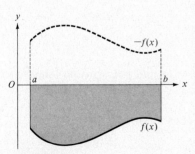

Figure 6.8
$\int_a^b f(x)$ is thought of as minus the shaded area; it is also called a *signed* area.

160 INTEGRALS

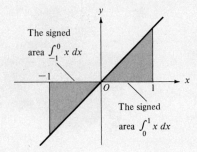

Figure 6.9

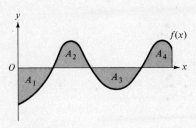

Figure 6.10

The two signed areas

$$\int_{-1}^{0} x\, dx = -\tfrac{1}{2} \quad \text{and} \quad \int_{0}^{1} x\, dx = \tfrac{1}{2}$$

cancel each other (see Figure 6.9). In a more general situation, when $f(x)$ is as pictured in Figure 6.10, then

$$\int_{a}^{b} f(x)\, dx = A_1 + A_2 + A_3 + A_4$$

where A_1, A_2, A_3, and A_4 are signed areas. Observe that A_1 and A_3 are negative, A_2 and A_4 are positive.

EXERCISES

Referring to Example 2, sketch the integrands and evaluate the definite integrals in Exercises 1–7.

1. $\int_{0}^{2} x^2\, dx$

2. $\int_{-2}^{0} x^2\, dx$

3. $\int_{-2}^{2} x^2\, dx$

4. $\int_{0}^{2} -x^2\, dx$

5. $\int_{0}^{2} 4x^2\, dx$

6. $\int_{0}^{2} (x-1)^2\, dx$

7. $\int_{0}^{2} (x+1)^2\, dx$

Evaluate the following limits.

8. $\lim_{n \to \infty} 55$

9. $\lim_{n \to \infty} \left(2 + \dfrac{3}{n}\right)$

10. $\lim_{n \to \infty} \left(1 + \dfrac{1}{n}\right)\left(2 + \dfrac{3}{n}\right)$

Sketch the integrand in each case below and guess which integral is zero. Give a reason for your guess.

11. $\int_{-4}^{4} x^3\, dx$

12. $\int_{0}^{3} x\, dx$

13. $\int_{0}^{4} (x-2)^3\, dx$

14. $\int_{-1}^{1} x^4\, dx$

15. $\int_{-1}^{1} -x^4\, dx$

16. $\int_{-4}^{4} |x|^3\, dx$

6.2 EXISTENCE OF THE DEFINITE INTEGRAL

The definite integral was defined as a limit; it was assumed throughout the preceding section that the limit exists. This can be proved to be the case for a wide class of functions (which contains all continuous functions). The conditions below are not the most general, but they are easy to state and they apply to all functions that appear in this text and that are likely to come up in applications.

CONDITIONS FOR THE EXISTENCE OF INTEGRALS
Let f be defined on $[a, b]$ and let $x_k = a + k \Delta x$ and $\Delta x = (b - a)/n$. Then

$$\int_a^b f(x)\ dx = \lim_{n \to \infty} [f(x_1)\ \Delta x + f(x_2)\ \Delta x + \cdots + f(x_n)\ \Delta x] \quad (1)$$

exists if

1. f is bounded on $[a, b]$,
2. f is continuous on $[a, b]$ except possibly at a finite number of points. Such a function is called *piecewise continuous*.

A proof of this statement involves arguments that would lead us too far afield, and is therefore omitted. Integrals of functions having discontinuities are discussed in Section 6.8.

OTHER DEFINITIONS OF INTEGRALS
We have defined the integral by using a specific approximation (limiting) scheme. There are other, more general approximation schemes that are often used. They offer no advantage at this introductory level, but they should at least be described. The procedure is this:

Step 1
Divide the interval $[a, b]$ into n parts (see Figure 6.11), *not necessarily equal,* by means of the points

$$x_0 = a < x_1 < x_2 < x_3 < \cdots < x_n = b$$

Designate the length of the kth interval $[x_{k-1}, x_k]$ by

$$(\Delta x)_k = x_k - x_{k-1}$$

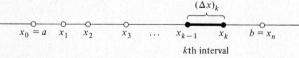

Figure 6.11
The division of $[a, b]$ into intervals of lengths $(\Delta x)_k$.

Step 2
Let t_k be *any* point in the closed interval $[x_{k-1}, x_k]$ (see Figure 6.12).

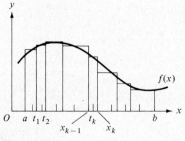

Figure 6.12

Step 3
Form the sum
$$\tilde{S}_n = f(t_1)(\Delta x)_1 + f(t_2)(\Delta x)_2 + \cdots + f(t_n)(\Delta x)_n$$

If the limit of the sequence $\tilde{S}_1, \tilde{S}_2, \tilde{S}_3, \ldots$ exists as $n \to \infty$, then it gives the integral of $f(x)$ from a to b as previously defined. A fact sometimes used is this:

> If $f(x)$ is bounded and piecewise continuous on $[a, b]$, then for each error $\epsilon > 0$, no matter how small, a number $\delta > 0$ can be found such that
> $$\left|\tilde{S}_n - \int_a^b f(x)\,dx\right| < \epsilon \text{ whenever } (\Delta x)_k < \delta \text{ for all } k$$
> This is the same as saying that
> $$\lim_{n \to \infty} \tilde{S}_n = \int_a^b f(x)\,dx$$

SUMMATION

In summation processes, such as those used to define integrals, a convenient shorthand notation is often used. For example, the sum $f(x_1)\,\Delta x + f(x_2)\,\Delta x + \cdots + f(x_n)\,\Delta x$ is abbreviated as

$$f(x_1)\,\Delta x + f(x_2)\,\Delta x + \cdots + f(x_n)\,\Delta x = \sum_{k=1}^{n} f(x_k)\,\Delta x$$

The Greek letter Σ (*sigma*) stands for *sum*. The meaning of the symbol $\sum_{k=m}^{n}$ is best explained by means of some examples.

Example 1

(a) $1 + 2 + 3 + 4 + 5 = \sum_{k=1}^{5} k$

(b) $1^2 + 2^2 + 3^2 + 4^2 + 5^2 = \sum_{k=1}^{5} k^2$

(c) $2^3 + 2^4 + 2^5 + 2^6 + 2^7 = \sum_{k=3}^{7} 2^k$

(d) $x_1 + x_2 + x_3 + x_4 + x_5 + x_6 = \sum_{k=1}^{6} x_k$

The letter k in the above sums is called the *summation index*. Other letters are often used instead of k. Thus, changing k to j gives

$$\sum_{k=m}^{n} x_k = \sum_{j=m}^{n} x_j$$

We now state the following definition:

> When $m \leq n$, then $\sum_{k=m}^{n}$ specifies a summation in which k takes on the values $m, m+1, \ldots, n$.

Example 2

Consider the sum

$$\sum_{k=1}^{3} f(x_k)$$

in which $f(x_1) = f(x_2) = f(x_3) = 1$. Here we have

$$\sum_{k=1}^{3} 1 = 1 + 1 + 1 = 3$$

In general,

$$\sum_{k=1}^{n} 1 = \underbrace{1 + 1 + \cdots + 1}_{n \text{ times}} = n$$

and

$$\sum_{k=1}^{n} x = \underbrace{x + x + \cdots + x}_{n \text{ times}} = nx$$

Two computational properties of the Σ notation are these:

PROPERTIES OF SUMS

$$\sum_{i=m}^{n} a_i + \sum_{i=m}^{n} b_i = \sum_{i=m}^{n} (a_i + b_i) \tag{2}$$

$$\sum_{i=m}^{n} c a_i = c \sum_{i=m}^{n} a_i \quad \text{for any constant } c \tag{3}$$

These properties are verified with simple algebra, by writing out the sums.

Using the summation notation, we can write formula (1) as

$$\int_{a}^{b} f(x)\, dx = \lim_{n \to \infty} \sum_{k=1}^{n} f(x_k)\, \Delta x$$

EXERCISES

Write the following sums using the Σ notation.

164 INTEGRALS

$\sum_{i=1}^{6} y_i$

1. $y_1 + y_2 + y_3 + y_4 + y_5 + y_6$
2. $2 \cdot 1 + 2 \cdot 2 + 2 \cdot 3 + 2 \cdot 4 + 2 \cdot 5$
3. $\frac{1}{1} + \frac{1}{2} + \frac{1}{3} + \frac{1}{4} + \frac{1}{5} + \frac{1}{6}$
4. $1^1 + 2^2 + 3^3 + 4^4$
5. $k^1 + k^2 + k^3 + k^4 + k^5 + k^6 + k^7$
6. $1 + 1 + 1 + 1 + 1 + 1 + 1$
7. $x + x^2 + x^3 + x^4 + x^5 + x^6$
8. $1 + x + x^2 + x^3 + x^4 + x^5 + x^6$
9. $\frac{1}{x_1 + 1} \Delta x + \frac{1}{x_2 + 1} \Delta x + \frac{1}{x_3 + 1} \Delta x + \frac{1}{x_4 + 1} \Delta x$

Write out the following sums.

10. $\sum_{i=1}^{5} \frac{i+1}{i+2}$
11. $\sum_{i=7}^{10} x^i$
12. $\sum_{j=1}^{5} 2j$
13. $\sum_{i=1}^{5} (2i + 1)$
14. $\sum_{k=1}^{3} (x_k^2 + k)$
15. $\sum_{j=10}^{15} (j - 10)$
16. $\sum_{i=1}^{5} (-1)^i x_i$
17. $\sum_{k=2}^{8} (-1)^k x_k$

6.3 PROPERTIES OF THE INTEGRAL

The results of the previous section enable us now to develop basic properties of integrals. The integral

$$\int_a^b f(x) \, dx$$

was defined for limits of integration $a < b$. This restriction is now removed by extending the definition of integral.

DEFINITION 1

$$\int_b^a f(x) \, dx = -\int_a^b f(x) \, dx \quad \text{for } a < b$$

$$\int_a^a f(x) \, dx = 0$$

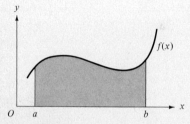

Figure 6.13
$\int_b^a f(x) \, dx = -$(shaded area).

The first part of the definition is explained in Figure 6.13.

Example 1

According to Section 6.1,

$$\int_0^3 x^2 \, dx = \tfrac{1}{3} 3^3 = 9$$

Hence,

$$\int_3^0 x^2 \, dx = -\int_0^3 x^2 \, dx = -9$$

THEOREM 1. PROPERTIES OF INTEGRALS

1. $\displaystyle\int_a^b c \, dx = c(b-a)$ for any constant c

2. $\displaystyle\int_a^b cf(x) \, dx = c \int_a^b f(x) \, dx$ for any constant c

3. $\displaystyle\int_a^b [f(x) + g(x)] \, dx = \int_a^b f(x) \, dx + \int_a^b g(x) \, dx$

4. $\displaystyle\int_a^b [f(x) - g(x)] \, dx = \int_a^b f(x) \, dx - \int_a^b g(x) \, dx$

5. $\displaystyle\int_a^c f(x) \, dx + \int_c^b f(x) \, dx = \int_a^b f(x) \, dx$

These properties are direct consequences of the definition of integral as a limit of sums. This will be demonstrated following some examples.

Example 2

Property (1) of Theorem 1 tells us that the integral of the constant function $f(x) = c$ from a to b gives the area of the rectangle $c(b-a)$ (see Figure 6.14). This is expected in view of the interpretation of the integral as area.

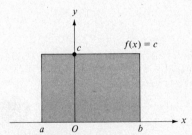

Figure 6.14
The area under $f(x) = c$ from a to b is $\int_a^b c \, dx = c(b-a)$.

Example 3

Find the area under $f(x) = x^2$ from a to b for $b > a \geq 0$.

Solution
Using the formula

$$\text{area from } a \text{ to } b = \int_a^b x^2 \, dx$$

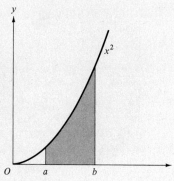

Figure 6.15
The area under $f(x) = x^2$ from a to b.

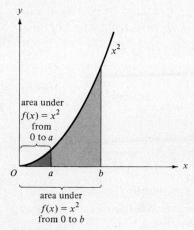

Figure 6.16

(see Figure 6.15), we proceed as follows:

$$\int_a^b x^2\, dx = \int_a^0 x^2\, dx + \int_0^b x^2\, dx \quad \text{(by Theorem 1, Property 5)}$$

$$= \int_0^b x^2\, dx + \int_a^0 x^2\, dx$$

$$= \int_0^b x^2\, dx - \int_0^a x^2\, dx \quad \text{(by Definition 1)}$$

According to Section 6.1,

$$\int_0^b x^2\, dx = \tfrac{1}{3} b^3 \quad \text{and} \quad \int_0^a x^2\, dx = \tfrac{1}{3} a^3$$

(see Figure 6.16). Thus,

$$\int_a^b x^2\, dx = \tfrac{1}{3} b^3 - \tfrac{1}{3} a^3$$

REMARKS ABOUT NOTATION
It is seen that $\int_a^b f(x)\, dx$ is a number. The presence of the variable x, often called a *dummy variable*, may thus be confusing. It does serve a useful purpose, however, as we shall subsequently see, and it is important to realize that the variable can be replaced by another without changing the integral. Thus,

$$\int_a^b f(x)\, dx = \int_a^b f(t)\, dt = \int_a^b f(s)\, ds$$

in particular,

$$\int_0^b x^2\, dx = \int_0^b t^2\, dt = \int_0^b s^2\, ds = \frac{b^3}{3}$$

The letters x, t, and s are *variables of integration*.

Now, rather than integrating from 0 to b, we could integrate from 0 to x ($x > 0$), thereby getting the result

$$\int_0^x t^2\, dt = \frac{x^3}{3}$$

Using the same variable as a dummy variable and as a limit of integration should be avoided.

Example 4

Evaluate $\int_{-1}^{2} (3x^2 - 5x)\, dx$.

Solution

$$\int_{-1}^{2} (3x^2 - 5x)\, dx = \int_{-1}^{2} 3x^2\, dx - \int_{-1}^{2} 5x\, dx \quad \text{(by Theorem 1, Property 4)}$$

$$= 3\int_{-1}^{2} x^2\, dx - 5\int_{-1}^{2} x\, dx \quad \text{(by Theorem 1, Property 2)}$$

$$= 3\{\tfrac{1}{3}[2^3 - (-1)^3]\} - 5\{\tfrac{1}{2}[2^2 - (-1)^2]\} = \tfrac{3}{2}$$

Example 5

Find the area bounded by the curves $f(x) = 1$, $g(x) = x^2$, and the y axis.

Solution

Since $f(1) = g(1) = 1$, the area asked for in the problem is the shaded area in Figure 6.17. This area equals the area under $f(x)$ minus the area under $g(x)$. Thus,

$$\text{area} = \int_0^1 f(x)\, dx - \int_0^1 g(x)\, dx$$

$$= \int_0^1 1\, dx - \int_0^1 x^2\, dx = 1 - \tfrac{1}{3} = \tfrac{2}{3}$$

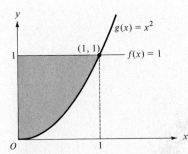

Figure 6.17
The area bounded by $f(x) = 1$, $g(x) = x^2$, and the y axis.

Formalizing the above reasoning, we have the following formula (see Figure 6.18).

THEOREM 2. AREA BETWEEN TWO CURVES

If $f(x) \geq g(x)$ for $a \leq x \leq b$, then the area A bounded by $f(x)$, $g(x)$, and the lines $x = a$ and $x = b$ is given by the formula

$$A = \int_a^b f(x)\, dx - \int_a^b g(x)\, dx$$

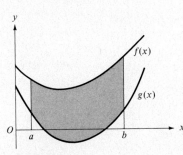

Figure 6.18
The area between $f(x)$ and $g(x)$ from a to b.

Example 6

Show that if $f(x) \geq g(x)$ for $a \leq x \leq b$, then

$$\int_a^b f(x)\, dx \geq \int_a^b g(x)\, dx$$

Solution

We shall use Theorem 2. Since A represents area, we have $A \geq 0$ and hence

$$0 \leq A = \int_a^b f(x)\, dx - \int_a^b g(x)\, dx$$

From this the desired inequality is readily obtained.

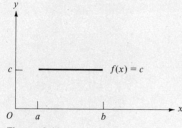

Figure 6.19
The graph of $f(x) = c$ for $a \leq x \leq b$.

Let us now turn to the proof of the properties listed earlier in this section. The notation will be that used in Section 6.2, and all integrals will be assumed to exist.

Proof of Property 1 of Theorem 1
Here the integral concerns the function $f(x) = c$ for $a \leq x \leq b$ (see Figure 6.19). Thus,

$$f(x_k) = c$$

and accordingly,

$$S_n = \sum_{k=1}^{n} f(x_k)\, \Delta x = \sum_{k=1}^{n} c\, \Delta x$$

$$= \underbrace{c\, \Delta x + c\, \Delta x + \cdots + c\, \Delta x}_{n \text{ times}} = nc\, \Delta x$$

But $\Delta x = (b - a)/n$, so

$$S_n = n c\, \Delta x = n c\, \frac{b-a}{n} = c(b - a)$$

Being true for each n, the last relation holds also in the limit as $n \to \infty$; that is,

$$\int_a^b c\, dx = \lim_{n \to \infty} S_n = c(b - a)$$

Proof of Property 2 of Theorem 1
Put $g(x) = cf(x)$ and let

$$S_n = \sum_{k=1}^{n} f(x_k)\, \Delta x$$

Then for each value of n,

$$S_n' = \sum_{k=1}^{n} g(x_k)\, \Delta x = \sum_{k=1}^{n} cf(x_k)\, \Delta x = c \sum_{k=1}^{n} f(x_k)\, \Delta x = cS_n$$

Since

$$\lim_{n \to \infty} S_n = \int_a^b f(x)\, dx \quad \text{and} \quad \lim_{n \to \infty} S_n' = \int_a^b g(x)\, dx$$

we have the relations

$$\int_a^b cf(x)\, dx = \int_a^b g(x)\, dx = \lim_{n \to \infty} S_n' = \lim_{n \to \infty} cS_n = c \lim_{n \to \infty} S_n$$

$$= c \int_a^b f(x)\, dx$$

Proof of Property 3 of Theorem 1

Put $h(x) = f(x) + g(x)$ and let

$$S_n = \sum_{k=1}^{n} h(x_k)\,\Delta x$$

$$S'_n = \sum_{k=1}^{n} f(x_k)\,\Delta x$$

$$S''_n = \sum_{k=1}^{n} g(x_k)\,\Delta x$$

Then

$$S_n = \sum_{k=1}^{n} h(x_k)\,\Delta x = \sum_{k=1}^{n} [f(x_k) + g(x_k)]\,\Delta x$$

$$= \sum_{k=1}^{n} [f(x_k)\,\Delta x + g(x_k)\,\Delta x] = \sum_{k=1}^{n} f(x_k)\,\Delta x + \sum_{k=1}^{n} g(x_k)\,\Delta x$$

$$= S'_n + S''_n$$

Since the limits of the sums S'_n and S''_n are assumed to exist, we can use Property 1 of limits given in Section 6.1. Hence

$$\int_a^b [f(x) + g(x)]\,dx = \int_a^b h(x)\,dx = \lim_{n \to \infty} S_n = \lim_{n \to \infty} [S'_n + S''_n]$$

$$= \lim_{n \to \infty} S'_n + \lim_{n \to \infty} S''_n = \int_a^b f(x)\,dx + \int_a^b g(x)\,dx$$

Proof of Property 4 of Theorem 1

$$\int_a^b [f(x) - g(x)]\,dx = \int_a^b [f(x) + (-1) \cdot g(x)]\,dx$$

$$= \int_a^b f(x)\,dx + \int_a^b (-1) \cdot g(x)\,dx$$

(by Property 3)

$$= \int_a^b f(x)\,dx + (-1) \int_a^b g(x)\,dx$$

(by Property 2)

$$= \int_a^b f(x)\,dx - \int_a^b g(x)\,dx$$

Proof of Property 5 of Theorem 1

The proof of this property is carried out for the case in which the lengths $|b - a|$ and $|c - a|$ are rational. The proof of the general case is technically involved, and it requires techniques that are not in the spirit of the introductory nature of this text. We further restrict the proof to the case

$$a < c < b$$

because the other cases follow from this one with a simple argument. The proof is carried out in several steps.

Step 1
Since $c - a$ and $b - a$ are rational (by assumption), so is their ratio; that is, there are integers $M < N$ such that

$$\frac{c-a}{b-a} = \frac{M}{N}$$

Solving this equation for c gives

$$c = a + M\left(\frac{b-a}{N}\right)$$

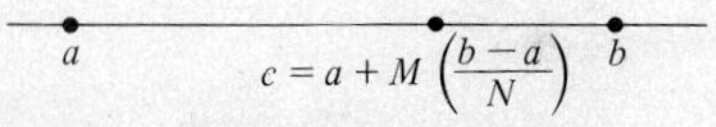

Figure 6.20

(see Figure 6.20).

Step 2
Suppose that the interval $[a, b]$ is divided into N equal parts. The points of partition are

$$x_k = a + k\left(\frac{b-a}{N}\right) \quad k = 1, 2, 3, \ldots, N-1$$

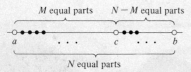

Figure 6.21

(see Figure 6.21) and we see that c is the Mth division point. From this we conclude that the interval $[a, c]$ is divided by the points x_k into M equal parts, and $[c, b]$ is divided by the points x_k into $N - M$ equal parts (see Figure 6.22).

Figure 6.22

Step 3
Consult Figure 6.23. From the fact that

$$\frac{c-a}{b-a} = \frac{M}{N} = \frac{nM}{nN} \quad \text{for } n = 1, 2, 3, 4, \ldots$$

and

$$c = a + nM\left(\frac{b-a}{nN}\right)$$

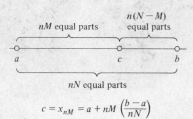

Figure 6.23

we conclude that if $[a, b]$ is divided into nN equal parts, then $[a, c]$ is divided into nM equal parts, and $[c, b]$ is divided into $nN - nM = n(N - M)$ equal parts. For each n, the length of the intervals $[x_{k-1}, x_k]$ of the partition is

$$\Delta x = \frac{b-a}{nN}$$

Step 4
Consult Figure 6.24. If we now put

$$S_{nN} = \sum_{k=1}^{nN} f(x_k)\, \Delta x$$

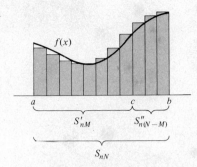

Figure 6.24

$$S_{nM} = \sum_{k=1}^{nM} f(x_k) \, \Delta x$$

$$S_{n(N-M)} = \sum_{k=nM+1}^{nN} f(x_k) \, \Delta x$$

then

$$S_{nN} = S_{nM} + S'_{n(N-M)}$$

and

$$\lim_{n \to \infty} S_{nN} = \int_a^b f(x) \, dx$$

$$\lim_{n \to \infty} S_{nM} = \int_a^c f(x) \, dx$$

$$\lim_{n \to \infty} S'_{n(N-M)} = \int_c^b f(x) \, dx$$

Step 5
Using Property 1 of Section 6.1, we conclude that

$$\int_a^b f(x) \, dx = \lim_{n \to \infty} S_{nN} = \lim_{n \to \infty} [S_{nM} + S'_{n(N-M)}]$$

$$= \lim_{n \to \infty} S_{nM} + \lim_{n \to \infty} S'_{n(N-M)} = \int_a^c f(x) \, dx + \int_c^b f(x) \, dx$$

EXERCISES

Evaluate the integrals in Exercises 1–9.

1. $\int_{-1}^{1} x^2 \, dx$
2. $\int_{-1}^{1} -x^2 \, dx$
3. $\int_{-1}^{1} 3t^2 \, dt$
4. $\int_{1}^{2} (x^2 - 1) \, dx$
5. $\int_{1}^{2} 3t \, dt$
6. $\int_{1}^{2} 5t^2 \, dt$
7. $\int_{1}^{2} (x^2 - 5) \, dx$
8. $\int_{1}^{x} (t^2 - 5) \, dt$
9. $\int_{0}^{x} 1 \, dt$

In Exercises 10–13, find the area bounded by the specified curves.

10. $f(x) = 1$, $g(x) = x^2$ (Figure 6.25)

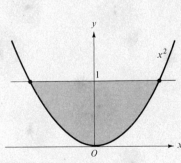

Figure 6.25

11. $f(x) = x^2$, $g(x) = 2x^2$, $x = a$ and $x = b$, $a < b$ (Figure 6.26)

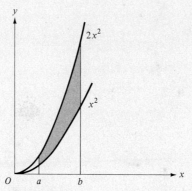

Figure 6.26

12. $f(x) = x^2$, $g(x) = \sqrt{x-1}$, $x = 1$ and $x = a > 1$ (Figure 6.27)

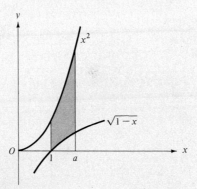

Figure 6.27

13. $f(x) = 2 - x^2$, $g(x) = 0$ (Figure 6.28)

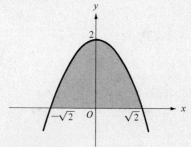

Figure 6.28

14. Show that

$$\int_0^b x^{3/2} \, dx \geq \tfrac{1}{3} b^3 \qquad \text{for } 0 < b < 1$$

15. Show that if $a < b$, then

$$\left|\int_a^b f(x)\,dx\right| \le \int_a^b |f(x)|\,dx$$

Hint: Consult Figures 6.29 and 6.30, and use the facts below in the given order:

(1) $-|f(x)| \le f(x) \le |f(x)|$ for all x

(2) $\int_a^b -|f(x)|\,dx = -\int_a^b |f(x)|\,dx$ by Property 2

(3) $-A \le B \le A$ is the same as $|B| \le A$

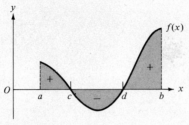

Figure 6.29
$\int_a^b f(x)\,dx =$ sum of signed areas.

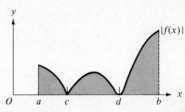

Figure 6.30
$\int_a^b |f(x)|\,dx =$ sum of areas.

6.4 THE EVALUATION OF DEFINITE INTEGRALS

The following method for evaluating integrals was discovered very early in the development of calculus, and it is one of its cornerstones. The method will be stated as a formal theorem following a description and some examples, but first we introduce the concept of antiderivative.

DEFINITION 1
Let $f(x)$ be defined on an open interval I. Then $F(x)$ is an *antiderivative* of $f(x)$ on I if $F'(x) = f(x)$ for all points x in I.

Example 1
If $f(x) = x^3$, then $F(x) = \tfrac{1}{4}x^4$ is an antiderivative of $f(x)$, since $F'(x) = x^3 = f(x)$. Notice that for any constant C, $G(x) = F(x) + C$ is also an antiderivative of $f(x)$, since $G'(x) = [F(x) + C]' = F'(x) = f(x)$. We thus see that $f(x)$ has infinitely many antiderivatives which differ by constants.

Example 2
If $f(x) = \dfrac{1}{x^2}$, then $F(x) = -\dfrac{1}{x}$ is an antiderivative.

With the concept of antiderivative, we can now formulate the following rule.

> **METHOD FOR EVALUATING DEFINITE INTEGRALS**
> To evaluate $\int_a^b f(x)\,dx$, find an antiderivative $F(x)$ of $f(x)$. Then
> $$\int_a^b f(x)\,dx = F(b) - F(a)$$

Example 3

Evaluate $\int_a^b x^3\,dx$.

Solution
Using Example 1, we have

$$\int_a^b x^3\,dx = \tfrac{1}{4}b^4 - \tfrac{1}{4}a^4$$

Example 4

Evaluate $\int_1^2 (x^4 - x^5)\,dx$.

Solution
Since $(x^5/5) - (x^6/6)$ is an antiderivative of $x^4 - x^5$, we have

$$\int_1^2 (x^4 - x^5)\,dx = \left(\frac{2^5}{5} - \frac{2^6}{6}\right) - \left(\frac{1^5}{5} - \frac{1^6}{6}\right) = -\frac{43}{10}$$

THEOREM 1
Let $f(x)$ be continuous on the closed interval $[a, b]$. If

1. $F(x)$ is continuous on $[a, b]$, and
2. $F'(x) = f(x)$ on (a, b),

then

$$\int_a^b f(x)\,dx = F(b) - F(a)$$

A rigorous proof of this fundamental theorem is given in Section 6.5. The following argument will give you an insight into this result and an indication of why it is true. From the definition of the integral,

$$\int_a^b f(x)\,dx = \lim_{n\to\infty} \sum_{k=1}^n f(x_k)\,\Delta x$$

where

$$\Delta x = \frac{b-a}{n} = x_k - x_{k-1}$$

$$x_0 = a \quad \text{and} \quad x_n = b$$

For sufficiently large values of n, therefore, the approximate equality

$$\int_a^b f(x)\, dx \approx \sum_{k=1}^n f(x_k)\, \Delta x$$

holds. When $\Delta x = x_k - x_{k-1}$ is sufficiently small, then

$$\frac{F(x_k) - F(x_{k-1})}{x_k - x_{k-1}} = \frac{F(x_k) - F(x_{k-1})}{\Delta x} \approx F'(x_k) = f(x_k)$$

that is,

$$f(x_k)\, \Delta x \approx F(x_k) - F(x_{k-1})$$

Hence,

$$\sum_{k=1}^n f(x_k)\, \Delta x \approx \sum_{k=1}^n [F(x_k) - F(x_{k-1})]$$

$$= [F(x_1) - F(x_0)] + [F(x_2) - F(x_1)] + \cdots$$
$$+ [F(x_{n-1}) - F(x_{n-2})] + [F(x_n) - F(x_{n-1})]$$

$$= F(x_n) - F(x_0) = F(b) - F(a)$$

Thus,

$$\int_a^b f(x)\, dx \approx F(b) - F(a)$$

The more exact proof given in Section 6.5 shows that the approximate equality is actually an exact equality.

According to Theorem 1, the problem of evaluating definite integrals is reduced to the problem of finding antiderivatives. The latter will be discussed in Section 6.6. In many cases, however, these can be guessed from the behavior of derivatives.

ANTIDERIVATIVES OF x^a
From the formula

$$\frac{d}{dx}(x^{a+1}) = (a+1)x^a$$

which is true for all numbers a, we see that a function whose derivative is x^a is $[1/(a+1)]x^{a+1}$, provided that $a + 1 \neq 0$. Thus:

> An antiderivative of $f(x) = x^a$ is $F(x) = \dfrac{1}{a+1} x^{a+1}$ $\quad (a \neq -1)$. (1)

In the next chapter we shall introduce the natural logarithm of x, designated $\ln x$, and we shall show that

$$\frac{d}{dx}(\ln x) = \frac{1}{x}$$

Thus, $F(x) = \ln x$ is an antiderivative of $f(x) = 1/x$, and this takes care of the case $a = -1$.

Example 5

Evaluate the integral $\int_0^2 x^5 \, dx$.

Solution
Use formula (1) with $a = 5$. Then $f(x) = x^5$, $F(x) = \frac{1}{6} x^6$ and hence

$$\int_0^2 x^5 \, dx = \tfrac{1}{6} \cdot 2^6 - \tfrac{1}{6} \cdot 0^6 = \tfrac{16}{3}$$

Example 6

Evaluate the integral $\int_{1/4}^{4} \sqrt{y} \, dy$.

Solution
The fact that y is used as the variable of integration does not change the problem of integration, and it should therefore cause no difficulty.
 Writing $\sqrt{y} = y^{1/2}$, we see that again formula (1) applies, this time with

$$f(y) = y^{1/2} \quad \text{and} \quad F(y) = \tfrac{2}{3} y^{3/2}$$

Thus,

$$\int_{1/4}^{4} \sqrt{y} \, dy = \int_{1/4}^{4} y^{1/2} \, dy = \tfrac{2}{3} 4^{3/2} - \tfrac{2}{3} (\tfrac{1}{4})^{3/2} = \tfrac{2}{3}[(4^{1/2})^3 - (4^{1/2})^{-3}]$$

$$= \tfrac{2}{3}[2^3 - 2^{-3}] = \tfrac{2}{3} \cdot \tfrac{63}{64} = \tfrac{21}{32}$$

Example 7

Evaluate the integral $\int_1^a [4x^3 - x^{-2}] \, dx$, where $a > 0$.

Solution
According to Section 6.3,

$$\int_1^a [4x^3 - x^{-2}]\, dx = \int_1^a 4x^3\, dx - \int_1^a x^{-2}\, dx$$

Since $(d/dx)(x^4) = 4x^3$, we have

$$\int_1^a 4x^3\, dx = a^4 - 1$$

Since $-x^{-1}$ is an antiderivative of x^{-2}, we have

$$\int_1^a x^{-2}\, dx = -a^{-1} - (-1) = 1 - \frac{1}{a}$$

Hence,

$$\int_1^a [4x^3 - x^{-2}]\, dx = (a^4 - 1) - \left(1 - \frac{1}{a}\right) = a^4 + \frac{1}{a} - 2$$

The following theorem is basic: It is used in the proof of Theorem 1 that appears in the next section. It also explains some frequently used notation introduced in Section 6.6 in connection with antiderivatives.

THEOREM 2
If $F(x)$ and $G(x)$ are both antiderivatives of $f(x)$ on the interval (a, b), then there is a constant C such that

$$G(x) = F(x) + C$$

for all x in (a, b).

Proof
Since

$$\frac{d}{dx} G(x) = f(x) \quad \text{and} \quad \frac{d}{dx} F(x) = f(x)$$

we conclude that

$$\frac{d}{dx}[G(x) - F(x)] = 0$$

It follows from Theorem 1 of Section 4.6 that for some constant C

$$G(x) - F(x) = C$$

for all x in (a, b). The theorem follows.

EXERCISES

In Exercises 1–12, evaluate the given integrals. You may have to use the properties of integrals discussed in Section 6.3.

1. $\int_{-1}^{1} (4x^2 - 2x + 1)\, dx$

2. $\int_{1}^{2} s^{1/3}\, ds$

3. $\int_{0}^{b} y^{100}\, dy$

4. $\int_{1}^{2} (x^5 + x^{-5})\, dx$

5. $\int_{a}^{b} t(t-1)(t+1)\, dt$

6. $\int_{0}^{x} [a_0 + a_1 t + a_2 t^2 + a_3 t^3 + a_4 t^4]\, dt$

7. $\int_{1}^{x} \left[\frac{a_2}{t^2} + \frac{a_3}{t^3} + \frac{a_4}{t^4}\right] dt \qquad (x > 0)$

8. $\int_{-2}^{2} (x + x^5)\, dx$

9. $\int_{-2}^{2} (x + x^5 + x^9)\, dx$

10. $\int_{-4}^{4} (x - x^5 + x^9)\, dx$

11. $\int_{0}^{x} \frac{1}{\sqrt{t}}\, dt$

12. $\int_{0}^{1} (x^2 + 1)^3\, dx$

6.5 THE FUNDAMENTAL THEOREM OF CALCULUS

We now present the fundamental theorem of the calculus, which states a new relation between integrals and derivatives. At the end of this section we give proofs of the fundamental theorem and the evaluation method of the last section. It is important for the student to understand what these theorems say; the proofs may be omitted.

We start with a function $f(t)$ that is continuous on a closed interval $[a, b]$; for each value x in $[a, b]$ we may consider the value of the integral

$$\int_{a}^{x} f(t)\, dt$$

(see Figure 6.31). As x changes, this integral changes; we thus have defined a new function, which we call $\phi(x)$:

$$\phi(x) = \int_{a}^{x} f(t)\, dt$$

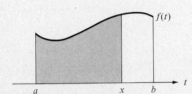

Figure 6.31
$\int_{a}^{x} f(t)\, dt =$ shaded area.

> **THE FUNDAMENTAL THEOREM OF THE CALCULUS**
> If $f(x)$ is continuous on $[a, b]$, then $\phi(x) = \int_a^x f(t) \, dt$ has a derivative at each point x_0 in (a, b) and
> $$\phi'(x_0) = f(x_0)$$

Before proving this theorem, we shall give some examples.

Example 1

Let $f(x) = x^2$, and let $[a, b] = [0, 1]$. Then for each value of x in $[0, 1]$, we have

$$\phi(x) = \int_0^x f(t) \, dt = \int_0^x t^2 \, dt = \tfrac{1}{3}x^3$$

Note, in particular, that $\phi'(x) = f(x)$.

Example 2

According to the fundamental theorem of calculus,

$$\frac{d}{dx} \int_1^x (t^3 - \tfrac{1}{4}t^2) \, dt = x^3 - \tfrac{1}{4}x^2$$

$$\frac{d}{dx} \int_{-4}^x \sqrt{s^2 + 1} \, ds = \sqrt{x^2 + 1}$$

$$\frac{d}{dy} \int_3^y \sqrt{x^2 + 1} \, dx = \sqrt{y^2 + 1}$$

As preliminary steps in the proof of the fundamental theorem, we establish two lemmas.

LEMMA 1
If

$$\phi(x) = \int_a^x f(t) \, dt$$

then

$$\phi(x) - \phi(x_0) = \int_{x_0}^x f(t) \, dt$$

Proof

Since $\phi(x_0) = \int_a^{x_0} f(t) \, dt$, we have

$$\phi(x) - \phi(x_0) = \int_a^x f(t)\,dt - \int_a^{x_0} f(t)\,dt$$

$$= \int_a^x f(t)\,dt + \int_{x_0}^a f(t)\,dt \qquad \text{(by Definition 1, Section 6.3)}$$

$$= \int_{x_0}^x f(t)\,dt \qquad \text{(by Theorem 1, Property 5, Section 6.3)}$$

LEMMA 2

If

$$\phi(x) = \int_a^x f(t)\,dt$$

then

$$\frac{\phi(x) - \phi(x_0)}{x - x_0} = f(x_0) + \frac{1}{x - x_0}\int_{x_0}^x [f(t) - f(x_0)]\,dt \qquad \text{for } x \neq x_0$$

Proof
According to Lemma 1,

$$\frac{\phi(x) - \phi(x_0)}{x - x_0} = \frac{1}{x - x_0}\int_{x_0}^x f(t)\,dt$$

Now, since $f(x_0)$ is constant, we have

$$\int_{x_0}^x f(x_0)\,dt = f(x_0)(x - x_0)$$

and so

$$f(x_0) = \frac{1}{x - x_0}\int_{x_0}^x f(x_0)\,dt$$

Hence, by Theorem 1, Property 4, Section 6.3,

$$\frac{\phi(x) - \phi(x_0)}{x - x_0} = f(x_0) - \frac{1}{x - x_0}\int_{x_0}^x f(x_0)\,dt + \frac{1}{x - x_0}\int_{x_0}^x f(t)\,dt$$

$$= f(x_0) + \frac{1}{x - x_0}\left[\int_{x_0}^x f(t)\,dt - \int_{x_0}^x f(x_0)\,dt\right]$$

$$= f(x_0) + \frac{1}{x - x_0}\int_{x_0}^x [f(t) - f(x_0)]\,dt$$

PROOF OF THE FUNDAMENTAL THEOREM
To show that $\phi'(x_0) = f(x_0)$, we have to show that

$$\lim_{x \to x_0} \frac{\phi(x) - \phi(x_0)}{x - x_0} = f(x_0)$$

6.5 THE FUNDAMENTAL THEOREM OF CALCULUS

According to Lemma 2, this is the same as demonstrating that

$$\lim_{x \to x_0} \frac{1}{x - x_0} \int_{x_0}^{x} [f(t) - f(x_0)] \, dt = 0 \tag{1}$$

To prove (1), in turn, we demonstrate that for each error $\epsilon > 0$, there is a number $\delta > 0$, such that

$$-\epsilon < \frac{1}{x - x_0} \int_{x_0}^{x} [f(t) - f(x_0)] \, dt < \epsilon \tag{2}$$

for all values of x, where $x_0 - \delta < x < x_0 + \delta$ and $x \neq x_0$.

Now, since $f(t)$ is continuous at x_0, it follows that for each error $\epsilon > 0$ there is a $\delta > 0$ such that

$$-\epsilon < f(t) - f(x_0) < \epsilon \tag{3}$$

for all values of t in the interval $x_0 - \delta < t < x_0 + \delta$. The inequalities in (3) are considered separately for the cases $x_0 < x < x_0 + \delta$ and $x_0 - \delta < x < x_0$. In the first case we integrate (3) from x_0 to x. This gives

$$\int_{x_0}^{x} (-\epsilon) \, dt < \int_{x_0}^{x} [f(t) - f(x_0)] \, dt < \int_{x_0}^{x} \epsilon \, dt$$

Since

$$\int_{x_0}^{x} (-\epsilon) \, dt = -\epsilon(x - x_0) \quad \text{and} \quad \int_{x_0}^{x} \epsilon \, dt = \epsilon(x - x_0)$$

we have

$$-\epsilon(x - x_0) < \int_{x_0}^{x} [f(t) - f(x_0)] \, dt < \epsilon(x - x_0)$$

The choice $x > x_0$ gives $x - x_0 > 0$, and dividing these inequalities by $x - x_0$ now yields the inequalities in (2). The second case is handled in a similar way.

We now return to Theorem 1 of Section 6.4, repeated here for convenience.

THEOREM 1
Let $f(x)$ be continuous on the closed interval $[a, b]$. If

1. $F(x)$ is continuous on $[a, b]$, and
2. $F'(x) = f(x)$ on (a, b),

then

$$\int_{a}^{b} f(x) \, dx = F(b) - F(a)$$

Proof
Let
$$H(x) = F(x) - \phi(x)$$
where
$$\phi(x) = \int_a^x f(t)\, dt$$
Now,
$$F'(x) = f(x) \quad \text{(by hypothesis)}$$
and
$$\phi'(x) = f(x) \quad \text{(by the fundamental theorem of calculus)}$$
Since $F(x)$ and $\phi(x)$ are both antiderivatives of $f(x)$, Theorem 2 of Section 6.4 tells us that a constant C can be found such that
$$F(x) = \phi(x) + C$$
or
$$F(x) = \int_a^x f(t)\, dt + C$$
Setting $x = a$, we have
$$F(a) = \int_a^a f(t)\, dt + C = C$$
and thus $F(a) = C$. Letting $x = b$ gives
$$F(b) = \int_a^b f(t)\, dt + F(a)$$
and the theorem follows.

EXERCISES

In Exercises 1–10, evaluate the given integrals.

1. $\displaystyle\int_{-2}^{2} x^7\, dx$

2. $\displaystyle\int_{-4}^{0} \sqrt[3]{x}\, dx$

3. $\displaystyle\int_{a}^{b} \frac{1}{x^2}\, dx$

4. $\displaystyle\int_{0}^{1} (x^4 + x^2 + 1)\, dx$

5. $\displaystyle\int_{-1}^{1} (x^4 + x)^2\, dx$

6. $\displaystyle\int_{0}^{c} x(x+1)(x+2)\, dx$

7. $\displaystyle\int_{0}^{10} 5\, dx$

8. $\displaystyle\int_{-1}^{1} x\sqrt{x^2 + 1}\, dx$

9. $\int_0^1 x(x^2+1)^3 \, dx$ **10.** $\int_1^2 \left(1 - \frac{1}{x^2}\right) dx$

Give formulas for the integrals in Exercises 11–13.

11. $\int_a^b f'(x) \, dx$

12. $\int_a^b [f'(x)g(x) + f(x)g'(x)] \, dx$

13. $\int_a^b [f(x)]^5 f'(x) \, dx$

6.6 INDEFINITE INTEGRALS

In Section 6.4 we introduced the concept of antiderivative, and we have shown how the problem of integration reduces to the problem of finding antiderivatives. We have also learned in Theorem 2 of Section 6.4 that any two antiderivatives of the same function differ only by a constant. In developing rules for finding antiderivatives, the following nomenclature will be found useful:

DEFINITION 1. INDEFINITE INTEGRALS
If $F'(x) = f(x)$, then we put

$$\int f(x) \, dx = F(x) + C$$

and call $\int f(x) \, dx$ the *indefinite integral* of $f(x)$.

The indefinite integral is seen to have the property

$$\frac{d}{dx} \int f(x) \, dx = f(x)$$

The idea of indefinite integral will become clear with a few examples.

Example 1

(a) $\int 5x^4 \, dx = x^5 + C$

(b) $\int x^4 \, dx = \frac{1}{5}x^5 + C$

(c) $\int (x^4 - x^5) \, dx = \frac{1}{5}x^5 - \frac{1}{6}x^6 + C$

These formulas are verified by differentiating the right-hand side.

In general:

> To verify that $\int f(x)\,dx = F(x) + C$, it suffices to show that $F'(x) = f(x)$.

Using letters other than x does not change the facts stated above. Thus,

$$\int y^4\,dy = \tfrac{1}{5}y^5 + C$$

$$\int t^4\,dt = \tfrac{1}{5}t^5 + C$$

and so on.

THEOREM 1. PROPERTIES OF INDEFINITE INTEGRALS

1. $\displaystyle\int F'(x)\,dx = F(x) + C$

2. $\displaystyle\int Af(x)\,dx = A\int f(x)\,dx$ for any constant A

3. $\displaystyle\int [f(x) + g(x)]\,dx = \int f(x)\,dx + \int g(x)\,dx$

4. $\displaystyle\int [f(x) - g(x)]\,dx = \int f(x)\,dx - \int g(x)\,dx$

5. $\displaystyle\int x^a\,dx = \frac{1}{a+1}x^{a+1} + C$ for $a \neq -1$

Proof

To prove these properties, it suffices to show that the derivative of the right side equals the integrand on the left side.

1. $\dfrac{d}{dx}[F(x) + C] = F'(x)$

2. $\dfrac{d}{dx}\left[A\int f(x)\,dx\right] = A\dfrac{d}{dx}\int f(x)\,dx = Af(x)$

3. $\dfrac{d}{dx}\left[\int f(x)\,dx + \int g(x)\,dx\right] = \dfrac{d}{dx}\int f(x)\,dx + \dfrac{d}{dx}\int g(x)\,dx$
$= f(x) + g(x)$

4. The proof is as (3) above, with $+$ replaced by $-$.

5. $\dfrac{d}{dx}\left[\dfrac{1}{a+1}x^{a+1}\right] = \dfrac{1}{a+1}\dfrac{d}{dx}[x^{a+1}] = x^a$

Example 2

Find a function $f(x)$ that satisfies the conditions
$$\begin{cases} f'(x) = 5 + 3x - 6x^2 \\ f(1) = -2 \end{cases}$$

Solution
The first condition tells us that we are to seek an indefinite integral of $5 + 3x - 6x^2$. We find that

$$f(x) = \int (5 + 3x - 6x^2) \, dx$$

$$= \int 5 \, dx + \int 3x \, dx - \int 6x^2 \, dx$$

$$= 5 \int 1 \, dx + 3 \int x \, dx - 6 \int x^2 \, dx$$

$$= 5x + \tfrac{3}{2} x^2 - 2x^3 + C$$

The second condition can now be used to determine C.

$$-2 = f(1) = 5 \cdot 1 + \tfrac{3}{2} \cdot 1^2 - 2 \cdot 1^3 + C = \tfrac{9}{2} + C$$

and hence $C = -2 - \tfrac{9}{2} = -\tfrac{13}{2}$.

The function
$$f(x) = 5x + \tfrac{3}{2} x^2 - 2x^3 - \tfrac{13}{2}$$
satisfies both conditions of the problem.

Example 3

Find the indefinite integral
$$\int (t + 2)^{10} \, dt$$

Solution
It is easy to verify that
$$\frac{d}{dt} \frac{1}{11} (t + 2)^{11} = (t + 2)^{10}$$

and hence
$$\int (t + 2)^{10} \, dt = \frac{1}{11} (t + 2)^{11} + C$$

REMARK 1
Every continuous function on an open interval has an indefinite integral.

Suppose that $f(x)$ is continuous on the open interval (a, b). Fixing an arbitrary number c, $a < c < b$, put

$$F(x) = \int_c^x f(t) \, dt$$

Then $F'(x) = f(x)$, and hence $F(x)$ is an indefinite integral. This is the origin of this nomenclature.

REMARK 2
The examples and exercises in this chapter have been chosen in such a way that the indefinite integrals can be "evaluated" in terms of known functions. The integral table at the end of the book gives many more examples of such evaluations. It is not always possible, however, to evaluate indefinite integrals. For example, the indefinite integral $\int 2^{x^2} \, dx$ exists, yet there is no function $F(x)$, other than those of the form $\int_a^x 2^{t^2} \, dt$, such that $F'(x) = 2^{x^2}$. Thus, there is no simple way for computing the definite integral $\int_a^b 2^{x^2} \, dx$.

EXERCISES

1. Find a formula for the indefinite integral of an arbitrary polynomial:

$$\int (a_0 + a_1 x + a_2 x^2 + \cdots + a_k x^k + \cdots + a_n x^n) \, dx =$$

2. Verify that

$$\int \frac{1}{(x+1)^2} \, dx = \frac{1}{2}\left(\frac{x-1}{x+1}\right) + C \quad \text{and} \quad \int \frac{1}{(x+1)^2} \, dx = -\frac{1}{x+1} + C$$

Find the indefinite integrals in Exercises 3–10.

3. $\int x^{-5} \, dx$

4. $\int t^{-5} \, dt$

5. $\int \frac{1}{(x+2)^7} \, dx$

6. $\int \sqrt{s+1} \, ds$

7. $\int \frac{1}{t^2} \, dt$

8. $\int \frac{1}{(t-1)^2} \, dt$

9. $\int (2x+1)^{10} \, dx$

10. $\int \sqrt[3]{2x+1} \, dx$

In Exercises 11–14, find a function $f(x)$ that satisfies the given conditions.

11. $\begin{cases} f'(x) = x^2 - 2x + 1 \\ f(0) = -1 \end{cases}$

12. $\begin{cases} f'(x) = x^3 - 2 \\ f(-1) = 0 \end{cases}$

13. $\begin{cases} f'(x) = -\dfrac{3}{x^2}\left(1 + \dfrac{1}{x}\right)^2 \\ f(1) = 0 \end{cases}$

14. $\begin{cases} f'(x) = \dfrac{1}{(x+1)^2}\left(\dfrac{x-1}{x+1}\right)^{-1/2} \\ f(1) = 1 \end{cases}$

Evaluate the definite integrals in Exercises 15 and 16.

15. $\displaystyle\int_1^2 x^{-5}\, dx$

16. $\displaystyle\int_{-1}^2 \sqrt{x+1}\, dx$

6.7 INTEGRATION BY SUBSTITUTION

Using the chain rule for derivatives, we get a useful formula for obtaining many integrals. For any differentiable function $g(x)$, we have

$$\frac{d}{dx}[g(x)]^{n+1} = (n+1)[g(x)]^n g'(x)$$

or

$$\frac{d}{dx}\frac{1}{n+1}[g(x)]^{n+1} = [g(x)]^n g'(x)$$

This gives the following integral:

$$\int [g(x)]^n g'(x)\, dx = \frac{1}{n+1}[g(x)]^{n+1} + C \tag{1}$$

This holds for all $n \neq -1$.

Let us look at some applications.

Example 1

Find

$$\int \left(x + \frac{1}{x}\right)^{3/2}\left(1 - \frac{1}{x^2}\right) dx$$

Solution

We use equation (1), with $n = \frac{3}{2}$ and $g(x) = x + 1/x$. We note that $g'(x) = 1 - 1/x^2$ and $n + 1 = \frac{5}{2}$. Thus,

$$\int \left(x + \frac{1}{x}\right)^{3/2} \left(1 - \frac{1}{x^2}\right) dx = \frac{2}{5}\left(x + \frac{1}{x}\right)^{5/2} + C$$

Example 2

For $n \neq -1$, we have the general formula

$$\int (x - a)^n \, dx = \frac{1}{n+1}(x - a)^{n+1} + C$$

This follows from (1), using $g(x) = (x - a)$ and noting that $g'(x) = 1$.

In some examples a constant must be supplied to apply (1). This point is illustrated in the next example.

Example 3

Find

$$\int (x^2 + 2x + 4)^{50}(x + 1) \, dx$$

Solution

If we let $g(x) = (x^2 + 2x + 4)$, we have

$$g'(x) = 2x + 2 = 2(x + 1)$$

Formula (1) now gives

$$\int (x^2 + 2x + 4)^{50} 2(x + 1) \, dx = \tfrac{1}{51}(x^2 + 2x + 4)^{51} + C$$

Dividing by 2, we find that

$$\int (x^2 + 2x + 4)^{50}(x + 1) \, dx = \tfrac{1}{2} \cdot \tfrac{1}{51}(x^2 + 2x + 4)^{51} + \tfrac{1}{2}C$$

Since $\tfrac{1}{2}C$ can represent any number, it can be replaced by C, giving

$$\int (x^2 + 2x + 4)^{50}(x + 1) \, dx = \tfrac{1}{102}(x^2 + 2x + 4)^{51} + C$$

In using formula (1), it is often convenient to use the notation of differentials. Recalling from Section 5.6 that if $u = g(x)$, then $du = g'(x) \, dx$, we can write formula (1) as

$$\int u^n \, du = \frac{1}{n+1} u^{n+1} + C \tag{2}$$

The following example illustrates the use of differentials.

Example 4

Find

$$\int \frac{x^2}{(x^3-1)^{1/5}}\,dx$$

Solution
Let $u = x^3 - 1$, so that

$$du = 3x^2\,dx$$

and

$$\tfrac{1}{3}\,du = x^2\,dx$$

Hence,

$$\int \frac{x^2}{(x^3-1)^{1/5}}\,dx = \int \frac{1}{u^{1/5}}\frac{1}{3}\,du = \frac{1}{3}\int u^{-1/5}\,du$$

$$= \frac{1}{3}\left(\frac{5}{4}u^{4/5} + C\right) = \frac{5}{12}(x^3-1)^{4/5} + C$$

A far-reaching generalization of equations (1) and (2) is contained in the following theorem.

THEOREM 1
If

$$\int f(u)\,du = F(u) + C$$

then

$$\int f[g(x)]g'(x)\,dx = F[g(x)] + C$$

Proof
We are given that $F'(u) = f(u)$. Using the chain rule,

$$\frac{d}{dx}F[g(x)] = F'[g(x)]g'(x) = f[g(x)]g'(x)$$

This completes the proof.

Theorem 1 will be applied in coming chapters to a variety of functions $f(u)$, taken from the trigonometric, logarithmic, and exponential functions. We note the following special case of the theorem, obtained by taking, in turn, $g(x) = x - a$ and $g(x) = bx$.

COROLLARY 1
If
$$\int f(u)\ du = F(u) + C$$
then
$$\int f(x-a)\ dx = F(x-a) + C$$
and
$$\int f(bx)\ dx = \frac{1}{b}F(bx) + C \qquad (b \neq 0)$$

From Theorem 1, we now derive a formula for substitution in definite integrals. First, however, we introduce the following very useful notation.

$$\boxed{\,F(x)\Big|_a^b = F(b) - F(a)\,}$$

Example 5

(a) $\frac{1}{2}x^2\Big|_a^b = \frac{1}{2}b^2 - \frac{1}{2}a^2 = \frac{1}{2}(b^2 - a^2)$

(b) $\sqrt{4-x^2}\,\Big|_0^2 = \sqrt{4-2^2} - \sqrt{4-0} = 0 - 2 = -2$

With the above notation, our basic theorem on evaluation of definite integrals can be stated as follows:
If
$$\int f(x)\ dx = F(x) + C$$
then
$$\int_a^b f(x)\ dx = F(x)\Big|_a^b$$

THEOREM 2. SUBSTITUTION IN DEFINITE INTEGRALS
$$\int_a^b f[g(x)]g'(x)\ dx = \int_{g(a)}^{g(b)} f(u)\ du$$
provided that the integrals are meaningful.

Proof
Suppose that
$$\int f(u)\, du = F(u) + C$$
Then
$$\int_{g(a)}^{g(b)} f(u)\, du = F(u)\Big|_{g(a)}^{g(b)} = F[g(b)] - F[g(a)]$$
On the other hand, Theorem 1 tells us that
$$\int_a^b f[g(x)]g'(x)\, dx = F[g(x)]\Big|_a^b = F[g(b)] - F[g(a)]$$
and this completes the proof.

REMARK
The following conditions are sufficient for the validity of the equation of Theorem 2.
1. $g(x)$ is continuous on the closed interval $[a, b]$ and differentiable on the open interval (a, b).
2. $g'(x)$ is continuous and bounded on (a, b).
3. $f(u)$ is continuous and bounded on the range of $g(x)$ for $a < x < b$.

Example 6

Evaluate $\int_0^3 \sqrt{x^2 + 16} \cdot 2x\, dx$.

Solution
Let
$$g(x) = x^2 + 16$$
$$f(u) = \sqrt{u}$$
Then
$$g'(x) = 2x$$
$$f[g(x)]g'(x) = \sqrt{x^2 + 16} \cdot 2x$$
Since
$$g(0) = 0^2 + 16 = 16$$
$$g(3) = 3^2 + 16 = 25$$
we have
$$\int_0^3 \sqrt{x^2 + 16} \cdot 2x\, dx = \int_{16}^{25} \sqrt{u}\, du = \tfrac{2}{3} u^{3/2}\Big|_{16}^{25} = \tfrac{2}{3}(25^{3/2} - 16^{3/2}) = \tfrac{122}{3}$$

Example 7

Evaluate $\int_0^1 (2x+1)^9\, dx$.

Solution
Let
$$g(x) = 2x + 1$$
$$f(u) = u^9$$

Then
$$g'(x) = 2$$
$$f[g(x)]g'(x) = (2x+1)^9 \cdot 2$$

Since
$$g(0) = 1$$
$$g(1) = 3$$

we have

$$\int_0^1 (2x+1)^9 \cdot 2\, dx = \int_1^3 u^9\, du = \tfrac{1}{10} u^{10}\Big|_1^3 = \tfrac{1}{10}(3^{10} - 1)$$

Dividing by 2 gives the final answer:

$$\int_0^1 (2x+1)^9\, dx = \tfrac{1}{20}(3^{10} - 1)$$

Example 8

The following equality is an interesting application of Theorem 2.

$$\int_a^b \frac{1}{x}\, dx = \int_{ka}^{kb} \frac{1}{x}\, dx$$

In order to prove this, we let $f(u) = 1/u$, $g(x) = kx$. Since $g'(x) = k$, $g(a) = ka$, and $g(b) = kb$, we have

$$\int_a^b \frac{1}{kx} k\, dx = \int_{ka}^{kb} \frac{1}{u}\, du$$

Thus

$$\int_a^b \frac{1}{x}\, dx = \int_{ka}^{kb} \frac{1}{x}\, dx$$

EXERCISES

Find the indefinite integrals in Exercises 1–17.

1. $\int (x^2 + 1)\sqrt{2}\, 2x\, dx$ 2. $\int \sqrt{1 - 2x}\, dx$

3. $\displaystyle\int \sqrt{x^4 - x^2 + 1} \cdot (2x^3 - x)\, dx$
4. $\displaystyle\int \frac{1}{(x-1)^2} \sqrt{\frac{x+1}{x-1}}\, dx$

5. $\displaystyle\int \frac{t}{\sqrt[3]{t^2 + 1}}\, dt$
6. $\displaystyle\int s^3 \sqrt{s^4 + a^4}\, ds$

7. $\displaystyle\int x\sqrt[3]{(2^2 - x^2)^2}\, dx$
8. $\displaystyle\int \frac{(\sqrt[3]{x} + 5)^5}{\sqrt[3]{x^2}}\, dx$

9. $\displaystyle\int (x^2 + 2x + 4)^{50}(x+1)\, dx$
10. $\displaystyle\int \left(x + \frac{1}{x}\right)^{3/2} \frac{x^2 - 1}{x^2}\, dx$

11. $\displaystyle\int (x^{10} + x)^{10}(x^9 + \tfrac{1}{10})\, dx$
12. $\displaystyle\int \frac{x}{(x^2 + 1)^4}\, dx$

13. $\displaystyle\int \frac{3t^2 - 1}{(t^3 - t + 2)^2}\, dt$
14. $\displaystyle\int (2u - 3)^{4/7}\, du$

15. $\displaystyle\int \frac{x}{(x+1)^5}\, dx$
16. $\displaystyle\int \frac{1}{x^2 - 2x + 1}\, dx$

17. $\displaystyle\int \frac{1}{(x^2 - 2x + 1)^3}\, dx$

Evaluate the definite integrals in Exercises 18–20.

18. $\displaystyle\int_0^{1/2} \sqrt{1 - 2t}\, dt$

19. $\displaystyle\int_{1/2}^0 \sqrt{1 - 2t}\, dt$

20. $\displaystyle\int_0^1 \frac{x^3 + \tfrac{1}{2}}{(x^4 + 2x + 1)^3}\, dx$

21. Show that for any constant A,
$$\int_a^b f(x - A)\, dx = \int_{a-A}^{b-A} f(x)\, dx$$
(see Figure 6.32).

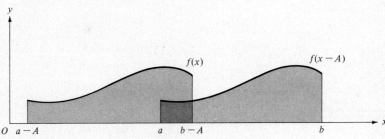

Figure 6.32
Illustrating the two integrals in Exercise 21.

22. Show that for any constant $B \neq 0$,
$$\int_a^b f(Bx)\, dx = \frac{1}{B} \int_{Ba}^{Bb} f(x)\, dx$$

23. Deduce from Exercise 22 that
$$\int_{-b}^{-a} f(-x)\, dx = \int_a^b f(x)\, dx$$

Hint: Express the integral in terms of integrals from $-a$ to 0, and from 0 to a.

24. Deduce from Exercise 23 the following fact:

If $f(x)$ is odd, then $\int_{-a}^{a} f(x)\, dx = 0$.

25. Deduce from Exercise 23 the following fact:

If $f(x)$ is even, then $\int_{-a}^{a} f(x)\, dx = 2 \int_0^a f(x)\, dx$.

6.8 PIECEWISE CONTINUOUS FUNCTIONS AND ABSOLUTE VALUES

Up to now the evaluation of integrals has depended on finding antiderivatives of functions over the entire interval of integration. This process must be modified when integrating piecewise continuous functions (see Section 6.2) or functions expressed in terms of absolute values. When integrating functions for which no antiderivative exists over the entire interval of integration, we divide that interval into subintervals in such a way that an antiderivative can be found on each one of the subintervals. This procedure is made clear in the following examples.

Example 1

Evaluate $\int_{-1}^{3} |x|\, dx$.

Solution
From the definition of absolute value (see Section 1.4), we have

$$|x| = \begin{cases} x & \text{for } -1 \leq x \leq 0 \\ -x & \text{for } 0 \leq x \leq 3 \end{cases}$$

(see Figure 6.33).
Thus, if we write

$$\int_{-1}^{3} |x|\, dx = \int_{-1}^{0} |x|\, dx + \int_{0}^{3} |x|\, dx$$

we know that

$$\int_{-1}^{0} |x|\, dx = \int_{-1}^{0} -x\, dx = -\tfrac{1}{2} x^2 \Big|_{-1}^{0} = \tfrac{1}{2}$$

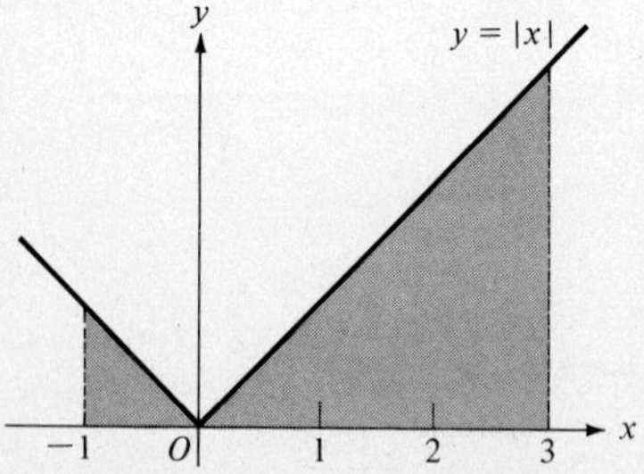

Figure 6.33
The area under $y = |x|$ from $x = -1$ to $x = 3$.

and
$$\int_0^3 |x|\, dx = \int_0^3 x\, dx = \tfrac{1}{2}x^2 \Big|_0^3 = \tfrac{9}{2}$$

Hence,
$$\int_{-1}^3 |x|\, dx = \tfrac{1}{2} + \tfrac{9}{2} = 5$$

Example 2

Evaluate $\int_{-1}^3 |x - 2|\, dx$.

Solution
Since
$$|x - 2| = \begin{cases} x - 2 & \text{for } x \geq 2 \\ -(x - 2) & \text{for } x \leq 2 \end{cases}$$

it appears appropriate to write
$$\int_{-1}^3 |x - 2|\, dx = \int_{-1}^2 |x - 2|\, dx + \int_2^3 |x - 2|\, dx$$

Since
$$\int_{-1}^2 |x - 2|\, dx = \int_{-1}^2 -(x - 2)\, dx = -\tfrac{1}{2}x^2 + 2x \Big|_{-1}^2 = \tfrac{9}{2}$$

and
$$\int_2^3 |x - 2|\, dx = \int_2^3 (x - 2)\, dx = \tfrac{1}{2}x^2 - 2x \Big|_2^3 = \tfrac{1}{2}$$

we find that
$$\int_{-1}^3 |x - 2|\, dx = \tfrac{9}{2} + \tfrac{1}{2} = 5$$

Example 3

Consider a function $f(x)$ defined on the interval $[0, 3]$ as follows (see Figure 6.34):
$$f(x) = \begin{cases} x^2 & \text{for } 0 \leq x \leq 2 \\ (x - 3)^2 & \text{for } 2 < x \leq 3 \end{cases}$$

Evaluate the integral $\int_0^3 f(x)\, dx$.

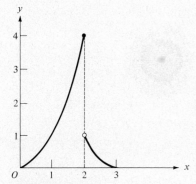

Figure 6.34
The graph of $y = f(x)$.

Solution
Since the integrand $f(x)$ is discontinuous at $x = 2$, it seems appropriate to write

$$\int_0^3 f(x)\,dx = \int_0^2 f(x)\,dx + \int_2^3 f(x)\,dx$$

Since $f(x) = x^2$ for $0 \le x \le 2$, we have

$$\int_0^2 f(x)\,dx = \int_0^2 x^2\,dx = \tfrac{1}{3}x^3\Big|_0^2 = \tfrac{8}{3} - \tfrac{0}{3} = \tfrac{8}{3}$$

Since $f(x) = (x-3)^2$ for $2 < x \le 3$, we have

$$\int_2^3 f(x)\,dx = \int_2^3 (x-3)^2\,dx = \tfrac{1}{3}(x-3)^3\Big|_2^3$$

$$= \tfrac{1}{3}(3-3)^3 - \tfrac{1}{3}(2-3)^3 = \tfrac{1}{3}$$

Hence,

$$\int_0^3 f(x)\,dx = \tfrac{8}{3} + \tfrac{1}{3} = \tfrac{9}{3} = 3$$

In evaluating

$$\int_2^3 f(x)\,dx = \int_2^3 (x-3)^2\,dx$$

we ignored the fact that $f(x) \ne (x-3)^2$ at $x = 2$, since $f(2) = 4$ whereas $(4-3)^2 = 1$. We justify this in the following theorem.

THEOREM 1
If $g(x)$ is continuous on $[a, b]$ and $f(x) = g(x)$ on (a, b), then

$$\int_a^b f(x)\,dx = \int_a^b g(x)\,dx$$

Thus, the value of a definite integral does not depend on the values of the integrand at the end points.

Proof
If we put

$$h(x) = f(x) - g(x)$$

then $h(x) = 0$ for all points x in (a, b). If we divide $[a, b]$ into n intervals of length

$$\Delta x = \frac{b-a}{n}$$

by means of the points $x_k = x_0 + k\,\Delta x$, where $x_0 = a$ and $x_n = b$, then

$$h(x_k) = 0 \quad \text{for} \quad k = 1, 2, \ldots, n-1,$$

and hence
$$\sum_{k=1}^{n} h(x_k)\, \Delta x = h(x_n)\, \Delta x = h(b)\, \Delta x$$

But
$$\int_a^b h(x)\, dx = \lim_{n\to\infty} \sum_{k=1}^{n} h(x_k)\, \Delta x = \lim_{n\to\infty} h(x_n)\, \Delta x$$
$$= \lim_{n\to\infty} h(b)\, \Delta x = 0$$

and hence
$$\int_a^b f(x)\, dx - \int_a^b g(x)\, dx = \int_a^b [f(x) - g(x)]\, dx$$
$$= \int_a^b h(x)\, dx = 0$$

and the theorem follows.

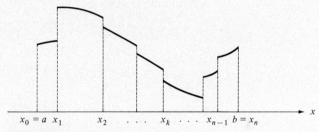

Figure 6.35
A piecewise continuous function.

Suppose that $f(x)$ is bounded and piecewise continuous on the closed interval $[a, b]$ (see Figure 6.35); then the definite integral of $f(x)$ from a to b can be written as

$$\int_a^b f(x)\, dx = \int_{x_0}^{x_1} f(x)\, dx + \int_{x_1}^{x_2} f(x)\, dx + \cdots$$
$$+ \int_{x_{k-1}}^{x_k} f(x)\, dx + \cdots + \int_{x_{n-1}}^{x_n} f(x)\, dx$$

where
$$x_0 = a < x_1 < x_2 < \cdots < x_{k-1} < x_k < \cdots < x_{n-1} < x_n = b$$

and $f(x)$ is continuous on each subinterval (x_{k-1}, x_k). Theorem 1 above lets us apply Theorem 1 of Section 6.5 to each integral on the right side. Thus, if for each value of k we let $F_k(x)$ have the properties

$F_k(x)$ is continuous on $[x_{k-1}, x_k]$, and
$F_k'(x) = f(x)$ on (x_{k-1}, x_k),

then
$$\int_{x_{k-1}}^{x_k} f(x)\,dx = F_k(x)\Big|_{x_{k-1}}^{x_k} = F_k(x_k) - F_k(x_{k-1})$$

It thus follows that
$$\int_a^b f(x)\,dx = F_1(x)\Big|_{x_0}^{x_1} + F_2(x)\Big|_{x_1}^{x_2} + \cdots + F_k(x)\Big|_{x_{k-1}}^{x_k} + \cdots + F_n(x)\Big|_{x_{n-1}}^{x_n}$$

Example 4

Let $f(x)$ be defined on $[0, 4]$ as follows (see Figure 6.36):

$f(x) = x^{1/k}$ for $k - 1 \le x < k$ and $k = 1, 2, 3, 4$;
$f(4) = 0$.

Evaluate $\int_0^4 f(x)\,dx$.

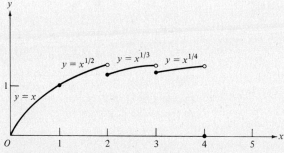

Figure 6.36
The graph of $y = f(x)$.

Solution
Breaking up the integral at the points 1, 2, and 3, we have

$$\int_0^4 f(x)\,dx = \int_0^1 x\,dx + \int_1^2 x^{1/2}\,dx + \int_2^3 x^{1/3}\,dx + \int_3^4 x^{1/4}\,dx$$

$$= \frac{1}{2}x^2\Big|_0^1 + \frac{2}{3}x^{3/2}\Big|_1^2 + \frac{3}{4}x^{4/3}\Big|_2^3 + \frac{4}{5}x^{5/4}\Big|_3^4$$

$$= \frac{1}{2}(1-0) + \frac{2}{3}(2^{3/2} - 1) + \frac{3}{4}(3^{4/3} - 2^{4/3}) + \frac{4}{5}(4^{5/4} - 3^{5/4})$$

$$= \frac{1}{2} + \frac{4 \cdot 2^{1/2}}{3} - \frac{2}{3} + \frac{9 \cdot 3^{1/3}}{4} - \frac{3 \cdot 2^{1/3}}{2} + \frac{16 \cdot 4^{1/4}}{5} - \frac{12 \cdot 3^{1/4}}{5}$$

$$= -\frac{1}{6} + \frac{4 \cdot 2^{1/2}}{3} - \frac{3 \cdot 2^{1/3}}{2} + \frac{9 \cdot 3^{1/3}}{4} - \frac{12 \cdot 3^{1/4}}{5} + \frac{16 \cdot 4^{1/4}}{5}$$

Note that in this example

$$F_k(x) = \frac{k}{k+1}x^{(k+1)/k} \quad \text{for} \quad k = 1, 2, 3, 4$$

Example 5

Evaluate $\int_0^2 [(|x-1|-x)^2 + 3]\,dx$.

Solution
Since

$$|x-1| = \begin{cases} 1-x & \text{for } x \leq 1 \\ x-1 & \text{for } x \geq 1 \end{cases}$$

we have

$$|x-1| - x = \begin{cases} 1-2x & \text{for } x \leq 1 \\ -1 & \text{for } x \geq 1 \end{cases}$$

and accordingly,

$$(|x-1|-x)^2 + 3 = \begin{cases} (1-2x)^2 + 3 & \text{for } x \leq 1 \\ 4 & \text{for } x \geq 1 \end{cases}$$

Hence,

$$\int_0^2 [(|x-1|-x)^2+3]\,dx = \int_0^1 [(1-2x)^2+3]\,dx + \int_1^2 4\,dx$$

$$= \int_0^1 (1-2x)^2\,dx + \int_0^1 3\,dx + \int_1^2 4\,dx = \tfrac{43}{6}$$

EXERCISES

In Exercises 1–11, plot $f(x)$ and evaluate $\int_a^b f(x)\,dx$.

1. $f(x) = \begin{cases} \dfrac{x}{|x|} & x \neq 0 \\ 0 & x = 0 \end{cases} \quad [a, b] = [-1, 1]$

2. $f(x) = \begin{cases} x & 0 \leq x \leq 1 \\ 1 & 1 \leq x \leq 2 \end{cases} \quad [a, b] = [0, 2]$

3. $f(x) = \begin{cases} 1 - x^2 & 0 \leq x \leq 1 \\ 1 - (x-2)^2 & 1 \leq x \leq 3 \\ 1 - (x-4)^2 & 3 \leq x \leq 4 \end{cases} \quad [a, b] = [0, 4]$

4. $f(x) = \begin{cases} x + \sqrt{x+1} & 0 \leq x \leq 1 \\ 0 & -1 \leq x \leq 0 \end{cases} \quad [a, b] = [-1, 1]$

5. $f(x) = \begin{cases} \dfrac{x - |x|}{x} & x \neq 0 \\ 1 & x = 0 \end{cases}$ $[a, b] = [-1, 1]$

6. $f(x) = \begin{cases} \dfrac{x^2 - 1}{x - 1} & x \neq 1 \\ 0 & x = 1 \end{cases}$ $[a, b] = [0, 2]$

7. $f(x) = |x^2 - 1|$ $[a, b] = [-2, 2]$

8. $f(x) = x|x|$ $[a, b] = [-1, 1]$

9. $f(x) = x|x|$ $[a, b] = [-2, 2]$

10. $f(x) = \begin{cases} \dfrac{x}{|x|} & x \neq 0 \\ 1 & x = 0 \end{cases}$ $[a, b] = [-1, 1]$

11. $f(x) = \begin{cases} x & \text{for } x \geq 0 \\ 0 & \text{for } x \leq 0 \end{cases}$ $[a, b] = [-1, 1]$

APPENDIX

1. Proof that $1 + 2 + \cdots + n = \tfrac{1}{2}n(n + 1)$.

   Let $G_n = \tfrac{1}{2}n(n + 1)$. Then

   $G_{n-1} = \tfrac{1}{2}(n - 1)n$

   and we have

   $G_n - G_{n-1} = n$

   Thus

   $1 + 2 + 3 + \cdots + n = (G_1 - G_0) + (G_2 - G_1) + \cdots + (G_n - G_{n-1}) = G_n - G_0$

   Thus, since $G_0 = 0$, we have

   $1 + 2 + \cdots + n = G_n = \tfrac{1}{2}n(n + 1)$

2. Proof that $1^2 + 2^2 + \cdots + n^2 = \tfrac{1}{6}n(n + 1)(2n + 1)$.

   Let $F_n = \tfrac{1}{6}n(n + 1)(2n + 1)$. Then

   $F_{n-1} = \tfrac{1}{6}(n - 1)n(2n - 1)$

   and we can verify that

   $F_n - F_{n-1} = n^2$

Thus

$$1^2 + 2^2 + \cdots + n^2 = (F_1 - F_0) + (F_2 - F_1) + \cdots$$
$$+ (F_n - F_{n-1}) = F_n - F_0$$

Thus, since $F_0 = 0$, we have

$$1^2 + 2^2 + \cdots + n^2 = F_n = \tfrac{1}{6}n(n+1)(2n+1)$$

QUIZ 1

1. Find the following indefinite integrals.

   (a) $\displaystyle\int \frac{1}{\sqrt{2x+5}}\, dx$ (b) $\displaystyle\int x^2\sqrt{3x^3+1}\, dx$

   (c) $\displaystyle\int (x-3)\,\sqrt[3]{(x-3)^2 - 3}\, dx$ (d) $\displaystyle\int (4t^3 - 1)(t^4 - t)^{-4}\, dt$

2. Evaluate the following definite integrals.

   (a) $\displaystyle\int_{-1}^{1} \frac{x}{(x^2+1)^2}\, dx$ (b) $\displaystyle\int_{0}^{1} (3x+4)^{-2}\, dx$

   (c) $\displaystyle\int_{0}^{8} |x-4|\, dx$ (d) $\displaystyle\int_{-1}^{1} x|x|\, dx$

3. Find the differential du for the following functions.

   (a) $u = \dfrac{x^2 - 1}{x^2 + 1}$

   (b) $u = \sqrt{x + \sqrt{x}}$

4. Find the area bounded by the curve $y^2 = x$ and the line $x = 1$.

5. Insert one of the symbols $=$ or $\neq$ into each box to produce a true statement.

   (a) $\displaystyle\int xf(x)\, dx \;\square\; \tfrac{1}{2}x^2 f(x) - \tfrac{1}{2}\int x^2 f'(x)\, dx$

   (b) $\displaystyle\int f(x)g(x)\, dx \;\square\; \left(\int f(x)\, dx\right)\left(\int g(x)\, dx\right)$

   (c) $\displaystyle\frac{d}{dx}\int_{a}^{x} f(t)\, dt \;\square\; \int_{a}^{x} f'(t)\, dt$

   (d) $\displaystyle\int f'(Ax + B)\, dx \;\square\; \frac{1}{A} f(Ax + B) + C$

   (e) $\displaystyle\int \sqrt{u(x)}\, u'(x)\, dx \;\square\; \tfrac{2}{3}[u(x)]^{3/2} + C$

QUIZ 2

1. Find the following indefinite integrals.

   (a) $\displaystyle\int \frac{x-2}{\sqrt{x^2-4x}}\, dx$

   (b) $\displaystyle\int x\sqrt{1+4x}\, dx$

   (c) $\displaystyle\int (x-1)\sqrt[4]{(x-1)^2-1}\, dx$

2. Evaluate the following definite integrals.

   (a) $\displaystyle\int_0^1 \frac{x}{(x^2+1)^{1/3}}\, dx$

   (b) $\displaystyle\int_{-2}^2 |x^2-1|\, dx$

3. If $F'(x)=f(x)$, find formulas for the following:

   (a) $\displaystyle\int_a^b f[F(x)]f(x)\, dx$

   (b) $\displaystyle\int_a^b f(x)[F(x)+1]^{10}\, dx$

   (c) $\displaystyle\int_a^b xf(x^2)\, dx$

4. Find the area bounded by the curve $y=\sqrt{x}$ and the line $y=x$.

5. Find the differential du for the following functions.

   (a) $u = [f(x)+1]^n$

   (b) $u = f\!\left(\dfrac{x^2-1}{x^2+1}\right)$

6. Insert one of the symbols $=$ or $\neq$ into each box to produce a true statement.

   (a) $\displaystyle\int_a^x f(t)\, dt\ \square\ \int_a^x f(u)\, du$

   (b) $\dfrac{d}{dx}\displaystyle\int f'(x)\, dx\ \square\ f''(x)$

   (c) $\displaystyle\int f'[f(x)]f'(x)\, dx\ \square\ f[f(x)]+C$

   (d) $\displaystyle\int \frac{u'(x)}{\sqrt{u(x)}}\, dx\ \square\ 2\sqrt{u(x)}+C$

CHAPTER 7

NATURAL LOGARITHMS AND EXPONENTIALS

Logarithms preceded the invention of calculus by several decades. They were invented by John Napier, Baron of Merchiston (1550–1617) to expedite computations. At the time of his death neither Newton (1643–1727) nor Leibnitz (1646–1716), the inventors of calculus, had yet been born. The first table of logarithms, to which Napier devoted 25 years of his life, was published in 1614. As an indication of the importance attached to logarithms at that time, we mention that Johannes Kepler (1571–1630), renowned for his laws of planetary motion, inscribed his third law to Napier.

Today, logarithms are still essential for the understanding of all areas of applied mathematics, but their use in computations has declined with the development of computers. Their use in many applications, however, is still widespread. Graph paper with logarithmic scales, for example, is commonly used in economics and in all applications where percentage changes are to be stressed; logarithmic scales are also used for measuring earthquakes, noise, and amount of information.

7.1 NATURAL LOGARITHMS

Logarithms are readily defined in terms of areas. Namely, we consider the problem of finding the area bounded by the x axis, the curve $y = 1/x$, and the vertical lines $x = 1$ and $x = a$. The numerical

NATURAL LOGARITHMS AND EXPONENTIALS

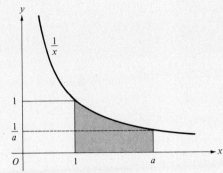

Figure 7.1
Natural logarithm of a.

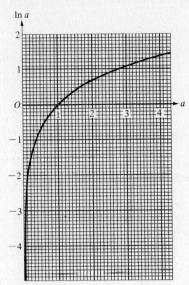

Figure 7.2
The graph of $y = \ln a$.

a	$\ln a$
0.01	−4.6052
0.05	−2.9958
0.10	−2.3026
0.20	−1.6094
0.30	−1.2040
0.40	−0.9163
0.50	−0.6931
0.60	−0.5108
0.70	−0.3567
0.80	−0.2231
0.90	−0.1054
1.00	0.0000
1.10	0.0953
1.20	0.1823
1.30	0.2624
1.40	0.3365
1.50	0.4055
1.60	0.4700
1.70	0.5306
1.80	0.5878
1.90	0.6419
2.00	0.6931
2.50	0.9163
3.00	1.0986
3.50	1.2528
4.00	1.3863

value of this area is given the special name *natural logarithm of a*. (see Figure 7.1 above).

DEFINITION 1

The *natural logarithm* of a, written $\ln a$, is defined for all numbers $a > 0$ by the integral

$$\ln a = \int_1^a \frac{1}{x}\, dx$$

It should be noticed that $\ln a$ is not defined for numbers $a \leq 0$. Several values of $\ln a$ have been tabulated in Figure 7.2, and these were used to plot the graph of $y = \ln a$.

The basic properties of natural logarithms are summarized in Theorem 1 below. Many of these properties will look familiar to those students who studied common logarithms (logarithms to the base 10, often designated $\log_{10} a$ or simply $\log a$).

THEOREM 1. PROPERTIES OF NATURAL LOGARITHMS

1. $\dfrac{d}{dx}(\ln x) = \dfrac{1}{x}$

2. $\ln 1 = 0$

3. $\ln(ab) = \ln a + \ln b$

4. $\ln \dfrac{1}{a} = -\ln a$

5. $\ln \dfrac{a}{b} = \ln a - \ln b$

6. $\ln a^r = r \ln a$

7. $\ln a < \ln b$ when $a < b$

Theorem 1 is proved later in this section. First we shall study some examples, which will provide some insight into the behavior of logarithms.

Example 1

Find a number N such that $\ln N > 10^6$.

Solution

Step 1
By Property 6,

$$\ln 10^r = r \ln 10$$

Step 2
From the table of natural logarithms at the end of the book, we have

$$\ln 10 > 2.3$$

and hence

$$r \ln 10 > 2.3r$$

By step 1 we thus have

$$\ln 10^r > 2.3r$$

Step 3
By step 2

$$\ln 10^r > 10^6 \quad \text{if} \quad 2.3r > 10^6$$

We can therefore take r to be any number greater than $10^6/2.3$; a convenient choice is

$$r = 5 \cdot 10^5 = 500{,}000$$

and

$$N = 10^r = 10^{500{,}000}$$

Step 4
With this choice

$$\ln N = \ln 10^{500{,}000} = 5 \cdot 10^5 \ln 10 = (5 \ln 10) \cdot 10^5 > 10^6$$

Notice that the number N is very large.

REMARK
By Property 7, $\ln a > \ln N$ when $a > N$. Hence, $\ln a > 10^6$ for *all* numbers $a > N$.

Example 2

Find a number M such that

$$\ln M < -10^6$$

Solution
By Property 4,

$$\ln \frac{1}{M} = -\ln M$$

and multiplying both sides by -1 gives

$$\ln M = -\ln \frac{1}{M}$$

Hence,

$$\ln M < -10^6 \quad \text{is the same as} \quad -\ln \frac{1}{M} < -10^6$$

and multiplying both sides of the last inequality by -1 gives

$$\ln \frac{1}{M} > 10^6$$

Putting $1/M = N$, we can use Example 1 and let

$$N = \frac{1}{M} = 10^r \quad \text{where} \quad r = 5 \cdot 10^5$$

Hence

$$M = 10^{-r} = 10^{-500,000}$$

and this solves the problem. Notice how small M is.

Example 3

Look at the table of natural logarithms at the end of the book. In it you will find values of $\ln x$ only for values of x lying between 1 and 10. How can you find $\ln x$ when $0 < x < 1$ or when $x > 10$?

Solution

Step 1
Write x in scientific notation (see Section 1.1):

$$x = p \cdot 10^n \quad \text{where} \quad 1 \leq p < 10$$

Step 2
By Properties 3 and 6,

$$\ln x = \ln(p \cdot 10^n) = \ln p + \ln 10^n = \ln p + n \ln 10$$

Thus, ln x can be calculated with the given table. Several values of $n \ln 10$ are given with the table for your convenience in doing calculations.

Example 4

A graphical illustration of Property 6 is very revealing. In Figure 7.3 we plot the values $\ln 2^n = n \ln 2$ for $n = -3, -2, -1, 1, 2$, and 3, using the value $\ln 2 = 0.6931$. Notice that the numbers 2^n form a geometric progression, whereas the numbers $n \ln 2$ form an arithmetic progression.

n	$n \ln 2$
-3	-2.0793
-2	-1.3862
-1	-0.6931
1	0.6931
2	1.3862
3	2.0793

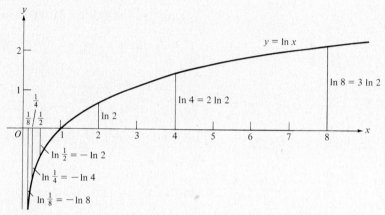

Figure 7.3
The graph of $y = \ln x$.

Proof of Theorem 1

1. $\dfrac{d}{dx}(\ln x) = \dfrac{1}{x}$

By definition,

$$\ln x = \int_1^x \frac{1}{t}\, dt$$

and the assertion thus follows from the Fundamental Theorem of Calculus.

2. $\ln 1 = 0$

$$\ln 1 = \int_1^1 \frac{1}{t}\, dt = 0$$

3. $\ln(ab) = \ln a + \ln b$

By the chain rule,

$$\frac{d}{dx}[\ln(ax)] = \frac{1}{ax} \cdot a = \frac{1}{x}$$

and hence

$$\frac{d}{dx}[\ln(ax)] = \frac{d}{dx}[\ln x]$$

for all $x > 0$. Hence, $\ln(ax)$ and $\ln x$ differ only by a constant:

$$\ln(ax) = \ln x + c$$

Taking $x = 1$ gives, by Property 2,

$$\ln a = \ln 1 + c = c$$

and so

$$\ln(ax) = \ln x + \ln a$$

Taking now $x = b$ gives the desired result.

4. $\ln \frac{1}{a} = -\ln a$

Writing $1 = a \cdot 1/a$, we have

$$0 = \ln 1 = \ln\left(a \cdot \frac{1}{a}\right) = \ln a + \ln \frac{1}{a}$$

and the result follows.

5. $\ln \frac{a}{b} = \ln a - \ln b$

$$\ln \frac{a}{b} = \ln\left(a \cdot \frac{1}{b}\right) = \ln a + \ln \frac{1}{b} = \ln a - \ln b$$

6. $\ln a^r = r \ln a$

This property is proved only for rational values of r; general exponents will be defined later, in Section 7.4. Using the chain rule, we note that

$$\frac{d}{dx}(\ln x^r) = \frac{1}{x^r} \cdot rx^{r-1} = r\frac{1}{x}$$

Also,

$$\frac{d}{dx}(r \ln x) = r \frac{d}{dx}(\ln x) = r \frac{1}{x}$$

Hence, $\ln x^r$ and $r \ln x$ differ only by a constant:

$$\ln x^r = r \ln x + c$$

Taking $x = 1$ shows that $c = 0$, and hence

$$\ln x^r = r \ln x$$

7. $\ln a < \ln b$ when $a < b$

From the Fundamental Theorem of Calculus, we know that

$$\int_a^b \frac{1}{x} \, dx = \ln b - \ln a$$

When $a < b$, this integral is positive and the result follows.

EXERCISES

Find the logarithms in Exercises 1–9. (Use the table at the end of the book.)

1. $\ln 0.0005$
2. $\ln 0.005$
3. $\ln 0.05$
4. $\ln 23$
5. $\ln 230$
6. $\ln 2300$
7. $\ln \dfrac{1}{10^7}$
8. $\ln \dfrac{5}{10^7}$
9. $\ln (5 \cdot 10^7)$

Here are some computational exercises involving logarithms. If $a = \ln 2$ and $b = \ln 5$, then $\ln 0.4 = \ln \frac{2}{5} = \ln 2 - \ln 5 = a - b$. We have thus expressed the logarithm of 0.4 in terms of a and b. Likewise, express each of the following logarithms in terms of a and b only.

10. $\ln 10$
11. $\ln 25$
12. $\ln \sqrt[2]{5}$
13. $\ln \sqrt[3]{10}$
14. $\ln 400$
15. $\ln 625$
16. $\ln 0.05$
17. $\ln (0.05)^3$
18. $\ln \dfrac{5^a}{2}$
19. $\ln \dfrac{1}{25}$

7.2 THE DERIVATIVE OF ln x

In Theorem 1 of the preceding section, we have shown that

$$\frac{d}{dx}(\ln x) = \frac{1}{x} \tag{1}$$

When we have a function $v = f(x)$, such that $v > 0$, then by the chain rule we have

$$\frac{d}{dx}(\ln v) = \frac{1}{v}\frac{dv}{dx} \tag{2}$$

We assume, of course, that $f(x)$ is differentiable.

Example 1

Find

$$\frac{d}{dx}[\ln(3x^2 + 1)]$$

Solution
Put $v = 3x^2 + 1$. Then $dv/dx = 6x$ and formula (2) gives

$$\frac{d}{dx}[\ln(3x^2 + 1)] = \frac{6x}{3x^2 + 1}$$

Example 2

Find

$$\frac{d}{dx}\left(\ln \frac{x-1}{x+1}\right)$$

Solution
Using the fact that

$$\ln \frac{x-1}{x+1} = \ln(x-1) - \ln(x+1)$$

we have [again using formula (2)]

$$\frac{d}{dx}[\ln(x-1)] = \frac{1}{x-1}$$

and

$$\frac{d}{dx}[\ln(x+1)] = \frac{1}{x+1}$$

and hence

$$\frac{d}{dx}\left(\ln\frac{x-1}{x+1}\right) = \frac{1}{x-1} - \frac{1}{x+1} = \frac{2}{x^2-1}$$

Example 3

Graph the function

$$f(x) = \ln(1+x^2)$$

Solution
Follow the procedure outlined in Sections 4.1–4.2. First, observe that this function has no asymptotes. Next, compute $f'(x)$ and $f''(x)$:

$$f'(x) = \frac{d}{dx}[\ln(1+x^2)] = \frac{1}{1+x^2}\frac{d}{dx}(1+x^2) = \frac{2x}{1+x^2}$$

$$f''(x) = \frac{d}{dx}\left(\frac{2x}{1+x^2}\right) = \frac{2(1-x^2)}{(1+x^2)^2}$$

Now tabulate the necessary information.

	$f'(x)$	$f''(x)$	$f(x)$
$x=-1, 1$		$=0$	inflection points
$-1 < x < 1$		>0	concave up
$x<-1, x>1$		<0	concave down
$x<0$	<0		decreasing
$x>0$	>0		increasing
$x=0$	$=0$		local minimum

The graph can now be obtained with a suitable table of values (Figure 7.4).

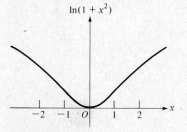

x	$\ln(1+x^2)$
0	0
$\pm\frac{1}{2}$	0.22
± 1	0.69
± 2	1.61
± 3	2.30

Figure 7.4
The graph of $y = \ln(1+x^2)$.

From formula (1) we have

$$\int \frac{1}{x}\,dx = \ln x + C \quad \text{for} \quad x > 0$$

When $x < 0$, then $-x > 0$, so that $\ln(-x)$ is defined, and

$$\frac{d}{dx}[\ln(-x)] = \frac{1}{-x}(-1) = \frac{1}{x}$$

This shows that

$$\frac{d}{dx}(\ln|x|) = \frac{1}{x} \quad \text{for} \quad x \neq 0$$

and we arrive at the formula

$$\int \frac{1}{x}\,dx = \ln|x| + C \quad \text{for} \quad x \neq 0 \tag{3}$$

Example 4

Evaluate

$$\int \frac{x^2}{x^3 + 2}\,dx$$

Solution
Letting $u = x^3 + 2$, we have $\frac{1}{3}\,du = x^2\,dx$, and thus

$$\int \frac{x^2}{x^3 + 2}\,dx = \frac{1}{3}\int \frac{1}{u}\,du = \frac{1}{3}\ln|u| + C = \frac{1}{3}\ln|x^3 + 2| + C$$

EXERCISES

Find the derivative in Exercises 1–14.

1. $y = x + \ln x$
2. $y = x \ln x$
3. $y = \dfrac{\ln x}{x}$
4. $y = \ln x^2$
5. $y = (\ln x)^2$
6. $y = \ln 2x$
7. $y = \sqrt{1 + \ln x}$
8. $s = t + \ln t$
9. $s = \ln \dfrac{t-1}{t+1}$
10. $u = \dfrac{1}{\ln t}$
11. $w = \ln(ax + b)$
12. $y = \ln(x + \sqrt{1 + x^2})$
13. $t = \ln \dfrac{\sqrt{x^2 + 1} - x}{\sqrt{x^2 - 1} + x}$
14. $y = \ln\!\left(x^2 + \dfrac{1}{x^2}\right)$

Find a formula for dy/dx in Exercises 15–18. u and v are positive functions of x.

Ans. $\dfrac{dy}{dx} = \dfrac{n}{v}\dfrac{dv}{dx}$

15. $y = \ln v^n$
16. $y = \ln(uv)$

17. $y = u \ln u$

18. $y = \ln \dfrac{u}{v}$

19. Find the value of $\ln x$ when $\ln' x = 2$ ($\ln' x = dy/dx$).

20. Find the point (x_0, y_0) if $y - y_0 = \tfrac{1}{3}(x - x_0)$ is tangent to the curve $y = \ln x$ at (x_0, y_0).

21. Find the point of intersection of the tangent lines to $y = \ln x$ at $(1, 0)$ and at $(5, \ln 5)$.

22. Find the line through the point $(3, \ln 3)$ that is perpendicular to the tangent to $y = \ln x$ at $(3, \ln 3)$ (see Exercise 7, Section 2.3).

23. Does the curve $y = \ln x$ have two tangents that are mutually perpendicular? Prove your answer (see Exercise 7, Section 2.3).

24. Find that tangent to $y = \ln x$ which is parallel to the line $y = x$.

25. Find that tangent to $y = \ln x$ which is perpendicular to the line $y = -x$ (see Exercise 7, Section 2.3).

26. Sketch the following curves.

    (a) $y = -\ln x$
    (b) $y = \ln x - x$
    (c) $y = \dfrac{\ln x}{\log_{10} x}$
    (d) $y = \ln(\ln x)$ $\quad (x > 1)$
    (e) $y = \ln(3x)$
    (f) $y = \ln(x + 2)$ $\quad (x > -2)$
    (g) $y = \ln(x - 2)$ $\quad (x > 2)$
    (h) $y = \ln x + 2$

27. Compare the graphs of $y = \ln x$ and $y = \ln(x - 2)$. Find the tangent to the curve $y = \ln(x - 2)$ at $(3, \ln 1)$ and at $(7, \ln 5)$.

28. Compare the graphs of $y = \ln x$ and $y = \ln x + 2$. Find the tangent to the curve $y = \ln x + 2$ at $(3, \ln 3 + 2)$ and $(1, 2)$.

7.3 THE EXPONENTIAL FUNCTION

Let us consider the problem of solving the equation $\ln b = a$, in which a is a known number.

Example 1

What value of b gives $\ln b = 2$?

Approximate Solution
An approximate numerical solution can be found by consulting a table of natural logarithms (see Table 7.1) or from a reasonably

214 NATURAL LOGARITHMS AND EXPONENTIALS

Table 7.1

x	$\ln x$
7.0	1.9459
7.1	1.9601
7.2	1.9741
7.3	1.9879
7.4	2.0015
7.5	2.0149
7.6	2.0282
7.7	2.0412
7.8	2.0541
7.9	2.0669
8.0	2.0794
8.1	2.0919
8.2	2.1041
8.3	2.1163
8.4	2.1282
8.5	2.1401
8.6	2.1518
8.7	2.1633
8.8	2.1748
8.9	2.1861

Table 7.2

x	$\ln x$
7.380	1.998774
7.381	1.998909
7.382	1.999045
7.383	1.999180
7.384	1.999315
7.385	1.999451
7.386	1.999586
7.387	1.999722
7.388	1.999857
7.389	1.999992
7.390	2.000128

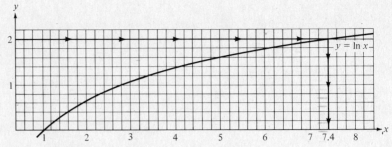

Figure 7.5
Finding the approximate solution of $\ln b = 2$ from the graph of $y = \ln x$.

accurate graph of $\ln x$ (see Figure 7.5). From the graph we could guess that $b \approx 7.4$. Looking at the table, we see that

$\ln 7.3 = 1.9879$
$\ln 7.4 = 2.0015$

From the fact that the function $f(x) = \ln x$ is increasing, we know that

$$\ln 7.3 < \ln b < \ln 7.4 \quad \text{when} \quad 7.3 < b < 7.4 \qquad (1)$$

Since 2.0015 is closer to 2 than 1.9879, we take as our approximate solution

$b = 7.4$

We could do better with a more accurate table. From the excerpt of the six-place table given in Table 7.2 we find that

$\ln 7.389 = 1.999992$
$\ln 7.390 = 2.000128$

Since 1.999992 is closer to 2 than 2.000128, we take

$b = 7.389$

The exact solution of the equation $\ln b = 2$ is called *exponential* 2 and designated exp 2. Thus, exp 2 is that unique number whose natural logarithm is 2; according to the above,

$\exp 2 \approx 7.389$

This leads us to the general definition of exponentials.

> **DEFINITION 1**
> *exp a*, read "exponential *a*," is the unique number whose natural logarithm is a.

Alternatively, this definition can be stated as follows:

$\exp a = b \quad \text{is the same as} \quad \ln b = a$

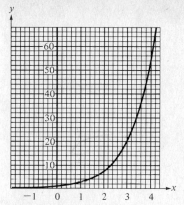

Figure 7.6
The graph of $y = \exp x$.

x	$\exp x$
−2.0	0.1353
−1.0	0.3679
0.0	1.0000
0.1	1.1052
0.2	1.2214
0.3	1.3499
0.4	1.4918
0.5	1.6487
0.6	1.8221
0.7	2.0138
0.8	2.2255
0.9	2.4596
1.0	2.7183
1.1	3.0042
1.2	3.3201
1.3	3.6693
1.4	4.0552
1.5	4.4817
1.6	4.9530
1.7	5.4739
1.8	6.0496
1.9	6.6859
2.0	7.3891
2.1	8.1662
2.2	9.0250
2.3	9.9742
2.4	11.0231
2.5	12.1824
2.6	13.4637
2.7	14.8797
2.8	16.4446
2.9	18.1741
3.0	20.0855
4.0	54.5981

In Figure 7.6 we present the graph of the function $y = \exp x$ together with a table of values.

Example 2

Use the table of exponentials at the end of the book to find $\exp 2$.

Solution

$$\exp 2 = 7.3891$$

This is correct to five significant figures.

The number $\exp 1$ is of special significance; it appears in all branches of pure and applied mathematics.

DEFINITION 2
The number $\exp 1$ is designated with the letter e:

$$e = \exp 1$$

The value of e to five significant figures is

$$e \approx 2.7183$$

It follows from Definition 2 that

$$\ln e = 1$$

REMARK

The number e is *transcendental*. This means that it is an irrational number that is not the root of any polynomial equation with rational coefficients. We recall that an *irrational* number is any number not expressible as a fraction m/n, where m and n are integers, and $n \neq 0$. Another familiar transcendental number is π.

We now present a list of the basic properties of the exponential function. It is followed by proofs, discussion, and examples.

THEOREM 1. PROPERTIES OF EXPONENTIALS
1. $\ln(\exp a) = a$
   $\exp(\ln b) = b$
2. $\exp 0 = 1$
3. $\exp a = e^a$
4. $\exp(a + b) = (\exp a)(\exp b)$
5. $\exp(-a) = \dfrac{1}{\exp a}$
6. $\exp a > 0$
7. $\exp a < \exp b$ whenever $a < b$

Proof of Property 1
Since $\exp a = b$ and $a = \ln b$ are equivalent, put $a = \ln b$ in the first relation, and $b = \exp a$ in the second relation.

Proof of Property 2
Since $\ln 1 = 0$, it follows from Property 1 that

$$\exp(\ln 1) = \exp 0 = 1$$

Proof of Property 3
We have proved in Section 7.1 that

$$\ln e^r = r \ln e = r$$

when r is rational. By the definition, therefore,

$$\exp r = e^r$$

Looking back at Definition 2 in Section 1.1, we notice that powers e^r were never defined for irrational values of r. In light of the above, it is now reasonable to lay down the following definition.

DEFINITION 3

$e^a = \exp a$ for irrational numbers a.

From the above we have $e^x = \exp x$ for all numbers x. The notations, e^x and $\exp x$, will be used interchangeably.

Proof of Property 4

Step 1
Put

$$u = \exp(a + b) \qquad v = \exp a \qquad w = \exp b$$

Step 2
By definition

$$\ln u = a + b \qquad \ln v = a \qquad \ln w = b$$

Step 3
By Section 7.1,

$$\ln v + \ln w = \ln(v \cdot w)$$

hence

$$a + b = \ln(v \cdot w)$$

Step 4
Steps 2 and 3 give

$$\ln u = \ln(vw)$$

This implies that

$$u = vw$$

Step 5
By step 1,

$$\exp(a + b) = (\exp a)(\exp b)$$

Proof of Property 5

Step 1
By Property 4,

$$\exp[a + (-a)] = (\exp a)[\exp(-a)]$$

Step 2
By Property 2, $\exp 0 = 1$; since $a + (-a) = 0$, this gives

$$(\exp a)[\exp(-a)] = 1$$

Step 3
Dividing both sides in step 2 by $\exp a$ yields the desired formula.

Proof of Property 6
By definition, $\exp a = b$ is the same as $a = \ln b$. Since $\ln b$ is defined only for $b > 0$, it follows that $b = \exp a > 0$. This is also clear from the fact that the graph of $y = \exp x$ is the reflection (mirror image) about the line $y = x$ of the graph of $y = \ln x$ (see Figure 7.7).

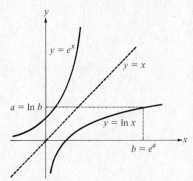

Figure 7.7
The graphs of $y = e^x$ and $y = \ln x$.

Proof of Property 7
The method of proof used here is called *proof by contradiction*. This method consists of assuming the contrary of what is to be proved, and deriving from this assumption a false conclusion.

Step 1
Consider two numbers a and b such that $a < b$. If $\exp a \not< \exp b$ (exponent a is *not* less than exponent b) then $\exp a \geq \exp b$.

Step 2
The property $\ln u \geq \ln v$ whenever $u \geq v$ gives

$$\ln(\exp a) \geq \ln(\exp b)$$

Step 3
Property 1 shows that

$$a \geq b$$

Step 4
Since the conclusion $a \geq b$ contradicts the assumption $a < b$, it follows that the assumption $\exp a \not< \exp b$ was false. Hence, $\exp a < \exp b$ when $a < b$.

Example 3

Simplify the expression

$$\exp(\ln 10 - \ln e^2)$$

as far as possible.

Solution
Since $\ln 10 - \ln e^2 = \ln(10/e^2)$, we have

$$\exp(\ln 10 - \ln e^2) = \exp\left(\ln \frac{10}{e^2}\right) = \frac{10}{e^2} = 10e^{-2}$$

Example 4

Find x if

$$\ln e^{-\ln x^2} = 5$$

Solution
The relation $\ln e^a = a$ for all a shows that $\ln e^{-\ln x^2} = -\ln x^2 = -2 \ln x$. Hence $-2 \ln x = 5$, or

$$\ln x = -\tfrac{5}{2} = -2.5$$

This is the same as (using an exponential table)

$$x = e^{-2.5} = 0.0821$$

REMARK
Exercises for this section are found with those of Section 7.4.

7.4 GENERAL EXPONENTS AND LOGARITHMS

We have yet to give meaning to expressions of the form c^a when a is an irrational number. We consider only numbers $c > 0$ to avoid the use of complex numbers. For instance, $(-1)^{1/2}$ is an example of a complex number that corresponds to no point on the number line.

Let us first show that

$$c^r = \exp[r \ln c] \quad \text{for rational numbers } r \tag{1}$$

We proved in Section 7.2 that

$$\ln c^r = r \ln c$$

It follows that

$$\exp(\ln c^r) = \exp(r \ln c)$$

But

$$\exp(\ln c^r) = c^r$$

and formula (1) follows.

As with Definition 3 of Section 7.3, we state

DEFINITION 1

$$c^a = \exp(a \ln c) \quad \text{for } c > 0 \text{ and irrational numbers } a.$$

Thus, whether x is rational or irrational,

$$\boxed{c^x = \exp(x \ln c) \quad \text{for all } c > 0} \tag{2}$$

Theorem 1 below extends the properties in Section 1.1 from rational exponents to all exponents.

Example 1

Find a decimal representation of $3^{3.1}$.

Solution

Step 1
According to (2),

$$3^{3.1} = \exp[3.1 \ln 3]$$

Step 2
Consulting a logarithms table,

$$3.1 \ln 3 = 3.1 \cdot 1.0986 = 3.40566 \approx 3.4$$

Step 3
Steps 1 and 2 give (with the use of an exponential table)

$$3^{3.1} = e^{3.4} = 29.964$$

THEOREM 1. PROPERTIES OF EXPONENTS

1. $c^0 = 1$
2. $\ln c^a = a \ln c$
3. $c^a \cdot c^b = c^{a+b}$
4. $\dfrac{c^a}{c^b} = c^{a-b}$
5. $(c^a)^b = c^{a \cdot b}$
6. $c^a \cdot d^a = (c \cdot d)^a$
7. $c^a < d^a$ when $0 < c < d$ and $a > 0$

Proof

The proofs below are based on formula (2) and on Theorem 1, Section 7.3.

1. Put $x = 0$ in formula (2). Then
$$c^0 = \exp(0 \cdot \ln c) = \exp 0 = 1$$

2. Since $c^a = \exp(a \ln c)$, we have
$$\ln c^a = \ln[\exp(a \ln c)] = a \ln c$$

3. Put $x = a + b$ in formula (2). Then
$$c^{a+b} = \exp[(a+b) \ln c] = \exp(a \ln c + b \ln c)$$
$$= [\exp(a \ln c)] \cdot [\exp(b \ln c)]$$

Since $\exp(a \ln c) = c^a$ and $\exp(b \ln c) = c^b$, we have
$$c^{a+b} = c^a \cdot c^b$$

4. $1/c^b = c^{-b}$ because $c^{-b} = \exp(-b \ln c)$ and $c^b = \exp(b \ln c)$. Hence,
$$\frac{c^a}{c^b} = c^a \frac{1}{c^b} = c^a c^{-b} = c^{a+(-b)} = c^{a-b}$$

5. Put $x = ab$ in formula (2). Then
$$c^{a \cdot b} = \exp(ab \ln c) = \exp(b \ln c^a)$$

But $\exp(b \ln c^a) = (c^a)^b$; hence,
$$c^{a \cdot b} = (c^a)^b$$

6. Replace c by cd in formula (2). Then
$$(cd)^a = \exp[a \ln(cd)] = \exp[a(\ln c + \ln d)]$$
$$= \exp(a \ln c) \cdot \exp(a \ln d) = c^a \cdot d^a$$

7. When $0 < c < d$, then $\ln c < \ln d$ and, since $a > 0$, $a \ln c < a \ln d$. By property 7 of Theorem 1 in Section 7.3,

$$\exp(a \ln c) < \exp(a \ln d)$$

and owing to formula (2) this is the same as $c^a < c^b$.

LOGARITHMS TO OTHER BASES

We now define a general logarithm

$$\log_c b$$

which is the logarithm of b to the base c. Here b and c are any positive numbers. When $c = 10$, we have logarithms to the base 10, which are called *common logarithms;* they are used for a wide variety of computations. Logarithms to the base 2 are used in information theory. We shall see below that logarithms to the base e coincide with the natural logarithm ln.

DEFINITION 2

$\log_c b$ is the unique number a that satisfies the equation $b = c^a$. Thus:

$\log_c b = a$ is the same as $b = c^a$.

The close connection between the functions $\log_c x$ and $\ln x$ is given in the following theorem.

THEOREM 2

$$\log_c x = \frac{1}{\ln c} \ln x$$

Proof
Suppose that

$$a = \log_c x$$

Then

$$c^a = x$$

Taking natural logarithms, we have

$$a \ln c = \ln x$$

or

$$a = \frac{1}{\ln c} \ln x$$

Theorem 2 follows.

COROLLARY 1

$$\log_e x = \ln x$$

This follows from Theorem 2, using the fact that $\ln e = 1$.

COROLLARY 2

$$\log_c b = \frac{1}{\log_b c}$$

This follows from Theorem 2, since

$$\log_c b = \frac{\ln b}{\ln c} \quad \text{and} \quad \log_b c = \frac{\ln c}{\ln b}$$

Example 2

Show that

$$\log_x e = \frac{1}{\ln x}$$

This follows from

$$\log_x e = \frac{1}{\log_e x} = \frac{1}{\ln x}$$

Example 3

Show that $\ln a$ can be found from $\log_{10} a$ by the formula

$$\ln a = \frac{\log_{10} a}{\log_{10} e}$$

Solution
Since

$$\log_{10} a = \frac{\ln a}{\ln 10}$$

we have

$$\ln a = \ln 10 \cdot \log_{10} a = \frac{1}{\log_{10} e} \log_{10} a$$

EXERCISES

Simplify each of the expressions in Exercises 1 and 2 as far as possible.

1. (a) $\exp(\ln 5 - \ln 2) =$

    (b) $\exp\{\ln[\exp(\ln 2)]\} =$

    (c) $\ln[\exp(\ln e + \ln e^2)] =$

2. (a) $\ln(e^{a-a^2}) =$

    (b) $\exp\left(-\ln \frac{1}{a}\right) =$

    (c) $e^{\ln 1/x} =$

(d) $\ln\left(\dfrac{1}{e^{x^2}}\right) =$

(e) $\exp[-\ln(x \cdot y^{-2})] =$

(f) $e^{\ln x^2} =$

(g) $\ln[\exp(e^{\ln 2})] =$

(d) $\ln \dfrac{1}{e^{-x}} =$

(e) $\exp(x + \ln x) =$

(f) $\exp(\ln a - 10 \ln b) =$

(g) $e^{-\ln(1/x)} =$

In Exercises 3 and 4, use tables of exponentials and natural logarithms to find the decimal representation of each of the following numbers. Your answers will, of course, be approximate.

3. (a) $2^{\sqrt{2}} =$

(b) $(\sqrt{2})^{\sqrt{2}} =$

(c) $e^e =$

(d) $2^{-\sqrt{2}} =$

(e) $(\sqrt{2})^e =$

4. (a) $3^{\ln 25} =$

(b) $\pi^{10} =$

(c) $5^{1/\sqrt{2}} =$

(d) $7^{3/5} =$

(e) $5^{\pi} =$

Mark which are true and which are false.

5. (a) $(e^e)^e = (\exp e)^e$

(b) $(2^3)^2 = 2^{3+2}$

(c) $(2^2)^2 = 4^2$

(d) $2^{(2^2)} = 4^2$

(e) $e^{(e^e)} = \exp(\exp e)$

6. (a) $\ln(e^e)^e = e^e$

(b) $\ln[\ln(e^e)^e] = 2$

(c) $(e^e)^e = e \cdot \exp e$

(d) $\log_2 2^2 = 2$

(e) $\ln(e^e)^e = e^2$

Evaluate each of the following logarithms.

7. (a) $\log_5 25 =$

(b) $2 \log_5 5 =$

(c) $\log_{25} 5 =$

(d) $\log_e e =$

8. (a) $\log_8 16 =$

(b) $\log_{0.5} 1 =$

(c) $\log_8 1 =$

(d) $\log_2 128 =$

Find x in each of the following equations.

9. (a) $5^{2x} = 10^x$

(b) $2^{x+1} = 3^{x-1}$

(c) $5^{\log_5 2} = 5^{\log_5 x^2}$

(d) $5^{\log_5 2} = 7^{\log_7 x^2}$

10. (a) $2^x = 7$

(b) $10^x = 5$

(c) $(\tfrac{1}{2})^x = 10$

224 NATURAL LOGARITHMS AND EXPONENTIALS

11. Is there a value x such that $e^x = x$? Look at the superimposed graphs of $y = e^x$ and $y = x$.

12. Is there a value x such that $e^x = 1/x$? Look at the superimposed graphs of $y = e^x$ and $y = 1/x$.

13. Is there a value x such that $\ln x = \log_{10} x$?

14. Graph the function $y = x^x$ on Figure 7.8.

7.5 DERIVATIVES AND INTEGRALS OF EXPONENTIALS

The exponential function e^x is characterized by the property that it equals its own derivative:

$$\frac{d}{dx}(e^x) = e^x \tag{1}$$

To verify this fact, we assume for the moment that a derivative exists. Since

$$\ln(e^x) = x$$

we have, by the chain rule,

$$\frac{1}{e^x} \frac{d}{dx}(e^x) = 1$$

and formula (1) follows. The existence of the derivative will follow from Theorem 1 in Section 9.4.

If $v = f(x)$ is a differentiable function, then we have by the chain rule

$$\frac{d}{dx}(e^v) = e^v \frac{dv}{dx} \tag{2}$$

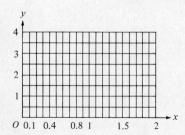
Figure 7.8

x	x^x
0.01	
0.05	
0.10	
0.20	
0.40	
0.60	
0.80	
1.00	
1.50	
2.00	

Example 1

Find dy/dx when $y = e^{\alpha x^2}$.

Solution
Using formula (2) with $v = \alpha x^2$ gives

$$\frac{dy}{dx} = e^{\alpha x^2} \frac{d}{dx}(\alpha x^2) = 2\alpha x e^{\alpha x^2}$$

With the notation $y = \exp(\alpha x^2)$, we have

$$\frac{dy}{dx} = 2\alpha x \exp(\alpha x^2)$$

THE DERIVATIVE OF a^x

The derivatives of functions $y = a^x$ for $a > 0$ can now be found from formula (1) and the chain rule. We have

$$\frac{d}{dx}(a^x) = a^x \ln a \quad \text{when } a > 0 \tag{3}$$

Proof

To prove this formula, we use the definition

$$a^x = \exp(x \ln a)$$

As in Example 1, we have

$$\frac{d}{dx}(a^x) = \frac{d}{dx}[\exp(x \ln a)] = \exp(x \ln a)\frac{d}{dx}(x \ln a)$$

$$= \exp(x \ln a)\ln a = a^x \ln a$$

If we have a differentiable function $v = f(x)$, then the chain rule gives

$$\frac{d}{dx}(a^v) = a^v \ln a \frac{dv}{dx} \quad \text{when } a > 0 \tag{4}$$

Example 2

Find dy/dx when $y = 10^{x^3}$.

Solution

Using formula (4) with $a = 10$ and $v = x^3$ gives

$$\frac{dy}{dx} = 10^{x^3} \ln 10 \frac{d}{dx}(x^3) = 3 \ln 10 \cdot x^2 \cdot 10^{x^3}$$

THE DERIVATIVE OF u^v

The following result is often useful.

THEOREM 1

If both u and v are functions of x, then

$$\frac{d}{dx}(u^v) = vu^{v-1}\frac{du}{dx} + u^v \ln u \frac{dv}{dx} \tag{5}$$

when the derivatives du/dx and dv/dx exist.

Proof

This formula is verified with a technique called *logarithmic differentiation*. The following steps are easy to justify.

Step 1
Put $y = u^v$. Then

$$\ln y = v \ln u$$

Step 2
Differentiate both sides:

$$\frac{d}{dx}(\ln y) = \frac{1}{y}\frac{dy}{dx}$$

$$\frac{d}{dx}(v \ln u) = v \frac{d}{dx}(\ln u) + \ln u \frac{dv}{dx} = \frac{v}{u}\frac{du}{dx} + \ln u \frac{dv}{dx}$$

and hence

$$\frac{1}{y}\frac{dy}{dx} = \frac{v}{u}\frac{du}{dx} + \ln u \frac{dv}{dx}$$

Step 3
Multiply both sides of the last equation by $y = u^v$ and simplify:

$$\frac{dy}{dx} = u^v\left(\frac{v}{u}\frac{du}{dx} + \ln u \frac{dv}{dx}\right)$$

$$= vu^{v-1}\frac{du}{dx} + u^v \ln u \frac{dv}{dx}$$

This is the formula we want.

Example 3

Find dy/dx when $y = x^{\sqrt{x}}$.

Solution
Put $u = x$ and $v = \sqrt{x}$. Then

$$\frac{du}{dx} = 1 \quad \text{and} \quad \frac{dv}{dx} = \frac{1}{2\sqrt{x}}$$

and according to the formula,

$$\frac{dy}{dx} = \sqrt{x} \cdot x^{\sqrt{x}-1} \cdot 1 + x^{\sqrt{x}} \ln x \frac{1}{2\sqrt{x}}$$

Writing

$$\frac{1}{2\sqrt{x}} = \frac{\sqrt{x}}{2\sqrt{x}\sqrt{x}} = \frac{\sqrt{x}}{2}x^{-1}$$

gives

$$\frac{dy}{dx} = \left[\sqrt{x}\,x^{-1} + \ln x \frac{\sqrt{x}}{2} x^{-1}\right] x^{\sqrt{x}} = (1 + \tfrac{1}{2} \ln x) \sqrt{x}\, x^{\sqrt{x}-1}$$

INTEGRALS INVOLVING EXPONENTIALS
From formula (1), we know that

$$\int e^x \, dx = e^x + C \qquad (6)$$

More generally, formula (3) gives

$$\int a^x \, dx = \frac{1}{\ln a} a^x + C \qquad (7)$$

Example 4

Find $\int x e^{3x^2} \, dx$.

Solution
Let $u = 3x^2$ so that $du = 6x\, dx$ and $\tfrac{1}{6} du = x\, dx$. We thus have

$$\int x e^{3x^2} \, dx = \tfrac{1}{6} \int e^u \, du = \tfrac{1}{6} e^u + C = \tfrac{1}{6} e^{3x^2} + C$$

Example 5

Find

$$\int 3x^2 2^{4x^3} \, dx$$

Solution
Let $u = 4x^3$; then $du = 12x^2 dx$ and $\tfrac{1}{4} du = 3x^2 \, dx$. Hence

$$\int 3x^2 2^{4x^3} dx = \frac{1}{4} \int 2^u \, du = \frac{1}{4}\left(\frac{1}{\ln 2}\right) 2^u + C = \left(\frac{1}{4 \ln 2}\right) 2^{4x^3} + C$$

EXERCISES

Find dy/dx in Exercises 1–24.

1. $y = (x^2 + 1) e^x$
2. $y = e^{-x}$
3. $y = e^{-x^2}$
4. $y = (e^{-x})^2$
5. $y = \exp(e^x + 1)$
6. $y = \ln(e^x + 1)$

7. $y = \dfrac{e^x - 1}{e^x + 1}$

8. $y = \dfrac{e^x + e^{-x}}{e^x - e^{-x}}$

9. $y = \ln 10^x$

10. $y = 10^{\ln x}$

11. $y = \sqrt{1 + 10^x}$

12. $y = \exp(10^{\ln x})$

13. $y = \dfrac{\exp(x + \ln x)}{\ln(x \exp x)}$

14. $y = 2^{x^2}$

15. $y = (2^x)^2$

16. $y = x^x$

17. $y = (x + 1)^{(x+1)}$

18. $y = x^{x^2}$

19. $y = (x^x)^2$

20. $y = e^{x^x}$

21. $y = a^{\sqrt{x}}$

22. $y = x^{1/x}$

23. $y = 5^x + x^5$

24. $y = e^{4x} + (4x)^e$

Find the derivative of each of the following functions at the indicated points.

25. $e^{0.5x}$ at $(0, 1)$

26. e^{7x} at $(1, e^7)$

27. 10^x at $(1, 10)$

28. 5^x at $(1, 5)$

29. 21^x at $(-1, \tfrac{1}{21})$

30. 5^{2x} at $(1, 5^2)$

31. 5^{3x} at $(1, 5^3)$

32. 5^{ax} at $(1, 5^a)$

33. Find the equation of the tangent to the curve $y = e^x$ at

   (a) $(-10, e^{-10})$

   (b) $(1, e)$

34. Find that tangent to $y = e^x$ which is parallel to the line $y = x$.

35. Find that tangent to $y = e^{3x}$ which is perpendicular to the line $y = -\tfrac{1}{2}x$.

36. Does the curve $y = e^x$ have two mutually perpendicular tangents?

37. Does the curve $y = e^x$ have a tangent line whose slope is undefined?

38. Find the tangent to $y = 5^x$ at $(0, 1)$.

39. Find the tangent to $y = e^{3x}$ that is parallel to the tangent in Exercise 38.

Graph the functions in Exercises 40–47.

40. $f(x) = \exp(x^2)$ **41.** $g(x) = xe^x$

42. $h(x) = e^x + e^{-x}$ **43.** $k(x) = e^x - e^{-x}$

44. $p(x) = e^{-x^2}$ **45.** $q(x) = e^{1/x}$

46. $r(x) = e^{-1/x}$ **47.** $s(x) = x \ln x$

Find the integrals in Exercises 48–57.

48. $\int e^{x/6} \, dx$ **49.** $\int e^{-x} \, dx$

50. $\int \dfrac{e^x - e^{-x}}{e^x + e^{-x}} \, dx$ **51.** $\int x^{n-1} e^{x^n} \, dx \quad (n \geq 1)$

52. $\int_0^a e^x \, dx$ **53.** $\int_{-a}^{a} (e^x + e^{-x}) \, dx$

54. $\int_0^a 2^{-x} \, dx$ **55.** $\int_0^3 e^{|x-1|-x} \, dx$

56. $\int_{-2}^{2} e^{|x|+x} \, dx$ **57.** $\int_a^b 2x e^{x^2} \, dx$

7.6 INTEGRATION BY PARTS

Our procedure in Section 6.6 for finding indefinite integrals relied on our ability to recognize antiderivatives. The substitution formula in Section 6.7 enabled us to apply this procedure in more complicated situations. Still, there are many integrals beyond our reach, a case in point being

$$\int x e^{ax} \, dx \quad (a \neq 0)$$

From our experience we know of no function $F(x)$ such that $F'(x) = xe^{ax}$, and we see at once that the substitution formula does not work here. This integral, and many others, can be evaluated with a formula derived from the product formula for derivatives. Namely, recalling that

$$\frac{d}{dx}(xe^{ax}) = x \frac{d}{dx}(e^{ax}) + e^{ax} \frac{d}{dx}(x) = axe^{ax} + e^{ax}$$

we write

$$xe^{ax} = \frac{1}{a} \frac{d}{dx}(xe^{ax}) - \frac{1}{a} e^{ax}$$

and find that

$$\int xe^{ax}\,dx = \int \left[\frac{1}{a}\frac{d}{dx}(xe^{ax}) - \frac{1}{a}e^{ax}\right]dx$$

$$= \int \frac{1}{a}\frac{d}{dx}(xe^{ax})\,dx - \int \frac{1}{a}e^{ax}\,dx$$

$$= \frac{1}{a}\int \frac{d}{dx}(xe^{ax})\,dx - \frac{1}{a}\int e^{ax}\,dx$$

$$= \frac{1}{a}xe^{ax} - \frac{1}{a}\int e^{ax}\,dx$$

We thus succeeded in replacing the original integral by an expression involving a simpler integral. We know, in fact, that

$$\int e^{ax}\,dx = \frac{1}{a}e^{ax} + C$$

and hence

$$\int xe^{ax}\,dx = \frac{1}{a}\left(x - \frac{1}{a}\right)e^{ax} + C \qquad a \neq 0 \tag{1}$$

This formula can easily be verified by differentiating the right side. In general, the following formula is valid:

THEOREM 1. INTEGRATION BY PARTS

$$\int g(x)f'(x)\,dx = g(x)f(x) - \int f(x)g'(x)\,dx$$

Proof
By Section 3.3,

$$[g(x)f(x)]' = g(x)f'(x) + f(x)g'(x)$$

This formula can be written as

$$g(x)f'(x) = [g(x)f(x)]' - f(x)g'(x)$$

and hence

$$\int g(x)f'(x)\,dx = \int \{[g(x)f(x)]' - f(x)g'(x)\}\,dx$$

$$= \int [g(x)f(x)]'\,dx - \int f(x)g'(x)\,dx$$

$$= g(x)f(x) - \int f(x)g'(x)\,dx$$

The formula for integration by parts is conveniently expressed in terms of differentials. Putting

$$u = g(x) \quad \text{and} \quad v = f(x)$$

we have

$$du = g'(x)\, dx \quad \text{and} \quad dv = f'(x)\, dx$$

and our formula becomes

$$\int u\, dv = uv - \int v\, du \qquad (2)$$

REMARK
The formula $dv = h(x)dx$ means that $dv/dx = h(x)$ and hence $v = \int h(x)\,dx + C$. When integrating by parts, we usually take $C = 0$.

The examples to be given involve logarithms and exponentials. We shall find other applications of integration by parts that involve trigonometric functions (Chapter 9) and rational functions (Chapter 10).

Example 1
Find $\int x \ln x\, dx$.

Solution
Our choice of u and dv has to be such that $u\, dv = x \ln x\, dx$. For this, we let

$$u = \ln x \quad \text{and} \quad dv = x\, dx$$

Then

$$du = \frac{1}{x}\, dx \quad \text{and} \quad v = \int x\, dx = \tfrac{1}{2} x^2$$

and hence

$$\int x \ln x\, dx = \tfrac{1}{2} x^2 \ln x - \int \tfrac{1}{2} x^2 \frac{1}{x}\, dx$$

$$= \tfrac{1}{2} x^2 \ln x - \tfrac{1}{2} \int x\, dx = \tfrac{1}{2} x^2 \ln x - \tfrac{1}{4} x^2 + C$$

$$= \tfrac{1}{2} x^2 (\ln x - \tfrac{1}{2}) + C$$

REMARK
Should you get stuck at any step in the process, start over with a different choice for u and dv. It goes without saying that dv has to

be chosen in such a way that v can be evaluated. For example, the choice $dv = \ln x \, dx$ in Example 1 would have resulted in the integral $\int \ln x \, dx$. We do not know yet how to find this integral.

Example 2

Find $\int \ln x \, dx$.

Solution
Let $u = \ln x$ and $dv = dx$. Then

$$du = \frac{1}{x} dx \quad \text{and} \quad v = \int dx = x$$

and hence

$$\int \ln x \, dx = (\ln x)x - \int x \frac{1}{x} dx = x \ln x - \int dx$$

$$= x \ln x - x + C = x(\ln x - 1) + C$$

That is,

$$\int \ln x \, dx = x(\ln x - 1) + C \tag{3}$$

Example 3

At times it is necessary to apply the technique of integration by parts several times before the final result is obtained. To illustrate this, consider the integral

$$\int x^n e^{ax} \, dx$$

where n is a positive integer. Putting

$$u = x^n \quad \text{and} \quad dv = e^{ax}$$

we get

$$du = nx^{n-1} \, dx \quad \text{and} \quad v = \frac{1}{a} e^{ax}$$

Formula (2) now becomes

$$\int x^n e^{ax} \, dx = x^n \left(\frac{1}{a}\right) e^{ax} - \int \frac{1}{a} e^{ax} nx^{n-1} \, dx$$

or

$$\int x^n e^{ax}\, dx = \frac{1}{a} x^n e^{ax} - \frac{n}{a} \int x^{n-1} e^{ax}\, dx \qquad (4)$$

This formula is an example of a *reduction formula*. It expresses an integral involving an nth power in terms of a similar integral involving the $(n-1)$st power. Observe that formula (1) is obtained from formula (4) by putting $n = 1$.

Example 4

Find $\int x^2 e^{ax}\, dx$.

Solution
With $n = 2$, formula (4) becomes

$$\int x^2 e^{ax}\, dx = \frac{1}{a} x^2 e^{ax} - \frac{2}{a} \int x e^{ax}\, dx$$

$$= \frac{1}{a} x^2 e^{ax} - \frac{2}{a}\left[\frac{1}{a}\left(x - \frac{1}{a}\right) e^{ax}\right] + C \qquad \text{by Formula (1)}$$

$$= \frac{1}{a}\left(x^2 - \frac{2}{a} x + \frac{2}{a^2}\right) e^{ax} + C$$

That is,

$$\int x^2 e^{ax}\, dx = \frac{1}{a}\left(x^2 - \frac{2}{a} x + \frac{2}{a^2}\right) e^{ax} + C \qquad (5)$$

Using this formula you can employ formula (4) again to find $\int x^3 e^{ax}\, dx$ and so on.

Integration by parts can also be used to derive a formula for evaluating definite integrals. For this, we observe that

$$\int_a^b g(x) f'(x)\, dx = \int_a^b [g(x) f(x)]'\, dx - \int_a^b f(x) g'(x)\, dx$$

With the fact that

$$\int_a^b [g(x) f(x)]'\, dx = g(x) f(x) \Big|_a^b = g(b) f(b) - g(a) f(a)$$

we arrive at the following result:

THEOREM 2. INTEGRATION BY PARTS FOR DEFINITE INTEGRALS
Let $f(x)$ and $g(x)$ satisfy the following conditions:

1. $f(x)$ and $g(x)$ are continuous on the interval $[a, b]$
2. $f'(x)$ and $g'(x)$ are continuous and bounded on the interval (a, b)

Then

$$\int_a^b g(x)f'(x)\, dx = [g(b)f(b) - g(a)f(a)] - \int_a^b g(x)f'(x)\, dx$$

With the notation $u = g(x)$ and $v = f(x)$, we get the formula

$$\int_a^b u\, dv = uv\Big|_a^b - \int_a^b v\, du$$

This is the analogous formula to (2).

Example 5

Evaluate $\int_0^{10} x(x+1)^9\, dx$.

Solution
Put $g(x) = x$ and $f'(x) = (x+1)^9$. Then

$$g'(x) = 1 \quad \text{and} \quad f(x) = \tfrac{1}{10}(x+1)^{10}$$

and

$$g(10)f(10) - g(0)f(0) = 10\tfrac{1}{10}(10+1)^{10} - 0 \cdot \tfrac{1}{10}(0+1)^{10} = 11^{10}$$

Since

$$\int_0^{10} f(x)g'(x)\, dx = \frac{1}{10}\int_0^{10}(x+1)^{10}\, dx = \frac{1}{10}\frac{1}{11}(x+1)^{11}\Big|_0^{10}$$

$$= \frac{1}{10}\frac{1}{11}(10+1)^{11} - \frac{1}{10}\frac{1}{11}(0+1)^{11}$$

$$= \frac{1}{10}11^{10} - \frac{1}{10 \cdot 11}$$

we find that

$$\int_0^{10} x(x+1)^9\, dx = 11^{10} - \left(\frac{1}{10}11^{10} - \frac{1}{10 \cdot 11}\right) = \frac{9}{10}11^{10} + \frac{1}{10 \cdot 11}$$

EXERCISES

Find the indefinite integrals in Exercises 1–7.

1. $\int x^3 e^x\, dx$ (Use equations (4) and (5))

2. $\int x^3 e^{-x}\, dx$

3. $\int x(x+5)^{20}\, dx$

4. $\int x^2(x+5)^{20}\, dx$

5. $\int \ln^2 x\, dx$ (Use equation (3))

6. $\int x \ln^2 x\, dx$

7. $\int x\sqrt{1-x}\, dx$

Evaluate the definite integrals in Exercises 8–12.

8. $\int_0^{10^6} xe^{-x}\, dx$

9. $\int_1^2 x(x-1)^5\, dx$

10. $\int_1^e x^2 \ln x\, dx$

11. $\int_{-1}^1 x(e^x - e^{-x})\, dx$

12. $\int_{-1}^1 x^4\sqrt{1-x^2}\, dx$

13. Obtain a reduction formula for $\int \ln^n x\, dx$ $(n \neq -1)$

14. Use Exercises 5 and 13 to find $\int \ln^3 x\, dx$.

15. Verify that for any polynomial $p(x)$,
$$\int p(x)e^x\, dx = [p(x) - p'(x) + p''(x) - \cdots]e^x + C$$

16. Use Exercise 15 to evaluate
$$\int_0^2 (5x^3 + 15x^2 - 4x + 3)e^x\, dx$$

17. Find a reduction formula for $\int x^{2n}\sqrt{1-x^2}\, dx$.

18. Find a reduction formula for $\int x^{2n} e^{-x^2}\, dx$.

QUIZ 1

1. Find dy/dx in each of the following cases.

   (a) $y = \ln \dfrac{x-1}{x+1}$

   (b) $y = \ln(x + \sqrt{x^2 + 4})$

   (c) $y = \dfrac{1}{x} \ln x$

   (d) $y = e^{2x^2}$

   (e) $y = \ln(e^x + 1)$

   (f) $y = 4^x + x^4$

   (g) $y = \exp(\ln x + \exp x)$

2. Simplify the following expressions as far as possible.

   (a) $\exp(\ln x^2)$

   (b) $\ln[\exp(x+1)]$

   (c) $e^{1+\ln 2x}$

3. Using tables of exponentials and logarithms, express the following numbers as decimals.

   (a) $3^{1/6}$

   (b) $4^{\sqrt{2}}$

   (c) $5^{-\sqrt{e}}$

4. Find the following integrals.

   (a) $\displaystyle\int 4x \exp(4x^2 + 1)\, dx$

   (b) $\displaystyle\int x e^{-5x}\, dx$

   (c) $\displaystyle\int x^2 \ln 4x\, dx$

5. Find a formula for $\displaystyle\int_a^b f(x) \ln[F(x) + 1]\, dx$ if $F'(x) = f(x)$.

QUIZ 2

1. Find dy/dx in each of the following cases.

   (a) $y = x^2 e^{-x}$

   (b) $y = e^{x \ln x}$

   (c) $y = x + \ln(xy)$

   (d) $y = 5^{5x}$

   (e) $y = \ln[\ln(e^x + 1)]$

   (f) $y = \exp(x - \sqrt{x^4 - 4})$

   (g) $y = 6^{\ln x}$

2. Simplify the following expressions as far as possible.
   (a) $\ln(e^e)^2$
   (b) $\exp(\ln 10 - \ln 2)$
   (c) $\ln(\ln e^e)$

3. Using tables of exponentials and natural logarithms, express the following numbers as decimals.
   (a) 2^e
   (b) π^7
   (c) e^π

4. Plot the graph of $g(x) = x + \ln|x|$.

5. Find the following integrals:
   (a) $\displaystyle\int \frac{e^{2x} - e^{-2x}}{e^{2x} + e^{-2x}}\, dx$
   (b) $\displaystyle\int_{-1}^{1} e^{-|x|-x}\, dx$
   (c) $\displaystyle\int_{-1}^{1} (x+2)\ln(x+2)\, dx$

6. Find a formula for $\displaystyle\int_a^b f(x) e^{F(x)}\, dx$ if $F'(x) = f(x)$.

CHAPTER 8

APPLICATIONS OF INTEGRATION

8.1 INTEGRATION OF RATES In this section we exploit the following facts:

> $f'(t)$ = rate of change of $f(t)$ with respect to t;
> $\int f'(t)\, dt = f(t) + C$

Rates were previously discussed in Section 5.3. The general procedure for solving the problems of this section is given following the example below.

Example 1

The size $P(t)$ of a population of mosquitos is an unknown function of time, t being measured in months. We wish to find $P(t)$ when it is known that

$P'(t) = 432t^2 - 5t^4$ — This represents the rate of change of the population with respect to time

$P(0) = 20$ — $t = 0$ represents the beginning of the year and $P(0)$ gives the population size at this time

Solution

We know that

$$P(t) = \int (432t^2 - 5t^4)\, dt = 144t^3 - t^5 + C$$

The constant of integration C can be determined from the condition $P(0) = 20$. Namely, for $t = 0$ we get

$$P(0) = 144 \cdot 0 - 0 + C = C = 20$$

Hence,

$$P(t) = 144t^3 - t^5 + 20$$

(see Figure 8.1).

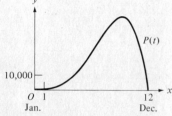

Figure 8.1
The population function $P(t) = 144t^3 - t^5$.

TERMINOLOGY

It is customary to use the time $t = 0$ as the *present* or *reference time*. The condition $P(0) = 0$ is called an *initial condition*, and Example 1 is an example of an *initial value problem*.

The preceding problem combines all the ingredients of the rate problems discussed in this section, and the following general procedure can be prescribed:

Initial Value Problem
Find the solution $P(t)$ of the problem

$$\left. \begin{array}{l} P'(t) = r(t) \\ P(t_0) = P_0 \end{array} \right\} \quad \begin{array}{l} r(t) \text{ is the given rate of change of } P(t) \\ t_0 \text{ and } P_0 \text{ are given numbers.} \end{array}$$

Solution

Step 1
Compute $\int r(t)\, dt = R(t) + C$ and put

$$P(t) = R(t) + C$$

Step 2
Substitute $t = t_0$ in the last equation to find C:

$$P(t_0) = P_0 = R(t_0) + C$$

and hence

$$C = P_0 - R(t_0)$$

Combining steps 1 and 2 gives

$$P(t) = R(t) + [P_0 - R(t_0)]$$

Example 2

A weather balloon of volume $v_0 = 400$ ft^3 develops a leak at time 10:50 a.m. The rate at which the balloon is deflating is measured to be

$$v'(t) = -0.01t \text{ ft}^3/\text{min}$$

Time is measured in minutes from the time the leak developed, and the minus sign is used to indicate that the volume is decreasing.

(a) Find the time at which the balloon is completely deflated.
(b) Express the volume $v(t)$ as a function of time.

Solution
Using the reference time $t_0 = 0$, we have the initial value problem

$$\left.\begin{array}{l} v'(t) = -0.01t \\ v(0) = 400 \end{array}\right\}$$

Step 1
Here $r(t) = -0.01t$, and hence

$$v(t) = \int -0.01t \, dt = -0.01 \frac{t^2}{2} + C$$

Step 2
Putting $R(t) = -0.01(t^2/2)$, we have

$$C = P_0 - R(0) = 400 - 0 = 400$$

The formula

$$v(t) = -0.01 \frac{t^2}{2} + 400$$

gives the volume of the balloon until it is completely deflated. To find the time it takes for this to happen, we set $v(t) = 0$ and solve for t:

$$0 = -0.01 \frac{t^2}{2} + 400$$

$$t^2 = \frac{2 \cdot 400}{0.01} = 2 \cdot 4 \cdot 10^4$$

and since we do not consider negative time,

$$t = \sqrt{2 \cdot 4 \cdot 10^4} = 200\sqrt{2} \approx 200 \cdot 1.414 = 282.8 \text{ min}$$

Thus, it takes about 4 hr and 42.8 min for the balloon to deflate completely. Since it started deflating at 10:50 a.m., we find that

(a) the balloon is completely deflated at about 3:33 p.m.
(b) $v(t)$ can now be expressed as follows:

$$v(t) = \begin{cases} -0.01\dfrac{t^2}{2} + 400 & \text{for } 0 \leq t \leq 200\sqrt{2} \\ 0 & \text{for } t > 200\sqrt{2} \end{cases}$$

The graph of $y = v(t)$ is given in Figure 8.2.

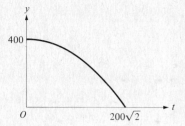

Figure 8.2
The graph of $y = v(t)$.

Example 3

A parachutist is falling from a height of $h_0 = 10{,}000$ ft with velocity

$$v(t) = 32(e^{-t} - 1) \text{ ft/sec} \qquad (1)$$

When will he hit the ground?

Solution

Let $h(t)$ be the unknown function expressing height as a function of time. Using the reference time $t_0 = 0$, we have the initial value $h_0 = h(t_0) = h(0) = 10{,}000$. Since velocity is rate of change of distance (height in our case) with respect to time, we have the relation $v(t) = h'(t)$. To solve the problem, we first have to solve the initial value problem

$$\left. \begin{array}{l} h'(t) = 32(e^{-t} - 1) \\ h(0) = 10{,}000 \end{array} \right\} \qquad (2)$$

For this we follow the procedure described above.

Step 1

$$h(t) = \int 32(e^{-t} - 1)\, dt = 32(-e^{-t} - t) + C$$
$$= -32(e^{-t} + t) + C$$

Step 2

Using the initial condition, we have

$$h(0) = 10{,}000 = -32(e^0 + 0) + C = -32 + C$$

and hence

$$C = 10{,}032$$

The solution of the initial value problem (2) is

$$h(t) = -32(e^{-t} + t) + 10{,}032 \qquad (3)$$

Step 3

To solve the problem we now set $h(t) = 0$ to find the time t at which the parachutist hits the ground. From (3) we find that when $h(t)=0$,

$$e^{-t}+t=\frac{10{,}032}{32}=313.5$$

To find an exact solution of this equation is difficult. We observe, however, that when $t \geq 313$, e^{-t} is negligible in comparison with t. We can thus take for our answer

$$t \approx 313.5 \text{ sec}$$

Thus, the parachutist will hit the ground after approximately 5 min and 13.5 sec.

Example 4

An experimental rocket is launched from the ground vertically with an initial velocity $v_0 = 7$ miles/sec. If the rocket's acceleration is given by the formula

$$a(t) = 6 \cdot 10^{-2} \cdot e^{-0.02t}$$

express its velocity $v(t)$ and height $h(t)$ in terms of t.

REMARK
Velocity is the rate of change of distance with respect to time,

$$v = h'(t)$$

Acceleration is the rate of change of velocity with respect to time,

$$v'(t) = h''(t)$$

If velocity is measured in miles per second (miles/sec), then acceleration is measured in miles per second per second, abbreviated miles/sec² and read "miles per second squared."

Solution
Using the relation $a(t) = v'(t)$, we obtain the velocity $v(t)$ as the solution of the initial value problem

$$\left. \begin{array}{l} v'(t) = 6 \cdot 10^{-2} \cdot e^{-0.02t} \\ v(0) = 7 \end{array} \right\}$$

Step 1

$$v(t) = \int 6 \cdot 10^{-2} \cdot e^{-0.02t}\, dt = 6 \cdot 10^{-2} \frac{e^{-0.02t}}{-0.02} + C$$

$$= -3e^{-0.02t} + C$$

Step 2
Since $v(0) = 7$, we have

$$7 = -3 \cdot e^{-0} + C = -3 + C$$

Hence

$$C = 10$$

and we find that

$$v(t) = -3 \cdot e^{-0.02t} + 10 \text{ ft/sec}$$

To find the height $h(t)$ we use the relation $v = h'(t)$ and the initial condition $h(0) = 0$. This yields the initial value problem

$$\left. \begin{array}{l} h'(t) = -3 \cdot e^{-0.02t} + 10 \\ h(0) = 0 \end{array} \right\}$$

Step 1'

$$h(t) = \int (-3e^{-0.02t} + 10) \, dt = -3\frac{e^{-0.02t}}{-0.02} + 10t + C$$

$$= 150 \cdot e^{-0.02t} + 10t + C$$

Step 2'
The initial condition gives

$$0 = 150 \cdot e^{-0} + 0 + C = 150 + C$$

and hence

$$C = -150$$

Thus,

$$h(t) = 150e^{-0.02t} + 10t - 150 \text{ ft}$$

Example 5

When a dairy produces $Q(t)$ pounds of cottage cheese in t days, its *production rate* is defined to be $Q'(t)$. We get insight into this interpretation of the derivative by noting that

$$Q'(t) \approx \frac{Q(t + \Delta t) - Q(t)}{\Delta t}$$

$$= \frac{\text{tons of cottage cheese produced in the period from } t \text{ to } t + \Delta t}{\Delta t \text{ days}}$$

Find $Q(t)$ and the average daily production when the production rate is

$$Q'(t) = 1000 + 0.4t \text{ lb/day}$$

Solution
The total production is

$$Q(t) = \int Q'(t) \, dt = \int (1000 + 0.4t) \, dt = 1000t + 0.2t^2 + C$$

and, observing that $Q(0) = 0$, we have

$$Q(t) = 1000t + 0.2t^2 \text{ lb}$$

The graph of $y = Q(t)$ is given in Figure 8.3.

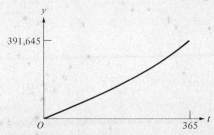

Figure 8.3
The production function $Q(t) = 1000t + 0.2t^2$.

In one year ($t = 365$), the production of our dairy is

$$Q(365) = 1000 \cdot 365 + 0.2 \cdot 365^2 \, (= 391{,}645) \text{ lb}$$

and hence the daily average is

$$\frac{Q(365)}{365} = 1000 + 0.2 \cdot 365 = 1073 \text{ lb}$$

EXERCISES

1. The acceleration of a rocket launched from the ground in an upward direction is $a = 3$ miles/sec^2; the rocket's velocity after 5 sec is 65 miles/sec. Find the height of the rocket after 10 sec.

2. A certain car has acceleration $a = \frac{1}{3}t$ miles/sec, t being time measured in seconds. Find the velocity of the car at time $t = 0$ in miles per hour if its velocity at time $t = 15$ sec is 50 mph.

3. A spherical balloon is being inflated at the rate of 3 in.^{3}/sec. What is the diameter of the balloon after 36 sec?

4. A water reservoir is being filled at the rate of 10,000 gallons/hr. Due to increased consumption, the water in the reservoir is decreasing at the rate of $\frac{5}{2}t$ gallons per hour at time t. When will

the reservoir be empty if the initial water volume was 10^5 gallons?

5. At t sec after an alpha particle enters an accelerator, its velocity is $v = 10^7 \, t^2$ ft/sec. How far does the particle travel during the first $\frac{1}{100}$ th of a second?

6. A car is traveling at 60 mph when the brakes are applied; t sec thereafter the car's velocity is $v = 60 - 24t$ mph. How far will the car travel before coming to a complete halt?

7. Starting from a dead stop, a car is driven 1 mile in 1 min. Find the constant acceleration a that is needed to do this.

8. A car is traveling at velocity v_0 when the brakes are applied; before stopping, it continues to travel 240 ft in 8 sec. Assuming that the deceleration a of the car was constant, find a and v_0.

9. A projectile is fired upward from a 100-ft tower with an initial velocity of 120 ft/sec; it is assumed that the deceleration acting on the body through gravity is $a = -32$ ft/sec^2.

   (a) Find the height reached by the projectile.

   (b) When will the projectile hit the ground?

10. Due to seasonal fluctuation, the price of oranges changes. Let $P(t)$ represent the price in dollars per ton as a function of time, where t is measured in weeks. If

    $$P'(t) = -\tfrac{2}{5}t + 16$$
    $$P(0) = 120$$

    $P'(t)$ represents the rate of change of price with respect to time
    $t = 0$ represents the beginning of the year

    find $P(t)$.

11. Let $q'(t)$ be the rate at which the rate of production $q(t) = Q'(t)$ is changing. Find and graph $q(t)$ and $Q(t)$ when it is known that

    $$q'(t) = 0.02 + 6t$$
    $$Q(0) = 5$$
    $$q(0) = 1$$

Exercises 12–15 use the terminology of *marginal analysis*. We explained in Section 5.1 that marginal analysis in economics refers to the application of derivatives to problems in economics. We recall from the same section the following basic definitions:

> If $f(q)$ is the *cost of production* of q units of a commodity, then $f'(q)$ is the *marginal cost*. If $m(q)$ is the *total revenue* (money received), then $m'(q)$ is the *marginal revenue*, and $m(q)/q$ is the *price*.

246 APPLICATIONS OF INTEGRATION

12. Suppose that the marginal cost of producing q tons of steel is $(q+1)^{-1/2}$ dollars per ton. In addition, it is known that the cost of producing the first 99 tons of steel is 2020 dollars. Find the cost function $f(q)$.

13. A steel manufacturer produces q tons of steel per week. He figures that his marginal revenue is $28/(q+1)^2$ dollars/ton. Find the price as a function of q when the revenue is fixed to be 0 when $q=0$.

14. The marginal cost of producing q gallons of lubricant is $\frac{10}{3}q$ dollars/gallon. If, due to overhead, the cost of producing 0 gallons is 300 dollars, find the cost function $f(q)$. The overhead does not depend on q.

15. The marginal revenue of producing q tons of Portland cement is $45q-5$ dollars/ton. Find the revenue and price functions when it is known that revenue for the first 1000 tons is 15,000,000 dollars.

8.2 GROWTH AND DECAY

Let $P(t)$ give the size of a population at time t. As time goes on, the population grows; one of the simplest examples of population growth is when the *rate of growth* $P'(t)$ is proportional to the size $P(t)$:

$$P'(t) = \alpha P(t)$$

where α is the *constant of proportionality*. The only function satisfying this condition is of the form

$$P(t) = A e^{\alpha t}$$

Since $P(0) = Ae^0 = A$, we can actually write

$$P(t) = P(0) e^{\alpha t}$$

THEOREM 1
If $f(t)$ satisfies the condition

$$f'(t) = \alpha f(t)$$

then

$$f(t) = f(0) e^{\alpha t} \tag{1}$$

We speak of *exponential growth* when $\alpha > 0$, *exponential decay* when $\alpha < 0$.

Proof

To simplify the proof, we assume that $f(t) > 0$ for all values of t. The condition $f'(t) = \alpha f(t)$ can be written

$$\frac{1}{f(t)} f'(t) = \alpha$$

and from formula (2) of Section 7.2 we have

$$\frac{d}{dt}[\ln f(t)] = \frac{1}{f(t)} f'(t)$$

Hence

$$\frac{d}{dt}[\ln f(t)] = \alpha$$

and accordingly,

$$\ln f(t) = \int \alpha \, dt = \alpha t + C$$

We now see that

$$e^{\ln f(t)} = f(t) = e^{\alpha t + C} = e^{\alpha t} e^C$$

but since $f(0) = e^C$, we arrive at the formula

$$f(t) = f(0) e^{\alpha t}$$

and this completes the proof.

REMARK 1

The logarithmic derivative of $f(t)$,

$$\frac{d}{dt}[\ln f(t)] = \frac{f'(t)}{f(t)}$$

is often called *rate of proportional change*. The reason for this terminology becomes clear when we write

$$\frac{f'(t)}{f(t)} \approx \frac{1}{f(t)} \cdot \frac{f(t + \Delta t) - f(t)}{\Delta t} = \frac{[f(t + \Delta t) - f(t)]/f(t)}{\Delta t}$$

The quantity $f(t + \Delta t) - f(t)$ represents *change*, and the quotient $[f(t + \Delta t) - f(t)]/f(t)$ is therefore called *proportional change*.

Example 1

The rate of proportional increase of a population is 5 percent per year. If the initial size is 10^6, what is the population after 1 year? After 2 years?

Solution
Here $\alpha = \frac{5}{100} = 0.05$ and, if the population is $P(t)$, where t is measured in years, then

$$\frac{P'(t)}{P(t)} = 0.05 \quad \text{and} \quad P(0) = 10^6$$

According to Theorem 1, therefore,

$$P(t) = 10^6 \cdot e^{0.05t}$$

Thus,

$$P(1) = 10^6 \cdot e^{0.05} \approx 1.0513 \cdot 10^6$$
$$P(2) = 10^6 \cdot e^{0.10} \approx 1.1052 \cdot 10^6$$

Observe that the population increase was 51,300 over the first year, 53,900 over the second year.

Example 2

The mass of a radioactive material decreases with the passage of time. The law that governs this change is

$$M'(t) = -kM(t)$$

where $M(t)$ is the mass at time t. Time t is measured in years, and k is the *rate of proportional decrease* of the mass $M(t)$. According to Theorem 1,

$$M(t) = M(0) \cdot e^{-kt}$$

The half-life of the material, $T_{1/2}$, is the length of time for half of the mass to disintegrate; that is,

$$M(T_{1/2}) = \tfrac{1}{2} M(0)$$

We now show that k is determined by $T_{1/2}$. Since

$$M(T_{1/2}) = M(0)e^{-kT_{1/2}} = \tfrac{1}{2} M(0)$$

we have

$$e^{-kT_{1/2}} = \tfrac{1}{2}$$

Taking logarithms, we arrive at

$$kT_{1/2} = \ln 2$$

Thus

$$k = \frac{\ln 2}{T_{1/2}}$$

For radium, $T_{1/2} \approx 1590$ years and thus $k \approx 4.4 \cdot 10^{-4}$.

Example 3

The rate of proportional decrease of a population of bacteria is 5 percent/day. If the initial size of the population is 10^6, when will the population reduce to 1 percent of this size?

Solution

As in Example 1, $\alpha = 0.05$, and if the population is $P(t)$, where t is measured in days, then $P(0) = 10^6$. Hence

$$P(t) = 10^6 e^{-0.05t}$$

Since 1 percent of 10^6 is $\frac{1}{100} \cdot 10^6 = 10^4$, we have to solve the equation

$$P(t) = 10^6 e^{-0.05t} = 10^4$$

This equation simplifies to

$$e^{-0.05t} = 10^{-2}$$

or

$$0.05t = 2 \ln 10$$

Solving for t gives

$$t = \frac{2 \ln 10}{0.05} = 40 \ln 10 \approx 40 \cdot 2.3026 = 92.104$$

Hence the population will reduce to 10^4 in 92.104 days.

The preceding problems dealt with situations in which the rate of proportional increase or decrease is constant. More general problems can be solved with our method.

Example 4

The rate of proportional change of a given population is

$$\alpha = 0.05 - 0.01t$$

where α represents a *proportional rate of increase* for $0 \leq t \leq 5$ since then $\alpha \geq 0$, and a *proportional rate of decrease* for $t > 5$ since then $\alpha < 0$. Find the number of days it takes the population to reduce to half its size if the initial size is 10^6.

Solution

Let the population be $P(t)$, where t is measured in days. Then $P(0) = 10^6$, and we have

$$\left. \begin{array}{l} \dfrac{P'(t)}{P(t)} = 0.05 - 0.01t \\ P(0) = 10^6 \end{array} \right\}$$

To find $P(t)$, we use the logarithmic derivative

$$\frac{d}{dt}[\ln P(t)] = \frac{P'(t)}{P(t)}$$

Using this fact gives

$$\frac{d}{dt}[\ln P(t)] = 0.05 - 0.01t$$

and hence

$$\ln P(t) = 0.05t - 0.01 \cdot \frac{t^2}{2} + C$$

This is the same as

$$P(t) = e^{0.05t - 0.01(t^2/2) + C} = e^C e^{0.05t - 0.005t^2} = 10^6 e^{0.05t - 0.005t^2}$$

To find the time it takes the population to reduce to half its size, $(\frac{1}{2})10^6$, we put $P(t) = (\frac{1}{2})10^6$ and solve for t. This gives upon simplification

$$e^{0.05t - 0.005t^2} = \tfrac{1}{2}$$

and hence

$$0.05t - 0.005t^2 = \ln \tfrac{1}{2} = -\ln 2$$

This equation can be written in the form

$$5 \cdot 10^{-3} t^2 - 5 \cdot 10^{-2} t - \ln 2 = 0$$

or

$$t^2 - 10t - 200 \ln 2 = 0$$

Solving for t and taking $t > 0$ gives

$$t = \frac{10 + \sqrt{10^2 + 4 \cdot 200 \cdot \ln 2}}{2} = 5 + \sqrt{\frac{100}{4} + \frac{4 \cdot 200 \cdot \ln 2}{4}}$$

$$= 5 + \sqrt{25 + 200 \cdot \ln 2} \approx 5 + \sqrt{25 + 200 \cdot 0.6931}$$

$$= 5 + \sqrt{163.62} \approx 17.79$$

Hence, the population will reduce to half its size in 17.79 days.

EXERCISES

1. The rate of proportional increase of a population is 2 percent/year. If the initial population numbers 10^6, how long will it take the population to double in size? (See Example 1.)

2. It is observed that a population of bacteria grows from 1000 to 4000 in 24 hr. Assume that the population $P(t)$ satisfies the condition $P'(t) = \alpha P(t)$.

(a) Find α.

(b) Find the size of the population after 10 hr.

3. Suppose that the function $f(t)$ satisfies the condition $f'(t) = \alpha f(t)$ for some number $\alpha \neq 0$. If T is any fixed number, what can you say about the sequences

   (a) $\dfrac{f(T)}{f(0)}, \dfrac{f(2T)}{f(T)}, \dfrac{f(3T)}{f(2T)}, \ldots, \dfrac{f(nT)}{f[(n-1)T]}, \ldots$

   (b) $f(T) - f(0), f(2T) - f(T), f(3T) - f(2T), \ldots,$
   $f(nT) - f[(n-1)T], \ldots$

4. Sugar in solution decomposes at a rate that is proportional to the amount of undecomposed sugar. It is found that 12 oz diminish to 4 oz in 3 hr. Find the length of time required to reduce the remaining 4 ounces of undecomposed sugar to 1 oz.

5. You make an investment of 1000 dollars which, in t years, has a rate of proportional increase equal to $0.05t$. Due to inflation and taxation, your investment has a constant rate of proportional decrease equal to 0.1.

   (a) At what times is the investment equal to 1000 dollars?

   (b) How long will it take for your investment to double?

6. The size $P(t)$ of a population of bacteria is a function of time, where t is measured in hours. Find $P(t)$ when it is known that

   $$\left. \begin{array}{l} P'(t) = 5 \cdot 10^4 \cdot e^{0.05t} \\ P(0) = 1.6 \cdot 10^6 \end{array} \right\}$$

7. If money $M(t)$ in the bank earns interest compounded daily at a rate of 6 percent/year, then a very small error is involved in assuming that interest is compounded "continuously," that is, in assuming that

   $$\frac{dM}{dt} = 0.06M$$

   where t is measured in years. How many years will it take to double your money? Treble your money?

8.3 VOLUMES OF SOLIDS OF REVOLUTION

Solids of revolution are solids generated by rotating a planar area about a fixed line. The two examples in Figure 8.4 will explain this idea. Our fixed line in both cases is the x axis.

To find volumes of solids of revolution is surprisingly simple,

252 APPLICATIONS OF INTEGRATION

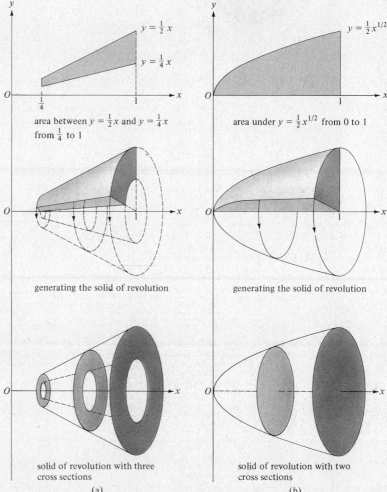

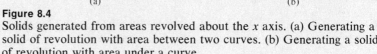

Figure 8.4
Solids generated from areas revolved about the x axis. (a) Generating a solid of revolution with area between two curves. (b) Generating a solid of revolution with area under a curve.

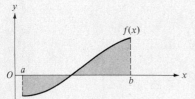

Figure 8.5
Area bounded between $f(x)$ and the x axis.

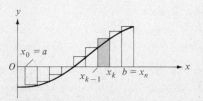

Figure 8.6
Approximation to area bounded between $f(x)$ and the x axis.

and we shall now derive a general formula. Consider the area bounded by the curve $y = f(x)$, the x axis, and the lines $x = a$ and $x = b$ (Figure 8.5). We divide the interval $[a, b]$ into subintervals $[x_{k-1}, x_k]$, $k = 1, 2, \ldots, n$, where $x_k = a + k\,\Delta x$ and $\Delta x = (b - a)/n$ (see Figure 8.6). On the interval $[x_{k-1}, x_k]$ we construct the rectangle of base Δx and height $|f(x_k)|$; when revolved about the x axis, this rectangle generates a shallow cylinder of radius $|f(x_k)|$ and height Δx, and its volume is, therefore, $\pi[f(x_k)]^2\,\Delta x$. The volume V_n of the solid generated from all the rectangles is

$$V_n = \sum_{k=1}^{n} \pi[f(x_k)]^2\,\Delta x = \pi \sum_{k=1}^{n} [f(x_k)]^2\,\Delta x$$

8.3 VOLUMES OF SOLIDS OF REVOLUTION

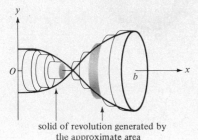

solid of revolution generated by the approximate area

third disk | kth disk

solid of revolution

Figure 8.7

(see Figure 8.7). If V designates the volume of the solid generated from the area bounded between $f(x)$ and the x axis, then quite clearly $V_n \to V$ as $n \to \infty$ (Figure 8.7). At the same time, however,

$$\lim_{n \to \infty} \pi \sum_{k=1}^{n} [f(x_k)]^2 \, \Delta x = \pi \int_a^b [f(x)]^2 \, dx$$

when $f(x)$ is bounded and piecewise continuous on $[a, b]$. This leads us to the following fact:

VOLUME OF SOLID OF REVOLUTION
If $f(x)$ is bounded and piecewise continuous on the closed interval $[a, b]$, then the solid of revolution generated by revolving the area bounded between $f(x)$, the x axis, and the lines $x = a$ and $x = b$ about the x axis has volume

$$V = \pi \int_a^b [f(x)]^2 \, dx \qquad (1)$$

This is illustrated for a piecewise continuous function in Figures 8.8 and 8.9. Before looking at complicated solids, we use this formula to obtain some familiar formulas for volumes in Table 8.1.

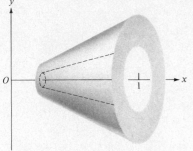

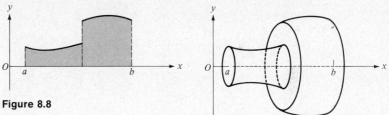

Figure 8.8

Figure 8.9

Example 1

Compute the volume of the hollow truncated cone given in Figure 8.4(a).

Solution
Consult Figure 8.10. The area generating this solid is bounded by the lines $f_1(x) = \frac{1}{2}x$, $f_2(x) = \frac{1}{4}x$, $x = \frac{1}{4}$, and $x = 1$. The volume of this solid is quite clearly given by the formula

$$V = \pi \int_{1/4}^{1} [f_1(x)]^2 \, dx - \pi \int_{1/4}^{1} [f_2(x)]^2 \, dx$$

$$= \pi \int_{1/4}^{1} \{[f_1(x)]^2 - [f_2(x)]^2\} \, dx$$

$$= \pi \int_{1/4}^{1} \left[\left(\frac{1}{2}x\right)^2 - \left(\frac{1}{4}x\right)^2\right] dx = \pi \int_{1/4}^{1} \left(\frac{1}{4}x^2 - \frac{1}{4^2}x^2\right) dx$$

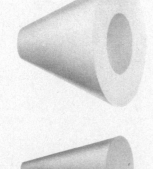

Figure 8.10
Subtracting two volumes to get the volume of a hollow solid.

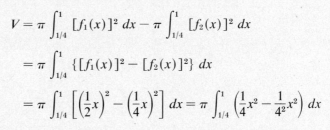

Table 8.1
Volumes of Some Familiar Solids

Solid	Generating area	Volume
cylinder with base of radius r and height h	$f(x) = r$ $a = 0$ $b = h$	$\pi \int_0^h r^2 \, dx = \pi r^2 h$
cone with base of radius r and height h	$f(x) = \frac{r}{h} x$ $a = 0$ $b = h$	$\pi \int_0^h \left(\frac{r}{h}x\right)^2 dx = \frac{1}{3}\pi r^2 h$
sphere of radius r	$f(x) = \sqrt{r^2 - x^2}$ $a = -r$ $b = r$	$\pi \int_{-r}^{r} (\sqrt{r^2 - x^2})^2 \, dx = \frac{4}{3}\pi r^3$
parabolic cone with base of radius $\sqrt{h}$ and height h	$f(x) = x^{1/2}$ $a = 0$ $b = h$	$\pi \int_0^h (x^{1/2})^2 \, dx = \frac{1}{2}\pi h^2$

Volumes of some familiar solids

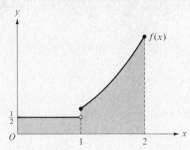

Figure 8.11
The generating area.

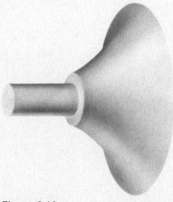

Figure 8.12
The solid of revolution resembles a door knob.

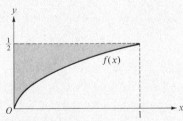

Figure 8.13
The generating area.

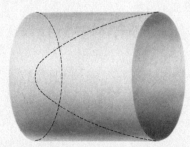

Figure 8.14
The solid of revolution resembles a drinking glass.

$$= \frac{3}{16}\pi \int_{1/4}^{1} x^2 \, dx = \frac{3}{16}\pi \frac{1}{3} x^3 \Big|_{1/4}^{1} = \frac{\pi}{16}\left(1 - \frac{1}{4^3}\right)$$

Example 2

Find the volume of the solid of revolution generated by the area bounded by the curve

$$f(x) = \begin{cases} \frac{1}{2} & \text{for } 0 \leq x < 1 \\ \frac{3}{4}x^2 & \text{for } 1 \leq x \leq 2 \end{cases}$$

the x axis, and the lines $x = 0$, $x = 1$, and $x = 2$ (see Figure 8.11).

Solution
By formula (1),

$$V = \pi \int_0^2 [f(x)]^2 \, dx = \pi \int_0^1 [f(x)]^2 \, dx + \pi \int_1^2 [f(x)]^2 \, dx$$

$$= \pi \int_0^1 \left(\frac{1}{2}\right)^2 dx + \pi \int_1^2 \left(\frac{3}{4}x^2\right)^2 dx$$

$$= \frac{1}{4}\pi x \Big|_0^1 + \frac{9}{16}\pi \cdot \frac{1}{5}x^5 \Big|_1^2 = \frac{1}{4}\pi + \frac{9}{5 \cdot 16}\pi(2^5 - 1)$$

$$= \frac{1}{4}\pi + \frac{18}{5}\pi - \frac{9}{80}\pi = \frac{299\pi}{80}$$

The solid of revolution is given in Figure 8.12.

Example 3

The area bounded by $f(x) = \frac{1}{2}x^{1/2}$, the lines $y = \frac{1}{2}$ and $x = 0$, is revolved about the x axis (see Figure 8.13). Find the volume of the solid of revolution.

Solution
The solid of revolution (see Figure 8.14) is cylindrical in shape on the outside, and the hollow interior is the parabolic cone described in Figure 8.4(b). The volume V is therefore seen to be given by the formula

$$V = \pi \underbrace{\int_0^1 \left(\frac{1}{2}\right)^2 dx}_{\text{volume of cylinder}} - \pi \underbrace{\int_0^1 \left(\frac{1}{2}x^{1/2}\right)^2 dx}_{\text{volume of parabolic cone}}$$

$$= \frac{\pi}{4} \int_0^1 (1 - x) \, dx = \frac{\pi}{8}$$

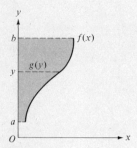

Figure 8.15
$g(x)$ is the inverse of $f(x)$ if $g(f(x)) = x$.

Figure 8.16

It is also easy to obtain a formula for the volume of a solid of revolution generated by a planar area revolved about the y axis (see Figure 8.15). Namely, let the area be bounded by a curve $y = f(x)$, the y axis, $y = a$, and $y = b$. If $f(x)$ has an inverse, $x = g(y)$, then the area of the solid is evidently given by the formula

$$V = \pi \int_a^b [g(y)]^2 \, dy$$

Such a solid of revolution is shown in Figure 8.16.

Example 4

The area bounded by $f(x) = e^x$, the y axis, and the line $y = e$ is revolved about the y axis (see Figure 8.17). Find the volume of the solid of revolution so generated.

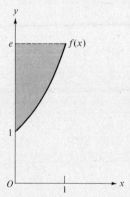

Figure 8.17

Solution
The function $y = e^x$ is the inverse of $x = \ln y$, and we put $g(y) = \ln y$. Since this curve intersects the y axis at $y = 1$, we have the formula

$$V = \pi \int_1^e (\ln y)^2 \, dy$$

Writing $(\ln y)^2 = \ln y \cdot \ln y$ and using integration by parts leads to the following (see Example 2 and Exercise 13 in Section 7.6):

$$V = \pi (y \ln y - y) \ln y \Big|_1^e - \pi \int_1^e (\ln y - 1) \, dy$$

$$= \pi y (\ln^2 y - 2 \ln y + 2) \Big|_1^e = e\pi (\ln^2 e - 2 \ln e + 2)$$
$$- \pi (\ln^2 1 - \ln 1 + 2)$$

$$= e\pi (1 - 2 + 2) - 2\pi = (e - 2)\pi$$

The solid of revolution is shown in Figure 8.18.

Figure 8.18

EXERCISES

Find the volumes of the solids described in Exercises 1–13.

	Solid	Generating area	Generating area bounded by
1.			$f_1(x) = 1.1$ $x = 0$ $f_2(x) = 1$ $x = h$
2.			$f_1(x) = 1.1$ y axis $f_2(x) = 1$ $x = 0.1$ x axis $x = h$
3.			$f(x) = x + \dfrac{1}{x}$ $x = \dfrac{1}{2}$ $x = 3$ x axis
4.			$f_1(x) = \sqrt{r^2 - x^2}$ $f_2(x) = \dfrac{r}{2}$

	Solid	Generating area	Generating area bounded by
5.		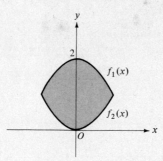	$f_1(x) = 2 - x^2$ $f_2(x) = x^2$
6.		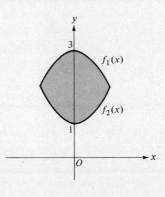	$f_1(x) = 3 - x^2$ $f_2(x) = 1 + x^2$
7.	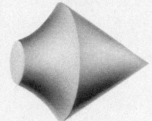	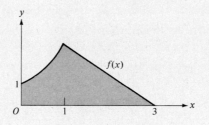	$f(x) = \begin{cases} e^x & \text{for } 0 \leq x \leq 1 \\ \dfrac{e}{2}(3-x) & \text{for } 1 \leq x \leq 3 \end{cases}$ x axis y axis
8.		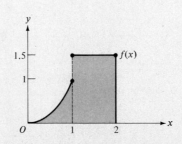	$f(x) = \begin{cases} x^2 & \text{for } 0 \leq x < 1 \\ 1.5 & \text{for } 1 \leq x \leq 2 \end{cases}$ x axis $x = 1 \quad x = 2$

	Solid	Generating area	Generating area bounded by
9.		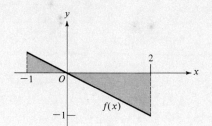	$f(x) = -\tfrac{1}{2}x$ $x = -1$ x axis $x = 2$
10.		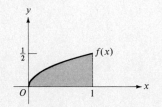	$f(x) = \tfrac{1}{2}x^{1/2}$ $x = 1$ x axis
11.		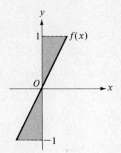	$f(x) = 2x$ $y = 1$ y axis $y = -1$
12.		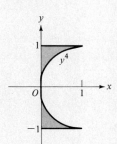	$y^4 = x$ $y = 1$ y axis $y = -1$
13.		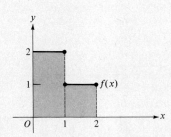	$f(x) = \begin{cases} 2 & \text{for } 0 \leq x \leq 1 \\ 1 & \text{for } 1 < x \leq 2 \end{cases}$ x axis $x = 1$ y axis $x = 2$

14. The area bounded between $f_1(x) = x^2$, $f_2(x) = x^3$, $x = 0$, and $x = 2$ is revolved about the y axis. Find the volume of the solid of revolution.

15. The rectangle with vertices $(1, 0)$, $(1.5, 0)$, $(1.5, 2)$, and $(1, 2)$ is revolved about the y axis. Find the volume of the generated solid.

16. Find the volume of the solid generated by revolving the triangle with vertices $(1, 1)$, $(3, 1)$, and $(2, 4)$ about the x axis.

8.4 VOLUME BY SLICING

Consider once more the formula

$$V = \int_a^b \pi [f(x)]^2 \, dx \tag{1}$$

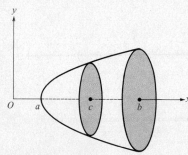

Figure 8.19
Solid of revolution with cross section perpendicular to the x axis at $x = c$. The cross section has radius $|f(c)|$ and area $A(c) = \pi [f(c)]^2$.

for the volume of a solid of revolution (see Figure 8.19). We observe that, for $x = c$, $\pi [f(c)]^2$ represents the area of the cross section of the solid formed by slicing it with the plane perpendicular to the x axis at $x = c$. Putting for each x

$$A(x) = \pi [f(x)]^2$$

formula (1) becomes

$$V = \int_a^b A(x) \, dx \tag{2}$$

We shall see that the latter formula gives the volume of an arbitrary solid with cross sections $A(x)$. We note that, in three-dimensional space,

$x = c$ is the equation of a plane perpendicular to the x axis at $x = c$ (see Figure 8.20).

Now consider an arbitrary solid S contained between the planes $x = a$ and $x = b$. We slice S into n parallel sections by means of the planes

$$x = x_0 = a, \ldots, x = x_k, \ldots, x = x_n = b$$

where, as before, $x_k = a + k \Delta x$ and $\Delta x = (b - a)/n$.

Let the area of the cross section determined by the plane $x = x_k$ be $A(x_k)$ (see Figure 8.21). Then the volume of that portion of S contained between the planes $x = x_{k-1}$ and $x = x_k$ is approximately $A(x_k) \Delta x$ (see Figure 8.22), which is the volume of a cylinder with base $A(x_k)$ and height Δx. The volume of the solid S is therefore given approximately by the expression

$$V_n = \sum_{k=1}^{n} A(x_k) \Delta x$$

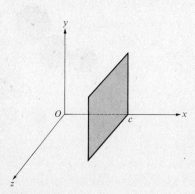

Figure 8.20
Section of the plane $x = c$.

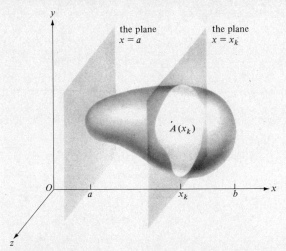

Figure 8.21
The cross section of a solid S determined by the plane $x = x_k$.

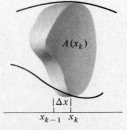

Figure 8.22
A cylinder with base of area $A(x_k)$ and height Δx.

(see Figure 8.23). We observe that V_n approaches the volume V as $n \to \infty$, but also $V_n \to \int_a^b A(x)\,dx$ when $A(x)$ is a bounded and piecewise continuous function of x. This fact is summarized in the following theorem.

THEOREM 1. VOLUME BY SLICING
Let the solid S be contained between the planes $x = a$ and $x = b$. For each c, $a \leq c \leq b$, let $A(c)$ be the area of the cross section determined by the plane $x = c$. Then the volume of S is

$$V = \int_a^b A(x)\,dx$$

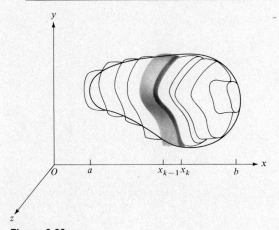

Figure 8.23
The approximation of the value of S by cylinders $A(x_k)\,\Delta x$ for $k = 1, 2, \cdots, n$.

REMARK
Observe that in this formula we need neither the shape of the solid nor the formula for the surface bounding it; the only thing required is a formula for the *area* of the cross sections.

Example 1

A walrus tusk of length 10 in. is found to have circular cross sections of radius

$$r(x) = \tfrac{1}{6}\sqrt{x}$$

where $r(x)$ is measured x in. from the tip of the tusk (see Figure 8.24). Find the volume.

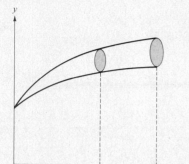

Figure 8.24
The walrus tusk with cross section $r(x)$ at x.

Solution
For each x the area of the cross section is

$$A(x) = \pi[r(x)]^2 = \pi\left(\tfrac{1}{6}\sqrt{x}\right)^2 = \frac{\pi}{36}x$$

Hence

$$V = \int_0^{10} A(x)\, dx = \frac{\pi}{36}\int_0^{10} x\, dx = \frac{\pi}{2\cdot 36}x^2\bigg|_0^{10} = \frac{100\pi}{2\cdot 36} \approx 1.39\pi \text{ in.}^3$$

Example 2

Consider the three solids pictured in Figure 8.25, each having circular cross sections of radius r and height h. According to Remark 1, all of these solids have the same volume. Since the value of the circular cone is $\pi r^2 h$, this is also the volume of the two solids to its right.

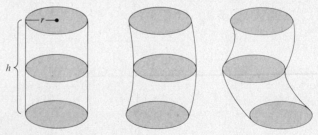

Figure 8.25
Volumes with circular cross sections of radius r and height h.

We further elaborate on this fact below.

VOLUME OF GENERAL CONES
In the previous section we calculated the volume of a right circular

cone having base of radius r and height h to be

$$V = \tfrac{1}{3}\pi r^2 h$$

(see Figure 8.26). The base has area $A = \pi r^2$ and we can put

$$V = \tfrac{1}{3}Ah \qquad (3)$$

This formula will be shown to give the volume of a general cone as well (see Figure 8.27).

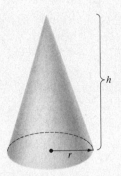

Figure 8.26

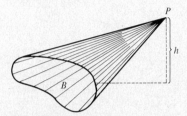

Figure 8.27
A general cone with base B of area A and height h.

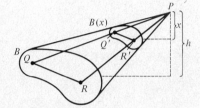

Figure 8.28

A *general cone* is the solid formed when all points of a planar region B are joined with line segments to a single point P lying outside the plane containing B.

To demonstrate that formula (3) gives the desired volume, consider Figure 8.28 and let $B(x)$ be the cross section of the cone determined by the plane parallel to B and x units from the vertex P; let the area of B be A, that of $B(x)$ be $A(x)$. If we consider arbitrary rays from P intersecting B in points Q and R, and $B(x)$ in points Q' and R', then

$$\frac{\overline{Q'R'}}{\overline{QR}} = \frac{x}{h}$$

From this we conclude that

$$\frac{A(x)}{A} = \left(\frac{x}{h}\right)^2$$

or

$$A(x) = A\frac{x^2}{h^2}$$

Hence, the volume V of the general cone is

$$V = \int_0^h A(x)\, dx = \frac{A}{h^2}\int_0^h x^2\, dx = \frac{A}{h^2}\frac{1}{3}x^3\bigg|_0^h = \frac{1}{3}Ah$$

Example 3

As illustrations of formula (3), consider the volumes in Figure 8.29.

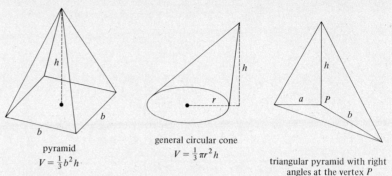

pyramid
$V = \frac{1}{3}b^2h$

general circular cone
$V = \frac{1}{3}\pi r^2 h$

triangular pyramid with right angles at the vertex P
$V = \frac{1}{3}(\frac{1}{2}ab)h = \frac{1}{6}abh$

Figure 8.29

EXERCISES

1. A cone of height h is fitted between the x axis and the line $y = x$; the cross sections perpendicular to the x axis are circles (Figure 8.30). Find the volume of the cone.

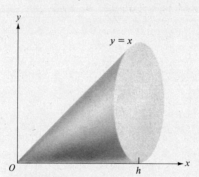

Figure 8.30

2. A solid with a circular base of radius r has vertical parallel cross sections that are squares (Figure 8.31). What is the volume of this solid?

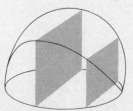

Figure 8.31

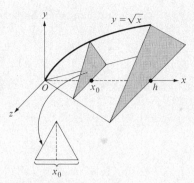

Figure 8.32

3. Consider the solid lying on the xz plane as pictured in Figure 8.32. Its cross section perpendicular to the x axis at $x = x_0$ is a triangle with base of length x_0. Find the volume.

4. Find the volume of the solid bounded by the coordinate planes and the surface

$$z = \frac{3}{x+1} - y - 1$$

(see Figure 8.33).

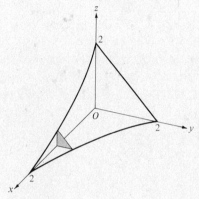

Figure 8.33

5. Find the volume of the tetrahedron having vertices $(0, 0, 0)$, $(3, 0, 4)$, $(0, 0, 4)$, and $(0, 3, 4)$.

Hint: Compute the cross sections $A(x)$.

6. The tent in Figure 8.34 has height h, rectangular base of length b and width $\frac{1}{4}b$, and its top has length $\frac{1}{2}b$. Find the tent's volume.

7. A solid of height h is contained between the two surfaces $x = 2y^2$ and $x = y^4 + 1$ (Figure 8.35). Find the volume of this solid.

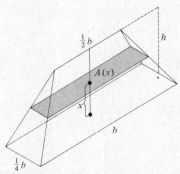

Figure 8.34

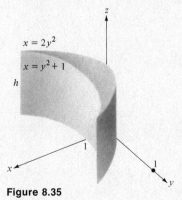

Figure 8.35

8. A hole with circular cross section of radius $\frac{1}{2}$ is bored through the middle of a sphere of radius 1 (Figure 8.36). Find the volume of the remaining solid.

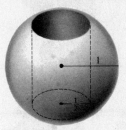

Figure 8.36

9. A solid of height 10 units is fitted between the x axis and the curve $y = \sqrt{x}$; the cross sections perpendicular to the x axis are circles (Figure 8.37). Find the volume of the cone.

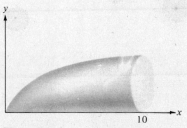

Figure 8.37

10. A solid with a square base of sides of length a has vertical cross sections parallel to a fixed diagonal that are half-circles (Figure 8.38). Find the volume of the solid.

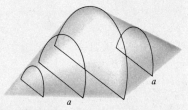

Figure 8.38

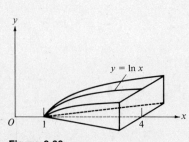

Figure 8.39

11. A solid of height 3 in. is fitted between the x axis and the curve $y = \ln x$; the cross sections perpendicular to the x axis are squares (Figure 8.39). Find the volume of the solid.

8.5 IMPROPER INTEGRALS OVER INFINITE INTERVALS

Improper integrals with infinite limits of integration result when in an integral $\int_a^b f(x)\, dx$ we let $b \to \infty$ or $a \to -\infty$.

8.5 IMPROPER INTEGRALS OVER INFINITE INTERVALS

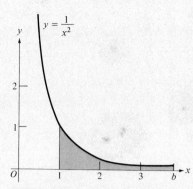

Figure 8.40
The area under $y = 1/x^2$ from 1 to b.

Example 1
Consider the integral
$$\int_1^b \frac{1}{x^2}\, dx = 1 - \frac{1}{b}$$
(see Figure 8.40). Since
$$\lim_{b \to \infty} \left(1 - \frac{1}{b}\right) = 1$$
we have
$$\lim_{b \to \infty} \int_1^b \frac{1}{x^2}\, dx = 1 \tag{1}$$

Geometrically, this means that the area under the curve $y = 1/x^2$ lying to the right of the line $x = 1$ approaches 1 as $b \to \infty$. In view of (1), we define the improper integral $\int_1^\infty (1/x^2)\, dx$ as the limit
$$\int_1^\infty \frac{1}{x^2}\, dx = \lim_{b \to \infty} \int_1^b \frac{1}{x^2}\, dx$$
and thus have
$$\int_1^\infty \frac{1}{x^2}\, dx = 1$$

This is an example of the first of three kinds of improper integrals with infinite limits of integration.

DEFINITION 1
The *improper integrals*
$$\int_a^\infty f(x)\, dx \quad \text{and} \quad \int_{-\infty}^b f(x)\, dx$$
are defined as the limits

$$\int_a^\infty f(x)\, dx = \lim_{b \to \infty} \int_a^b f(x)\, dx$$
$$\int_{-\infty}^b f(x)\, dx = \lim_{a \to -\infty} \int_a^b f(x)\, dx$$

if the limits exist, and they are said to *converge*. If the limits do not exist, they are said to *diverge*.

Example 2
Show that the improper integral $\int_2^\infty \frac{1}{x}\, dx$ diverges.

Solution

$$\int_2^b \frac{1}{x}\,dx = \ln b - \ln 2 \quad \text{for } b > 0$$

and since $\lim_{b\to\infty}(\ln b - \ln 2) = \infty$, the improper integral diverges.

Example 3

Evaluate the improper integral $\int_{-\infty}^{0} e^x\,dx$.

Solution

$$\int_a^0 e^x\,dx = e^0 - e^a = 1 - e^a$$

and thus

$$\lim_{a\to-\infty} \int_a^0 e^x\,dx = \lim_{a\to-\infty}(1 - e^a) = 1$$

This shows that

$$\int_{-\infty}^{0} e^x\,dx = 1$$

The third kind of improper integrals, where both limits of integration are infinite, is introduced in the following definition.

DEFINITION 2

$$\int_{-\infty}^{\infty} f(x)\,dx = \lim_{b\to\infty} \int_{-\infty}^{b} f(x)\,dx$$

if the limit exists. The improper integral $\int_{-\infty}^{\infty} f(x)\,dx$ is said to *converge* when

1. the improper integral $\int_{-\infty}^{b} f(x)\,dx$ converges;

2. the limit $\lim_{b\to\infty} \int_{-\infty}^{b} f(x)\,dx$ exists.

The improper integral is said to *diverge* when (1) or (2) is false.

Example 4

Evaluate $\int_{-\infty}^{\infty} xe^{-x^2}\,dx$.

Solution

We begin with the following integration:

$$\int_a^b xe^{-x^2}\,dx = -\tfrac{1}{2}e^{-x^2}\Big|_a^b = -\tfrac{1}{2}(e^{-b^2} - e^{-a^2})$$

The interpretation of this integral in terms of signed area (see Section 6.1) is given in Figure 8.41. Observe that, due to cancellation of the signed areas, this integral is zero when $a = b$.

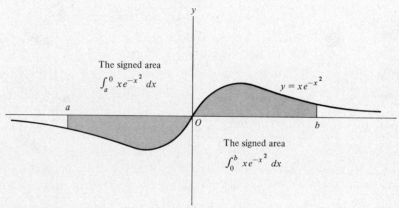

Figure 8.41
$\int_a^b xe^{-x^2}\,dx$ interpreted in terms of signed area.

Since $\lim\limits_{a \to -\infty} -\tfrac{1}{2}(e^{-b^2} - e^{-a^2}) = -\tfrac{1}{2}e^{-b^2}$, we find that

$$\int_{-\infty}^b xe^{-x^2}\,dx = -\tfrac{1}{2}e^{-b^2}$$

Since $\lim\limits_{b \to \infty} -\tfrac{1}{2}e^{-b^2} = 0$, we now find that

$$\int_{-\infty}^{\infty} xe^{-x^2}\,dx = 0$$

Example 5

Evaluate $\int_{-\infty}^{\infty} e^{-|x|}\,dx$.

Solution

In evaluating this integral we use the fact that

$$\int_{-\infty}^{b} f(x)\,dx = \int_{-\infty}^{c} f(x)\,dx + \int_{c}^{b} f(x)\,dx$$

whenever the improper integrals exist.

From the definition of absolute value, we know that

$$e^{-|x|} = \begin{cases} e^{-x} & \text{for } x \geq 0 \\ e^{x} & \text{for } x \leq 0 \end{cases}$$

From Example 3 we have

$$\int_{-\infty}^{0} e^{-|x|}\, dx = \int_{-\infty}^{0} e^{x}\, dx = 1$$

and hence, for each number $b > 0$,

$$\int_{-\infty}^{b} e^{-|x|}\, dx = \int_{-\infty}^{0} e^{-|x|}\, dx + \int_{0}^{b} e^{-|x|}\, dx$$

$$= 1 + \int_{0}^{b} e^{-x}\, dx$$

Since

$$\int_{0}^{b} e^{-x}\, dx = -(e^{-b} - e^{-0}) = 1 - e^{-b}$$

we see that

$$\int_{-\infty}^{b} e^{-|x|}\, dx = 1 + (1 - e^{-b}) = 2 - e^{-b}$$

and consequently,

$$\lim_{b \to \infty} \int_{-\infty}^{b} e^{-|x|}\, dx = \lim_{b \to \infty} (2 - e^{-b}) = 2$$

This gives us the final result

$$\int_{-\infty}^{\infty} e^{-|x|}\, dx = 2$$

EXERCISES

Decide which of the following improper integrals converge, and evaluate those that do.

1. $\displaystyle\int_{1}^{\infty} \frac{1}{x^3}\, dx$

2. $\displaystyle\int_{0}^{\infty} xe^{-x}\, dx$

3. $\displaystyle\int_{0}^{\infty} \frac{e^{-x}}{e^{-x} + 1}\, dx$

4. $\displaystyle\int_{1}^{\infty} (2x + 6)^{-4}\, dx$

Hint: $\dfrac{1}{x^2 - x} = \dfrac{1}{x - 1} - \dfrac{1}{x}$

5. $\displaystyle\int_{2}^{\infty} \frac{1}{x^2 - x}\, dx$

Hint: $\dfrac{1}{x^2 - 1} = \dfrac{1}{2}\left(\dfrac{1}{x - 1} - \dfrac{1}{x - 1}\right)$

6. $\displaystyle\int_{-\infty}^{-2} \frac{1}{x^2 - 1}\, dx$

7. $\int_{-\infty}^{-1} \frac{1}{x^2} \, dx$

8. $\int_{-\infty}^{-2} \frac{1}{(x+1)^4} \, dx$

9. $\int_{-\infty}^{0} x^3 e^x \, dx$

10. $\int_{-\infty}^{4} x^{2/3} \, dx$

11. $\int_{-\infty}^{\infty} xe^{-x} \, dx$

12. $\int_{-\infty}^{\infty} \frac{x}{x^2+1} \, dx$

13. $\int_{-\infty}^{\infty} e^{-|x-3|} \, dx$

14. $\int_{-\infty}^{\infty} \frac{1}{|x|+1} \, dx$

15. Show that $\int_{0}^{\infty} x^n e^{-x} \, dx = n!$ for any positive integer n.

16. Using the fact that $\int_{0}^{\infty} e^{-x^2} \, dx = \frac{1}{2}\sqrt{\pi}$, evaluate $\int_{0}^{\infty} x^2 e^{-x^2} \, dx$.

17. Show that

$$\int_{0}^{\infty} x^{2n} e^{-x^2} \, dx = \frac{2n-1}{2} \int_{0}^{\infty} x^{2n-2} e^{-x^2} \, dx$$

for any positive integer n.

18. Using Exercises 16 and 17, show that

$$\int_{0}^{\infty} x^{2n} e^{-x^2} \, dx = \frac{1 \cdot 3 \cdot 5 \cdot \cdots \cdot (2n-1)}{2^{n+1}} \sqrt{\pi}$$

8.6 IMPROPER INTEGRALS WITH UNBOUNDED INTEGRANDS

The second class of improper integrals concerns integrands that are not bounded in the interval of integration.

Example 1

Consider the integral

$$\int_{0}^{1} \frac{1}{\sqrt{x}} \, dx \tag{1}$$

This integral is undefined, since $1/\sqrt{x}$ is undefined at 0. To get around this difficulty, consider instead the integral

$$\int_{\epsilon}^{1} \frac{1}{\sqrt{x}} \, dx \quad \text{for } \epsilon > 0$$

(see Figure 8.42). Then

$$\int_{\epsilon}^{1} \frac{1}{\sqrt{x}} \, dx = 2\sqrt{x} \Big|_{\epsilon}^{1} = 2 - 2\sqrt{\epsilon}$$

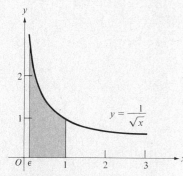

Figure 8.42
The area under $y = 1/\sqrt{x}$ from ϵ to 1.

and since $\lim_{\epsilon \to 0+}(2 - 2\sqrt{\epsilon}) = 2$, we see that also

$$\lim_{\epsilon \to 0+} \int_\epsilon^1 \frac{1}{\sqrt{x}}\, dx = 2 \tag{2}$$

Thus, we define integral (1) as the limit

$$\int_0^1 \frac{1}{\sqrt{x}}\, dx = \lim_{\epsilon \to 0+} \int_\epsilon^1 \frac{1}{\sqrt{x}}\, dx$$

and call (1) an improper integral; according to (2), we can now write

$$\int_0^1 \frac{1}{\sqrt{x}}\, dx = 2$$

DEFINITION 1
If

1. $f(x)$ is unbounded on the interval $(a, b]$,

2. $\int_{a+\epsilon}^b f(x)\, dx$ is defined for each $\epsilon > 0$ $(a + \epsilon < b)$,

then the improper integral $\int_a^b f(x)\, dx$ is defined as the limit

$$\int_a^b f(x)\, dx = \lim_{\epsilon \to 0+} \int_{a+\epsilon}^b f(x)\, dx$$

when the limit exists, and it is said to *converge*. When the limit does not exist, the improper integral is said to *diverge*.

In the example below we show that not every improper integral converges.

Example 2
Show that the improper integral $\int_0^2 \frac{1}{x}\, dx$ diverges.

Solution
We have

$$\int_\epsilon^2 \frac{1}{x}\, dx = \ln 2 - \ln \epsilon$$

and since $\lim_{\epsilon \to 0+}(\ln 2 - \ln \epsilon) = \infty$, we conclude that the integral diverges.

Improper integrals can also, of course, be defined for integrands that are unbounded at the right end point of the interval of integration. Such a situation is considered in the following example.

Example 3

Evaluate the improper integral

$$\int_0^1 \frac{1}{(x-1)^{2/3}}\,dx$$

Solution
Since the integrand this time is undefined at $x = 1$ (see Figure 8.43), we consider the integral

$$\int_0^{1-\epsilon} \frac{1}{(x-1)^{2/3}}\,dx$$

and put

$$\int_0^1 \frac{1}{(x-1)^{2/3}}\,dx = \lim_{\epsilon \to 0+} \int_0^{1-\epsilon} \frac{1}{(x-1)^{2/3}}\,dx$$

Now,

$$\int_0^{1-\epsilon} \frac{1}{(x-1)^{2/3}}\,dx = 3(x-1)^{1/3}\Big|_0^{1-\epsilon} = 3(-\epsilon)^{1/3} - 3(-1)^{1/3}$$
$$= 3(-\epsilon)^{1/3} + 3$$

and since $\lim_{\epsilon \to 0+} [3(-\epsilon)^{1/3} + 3] = 3$, we have

$$\int_0^1 \frac{1}{(x-1)^{2/3}}\,dx = 3$$

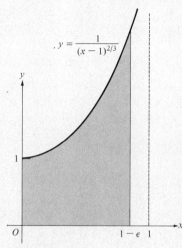

Figure 8.43
The area under $y = 1/(x-1)^{2/3}$ from 0 to $1 - \epsilon$.

Example 4

Evaluate the improper integral

$$\int_0^4 \frac{1}{(x-2)^{2/3}}\,dx$$

Solution
This situation differs from those considered above in that the integrand is unbounded *inside* the interval of integration rather than at an end point (see Figure 8.44). This integral is evaluated by writing

$$\int_0^4 \frac{1}{(x-2)^{2/3}}\,dx = \int_0^2 \frac{1}{(x-2)^{2/3}}\,dx + \int_2^4 \frac{1}{(x-2)^{2/3}}\,dx$$

and evaluating the two improper integrals on the right side. With the procedure employed in the preceding examples, we find that

$$\int_0^2 \frac{1}{(x-2)^{2/3}}\,dx = \lim_{\epsilon \to 0+} \int_0^{2-\epsilon} \frac{1}{(x-2)^{2/3}}\,dx = \lim_{\epsilon \to 0+} 3(x-2)^{1/3}\Big|_0^{2-\epsilon}$$
$$= \lim_{\epsilon \to 0+} [3(-\epsilon)^{1/3} - 3(-2)^{1/3}] = 3 \cdot 2^{1/3}$$

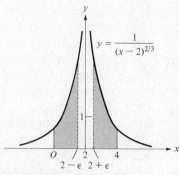

Figure 8.44
The area under $y = 1/(x-2)^{2/3}$ from 0 to $2 - \epsilon$ and from $2 + \epsilon$ to 4.

$$\int_2^4 \frac{1}{(x-2)^{2/3}}\,dx = \lim_{\epsilon \to 0+} \int_{2+\epsilon}^4 \frac{1}{(x-2)^{2/3}}\,dx = \lim_{\epsilon \to 0+} 3(x-2)^{1/3} \Big|_{2+\epsilon}^4$$

$$= \lim_{\epsilon \to 0+} [3 \cdot 2^{1/3} - 3\epsilon^{1/3}] = 3 \cdot 2^{1/3}$$

Hence

$$\int_0^3 \frac{1}{(x-2)^{2/3}}\,dx = 3 \cdot 2^{1/3} + 3 \cdot 2^{1/3} = 6 \cdot 2^{1/3}$$

EXERCISES

Decide which of the following improper integrals converge and evaluate those that do.

1. $\displaystyle\int_0^1 x^{-1/3}\,dx$
2. $\displaystyle\int_0^1 x^{-4/3}\,dx$
3. $\displaystyle\int_0^1 \frac{1}{x^2 - 1}\,dx$
4. $\displaystyle\int_0^2 \frac{1}{(x-1)^{1/5}}\,dx$
5. $\displaystyle\int_{-1}^1 |x|^{-3/4}\,dx$
6. $\displaystyle\int_0^1 \frac{1}{x-1}\,dx$
7. $\displaystyle\int_{-1/2}^4 \frac{1}{\sqrt{2x+1}}\,dx$
8. $\displaystyle\int_0^{10} \frac{x}{|x|}\,dx$
9. $\displaystyle\int_{-10}^{10} \frac{x}{|x|}\,dx$
10. $\displaystyle\int_0^1 \frac{e^x}{e^x - 1}\,dx$
11. $\displaystyle\int_{-1}^1 \frac{|x|}{\sqrt{1-x^2}}\,dx$
12. $\displaystyle\int_{-1}^1 \frac{x}{\sqrt{1-x^2}}\,dx$

8.7 NUMERICAL INTEGRATION

The methods we developed for evaluating definite integrals are based on the fundamental theorem of calculus. As such, they depend on our ability to find indefinite integrals that can be used in numerical calculations. We have remarked in Section 6.6 that this is not always possible. It can be proved with considerable difficulty, for example, that the function $f(x) = Ae^{-ax^2}$ has no usable indefinite integral when A and a are nonzero constants. Being a little more precise, this means that $\int f(x)\,dx$ cannot be expressed in terms of the functions ordinarily encountered in calculus, such as polynomials, rational functions, algebraic functions, trigonometric functions, and the like. The particular definite integrals

$$\int_a^b \frac{1}{\sqrt{2\pi}} e^{-x^2/2}\,dx$$

are of importance in engineering, physics, and probability.

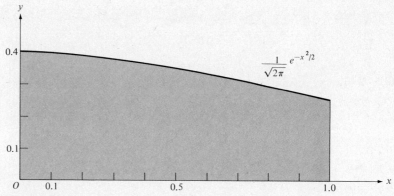
Figure 8.45

How are we, then, to find values of a definite integral when no indefinite integral is available? In general, we have to be content with approximate values, and methods for finding these are derived from the definition of the definite integral. The three methods we discuss are presented in increasing order of sophistication and degree of accuracy. For comparison purposes, all three methods will be applied to the evaluation of the definite integral (see Figure 8.45).

$$I = \int_0^1 \frac{1}{\sqrt{2\pi}} e^{-x^2/2} \, dx \tag{1}$$

The value of I to six significant figures is known from other sources to be

$$I \approx 0.341345 \tag{2}$$

We shall use this value to check the degree of accuracy of our methods.

METHOD I. APPROXIMATION WITH RECTANGLES
Recall from Section 6.1 the definition of definite integrals:

$$\int_a^b f(x) \, dx = \lim_{n \to \infty} \sum_{k=1}^n f(x_k) \, \Delta x$$

where $\Delta x = (b - a)/n$ and $x_k = a + k \Delta x$. When n is appropriately chosen, the sum on the right side will give a good approximation to the integral, a fact expressed by writing

$$\int_a^b f(x) \, dx \approx \sum_{k=1}^n f(x_k) \, \Delta x$$

Using the convenient notation

$$y_k = f(x_k)$$

we can present our first method as follows (see Figure 8.46):

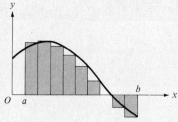

Figure 8.46
The interpretation of $\Delta x(y_1 + y_2 + \cdots + y_n)$ as the sum of signed areas (Section 7.1).

APPROXIMATION WITH RECTANGLES

$$\int_a^b f(x)\, dx \approx \Delta x(y_1 + y_2 + \cdots + y_n)$$

Using $n = 10$, let us apply this formula to the approximate evaluation of the integral in (1). We thus have the following data:

$$f(x) = \frac{1}{\sqrt{2\pi}} e^{-x^2/2}$$

$$a = 0 \qquad b = 1$$

$$\Delta x = \frac{b - a}{n} = \frac{1 - 0}{10} = \frac{1}{10} = 0.1$$

$$x_k = a + k\,\Delta x = 0 + 0.1 \cdot k = 0.1 \cdot k$$

$$y_k = f(x_k) = \frac{1}{\sqrt{2\pi}} e^{-(0.1 \cdot k)^2/2}$$

Table 8.2

x_k	$f(x_k) = y_k$
0.1	0.39695
0.2	0.39104
0.3	0.38139
0.4	0.36827
0.5	0.35207
0.6	0.33322
0.7	0.31225
0.8	0.28969
0.9	0.26609
1.0	0.24197
	3.33294

The values y_k are correct to five significant figures.

From Table 8.2 we find that

$$\Delta x(y_1 + y_2 + \cdots + y_{10}) = 0.1 \cdot 3.33294 = 0.333294$$

Comparing this figure with our reference value in (2) shows that the error this method introduces is

$$0.341345 - 0.333294 = 0.008051 \approx 8 \cdot 10^{-3}$$

METHOD II. THE TRAPEZOIDAL RULE

A significant improvement in the approximation is obtained when rectangles are replaced by trapezoids. This is clear from an inspection of Figure 8.47. When the definite integral from a to b of $f(x)$ is approximated with n trapezoids, then each trapezoid has base $\Delta x = (b - a)/n$ and the kth trapezoid has signed area

$$\Delta x \frac{y_{k-1} + y_k}{2}$$

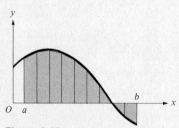

Figure 8.47
Approximations with trapezoids.

(see Figure 8.48). The sum of the n trapezoids is

$$\Delta x \frac{y_0 + y_1}{2} + \Delta x \frac{y_1 + y_2}{2} + \cdots + \Delta x \frac{y_{n-2} + y_{n-1}}{2} + \Delta x \frac{y_{n-1} + y_n}{2}$$

$$= \Delta x \left(\frac{y_0 + y_1}{2} + \frac{y_1 + y_2}{2} + \cdots + \frac{y_{n-2} + y_{n-1}}{2} + \frac{y_{n-1} + y_n}{2} \right)$$

$$= \Delta x \left(\frac{1}{2} y_0 + y_1 + y_2 + \cdots + y_{n-1} + \frac{1}{2} y_n \right)$$

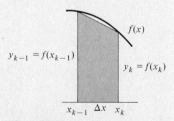

Figure 8.48
The area of the trapezoid is $\Delta x(y_{k-1} + y_k)/2$.

This brings us to our second method.

> **TRAPEZOIDAL RULE**
>
> $$\int_a^b f(x)\, dx \approx \Delta x(\tfrac{1}{2}y_0 + y_1 + y_2 + \cdots + y_{n-1} + \tfrac{1}{2}y_n)$$

Observe that

This formula gives a relation between Method I and Method II.

$$\Delta x\left(\frac{1}{2}y_0 + y_1 + y_2 + \cdots + y_{n-1} + \frac{1}{2}y_n\right)$$
$$= \Delta x(y_1 + y_2 + \cdots + y_{n-1} + y_n) + \Delta x \frac{y_0 - y_n}{2}$$

In our example $x_0 = 0$ and $y_0 = f(x_0) = f(0) = (1/\sqrt{2\pi})e^0 = 1/\sqrt{2\pi}$ and this is found to be 0.39894 to five significant figures. Thus,

$$\Delta x \frac{y_0 - y_{10}}{2} = 0.1\frac{0.39894 - 0.24197}{2} = 0.1\frac{0.15697}{2} = 0.0078485$$

Thus,

$$\Delta x(\tfrac{1}{2}y_0 + y_1 + \cdots + y_{n-1} + \tfrac{1}{2}y_n) = 0.333294 + 0.0078485$$
$$= 0.3411425$$

Compared with (2), the error introduced is

$$0.341345 - 0.3411425 = 0.0002025 = 2.025 \cdot 10^{-4} \approx 2 \cdot 10^{-4}$$

Thus, the error has decreased from $8 \cdot 10^{-3}$ to $2 \cdot 10^{-4}$.

Example 1

Use Method II with $n = 10$ to approximate $\int_0^1 x^2\, dx$.

Solution
We have the following data:

$$f(x) = x^2$$
$$a = 0 \quad b = 1$$
$$\Delta x = \frac{b - a}{n} = \frac{1 - 0}{10} = \frac{1}{10}$$
$$x_k = a + k\,\Delta x = 0 + k \cdot \frac{1}{10} = \frac{k}{10}$$
$$y_k = f(x_k) = x_k^2 = \left(\frac{k}{10}\right)^2$$

To use formula (2), we compute the following:

$$\tfrac{1}{2}y_0 + y_1 + y_2 + \cdots + y_9 + \tfrac{1}{2}y_{10}$$
$$= \tfrac{1}{2} \cdot 0 + (\tfrac{1}{10})^2 + (\tfrac{2}{10})^2 + \cdots + (\tfrac{9}{10})^2 + \tfrac{1}{2}(\tfrac{10}{10})^2$$
$$= (\tfrac{1}{10})^2(1 + 2^2 + \cdots + 9^2) + \tfrac{1}{2}$$

According to the appendix to Chapter 6,

$$1 + 2^2 + \cdots + 9^2 = \frac{9 \cdot 10 \cdot 19}{6} = 285$$

and hence

$$\Delta x \left(\frac{1}{2} y_0 + y_1 + y_2 + \cdots + y_9 + \frac{1}{2} y_{10} \right)$$
$$= \frac{1}{10}\left(\frac{1}{10^2} \cdot 285 + \frac{1}{2}\right) = \frac{1}{10}(2.85 + 0.5) = 0.335$$

Thus,

$$\int_0^1 x^2 \, dx \approx 0.335$$

Since

$$\int_0^1 x^2 \, dx = \tfrac{1}{3}x^3 \Big|_0^1 = \tfrac{1}{3} = 0.333 \ldots$$

We see that the error is

$$0.335 - 0.333 \ldots < 0.002 = 2 \cdot 10^{-3}$$

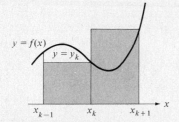

Figure 8.49
Approximation with rectangles.

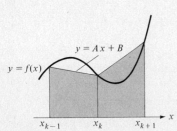

Figure 8.50
Approximation with trapezoids.

A *parabola* is any curve given by a second degree equation.

METHOD III. SIMPSON'S RULE

Thus far our methods for approximating definite integrals numerically have seemed like logical consequences of the definition of definite integrals. To provide an explanation for what we are about to do, let us briefly review what we have done.

In our first method we replaced the curve $y = f(x)$ over each subinterval $[x_{k-1}, x_k]$ by the constant function $y = y_k$ (Figure 8.49). In the second method we obtained a better result by replacing the curve over each subinterval $[x_{k-1}, x_k]$ by the line segment

$$y = Ax + B$$

where A and B were chosen so that the line segment joined the points (x_{k-1}, y_{k-1}) and (x_k, y_k) (Figure 8.50). It thus seems reasonable to try our luck by replacing the curve over subintervals by an appropriately chosen parabola

$$y = Ax^2 + Bx + C$$

Since now there are three arbitrary coefficients, the parabola may be chosen to pass through three preassigned points on the curve $y = f(x)$. This suggests that the interval $[a, b]$ should be divided into an *even* number of subintervals $[x_{k-1}, x_k]$, $k = 1, 2, \ldots, n$. We shall show at the end of this section that if A, B, and C are chosen so that the parabola $y = Ax^2 + Bx + C$ will pass through the points

$$(x_{k-1}, y_{k-1}), \qquad (x_k, y_k), \qquad (x_{k+1}, y_{k+1})$$

then

$$\int_{x_{k-1}}^{x_{k+1}} (Ax^2 + Bx + C)\, dx = \frac{\Delta x}{3}(y_{k-1} + 4y_k + y_{k+1}) \qquad (3)$$

where, as before, $y_k = f(x_k)$ and $\Delta x = (b - a)/n$. We thus have the formula (see Figure 8.51)

$$\int_{x_{k-1}}^{x_{k+1}} f(x)\, dx \approx \frac{\Delta x}{3}(y_{k-1} + 4y_k + y_{k+1}) \qquad (4)$$

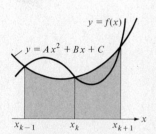

Figure 8.51
Because of formula (3), the shaded area is given by the right side of formula (4).

This is the first version of *Simpson's formula*.

Example 2

Use formula (4) to approximate $\int_0^1 x^4\, dx$, dividing the interval $[0, 1]$ into two parts.

Solution
We have the following data:

$$f(x) = x^4$$
$$a = 0 \qquad b = 1$$
$$n = 2$$
$$\Delta x = \frac{b - a}{2} = \frac{1 - 0}{2} = \frac{1}{2}$$

Taking $k = 1$,

$$x_{k-1} = x_0 = 0 \qquad x_k = x_1 = \tfrac{1}{2} \qquad x_{k+1} = x_2 = \tfrac{2}{2} = 1$$
$$y_{k-1} = x_0^4 = 0^4 = 0 \qquad y_k = x_1^4 = (\tfrac{1}{2})^4 = \tfrac{1}{16} \qquad y_{k+1} = x_2^4 = 1^4 = 1$$

Hence,

$$\frac{\Delta x}{3}(y_{k-1} + 4y_k + y_{k+1}) = \frac{\Delta x}{3}(y_0 + 4y_1 + y_2)$$
$$= \tfrac{1}{2} \cdot \tfrac{1}{3}(0 + 4 \cdot \tfrac{1}{16} + 1)$$
$$= \tfrac{1}{6}(\tfrac{1}{4} + 1) = \tfrac{5}{24}$$

Since

$$\int_0^1 x^4 \, dx = \tfrac{1}{5} x^5 \Big|_0^1 = \tfrac{1}{5}$$

we see that the error this method introduces here is

$$\frac{5}{24} - \frac{1}{5} = \frac{25 - 24}{120} = \frac{1}{120} = 0.00833 \ldots \approx 0.008 = 8 \cdot 10^{-3}$$

To obtain Simpson's formula for $\int_a^b f(x) \, dx$, divide $[a, b]$ into an even number of subintervals $[x_{k-1}, x_k]$, and apply formula (4) to successive pairs of intervals. This results in the following:

$$\int_a^b f(x) \, dx = \int_{x_0}^{x_2} f(x) \, dx + \int_{x_2}^{x_4} f(x) \, dx + \cdots + \int_{x_{n-2}}^{x_n} f(x) \, dx$$

$$\approx \frac{\Delta x}{3}(y_0 + 4y_1 + y_2) + \frac{\Delta x}{3}(y_2 + 4y_3 + y_4) + \cdots$$

$$+ \frac{\Delta x}{3}(y_{n-2} + 4y_{n-1} + y_n)$$

$$= \frac{\Delta x}{3}(y_0 + 4y_1 + y_2 + y_2 + 4y_3 + y_4 + \cdots$$

$$+ y_{n-2} + 4y_{n-1} + y_n)$$

$$= \frac{\Delta x}{3}[y_0 + y_n + 4(y_1 + y_3 + \cdots + y_{n-1})$$

$$+ 2(y_2 + y_4 + \cdots + y_{n-2})]$$

SIMPSON'S RULE

$$\int_a^b f(x) \, dx \approx \frac{\Delta x}{3}[y_0 + y_n + 4(y_1 + y_3 + \cdots + y_{n-1})$$

$$+ 2(y_2 + y_4 + \cdots + y_{n-2})]$$

where n is an *even* integer.

When this formula is applied to

$$\int_0^1 \frac{1}{\sqrt{2\pi}} e^{-x^2/2} \, dx$$

with $n = 10$, it gives the following:

$$y_0 + y_{10} = 0.64091$$
$$y_1 + y_3 + \cdots + y_9 = 1.70875$$
$$y_2 + y_4 + \cdots + y_8 = 1.38222$$

so that the right side in Simpson's rule is

$$\tfrac{1}{3} \cdot \tfrac{1}{10} \cdot (0.64091 + 4 \cdot 1.70875 + 2 \cdot 1.38222)$$
$$= \tfrac{1}{30}(0.64091 + 6.83500 + 2.76444)$$
$$= \tfrac{1}{30} \cdot 10.24035 = 0.341345$$

Hence

$$\int_0^1 \frac{1}{\sqrt{2\pi}} e^{-x^2/2} \, dx \approx 0.341345$$

This agrees with our reference number in (2). Simpson's rule has thus given the correct answer to six significant figures, and the error is therefore less than $5 \cdot 10^{-7}$. This compares favorably with the errors resulting from the other two methods, which were $8 \cdot 10^{-3}$ and $2 \cdot 10^{-4}$, respectively.

DERIVATION OF FORMULA (4)
If the parabola

$$y = Ax^2 + Bx + C$$

passes through the points

$$(x_{k-1}, y_{k-1}), \quad (x_k, y_k), \quad (x_{k+1}, y_{k+1})$$

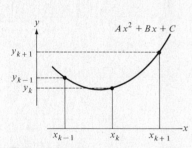

Figure 8.52

(see Figure 8.52), then the following three equations hold:

$$y_{k-1} = Ax_{k-1}^2 + Bx_{k-1} + C$$
$$y_k \;\;\, = Ax_k^2 \;\;\, + Bx_k \;\;\, + C$$
$$y_{k+1} = Ax_{k+1}^2 + Bx_{k+1} + C$$

Let us use the abbreviated notation

$$I = \int_{x_{k-1}}^{x_{k+1}} (Ax^2 + Bx + C) \, dx$$

Then

$$I = (\tfrac{1}{3}Ax^3 + \tfrac{1}{2}Bx^2 + Cx)\Big|_{x_{k-1}}^{x_{k+1}}$$
$$= (\tfrac{1}{3}Ax_{k+1}^3 + \tfrac{1}{2}Bx_{k+1}^2 + Cx_{k+1}) - (\tfrac{1}{3}Ax_{k-1}^3 + \tfrac{1}{2}Bx_{k-1}^2 + Cx_{k-1})$$
$$= \tfrac{1}{3}A(x_{k+1}^3 - x_{k-1}^3) + \tfrac{1}{2}B(x_{k+1}^2 - x_{k-1}^2) + C(x_{k+1} - x_{k-1})$$

We now use the identities

$$x_{k+1}^3 - x_{k-1}^3 = (x_{k+1} - x_{k-1})(x_{k+1}^2 + x_{k-1}x_{k+1} + x_{k-1}^2)$$
$$x_{k+1}^2 - x_{k-1}^2 = (x_{k+1} - x_{k-1})(x_{k+1} + x_{k-1})$$
$$x_{k+1} - x_{k-1} = 2\,\Delta x$$

With these, we see that

$$I = \frac{1}{3}\Delta x[2A(x_{k+1}^2 + x_{k-1}x_{k+1} + x_{k-1}^2) + 3B(x_{k+1} + x_{k-1}) + 6C]$$

$$= \frac{\Delta x}{3}[(Ax_{k+1}^2 + Bx_{k+1} + C) + (Ax_{k-1}^2 + Bx_{k-1} + C)$$

$$+ A(x_{k+1}^2 + 2x_{k-1}x_{k+1} + x_{k-1}^2) + 2B(x_{k+1} + x_{k-1}) + 4C]$$

$$= \frac{\Delta x}{3}[y_{k+1} + y_{k-1} + A(x_{k+1} + x_{k-1})^2 + 2B(x_{k+1} + x_{k-1}) + 4C]$$

Since

$$x_{k+1} + x_{k-1} = (x_{k-1} + 2\Delta x) + x_{k-1} = 2(x_{k-1} + \Delta x) = 2x_k$$

we have

$$A(x_{k+1} + x_{k-1})^2 + 2B(x_{k+1} + x_{k-1}) + 4C = A(2x_k)^2 + 2B(2x_k) + 4C$$
$$= 4(Ax_k^2 + Bx_k + C) = 4y_k$$

This gives

$$I = \frac{\Delta x}{3}(y_{k+1} + y_{k-1} + 4y_k) = \frac{\Delta x}{3}(y_{k-1} + 4y_k + y_{k+1})$$

as was to be shown.

EXERCISES

Use Method I (approximation with rectangles) with $n = 4$ to approximate the integrals in Exercises 1–3.

1. $\int_0^4 2\, dx$

2. $\int_0^1 (3x + 1)\, dx$

3. $\int_1^5 e^x\, dx$

Use Method II (trapezoidal rule) with $n = 4$ to approximate the integrals in Exercises 4–7.

4. $\int_0^1 (ax + b)\, dx$ 5. $\int_1^5 e^x\, dx$

6. $\int_1^2 \ln x\, dx$ 7. $\int_0^1 \frac{1}{1 + x^2}\, dx$

Use Method III (Simpson's rule) with $n = 4$ to approximate the integrals in Exercises 8–11.

8. $\int_a^b (\alpha x^2 + \beta x + \gamma)\, dx$ 9. $\int_1^5 e^x\, dx$

10. $\int_0^1 \dfrac{1}{1+x^2}\, dx$ 11. $\int_0^4 \sqrt{1+x^2}\, dx$

12. Use the trapezoidal rule to approximate the area between the curves $y = e^x$ and $y = e^{2x}$ for $0 \le x \le 1$ and $n = 5$.

QUIZ 1

1. The size of a bacteria population is given by a function $P(t)$ of time, where t is measured in days. It is known that

$$P'(t) = 3 \cdot 10^4 t^2 - 20 t^4$$
$$P(0) = 10^6$$

   (a) Find the function $P(t)$.

   (b) After how many days will the population reach its maximum size?

   (c) After how many days will the population reduce to its original size at time $t = 0$?

2. A steel manufacturer produces q tons of steel per week with a marginal revenue of $7/[2(q+2)^2]$ dollars/ton. What is the price as a function of q when revenue is fixed at 0 when $q = 0$?

3. Find the volumes of the solids generated when the areas given below are revolved about the x axis.

Solid	Generating area	Generating area bounded by
	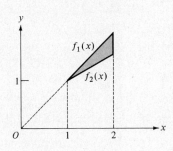	$f_1(x) = x$ $f_2(x) = \tfrac{1}{2}x + \tfrac{1}{2}$ $x = 2$

Solid	Generating area	Generating area bounded by
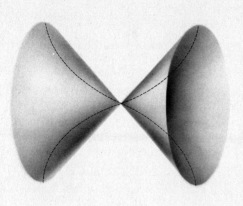	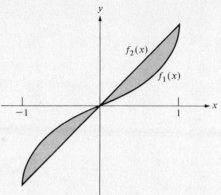	$f_1(x) = x^3$ $f_2(x) = x$

4. A horn, ten feet long, has elliptical cross sections of area $A(x) = e^{0.01x}$, where x is the distance from one end. Find the volume of the horn.

5. Evaluate the following improper integrals.

   (a) $\int_1^\infty \dfrac{\ln x}{x^3}\, dx$ (b) $\int_{-\infty}^\infty xe^{-4x}\, dx$

   (c) $\int_1^2 \dfrac{1}{x\sqrt{\ln x}}\, dx$ (d) $\int_{-4/3}^2 \dfrac{1}{\sqrt{3x+4}}\, dx$

6. Use the trapezoidal rule (Method II) with $n = 4$ to approximate $\int_0^4 \sqrt{1+x^2}\, dx$.

QUIZ 2

1. A missile is falling from a height of 20,000 ft with velocity $v(t) = 32(e^{-0.25t} - 1)$ ft/sec.

   (a) What is the missile's acceleration at time $t = 100$ sec?

   (b) When will the missile hit the ground?

2. A speed boat is traveling at a velocity of 125 mph when the motor is shut off. At t sec later, the boat's velocity has reduced to $v = 125 - 15t^{3/2}$ mph.

   (a) What is the boat's deceleration 8 sec after the motor was shut off?

   (b) How far will the boat travel before coming to a halt?

3. Find the volumes of the solids generated when the areas given below are revolved about the x axis.

Solid	Generating area	Generating area bounded by
		$f_1(x) = \sqrt{x}$ $f_2(x) = \ln x$ $x = 3$
		$f_1(x) = e^x$ $f_2(x) = e^{2-x}$ $f_3(x) = (1-x)^2$

4. Find the volume of the tetrahedron with vertices $(1, 0, 0)$, $(0, 2, 0)$, $(0, 0, 3)$, and $(0, 0, 0)$.

5. Evaluate the following improper integrals.

(a) $\int_{-\infty}^{0} \dfrac{1}{(2x-1)^{4/3}} \, dx$

(b) $\int_{-\infty}^{\infty} x^3 e^{-5x} \, dx$

(c) $\int_{-5}^{5} \dfrac{x+1}{|x+1|} \, dx$

(d) $\int_{0}^{1} \dfrac{1}{x^3} e^{-1/x} \, dx$

6. Use the trapezoidal rule (Method II) with $n = 4$ to approximate $\int_0^1 \ln(1+x^2) \, dx$.

7. Use Simpson's rule (Method III) with $n = 4$ to approximate $\int_0^1 \ln(1+x^2) \, dx$.

CHAPTER 9

TRiGONOMETRiC FUNCTiONS

9.1 SINE AND COSINE

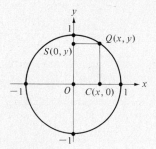

Figure 9.1
The unit circle $x^2 + y^2 = 1$.

Figure 9.2

As a point $Q(x, y)$ moves around the unit circle $x^2 + y^2 = 1$ (see Figure 9.1), its projection (shadow) $C(x, 0)$ on the x axis moves back and forth in the interval $-1 \leq x \leq 1$, and its projection $S(0, y)$ on the y axis moves back and forth in the interval $-1 \leq y \leq 1$. Below we describe the cosine and sine functions that relate $Q(x, y)$ to its projections.

Let $\theta > 0$ be any number. Then $P(\theta)$ will stand for the point on the unit circle $x^2 + y^2 = 1$ that is reached by tracing an arc of length θ on the circle from $(1, 0)$ in a counterclockwise direction (see Figure 9.2). Intuitively you can imagine that a line of length θ is attached at one end to the point $(1, 0)$ and wrapped around the circle in a counterclockwise direction. $P(-\theta)$ will stand for the point on

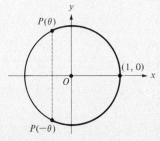

Figure 9.3

the circle that is reached by tracing an arc of length θ from $(1, 0)$ in a clockwise direction (see Figure 9.3).

Thus, if we put $P(0) = (1, 0)$, then we conclude that:

> Every number θ, positive, zero, or negative, determines a unique point $P(\theta)$ on the circle $x^2 + y^2 = 1$.

Since the circle $x^2 + y^2 = 1$ has circumference of length 2π, it follows that θ, $\theta + 2\pi$, and $\theta - 2\pi$ determine the same point. That is,

$$P(\theta \pm 2\pi) = P(\theta) \quad \text{for all numbers } \theta \tag{1}$$

Example 1

Plot the points $P(k\pi/2)$ for $k = \pm 1, \pm 2, \pm 3, \pm 4$.

Solution
Owing to formula (1),

$$P\left(\frac{\pi}{2}\right) = P\left(\frac{\pi}{2} - 2\pi\right) = P\left(-\frac{3\pi}{2}\right)$$

$$P\left(-\frac{\pi}{2}\right) = P\left(-\frac{\pi}{2} + 2\pi\right) = P\left(\frac{3\pi}{2}\right)$$

$$P(\pi) = P(\pi - 2\pi) = P(-\pi)$$

$$P(2\pi) = P(-2\pi) = P(0)$$

These four points are plotted in Figure 9.4.

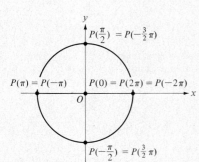

Figure 9.4

Now, a number θ determines the unique point $P(\theta)$ on the unit circle and this point determines coordinates (x, y) such that $P(\theta) = (x, y)$. The function that converts θ into x is called the *cosine* function, written

$$x = \cos \theta$$

The function that converts θ into y is called the *sine* function, written

$$y = \sin \theta$$

Observe that $\cos \theta$ is the x coordinate of $P(\theta)$, and $\sin \theta$ is the y coordinate of $P(\theta)$ (see Figure 9.5).

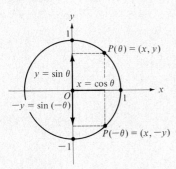

Figure 9.5

The sine and cosine functions are noted to have the following properties:

1. $-1 \leq \sin \theta \leq 1$
2. $-1 \leq \cos \theta \leq 1$

288 TRIGONOMETRIC FUNCTIONS

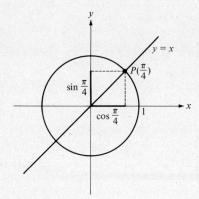

Figure 9.6

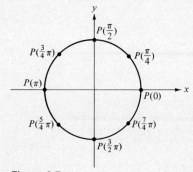

Figure 9.7

θ	$\sin \theta$	$\cos \theta$
0	0	1
$\dfrac{\pi}{4}$	$\dfrac{\sqrt{2}}{2}$	$\dfrac{\sqrt{2}}{2}$
$\dfrac{\pi}{2}$	1	0
$\dfrac{3\pi}{4}$	$\dfrac{\sqrt{2}}{2}$	$-\dfrac{\sqrt{2}}{2}$
π	0	-1
$\dfrac{5\pi}{4}$	$-\dfrac{\sqrt{2}}{2}$	$-\dfrac{\sqrt{2}}{2}$
$\dfrac{3\pi}{2}$	-1	0
$\dfrac{7\pi}{4}$	$-\dfrac{\sqrt{2}}{2}$	$\dfrac{\sqrt{2}}{2}$

3. $\sin^2 \theta + \cos^2 \theta = 1$
4. $\sin(-\theta) = -\sin \theta$
5. $\cos(-\theta) = \cos \theta$
6. $\sin(\theta \pm 2\pi) = \sin \theta$
7. $\cos(\theta \pm 2\pi) = \cos \theta$

It is customary to write $\sin^2 \theta$ (read "sine square theta") instead of $(\sin \theta)^2$, and $\cos^2 \theta$ (read "cosine square theta") instead of $(\cos \theta)^2$.

Property (4) tells us that the sine function is *odd,* and Property (5) tells us that the cosine function is *even* (see Figure 9.5). Properties (6) and (7) follow from the fact that

$$P(\theta \pm 2\pi) = P(\theta)$$

Example 2

Find the values of $\sin \pi/4$ and $\cos \pi/4$.

Solution

Consult Figure 9.6. Since $P(\pi/4)$ is situated halfway between the points $P(0)$ and $P(\pi/2)$, it follows that it lies on the line $y = x$. This tells us that

$$\sin \frac{\pi}{4} = \cos \frac{\pi}{4}$$

Let the common value be a. Using the relation $(\sin \pi/4)^2 + (\cos \pi/4)^2 = 1$, we get $2a^2 = 1$, or $a^2 = \frac{1}{2}$. Since $a > 0$, we get $a = 1/\sqrt{2} = \sqrt{2}/2 \approx 0.707$. Thus, $\sin \pi/4 = \cos \pi/4 = \sqrt{2}/2$.

Let us tabulate the values of $\sin k\pi/4$ and $\cos k\pi/4$ for $k = 0, 1, 2, \ldots, 7$. These are read off in Figure 9.7 with the help of the fact $P(\theta) = (x, y) = (\cos \theta, \sin \theta)$.

THE GRAPHS OF SINE AND COSINE

In Figures 9.8 and 9.9, the curves for $\sin \theta$ and $\cos \theta$ have been plotted with the help of the table with Figure 9.7 and the table of trigonometric functions at the end of the book. With the relations $\sin(\theta \pm 2\pi) = \sin \theta$ and $\cos(\theta \pm 2\pi) = \cos \theta$, which are true for all θ, we can graph $\sin \theta$ and $\cos \theta$ on any interval using translations. The result is pictured in Figures 9.10 and 9.11. Observe that the curves $y = \sin \theta$ and $y = \cos \theta$ look as though they can be obtained from each other through a translation. This is a correct observation, which is a consequence of the relation

$$\sin \theta = \cos\left(\frac{\pi}{2} - \theta\right)$$

This relation will be justified in Exercise 20 at the end of this section. Observe further that $\sin \theta$ as well as $\cos \theta$ look the same on

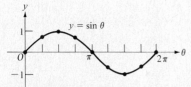

Figure 9.8
The graph of $y = \sin \theta$.

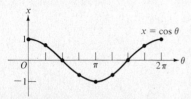

Figure 9.9
The graph of $x = \cos \theta$. The points on the graphs correspond to the values tabulated above.

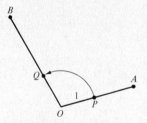

Figure 9.12
The angle from $\overline{OA}$ to $\overline{OB}$.

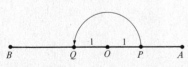

Figure 9.13
The straight angle from $\overline{OA}$ to $\overline{OB}$.

Figure 9.10
The graph of $y = \sin \theta$.

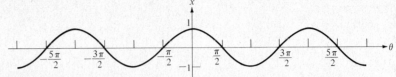

Figure 9.11
The graph of $x = \cos \theta$.

each interval of length 2π. These functions are examples of periodic functions. In general,

> $f(\theta)$ is *periodic with period* p if $f(\theta + p) = f(\theta)$ for all θ and if p is the smallest positive number for which this is true.

ANGLES, DEGREES, AND RADIANS

The functions $\sin \theta$ and $\cos \theta$ have been defined for numbers θ, where $|\theta|$ denotes the length of a circular arc of radius 1. This circular arc is conveniently used to measure angles. To measure the angle from $\overline{OA}$ to $\overline{OB}$, we draw a circular arc with center at O and radius 1 from the point P on $\overline{OA}$ in a *counterclockwise* direction to the point Q on $\overline{OB}$ (see Figure 9.12). The length of this arc is called the *radian measure* of the angle. Thus,

> measure of angle from $\overline{OA}$ to $\overline{OB}$ in radians = length of arc PQ

The circumference of a unit circle is 2π. If the angle from $\overline{OA}$ to $\overline{OB}$ is a straight angle (Figure 9.13), then the angle has a measure of π radians, the arc from P to Q being one-half of a unit circle. We know that a straight angle has 180°. Hence

$$\pi \text{ radians} = 180°$$

The following rules for conversion between radians and degrees hold:

> $$1 \text{ radian} = \left(\frac{180}{\pi}\right)^\circ \approx 57.296°$$
>
> $$1° = \left(\frac{\pi}{180}\right) \text{ radians} \approx 0.0174 \text{ radians}$$
>
> $$\pi \approx 3.14159$$

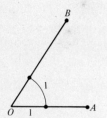

Figure 9.14
An angle of measure 1 radian.

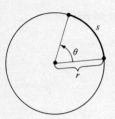

Figure 9.15

In Figure 9.14, we have an angle whose measure is 1 radian.

> 1 *radian* is the measure of a central angle from OA to OB corresponding to a circular arc of radius 1 and length 1.

Degrees are often used in applications, whereas radians are used in theoretical discussions. Reference to radians is usually omitted.

Example 3

In a circle of radius r, consider a sector with central angle of θ radians (see Figure 9.15). What is the length s of the arc subtended in terms of r and θ?

Solution
The following statement of proportionality is intuitively clear:

$$\frac{\text{length of subtended arc}}{\text{circumference}} = \frac{\theta \text{ radians}}{2\pi \text{ radians}}$$

Since the circumference of the circle is $2\pi r$, we get

$$\frac{s}{2\pi r} = \frac{\theta}{2\pi}$$

or

$$s = r\theta$$

Example 4

What is the area A of the sector in Example 3?

Solution
Again we use an intuitive proportionality:

$$\frac{\text{area of sector}}{\text{area of circle}} = \frac{\theta \text{ radians}}{2\pi \text{ radians}}$$

Remembering that the area of the circle is πr^2, this gives

$$\frac{A}{\pi r^2} = \frac{\theta}{2\pi}$$

and thus

$$A = \tfrac{1}{2} r^2 \theta$$

AN ALTERNATIVE DEFINITION OF SINE AND COSINE

In elementary trigonometry, the sine and cosine were defined by looking at triangles. Let us see how this compares with our approach. As the point (x, y) varies over a circle $x^2 + y^2 = r^2$ for some $r > 0$, then the point $(x/r, y/r)$ lies on the unit circle since $(x/r)^2 + (y/r)^2 = 1$, so that

$$\frac{x}{r} = \cos \theta$$

$$\frac{y}{r} = \sin \theta$$

where θ is the length of the circular arc joining $(1, 0)$ to $(x/r, y/r)$ (see Figure 9.16); or, what is the same thing, θ is the radian measure of the angle from the positive x axis to the line joining $(0, 0)$ to (x, y). If the point (x, y) is in the first quadrant, then we have the familiar triangle pictured in Figure 9.17.

We terminate this section with some important trigonometric identities. The most important ones are the so-called addition formulas for trigonometric functions.

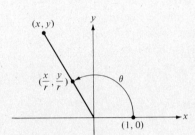

Figure 9.16

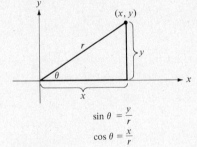

Figure 9.17

ADDITION FORMULAS FOR SINE AND COSINE
The following identities are true for all numbers θ_1 and θ_2:

$$\cos(\theta_1 - \theta_2) = \cos \theta_1 \cos \theta_2 + \sin \theta_1 \sin \theta_2 \qquad (2)$$
$$\cos(\theta_1 + \theta_2) = \cos \theta_1 \cos \theta_2 - \sin \theta_1 \sin \theta_2 \qquad (3)$$
$$\sin(\theta_1 + \theta_2) = \sin \theta_1 \cos \theta_2 + \sin \theta_2 \cos \theta_1 \qquad (4)$$
$$\sin(\theta_1 - \theta_2) = \sin \theta_1 \cos \theta_2 - \sin \theta_2 \cos \theta_1 \qquad (5)$$

The first formula is proved below; the other three formulas, which are easy consequences thereof, are assigned to the exercises.

Proof of Formula (2)
Consider arbitrary points $P(\theta_1)$ and $P(\theta_2)$ on the circle $x^2 + y^2 = 1$. Then the circular arc joining $P(\theta_1)$ to $P(\theta_2)$ has the same length as the arc joining $P(\theta_1 - \theta_2)$ to $(1, 0)$ (see Figure 9.18). The chords (straight line segments) joining these pairs of points also have the same length:

$$\text{dist}[P(\theta_1), P(\theta_2)] = \text{dist}[P(\theta_1 - \theta_2), (1, 0)]$$

and hence also

$$\{\text{dist}[P(\theta_1), P(\theta_2)]\}^2 = \{\text{dist}[P(\theta_1 - \theta_2), (1, 0)]\}^2 \qquad (6)$$

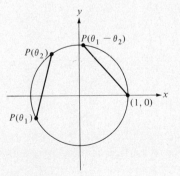

Figure 9.18

(see Section 1.6). Using the definition of the cosine and sine functions

$$P(\theta_1) = (\cos\theta_1, \sin\theta_1)$$
$$P(\theta_2) = (\cos\theta_2, \sin\theta_2)$$
$$P(\theta_1 - \theta_2) = [\cos(\theta_1 - \theta_2), \sin(\theta_1 - \theta_2)]$$

By the distance formula

$$\begin{aligned}\{\text{dist}[P(\theta_1), P(\theta_2)]\}^2 \\
&= (\cos\theta_1 - \cos\theta_2)^2 + (\sin\theta_1 - \sin\theta_2)^2 \\
&= (\cos^2\theta_1 - 2\cos\theta_1\cos\theta_2 + \cos^2\theta_2) \\
&\quad + (\sin^2\theta_1 - 2\sin\theta_1\sin\theta_2 + \sin^2\theta_2) \\
&= (\cos^2\theta_1 + \sin^2\theta_1) + (\cos^2\theta_2 + \sin^2\theta_2) \\
&\quad - 2(\cos\theta_1\cos\theta_2 + \sin\theta_1\sin\theta_2) \\
&= 2 - 2(\cos\theta_1\cos\theta_2 + \sin\theta_1\sin\theta_2)\end{aligned}$$

$$\begin{aligned}\{\text{dist}[P(\theta_1 - \theta_2), (1, 0)]\}^2 \\
&= [\cos(\theta_1 - \theta_2) - 1]^2 + [\sin(\theta_1 - \theta_2) - 0]^2 \\
&= [\cos^2(\theta_1 - \theta_2) - 2\cos(\theta_1 - \theta_2) + 1] + \sin^2(\theta_1 - \theta_2) \\
&= [\cos^2(\theta_1 - \theta_2) + \sin^2(\theta_1 - \theta_2)] + 1 - 2\cos(\theta_1 - \theta_2) \\
&= 2 - 2\cos(\theta_1 - \theta_2)\end{aligned}$$

After a substitution and a simplification, formula (6) becomes

$$\cos(\theta_1 - \theta_2) = \cos\theta_1\cos\theta_2 + \sin\theta_1\sin\theta_2$$

REMARK

From the addition formulas we draw an important conclusion. The relation

$$\sin(\pi + \theta) = \sin\pi\cos\theta + \sin\theta\cos\pi = -\sin\theta$$

tells us that the graph of $y = \sin\theta$ for $\pi \leq \theta \leq 2\pi$ is the inverted image (mirror image) of the graph for $0 \leq \theta \leq \pi$ (see Figure 9.19). In addition, with $\theta_1 = \pi/2$, $\theta_2 = \theta$ we get

$$\sin\left(\frac{\pi}{2} - \theta\right) = \sin\frac{\pi}{2}\cos\theta - \sin\theta\cos\frac{\pi}{2} = \cos\theta$$

$$\sin\left(\frac{\pi}{2} + \theta\right) = \sin\frac{\pi}{2}\cos\theta + \sin\theta\cos\frac{\pi}{2} = \cos\theta$$

so that

$$\sin\left(\frac{\pi}{2} - \theta\right) = \sin\left(\frac{\pi}{2} + \theta\right)$$

This tells us that *all* the values of $\sin\theta$ can be obtained from the values of $\sin\theta$ for $0 \leq \theta \leq \pi/2$. We see, in particular, that the graph of $\sin\theta$ is symmetric with respect to the line $\theta = \pi/2$ (see Figure 9.20).

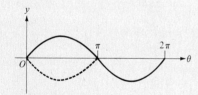

Figure 9.19

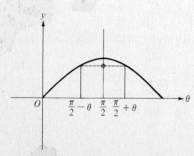

Figure 9.20

Example 5

Verify the formulas

$$\sin 2\theta = 2 \sin \theta \cos \theta$$
$$\cos 2\theta = \cos^2\theta - \sin^2\theta$$

Solution
Put $\theta_1 = \theta_2 = \theta$ in formula (4). Therefore $\sin(\theta + \theta) = \sin \theta \cos \theta + \sin \theta \cos \theta$, and the first formula follows. To get the second formula, put $\theta_1 = \theta_2 = \theta$ in formula (3).

Throughout this section, we have used the notation

$$x = \cos \theta \quad \text{and} \quad y = \sin \theta$$

In the exercises and throughout the rest of the book we most often use the notation

$$y = \cos x \quad \text{and} \quad y = \sin x$$

Other variables may appear from time to time.

EXERCISES

1. Complete the following table:

degrees	270	30	295	5	π	400	$\frac{1}{60}$				
radians								$\frac{1}{2\pi}$	$\frac{\pi}{3}$	5π	5	400

2. Plot the following points on the unit circle $x^2 + y^2 = 1$.

$$P(8\pi) \quad P\left(\frac{33\pi}{4}\right) \quad P(-8\pi) \quad P\left(-\frac{33\pi}{4}\right)$$

In Exercises 3–8, find which functions are even, which are odd, and determine the period if they are periodic.

3. $\sin x + \cos x$

4. $\sin 3x$

5. $\cos 2x + \cos 4x$

6. $x \sin x$

7. $\sin \frac{x}{2} \cdot \cos \frac{x}{2}$

8. $\sin^2 x$

Aided by the table of trigonometric functions at the end of the book, plot the graphs in Exercises 9–16.

9. $y = \sin 3x$

10. $y = \sin x + \cos x$

11. $y = \sin x + \sin \frac{x}{2}$

12. $y = \sin^2 x$

13. $y = |\cos x|$

14. $y = \sin x + 2\sin \dfrac{x}{2}$

15. $y = \sin 2\pi x$

16. $y = \cos\left(\theta - \dfrac{\pi}{4}\right)$

Establish the trigonometric identities in Exercises 17–26:

17. $\cos 2\theta = 2\cos^2\theta - 1 = 1 - 2\sin^2\theta$ (Use Example 5)

18. $\sin^2 \dfrac{\theta}{2} = \dfrac{1 - \cos\theta}{2}$ (Use Exercise 17)

19. $\cos^2 \dfrac{\theta}{2} = \dfrac{1 + \cos\theta}{2}$ (Use Exercise 17)

20. (a) $\cos\left(\dfrac{\pi}{2} - \theta\right) = \sin\theta$ $\left(\text{Use formula (2) with } \theta_1 = \dfrac{\pi}{2} \text{ and } \theta_2 = \theta\right)$

    (b) $\sin\left(\dfrac{\pi}{2} - \theta\right) = \cos\theta$ $\left(\text{Use (a) with } \theta \text{ replaced by } \dfrac{\pi}{2} - \theta\right)$

21. $\cos(\theta_1 + \theta_2) = \cos\theta_1 \cos\theta_2 - \sin\theta_1 \sin\theta_2$ (This is formula (3); use formula (2) with θ_2 replaced by $-\theta_2$.)

22. $\sin(\theta_1 + \theta_2) = \sin\theta_1 \cos\theta_2 + \sin\theta_2 \cos\theta_1$ (This is formula (4). Putting $\theta = \theta_1 + \theta_2$ in Exercise 20 gives $\sin(\theta_1 + \theta_2) = \cos\{[(\pi/2) - \theta_1] - \theta_2\}$: Apply formula (2) to the right side, then use Exercise 20 once more.)

23. $\sin(\theta_1 - \theta_2) = \sin\theta_1 \cos\theta_2 - \sin\theta_2 \cos\theta_1$ (This is formula (5); use formula (4) with θ_2 replaced by $-\theta_2$.)

24. $\cos 3\theta = \cos^3\theta - 3\sin^2\theta \cos\theta$ (Apply formula (3) to $\cos(2\theta + \theta)$, then use Example 5.)

25. $\sin 3\theta = 3\sin\theta \cos^2\theta - \sin^3\theta$ (Apply formula (4) to $\sin(2\theta + \theta)$, then use Example 5.)

26. $\cos\theta_1 \cos\theta_2 = \tfrac{1}{2}[\cos(\theta_1 - \theta_2) + \cos(\theta_1 + \theta_2)]$

27. Give a formula for the values of x at which $\cos 5x = k$ for $k = -1, 0, 1$.

28. From Exer. 24 derive the formula $\cos 3\theta = \cos\theta[4\cos^2\theta - 3]$.

With $\theta = \pi/6$, use this formula to compute $\cos 30° = \cos \pi/6$. From this, find the values of $\cos 60°$, $\sin 30°$, and $\sin 60°$.

29. How often does $y = \sin(2\pi/3)x$ cross the x axis for $-6 \le x \le 6$?

9.2 THE DERIVATIVES OF SINE AND COSINE

In this section we establish two basic differentiation formulas.

$$\frac{d}{dx}(\sin x) = \cos x \tag{1}$$

$$\frac{d}{dx}(\cos x) = -\sin x \tag{2}$$

The proof of the first formula requires many steps, and it is given following the examples and comments below. The second formula is a consequence of the first, as we now demonstrate.

Using the formula

$$\cos x = \sin\left(\frac{\pi}{2} - x\right)$$

we put $u = (\pi/2) - x$ and appeal to the chain rule:

$$\frac{d}{dx}(\cos x) = \frac{d}{dx}(\sin u) = \frac{d}{du}(\sin u)\frac{du}{dx} = \cos u \frac{du}{dx}$$

But $\cos u = \cos(\pi/2 - x) = \sin x$ and $du/dx = -1$, so

$$\frac{d}{dx}(\cos x) = -\sin x$$

Example 1

Find

$$\frac{d}{dx}[\ln(\sin x)]$$

Solution
Put $u = \sin x$ and apply the chain rule:

$$\frac{d}{dx}\ln u = \frac{1}{u}\frac{du}{dx} = \frac{1}{\sin x}\frac{du}{dx} = \frac{1}{\sin x} \cdot \cos x = \frac{\cos x}{\sin x}$$

Notice that $\ln(\sin x)$ is undefined when $\sin x \le 0$.

As a direct result of the chain rule, formulas (1) and (2) have the following generalization.

THEOREM 1
If u is a differentiable function of x, then

$$\frac{d}{dx}(\sin u) = \cos u \cdot \frac{du}{dx}$$

$$\frac{d}{dx}(\cos u) = -\sin u \cdot \frac{du}{dx}$$

The first of these formulas was already used with $u = (\pi/2) - x$.

Example 2

Find the first two derivatives of $y = \sin kx$ and $y = \cos kx$, where k is a constant.

Solution

$y = \sin kx \qquad \dfrac{dy}{dx} = k \cos kx \qquad \dfrac{d^2y}{dx^2} = -k^2 \sin kx$

$y = \cos kx \qquad \dfrac{dy}{dx} = -k \sin kx \qquad \dfrac{d^2y}{dx^2} = -k^2 \cos kx$

Comparing y with d^2y/dx^2, we notice that both $\sin kx$ and $\cos kx$ satisfy the equation

$$\frac{d^2y}{dx^2} = -k^2 y \quad \text{or} \quad \frac{d^2y}{dx^2} + k^2 y = 0$$

This equation is an example of a *differential equation,* that is, an equation involving an unknown function and its derivatives. We also observe the following interesting facts:

$y = \sin x \qquad\qquad y = \cos x$

$\dfrac{dy}{dx} = \cos x \qquad\qquad \dfrac{dy}{dx} = -\sin x$

$\dfrac{d^2y}{dx^2} = -\sin x = -y \qquad \dfrac{d^2y}{dx^2} = -\cos x = -y$

$\dfrac{d^3y}{dx^3} = -\cos x \qquad\qquad \dfrac{d^3y}{dx^3} = \sin x$

$\dfrac{d^4y}{dx^4} = \sin x = y \qquad\qquad \dfrac{d^4y}{dx^4} = \cos x = y$

THE DERIVATIVE OF $\sin x$

To find the derivative of $\sin x$, we investigate the quotient

$$\frac{\Delta y}{\Delta x} = \frac{\sin(x + \Delta x) - \sin x}{\Delta x}$$

9.2 THE DERIVATIVES OF SINE AND COSINE

By the addition formula,
$$\sin(x + \Delta x) = \sin x \cos \Delta x + \sin \Delta x \cos x$$
and hence
$$\Delta y = \sin(x + \Delta x) - \sin x = \sin x \cos \Delta x + \sin \Delta x \cos x - \sin x$$
$$= \sin x (\cos \Delta x - 1) + \sin \Delta x \cos x$$

We thus find that
$$\frac{\Delta y}{\Delta x} = \frac{\sin x (\cos \Delta x - 1) + \sin \Delta x \cos x}{\Delta x}$$
$$= \sin x \cdot \frac{\cos \Delta x - 1}{\Delta x} + \cos x \cdot \frac{\sin \Delta x}{\Delta x} \quad (3)$$

To find the derivative of $\sin x$ means to evaluate the limit of $\Delta y / \Delta x$ as $\Delta x \to 0$. Since x is held fixed, this limit is found by evaluating the limits of $(\cos \Delta x - 1)/\Delta x$ and $(\sin \Delta x)/\Delta x$ as $\Delta x \to 0$. When we compare $\Delta y / \Delta x$ with the derivative of $\sin x$, we see what these limits should be.

The numerical evidence in Table 9.1 indicates that the following relations hold.

$$\lim_{\Delta x \to 0} \frac{\sin \Delta x}{\Delta x} = 1 \quad (4)$$

$$\lim_{\Delta x \to 0} \frac{\cos \Delta x - 1}{\Delta x} = 0 \quad (5)$$

These are proved below; note that in both cases the numerator and denominator vanish when $\Delta x = 0$.

From (3), (4), and (5), we conclude that
$$\lim_{\Delta x \to 0} \frac{\Delta y}{\Delta x} = \sin x \cdot \lim_{\Delta x \to 0} \frac{\cos \Delta x - 1}{\Delta x} + \cos x \cdot \lim_{\Delta x \to 0} \frac{\sin \Delta x}{\Delta x}$$
$$= \sin x \cdot 0 + \cos x \cdot 1 = \cos x$$

which is the result we are seeking.

We now proceed to establish the limits in (4) and (5). The proofs depend on the following two lemmas; a *lemma* is a preparatory statement to a theorem.

Table 9.1
Numerical evidence to support the claim made in (4) and (5).

Δx	$\dfrac{\sin \Delta x}{\Delta x}$	$\dfrac{\cos \Delta x - 1}{\Delta x}$
0.5	0.95886	−0.24484
0.4	0.97355	−0.19735
0.3	0.98507	−0.14887
0.2	0.99335	−0.09965
0.1	0.99830	−0.05000
0.08	0.99887	−0.04000
0.06	0.99933	−0.03000
0.04	0.99975	−0.02000

LEMMA 1
If $0 < \theta < \pi/2$, then
$$\sin \theta < \theta \quad \text{and} \quad 1 - \cos \theta < \theta$$

Proof
Consult Figure 9.21. The arc from B to C is a circular arc of length θ

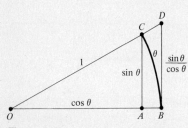

Figure 9.21

from a circle of radius 1. An inspection of the figure shows the following to be true:

$$\text{dist}(B, C) < \theta$$

because the shortest distance between two points is a straight line;

$$\sin \theta = \text{dist}(A, C) \leq \text{dist}(B, C) < \theta$$

and

$$1 - \cos \theta = \text{dist}(A, B) \leq \text{dist}(B, C) < \theta$$

because the sides of a right triangle are less than the length of the hypotenuse.

COROLLARY 1

$$\lim_{\theta \to 0} \cos \theta = 1$$

Proof
The fact that $\cos(-\theta) = \cos \theta$ permits us to limit the proof to values $\theta > 0$; being interested in small values of θ, we may suppose that $0 < \theta < \pi/2$. Then by the lemma, $1 - \theta < \cos \theta$ and, of course, $\cos \theta \leq 1$. Thus, $\cos \theta$ lies in the closed interval $[1 - \theta, 1]$. As $\theta \to 0$, these intervals narrow down to the single number 1, and it follows that $\lim_{\theta \to 0} \cos \theta = 1$, as was to be shown.

LEMMA 2
If $0 < \theta < \pi/2$, then

$$\cos \theta < \frac{\sin \theta}{\theta}$$

Figure 9.22

Proof
As in the proof of Lemma 1, the arc from B to C is a circular arc of length θ and radius 1. We proceed with the following steps (consult Figure 9.22).

Step 1
The triangles OBD and OAC are similar, and hence

$$\frac{\text{dist}(B, D)}{\text{dist}(O, B)} = \frac{\text{dist}(A, C)}{\text{dist}(O, A)}$$

Noting that $\text{dist}(O, B) = 1$, $\text{dist}(A, C) = \sin \theta$, and $\text{dist}(O, A) = \cos \theta$, we have

$$\text{dist}(B, D) = \frac{\sin \theta}{\cos \theta}$$

Step 2
We observe that

   area of circular sector OBC < area of triangle OBD

These areas are

$$\frac{1}{2} \cdot 1^2 \cdot \theta \quad \text{and} \quad \frac{1}{2} \cdot 1 \cdot \frac{\sin \theta}{\cos \theta}$$

respectively, and hence we have

$$\theta < \frac{\sin \theta}{\cos \theta} \quad \text{or} \quad \cos \theta < \frac{\sin \theta}{\theta}$$

We are now ready to prove the assertions made in (4) and (5).

THEOREM 2

$$\lim_{\theta \to 0} \frac{\sin \theta}{\theta} = 1$$

Proof
Again we restrict the proof to values $\theta > 0$, since

$$\frac{\sin(-\theta)}{-\theta} = \frac{-\sin \theta}{-\theta} = \frac{\sin \theta}{\theta}$$

The argument runs as follows:

$$\frac{\sin \theta}{\theta} < 1 \qquad \text{by Lemma 1}$$

$$\cos \theta < \frac{\sin \theta}{\theta} \qquad \text{by Lemma 2}$$

$$1 - \theta < \cos \theta \qquad \text{by Lemma 1}$$

and hence

$$1 - \theta < \frac{\sin \theta}{\theta} < 1$$

We can thus repeat the argument in the proof of Corollary 1. Namely, $\sin \theta/\theta$ lies in the closed interval $[1 - \theta, 1]$, and as $\theta \to 0$ these intervals narrow down to the unique number 1. This implies the theorem.

THEOREM 3

$$\lim_{\theta \to 0} \frac{1 - \cos \theta}{\theta} = 0$$

Proof

We observe that

$$\frac{1-\cos(-\theta)}{-\theta} = \frac{1-\cos\theta}{-\theta} = -\frac{1-\cos\theta}{\theta}$$

hence we can once more restrict the discussion to values $\theta > 0$. We proceed with the following steps.

Step 1

By Exercise 17 in Section 9.1, we have $\cos 2x = 1 - 2\sin^2 x$ or, equivalently, $1 - \cos 2x = 2\sin^2 x$. With $x = \theta/2$, this becomes $1 - \cos \theta = 2 \sin^2 \theta/2$.

Step 2

By Lemma 1, $\sin \theta < \theta$ and so also $\sin(\theta/2) < \theta/2$ for $\theta > 0$; squaring both sides gives $\sin^2(\theta/2) < \theta^2/4$ and hence $1 - \cos \theta < \theta^2/2$ by step 1. Dividing by θ gives

$$\frac{1-\cos\theta}{\theta} < \frac{\theta}{2}$$

Step 3

For $0 < \theta < \pi/2$, $(1-\cos\theta)/\theta$ is positive, and hence $0 < (1-\cos\theta)/\theta < \theta/2$. It follows that $(1-\cos\theta)/\theta$ lies in the closed interval $[0, \theta/2]$, which narrows down to the number 0 as $\theta \to 0$. This establishes the result we are after.

EXERCISES

Find dy/dx in Exercises 1–16.

1. $y = \sin(e^x)$
2. $y = \dfrac{\sin x}{\cos x}$
3. $y = \dfrac{\cos x}{\sin x}$
4. $y = \dfrac{1}{\sin x}$
5. $y = \dfrac{1}{\cos x}$
6. $y = \sin(\ln x)$
7. $y = \sin x \cdot \cos x$
8. $y = e^{\sin x}$
9. $y = \sin(x^4)$
10. $y = \sin^4 x$
11. $y = \cos 5x$
12. $y = \sin(x^x)$
13. $y = (\sin x)^x$
14. $y = x^{\cos x}$
15. $y = x \sin \dfrac{1}{x}$
16. $y = a^{1/\cos x}$

Find dy/dx in Exercises 17–22 using implicit differentiation.

17. $\sin y + \cos x = 0$
18. $y + \sin xy = 1$
19. $x = \sin y$
20. $\cos xy = 1$
21. $\sin^2 x + \sin^2 y = c$
22. $\sin(x+y) + \sin(x-y) = k$, $k = $ const.

Find the indicated limits in Exercises 23–30.

Hint: $1 - \cos^2 \theta = (1 - \cos \theta)(1 + \cos \theta)$

23. $\lim\limits_{\theta \to 0} \dfrac{1 - \cos^2 \theta}{\theta}$

24. $\lim\limits_{x \to 0} \dfrac{\sin x \cos x}{x}$

Hint: Exercise 25 of Section 9.1.

25. $\lim\limits_{x \to 0} \dfrac{\sin 3x}{x}$

26. $\lim\limits_{\theta \to \pi/2} \dfrac{\cos \theta}{\theta}$

27. $\lim\limits_{x \to 0} \dfrac{\sin x}{\sin 2x}$

28. $\lim\limits_{x \to \pi/2} \dfrac{\cos x}{(\pi/2) - x}$

Hint: Exercise 25 of Section 9.1.

29. $\lim\limits_{x \to 0} \dfrac{\sin 3x}{\sin 2x}$

Hint: Take logarithms.

30. $\lim\limits_{x \to 0} (e^{1/x})^{\sin x}$

Graph the functions in Exercises 31–36.

31. $y = |\sin x|$
32. $y = \cos(x + \tfrac{1}{2})$
33. $y = 2 \cos x$
34. $y = \sin 2\pi x$
35. $y = \sin \dfrac{x}{\pi}$
36. $y = \sin x + \sin \dfrac{x}{2}$

37. Find $y^{(8)}$ when $y = \cos x$
38. Find $y^{(7)}$ when $y = \sin 2x$
39. Find $y^{(2)} + y$ when $y = \sin x$
40. Find $y^{(2)} + y$ when $y = x \sin x$
41. Find explicit formulas for the points x at which

Ans. $x = 2k\pi$ for $k = 0, \pm 1, \pm 2, \pm 3, \ldots$

(a) $\sin x = 0$
(b) $\sin x = 1$
(c) $\sin x = -1$
(d) $\sin x$ is concave up

Ans. $2k\pi < x < (2k+1)\pi$ for $k = 0, \pm 1, \pm 2, \pm 3, \ldots$

(e) $\sin x$ is concave down

Ans. $(2k - \frac{1}{2})\pi < x < (2k + \frac{1}{2})\pi$
for $k = 0, \pm 1, \pm 2, \pm 3, \ldots$.

(f) $\sin x$ is increasing

(g) $\sin x$ is decreasing

42. Find explicit formulas for the points x at which

(a) $\cos 2x = 0$

(b) $\sin x = \frac{1}{2}$

(c) $\sin 3x = \frac{1}{2}$

(d) $\cos\left(\frac{\pi}{4} - x\right) = \frac{1}{2}$

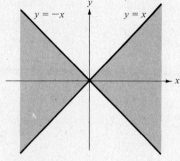

Figure 9.23

Graph the functions in Exercises 43–48.

43. $y = x \sin x$

44. $y = \frac{1}{x} \sin x$

45. $y = x \sin \frac{1}{x}$

Hint: (1) $|x \sin 1/x| \leq |x|$, and this tells you that the graph lies in the shaded area in Figure 9.23 above.
(2) $\sin 1/x = 0$ when $1/x = k\pi$, or $x = 1/k\pi$ for $k = \pm 1, \pm 2, \pm 3, \ldots$. Similarly, find the points where $\sin 1/x = 1$ and $\sin 1/x = -1$.

Hint: $|x^2 \sin 1/x| \leq x^2$. From this you can determine the region in the plane where the graph lies.

46. $y = x^2 \sin \frac{1}{x}$

47. $y = \frac{1}{x} \sin \frac{1}{x}$

48. $y = \frac{1}{2}x - \sin x$ (see Figure 9.24).

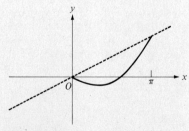

Figure 9.24
The graph of $y = \frac{1}{2}x - \sin x$ for $0 \leq x \leq \pi$.

9.3 FURTHER TRIGONOMETRIC FUNCTIONS

The sine and cosine functions defined in Section 9.1 give rise to four more trigonometric functions. They are listed in Table 9.2. A geometric interpretation of the various trigonometric functions is given in Figure 9.25. It goes without saying that angles are measured in radians.

Table 9.2

Name of the function	Definition and notation	Remarks	
tangent	$\tan x = \dfrac{\sin x}{\cos x}$	not defined for $x = (k + \tfrac{1}{2})\pi$	periodic with period π
secant	$\sec x = \dfrac{1}{\cos x}$		periodic with period 2π
cotangent	$\cot x = \dfrac{\cos x}{\sin x} = \dfrac{1}{\tan x}$	not defined for $x = k\pi$	periodic with period π
cosecant	$\csc x = \dfrac{1}{\sin x}$		periodic with period 2π

$k = 0, \pm 1, \pm 2, \pm 3, \ldots$

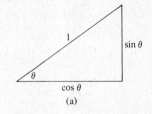

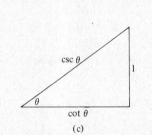

Figure 9.25

The periodicity of these functions is easily established.

Example 1

Show that $\tan x$ is periodic with period π.

Solution
It has to be shown that

$$\tan(\pi + x) = \tan x$$

Using the definition, this is the same as showing that

$$\frac{\sin(\pi + x)}{\cos(\pi + x)} = \frac{\sin x}{\cos x}$$

By the addition formulas,

$$\sin(\pi + x) = \sin \pi \cos x + \sin x \cos \pi = -\sin x$$

$$\cos(\pi + x) = \cos \pi \cos x - \sin \pi \sin x = -\cos x$$

and the result follows.

The derivatives of these new functions are obtained from the formula for differentiating a quotient.

Example 2

Find the derivative of $\tan x$.

Solution

$$\frac{d}{dx}(\tan x) = \frac{d}{dx}\left(\frac{\sin x}{\cos x}\right) = \frac{\cos x (d/dx)(\sin x) - \sin x (d/dx)(\cos x)}{\cos^2 x}$$

$$= \frac{\cos^2 x + \sin^2 x}{\cos^2 x} = \frac{1}{\cos^2 x} = \sec^2 x$$

In general, the following formulas are true.

	If u is a differentiable function of x, then
$\dfrac{d}{dx}(\tan x) = \sec^2 x$	$\dfrac{d}{dx}(\tan u) = \sec^2 u \dfrac{du}{dx}$
$\dfrac{d}{dx}(\cot x) = -\csc^2 x$	$\dfrac{d}{dx}(\cot u) = -\csc^2 u \dfrac{du}{dx}$
$\dfrac{d}{dx}(\sec x) = \sec x \tan x$	$\dfrac{d}{dx}(\sec u) = \sec u \tan u \dfrac{du}{dx}$
$\dfrac{d}{dx}(\csc x) = -\csc x \cot x$	$\dfrac{d}{dx}(\csc u) = -\csc u \cot u \dfrac{du}{dx}$

Example 3

Find dy/dx if $y + \cot xy = 1$.

Solution
Using implicit differentiation and the chain rule,

$$\frac{dy}{dx} - \csc^2 xy \frac{d}{dx}(xy) = \frac{dy}{dx} - \csc^2 xy \cdot \left(y + x\frac{dy}{dx}\right) = 0$$

Solving for dy/dx, we get

$$\frac{dy}{dx} = \frac{y \csc^2 xy}{1 - x \csc^2 xy} = \frac{y}{\sin^2 xy - x}$$

THE GRAPHS OF THE NEW TRIGONOMETRIC FUNCTIONS
The graphs of the trigonometric functions introduced in Figure 9.26 are obtained with the procedure described in Section 4.3.

EXERCISES

1. Show that $\cot x$ is periodic with period π.

2. Show that $\sec x$ and $\csc x$ are periodic with period 2π.

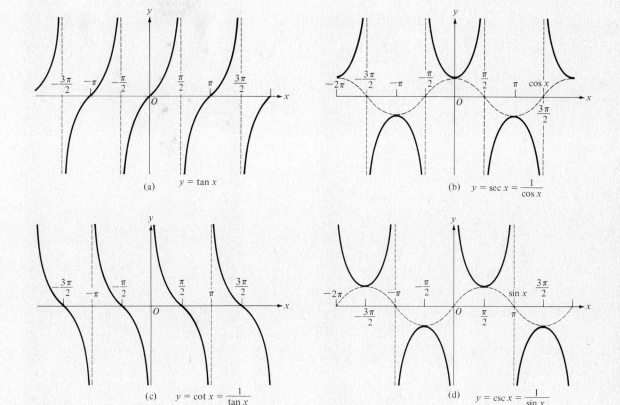

Figure 9.26
(a) The graph of $y = \tan x$. (b) The graph of $y = \sec x = 1/\cos x$. (c) The graph of $y = \cot x = 1/\tan x$. (d) The graph of $y = \csc x = 1/\sin x$. Since $\sin x = \cos[(\pi/2) - x]$, we have $\csc x = \sec[(\pi/2) - x)]$. The graph of $y = \csc x$ is therefore a translation to the right of $y = \sec x$ by a distance $\pi/2$.

3. Verify the differentiation formulas

   (a) $\dfrac{d}{dx}(\cot x) = -\csc^2 x$

   (b) $\dfrac{d}{dx}(\sec x) = \sec x \tan x$

   (c) $\dfrac{d}{dx}(\csc x) = -\csc x \cot x$

Evaluate the derivatives in Exercises 4–17.

4. $y = \tan^2 x$

5. $y = \tan\left(\dfrac{1}{x}\right)$

6. $y = \tan 5x$

7. $y = x \cot x$

8. $y = \sec \dfrac{1}{x}$ 9. $y = \ln(\tan x)$

10. $y = \cos x \tan x$ 11. $y = \cot\sqrt{x^2 - 1}$

12. $y = \dfrac{\tan x}{\cot x}$ 13. $y = x^{\sec x}$

14. $y + \tan xy = 0$ 15. $x = \cot(x + y)$

16. $y = \sec y$ 17. $\tan x + \tan y = 1$

18. Prove from the identity $\sin^2 \theta + \cos^2 \theta = 1$ that

$$\tan^2 \theta + 1 = \sec^2 \theta$$
$$\cot^2 \theta + 1 = \csc^2 \theta$$

Use the addition formulas for sine and cosine to verify the identities in Exercises 19 and 20.

19. $\tan(\theta_1 + \theta_2) = \dfrac{\tan \theta_1 + \tan \theta_2}{1 - \tan \theta_1 \tan \theta_2}$

20. $\tan(\theta_1 - \theta_2) = \dfrac{\tan \theta_1 - \tan \theta_2}{1 + \tan \theta_1 \tan \theta_2}$

21. Show that

$$\tan 2\theta = \dfrac{2 \tan \theta}{1 - \tan^2 \theta}$$

22. Show that

$$\tan \dfrac{\theta}{2} = \dfrac{\sin \theta}{1 + \cos \theta}$$

23. Show that

$$\tan \dfrac{\theta}{2} = \dfrac{1 - \cos \theta}{\sin \theta}$$

Plot the graphs in Exercises 24–27.

24. $y = -\tan x$ 25. $y = \tan 3x$

26. $y = x - \tan x$ 27. $y = \tfrac{1}{2} \sec 2x$

9.4 INVERSE FUNCTIONS

A *function* $y = f(x)$ was defined as a relation that converts x into y (Section 2.1). By an *inverse function* of $y = f(x)$ we mean a function $x = g(y)$ that converts y back into x. This is illustrated schematically in Figure 9.27. Such a situation was encountered in Chapter 7, where we introduced the exponential function $x = \exp y$ as the inverse of the logarithmic function $y = \ln x$ (see Figure 9.28). The definition of $x = \exp y$ was

$y = \ln x$ is the same as $x = \exp y$

To summarize, a function $f(x)$ has an inverse when the equation $y = f(x)$ has a *unique* solution x for each y in its range. The symbol f^{-1} is often used for the inverse of f. Thus:

DEFINITION 1
$f^{-1}(y)$ is the unique x for which $y = f(x)$, and $x = f^{-1}(y)$ is the same as $y = f(x)$.

A function $f(x)$ for which the equation $y = f(x)$ has more than one solution for some y has no inverse. This cannot happen with functions that are always increasing or always decreasing. Some examples will make this clear.

Example 1

Examine the function $y = x^2 - \frac{1}{4}$ (see Figure 9.29). All horizontal lines $y = y_0$ for $y_0 > -\frac{1}{4}$ intersect the curve in *two* points. Indeed, the equation $y = x^2 - \frac{1}{4}$ does not have a unique solution for x. Since $x^2 = y + \frac{1}{4}$, there are two solutions,

$$x = \sqrt{y + \tfrac{1}{4}} \quad \text{and} \quad x = -\sqrt{y + \tfrac{1}{4}}$$

The function has, therefore, *no* inverse.

On the other hand, the function $y = x^2 - \frac{1}{4}$ restricted to values $x \geq 0$ does have an inverse, namely, $x = \sqrt{y + \frac{1}{4}}$ (Figure 9.30). When we put $f(x) = x^2 - \frac{1}{4}$, then $f^{-1}(y) = \sqrt{y + \frac{1}{4}}$ (Figure 9.31).

The restriction of this function to values $x \leq 0$ (Figure 9.32) also has an inverse, namely, $x = -\sqrt{y + \frac{1}{4}}$ (Figure 9.33).

There are many important cases in which inverses can be obtained by restricting functions to suitable intervals (see Section 9.5).

Figure 9.27

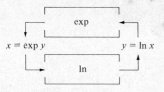

Figure 9.28

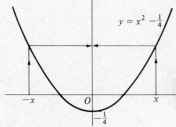

Figure 9.29
The graph of $y = x^2 - \frac{1}{4}$.

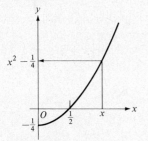

Figure 9.30
The graph of $f(x) = x^2 - \frac{1}{4}$ for $x \geq 0$.

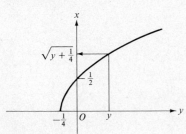

Figure 9.31
The graph of $f^{-1}(y) = \sqrt{y + \frac{1}{4}}$.

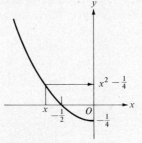

Figure 9.32
The graph of $f(x) = x^2 - \frac{1}{4}$ for $x \leq 0$.

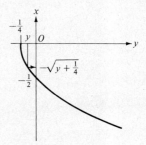

Figure 9.33
The graph of $f^{-1}(y) = -\sqrt{y + \frac{1}{4}}$.

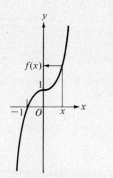

Figure 9.34
The graph of $y = x^3 + 1$.

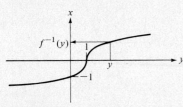

Figure 9.35
The graph of $x = (y - 1)^{1/3}$.

Example 2

Consider the function $y = x^3 + 1$ (see Figure 9.34). This function is increasing, so each horizontal line intersects it in exactly one point. Each y therefore determines a unique number x which can be found by solving for x in terms of y:

$$x = (y - 1)^{1/3} \quad \text{is the same as} \quad y = x^3 + 1$$

(see Figure 9.35). Thus, when $f(x) = x^3 + 1$, then $f^{-1}(y) = (y - 1)^{1/3}$. We observe that

$$\frac{dy}{dx} = 3x^2$$

and

$$\frac{dx}{dy} = \frac{1}{3}(y - 1)^{-2/3} = \frac{1}{3}[(y - 1)^{1/3}]^{-2} = \frac{1}{3}x^{-2} = \frac{1}{3x^2}$$

so that the formula

$$\frac{dx}{dy} = \frac{1}{dy/dx}$$

ensues.

Example 3

For the previous examples we were able to solve the functions explicitly for x in terms of y. This is not always possible, as in the following case. Consider the function

$$y = \tfrac{1}{5}x^5 + x$$

The derivative of this function is always positive,

$$\frac{dy}{dx} = x^4 + 1 > 0$$

and this says that the function is increasing. Accordingly, the function has an inverse, but there is no formula for it. Despite this difficulty we observe this: Treating x as a function of y and using implicit differentiation,

$$1 = x^4 \frac{dx}{dy} + \frac{dx}{dy} = (x^4 + 1) \frac{dx}{dy}$$

and you may be surprised to discover that

$$\frac{dx}{dy} = \frac{1}{x^4 + 1} = \frac{1}{dy/dx}$$

In words, we were able to ascertain the existence of an inverse function and to calculate its derivative dx/dy without having a formula for it. This is, of course, only a special case of our results on implicit functions.

THEOREM 1
Let $y = f(x)$ have domain $[x_1, x_2]$ and range $[y_1, y_2]$. If $f'(x) > 0$ for all x in (x_1, x_2), or $f'(x) < 0$ for all x in (x_1, x_2), then

1. $y = f(x)$ has an inverse $x = g(y)$,
2. $x = g(y)$ is differentiable on (y_1, y_2), and

$$g'(y) = \frac{1}{f'(x)} = \frac{1}{f'[g(y)]}$$

Proof
We first remark that the conditions on $f'(x)$ guarantee that $f(x)$ is always increasing or always decreasing. This fact tells us that the inverse function $g(y)$ exists. To show that $g'(y)$ exists at y_0, $y_1 < y_0 < y_2$, and is of the asserted form, we proceed as follows.

Step 1
Form the difference quotient

$$\frac{g(y_0 + \Delta y) - g(y_0)}{\Delta y}$$

Step 2
Putting $x_0 = g(y_0)$ and $x_0 + \Delta x = g(y_0 + \Delta y)$, then $y_0 = f(x_0)$ and $y_0 + \Delta y = f(x_0 + \Delta x)$. This leads to the relations

$$\frac{g(y_0 + \Delta y) - g(y_0)}{\Delta y} = \frac{(x_0 + \Delta x) - x_0}{\Delta y} = \frac{\Delta x}{\Delta y} = \frac{1}{\Delta y / \Delta x}$$

$$= \frac{1}{[f(x_0 + \Delta x) - f(x_0)]/\Delta x}$$

Since $\Delta x \to 0$ as $\Delta y \to 0$, and

$$\lim_{\Delta x \to 0} \frac{f(x_0 + \Delta x) - f(x_0)}{\Delta x} = f'(x_0)$$

we see that

$$\lim_{\Delta y \to 0} \frac{g(y_0 + \Delta y) - g(y_0)}{\Delta y} = \frac{1}{f'(x_0)} = \frac{1}{f'[g(y_0)]}$$

This tells us that $g'(y_0)$ exists and is as asserted.

REMARK
Observe that if $x = f^{-1}(y)$ is the inverse of $y = f(x)$, then

$$f^{-1}[f(x)] = x \quad \text{for all } x \text{ in the domain of } f(x)$$

$$f[f^{-1}(y)] = y \quad \text{for all } y \text{ in the range of } f(x) \text{ [the domain of } f^{-1}(y)\text{]}$$

REMARKS ON NOTATION

1. Once the inverse function has been constructed, it comes into its own. For example, the function $x = \sqrt[3]{y}$ can be treated as a function without regard to the fact that it is the inverse of $y = x^3$. We can therefore use the notation $y = \sqrt[3]{x}$.
2. Be forewarned that $f^{-1}(x)$ is *not* the same as $1/f(x)$. We shall use the notation

$$\frac{1}{f(x)} = [f(x)]^{-1}$$

3. In the formula

$$\frac{dx}{dy} = \frac{1}{dy/dx}$$

the roles of x and y are interchanged on the left and right sides: On the left side y is the independent variable, x the dependent variable; the opposite is the case on the right side.

Example 4

Show that

$$y = \frac{e^x - e^{-x}}{e^x + e^{-x}}$$

has an inverse, and find its derivative.

Solution
We use the fact that the function is differentiable to find out if it is increasing or decreasing:

$$\frac{dy}{dx} = \frac{(e^x + e^{-x})^2 - (e^x - e^{-x})^2}{(e^x + e^{-x})^2} = 1 - \left(\frac{e^x - e^{-x}}{e^x + e^{-x}}\right)^2$$

and it is easy to see that $dy/dx \geq 0$ since $(e^x + e^{-x})^2 \geq (e^x - e^{-x})^2$. Hence the function is increasing and thus has an inverse. By Theorem 1, the derivative of the inverse is

$$\frac{dx}{dy} = \frac{1}{dy/dx} = \frac{1}{1 - [(e^x - e^{-x})/(e^x + e^{-x})]^2} = \frac{1}{1 - y^2}$$

EXERCISES

Find explicit formulas for the inverse functions in Exercises 1–5.

1. $y = \dfrac{x-1}{x+1}$ $(x \neq -1)$

2. $x = 3y^7 - 2$

3. $y = \sqrt{\dfrac{x-1}{x+1}}$ $(x \geq 1)$

4. $y = x + \sqrt{x}$ $(x \geq 0)$

5. $x = \dfrac{\sqrt{y}+1}{\sqrt{y}-1}$ $(y > 1)$

6. Let $f(x) = x^3 + 3$ and let $f^{-1}(x)$ be its inverse. Find the tangents to $y = f(x)$ and $y = f^{-1}(x)$ that pass through the point $(0, 0)$.

Decide which of the functions $y = f(x)$ in Exercises 7–14 has an inverse $x = f^{-1}(y)$. Find dx/dy, the derivative of the inverse, when it exists.

7. $y = \cos x$ $(0 \leq x \leq \pi)$

8. $y = x^4$

9. $y = \sqrt{x^2 + 1}$

10. $y = x^3 - 3x$ $(0 \leq x \leq 2)$

11. $y = x \cos x$ $\left(0 \leq x \leq \dfrac{\pi}{4}\right)$

12. $y = \dfrac{1}{2}x^2 - \cos x$ $\left(0 \leq x \leq \dfrac{\pi}{2}\right)$

13. $y = x + \ln x$ $(x > 0)$

14. $y = \dfrac{1 - e^x}{1 + e^x}$

15. In Example 3 we saw that the inverse $x = g(y)$ of $y = f(x) = \frac{1}{5}x^5 + x$ has a derivative $g'(y) = 1/(x^4 + 1)$.

   (a) Taking into account that $f(1) = \frac{6}{5}$, find $g'(\frac{6}{5})$.

   (b) Find a formula for $g''(y)$ and calculate $g''(\frac{6}{5})$.

9.5 THE INVERSE TRIGONOMETRIC FUNCTIONS

It is clear from the outset that the trigonometric functions have no inverses because they are neither increasing nor decreasing. The inverses of these functions are defined for their restrictions to suitable intervals. This idea was already used in Section 9.4, Example 1, in discussing the inverse of $y = x^2 - \frac{1}{4}$.

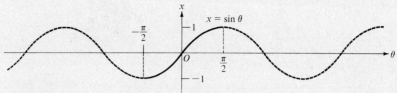

Figure 9.36
The restriction of $x = \sin \theta$ to the interval $-\pi/2 \leq \theta \leq \pi/2$. Note that as θ goes from $-\pi/2$ to $\pi/2$, $\sin \theta$ gives from -1 to 1.

THE FUNCTION $\sin^{-1} x$

For any x in the interval $[-1, 1]$, the equation $x = \sin \theta$ has infinitely many solutions. The solution θ that lies in the interval $[-\pi/2, \pi/2]$ is designated $\theta = \sin^{-1} x$. This solution is indeed unique, since $\sin \theta$ is increasing in this interval (see Figure 9.36). Thus (see Figure 9.37):

> When $-\pi/2 \leq \theta \leq \pi/2$, then $-1 \leq x \leq 1$ and $\theta = \sin^{-1} x$ is the same as $x = \sin \theta$. The function $\theta = \sin^{-1} x$ is the inverse function of $x = \sin \theta$.

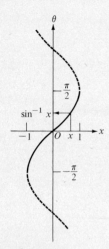

Figure 9.37
The graph of $\theta = \sin^{-1} x$.

WARNING

If you think of θ as the measure of an angle, then in the formula $\theta = \sin^{-1} x$, θ is the measure in *radians*.

Example 1

Find the values of $\sin^{-1} x$ when $x = -1, -\tfrac{1}{2}, 0, \tfrac{1}{2}, 1$.

Solution

The solution is obtained by reading a table of values of $\sin \theta$ "backwards," as illustrated in Figure 9.38.

Figure 9.38

Observe that $\sin^{-1} x$ has a derivative at each x in the interval $(-1, 1)$, but the derivative is undefined at $x = -1$ and at $x = 1$.

THE DERIVATIVE OF $\sin^{-1} x$

If $\theta = \sin^{-1} x$, then $x = \sin \theta$ and, making use of Theorem 1 in the last section, we have $d\theta/dx = 1/(dx/d\theta)$. We shall show that

$$\frac{d}{dx}(\sin^{-1} x) = \frac{1}{\sqrt{1 - x^2}}$$

This is done as follows:

Step 1

$$\frac{dx}{d\theta} = \cos \theta$$

Step 2

$$\cos^2 \theta = 1 - \sin^2 \theta \quad \text{because } \sin^2 \theta + \cos^2 \theta = 1$$
$$= 1 - x^2 \quad \text{because } x = \sin \theta$$

Step 3

$$\cos \theta = \pm\sqrt{1 - x^2}$$

but since $\cos \theta \geq 0$ for $-\pi/2 \leq \theta \leq \pi/2$,

$$\cos \theta = \sqrt{1 - x^2}$$

Hence,

$$\frac{d}{dx}(\sin^{-1} x) = \frac{d\theta}{dx} = \frac{1}{\cos \theta} = \frac{1}{\sqrt{1 - x^2}}$$

A direct application of the chain rule shows the following formula to be true.

> If u is a differentiable function of x, then
> $$\frac{d}{dx}(\sin^{-1} u) = \frac{1}{\sqrt{1 - u^2}} \frac{du}{dx}.$$

Example 2

Let a be a positive constant. Find where $\sin^{-1}(x/a)$ is defined and calculate its derivative.

Solution
The function $\sin^{-1}(x/a)$ is defined when $-1 \leq x/a \leq 1$, that is, when $-a \leq x \leq a$. Using the chain rule with $u = x/a$, we get

$$\frac{d}{dx}\left(\sin^{-1} \frac{x}{a}\right) = \frac{1}{\sqrt{1 - (x/a)^2}} \cdot \frac{1}{a} = \frac{1}{\sqrt{a^2 - x^2}}$$

THE FUNCTION $\cos^{-1} x$
For a given x in the interval $[-1, 1]$, the equation $x = \cos \theta$ has infinitely many solutions (see Figure 9.39). The solution θ which lies

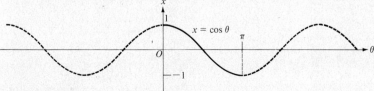

Figure 9.39
The restriction of $x = \cos \theta$ to the interval $0 \leq \theta \leq \pi$. Note that as θ goes from 0 to π, $\cos \theta$ goes from 1 to -1.

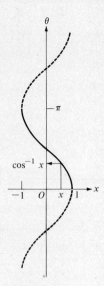

Figure 9.40
The graph of $\theta = \cos^{-1} x$.

in the interval $[0, \pi]$ is designated $\theta = \cos^{-1} x$ (see Figure 9.40). Thus:

> When $0 \leq \theta \leq \pi$, then $-1 \leq x \leq 1$ and $\theta = \cos^{-1} x$ is the same as $x = \cos \theta$. The function $\theta = \cos^{-1} x$ is the inverse function of $x = \cos \theta$.

THE DERIVATIVE OF $\cos^{-1} x$

The functions $\sin^{-1} x$ and $\cos^{-1} x$ are related through the equation

$$\cos^{-1} x = \frac{\pi}{2} - \sin^{-1} x \qquad \text{for } -1 \leq x \leq 1 \tag{1}$$

From this equation we obtain the derivative of $\cos^{-1} x$.

Proof of Equation (1)
Put $\theta = \sin^{-1} x$. Then $x = \sin \theta$, where $-1 \leq x \leq 1$ and $-\pi/2 \leq \theta \leq \pi/2$. Recall the relation $\sin \theta = \cos[(\pi/2) - \theta]$; when $-\pi/2 \leq \theta \leq \pi/2$, then $0 \leq \pi/2 - \theta \leq \pi$. Hence the inverse of $x = \cos[(\pi/2) - \theta]$ is defined, and $(\pi/2) - \theta = \cos^{-1} x$. Since $\theta = (\pi/2) - \cos^{-1} x$ and $\theta = \sin^{-1} x$, it follows that $(\pi/2) - \cos^{-1} x = \sin^{-1} x$, which is the same as equation (1).

Differentiating both sides of equation (1) yields the formula

$$\frac{d}{dx}(\cos^{-1} x) = \frac{-1}{\sqrt{1-x^2}} \qquad \text{for } -1 < x < 1$$

THE FUNCTION $\tan^{-1} x$
We recall that the graph of $x = \tan \theta$ consists of separate branches one of which passes through the origin (see Figure 9.41). This

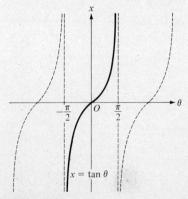

Figure 9.41
The restriction of $x = \tan \theta$ to the interval $-\pi/2 < \theta < \pi/2$. Note that as θ goes from $-\pi/2$ to $\pi/2$, $\tan \theta$ goes from $-\infty$ to ∞.

branch is defined for $-\pi/2 < \theta < \pi/2$, and for these values of θ the equation $x = \tan \theta$ has a unique solution $\theta = \tan^{-1} x$ for all x. Hence (see Figure 9.42):

> When $-\pi/2 < \theta < \pi/2$, then $-\infty < x < \infty$ and $\theta = \tan^{-1} x$ is the same as $x = \tan \theta$. The function $\theta = \tan^{-1} x$ is the inverse function of $x = \tan \theta$.

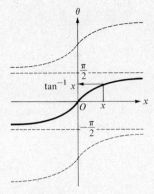

Figure 9.42
The graph of $\theta = \tan^{-1} x$.

$\theta \longrightarrow \tan \theta$	
$-\dfrac{\pi}{3}$	$-\sqrt{3}$
$-\dfrac{\pi}{6}$	$-\dfrac{1}{\sqrt{3}}$
0	0
$\dfrac{\pi}{6}$	$\dfrac{1}{\sqrt{3}}$
$\dfrac{\pi}{3}$	$\sqrt{3}$

$\tan^{-1} x \longleftarrow x$

Figure 9.43

Example 3

Find the values of $\tan^{-1} x$ when $x = -\sqrt{3}, -1/\sqrt{3}, 0, 1/\sqrt{3}, \sqrt{3}$.

Solution
As in Example 1, we are aided by the table of values for $\tan \theta$, simply reading it "backwards," as illustrated in Figure 9.43.

THE DERIVATIVE OF $\tan^{-1} x$
The formula

$$\frac{d}{dx}(\tan^{-1} x) = \frac{1}{1 + x^2} \qquad \text{for all values of } x$$

is verified as follows. Putting $\theta = \tan^{-1} x$, we have $x = \tan \theta$, and by Theorem 1 of the last section

$$\frac{d\theta}{dx} = \frac{1}{dx/d\theta} = \frac{1}{\sec^2 \theta}$$

Since

$$\sec^2 \theta = 1 + \tan^2 \theta = 1 + x^2$$

we have the desired formula.

The following is a summary of the differentiation formulas of the inverse trigonometric functions discussed thus far.

	If u is a differentiable function of x, then
$\dfrac{d}{dx}(\sin^{-1} x) = \dfrac{1}{\sqrt{1 - x^2}}$	$\dfrac{d}{dx}(\sin^{-1} u) = \dfrac{1}{\sqrt{1 - u^2}} \dfrac{du}{dx}$
$\dfrac{d}{dx}(\cos^{-1} x) = \dfrac{-1}{\sqrt{1 - x^2}}$	$\dfrac{d}{dx}(\cos^{-1} u) = \dfrac{-1}{\sqrt{1 - u^2}} \dfrac{du}{dx}$
$\dfrac{d}{dx}(\tan^{-1} x) = \dfrac{1}{1 + x^2}$	$\dfrac{d}{dx}(\tan^{-1} u) = \dfrac{1}{1 + u^2} \dfrac{du}{dx}$

316 TRIGONOMETRIC FUNCTIONS

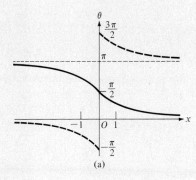

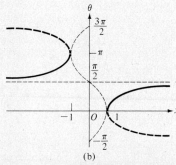

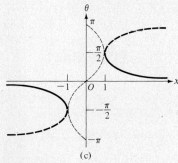

Figure 9.44
(a) The graph of $\theta = \cot^{-1} x$.

$$\cot^{-1} x = \frac{\pi}{2} - \tan^{-1} x$$

(b) The graph of $\theta = \sec^{-1} x$.

$$\sec^{-1} x = \cos^{-1} \frac{1}{x}$$

Observe that $\sec^{-1} x$ is undefined for values $-1 < x < 1$,

$$0 \leq \sec^{-1} x < \frac{\pi}{2} \quad \text{for } x \geq 1$$
$$\frac{\pi}{2} < \sec^{-1} x \leq \pi \quad \text{for } x \leq -1$$

Example 4

Find $d\theta/dx$ when

$$\theta = \tan^{-1} \frac{x-1}{x+1}$$

Solution
With $u = (x-1)/(x+1)$, we get by the chain rule

$$\frac{d}{dx}\left(\tan^{-1} \frac{x-1}{x+1}\right) = \frac{1}{1 + [(x-1)/(x+1)]^2} \cdot \frac{d}{dx}\left(\frac{x-1}{x+1}\right)$$

Since

$$\frac{d}{dx}\left(\frac{x-1}{x+1}\right) = \frac{2}{(x+1)^2}$$

we find, upon simplification, that

$$\frac{d}{dx}\left(\tan^{-1} \frac{x-1}{x+1}\right) = \frac{1}{1 + [(x-1)/(x+1)]^2} \cdot \frac{2}{(x+1)^2}$$

$$= \frac{2}{(x+1)^2 + (x-1)^2} = \frac{1}{1 + x^2}$$

THE OTHER INVERSE TRIGONOMETRIC FUNCTIONS
The inverses of $\cot \theta$, $\sec \theta$, and $\csc \theta$ are included here for the sake of completeness, but these functions will have no applications in this course. For this reason we present the defining relations and graphs in Figure 9.44 without discussion. The differentiation formulas of these inverse trigonometric functions are easily obtained with the help of the chain rule.

$$\frac{d}{dx}(\cot^{-1} x) = \frac{-1}{1 + x^2}$$

$$\frac{d}{dx}(\sec^{-1} x) = \begin{cases} \dfrac{1}{x\sqrt{x^2 - 1}} & \text{for } x > 1 \\[2ex] \dfrac{-1}{x\sqrt{x^2 - 1}} & \text{for } x < -1 \end{cases}$$

$$\frac{d}{dx}(\csc^{-1} x) = \begin{cases} \dfrac{-1}{x\sqrt{x^2 - 1}} & \text{for } x > 1 \\[2ex] \dfrac{1}{x\sqrt{x^2 - 1}} & \text{for } x < -1 \end{cases}$$

Figure 9.44 (continued)

(c) The graph of $\theta = \csc^{-1} x$.

$$\csc^{-1} x = \sin^{-1} \frac{1}{x}$$

Observe that $\csc^{-1} x$ is undefined for values $-1 < x < 1$,

$$0 < \csc^{-1} x \leq \frac{\pi}{2} \quad \text{for } x \geq 1$$
$$-\frac{\pi}{2} \leq \csc^{-1} x < 0 \quad \text{for } x \leq -1$$

Ans. $\dfrac{\sqrt{a^2 - x^2}}{b^2 + x^2}$

Ans. $\dfrac{1}{a + b \sin x}$

EXERCISES

1. Evaluate

(a) $\sin^{-1}\left(\dfrac{1}{\sqrt{2}}\right)$

(b) $\cos^{-1}\left(\dfrac{1}{\sqrt{2}}\right)$

(c) $\sin^{-1}(1) - \sin^{-1}(-1)$

(d) $\tan^{-1}\left(\tan \dfrac{\pi}{4}\right)$

2. Evaluate

(a) $\tan^{-1}(2) + \tan^{-1}(-2)$

(b) $\sin^{-1}\left(\sin \dfrac{\pi}{6}\right)$

(c) $\sin^{-1}\left[\sin\left(2\pi + \dfrac{\pi}{6}\right)\right]$

(d) $\sin^{-1}\left[\sin\left(\pi + \dfrac{\pi}{6}\right)\right]$

Find $d\theta/dx$ in Exercises 3–16 and indicate for which values of x the functions are defined.

3. $\theta = \tan^{-1} \dfrac{a}{x}$

4. $\theta = \sin^{-1} \dfrac{x-1}{x+1}$

5. $\theta = \sin^{-1} \dfrac{x}{2} + \cos^{-1} \dfrac{x}{2}$

6. $\theta = x \sin^{-1} x$

7. $\theta = \sin^{-1} \dfrac{1}{1+x^2}$

8. $\theta = \cos^{-1}(\cos x)$

9. $\theta = \cos^{-1}(\sin x)$

10. $\theta = \tan^{-1} e^x$

11. $\theta = \sin(\cos^{-1} x)$

12. $\theta = \cos(\sin^{-1} x)$

13. $\theta = \sin^{-1}(1 + \cot x)$

14. $\theta = \dfrac{\sqrt{a^2 + b^2}}{b} \sin^{-1}\left(\dfrac{x\sqrt{a^2 + b^2}}{a\sqrt{x^2 + b^2}}\right) - \sin^{-1} \dfrac{x}{a}$

15. $\theta = \dfrac{2}{\sqrt{a^2 - b^2}} \tan^{-1}\left(\dfrac{a \tan(x/2) + b}{\sqrt{a^2 - b^2}}\right) \quad (a > b)$

16. $\theta = \dfrac{1}{ab} \tan^{-1} \dfrac{b \tan x}{a}$

Graph the functions in Exercises 17–20.

17. $\theta = 2 \sin^{-1} \dfrac{x}{2}$
18. $\theta = \sin^{-1}(1 - x)$
19. $\theta = \tan^{-1}(x + 5)$
20. $\theta = \cos^{-1}(2x + 1)$

9.6 APPLICATIONS

In this section we consider related rate and maximum-minimum problems involving trigonometric functions. These problems are of the types investigated in Sections 5.2 and 5.3, and the following examples should serve as adequate preparation for tackling the exercises.

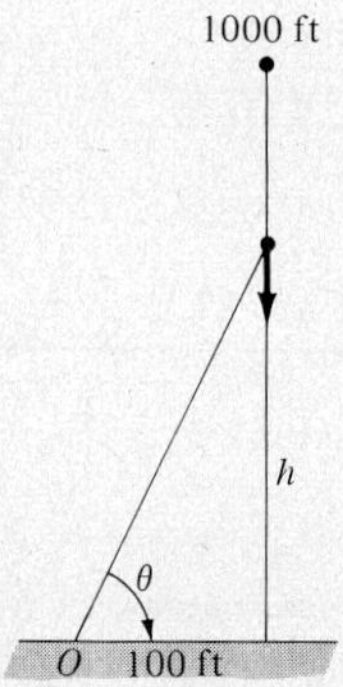

Example 1

Consider Figure 9.45. A camera mounted at O is tracking a free-falling object dropped from a height of 1000 ft. At what rate does the angle of inclination θ change when the elapsed time is 5 sec and the height of the object at time t is $h = 1000 - 16t^2$?

Figure 9.45

Solution
We shall present here two methods for solving this problem, the first involving inverse trigonometric functions, and the second involving implicit differentiation.

Method 1
(a) From the diagram we deduce that $\tan \theta = h/100$. The variables θ and t are thus seen to be related through the equations

$$\theta = \tan^{-1} \frac{h}{100}$$

$$h = 1000 - 16t^2$$

We want to find $d\theta/dt$, for which the chain rule is used:

$$\frac{d\theta}{dt} = \frac{d\theta}{dh} \cdot \frac{dh}{dt}$$

(b) This gives

$$\frac{d\theta}{dh} = \frac{1}{1 + (h/100)^2} \frac{d}{dh}\left(\frac{h}{100}\right) = \frac{100}{100^2 + h^2}$$

$$\frac{dh}{dt} = -32t$$

and we find that

$$\frac{d\theta}{dt} = -\frac{3200t}{100^2 + h^2}$$

(c) For $t = 5$, we have

$$h = (1000 - 16 \cdot 5^2) = 600$$

and hence

$$\frac{d\theta}{dt} = -\frac{3200 \cdot 5}{100^2 + 600^2} = -\frac{16}{370} \approx -0.0432 \text{ radians/sec}$$

In terms of degrees the answer is

$$\frac{16}{370} \text{ radians} = \left(\frac{16}{370} \cdot \frac{180}{\pi}\right)^\circ \approx (0.0432 \cdot 57.3)^\circ \approx 2.475^\circ$$

and hence

$$\frac{d\theta}{dt} \approx -2.475 \text{ deg/sec}$$

Remark: $d\theta/dt$ is negative because θ is decreasing.

Method 2
(a') Here we work with the equations

$$\tan \theta = \frac{h}{100}$$

$$h = 1000 - 16t^2$$

(b') We want to find $d\theta/dt$. Using implicit differentiation,

$$\sec^2 \theta \frac{d\theta}{dt} = \frac{1}{100} \frac{dh}{dt}$$

or

$$\frac{d\theta}{dt} = \frac{1}{100} \frac{1}{\sec^2 \theta} \frac{dh}{dt} = \frac{1}{100} \cos^2 \theta \frac{dh}{dt}$$

We also have

$$\frac{dh}{dt} = -32t$$

and from the diagram we find that

$$\cos \theta = \frac{100}{\sqrt{100^2 + h^2}}$$

A substitution and simplification leads to the equation

$$\frac{d\theta}{dt} = -\frac{3200t}{100^2 + h^2}$$

This is the formula obtained at the end of (b) in Method 1, and from this point we proceed as in (c).

Example 2

Examine Figure 9.46. A radar antenna mounted at O is tracking a 10-ft rocket going straight up with velocity $v = 500$ ft/min. At what rate is the angle θ changing when $\phi = 45°$?

Solution
(a) Simple geometry gives

$$\tan \phi = \frac{h}{1000}$$

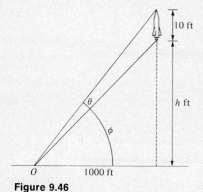

Figure 9.46

$$\tan(\phi + \theta) = \frac{h + 10}{1000}$$

Taking inverse tangents and eliminating ϕ results in the equation

$$\theta = \tan^{-1} \frac{h + 10}{1000} - \tan^{-1} \frac{h}{1000}$$

(b) We use the chain rule

$$\frac{d\theta}{dt} = \frac{d\theta}{dh} \frac{dh}{dt}$$

First we find that

$$\frac{d\theta}{dh} = \frac{1}{1 + [(h+10)/1000]^2} \frac{d}{dh}\left(\frac{h+10}{1000}\right) - \frac{1}{1 + (h/1000)^2} \frac{d}{dh}\left(\frac{h}{1000}\right)$$

$$= \left\{\frac{1}{1 + [(h+10)/1000]^2} - \frac{1}{1 + (h/1000)^2}\right\} \frac{1}{1000}$$

and

$$\frac{dh}{dt} = v = 500 \text{ ft/min}$$

Thus

$$\frac{d\theta}{dt} = \left\{\frac{1}{1 + [(h+10)/1000]^2} - \frac{1}{1 + (h/1000)^2}\right\} \cdot \frac{1}{2}$$

(c) Recalling that $45° = \pi/4$, we see from the diagram that

$$\frac{h}{1000} = \tan \frac{\pi}{4} = 1$$

$$\frac{h+10}{1000} = 1 + 0.01 = 1.01$$

and simple arithmetic leads to the final result

$$\frac{d\theta}{dt} = \left(\frac{1}{1+1.01^2} - \frac{1}{1+1}\right) \cdot \frac{1}{2} \approx \left(\frac{1}{2.02} - \frac{1}{2}\right) \cdot \frac{1}{2} = -\frac{1}{404} \text{ radians/min}$$

Again the sign of $d\theta/dt$ is negative because θ is decreasing.

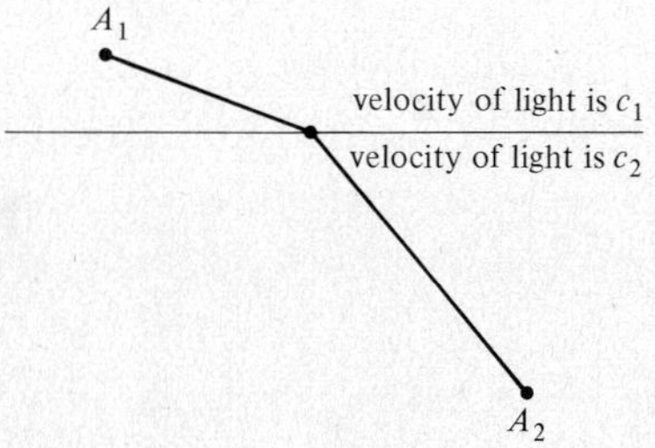

Example 3

It is known that light travels at different velocities in different substances (such as air, water, glass, and so on). Suppose that light travels from a point A_1 in a substance where its velocity is c_1 to a point A_2 in a substance where its velocity is c_2 (see Figure 9.47). *Fermat's principle* states that light always travels along the path that requires the least time. Describe the path from A_1 to A_2.

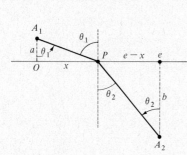

Figure 9.47

Solution

(a) Let the light cross from one substance to the other at a point P. Then paths taken by the light from A_1 to P and from P to A_2 are straight line segments because these are the shortest distances.

(b) Using the formula

$$\text{time} = \frac{\text{distance}}{\text{velocity}}$$

we see from Figure 9.48 that the travel time from A_1 to P is

$$t_1 = \frac{\sqrt{x^2 + a^2}}{c_1}$$

and the travel time from P to A_2 is

$$t_2 = \frac{\sqrt{(e-x)^2 + b^2}}{c_2}$$

Figure 9.48

The total time is therefore given by the formula

$$t = t_1 + t_2 = \frac{\sqrt{x^2 + a^2}}{c_1} + \frac{\sqrt{(e-x)^2 + b^2}}{c_2}$$

(c) To find the least time, we evaluate dt/dx and let it equal zero:

$$\frac{dt}{dx} = \frac{x}{c_1\sqrt{x^2 + a^2}} - \frac{e-x}{c_2\sqrt{(e-x)^2 + b^2}} = 0$$

From the diagram we see that

$$\frac{x}{\sqrt{x^2 + a^2}} = \sin \theta_1$$

$$\frac{e-x}{\sqrt{(e-x)^2 + b^2}} = \sin \theta_2$$

and hence

$$\frac{dt}{dx} = \frac{\sin \theta_1}{c_1} - \frac{\sin \theta_2}{c_2} = 0$$

or

$$\frac{\sin \theta_1}{c_1} = \frac{\sin \theta_2}{c_2}$$

This is *Snell's law of refraction*.

(d) To see that a minimum does, indeed, occur when $dt/dx = 0$, we shall demonstrate that $d^2t/dx^2 > 0$, thereby showing that the graph of

$$t = \frac{1}{c_1}\sqrt{x^2 + a^2} + \frac{1}{c_2}\sqrt{(e-x)^2 + b^2}$$

is concave up. Now, using the first formula in (c),

$$\frac{d^2}{dx^2}\left(\sqrt{x^2 + a^2}\right) = \frac{a^2}{(x^2 + a^2)^{3/2}}$$

$$\frac{d^2}{dx^2}\left(\sqrt{(e-x)^2 + b^2}\right) = \frac{-b^2}{[(e-x)^2 + b^2]^{3/2}}$$

and accordingly,

$$\frac{d^2t}{dx^2} = \frac{a^2}{c_1(x^2 + a^2)^{3/2}} + \frac{b^2}{c_2[(e-x)^2 + b^2]^{3/2}} > 0$$

We recall that the conditions $dt/dx = 0$, $d^2t/dx^2 > 0$ guarantee a minimum.

EXERCISES

1. Consult Figure 9.49. A camera mounted at O is tracking a 10-ft rocket going straight up with velocity $v = 500$ ft/min. At what rate is the angle θ changing when 1 min has elapsed?

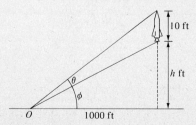

Figure 9.49

2. Consult Figure 9.50. The hands of a clock have lengths 2 and 3 in. As the hands move around the clock, they sweep out the triangle OAB. At what rate is the area of the triangle changing at time 12:10?

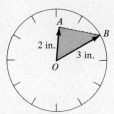

Figure 9.50

3. Consult Figure 9.51. A searchlight is following a plane flying at an altitude of 3000 ft in a straight line over the light; the plane's velocity is 500 mph. At what rate is the searchlight turning when the distance between light and plane is 5000 ft?

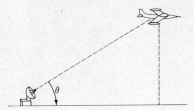

Figure 9.51

4. A trough of the dimensions shown in Figure 9.52 contains 20 ft³ of water. One of its sides is hinged along the lower edge, and as this side is swung out the water level is decreasing. If the angle θ is decreasing at the rate of 1 deg/sec, at what rate is the water level x decreasing when $\theta = 30°$?

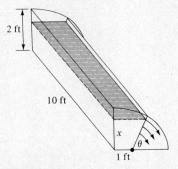

Figure 9.52

5. A weight hangs on a 2-ft-long spring (see Figure 9.53). The weight is pulled down and then released. The weight oscillates

up and down, and the length L of the spring when t seconds elapsed is given by the formula

$$L = 2 + \cos 2\pi t$$

Find

(a) the length of the spring at times $t = 0, \frac{1}{2}, 1, \frac{3}{2}, \frac{5}{8}$;

(b) the velocity of the weight at time $t = \frac{1}{4}$;

(c) the acceleration of the weight at time $t = \frac{1}{4}$;

(d) the time intervals during which the weight is moving down.

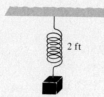

Figure 9.53

6. The sides of an isosceles triangle (see Figure 9.54) are sliding in an outward direction with velocity 1 in./min. At what rate is the area bounded by the triangle changing when $\theta = 20°$?

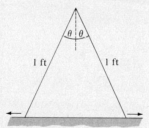

Figure 9.54

7. An observer (Figure 9.55) is viewing a picture 5 ft tall hanging 10 ft high; the observer's eyes are 6 ft from the ground. At what distance x from the picture will the observation angle θ be largest?

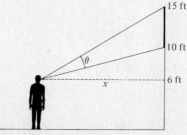

Figure 9.55

8. What is the longest beam that can be moved horizontally around the corner pictured at right (see Figure 9.56)?

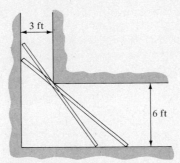

Figure 9.56

9. Consult Figure 9.57. A 6-ft fence is erected 4 ft from the house. What is the shortest ladder that can stand outside the fence and lean against the wall? What is its angle of inclination θ?

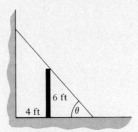

Figure 9.57

10. A sector is removed from a circular metal sheet of radius 4 ft (Figure 9.58a) and made into a cone by joining its edges together without overlap (Figure 9.58b). What is the angle θ of the sector that yields a cone of maximum volume? The volume of a circular cone of height h and base of radius r is $v = \frac{1}{3}\pi r^2 h$.

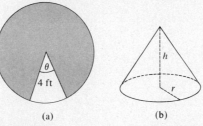

Figure 9.58

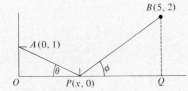

Figure 9.59

11. Find the shortest distance between the points $A(0, 1)$ and $B(5, 2)$ along a path that must touch the x axis (see Figure 9.59 at left). Also, find the angles θ and ϕ.

9.7 INTEGRATION OF TRIGONOMETRIC FUNCTIONS

The interrelations existing between the trigonometric functions and their derivatives allow us to extend the class of functions that we can integrate. A basic table of integrals is this:

INDEFINITE INTEGRALS OF TRIGONOMETRIC FUNCTIONS

1. $\int \sin x \, dx = -\cos x + C$

2. $\int \cos x \, dx = \sin x + C$

3. $\int \sec^2 x \, dx = \tan x + C$

4. $\int \csc^2 x \, dx = -\cot x + C$

5. $\int \sec x \tan x \, dx = \sec x + C$

6. $\int \csc x \cot x \, dx = -\csc x + C$

7. $\int \tan x \, dx = -\ln|\cos x| + C$

8. $\int \cot x \, dx = \ln|\sin x| + C$

Except for the last two formulas, these are mere restatements of the differentiation formulas of Section 9.2 and Section 9.3. The last two formulas are consequences of the formula

$$\int \frac{f'(x)}{f(x)} \, dx = \ln|f(x)| + C$$

and the facts

$$\tan x = \frac{\sin x}{\cos x} = \left[\frac{d}{dx}(-\cos x)\right] \bigg/ \cos x$$

$$\cot x = \frac{\cos x}{\sin x} = \left[\frac{d}{dx}(\sin x)\right] \bigg/ \sin x$$

Example 1

Find $\int \sin^a x \cos x \, dx$ for $a \neq -1$.

Solution

We use the method of substitution (Section 6.7). Let $u = \sin x$, so that $du = \cos x \, dx$. Then

$$\sin^a x \cos x \, dx = u^a \, du$$

Since

$$\int u^a \, du = \frac{u^{a+1}}{a+1} + C$$

we see that

$$\int \sin^a x \cos x \, dx = \frac{1}{a+1} \sin^{a+1} x + C$$

Example 2

Find $\int \cos^2 x \, dx$ and $\int \sin^2 x \, dx$.

Solution

Using the identity (Section 9.1, Exercise 17)

$$\cos^2 x = \tfrac{1}{2}(1 + \cos 2x)$$

we have

$$\int \cos^2 x \, dx = \tfrac{1}{2} \int (1 + \cos 2x) \, dx$$

$$= \tfrac{1}{2} \int 1 \, dx + \tfrac{1}{2} \int \cos 2x \, dx = \tfrac{1}{2}x + \tfrac{1}{4} \sin 2x + C$$

With the identity

$$\sin 2x = 2 \sin x \cos x$$

the final answer can be put in the form

$$\int \cos^2 x \, dx = \tfrac{1}{2}(x + \sin x \cos x) + C$$

For the second integral we now have

$$\int \sin^2 x \, dx = \int (1 - \cos^2 x) \, dx$$

$$= \int 1 \, dx - \int \cos^2 x \, dx$$

$$= x - \tfrac{1}{2}(x + \sin x \cos x) + C$$
$$= \tfrac{1}{2}(x - \sin x \cos x) + C$$

That is,

$$\int \sin^2 x \, dx = \tfrac{1}{2}(x - \sin x \cos x) + C$$

While the indefinite integral of $\cos^2 x$ was easy to find, this is not so for $\cos^n x$, when n is large. For this case we use a reduction formula (Section 7.6):

$$\int \cos^n x \, dx = \frac{1}{n} \cos^{n-1} x \cdot \sin x + \frac{n-1}{n} \int \cos^{n-2} x \, dx \quad (1)$$

It is assumed, of course, that $n \geq 2$. The formula is verified using integration by parts as follows:

We set $u = \cos^{n-1} x$ and $dv = \cos x$, so that

$$du = -(n-1) \cos^{n-2} x \sin x \, dx \quad \text{and} \quad v = \sin x$$

Hence,

$$\int \cos^n x \, dx = \cos^{n-1} x \cdot \sin x$$
$$\qquad - \int \sin x \cdot [-(n-1)\cos^{n-2} x \cdot \sin x] \, dx$$
$$= \cos^{n-1} x \cdot \sin x + (n-1) \int \cos^{n-2} x \cdot \sin^2 x \, dx$$

The substitution $\sin^2 x = 1 - \cos^2 x$ gives for the last integral

$$\int \cos^{n-2} x \cdot \sin^2 x \, dx = \int \cos^{n-2} x (1 - \cos^2 x) \, dx$$
$$= \int \cos^{n-2} x \, dx - \int \cos^n x \, dx$$

and accordingly,

$$\int \cos^n x \, dx = \cos^{n-1} x \cdot \sin x + (n-1) \int \cos^{n-2} x \, dx$$
$$\qquad - (n-1) \int \cos^n x \, dx$$

Solving this equation for $\int \cos^n x \, dx$ gives formula (1).

Example 3

Find $\int \cos^3 x \, dx$ and $\int \cos^4 x \, dx$.

Solution
Formula (1) becomes, with $n = 3$,

$$\int \cos^3 x \, dx = \tfrac{1}{3} \cos^2 x \sin x + \tfrac{2}{3} \int \cos x \, dx$$

$$= \tfrac{1}{3} \cos^2 x \sin x + \tfrac{2}{3} \sin x + C$$

For $n = 4$ we have, by Example 2,

$$\int \cos^4 x \, dx = \tfrac{1}{4} \cos^3 x \sin x + \tfrac{3}{4} \int \cos^2 x \, dx$$

$$= \tfrac{1}{4} \cos^3 x \sin x + \tfrac{3}{8}(x + \sin x \cos x) + C$$

To find indefinite integrals of $\tan^n x$ for large integers n, we again use a reduction formula:

$$\int \tan^n x \, dx = \frac{1}{n-1} \tan^{n-1} x - \int \tan^{n-2} x \, dx \qquad (2)$$

To verify this formula we write

$$\int \tan^n x \, dx = \int \tan^{n-2} x \tan^2 x \, dx$$

and use the identity (Section 9.3, Exercise 18)

$$\tan^2 x = \sec^2 x - 1$$

This gives

$$\int \tan^n x \, dx = \int \tan^{n-2} x (\sec^2 x - 1) \, dx$$

$$= \int \tan^{n-2} x \sec^2 x \, dx - \int \tan^{n-2} x \, dx$$

Using the method of substitution (see Exercise 18 below),

$$\int \tan^{n-2} x \sec^2 x \, dx = \frac{1}{n-1} \tan^{n-1} x + C$$

and hence

$$\int \tan^n x \, dx = \frac{1}{n-1} \tan^{n-1} x - \int \tan^{n-2} x \, dx$$

which is formula (2).

Example 4

Find $\int \tan^3 x \, dx$.

Solution
By formula (2) and the indefinite integral for $\int \tan x \, dx$,

$$\int \tan^3 x \, dx = \tfrac{1}{2} \tan^2 x - \int \tan x \, dx$$

$$= \tfrac{1}{2} \tan^2 x + \ln|\cos x| + C$$

EXERCISES

Find the integrals in Exercises 1–14.

1. $\int \sin 7x \, dx$

2. $\int x^2 \sin x^3 \, dx$

3. $\int \cos(2x + 1) \, dx$

4. $\int \dfrac{1}{x} \cos(\ln x) \, dx$

5. $\int (\sin x + \cos x)^2 \, dx$

6. $\int (\sin^2 x + \cos^2 x) \, dx$

7. $\int x \sec^2(2x^2 + 1) \, dx$

8. $\int \tan \tfrac{1}{5} x \, dx$

9. $\int \cot \sqrt{x+1} \, dx$

10. $\int e^x \tan e^x \, dx$

11. $\int (\sin^4 x - \cos^4 x) \, dx$

Hint: You could use the identity $a^3 - b^3 = (a - b)(a^2 + ab + b^2)$.

12. $\int (\sin^3 x - \cos^3 x) \, dx$

13. $\int (\cos^3 x - 3 \sin^2 x \cos x) \, dx$

14. $\int (\sin x \cos 4x + \sin 4x \cos x) \, dx$

15. Verify the reduction formula

$$\int \sin^n x \, dx = -\frac{1}{n} \sin^{n-1} x \cdot \cos x + \frac{n-1}{n} \int \sin^{n-2} x \, dx$$

16. Use Exercise 15 to find $\int \sin^3 x \, dx$ and $\int \sin^4 x \, dx$.

17. Find a reduction formula for $\int \cot^n x \, dx$.

18. Verify the formula

$$\int \tan^m x \cdot \sec^2 x \, dx = \frac{1}{m+1} \tan^{m+1} x + C$$

by using the method of substitution.

Find the integrals in Exercises 20–23 using integration by parts.

19. $\displaystyle\int x \cos x \, dx$

20. $\displaystyle\int x \sin x \, dx$

21. $\displaystyle\int x^2 \cos x \, dx$

Hint: Use integration by parts twice, then solve for $\int e^x \sin x \, dx$.

22. $\displaystyle\int e^x \sin x \, dx$

23. $\displaystyle\int x \sec^2 x \, dx$

24. (a) Verify the following formulas with the help of the addition formulas in Section 9.1.

$$\sin kx \sin nx = \tfrac{1}{2}[\cos(k-n)x - \cos(k+n)x]$$
$$\cos kx \cos nx = \tfrac{1}{2}[\cos(k-n)x + \cos(k+n)x]$$
$$\sin kx \cos nx = \tfrac{1}{2}[\sin(k-n)x + \sin(k+n)x]$$

Use these identities to verify the following formulas:

(b) $\displaystyle\int_{-\pi}^{\pi} \cos kx \cos nx \, dx = 0 \quad$ for $k \neq n$

(c) $\displaystyle\int_{-\pi}^{\pi} \sin kx \sin nx \, dx = 0 \quad$ for $k \neq n$

(d) $\displaystyle\int_{-\pi}^{\pi} \sin kx \cos nx \, dx = 0 \quad$ for all integers k and n

(e) Evaluate the integral in (b) when $k = n$.

(f) Evaluate the integral in (c) when $k = n = 0$ and when $k = n \neq 0$.

9.8 TECHNIQUES OF INTEGRATION

Among the many integrals yet beyond our reach are integrals of the general form

$$\int \frac{Ax + B}{\sqrt{Cx^2 + Dx + E}} \, dx \quad \text{and} \quad \int \frac{Ax + B}{Cx^2 + Dx + E} \, dx$$

We shall show here how some of these integrals can be reduced to one of the following forms:

INTEGRATION FORMULAS

$$\int \frac{1}{\sqrt{a^2 - x^2}}\, dx = \sin^{-1}\frac{x}{a} + C \qquad (1)$$

$$\int \frac{1}{x^2 + a^2}\, dx = \frac{1}{a}\tan^{-1}\frac{x}{a} + C \qquad (2)$$

$$\int \frac{1}{x^2 - a^2}\, dx = \frac{1}{2a}\ln\left|\frac{x-a}{x+a}\right| + C \qquad (3)$$

$$\int \frac{x}{x^2 + a}\, dx = \frac{1}{2}\ln|x^2 + a| + C \qquad (4)$$

Further integration formulas are described in Chapter 10. Formulas (1) and (2) are just restatements of differentiation formulas in Section 9.5. From that section we also know that

$$\frac{d}{dx}\left(-\cos^{-1}\frac{x}{a}\right) = \frac{1}{\sqrt{a^2 - x^2}}$$

so that (1) can also be written in the form

$$\int \frac{1}{\sqrt{a^2 - x^2}}\, dx = -\cos^{-1}\frac{x}{a} + C$$

Formula (3) is derived from the algebraic identity

$$\frac{1}{x^2 - a^2} = \frac{1}{2a}\left(\frac{1}{x-a} - \frac{1}{x+a}\right)$$

From this identity we see that

$$\int \frac{1}{x^2 - a^2}\, dx = \int \frac{1}{2a}\left(\frac{1}{x-a} - \frac{1}{x+a}\right) dx$$

$$= \frac{1}{2a}\int \frac{1}{x-a}\, dx - \frac{1}{2a}\int \frac{1}{x+a}\, dx \qquad (5)$$

We now use formula (3) from Section 7.2,

$$\int \frac{1}{x}\, dx = \ln|x| + C$$

and Corollary 1 from Section 6.7, which states:

If $\int f(x)\, dx = F(x) + C$, then $\int f(x - A)\, dx = F(x - A) + C$ \quad (6)

With $A = a$, we thus get

$$\int \frac{1}{x-a} dx = \ln|x-a| + C$$

and with $A = -a$, we get

$$\int \frac{1}{x+a} dx = \ln|x+a| + C$$

Using these results in (5), we find that

$$\int \frac{1}{x^2 - a^2} dx = \frac{1}{2a} \ln|x-a| - \frac{1}{2a} \ln|x+a| + C$$

$$= \frac{1}{2a}(\ln|x-a| - \ln|x+a|) + C = \frac{1}{2a} \ln\frac{|x-a|}{|x+a|} + C$$

$$= \frac{1}{2a} \ln\left|\frac{x-a}{x+a}\right| + C$$

The derivation of formula (4) is relegated to the exercises.

Example 1

Find

$$\int \frac{1}{x^2 + 2x + 4} dx$$

Solution
We use the method of *completing the square*, which in this case consists of writing

$$x^2 + 2x + 4 = (x^2 + 2x + 1) + 3 = (x+1)^2 + (\sqrt{3})^2$$

Thus,

$$\int \frac{1}{x^2 + 2x + 4} dx = \int \frac{1}{(x+1)^2 + (\sqrt{3})^2} dx$$

and the right integral resembles the integral in (2). Using formula (6) with $A = -1$, we find that

$$\int \frac{1}{x^2 + 2x + 4} dx = \frac{1}{\sqrt{3}} \tan^{-1} \frac{x+1}{\sqrt{3}} + C$$

Here formula (2) was used with $a = \sqrt{3}$.

Example 2

Find

$$\int \frac{1}{x^2 + 2x - 4} dx$$

Solution
Completing the square in this case gives
$$x^2 + 2x - 4 = (x^2 + 2x + 1) - 5 = (x+1)^2 - (\sqrt{5})^2$$
and hence
$$\int \frac{1}{x^2 + 2x - 4} dx = \int \frac{1}{(x+1)^2 - (\sqrt{5})^2} dx$$
Using formula (3) with $a = \sqrt{5}$, and formula (6) with $A = -1$, we find that
$$\int \frac{1}{x^2 + 2x - 4} dx = \frac{1}{2\sqrt{5}} \ln \left| \frac{(x+1) - \sqrt{5}}{(x+1) + \sqrt{5}} \right| + C$$

Example 3

Find
$$\int \frac{3x + 1}{2x^2 - 3x + 5} dx$$

Solution
We again begin by completing the square, as follows:
$$2x^2 - 3x + 5 = 2\left(x^2 - \frac{3}{2}x + \frac{5}{2}\right)$$
$$= 2\left[\left(x^2 - 2 \cdot \frac{3}{4}x + \frac{3^2}{4^2}\right) + \left(\frac{5}{2} - \frac{3^2}{4^2}\right)\right]$$
$$= 2\left[\left(x - \frac{3}{4}\right)^2 + \left(\sqrt{\frac{31}{16}}\right)^2\right]$$

At this point it is useful to introduce the substitution
$$u = x - \tfrac{3}{4}$$
which gives
$$2x^2 - 3x + 5 = 2(u^2 + \tfrac{31}{16})$$
In addition, we have
$$x = u + \tfrac{3}{4}$$
and thus
$$3x + 1 = 3(u + \tfrac{3}{4}) + 1 = 3u + \tfrac{13}{4}$$

Since $du = dx$,
$$\int \frac{3x+1}{2x^2 - 3x + 5} dx = \int \frac{3u + \tfrac{13}{4}}{2(u^2 + \tfrac{31}{16})} du = \frac{3}{2} \frac{u}{u^2 + \tfrac{31}{16}} + \frac{13}{8} \frac{1}{u^2 + \tfrac{31}{16}} du$$

9.8 TECHNIQUES OF INTEGRATION

We have, using formulas (2) and (4),

$$\int \frac{3u + \frac{13}{4}}{2(u^2 + \frac{31}{16})} \, du = \frac{3}{2} \int \frac{u}{u^2 + \frac{31}{16}} \, du + \frac{13}{8} \int \frac{1}{u^2 + \frac{31}{16}} \, du$$

$$= \frac{3}{2} \cdot \frac{1}{2} \ln \left| u^2 + \frac{31}{16} \right| + \frac{13}{8} \frac{1}{\sqrt{\frac{31}{16}}} \tan^{-1} \frac{u}{\sqrt{\frac{31}{16}}}$$

Substituting $u = x - \frac{3}{4}$ and simplifying, we have the final result

$$\int \frac{3x+1}{2x^2 - 3x + 5} \, dx = \frac{3}{4} \ln|2x^2 - 3x + 5| + \frac{13}{2\sqrt{31}} \tan^{-1} \frac{4x - 3}{\sqrt{31}} + C$$

REMARK
You may find at times that your answer to an exercise does not agree with the answer in the book. You should not assume automatically that either your answer or the book's answer is wrong: The two may be related through some algebraic manipulation that is not apparent. In such a case it is best to settle the question by verifying the answer through differentiation.

Example 4
Find

$$\int \frac{x}{\sqrt{5 + 4x - x^2}} \, dx$$

Solution
We again begin by completing the square:

$$5 + 4x - x^2 = 5 - (x^2 - 2 \cdot 2x) = (5 + 4) - (x^2 - 2 \cdot 2x + 4)$$
$$= 3^2 - (x - 2)^2$$

Hence,

$$\int \frac{x}{\sqrt{5 + 4x - x^2}} \, dx = \int \frac{(x-2) + 2}{\sqrt{3^2 - (x-2)^2}} \, dx$$

$$= \int \frac{x - 2}{\sqrt{3^2 - (x-2)^2}} \, dx + 2 \int \frac{1}{\sqrt{3^2 - (x-2)^2}} \, dx$$

$$= I_1 + 2I_2$$

The integral I_2 is obtained from formula (1) with $a = 3$ and formula (6) with $A = 2$:

$$I_2 = \int \frac{1}{\sqrt{3^2 - (x-2)^2}} \, dx = \sin^{-1} \frac{x - 2}{3} + C$$

To find I_1, we let $u = 3^2 - (x-2)^2$. Since $du = -2(x-2) \, dx$, we have

$$(x - 2) \, dx = -\tfrac{1}{2} \, du$$

We can thus write I_1 as follows:

$$I_1 = \int \frac{x-2}{\sqrt{3^2 - (x-2)^2}} dx = \int \frac{-\frac{1}{2} du}{u^{1/2}} = -\frac{1}{2} \int u^{-1/2} du = -u^{1/2} + C$$

Thus,

$$I_1 = -\sqrt{3^2 - (x-2)^2} + C = -\sqrt{5 + 4x - x^2} + C$$

Hence,

$$\int \frac{x}{\sqrt{5 + 4x - x^2}} dx = -\sqrt{5 + 4x - x^2} + 2 \sin^{-1} \frac{x-2}{3} + C$$

EXERCISES

Find the indefinite integrals in Exercises 1–12.

1. $\displaystyle\int \frac{1}{\sqrt{1 - 2x^2}} dx$

2. $\displaystyle\int \frac{1}{\sqrt{12 - x - \frac{1}{2}x^2}} dx$

3. $\displaystyle\int \frac{1}{2x^2 + 7} dx$

4. $\displaystyle\int \frac{1}{\frac{1}{2}x^2 + x - \frac{1}{2}} dx$

5. $\displaystyle\int_0^1 \frac{x}{x^2 - 2x + 6} dx$

6. $\displaystyle\int_{-5}^{-3} [5 - (x + 5)^2]^{-1/2} dx$

7. $\displaystyle\int \frac{1}{(x-5)(x+5)} dx$

8. $\displaystyle\int \frac{x-1}{x^2 - 2x + 2} dx$

9. $\displaystyle\int_1^2 \frac{1}{x^2 + 6x} dx$

10. $\displaystyle\int_0^1 \frac{1}{2 - x^2} dx$

11. $\displaystyle\int \frac{1}{2x^2 + 4x + 4} dx$

12. $\displaystyle\int \frac{1}{\frac{1}{4}x^2 + x + 2} dx$

The indefinite integrals in Exercises 13–20 require more complicated substitutions than the preceding integrals.

13. $\displaystyle\int \frac{e^x}{\sqrt{1 - e^{2x}}} dx$

14. $\displaystyle\int \frac{2x}{(x^2 + 1)^2 + 1} dx$

15. $\displaystyle\int \frac{x^2}{x^3 + 1} dx$

16. $\displaystyle\int \frac{e^x}{(e^x + 1)(e^x - 1)} dx$

Hint: Try $u = \sqrt{x}$.

17. $\displaystyle\int \frac{1}{\sqrt{1 - x\sqrt{x}}} dx$

18. $\displaystyle\int \frac{2x^2}{x^4 - 1} dx$

19. $\displaystyle\int \frac{-2}{x^4 - 1} dx$

20. $\displaystyle\int \frac{(x+1)^2}{x^2 + 1} dx$

21. Derive formula (4).

QUIZ 1

1. Plot the graphs of $y = 2 \sin 3x$ and $y = \frac{1}{2} \sin^{-1} 2x$.
2. Find the period of $g(x) = \tan x \sec 4x$.
3. Find a formula for the points x at which

    (a) $\dfrac{d}{dx}\left(\sin \dfrac{x}{4}\right) = 0$

    (b) $\cos 8x = -1$

4. Verify the formula
$$\tan(\theta_1 + \theta_2) = \frac{\tan \theta_1 + \tan \theta_2}{1 - \tan \theta_1 \tan \theta_2}$$

5. Find dy/dx in each case.

    (a) $y = \sin 4x \cos \dfrac{x}{4}$ (b) $y = \ln \sin^2 x$

    (c) $y = \dfrac{\tan^2 x}{x}$ (d) $y = \sin^{-1} \dfrac{x}{4}$

    (e) $y = \tan^{-1} \dfrac{x-1}{x+1}$ (f) $y = \cos^{-1}(12 \cos x)$

    (g) $\sin x + \cos y = x$ (h) $x - y = \cot xy$

6. Find the following indefinite integrals.

    (a) $\displaystyle\int \cot(2x + 3)\, dx$ (b) $\displaystyle\int \sin^2 \dfrac{x}{5}\, dx$

    (c) $\displaystyle\int \dfrac{1}{x} \tan(\ln x)\, dx$ (d) $\displaystyle\int x^3 \sin 2x\, dx$

    (e) $\displaystyle\int \dfrac{1}{\sqrt{5 - 2x^2}}\, dx$ (f) $\displaystyle\int \dfrac{1}{\frac{1}{2}x^2 + 2}\, dx$

    (g) $\displaystyle\int \dfrac{x}{x^2 - x + 1}\, dx$ (h) $\displaystyle\int \dfrac{1}{\sqrt{1 - 4x^2}}\, dx$

QUIZ 2

1. Plot the graph of $f(x) = x \cos x$.
2. Evaluate the following expressions.

    (a) $\sin^{-1}\left(\sin \dfrac{\pi}{6}\right)$

    (b) $\cos^{-1}\left(\dfrac{\sqrt{2}}{2}\right)$

(c) $\tan^{-1}\left(\tan \dfrac{\pi}{4}\right)$

3. Find dy/dx in each case.

   (a) $y = e^{\sin x}$ \hspace{2em} (b) $y = \tan^5 x$

   (c) $y = \sec \dfrac{x^2 - 1}{x^2 + 1}$ \hspace{2em} (d) $y = x^{\cos x}$

   (e) $y = \sin^{-1}(\cos x)$ \hspace{2em} (f) $x^3 + y^3 = \sin(x + y)$

4. Evaluate the following definite integrals.

   (a) $\displaystyle\int_{-\pi}^{\pi} \sin x \cos 2x \, dx$

   (b) $\displaystyle\int_{-\pi}^{\pi} \sin^2 x \, dx$

   (c) $\displaystyle\int_{0}^{\pi} x \sin 3x \, dx$

5. Find the following indefinite integrals.

   (a) $\displaystyle\int \tan^3 x \, dx$ \hspace{2em} (b) $\displaystyle\int \dfrac{1}{\tfrac{1}{2}x^2 - 2} \, dx$

   (c) $\displaystyle\int \dfrac{1}{x^2 + 5x} \, dx$ \hspace{2em} (d) $\displaystyle\int e^x \cos e^x \, dx$

   (e) $\displaystyle\int \dfrac{1}{x^2 + 4x + 13} \, dx$

6. An observer at O is watching a 12-ft long car going by on a straight road (see Figure 9.60) with a velocity $v = 150$ ft/sec. At what rate is the angle of vision θ changing 1 min after the car passed the observer?

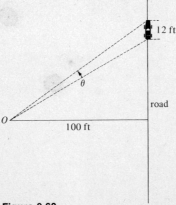

Figure 9.60

CHAPTER 10

TECHNIQUES OF INTEGRATION

Following a brief review of the method of substitution with differentials (see Sections 5.6 and 6.7), we shall discuss four new methods that will substantially increase the class of functions we can integrate. This chapter is not essential for the continuity of the material, and one can proceed directly to Chapter 11.

10.1 SUBSTITUTION WITH DIFFERENTIALS

As a review, we shall show how to find the integral

$$\int \sqrt{9 - x^2}\, dx$$

Example 1

Express the differential $(9 - x^2)\, dx$ in terms of t and dt if $x = 3 \sin t$ and $-\pi/2 < t < \pi/2$.

Solution
We have

$$dx = 3 \sin t\, dt$$

and

$$\sqrt{9 - x^2} = \sqrt{9 - 9 \sin^2 t} = 3\sqrt{1 - \sin^2 t} = 3\sqrt{\cos^2 t}$$

339

Since $\cos t > 0$ for $-\pi/2 < t < \pi/2$, we can eliminate the last radical and write $3\sqrt{\cos^2 t} = 3 \cos t$. Hence

$$\sqrt{9 - x^2} = 3 \cos t$$

and therefore

$$\sqrt{9 - x^2}\, dx = (3 \cos t)(3 \cos t\, dt) = 9 \cos^2 t\, dt$$

Example 2

Find $\int \sqrt{9 - x^2}\, dx$.

Solution
In this problem we substitute directly for x in terms of t, the function being

$$x = 3 \sin t \qquad (1)$$

Since $\sqrt{9 - x^2}$ must be defined, x must be restricted to the interval $-3 < x < 3$, and this confines t to the interval $-\pi/2 < t < \pi/2$. We observe that (1) is the same as $t = \sin^{-1} x/3$, so (1) is equivalent to setting $t = u(x)$, where $u(x) = \sin^{-1} x/3$. We now proceed with the following steps.

Step 1
According to Example 1,

$$\sqrt{9 - x^2}\, dx = 9 \cos^2 t\, dt$$

Step 2

$$\int 9 \cos^2 t\, dt = \tfrac{9}{2}(t + \sin t \cos t) + C$$

(see Example 2 in Section 9.7).

Step 3

$$\int \sqrt{9 - x^2}\, dx = \frac{9}{2}\left[\sin^{-1}\frac{x}{3} + \sin\left(\sin^{-1}\frac{x}{3}\right)\cos\left(\sin^{-1}\frac{x}{3}\right)\right] + C$$

This can be simplified by noting that

$$\sin\left(\sin^{-1}\frac{x}{3}\right) = \frac{x}{3}$$

and

$$\cos t = \sqrt{1 - \sin^2 t}$$

so that
$$\cos\left(\sin^{-1}\frac{x}{3}\right) = \sqrt{1 - \sin^2\left(\sin^{-1}\frac{x}{3}\right)} = \sqrt{1 - \left(\frac{x}{3}\right)^2} = \frac{\sqrt{9 - x^2}}{3}$$

Hence
$$\int \sqrt{9 - x^2}\, dx = \frac{9}{2}\left[\sin^{-1}\frac{x}{3} + \frac{x}{3}\frac{\sqrt{9 - x^2}}{3}\right] + C$$
$$= \frac{9}{2}\sin^{-1}\frac{x}{3} + \frac{x\sqrt{9 - x^2}}{2} + C$$

EXERCISES

Find the integrals in Exercises 1–14 using differentials.

Hint: Let $t = x + \frac{1}{x}$.

1. $\int \left(x + \frac{1}{x}\right)^{3/2}\left(1 - \frac{1}{x^2}\right) dx$

2. $\int \frac{x}{x^2 + 1}\, dx$

3. $\int \frac{x}{\sqrt{x^2 + 4}}\, dx$

4. $\int (2x + 1)^5\, dx$

5. $\int x e^{4x^2 + 2}\, dx$

Hint: Let $t = \frac{x + 1}{x - 1}$.

6. $\int \frac{1}{(x - 1)^2}\sqrt{\frac{x + 1}{x - 1}}\, dx$

7. $\int 5s^3 \sqrt{s^4 + 1}\, ds$

8. $\int \sqrt{a^2 - x^2}\, dx \quad (a > 0)$

9. $\int \frac{1}{x^2 - 2x - 2}\, dx$

10. $\int \frac{3x^2 - 1}{x^3 - x + 2}\, dx$

11. $\int \frac{\sqrt{1 - x^2}}{x^4}\, dx$, using the substitution $x = \frac{1}{t} \quad (0 < x < 1)$

12. $\int \frac{x^3}{\sqrt{4 - x^2}}\, dx$, using the substitution $x = 2\sin t$ for $-\frac{\pi}{2} < t < \frac{\pi}{2}$

13. $\int \frac{1}{(x^2 + 4)^2}\, dx$, using the substitution $x = 2\tan t$ for $-\frac{\pi}{2} < t < \frac{\pi}{2}$

14. $\int x^5 \sqrt{x^2 + 1}\, dx$, using the substitution $t = \sqrt{x^2 + 1}$

10.2 SOME TRIGONOMETRIC INTEGRALS

In this section, which is a continuation of Section 9.7, we discuss integrals of the form

$$\int \sin^p x \cos^q x\, dx \quad \text{and} \quad \int \tan^p x \sec^q x\, dx$$

where p and q are nonnegative numbers. The substitutions required to find these integrals depend on p and q, and while we do not discuss all cases, we discuss enough cases to enable you to continue on your own.

$\int \sin^p x \cos^q x\, dx$ WHEN p IS AN ODD INTEGER
The substitution

$$t = \cos x$$

gives

$$dt = -\sin x\, dx$$

Noting that, being odd, p can be written as $p = 2k + 1$, we have

$$\sin^p x \cos^q x\, dx = \sin^{2k+1} x \cos^q x\, dx = (\sin^2 x)^k \cos^q x (\sin x\, dx)$$
$$= -(1 - \cos^2 x)^k \cos^q x (-\sin x\, dx)$$
$$= -(1 - t^2)^k t^q\, dt$$

The integral $\int (1 - t^2)^k t^q\, dt$ can be evaluated by expanding $(1 - t^2)^k$ using the binomial theorem; when q is an integer, we can use integration by parts instead.

Example 1

Find $\int \sin^5 x \cos^{3/2} x\, dx$.

Solution
For $\cos^{3/2} x$ to be defined, we take $\cos x \geq 0$. Setting $t = \cos x$, we have

$$\sin^5 x \cos^{3/2} x\, dx = -(1 - t^2)^2 t^{3/2}\, dt$$
$$= -(1 - 2t^2 + t^4) t^{3/2}\, dt$$
$$= (-t^{3/2} + 2t^{7/2} - t^{11/2})\, dt$$

Since

$$\int (-t^{3/2} + 2t^{7/2} - t^{11/2})\, dt = -\tfrac{2}{5} t^{5/2} + \tfrac{4}{9} t^{9/2} - \tfrac{2}{13} t^{13/2} + C$$

we have

$$\int \sin^5 x \cos^{3/2} x \, dx = -\tfrac{2}{5} \cos^{5/2} x + \tfrac{4}{9} \cos^{9/2} x - \tfrac{2}{13} \cos^{13/2} x + C$$

∫ $\sin^p x \cos^q x \, dx$ WHEN q IS AN ODD INTEGER
Proceed as in the case above, with the substitution

$$t = \sin x$$

using the facts $\cos^2 x = 1 - \sin^2 x = 1 - t^2$ and $dt = \cos x \, dx$.

Example 2

Find $\int \sin^{1/6} x \cos^3 x \, dx$.

Solution
Using the substitution $t = \sin x$ gives $dt = \cos x \, dx$ and accordingly,

$$\begin{aligned}\sin^{1/6} x \cos^3 x \, dx &= \sin^{1/6} x \cos^2 x (\cos x \, dx) \\ &= \sin^{1/6} x (1 - \sin^2 x)(\cos x \, dx) \\ &= t^{1/6}(1 - t^2) \, dt = (t^{1/6} - t^{13/6}) \, dt\end{aligned}$$

Since

$$\int (t^{1/6} - t^{13/6}) \, dt = \tfrac{6}{7} t^{7/6} - \tfrac{6}{19} t^{19/6} + C$$

we find that

$$\int \sin^{1/6} x \cos^3 x \, dx = \tfrac{6}{7} \sin^{7/6} x - \tfrac{6}{19} \sin^{19/6} x + C$$

∫ $\sin^p x \cos^q x \, dx$ WHEN p AND q ARE EVEN INTEGERS
Here we use the identities

$$\begin{aligned}\sin^2 x &= \tfrac{1}{2}(1 - \cos 2x) \\ \cos^2 x &= \tfrac{1}{2}(1 + \cos 2x)\end{aligned} \tag{1}$$

They were used in Section 9.7, Example 2, to give the integrals

$$\begin{aligned}\int \sin^2 x \, dx &= \tfrac{1}{2}(x - \tfrac{1}{2} \sin 2x) + C \\ \int \cos^2 x \, dx &= \tfrac{1}{2}(x + \tfrac{1}{2} \sin 2x) + C\end{aligned} \tag{2}$$

Example 3

Find $\int \sin^4 x \cos^2 x \, dx$.

Solution
Using the above identities, we have

$$\int \sin^4 x \cos^2 x \, dx = \int [\tfrac{1}{2}(1-\cos 2x)]^2 [\tfrac{1}{2}(1+\cos 2x)] \, dx$$

$$= \tfrac{1}{8} \int (1 - 2\cos 2x + \cos^2 2x)(1 + \cos 2x) \, dx$$

$$= \tfrac{1}{8} \int (1 - \cos 2x - \cos^2 2x + \cos^3 2x) \, dx$$

$$= \tfrac{1}{8}(x - \int \cos 2x \, dx - \int \cos^2 2x \, dx + \int \cos^3 2x \, dx)$$

We now look at these integrals one at a time. The first is immediate:

$$\int \cos 2x \, dx = \tfrac{1}{2} \sin 2x + C_1$$

The second integral is found by letting $t = 2x$ and using the second identity in (2) with x replaced by t. Thus, $dt = 2 \, dx$, so that $dx = \tfrac{1}{2} dt$, $\cos^2 2x \, dx = \tfrac{1}{2} \cos^2 t \, dt$, and hence

$$\int \tfrac{1}{2} \cos^2 t \, dt = \tfrac{1}{4}(t + \tfrac{1}{2} \sin 2t) + C_2$$

This tells us that

$$\int \cos^2 2x \, dx = \tfrac{1}{4}(2x + \tfrac{1}{2} \sin 4x) + C_2 = \tfrac{1}{2} x + \tfrac{1}{8} \sin 4x + C_2$$

The last integral is found by letting $t = \sin 2x$. Thus,

$$\int \cos^3 2x \, dx = \int \cos^2 2x \cos 2x \, dx$$

$$= \tfrac{1}{2} \int (1 - t^2) \, dt = \tfrac{1}{2}\left(t - \frac{t^3}{3}\right) + C_3$$

$$= \tfrac{1}{2}(\sin 2x - \tfrac{1}{3} \sin^3 2x) + C_3$$

$$= \tfrac{1}{2} \sin 2x - \tfrac{1}{6} \sin^3 2x + C_3$$

Adding these integrals and writing $C_1 + C_2 + C_3 = C$ shows that

$$\int \sin^4 x \cos^2 x \, dx = \tfrac{1}{8}[x - \tfrac{1}{2} \sin 2x - (\tfrac{1}{2} x + \tfrac{1}{8} \sin 4x)$$
$$+ (\tfrac{1}{2} \sin 2x - \tfrac{1}{6} \sin^3 2x)] + C$$
$$= \tfrac{1}{8}(\tfrac{1}{2} x - \tfrac{1}{8} \sin 4x - \tfrac{1}{6} \sin^3 2x) + C$$

We now consider the integrals of $\tan^p x \sec^q x$. When $p = 0$ and q is an integer, we have the following:

> **THEOREM 1.** $\int \sec^q x \, dx$ **WHEN** q **IS AN INTEGER**
>
> $$\int \sec x \, dx = \ln|\sec x + \tan x| + C \tag{3}$$
>
> $$\int \sec^2 x \, dx = \tan x + C \tag{4}$$
>
> $$\int \sec^q x \, dx = \frac{1}{q-1} \sin x \cos^{1-q} x$$
>
> $$+ \frac{q-2}{q-1} \int \sec^{q-2} x \, dx \qquad (q \geq 2) \tag{5}$$

Proof
Using the trivial relation

$$\sec x \, dx = \frac{\sec x (\sec x + \tan x)}{\sec x + \tan x} \, dx$$

we put $t = \sec x + \tan x$ and find that

$$dt = (\sec^2 x + \sec x \tan x) \, dx = \sec x (\sec x + \tan x) \, dx$$
$$= (\sec x) \, t \, dx$$

so that

$$\sec x \, dx = \frac{1}{t} \, dt$$

From the integral

$$\int \frac{1}{t} \, dt = \ln|t| + C$$

we get (3) with the substitution $t = \sec x + \tan x$. The integral (4) was obtained in Section 9.7, so it remains only to verify (5). This is done by integration by parts (see Exercise 15 at the end of this section).

$\int \tan^p x \sec^q x \, dx$ **WHEN** q **IS AN EVEN INTEGER**
The substitution

$$t = \tan x$$

gives

$$dt = \sec^2 x \, dx$$

Since q is even, it can be written as $q = 2k$, and with the identity

$$\sec^2 x = 1 + \tan^2 x$$

(Section 9.3, Exercise 18) we find that

$$\tan^p x \sec^q x \, dx = \tan^p x \sec^{2k} x \, dx = \tan^p x (\sec^2 x)^{k-1} (\sec^2 x) \, dx$$
$$= \tan^p x (1 + \tan^2 x)^{k-1} (\sec^2 x) \, dx$$
$$= t^p (1 + t^2)^{k-1} \, dt$$

The integral $\int t^p (1 - t^2)^{k-1} \, dt$ is one we can handle.

Example 4

Find $\int \tan^{3/2} x \sec^4 x \, dx$.

Solution
By the above,

$$\tan^{3/2} x \sec^4 x \, dx = \tan^{3/2} x \sec^2 x (\sec^2 x \, dx)$$
$$= t^{3/2} (1 + t^2) \, dt = (t^{3/2} + t^{7/2}) \, dt$$

and we have

$$\int (t^{3/2} + t^{7/2}) \, dt = \tfrac{2}{5} t^{5/2} + \tfrac{2}{9} t^{9/2} + C$$

Hence

$$\int \tan^{3/2} x \sec^4 x \, dx = \tfrac{2}{5} \tan^{5/2} x + \tfrac{2}{9} \tan^{9/2} x + C$$

$\int \tan^p x \sec^q x \, dx$ WHEN p IS AN ODD INTEGER
The substitution

$$t = \sec x$$

gives

$$dt = \tan x \sec x \, dx$$

and it is used with the identity

$$\tan^2 x = \sec^2 x - 1$$

Example 5

Find $\int \tan^3 x \sec^{2/3} x \, dx$.

Solution
By the above,

$$\tan^3 x \sec^{2/3} x \, dx = \tan^2 x \sec^{-1/3} x (\tan x \sec x \, dx)$$
$$= (\sec^2 x - 1) \sec^{-1/3} x (\tan x \sec x \, dx)$$
$$= (t^2 - 1) t^{-1/3} \, dt = (t^{5/3} - t^{-1/3}) \, dt$$

and we have

$$\int (t^{5/3} - t^{-1/3})\, dt = \tfrac{3}{8} t^{8/3} - \tfrac{3}{2} t^{2/3} + C$$

Hence,

$$\int \tan^3 x \, \sec^{2/3} x \, dx = \tfrac{3}{8} \sec^{8/3} x - \tfrac{3}{2} \sec^{2/3} x + C$$

EXERCISES

Find the integrals in Exercises 1–14.

1. $\displaystyle\int \sin^5 x \sqrt{\cos x}\, dx$
2. $\displaystyle\int \sin^7 x \, \cos^{\sqrt{2}} x \, dx$
3. $\displaystyle\int \sin^5 x \, \cos^5 x \, dx$
4. $\displaystyle\int \sin^{1/3} x \, \cos^3 x \, dx$
5. $\displaystyle\int \sin^3 x \, \cos x \, dx$
6. $\displaystyle\int \sin^3 x \, \cos^2 x \, dx$
7. $\displaystyle\int \sin^2 x \, \cos^3 x \, dx$
8. $\displaystyle\int \sec^4 x \, dx$
9. $\displaystyle\int \sec^6 x \, dx$
10. $\displaystyle\int \sec^5 x \, dx$
11. $\displaystyle\int \tan^2 x \, \sec^2 x \, dx$
12. $\displaystyle\int \sqrt{\tan x} \, \sec^6 x \, dx$
13. $\displaystyle\int \tan^3 x \, \sec^{\sqrt{2}} x \, dx$
14. $\displaystyle\int \tan^7 x \, \sec^2 x \, dx$

15. Verify formula (5) in Theorem 1.

$$\int \sec^q x \, dx = \frac{1}{q-1} \sin x \cos^{1-q} x + \frac{q-2}{q-1} \int \sec^{q-2} x \, dx \qquad (q \geq 2)$$

Hints: Integrate $\int \sec^{q-2} x \, dx$ by parts by using the following:

(1) $\sec^{q-2} x \, dx = \dfrac{1}{\cos^{q-2} x}\, dx = \dfrac{1}{\cos^{q-1} x} \cos x \, dx$

(2) Put $u = 1/\cos^{q-1} x$, $dv = \cos x \, dx$ and find du and v.
(3) Use the formula for integration by parts, which can be put in the form

$$\int u \, dv = uv - \int v \, du$$

(4) You ended up with an equation involving the integral

$$\int \frac{1}{\cos^{q-2} x}\, dx$$

in the left and right sides. Combine these two integrals and solve the equation for

$$\int \frac{1}{\cos^q x}\, dx$$

10.3 TRIGONOMETRIC SUBSTITUTIONS

Continuing the discussion of techniques of integration begun in Section 9.8, we consider here integrals involving $\sqrt{a^2 - x^2}$ and $a^2 + x^2$.

> **INTEGRALS INVOLVING $\sqrt{a^2 - x^2}$**
> Try the substitution
> $$x = a \sin t$$
> which gives
> $$dx = a \cos t\, dt$$
> and use the identities
> $$\sqrt{a^2 - x^2} = a \cos t \quad \text{and} \quad t = \sin^{-1} \frac{x}{a}$$

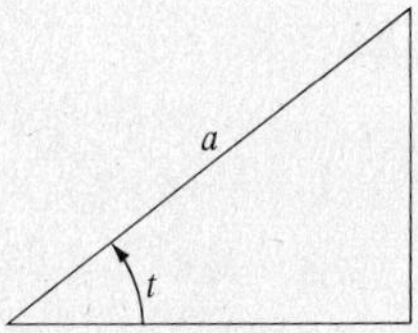

It is being assumed that $|x| < a$ and $-\pi/2 < t < \pi/2$.

In using the above substitution, it is convenient to read off trigonometric relations from Figure 10.1. Thus, we find that

$$\left.\begin{aligned} \sin t &= \frac{x}{a} & \cos t &= \frac{\sqrt{a^2 - x^2}}{a} \\ \sec t &= \frac{a}{\sqrt{a^2 - x^2}} & \csc t &= \frac{a}{x} \\ \tan t &= \frac{x}{\sqrt{a^2 - x^2}} & \cot t &= \frac{\sqrt{a^2 - x^2}}{x} \end{aligned}\right\} \quad (1)$$

Figure 10.1

Example 1

Find

$$\int \frac{1}{x\sqrt{a^2 - x^2}}\, dx$$

Solution
Let $x = a \sin t$; then

$$\frac{1}{x\sqrt{a^2-x^2}}\,dx = \frac{1}{a\sin t \cdot a\cos t}\,a\cos t\,dt = \frac{1}{a}\csc t\,dt$$

Since

$$\int \frac{1}{a}\csc t\,dt = \frac{1}{a}\ln|\csc t - \cot t| + C$$

we find from the relations in (1) that

$$\int \frac{1}{x\sqrt{a^2-x^2}}\,dx = \frac{1}{a}\ln\left|\frac{a}{x} - \frac{\sqrt{a^2-x^2}}{x}\right| + C$$

> INTEGRALS INVOLVING $a^2 + x^2$
> Try the substitution
>
> $$x = a\tan t$$
>
> which gives
>
> $$dx = a\sec^2 t\,dt$$
>
> and use the identities
>
> $$a^2 + x^2 = a^2\sec^2 t \quad \text{and} \quad t = \tan^{-1}\frac{x}{a}$$

For this substitution it is convenient to use the triangle in Figure 10.2. From it we read off the relations

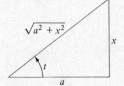

Figure 10.2

$$\sin t = \frac{x}{\sqrt{a^2+x^2}} \qquad \cos t = \frac{a}{\sqrt{a^2+x^2}}$$

$$\sec t = \frac{\sqrt{a^2+x^2}}{a} \qquad \csc t = \frac{\sqrt{a^2+x^2}}{x} \qquad (2)$$

$$\tan t = \frac{x}{a} \qquad \cot t = \frac{a}{x}$$

Example 2

Find

$$\int \frac{1}{\sqrt{a^2+x^2}}\,dx$$

Solution
With the substitution $x = a\tan t$, we have

$$\frac{1}{\sqrt{a^2+x^2}}\,dx = \frac{1}{a\sec t}\,a\sec^2 t\,dt = \sec t\,dt$$

Since

$$\int \sec t \, dt = \ln|\sec t + \tan t| + C$$

we find from the relations in (2) that

$$\int \frac{1}{\sqrt{a^2 + x^2}} \, dx = \ln\left|\frac{\sqrt{a^2 + x^2}}{a} + \frac{x}{a}\right| + C$$

$$= \ln\left|\frac{\sqrt{a^2 + x^2} + x}{a}\right| + C$$

Example 3

Find

$$\int \frac{1}{x\sqrt{a^2 + x^2}} \, dx$$

Solution
Again we use the substitution $x = a \tan t$ to get

$$\frac{1}{x\sqrt{a^2 + x^2}} \, dx = \frac{1}{a \tan t \cdot a \sec t} a \sec^2 t \, dt$$

$$= \frac{1}{a} \frac{\sec t}{\tan t} \, dt = \frac{1}{a} \csc t \, dt$$

Since

$$\int \frac{1}{a} \csc t \, dt = \frac{1}{a} \ln|\csc t - \cot t| + C$$

we find from the relations in (2) that

$$\int \frac{1}{x\sqrt{a^2 + x^2}} \, dx = \frac{1}{a} \ln\left|\frac{\sqrt{a^2 + x^2}}{x} - \frac{a}{x}\right| + C$$

$$= \frac{1}{a} \ln\left|\frac{\sqrt{a^2 + x^2} - a}{x}\right| + C$$

Example 4

Find

$$\int \frac{1}{(a^2 + x^2)^2} \, dx$$

Solution
Once more the substitution $x = a \tan t$ is used; we get

$$\frac{1}{(a^2 + x^2)^2} \, dx = \frac{1}{(a^2 + a^2 \tan^2 t)^2} a \sec^2 t \, dt$$

$$= \frac{1}{a^3} \frac{1}{(\tan^2 t + 1)^2} \sec^2 t \, dt$$

$$= \frac{1}{a^3} \frac{1}{\sec^4 t} \sec^2 t \, dt$$

$$= \frac{1}{a^3} \frac{1}{\sec^2 t} \, dt = \frac{1}{a^3} \cos^2 t \, dt$$

From Section 9.7 we know that

$$\int \cos^2 t \, dt = \tfrac{1}{2}(t + \sin t \cos t) + C$$

and hence we see with the help of (2) that

$$\int \frac{1}{(a^2 + x^2)^2} \, dx = \frac{1}{2a^3}\left(\tan^{-1} \frac{x}{a} + \frac{ax}{a^2 + x^2}\right) + C$$

EXERCISES

Find the integrals in Exercises 1–12.

1. $\int \dfrac{1}{x\sqrt{1-x^2}} \, dx$
2. $\int \dfrac{1}{(5-x^2)^{3/2}} \, dx$
3. $\int \dfrac{x^3}{\sqrt{9-x^2}} \, dx$
4. $\int \sqrt{16+x^2} \, dx$
5. $\int \dfrac{1}{x^2\sqrt{1-x^2}} \, dx$
6. $\int \dfrac{1}{x\sqrt{100+x^2}} \, dx$
7. $\int \dfrac{1}{x^4\sqrt{9+x^2}} \, dx$
8. $\int \dfrac{5}{(2+x^2)^2} \, dx$
9. $\int \dfrac{1}{(x+2)^3\sqrt{1-(x+2)^2}} \, dx$
10. $\int \dfrac{1}{x\sqrt{1+4x^2}} \, dx$
11. $\int \dfrac{x^3}{\sqrt{9-16x^2}} \, dx$
12. $\int \sqrt{4+5x^2} \, dx$

In Exercises 14–18, use the substitution $x = a \sec t$ and the relations obtained in Exercise 13.

13. Express the trigonometric functions in terms of a and x by using Figure 10.3.

14. $\int \dfrac{1}{\sqrt{x^2-4}} \, dx$
15. $\int \dfrac{1}{x\sqrt{x^2-1}} \, dx$
16. $\int \dfrac{3}{\sqrt{x^2-x^4}} \, dx$
17. $\int \sqrt{x^2-9} \, dx$
18. $\int \dfrac{1}{x^2\sqrt{x^2-5}} \, dx$

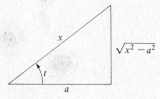

Figure 10.3

10.4 PARTIAL FRACTIONS

If $P(x)$ and $Q(x)$ are polynomials, then the function $f(x) = P(x)/Q(x)$ is called a *rational function*. In this section we describe a method whereby all rational functions can be integrated. The method depends on breaking down a rational function into simpler rational functions, the so-called *partial fractions*, which we now define.

> **DEFINITION 1**
> A *partial fraction* is a rational function expressible in one of the following forms:
>
> 1. $\dfrac{A}{x-a}$
>
> 2. $\dfrac{A}{(x-a)^k}$ $(k = 2, 3, 4, 5, \ldots)$
>
> 3. $\dfrac{Ax+B}{x^2+px+q}$ $\left(\dfrac{p^2}{4} < q\right)$
>
> 4. $\dfrac{Ax+B}{(x^2+px+q)^k}$ $\left(\dfrac{p^2}{4} < q \text{ and } k = 2, 3, 4, 5, \ldots\right)$

REMARK
The condition $p^2/4 < q$ implies that $x^2 + px + q$ cannot be written in the form $(x+a)(x+b)$, where a and b are real numbers. In fact, since by completing the square

$$x^2 + px + q = \left(x + \frac{p}{2}\right)^2 + \left(q - \frac{p^2}{4}\right) \tag{1}$$

it follows that $p^2/4 < q$ forces the graph of $y = x^2 + px + q$ to lie in its entirety *above* the x axis.

We now state the following result:

PARTIAL FRACTION DECOMPOSITION
Every rational function $P(x)/Q(x)$ can be expressed (uniquely) as the sum of a polynomial $M(x)$ and partial fractions:

$$\frac{P(x)}{Q(x)} = M(x) + \text{sum of partial fractions}$$

We have $M(x) = 0$ when the degree of $P(x)$ is less than the degree of $Q(x)$; otherwise, $M(x)$ can be determined by long division.

The process of obtaining partial fraction decompositions is best explained with specific examples. Later in this section we describe the general procedure.

Example 1

Decompose $(2x + 1)/x(x + 1)^2$ into partial fractions.

Solution
The method we use is called the *method of undetermined coefficients*. It works as follows: In a partial fraction decomposition of the rational function above, we encounter the denominators x, $(x + 1)$, and $(x + 1)^2$. We therefore write

$$\frac{2x + 1}{x(x + 1)^2} = \frac{A}{x} + \frac{B}{x + 1} + \frac{C}{(x + 1)^2} \tag{2}$$

where the coefficients A, B, and C, are to be determined. Adding the partial fractions gives

$$\frac{A}{x} + \frac{B}{x + 1} + \frac{C}{(x + 1)^2} = \frac{A(x + 1)^2 + Bx(x + 1) + Cx}{x(x + 1)^2}$$

Comparing this with (2) shows that we must have

$$\begin{aligned} 2x + 1 &= A(x + 1)^2 + Bx(x + 1) + Cx \\ &= A(x^2 + 2x + 1) + B(x + x^2) + Cx \\ &= (A + B)x^2 + (2A + B + C)x + A \end{aligned}$$

Thus, the following equations must be satisfied:

$$\begin{aligned} A + B &= 0 \\ 2A + B + C &= 2 \\ A &= 1 \end{aligned}$$

Solving this system of equations gives

$$A = 1 \quad B = -1 \quad C = 1$$

and hence

$$\frac{2x + 1}{x(x + 1)^2} = \frac{1}{x} - \frac{1}{x + 1} + \frac{1}{(x + 1)^2}$$

Example 2

Express $(2x + 1)/x^4(x + 1)$ in terms of partial fractions.

Solution
Here we expect to encounter the denominators x, x^2, x^3, x^4, and $x + 1$, and we therefore write

$$\frac{2x + 1}{x^4(x + 1)} = \frac{A}{x} + \frac{B}{x^2} + \frac{C}{x^3} + \frac{D}{x^4} + \frac{E}{x + 1} \tag{3}$$

For the right side we have

$$\frac{A}{x} + \frac{B}{x^2} + \frac{C}{x^3} + \frac{D}{x^4} + \frac{E}{x+1} =$$

$$\frac{Ax^3(x+1) + Bx^2(x+1) + Cx(x+1) + D(x+1) + Ex^4}{x^4(x+1)}$$

and a comparison with (3) results in the equation

$$2x + 1 = Ax^3(x+1) + Bx^2(x+1) + Cx(x+1) + D(x+1) + Ex^4$$
$$= (A+E)x^4 + (A+B)x^3 + (B+C)x^2 + (C+D)x + D$$

Thus, the following system of equations must be satisfied:

$$A + E = 0$$
$$A + B = 0$$
$$B + C = 0$$
$$C + D = 2$$
$$D = 1$$

The solution of this system is

$$A = 1 \quad B = -1 \quad C = 1 \quad D = 1 \quad E = -1$$

Hence,

$$\frac{2x+1}{x^4(x+1)} = \frac{1}{x} - \frac{1}{x^2} + \frac{1}{x^3} + \frac{1}{x^4} - \frac{1}{x+1}$$

Example 3

Decompose $6x^3/(x^4 - 1)$ into partial fractions.

Solution
Writing the denominator in the form

$$x^4 - 1 = (x^2 - 1)(x^2 + 1) = (x-1)(x+1)(x^2 + 1)$$

we expect to find the denominators $(x-1)$, $(x+1)$, and (x^2+1) in the partial fraction decomposition. We therefore put

$$\frac{6x^3}{x^4 - 1} = \frac{A}{x-1} + \frac{B}{x+1} + \frac{Cx+D}{x^2+1}$$

$$= \frac{A(x+1)(x^2+1) + B(x-1)(x^2+1) + (Cx+D)(x-1)(x+1)}{(x-1)(x+1)(x^2+1)}$$

Thus we must have

$$6x^3 = A(x^3 + x^2 + x + 1) + B(x^3 - x^2 + x - 1)$$
$$+ (Cx+D)(x^2 - 1)$$
$$= (A+B+C)x^3 + (A-B+D)x^2 + (A+B-C)x$$
$$+ (A-B-D)$$

and this yields the equations

$$A + B + C = 6$$
$$A - B + D = 0$$
$$A + B - C = 0$$
$$A - B - D = 0$$

The solution of this system of equations is

$$A = \tfrac{3}{2} \quad B = \tfrac{3}{2} \quad C = 3 \quad D = 0$$

and hence

$$\frac{6x^3}{x^4 - 1} = \frac{3/2}{x - 1} + \frac{3/2}{x + 1} + \frac{3x}{x^2 + 1}$$

Example 4

Decompose $(3x^4 - 3x^3 - x^2 + 3x - 3)/x(x - 1)$ into partial fractions.

Solution
Since the degree of the numerator is greater than that of the denominator, we start with long division, which gives

$$\frac{3x^4 - 3x^3 - x^2 + 3x - 3}{x(x - 1)} = (3x^2 - 1) + \frac{2x - 3}{x(x - 1)}$$

To break down $(2x - 3)/x(x - 1)$ into partial fractions, we put

$$\frac{2x - 3}{x(x - 1)} = \frac{A}{x} + \frac{B}{x - 1} = \frac{A(x - 1) + Bx}{x(x - 1)} = \frac{(A + B)x - A}{x(x - 1)}$$

and find that we must have

$$A + B = 2$$
$$-A = -3$$

Hence $A = 3$, $B = -1$, and consequently

$$\frac{3x^4 - 3x^3 - x^2 + 3x - 3}{x(x - 1)} = (3x^2 - 1) + \frac{3}{x} - \frac{1}{x - 1}$$

We now state the following general rule:

PROCEDURE FOR PARTIAL FRACTION DECOMPOSITION

Step 1
If degree of $P(x) \geq$ degree of $Q(x)$, use long division to find polynomials $M(x)$ and $R(x)$ such that

$$\frac{P(x)}{Q(x)} = M(x) + \frac{R(x)}{Q(x)}$$

where

> degree of $R(x)$ < degree of $Q(x)$

If degree of $P(x)$ < degree of $Q(x)$, take $M(x) = 0$ and $R(x) = P(x)$.

Step 2
Factor $Q(x)$ in the form
$$Q(x) = C(x-a_1)^{k_1}(x-a_2)^{k_2}\cdots(x^2+p_1x+q_1)^{m_1}(x^2+p_2x+q_2)^{m_2}\cdots$$

For each factor $(x-a)^k$, write
$$\frac{A_1}{x-a} + \frac{A_2}{(x-a)^2} + \cdots + \frac{A_k}{(x-a)^k}$$

and for each factor $(x^2 + px + q)^m$, write
$$\frac{B_1x+C_1}{x^2+px+q} + \frac{B_2x+C_2}{(x^2+px+q)^2} + \cdots + \frac{B_mx+C_m}{(x^2+px+q)^m}$$

Step 3
Using the common denominator $Q(x)$, add all the partial fractions to obtain a system of equations for determining the indeterminate coefficients $A_1, A_2, \ldots, A_k, B_1, B_2, \ldots, B_m$, and $C_1, C_2, \ldots, C_m$.

Turning now to the integration of partial fractions, we begin with the following facts:

THEOREM 1. INTEGRATION OF PARTIAL FRACTIONS

$$\int \frac{A}{x-a}\,dx = A\ln|x-a| + C \qquad (4)$$

$$\int \frac{A}{(x-a)^k}\,dx = -\frac{A}{k-1}\frac{1}{(x-a)^{k-1}} + C \qquad (k>1) \qquad (5)$$

$$\int \frac{Ax+B}{x^2+px+q}\,dx = \frac{A}{2}\ln|x^2+px+q|$$
$$+ \frac{B-Ap/2}{\sqrt{q-p^2/4}}\tan^{-1}\frac{x+p/2}{\sqrt{q-p^2/4}} + C \qquad (6)$$

The first two integrals were obtained earlier and need no further justification. The last integral is obtained by using (1) and the substitution $t = x + p/2$ (see Exercise 11 at the end of the section). Note that we did not give an integration formula for partial fraction 4 in Definition 1. This partial fraction occurs very seldom, and we therefore discuss it only in exercises (see Exercises 12–18).

Example 5
Find
$$\int \frac{2x+1}{x(x+1)^2}\, dx$$

Solution
By Example 1,
$$\int \frac{2x+1}{x(x+1)^2}\, dx = \int \left[\frac{1}{x} - \frac{1}{x+1} + \frac{1}{(x+1)^2}\right] dx$$
$$= \int \frac{1}{x}\, dx - \int \frac{1}{x+1}\, dx + \int \frac{1}{(x+1)^2}\, dx$$
$$= \ln|x| - \ln|x+1| - \frac{1}{x+1} + C$$
$$= \ln\left|\frac{x}{x+1}\right| - \frac{1}{x+1} + C$$

Example 6
Find
$$\int \frac{2x+1}{x^4(x+1)}\, dx$$

Solution
By Example 2,
$$\int \frac{2x+1}{x^4(x+1)}\, dx = \int \left(\frac{1}{x} - \frac{1}{x^2} + \frac{1}{x^3} + \frac{1}{x^4} - \frac{1}{x+1}\right) dx$$
$$= \ln|x| + \frac{1}{x} - \frac{1}{2}\frac{1}{x^2} - \frac{1}{3}\frac{1}{x^3} - \ln|x+1| + C$$
$$= \ln\left|\frac{x}{x+1}\right| + \frac{1}{x} - \frac{1}{2x^2} - \frac{1}{3x^3} + C$$

Example 7
Find
$$\int \frac{6x^3}{x^4-1}\, dx$$

Solution
By Example 3,
$$\int \frac{6x^3}{x^4-1}\, dx = \int \left(\frac{3/2}{x-1} + \frac{3/2}{x+1} + \frac{3x}{x^2+1}\right) dx$$

$$= \frac{3}{2}\ln|x-1| + \frac{3}{2}\ln|x+1| + \int \frac{3x}{x^2+1}\,dx$$

The last integral is obtained from Theorem 1, formula (6), by putting $A = 3$, $B = 0$, $p = 0$, and $q = 1$; we get

$$\int \frac{3x}{x^2+1}\,dx = \frac{3}{2}\ln|x^2+1| + C$$

This, together with the simplification

$$\frac{3}{2}\ln|x-1| + \frac{3}{2}\ln|x+1| = \frac{3}{2}(\ln|x-1| + \ln|x+1|)$$

$$= \frac{3}{2}\ln|x-1|\cdot|x+1| = \frac{3}{2}\ln|x^2-1|$$

gives the final result

$$\int \frac{6x^3}{x^4-1}\,dx = \frac{3}{2}[\ln|x^2-1| + \ln|x^2+1|] + C$$

$$= \frac{3}{2}\ln|x^4-1| + C$$

Observe that here you can check the answer by using integration by substitution.

EXERCISES

Find the integrals in Exercises 1–10.

1. $\displaystyle\int \frac{2x+5}{(x-1)(x+3)^2}\,dx$ 2. $\displaystyle\int \frac{3x-2}{x^2-7x+12}\,dx$

3. $\displaystyle\int \frac{-2x+4}{(x-1)^2(x^2+1)}\,dx$ 4. $\displaystyle\int \frac{x^3-34x+82}{x^2-7x+12}\,dx$

5. $\displaystyle\int \frac{x^2}{(x^2+1)(x+1)(x-1)}\,dx$ 6. $\displaystyle\int \frac{x}{x^4-1}\,dx$

7. $\displaystyle\int \frac{x^2}{(x-1)(x+2)}\,dx$ 8. $\displaystyle\int \frac{x+1}{x^2+2x+2}\,dx$

9. $\displaystyle\int \frac{x^2+1}{x(x^2+x+1)}\,dx$ 10. $\displaystyle\int \frac{x^3+1}{x^2+x}\,dx$

11. Verify formula (6) of Theorem 1 using the identity

$$x^2 + px + q = \left(x + \frac{p}{2}\right)^2 + \left(q - \frac{p^2}{4}\right)$$

and the substitution $t = x + p/2$.

12. (a) Show that

$$\int \frac{Ax+B}{(x^2+px+q)^k}\,dx = \frac{A}{2}\int \frac{2x+p}{(x^2+px+q)^k}\,dx$$
$$+ \left(B - \frac{Ap}{2}\right)\int \frac{1}{(x^2+px+q)^k}\,dx$$

(b) Find

$$\int \frac{2x+p}{(x^2+px+q)^k}\,dx$$

(c) Verify the following recursion formula by differentiation:

$$\boxed{\int \frac{1}{(t^2+\alpha)^k}\,dt = \frac{1}{2(k-1)\alpha}\frac{t}{(t^2+\alpha)^{k-1}} \\ + \frac{2k-3}{(2k-2)\alpha}\int \frac{1}{(t^2+\alpha)^{k-1}}\,dt \qquad (k\geq 2)}$$

The following exercises can be done with the help of Exercise 12.

13. (a) $\int \dfrac{1}{(x^2+x+1)^2}\,dx$ (b) $\int \dfrac{1}{(x^2+x+1)^3}\,dx$

14. $\int \dfrac{x}{(x^2+x+1)^2}\,dx$

15. $\int \dfrac{x}{(x^2+x+1)^3}\,dx$

16. $\int \dfrac{2x+5}{(x^2+3x+3)^2}\,dx$

10.5 RATIONALIZING SUBSTITUTIONS

This section is devoted to some examples of substitutions that lead to rational functions; the latter can be integrated with the method of the last section.

The first substitution we discuss works for integrals involving $\sin x$, $\cos x$, and combinations thereof. With the basic substitution

$$t = \tan \frac{x}{2}$$

we can express $\sin x$ and $\cos x$ in terms of t as follows:

$$\sin x = 2 \sin \frac{x}{2} \cos \frac{x}{2} = \frac{2 \sin(x/2) \cos(x/2)}{\sin^2(x/2) + \cos^2(x/2)}$$
$$= \frac{2 \tan(x/2)}{1 + \tan^2(x/2)} = \frac{2t}{1+t^2}$$

$$\cos x = \cos^2 \frac{x}{2} - \sin^2 \frac{x}{2} = \frac{\cos^2(x/2) - \sin^2(x/2)}{\cos^2(x/2) + \sin^2(x/2)}$$

$$= \frac{1 - \tan^2(x/2)}{1 + \tan^2(x/2)} = \frac{1 - t^2}{1 + t^2}$$

(see Section 9.1, Example 5, with $2\theta = x$). From the substitution $t = \tan(x/2)$ we also have

$$dt = \frac{1}{2} \sec^2 \frac{x}{2} \, dx = \frac{1}{2}\left(1 + \tan^2 \frac{x}{2}\right) dx = \frac{1}{2}(1 + t^2) \, dx$$

so that

$$dx = \frac{2}{1 + t^2} \, dt$$

Summarizing the relations just derived, we have the following:

> If
> $$t = \tan \frac{x}{2}$$
> then
> $$\sin x = \frac{2t}{1 + t^2}$$
> $$\cos x = \frac{1 - t^2}{1 + t^2}$$
> $$dx = \frac{2}{1 + t^2} \, dt$$

(1)

Example 1

Find $\int \csc x \, dx$.

Solution
With the substitution $t = \tan(x/2)$ we have, by (1),

$$\csc x \, dx = \frac{1}{\sin x} \, dx = \frac{1 + t^2}{2t} \cdot \frac{2}{1 + t^2} \, dt = \frac{1}{t} \, dt$$

Since

$$\int \frac{1}{t} \, dt = \ln|t| + C$$

we find that

$$\int \csc x \, dx = \ln\left|\tan \frac{x}{2}\right| + C$$

Example 2

Find

$$\int \frac{1}{4 - 5 \sin x} \, dx$$

Solution

Again using the substitution $t = \tan(x/2)$, we find from (1) that

$$\int \frac{1}{4 - 5 \sin x} \, dx = \int \frac{1}{4 - 5[2t/(1 + t^2)]} \frac{2}{1 + t^2} \, dt = \int \frac{2}{4 + 4t^2 - 10t} \, dt$$

$$= \int \frac{1}{2t^2 - 5t + 2} \, dt = \int \frac{1}{(2t - 1)(t - 2)} \, dt$$

Writing

$$\frac{1}{(2t - 1)(t - 2)} = \frac{A}{2t - 1} + \frac{B}{t - 2} = \frac{A(t - 2) + B(2t - 1)}{(2t - 1)(t - 2)}$$

leads to the equation

$$1 = A(t - 2) + B(2t - 1) = (A + 2B)t - (2A + B)$$

which, in turn, gives

$$A + 2B = 0$$
$$-(2A + B) = 1$$

The solution of this system of equations is

$$A = -\tfrac{2}{3} \qquad B = \tfrac{1}{3}$$

and hence

$$\frac{1}{(2t - 1)(t - 2)} = \frac{-\tfrac{2}{3}}{2t - 1} + \frac{\tfrac{1}{3}}{t - 2}$$

From this we get

$$\int \frac{1}{(2t - 1)(t - 2)} \, dt = -\frac{2}{3} \int \frac{1}{2t - 1} \, dt + \frac{1}{3} \int \frac{1}{t - 2} \, dt$$

$$= -\frac{1}{3} \ln|2t - 1| + \frac{1}{3} \ln|t - 2| + C$$

$$= \frac{1}{3} \ln \left| \frac{t - 2}{2t - 1} \right| + C$$

The final result is

$$\int \frac{1}{4 - 5 \sin x} \, dx = \frac{1}{3} \ln \left| \frac{\tan(x/2) - 2}{2 \tan(x/2) - 1} \right| + C$$

The next problem shows how to use two consecutive substitutions to reduce an integral to rational form.

Example 3

Find

$$\int \frac{1}{x^2\sqrt{(1+x^2)^3}}\, dx$$

Solution

We begin with the substitution $t = x^2$, whose purpose is to eliminate the square under the radical sign. Thus, $dt = 2x\, dx$ and hence $dx = (1/2x)\, dt = (1/2\sqrt{t})\, dt$, and we find that

$$\frac{1}{x^2\sqrt{(1+x^2)^3}}\, dx = \frac{1}{t\sqrt{(1+t)^3}}\, \frac{1}{2\sqrt{t}}\, dt = \frac{1}{2t^{3/2}(1+t)^{3/2}}\, dt$$

$$= \frac{1}{2}\frac{1}{t^3}\left(\frac{t}{1+t}\right)^{3/2}\, dt$$

We have thus reduced the original integral to

$$\int \frac{1}{2}\frac{1}{t^3}\left(\frac{t}{1+t}\right)^{3/2}\, dt$$

The second substitution we use is

$$z = \left(\frac{1+t}{t}\right)^{1/2} \tag{2}$$

which gives

$$\frac{1+t}{t} = z^2 \quad \text{or} \quad t = \frac{1}{z^2-1}$$

and hence

$$dt = -\frac{2z}{(z^2-1)^2}\, dz$$

Accordingly,

$$\frac{1}{2}\frac{1}{t^3}\left(\frac{t}{1+t}\right)^{3/2}\, dt = \frac{1}{2}(z^2-1)^3\frac{1}{z^3}\left[-\frac{2z}{(z^2-1)^2}\right]\, dz$$

$$= \frac{1-z^2}{z^2}\, dz = \left(\frac{1}{z^2}-1\right)\, dz$$

and clearly

$$\int \left(\frac{1}{z^2}-1\right)\, dz = -\frac{1}{z} - z + C$$

Using (2), we have

$$\int \frac{1}{2}\frac{1}{t^3}\left(\frac{t}{1+t}\right)^{3/2}\, dt = -\left(\frac{t}{1+t}\right)^{1/2} - \left(\frac{1+t}{t}\right)^{1/2} + C$$

and the substituting $t = x^2$ gives

$$\int \frac{1}{x^2\sqrt{(1+x^2)^3}} dx = -\left(\frac{x^2}{1+x^2}\right)^{1/2} - \left(\frac{1+x^2}{x^2}\right)^{1/2} + C$$

$$= -\frac{x}{\sqrt{1+x^2}} - \frac{\sqrt{1+x^2}}{x} + C$$

$$= -\frac{2x^2+1}{x\sqrt{1+x^2}} + C$$

Example 4

Find

$$\int \frac{\sec x}{1+\sin x} dx$$

Solution
The substitution $t = \tan(x/2)$ gives

$$\int \frac{\sec x}{1+\sin x} dx = \int \frac{1/\cos x}{1+\sin x} dx = \int \frac{(1+t^2)/(1-t^2)}{1+2t/(1+t^2)} \cdot \frac{2}{1+t^2} dt$$

$$= 2\int \frac{1+t^2}{(1-t^2)(1+t)^2} dt = 2\int \frac{1+t^2}{(1-t)(1+t)^3} dt$$

Using partial fraction decomposition, we put

$$\frac{1+t^2}{(1-t)(1+t)^3} = \frac{A}{1-t} + \frac{B}{1+t} + \frac{C}{(1+t)^2} + \frac{D}{(1+t)^3}$$

and, upon adding the right side, this leads to the equation

$$1+t^2 = A(1+t)^3 + B(1-t)(1+t)^2 + C(1-t)(1+t) + D(1-t)$$

From it we get the system of equations

$$A - B = 0$$
$$3A - B - C = 1$$
$$3A + B - D = 0$$
$$A + B + C + D = 1$$

and we find the solution to be

$$A = \tfrac{1}{4} \quad B = \tfrac{1}{4} \quad C = -\tfrac{1}{2} \quad D = 1$$

Thus,

$$\frac{1+t^2}{(1-t)(1+t)^3} = \frac{\tfrac{1}{4}}{1-t} + \frac{\tfrac{1}{4}}{1+t} - \frac{\tfrac{1}{2}}{(1+t)^2} + \frac{1}{(1+t)^3}$$

The reader should be able to complete this problem on his own.

EXERCISES

Find the integrals in Exercises 1–11.

1. $\int \sec x \, dx$

2. $\int \csc^2 x \, dx$

3. $\int \csc^3 x \, dx$

4. $\int \dfrac{1}{1 + \cos x} \, dx$

5. $\int \dfrac{\cos x}{1 - \sin x} \, dx$

6. $\int \dfrac{1}{x\sqrt{(1 + x)^3}} \, dx$

7. $\int \dfrac{x^3}{\sqrt{(1 + x^2)^3}} \, dx$

8. $\int \dfrac{4 + 5 \cos x}{4 - 5 \sin x} \, dx$

Hint: First, let $z = x^{2/3}$.

9. $\int \dfrac{1}{x^{2/3}(1 + x^{2/3})} \, dx$

Hint: First, let $z = 1 - x^2$.

10. $\int \dfrac{x^3}{\sqrt{1 - x^2}} \, dx$

Hint: Let $t = \tan x$.

11. $\int \dfrac{1}{2 - \sin^2 x} \, dx$

QUIZ 1

Find the integrals below.

1. $\int \sin^3 2x \, \cos^{3/2} 2x \, dx$

2. $\int \sin^{1/8} 5x \, \cos 5x \, dx$

3. $\int \sin^4 x \, dx$

4. $\int \tan^5 x \, \sec^3 x \, dx$

5. $\int \dfrac{1}{x\sqrt{4 - x^2}} \, dx$

6. $\int \dfrac{1}{\sqrt{16 + x^2}} \, dx$

7. $\int \dfrac{1}{x^4\sqrt{1 + x^2}} \, dx$

8. $\int \dfrac{x^2 + 4x + 8}{x(x + 2)(x + 4)} \, dx$

9. $\int \dfrac{2}{(1 + x)(1 + x^2)} \, dx$

10. $\int \dfrac{2x - 1}{x(x - 1)^2} \, dx$

QUIZ 2

Find the integrals below.

1. $\int \sin^5 5x \, \cos^{3/2} 5x \, dx$

2. $\int \sin^{1/6} x \, \cos^3 x \, dx$

3. $\int \sin^2 x \, \cos^2 x \, dx$

4. $\int \sec^3 x \, dx$

5. $\displaystyle\int \tan^2 x \sec^4 x \, dx$

6. $\displaystyle\int \frac{1}{x\sqrt{x^2-4}} \, dx$

7. $\displaystyle\int \frac{1}{x\sqrt{16+x^2}} \, dx$

8. $\displaystyle\int \frac{1}{(6^2+x^2)^4} \, dx$

9. $\displaystyle\int \frac{1}{x^2\sqrt{1+x^2}} \, dx$

10. $\displaystyle\int \frac{7x^5}{x^4-2^4} \, dx$

11. $\displaystyle\int \frac{12x^2+x-2}{(4x^2+x+1)(x-1)} \, dx$

12. $\displaystyle\int \frac{\csc x}{1+\cos x} \, dx$

Hint: Try $u = \sqrt{2+x^2}$

13. $\displaystyle\int \frac{x^5}{\sqrt{(2+x^2)^3}} \, dx$

CHAPTER 11

TAYLOR APPROXIMATIONS AND INFINITE SERIES

In Section 5.5, we discussed the formula

$$f(x) \approx f(a) + f'(a)(x - a) \qquad \text{when } x \approx a$$

which gives an approximation to $f(x)$ for values of x close to a. Following a discussion of sequences, we shall develop the more refined formula

$$f(x) \approx f(a) + f'(a)(x - a) + \frac{1}{2}f''(a)(x - a)^2 + \frac{1}{3!}f'''(a)(x - a)^3$$

$$+ \cdots + \frac{1}{n!}f^{(n)}(a)(x - a)^n \qquad \text{when } x \approx a$$

We shall lead to this general formula, called the *n*th-*order Taylor approximation*, through a study of second- and third-order approximations.

The study of Taylor approximations and the errors they introduce will lead us to a study of *infinite series*.

11.1 SEQUENCES

Sequences are very basic in mathematics. We have already encountered sequences in defining the definite integral as a limit of sums S_n of rectangles: The succession of numbers $S_1, S_2, S_3, \ldots$ is an example of a sequence. Thus, by a sequence we simply mean a list of numbers given in a definite order, such as

$$1, 2, 3, 4, 5, \ldots$$

$$\tfrac{1}{2}, \tfrac{2}{3}, \tfrac{2}{4}, \tfrac{4}{5}, \tfrac{5}{6}, \ldots$$

or, more generally,

$$a_1, a_2, a_3, a_4, a_5, \ldots$$

where the dots indicate that the list goes on and on, without terminating. We shall use the dual notation

$$\{a_n\} = \{a_1, a_2, a_3, \ldots\}$$

for a sequence.

Observe that the sequence $\{a_n\}$ assigns to each positive integer a unique number. A sequence is, therefore, a function whose domain is the positive integers.

The concept of limit of a sequence was also encountered in the definition of the integral. Before giving a formal definition, we look at some examples.

Example 1

Consider the sequence

$$\left\{\frac{1}{n}\right\} = \left\{1, \frac{1}{2}, \frac{1}{3}, \frac{1}{4}, \ldots\right\}$$

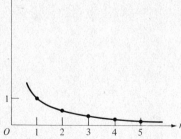

Figure 11.1
The sequence $\{1/n\}$.

This sequence is represented graphically in Figure 11.1 by means of the solid dots. This sequence is seen to approach 0 as n increases beyond all bounds, and we express this by writing

$$\lim_{n \to \infty} \frac{1}{n} = 0$$

We say that the sequence $\{1/n\}$ *converges* to the limit 0.

Example 2

The sequence

$$\left\{\frac{n}{n+1}\right\} = \left\{\frac{1}{2}, \frac{2}{3}, \frac{3}{4}, \frac{4}{5}, \ldots\right\}$$

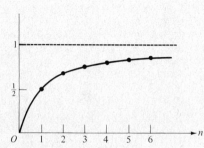

Figure 11.2
The sequence $\{n/(n+1)\}$.

is represented graphically in Figure 11.2. Writing

$$\frac{n}{n+1} = 1 - \frac{1}{n+1}$$

we observe that $1/(n+1) \to 0$ as $n \to \infty$, and so $1 - 1/(n+1) \to 1$ as $n \to \infty$. We thus write

$$\lim_{n \to \infty} \frac{n}{n+1} = 1$$

and say that the sequence $\{n/(n+1)\}$ converges to the limit 1.

Example 3

Consider the sequence

$$\{(-1)^n\} = \{-1, 1, -1, 1, -1, \ldots\}$$

(see Figure 11.3). As n increases, this sequence does not approach any particular number, since the terms alternate between -1 and 1. This sequence is said to *diverge*.

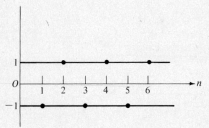

Figure 11.3
The sequence $\{(-1)^n\}$.

Example 4

The sequence

$$\{n\} = \{1, 2, 3, 4, \ldots\}$$

is seen to increase with n (see Figure 11.4). Also this sequence fails to have a limit and it is said to diverge.

With these examples the following definitions will be clear.

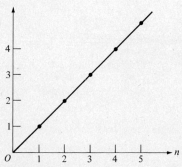

Figure 11.4
The sequence $\{n\}$.

DEFINITION 1

If the sequence $\{a_n\}$ approaches the number A as $n \to \infty$, we write

$$\lim_{n \to \infty} a_n = A$$

and say that the sequence *converges* to the limit A. A sequence that does not converge is said to *diverge*.

This intuitive definition suffices for most of our purposes. In proofs, however, a more formal definition is needed.

DEFINITION 1'

Let $\{a_n\}$ be a given sequence. Then

$$\lim_{n \to \infty} a_n = A$$

if, for each error $\epsilon > 0$, no matter how small, there is an integer N such that

$$|a_n - A| < \epsilon \quad \text{whenever } n > N$$

The use of this definition will be illustrated in the proofs of the facts below.

THEOREM 1. SOME SPECIAL LIMITS

1. If $a_n = c$ for $n = 1, 2, 3, \ldots$, then $\lim_{n \to \infty} a_n = c$;

2. $\lim_{n \to \infty} \dfrac{1}{n^\alpha} = 0$ for $\alpha > 0$;

3. $\lim_{n \to \infty} a^n = 0$ for $|a| < 1$.

Proof

1. If $a_n = c$, then $|a_n - c| = 0$ for $n = 1, 2, 3, \ldots$. Hence, given any error $\epsilon > 0$,

$$|a_n - c| < \epsilon \quad \text{for all values of } n.$$

2. Given an error $\epsilon > 0$, we must find an integer N such that

$$\left| \frac{1}{n^\alpha} - 0 \right| < \epsilon \quad \text{whenever } n > N$$

Since

$$\left| \frac{1}{n^\alpha} - 0 \right| = \frac{1}{n^\alpha}$$

we must simply have

$$\frac{1}{n^\alpha} < \epsilon \quad \text{whenever } n > N$$

This, however, is equivalent to asking that

$$n > \frac{1}{\sqrt[\alpha]{\epsilon}} \quad \text{whenever } n > N$$

Hence, given any $\epsilon > 0$, it is sufficient to let N be any integer greater than $1/\sqrt[\alpha]{\epsilon}$, and (2) will follow from definition 1'.

3. The statement

$$|a^n - 0| < \epsilon$$

is equivalent to

$$|a|^n < \epsilon$$

Taking logarithms gives

$$n \ln|a| < \ln \epsilon$$

and since $|a| < 1$ (by assumption), we have $\ln|a| < 0$. Hence the last inequality can be expressed as

$$n > \frac{\ln \epsilon}{\ln |a|}$$

Taking now N to be any integer exceeding $\ln \epsilon / \ln |a|$, then it will follow that, for a given error $\epsilon > 0$,

$$|a^n - 0| < \epsilon \quad \text{whenever } n > N$$

By Definition 1', therefore,

$$\lim_{n \to \infty} a^n = 0$$

Example 5

(a) $\lim_{n \to \infty} 5 = 5$

(b) $\lim_{n \to \infty} e = e$

(c) $\lim_{k \to \infty} \frac{1}{2^k} = 0$

(d) $\lim_{n \to \infty} (\frac{4}{5})^n = 0$

(e) $\lim_{k \to \infty} \frac{1}{\sqrt{k}} = 0$

The following properties of limits of sequences are exact analogues of the properties of limits of functions given in Theorem 1 of Section 2.6.

THEOREM 2. PROPERTIES OF LIMITS
If $\lim_{n \to \infty} a_n = A$ and $\lim_{n \to \infty} b_n = B$, then

1. $\lim_{n \to \infty} c a_n = cA$ for any constant c;

2. $\lim_{n \to \infty} (a_n + b_n) = A + B$;

3. $\lim_{n \to \infty} (a_n - b_n) = A - B$;

4. $\lim_{n \to \infty} (a_n \cdot b_n) = A \cdot B$;

5. $\lim_{n \to \infty} \frac{a_n}{b_n} = \frac{A}{B}$, provided that $B \neq 0$.

The proof of this theorem is omitted here because it is technical and it does not contribute to our understanding of limits.

Example 6

Find
$$\lim_{n\to\infty} \frac{n+1}{n}$$

Solution

$$\lim_{n\to\infty} \frac{n+1}{n} = \lim_{n\to\infty}\left(1 + \frac{1}{n}\right)$$
$$= \lim_{n\to\infty} 1 + \lim_{n\to\infty} \frac{1}{n} \quad \text{(by Property 2 in Theorem 2)}$$
$$= 1 + 0 = 1$$

Hence,
$$\lim_{n\to\infty} \frac{n+1}{n} = 1$$

Example 7

Find
$$\lim_{n\to\infty} \frac{2n^2 + 3n - 5}{5n^2 + 8n + 6}$$

Solution
Since the numerator approaches no limit as $n \to \infty$, we cannot use Property 5 in Theorem 2, and hence an algebraic manipulation is required before the limit can be found. Dividing both numerator and denominator by n^2 gives
$$\frac{2n^2 + 3n - 5}{5n^2 + 8n + 6} = \frac{2 + (3/n) - (5/n^2)}{5 + (8/n) + (6/n^2)}$$

Since
$$\lim_{n\to\infty}\left(2 + \frac{3}{n} - \frac{5}{n^2}\right) = 2 + 0 - 0 = 2$$

and
$$\lim_{n\to\infty}\left(5 + \frac{8}{n} + \frac{6}{n^2}\right) = 5 + 0 + 0 = 5$$

by Property 2, we can apply Property 5 to obtain
$$\lim_{n\to\infty} \frac{2 + (3/n) - (5/n^2)}{5 + (8/n) + (6/n^2)} = \frac{2}{5}$$

Hence,
$$\lim_{n\to\infty} \frac{2n^2 + 3n - 5}{5n^2 + 8n + 6} = \frac{2}{5}$$

Example 8

Let us show that

$$\lim_{n\to\infty} \frac{a^n}{n!} = 0$$

for any number a. This sequence will be of importance in later sections. The symbol $n!$ stands for the product of the first n integers. Thus,

$$1! = 1 \quad 2! = 1 \times 2 = 2 \quad 3! = 1 \times 2 \times 3 = 6 \quad 4! = 1 \times 2 \times 3 \times 4 = 24$$

and so on. $n!$ is read "n factorial."

Solution

Step 1
Let M be a fixed integer, $M \geq |a|$, and let $n > M$. Then

$$n! = 1 \cdot 2 \cdot 3 \cdot \cdots \cdot M \cdot (M+1) \cdot (M+2) \cdot \cdots \cdot n$$
$$= M!(M+1)(M+2) \cdot \cdots \cdot n$$
$$|a|^n = |a|^M |a|^{n-M}$$

and hence

$$\left|\frac{a^n}{n!}\right| = \frac{|a|^n}{n!} = \frac{|a|^M |a|^{n-M}}{M!(M+1)\cdots(n-1)n}$$

$$= \frac{|a|^M}{M!} \cdot \frac{\overbrace{|a| \cdot |a| \cdots |a| \cdot |a|}^{n-M}}{(M+1)(M+2)\cdots(n-1)n}$$

$$= \frac{|a|^M}{M!} \cdot \frac{|a|}{M+1} \cdot \frac{|a|}{M+2} \cdot \cdots \cdot \frac{|a|}{n-1} \cdot \frac{|a|}{n}$$

Step 2
Since $M \geq |a|$, it follows that

$$\frac{|a|}{M+1} < 1, \frac{|a|}{M+2} < 1, \ldots, \frac{|a|}{n-1} < 1$$

and hence

$$\frac{|a|}{M+1} \cdot \frac{|a|}{M+2} \cdot \cdots \cdot \frac{|a|}{n-1} \cdot \frac{|a|}{n} < \frac{|a|}{n}$$

Step 3
Combining steps 1 and 2 shows that

$$\left|\frac{a^n}{n!}\right| < \frac{|a|^M}{M!} \cdot \frac{|a|}{n} = \frac{|a|^{M+1}}{M!} \cdot \frac{1}{n}$$

Step 4
Since

$$\lim_{n\to\infty} \frac{|a|^{M+1}}{M!} \cdot \frac{1}{n} = \frac{|a|^{M+1}}{M!} \lim_{n\to\infty} \frac{1}{n} = 0$$

and

$$0 \le \left|\frac{a^n}{n!}\right| < \frac{|a|^{M+1}}{M!} \cdot \frac{1}{n}$$

it follows that

$$\lim_{n\to\infty} \left|\frac{a^n}{n!}\right| = 0$$

and our assertion is now obvious.

EXERCISES

Write out the first few terms of the sequences in Exercises 1–10.

1. $\left\{\dfrac{1}{n+5}\right\} =$
2. $\left\{\dfrac{n-1}{n+1}\right\} =$
3. $\{(-1)^n \cdot n\} =$
4. $\{5n\} =$
5. $\{n^2 - 2\} =$
6. $\left\{\dfrac{1}{n+2} - \dfrac{1}{4}\right\} =$
7. $\left\{\dfrac{2n}{2n+1}\right\} =$
8. $\{n^n\} =$
9. $\{n!\} =$
10. $\left\{\left(\dfrac{n}{2}\right)^n\right\} =$

Give the formula for the general (nth) term of the sequences in Exercises 11–16.

11. $\tfrac{1}{4}, \tfrac{1}{5}, \tfrac{1}{6}, \tfrac{1}{7}, \tfrac{1}{8}, \ldots$
12. $2, 4, 8, 16, 32, \ldots$
13. $\tfrac{1}{5}, \tfrac{1}{10}, \tfrac{1}{15}, \tfrac{1}{20}, \tfrac{1}{25}, \ldots$
14. $-4, -2, 0, 2, 4, \ldots$
15. $2, -2, 2, -2, 2, \ldots$
16. $-2, 2, -2, 2, -2, \ldots$

In Exercises 17–20, plot the first five terms of the given sequence.

17. $\left\{(-1)^n \cdot \dfrac{1}{n}\right\}$
18. $\{n^2 - 2\}$
19. $\left\{\dfrac{n}{n^2 + 10}\right\}$
20. $\left\{1 + \dfrac{1}{n}\right\}$

Find the limit of the sequences in Exercises 21–26.

21. $\lim_{n\to\infty} \dfrac{n}{n+1}$

22. $\lim_{n\to\infty} \dfrac{n}{n+1000}$

23. $\lim_{n\to\infty} \left(1 + \left(\dfrac{1}{2}\right)^n + \left(\dfrac{2}{3}\right)^n\right)$

24. $\lim_{n\to\infty} \dfrac{\sqrt{n}}{n+5}$

25. $\lim_{n\to\infty} \dfrac{n^2 - n + 1}{n^3 - 5n^2 + 7}$

26. $\lim_{n\to\infty} \dfrac{n-1}{n+1}$

11.2 PROPERTIES OF SEQUENCES

In the preceding section we introduced the concept of limit of a sequence and we examined some special limits. In this section we shall discuss basic conditions that guarantee convergence.

Consider the sequence

$$\left\{\frac{n}{n+1}\right\} = \frac{1}{2}, \frac{2}{3}, \frac{3}{4}, \ldots, \frac{n}{n+1}, \frac{n+1}{n+2}, \ldots$$

This sequence has the property that each term is less than its successor:

$$\frac{1}{2} < \frac{2}{3} < \frac{3}{4} < \cdots < \frac{n}{n+1} < \frac{n+1}{n+2} < \cdots$$

A sequence such as $\{n/(n+1)\}$ is said to be *increasing*. Considering the sequence

$$\left\{\frac{1}{n}\right\} = 1, \frac{1}{2}, \frac{1}{3}, \frac{1}{4}, \ldots, \frac{1}{n}, \frac{1}{n+1}, \ldots$$

we see that each term is bigger than its successor:

$$1 > \frac{1}{2} > \frac{1}{3} > \frac{1}{4} > \cdots > \frac{1}{n} > \frac{1}{n+1} > \cdots$$

Such a sequence is said to be *decreasing*. In general, we have the following definition.

DEFINITION 1

The sequence $\{a_n\}$ is *increasing* when $a_n < a_{n+1}$ for $n = 1, 2, 3, \ldots$, that is, when

$$a_1 < a_2 < a_3 < \ldots$$

The sequence $\{a_n\}$ is *decreasing* when $a_n > a_{n+1}$ for $n = 1, 2, 3, \ldots$, that is, when

$$a_1 > a_2 > a_3 > \ldots$$

A sequence that is increasing or decreasing is said to be *monotonic*.

Inspecting the sequences $\{n\}$ and $\{n/(n+1)\}$, we see that the first sequence becomes larger and larger: It increases beyond all bounds as n increases. What about the second sequence? We can write

$$\frac{n}{n+1} = 1 - \frac{1}{n+1}$$

and this tells us that no matter how large n is,

$$\frac{n}{n+1} < 1$$

The observations just made are often expressed by saying that the sequence $\{n\}$ is *unbounded,* the sequence $\{n/(n+1)\}$ is *bounded.* Formally, we have the following definition.

> **DEFINITION 2**
> A sequence is *bounded* when all its members are contained in some interval $[a, b]$. When this is not the case, then the sequence is *unbounded.*

Example 1

Decide if the sequence

$$\left\{\frac{n^n}{n!}\right\} = \frac{1}{1}, \frac{2^2}{2!}, \frac{3^3}{3!}, \frac{4^4}{4!}, \frac{5^5}{5!}, \ldots$$

is bounded.

Solution
We see that, regardless of the value of n,

$$\frac{n^n}{n!} = \frac{\overbrace{n \cdot n \cdot n \cdots n}^{n \text{ times}}}{1 \cdot 2 \cdot 3 \cdots n} = \frac{n}{1} \cdot \frac{n}{2} \cdot \frac{n}{3} \cdots \frac{n}{n} \geq n$$

since $n/2 \geq 1$, $n/3 \geq 1$, and so on. Thus, as n increases beyond all bounds, so does $n^n/n!$, and from this we conclude that the sequence in question is unbounded.

Example 2

Decide if the sequence

$$\left\{\frac{n}{n^2 + 100}\right\} = \frac{1}{101}, \frac{2}{104}, \frac{3}{109}, \frac{4}{116}, \frac{5}{125}, \ldots$$

is monotonic.

Solution
If the sequence is monotonic, then either

$$\frac{n}{n^2 + 100} < \frac{n+1}{(n+1)^2 + 100} \quad \text{for } n = 1, 2, 3, 4, \ldots \quad (1)$$

or else

$$\frac{n}{n^2 + 100} > \frac{n+1}{(n+1)^2 + 100} \quad \text{for } n = 1, 2, 3, 4, \ldots \quad (2)$$

The first inequality is the same as

$$n[(n+1)^2 + 100] < (n+1)[n^2 + 100]$$

which becomes upon simplification

$$n^3 + 2n^2 + 101n < n^3 + n^2 + 100n + 100$$

This, in turn, reduces to

$$n^2 + n < 100$$

We see that when $n \leq 9$,

$$n^2 + n \leq 9^2 + 9 = 90 < 100$$

whereas when $n \geq 10$,

$$n^2 + n \geq 10^2 + 10 = 110 > 100$$

Hence, (1) holds for $n = 1, 2, 3, \ldots, 9$, and (2) holds when $n = 10, 11, 12, 13, \ldots$. Hence, the sequence is *not* monotonic. This is shown graphically in Figure 11.5.

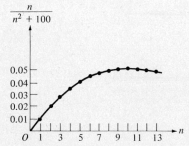

Figure 11.5
The sequence $\{n/(n^2 + 100)\}$.

Regarding bounded monotonic sequences we have the following theorem, stated here without proof:

THEOREM 1
Every bounded increasing sequence and every bounded decreasing sequence has a limit.

Example 3

Does the sequence

$$\left\{\frac{n}{n+1}\right\} = \frac{1}{2}, \frac{2}{3}, \frac{3}{4}, \ldots$$

converge?

Solution
At the beginning of this section we noted that this sequence is *increasing*. This is actually shown by observing that $n/(n+1) <$

$(n+1)/(n+2)$ is the same as $n(n+2) < (n+1)^2$ for all n, and the last inequality simplifies to $0 < 1$. Next, we note that the sequence is bounded, since $0 < n/(n+1) < 1$ for all n. Hence, by Theorem 1, the sequence converges.

REMARK

When the sequence $\{a_n\}$ is *increasing* but *unbounded*, we write

$$\lim_{n\to\infty} a_n = \infty$$

when $\{a_n\}$ is *decreasing* and *unbounded*, we write

$$\lim_{n\to\infty} a_n = -\infty$$

Thus, for example,

$$\lim_{n\to\infty} n = \infty$$

$$\lim_{n\to\infty} \sqrt{n} = \infty$$

$$\lim_{n\to\infty}(1-n) = -\infty$$

Example 4

Let us define a sequence $\{S_n\}$ as follows:

$S_1 = 1 + \frac{1}{3}$
$S_2 = 1 + \frac{1}{3} + (\frac{1}{3})^2$
$S_3 = 1 + \frac{1}{3} + (\frac{1}{3})^2 + (\frac{1}{3})^3$
$\vdots$
$S_n = 1 + \frac{1}{3} + (\frac{1}{3})^2 + \cdots + (\frac{1}{3})^n$

This sequence is evidently increasing since, for all n,

$$S_n = S_{n-1} + (\tfrac{1}{3})^n > S_{n-1}$$

On the other hand, $S_n < \frac{3}{2}$ for all values of n. We see this by observing that

$$S_n = 1 + \tfrac{1}{3} + (\tfrac{1}{3})^2 + \cdots + (\tfrac{1}{3})^n = \frac{1 - (\tfrac{1}{3})^{n+1}}{1 - \tfrac{1}{3}} \tag{3}$$

(This is readily verified by multiplying both sides by $1 - \frac{1}{3}$ and simplifying. See equation (4) below with $a = \frac{1}{3}$.) Since

$$\frac{1 - (\tfrac{1}{3})^{n+1}}{1 - \tfrac{1}{3}} = \tfrac{3}{2}[1 - (\tfrac{1}{3})^{n+1}] < \tfrac{3}{2}$$

we justified our claim. Hence, the sequence $\{S_n\}$ converges.

Generalizing formula (3), we have the following theorem.

THEOREM 2
If $a \neq 1$, then
$$1 + a + a^2 + \cdots + a^n = \frac{1 - a^{n+1}}{1 - a} \tag{4}$$
If $|a| < 1$, then
$$\lim_{n \to \infty} (1 + a + a^2 + \cdots + a^n) = \frac{1}{1 - a} \tag{5}$$

Proof

$$(1 + a + a^2 + \cdots + a^n)(1 - a)$$
$$= (1 + a + \cdots + a^n) - (a + a^2 + a^3 + \cdots + a^{n+1})$$
$$= 1 - a^{n+1}$$

Dividing by $1 - a$ gives equation (4).

To prove equation (5), we note that

$$\lim_{n \to \infty} (1 + a + a^2 + \cdots + a^n) = \lim_{n \to \infty} \frac{1 - a^{n+1}}{1 - a} \quad \text{by (4)}$$
$$= \lim_{n \to \infty} \left(\frac{1}{1 - a} - \frac{a^{n+1}}{1 - a} \right)$$
$$= \lim_{n \to \infty} \frac{1}{1 - a} - \lim_{n \to \infty} \frac{a^{n+1}}{1 - a}$$
$$= \frac{1}{1 - a} - \frac{1}{1 - a} \lim_{n \to \infty} a^{n+1}$$

According to Theorem 1 of Section 11.1,

$$\lim_{n \to \infty} a^{n+1} = a \lim_{n \to \infty} a^n = 0 \quad \text{for } |a| < 1$$

and formula (5) follows.

The following theorem, stated without proof, will allow us to evaluate many new kinds of limits. The reader will observe, however, that the theorem is a direct consequence of the concept of continuity (see Section 4.5).

THEOREM 3
Suppose that

1. $\lim_{n \to \infty} a_n = A$;
2. $f(a_n)$ is defined;
3. $f(x)$ is continuous at A.

Then
$$\lim_{n\to\infty} f(a_n) = f(A).$$

Example 5

(a) $\lim_{n\to\infty} e^{n/(n+1)} = e$

(b) $\lim_{n\to\infty} \ln\left(\dfrac{2n^2 + 3n - 5}{5n^2 + 8n + 6}\right) = \ln \dfrac{2}{5}$

(c) $\lim_{n\to\infty} \sin\left(\pi - \dfrac{1}{n}\right) = \sin \pi = 0$

EXERCISES

In Exercises 1–12, decide whether the sequence is increasing or decreasing, and whether it is bounded or unbounded.

1. $\{n^2 - n\}$
2. $\left\{(-1)^n \cdot \dfrac{1}{n}\right\}$
3. $\left\{\dfrac{n+1}{n+2}\right\}$
4. $\left\{\dfrac{n!}{n^n}\right\}$
5. $\{2^n - n\}$
6. $\{2^n - n^2\}$
7. $\{(-1)^n \cdot n\}$
8. $\{1 + \tfrac{3}{4} + (\tfrac{3}{4})^2 + \cdots + (\tfrac{3}{4})^n\}$
9. $\left\{\dfrac{\sin n}{n}\right\}$
10. $\{n^{1/n}\}$
11. $\{n - \sqrt{n}\}$
12. $\{n^2 - 100n\}$

Find the limits in Exercises 13–21.

13. $\lim_{n\to\infty} e^{-2n}$

14. $\lim_{n\to\infty} e^{1/n}$

15. $\lim_{n\to\infty} (a^2 + a^3 + a^4 + \cdots + a^{n+2})$ for $0 < a < 1$

16. $\lim_{n\to\infty} (-1)^n \dfrac{1}{n}$

17. $\lim_{n\to\infty} \dfrac{(n-1)n(n+1)}{n^3}$

18. $\lim_{n\to\infty} \sin \dfrac{2n+1}{n} \pi$

19. $\lim_{n\to\infty}\left(\dfrac{10}{7}+\dfrac{10}{7^2}+\dfrac{10}{7^3}+\cdots+\dfrac{10}{7^n}\right)$

20. $\lim_{n\to\infty}\left(1+\dfrac{1}{5^2}+\dfrac{1}{5^4}+\dfrac{1}{5^6}+\cdots+\dfrac{1}{5^{2n}}\right)$

21. $\lim_{n\to\infty} a_n \quad$ where $a_n = \displaystyle\int_1^n \dfrac{1}{x^2}\,dx$

11.3 THE ERROR IN LINEAR APPROXIMATIONS

The mean value theorem in Section 4.6 gave us the formula

$$f(x) = f(a) + f'(c)(x-a)$$

where c is a suitable point lying between a and x. Replacing c by a gives the approximate formula

$$f(x) \approx f(a) + f'(a)(x-a) \tag{1}$$

which is valid when $x \approx a$ (see Figure 11.6). This is the linear approximation formula discussed in Section 5.5, and the theorem below will tell us how good the approximation is when $f(x)$ is twice differentiable.

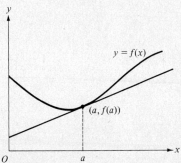

Figure 11.6
The linear approximation to $y = f(x)$ at $x = a$. The tangent at $(a, f(a))$ is $y = f(a) + f'(a)(x-a)$.

THEOREM 1
Let $f(x)$ be defined on the open interval (α, β) and have a second derivative at each point thereof. Then between any two points a and b in (α, β), there is a point c such that

$$f(b) = f(a) + f'(a)(b-a) + \dfrac{f''(c)}{2}(b-a)^2 \tag{2}$$

REMARKS
In contrast to Section 4.6, it is convenient here to consider functions defined on open rather than closed intervals. A geometric interpretation of formula (2) for the case $a < b$ is given in Figure 11.7;

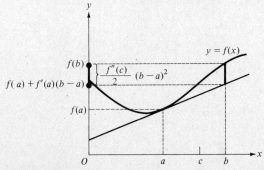

Figure 11.7
An illustration of formula (2).

it should be borne in mind, however, that the formula is true also when $b < a$.

Proof

For the sake of argument, let us suppose that $a < b$. Then both $f(x)$ and $f'(x)$ are continuous on the closed interval $[a, b]$ and differentiable on the open interval (a, b), and this reminds us of the Mean Value Theorem in Section 4.6. We thus try to reduce the present theorem to Rolle's theorem (see Section 4.6), and this is done with the function

$$G(x) = F(x) - F(a)\frac{(b-x)^2}{(b-a)^2} \qquad (3)$$

where

$$F(x) = f(b) - [f(x) + f'(x)(b-x)] \qquad (4)$$

[We offer no motivation for introducing the functions $G(x)$ and $F(x)$; regard them as an inspired choice.]

We now find that

1. $G(x)$ is continuous on $[a, b]$;
2. $G(x)$ is differentiable on (a, b);
3. $G(a) = 0$ and $G(b) = 0$.

According to Rolle's theorem, there is a point c satisfying $a < c < b$, and such that $G'(c) = 0$. To evaluate $G'(c)$, we first use equation (4) to get

$$F'(c) = 0 - [f'(c) + f''(c)(b-c) - f'(c)]$$
$$= -f''(c)(b-c)$$

Using this in (3) gives

$$G'(c) = F'(c) + 2F(a)\frac{(b-c)}{(b-a)^2}$$
$$= -f''(c)(b-c) + 2F(a)\frac{(b-c)}{(b-a)^2}$$
$$= (b-c)\left[-f''(c) + 2F(a)\frac{1}{(b-a)^2}\right] = 0$$

Since $c \neq b$ we have $b - c \neq 0$, and this tells us that

$$-f''(c) + 2F(a)\frac{1}{(b-a)^2} = 0$$

Solving for $F(a)$ shows that

$$F(a) = \frac{f''(c)}{2}(b-a)^2$$

By equation (4), however,
$$F(a) = f(b) - [f(a) + f'(a)(b-a)]$$
and comparing this equation with the preceding one gives formula (2), and this completes the proof.

Let us now replace b by x in formula (2), so that it becomes
$$f(x) = f(a) + f'(a)(x-a) + \frac{f''(c)}{2}(x-a)^2 \qquad (5)$$

We observe that if we omit the last term, then we get the approximate formula (1). This term, $[f''(c)/2](x-a)^2$, represents the *error* $E(x)$ of the linear approximation of a function having a second derivative. Thus,
$$E(x) = f(x) - [f(a) + f'(a)(x-a)] = \frac{f''(c)}{2}(x-a)^2 \qquad (6)$$

Example 1

Find the linear approximation to $\sin x$ for values $x \approx \pi/6$ and show that
$$|E(x)| \le \frac{1}{2}\left(x - \frac{\pi}{6}\right)^2$$

Solution
Consult Figure 11.8. Taking $a = \pi/6$ and $f(x) = \sin x$, we have the linear approximation
$$\sin x \approx \sin \frac{\pi}{6} + \cos \frac{\pi}{6} \cdot \left(x - \frac{\pi}{6}\right)$$
We recall that
$$\sin \frac{\pi}{6} = \frac{1}{2} \quad \text{and} \quad \cos \frac{\pi}{6} = \frac{\sqrt{3}}{2}$$
and this gives the formula
$$\sin x \approx \frac{1}{2} + \frac{\sqrt{3}}{2}\left(x - \frac{\pi}{6}\right) \quad \text{when } x \approx \frac{\pi}{6}$$

According to formula (6),
$$|E(x)| = \left|\frac{-\sin c}{2}\right|\left(x - \frac{\pi}{6}\right)^2 \le \frac{1}{2}\left(x - \frac{\pi}{6}\right)^2$$
since $|-\sin c| = |\sin c| \le 1$.

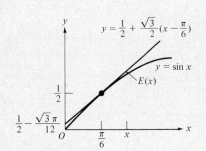

Figure 11.8
The linear approximation to $\sin x$ near $x = \pi/6$.

Example 2

In numerous important applications of formula (5) we take $a = 0$; the following formulas are readily verified:

$$e^x = 1 + x + \frac{e^c}{2}x^2$$

$$\sin x = x + \frac{(-\sin c)}{2}x^2$$

$$\cos x = 1 + \frac{(-\cos c)}{2}x^2$$

$$\ln(1 + x) = x + \left(-\frac{1}{2(1+c)^2}\right)x^2 \qquad (x > -1)$$

$$(1 + x)^r = 1 + rx + \left[\frac{1}{2}r(r-1)(1+c)^{r-2}\right]x^2 \qquad (x > -1)$$

In these formulas c lies between 0 and x. It must be remembered that c represents a different constant in each formula.

Example 3

Show that in the linear approximation

$$e^x \approx 1 + x$$

we have

$|E(x)| < 0.5x^2$ when $x \leq 0$
$|E(x)| < 0.83x^2$ when $0 < x < \frac{1}{2}$

Solution
Since the second derivative of e^x is e^x, we get from formula (6) with $a = 0$

$$E(x) = \frac{e^c}{2}x^2$$

When $x < 0$, then $x < c < 0$, and since $e^c < e^0 = 1$ we find that

$$|E(x)| < \tfrac{1}{2}x^2 = 0.5x^2$$

On the other hand, when $0 < x < \frac{1}{2}$, then $c < \frac{1}{2}$ and hence

$$|E(x)| < \frac{e^{1/2}}{2}x^2 < 0.83x^2$$

since $e^{1/2} < 1.6487$.

With formula (5) we can now tie up some loose ends. Specifically, in Section 4.2 we made the statement that a curve $y = f(x)$ lies *above*

Figure 11.9
A curve lying above its tangent lines.

Figure 11.10
A curve lying below its tangent lines.

its tangent lines for x in an interval (a, b) when $f''(x) > 0$ for all values of x in (a, b) (see Figure 11.9); $y = f(x)$ lies *below* its tangent lines when $f''(x) < 0$ on (a, b) (see Figure 11.10). These facts are now stated as a corollary.

COROLLARY 1
Let $f(x)$ be defined on the open interval (α, β) and let $f''(x)$ exist for each x in (α, β). Then

1. $f(x) > f(a) + f'(a)(x - a)$ for all values $x \neq a$ when $f''(x) > 0$ on (α, β);
2. $f(x) < f(a) + f'(a)(x - a)$ for all values $x \neq a$ when $f''(x) < 0$ on (α, β).

(See Figure 11.11.)

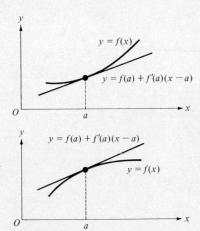

Figure 11.11
Illustrating conditions 1 and 2 of Corollary 1.

Proof
Conclusions 1 and 2 are direct consequences of formula (5), since $[f''(c)/2](x - a)^2$ is positive or negative depending on the sign of $f''(x)$.

As a further consequence of formula (5), we can also establish the first test for local minima and maxima given in Section 4.2.

COROLLARY 2
Let $f(x)$ be defined on an open interval (α, β) and suppose that $\alpha < a < \beta$.

1. If $f'(a) = 0$ and $f''(x) > 0$ on (α, β), then $f(x)$ has a *local minimum* at a;
2. If $f'(a) = 0$ and $f''(x) < 0$ on (α, β), then $f(x)$ has a *local maximum* at b.

Proof
The corollary is explained graphically in Figure 11.12, and it follows directly from Corollary 1. Namely, when $f'(a) = 0$, Corollary 1 states that

$f(x) > f(a)$ for all values $x \neq a$ when $f''(x) > 0$ on (α, β);
$f(x) < f(a)$ for all values $x \neq a$ when $f''(x) < 0$ on (α, β).

The first statement tells us that $f(x)$ has a local minimum at a, the second that $f(x)$ has a local maximum at a.

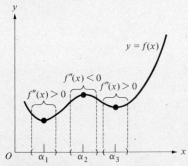

Figure 11.12
Illustrating Corollary 2. The function $f(x)$ has local minima at $x = \alpha_1$ and $x = \alpha_3$, and a local maximum at $x = \alpha_2$.

REMARK
If, in Corollary 2, we merely assume that $f''(a) > 0$ and $f''(x)$ is continuous, then $f''(x) > 0$ on an interval containing a, and the conclusion still holds.

EXERCISES

1. Referring to Example 3, what can you say about the error in the approximations

$$e^{0.000001} \approx 1.000001$$
$$e^{-0.000001} \approx 0.999999$$

In Exercises 2-7, find the linear approximation to the given function $f(x)$ for points $x \approx a$ and estimate the error $E(x)$ in the given interval.

2. $f(x) = \tan x \qquad a = \frac{\pi}{3}, \quad \left[\frac{\pi}{4}, \frac{2\pi}{5}\right]$

3. $F(x) = \sin^{-1} x \qquad a = 0, \quad [-0.1, 0.1]$
$\qquad\qquad\qquad\qquad a = \frac{1}{2}, \quad [0.45, 0.55]$

4. $f(x) = \tan^{-1} x \qquad a = 1, \quad [0.9, 1.1]$

5. $f(x) = x^{10} \qquad a = 1, \quad [0.98, 1.05]$

6. $f(x) = \dfrac{e^x - e^{-x}}{e^x + e^{-x}} \qquad a = 0, \quad [-0.2, 0.2]$

7. $f(x) = \displaystyle\int_0^x e^{-t^2}\, dt \qquad a = 0, \quad [-0.2, 0.2]$

8. Consider the linear approximation to $f(x) = x^2$ for points $x \approx 2$. Find the interval containing 2 inside of which the error $E(x)$ will be less than 0.1 in absolute value.

11.4 TAYLOR APPROXIMATIONS

The formula

$$f(x) = f(a) + f'(a)(x-a) + \frac{f''(c)}{2}(x-a)^2$$

was used in the previous section to determine the error in the linear approximation

$$f(x) \approx f(a) + f'(a)(x-a) \tag{1}$$

The number c, as we remarked, is generally unknown, but replacing it by a yields the useful approximate formula

$$f(x) \approx f(a) + f'(a)(x-a) + \frac{f''(a)}{2}(x-a)^2 \tag{2}$$

This formula is called the *second-degree Taylor approximation* to $f(x)$ at a. That this formula is an improvement of formula (1) is indicated in the following example.

Example 1

Consider the function $f(x) = \sin x$ evaluated near $\pi/6$, and let us denote the right sides of formulas (1) and (2) by $T_1(x)$ and $T_2(x)$, respectively. Since $f'(x) = \cos x$ and $f''(x) = -\sin x$, we find that

$$T_1(x) = \sin \frac{\pi}{6} + \cos \frac{\pi}{6} \cdot \left(x - \frac{\pi}{6}\right)$$

$$T_2(x) = \sin \frac{\pi}{6} + \cos \frac{\pi}{6} \cdot \left(x - \frac{\pi}{6}\right) - \frac{1}{2} \sin \frac{\pi}{6} \cdot \left(x - \frac{\pi}{6}\right)^2$$

Using the fact that

$$\sin \frac{\pi}{6} = \frac{1}{2} \quad \text{and} \quad \cos \frac{\pi}{6} = \frac{\sqrt{3}}{2}$$

(see Section 9.1) gives

$$T_1(x) = \frac{1}{2} + \frac{\sqrt{3}}{2}\left(x - \frac{\pi}{6}\right)$$

$$T_2(x) = \frac{1}{2} + \frac{\sqrt{3}}{2}\left(x - \frac{\pi}{6}\right) - \frac{1}{4}\left(x - \frac{\pi}{6}\right)^2$$

Carrying out the numerical calculation for $x = (\pi/6) + (\pi/60) = (11\pi/60)$ (note that $11\pi/60$ corresponds to 33°) to eight significant figures reveals the following:

$$\sin \frac{11\pi}{60} = 0.54463903$$

$$T_1\left(\frac{11\pi}{60}\right) = 0.54534498$$

$$T_2\left(\frac{11\pi}{60}\right) = 0.54465959$$

The respective errors resulting from the use of formulas (1) and (2) are thus the following:

$$E_1 = \sin \frac{11\pi}{60} - T_1\left(\frac{11\pi}{60}\right) = -0.00070595 = -7.0595 \times 10^{-4}$$

$$E_2 = \sin \frac{11\pi}{60} - T_2\left(\frac{11\pi}{60}\right) = -0.00002056 = -2.056 \times 10^{-5}$$

Since

$$E_2 \approx \tfrac{1}{34} E_1$$

we see that we have a marked improvement.

The success of the second-degree Taylor approximation suggests that we try "nth-degree" Taylor approximations. This is done below, but the real significance of this process will unveil itself in the next section, where it is shown that the errors resulting from the approximations often diminish to 0 as n approaches infinity.

DEFINITION 1
Let $f(x)$ be n-times differentiable at a. The polynomial $T_n(x)$, given by the formula

$$T_n(x) = f(a) + f'(a)(x-a) + \cdots + \frac{f^{(k)}(a)}{k!}(x-a)^k$$
$$+ \cdots + \frac{f^{(n)}(a)}{n!}(x-a)^n$$

is called the *nth-degree Taylor approximation* to $f(x)$ at a with an error

$$E_n(x) = f(x) - T_n(x)$$

Example 2

Find the third Taylor approximation to $f(x) = \sin x$ at $\pi/6$.

Solution
This is a continuation of Example 1. We have

$$f\left(\frac{\pi}{6}\right) = \sin\frac{\pi}{6} = \frac{1}{2}$$

$$f'\left(\frac{\pi}{6}\right) = \cos\frac{\pi}{6} = \frac{\sqrt{3}}{2}$$

$$f''\left(\frac{\pi}{6}\right) = -\sin\frac{\pi}{6} = -\frac{1}{2}$$

$$f'''\left(\frac{\pi}{6}\right) = -\cos\frac{\pi}{6} = -\frac{\sqrt{3}}{2}$$

and hence

$$T_3(x) = \frac{1}{2} + \frac{\sqrt{3}}{2}\left(x - \frac{\pi}{6}\right) - \frac{1}{4}\left(x - \frac{\pi}{6}\right)^2 - \frac{\sqrt{3}}{12}\left(x - \frac{\pi}{6}\right)^3$$

For $x = 11\pi/60$, a direct computation gives

$$T_3\left(\frac{11\pi}{60}\right) = 0.54463887$$

and hence

$$E_3\left(\frac{11\pi}{60}\right) = f\left(\frac{11\pi}{60}\right) - T_3\left(\frac{11\pi}{60}\right) = 0.54463903 - 0.54463887$$

$$= 1.6 \times 10^{-7}$$

Comparing this with the errors computed in Example 1, we see that

$$E_3\left(\frac{11\pi}{60}\right) \approx \frac{1}{128}\left|E_2\left(\frac{11\pi}{60}\right)\right| \approx \frac{1}{4412}\left|E_1\left(\frac{11\pi}{60}\right)\right|$$

Thus, the improvement is quite dramatic.

An important special case of Taylor approximations is $a = 0$; these approximations are also known under the name *Maclaurin approximations*, and the polynomial

$$T_n(x) = f(0) + f'(0)x + \cdots + \frac{f^{(k)}(0)}{k!}x^k + \cdots + \frac{f^{(n)}(0)}{n!}x^n$$

is called the *n*th-degree *Maclaurin approximation*.

Some of the Maclaurin approximations most used are presented below; their verification is straightforward and for the most part are relegated to the exercises.

MACLAURIN APPROXIMATIONS

$$e^x \approx 1 + x + \frac{1}{2!}x^2 + \frac{1}{3!}x^3 + \frac{1}{4!}x^4 + \cdots + \frac{1}{n!}x^n \tag{3}$$

$$\ln(1 + x) \approx x - \frac{1}{2}x^2 + \frac{1}{3}x^3 - \frac{1}{4}x^4 + \cdots$$

$$+ (-1)^{n+1}\frac{1}{n}x^n \quad (-1 < x \leq 1) \tag{4}$$

$$\sin x \approx x - \frac{1}{3!}x^3 + \frac{1}{5!}x^5 + \cdots$$

$$+ (-1)^{n+1}\frac{1}{(2n-1)!}x^{2n-1} \tag{5}$$

$$\cos x \approx 1 - \frac{1}{2!}x^2 + \frac{1}{4!}x^4 + \cdots + (-1)^n \frac{1}{(2n)!}x^{2n} \tag{6}$$

$$(1 + x)^r \approx 1 + rx + \frac{r(r-1)}{2!}x^2 + \frac{r(r-1)(r-2)}{3!}x^3$$

$$+ \cdots + \frac{r(r-1)(r-2)\cdots(r-n+1)}{n!}x^n$$

$$(|x| < 1) \tag{7}$$

$$\frac{1}{1-x} \approx 1 + x + x^2 + x^3 + \cdots + x^n \quad (|x| < 1) \tag{8}$$

Example 3

Find the fourth-degree Maclaurin approximation of $\sqrt{1 + x}$.

Solution
Using formula (7) with $r = \frac{1}{2}$ and $n = 4$ gives

$$\sqrt{1+x} \approx 1 + \frac{1}{2}x + \frac{\frac{1}{2}(\frac{1}{2}-1)}{2!}x^2 + \frac{\frac{1}{2}(\frac{1}{2}-1)(\frac{1}{2}-2)}{3!}x^3$$

$$+ \frac{\frac{1}{2}(\frac{1}{2}-1)(\frac{1}{2}-2)(\frac{1}{2}-3)}{4!}x^4$$

$$= 1 + \frac{1}{2}x - \frac{1}{8}x^2 + \frac{1}{16}x^3 - \frac{5}{128}x^4$$

Example 4

Verify formula (5) for the approximation of $f(x) = \sin x$.

Solution
Since

$$f(0) = \sin 0 = 0$$
$$f^{(1)}(0) = \cos 0 = 1$$
$$f^{(2)}(0) = -\sin 0 = 0$$
$$f^{(3)}(0) = -\cos 0 = -1$$
$$f^{(4)}(0) = \sin 0 = 0$$

we conclude from the periodic behavior of the derivatives of $\sin x$ that

$$f^{(k)}(0) = \begin{cases} 0 & \text{for } k = 2, 4, 6, 8, 10, \ldots \\ 1 & \text{for } k = 1, 5, 9, 13, \ldots \\ -1 & \text{for } k = 3, 7, 11, 15, \ldots \end{cases}$$

From this fact we see that

$$T_1(x) = 0 + 1 \cdot x = x$$

$$T_2(x) = T_1(x) + \frac{f^{(2)}(0)}{2!}x^2 = T_1(x) = x$$

$$T_3(x) = T_2(x) + \frac{f^{(3)}(0)}{3!}x^3 = x - \frac{1}{3!}x^3$$

and since $f^{(4)}(0) = 0$, we have

$$T_4(x) = T_3(x)$$

In general,

$$T_{2n-1}(x) = T_{2n}(x) = \sum_{k=1}^{n} (-1)^{k+1} \frac{1}{(2k-1)!} x^{2k-1}$$

EXERCISES

1. Using the convention $f^{(0)}(a) = f(a)$, write the Taylor and

Maclaurin approximations $T_n(x)$ with the summation notation.

2. Write the Maclaurin approximations in formulas (3)–(8) using the summation notation.

Find the fourth-degree Maclaurin approximations of the functions in Exercises 3–9.

3. $f(x) = \dfrac{1}{x+1}$

4. $f(x) = \dfrac{1}{x^2+1}$

5. $f(x) = e^{-x}$

6. $f(x) = e^{x^2}$

7. $f(x) = 5^x$

8. $f(x) = \ln \dfrac{1+x}{1-x}$

9. $f(x) = \dfrac{1}{\sqrt{x+1}}$

Give the nth term of the Taylor approximation $T_n(x)$ at a of the functions in Exercises 10–14.

10. $f(x) = \cos x$

11. $f(x) = e^{2x}$

12. $f(x) = 10^x$

13. $f(x) = \ln(2x+1)$

14. $f(x) = e^{-x}$

15. Referring to Definition 1, show that the nth-degree Taylor approximation $T_n(x)$ to $f(x)$ at a has the following properties:

$$T_n(a) = f(a)$$
$$T_n^{(1)}(a) = f^{(1)}(a)$$
$$\vdots$$
$$T_n^{(k)}(a) = f^{(k)}(a)$$
$$\vdots$$
$$T_n^{(n)}(a) = f^{(n)}(a)$$

16. Find the first, second, and third Taylor approximations of the polynomial $p(x) = 1 - 5x + 3x^2$ at a, and find the errors.

17. Find the $(n-1)$st, nth, and $(n+1)$st Maclaurin approximations of the polynomial

$$p(x) = A_0 + A_1 x + A_2 x^2 + \cdots + A_n x^n$$

What are the errors?

11.5 THE ERROR IN TAYLOR APPROXIMATIONS

In this section we consider the error $E_n(x) = f(x) - T_n(x)$ of the nth-degree Taylor approximation of $f(x)$. We shall show that in many important cases $E_n(x)$ diminishes to 0 as $n \to \infty$. Later (Theorem 1), we shall relate $E_n(x)$ to the $(n + 1)$st derivative of $f(x)$. By way of motivation, consider the following example.

Example 1

Consider the function

$$f(x) = \frac{1}{1-x} \qquad (x \neq 1)$$

and its nth-degree Maclaurin approximation (see Section 11.4)

$$T_n(x) = 1 + x + x^2 + \cdots + x^n$$

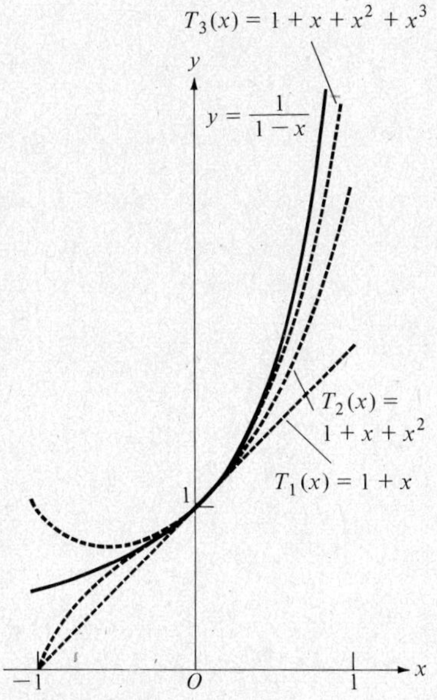

Figure 11.13
The first three Maclaurin approximations of $f(x) = 1/(1 - x)$ for $|x| < 1$.

Figure 11.13 shows the relation between $f(x)$, $T_1(x)$, $T_2(x)$, and $T_3(x)$. In view of the identity

$$\frac{1}{1-x} = 1 + x + x^2 + \cdots + x^n + \frac{x^{n+1}}{1-x} \qquad (x \neq 1)$$

(see Theorem 2, Section 11.2), we see that

$$E_n(x) = f(x) - T_n(x) = \frac{x^{n+1}}{1-x}$$

When $|x| < 1$, then we know that

$$\lim_{n \to \infty} |x|^{n+1} = 0$$

(see Section 11.1). Hence

$$\lim_{n\to\infty} |E_n(x)| = \lim_{n\to\infty} \left|\frac{x^{n+1}}{1-x}\right| = \frac{1}{|1-x|} \lim_{n\to\infty} |x^{n+1}| = 0$$

and this is also expressed by writing

$$\lim_{n\to\infty} T_n(x) = f(x) \quad \text{for } |x| < 1$$

A special symbol is used for this limit; it is

$$\lim_{n\to\infty} T_n(x) = 1 + \sum_{k=1}^{\infty} x^k$$

and we write

$$\frac{1}{1-x} = 1 + \sum_{k=1}^{\infty} x^k \tag{1}$$

Before explaining this further, we present the following definition.

DEFINITION 1

Suppose that all derivatives $f^{(k)}(a)$, $k = 1, 2, 3, \ldots$, exist. If the sequence of Taylor polynomials

$$T_n(x) = f(a) + \sum_{k=1}^{n} \frac{f^{(k)}(a)}{k!}(x-a)^k$$

has a limit as $n \to \infty$, we write

$$\lim_{n\to\infty} T_n(x) = f(a) + \sum_{k=1}^{\infty} \frac{f^{(k)}(a)}{k!}(x-a)^k$$

and call the right side a *Taylor series* about a. We write

$$f(x) = f(a) + \sum_{k=1}^{\infty} \frac{f^{(k)}(a)}{k!}(x-a)^k$$

and say that the Taylor series *converges* to $f(x)$ when

$$\lim_{n\to\infty} T_n(x) = f(x).$$

We remark that the statement

$$\lim_{n\to\infty} T_n(x) = f(x)$$

is the same as

$$\lim_{n\to\infty} E_n(x) = 0$$

where $E_n(x) = f(x) - T_n(x)$ represents the error for each value of n.

The reader will notice that nothing was said in Definition 1 about the values of x for which the Taylor series may converge. We shall see that Taylor series about a converge for all values of x in some interval containing a, but that the interval depends on $f(x)$. It may happen that the interval is the single point a, but in all important cases the interval contains other points as well.

Comparing Example 1 with the definition shows that the right side of equation (1) is the Taylor series of the function $f(x) = 1/(1-x)$ about 0 for all values $|x| < 1$. Taylor series about 0 are also called *Maclaurin series*.

To investigate the convergence of Taylor polynomials $T_n(x)$, we now relate the error $E_n(x)$ to the $(n+1)$st derivative of $f(x)$; the theorem that follows is an extension of Theorem 1, Section 11.3.

THEOREM 1
Let $f(x)$ be defined on the open interval (α, β) and, for a given positive integer n, let the derivatives $f^{(k)}(x)$ exist on that interval for $k = 1, 2, 3, \ldots, n+1$. Then between any two points a and b in (α, β), there is a point c such that

$$f(b) = \left[f(a) + \sum_{k=1}^{n} \frac{f^{(k)}(a)}{k!}(b-a)^k\right] + \frac{f^{(n+1)}(c)}{(n+1)!}(b-a)^{n+1} \qquad (2)$$

Like the proof of Theorem 1, Section 11.3, the proof of this theorem is very technical and offers no additional insight; it is presented at the end of this section for the sake of completeness. Replacing b by x shows that

$$E_n(x) = f(x) - T_n(x) = \frac{f^{(n+1)}(c)}{(n+1)!}(x-a)^{n+1} \qquad (3)$$

and with this formula we easily establish the convergence of the Taylor expansions of e^x, $\sin x$, $\cos x$, and other functions, for all values of x. It must be emphasized that c depends on a, x, and n.

Example 2

Show that the Taylor expansion of $f(x) = \sin x$ about a converges for each value of x.

Solution
It is readily computed that

$$T_n(x) = \sin a + \cos a(x-a) - \frac{\sin a}{2!}(x-a)^2$$

$$- \frac{\cos a}{3!}(x-a)^3 + \cdots + \frac{A}{n!}(x-a)^n$$

where $A = \pm\sin a$ or $A = \pm\cos a$, depending on the value of n; according to formula (3),

$$E_n(x) = \frac{f^{(n+1)}(c)}{(n+1)!}(x-a)^{n+1}$$

where $f^{(n+1)}(c) = \pm\sin c$ or $f^{(n+1)}(c) = \pm\cos c$, again depending on the value of n. In any case,

$$|f^{(n+1)}(c)| \leq 1$$

no matter what c and n are, and hence

$$|E_n(x)| = \left|\frac{f^{(n+1)}(c)}{(n+1)!}(x-a)^{n+1}\right|$$

$$= |f^{(n+1)}(c)|\frac{|x-a|^{n+1}}{(n+1)!}$$

$$\leq \frac{|x-a|^{n+1}}{(n+1)!}$$

From Example 8 in Section 11.1 we know that

$$\lim_{n\to\infty}\frac{|x-a|^{n+1}}{(n+1)!} = 0$$

for any choice of x and a. Hence

$$\lim_{n\to\infty}|E_n(x)| = 0$$

and consequently,

$$\lim_{n\to\infty}E_n(x) = 0$$

which is the same as saying

$$\lim_{n\to\infty}T_n(x) = \sin x$$

Example 3

In approximating $f(x) = e^x$ by

$$T_n(x) = 1 + x + \frac{1}{2!}x^2 + \cdots + \frac{1}{n!}x^n$$

at $x = 0.1$, how big must n be to have $|E_n(x)| < 10^{-10}$?

Solution
We use formula (3) with $a = 0$, and since $f^{(k)}(x) = e^x$ for all values of k, we find that

$$E_n(x) = \frac{e^c}{(n+1)!}x^{n+1}$$

where the number c lies between 0 and x. Now, taking $x = 0.1$, we find that

$$e^c < e^{0.1} \quad \text{since} \quad c < 0.1$$

and from the table of exponentials we see that

$$e^{0.1} < 1.1052$$

Hence

$$|E_n(x)| < \frac{1.1052}{(n+1)!} 0.1^{n+1} = \frac{1.1052}{(n+1)!} 10^{-n-1}$$

From Table 11.1 we see that

$$|E_5(x)| < \frac{1.1052}{6!} 10^{-6} = 1.1052 \times 0.14 \times 10^{-8} > 10^{-10}$$

whereas

$$|E_6(x)| < \frac{1.1052}{7!} 10^{-7} = 1.1052 \times 0.20 \times 10^{-10} < 10^{-10}$$

Hence, the error is less than 10^{-10} when $n = 6$.

Table 11.1
Reciprocals of Factorials

n	$1/n!$
5	0.83×10^{-2}
6	0.14×10^{-2}
7	0.20×10^{-3}
8	0.25×10^{-4}

Example 4

Show that

$$e^x = 1 + \sum_{k=1}^{\infty} \frac{1}{k!} x^k$$

Solution

The equation

$$E_n(x) = \frac{e^c}{(n+1)!} x^{n+1}$$

used in the solution of Example 3 is also used here; to solve the problem we must show that $E_n(x) \to 0$ as $n \to \infty$ for any choice of x.

Since different values of n give different values of c, we estimate $E_n(x)$ as follows: When $x > 0$, then $0 < c < x$, implying that $e^c < e^x$, and thus

$$|E_n(x)| \leq e^x \frac{x^{n+1}}{(n+1)!}$$

When $x < 0$, then $x < c < 0$, implying that $e^c < e^0 = 1$, and hence

$$|E_n(x)| \leq \frac{|x|^{n+1}}{(n+1)!}$$

In any case, we know that

$$\lim_{n\to\infty} |E_n(x)| = 0$$

and hence

$$\lim_{n\to\infty} E_n(x) = 0$$

Thus,

$$\lim_{n\to\infty} T_n(x) = e^x$$

for all values of x.

Corresponding to the Maclaurin approximations of Section 11.4, we have the following Maclaurin series.

MACLAURIN SERIES

$$e^x = 1 + \sum_{k=1}^{\infty} \frac{1}{k!} x^k \tag{4}$$

$$\ln(1+x) = \sum_{k=1}^{\infty} (-1)^{k+1} \frac{1}{k} x^k \quad \text{for } -1 < x \leq 1 \tag{5}$$

$$\sin x = \sum_{k=1}^{\infty} (-1)^{k+1} \frac{1}{(2k-1)!} x^{2k-1} \tag{6}$$

$$\cos x = 1 + \sum_{k=1}^{\infty} (-1)^k \frac{1}{(2k)!} x^{2k} \tag{7}$$

$$(1+x)^r = 1 + \sum_{k=1}^{\infty} \frac{r(r-1)\cdots(r-k+1)}{k!} x^k \tag{8}$$

$$\text{for } |x| < 1$$

We conclude this section with a proof of Theorem 1.

Proof of Theorem 1
Consider the functions

$$F(x) = f(b) - \left[f(x) + f^{(1)}(x)(b-x) \right.$$
$$\left. + \frac{f^{(2)}(x)}{2!}(b-x)^2 + \cdots + \frac{f^{(n)}(x)}{n!}(b-x)^n \right] \tag{9}$$

$$G(x) = F(x) - F(a) \frac{(b-x)^{n+1}}{(b-a)^{n+1}}$$

For the sake of argument, suppose that $a < b$. From the statement of the theorem, we deduce that $F(x)$ is continuous and differentiable on the closed interval $[a, b]$. From this we deduce that $G(x)$ is likewise continuous and differentiable on $[a, b]$. In addition, it is seen that

$$G(a) = F(a) - F(a)\frac{(b-a)^{n+1}}{(b-a)^{n+1}} = F(a) - F(a) = 0$$

$$G(b) = F(b) - F(a) \cdot 0 = F(b) = f(b) - f(b) = 0$$

Rolle's theorem (Section 4.6) can therefore be applied to the function $G(x)$ to give a point c, $a < c < b$, such that $G'(c) = 0$. This fact will lead to formula (2) in the statement of the theorem.

A direct differentiation shows the following:

$$G'(x) = F'(x) + (n+1)F(a)\frac{(b-x)^n}{(b-a)^{n+1}}$$

$$F'(x) = -\bigg[f^{(1)}(x) - f^{(1)}(x) + f^{(2)}(x)(b-x)$$

$$- 2\frac{f^{(2)}(x)}{2!}(b-x) + \frac{f^{(3)}(x)}{2!}(b-x)^2$$

$$- \cdots - n\frac{f^{(n)}(x)}{n!}(b-x)^{n-1} + \frac{f^{(n+1)}(x)}{n!}(b-x)^n\bigg]$$

It is not difficult to see that the terms on the right cancel pairwise, except for the last term; thus,

$$F'(x) = -\frac{f^{(n+1)}(x)}{n!}(b-x)^n$$

Substituting this into the formula for $G'(x)$ and setting $x = c$ gives

$$0 = G'(c) = -\frac{f^{(n+1)}(c)}{n!}(b-c)^n + (n+1)F(a)\frac{(b-c)^n}{(b-a)^{n+1}}$$

and from this we find that

$$F(a) = \frac{f^{(n+1)}(c)}{(n+1)n!}(b-a)^{n+1} = \frac{f^{(n+1)}(c)}{(n+1)!}(b-a)^{n+1}$$

Substituting a for x in formula (9) and using the result just computed shows that

$$\frac{f^{(n+1)}(c)}{(n+1)!}(b-a)^{n+1} = f(b) - \bigg[f(a) + f^{(1)}(a)(b-a)$$

$$+ \cdots + \frac{f^{(n)}(a)}{n!}(b-a)^n\bigg]$$

With simple algebra we obtain formula (2).

EXERCISES

1. Show that the Taylor expansion of $f(x) = \cos x$ about a converges for each value of x.

Find the Taylor series of the functions in Exercises 2–10 about the given points a.

2. $f(x) = \dfrac{1}{x}$ $\qquad a = 1$

3. $f(x) = \sin x$ $\qquad a = \dfrac{\pi}{2}$

4. $f(x) = e^x$ $\qquad a = \ln 2$

5. $f(x) = \tfrac{1}{2}(e^x + e^{-x})$ $\qquad a = 0$

6. $f(x) = \tfrac{1}{2}(e^x - e^{-x})$ $\qquad a = 0$

7. $f(x) = b^x$ $\qquad a$

8. $f(x) = \sin^2 x$ $\qquad a = 0$

9. $f(x) = \ln x$ $\qquad a$ (write out the first few terms)

10. $f(x) = x^5$ $\qquad a$

11. In approximating $f(x) = \tfrac{1}{2}(e^x + e^{-x})$ with a Maclaurin polynomial of degree n, how large must n be to have $|E_n(x)| < 10^{-6}$ for x in the interval $[-0.5, 0.5]$.

12. Show that

$$\ln(1+x) = \sum_{k=1}^{\infty} (-1)^{k+1} \frac{1}{k} x^k \qquad \text{for } 0 < x \leq 1$$

Procedure:

(a) In the identity

$$\frac{1}{1-x} = 1 + x + x^2 + \cdots + x^{n-1} + \frac{x^n}{1-x}$$

substitute $x = -t$ and integrate both sides of the resulting equation from 0 to x.

(b) From (a) you should get an equation

$$\ln(1+x) = T_n(x) + E_n(x)$$

where

$$E_n(x) = (-1)^n \int_0^x \frac{t^n}{1+t} dt$$

Write out $T_n(x)$.

(c) To show that
$$\lim_{n \to \infty} E_n(x) = 0$$
show that
$$|E_n(x)| \le \frac{1}{n+1}$$

This is done by using the fact that
$$\int_0^x \frac{t^n}{1+t}\, dt \le \int_0^x t^n\, dt$$

which is due to the inequality $1/(1+t) \le 1$ for $0 \le t \le x$.

(d) Show that the Taylor expansion converges to $\ln(1+x)$ for $-1 < x < 0$ by following a similar procedure.

11.6 INFINITE SERIES

Consider the Taylor series of $f(x) = 1/(1-x)$ about 0:

$$\frac{1}{1-x} = \sum_{k=0}^{\infty} x^k \qquad \text{for } |x| < 1 \tag{1}$$

Evaluating this series at $x = \frac{1}{2}$ gives

$$\sum_{k=0}^{\infty} \frac{1}{2^k} = 2 \tag{2}$$

This is an example of an *infinite series*. We call 2 the *sum* of the infinite series, but this is not to be understood as an arithmetic sum; rather, it is the *limit* of the sequence of partial sums

$$s_0 = 1 = 2 - 1$$

$$s_1 = 1 + \frac{1}{2} = 2 - 0.5$$

$$s_2 = 1 + \frac{1}{2} + \frac{1}{4} = 2 - 0.25$$

$$s_3 = 1 + \frac{1}{2} + \frac{1}{4} + \frac{1}{8} = 2 - 0.125 \tag{3}$$

$$s_4 = 1 + \frac{1}{2} + \frac{1}{4} + \frac{1}{8} + \frac{1}{16} = 2 - 0.0625$$

$$\vdots$$

$$s_n = 1 + \frac{1}{2} + \frac{1}{4} + \cdots + \frac{1}{2^n} = 2 - \frac{1}{2^n}$$

Infinite series are of considerable importance because they enable us to evaluate accurately logarithms, trigonometric functions, and so on. This and the next two sections are devoted, therefore, to a study of the basic properties of infinite series and techniques for finding their sums when such exist.

The basic nomenclature of this section is the following definition.

> **DEFINITION 1**
> Let $a_0, a_1, \ldots, a_k, \ldots$ be a given sequence of numbers. The symbol
> $$\sum_{k=0}^{\infty} a_k$$
> is called an *infinite series* or simply a *series*. For each n the sum
> $$s_n = \sum_{k=0}^{n} a_k$$
> is called the nth *partial sum* of the series.

In contrast to the previous section, it is now convenient to start the summation with 0 rather than with 1. We shall also use the notation

$$a_0 + a_1 + a_2 + a_3 + \ldots$$

or

$$a_0 + a_1 + a_2 + a_3 + \cdots + a_k + \ldots$$

to designate an infinite series.

Looking at the series in (2), we say that it *converges* to the sum 2 because, according to (3)

$$\lim_{n \to \infty} s_n = 2$$

In general, we shall say that an infinite series converges when its sequence of partial sums converges; we shall show now that the latter is not always the case.

Example 1

Consider the series

$$\sum_{k=0}^{\infty} (-1)^k = 1 - 1 + 1 - 1 + 1 - 1 + \ldots$$

Its partial sums are

$s_0 = 1$
$s_1 = 1 - 1 = 0$

$$s_2 = 1 - 1 + 1 = 1$$
$$s_3 = 1 - 1 + 1 - 1 = 0$$
$$\vdots$$
$$s_n = \begin{cases} 0 & \text{when } n \text{ is odd} \\ 1 & \text{when } n \text{ is even} \end{cases}$$

The sequence of partial sums is thus

$$1, 0, 1, 0, 1, 0, \ldots$$

This sequence does not converge and hence the corresponding series does not converge.

Formally, we put down the following definition.

DEFINITION 2
The infinite series

$$\sum_{k=0}^{\infty} a_k$$

with partial sums $s_0, s_1, s_2, s_3, \ldots$, is said to *converge* to the sum A, written

$$\sum_{k=0}^{\infty} a_k = A$$

when

$$\lim_{n \to \infty} s_n = A$$

The infinite series is said to *diverge* when its sequence of partial sums diverges.

Example 2

Using the convention $0! = 1$, we have

$$e = \sum_{k=0}^{\infty} \frac{1}{k!}$$

This is obtained from the formula

$$e^x = 1 + \sum_{k=1}^{\infty} \frac{x^k}{k!}$$

discussed in Example 4, Section 11.5, by taking $x = 1$. We know that

$$|e - s_n| \leq \frac{e}{(n+1)!}$$

where

$$s_n = \sum_{k=0}^{n} \frac{1}{k!} = 1 + 1 + \frac{1}{2!} + \cdots + \frac{1}{n!}$$

Example 3

Verify that

$$\ln 2 = \sum_{k=1}^{\infty} (-1)^{k+1} \frac{1}{k}$$

Solution
In Exercises 12 and 13 of Section 11.5, you were asked to establish the formula

$$\ln(1+x) = \sum_{k=1}^{\infty} (-1)^{k+1} \frac{1}{k} x^k \qquad \text{for } -1 < x \leq 1$$

Taking $x = 1$ yields the series converging to $\ln 2$.

In the last two examples we saw that a Taylor series, evaluated for a specific value of x, gives rise to an infinite series.

Example 4

Show that the series

$$\sum_{k=0}^{\infty} a_k$$

in which $a_k = 1$ for all values of k, diverges.

Solution
Since

$$\sum_{k=0}^{\infty} a_k = 1 + 1 + 1 + 1 + \ldots$$

the nth partial sum of this series is

$$s_n = \sum_{k=0}^{n} a_k = \underbrace{1 + 1 + 1 + \cdots + 1}_{n+1} = n + 1$$

and, according to Section 11.1,

$$\lim_{n \to \infty} s_n = \infty$$

Hence, the sequence of partial sums diverges and so does the series.

The second inequality in (4) therefore tells us that

$$\lim_{n\to\infty} s_{n-1} \geq \lim_{n\to\infty} \int_1^n f(x)\, dx = \infty$$

Hence the series diverges.

EXERCISES

In Exercises 1–12, decide which series converges and which series diverges. Always substantiate your answer.

1. $\sum_{k=1}^{\infty} \dfrac{1}{\sqrt{k}}$

2. $\sum_{k=1}^{\infty} \dfrac{1}{k^r} \quad (r > 1)$

3. $\sum_{k=1}^{\infty} \left(\dfrac{100}{k}\right)^2$

4. $\sum_{k=1}^{\infty} k e^{-k}$

5. $\sum_{k=1}^{\infty} k^2 e^{-k}$

6. $\sum_{k=1}^{\infty} \dfrac{k}{2^k}$

7. $\sum_{k=1}^{\infty} \dfrac{1}{k^3 + 1}$

8. $\sum_{k=1}^{\infty} \dfrac{k}{k^2 + 2}$

9. $\sum_{k=1}^{\infty} \dfrac{1}{k \ln k}$

10. $\sum_{k=1}^{\infty} \left(\dfrac{7}{9}\right)^{3k}$

11. $\sum_{k=1}^{\infty} (-1)^k \dfrac{1}{(2k)!} \pi^{2k}$

12. $\sum_{k=1}^{\infty} (-1)^{k+1} \dfrac{1}{(2k-1)!} \left(\dfrac{\pi}{6}\right)^{2k-1}$

13. In Example 7 we have shown that the harmonic series $\sum_{k=1}^{\infty} (1/k)$ diverges. What about the series $\sum_{k=10^6}^{\infty} 1/k$? Does it also diverge? Explain your answer.

14. Suppose that the series $\sum_{k=0}^{\infty} a_k$ diverges. Could you choose a large integer N for which the series $\sum_{k=N}^{\infty} a_k$ could converge? Explain your answer.

11.7 PROPERTIES OF INFINITE SERIES

Because convergent series are the limits of sequences (of partial sums), it is natural that sequences and series should share some common features. Corresponding to Theorem 2, Section 11.1, we have the following theorem.

THEOREM 1

If $\sum_{k=0}^{\infty} a_k = A$ and $\sum_{k=0}^{\infty} b_k = B$, then

$$\sum_{k=0}^{\infty} (a_k + b_k) = A + B \tag{1}$$

and

$$\sum_{k=0}^{\infty} ca_k = cA \quad \text{for any constant } c \tag{2}$$

Stated verbally, this theorem says that convergent series may be added term by term, and that if each term of a convergent series is multiplied by c, then so is its sum.

Proof

The proof of this theorem is very simple. Let

$$s_n = \sum_{k=0}^{n} a_k \quad \text{and} \quad t_n = \sum_{k=0}^{n} b_k$$

Then, by elementary algebra,

$$s_n + t_n = \sum_{k=0}^{n} a_k + \sum_{k=0}^{n} b_k = \sum_{k=0}^{n} (a_k + b_k)$$

Since, by Theorem 2 of Section 11.1,

$$\lim_{n \to \infty} (s_n + t_n) = \lim_{n \to \infty} s_n + \lim_{n \to \infty} t_n = A + B$$

we see that (1) is true; by the same theorem

$$\lim_{n \to \infty} cs_n = cA$$

and hence (2) is true.

Example 1

With Theorem 1 and previous results, you can easily verify the following facts:

(a) $\sum_{k=0}^{\infty} \left(\frac{1}{2^k} + \frac{1}{3^k} \right) = 2 + 1.5 = 3.5;$

(b) $\sum_{k=0}^{\infty} \left(\frac{1}{2^k} - \frac{1}{3^k} \right) = 2 - 1.5 = 0.5;$

(c) $\sum_{k=0}^{\infty} -\frac{3}{2^k} = -6.$

One of the most fundamental properties of series is given in the following theorem.

> **THEOREM 2**
>
> If the series $\sum_{k=0}^{\infty} a_k$ converges, then
>
> $$\lim_{k \to \infty} a_k = 0$$

WARNING

The statement "if $\lim_{k \to \infty} a_k = 0$, then $\sum_{k=0}^{\infty} a_k$ converges" is *false*. It is quite common for students to fall into this trap in logic. You already know that, for example, $\sum_{k=1}^{\infty} 1/k$ diverges, yet $\lim_{k \to \infty} (1/k) = 0$.

Proof of Theorem 2

Let the sum of the series $\sum_{k=0}^{\infty} a_k$ be A. Then

$$\lim_{n \to \infty} s_n = A$$

and

$$\lim_{n \to \infty} s_{n-1} = A$$

where s_n and s_{n-1} are, respectively, the nth and $(n-1)$st partial sums of the series. Since, for each value of n,

$$a_n = s_n - s_{n-1}$$

we see that

$$\lim_{n \to \infty} a_n = \lim_{n \to \infty}(s_n - s_{n-1}) = \lim_{n \to \infty} s_n - \lim_{n \to \infty} s_{n-1} = A - A = 0$$

Theorem 2 is often used in the following form as a criterion for divergence of a series: If $\lim_{k \to \infty} a_k = 0$ does not hold, then the series $\sum_{k=0}^{\infty} a_k$ diverges.

Example 2

(a) $\sum_{k=0}^{\infty} (-1)^k$ diverges because the sequence $1, -1, 1, -1, \ldots$ diverges.

(b) $\sum_{k=1}^{\infty}(1-1/k)$ diverges, because $\lim_{k\to\infty}(1-1/k)=1$.

A property of series that has ramifications throughout mathematics is given in the following theorem.

THEOREM 3

If the series $\sum_{k=0}^{\infty}|a_k|$ converges, then the series $\sum_{k=0}^{\infty} a_k$ also converges.

The proof of this result is given in the next section; here we consider some examples and some applications of this theorem.

Warning: No conclusion can be drawn about $\sum_{k=0}^{\infty} a_k$ when $\sum_{k=0}^{\infty}|a_k|$ diverges.

Example 3

(a) The series

$$\sum_{k=1}^{\infty}(-1)^{k+1}\frac{1}{k^2} = 1 - \frac{1}{4} + \frac{1}{9} - \frac{1}{16} + \frac{1}{25} - \cdots$$

converges because the series

$$\sum_{k=1}^{\infty}\left|(-1)^{k+1}\frac{1}{k^2}\right| = \sum_{k=1}^{\infty}\frac{1}{k^2}$$

converges (see Example 6, Section 11.6).

(b) Theorem 3 cannot be used to decide the question of convergence of the series

$$\sum_{k=1}^{\infty}(-1)^{k+1}\frac{1}{\sqrt{k}} = 1 - \frac{1}{\sqrt{2}} + \frac{1}{\sqrt{3}} - \frac{1}{\sqrt{4}} + \cdots$$

because the series

$$\sum_{k=1}^{\infty}\left|(-1)^{k+1}\frac{1}{\sqrt{k}}\right| = \sum_{k=1}^{\infty}\frac{1}{\sqrt{k}}$$

diverges. (See Example 4 below.)

It may happen that $\sum_{k=0}^{\infty}|a_k|$ diverges whereas $\sum_{k=0}^{\infty} a_k$ converges.

This is the case with the series

$$\sum_{k=1}^{\infty} (-1)^{k+1} \frac{1}{k}$$

(see Examples 3 and 7 in Section 11.6). We thus introduce the following terminology:

DEFINITION 1
The series $\sum_{k=1}^{\infty} a_k$ is said to converge *absolutely* when $\sum_{k=1}^{\infty} |a_k|$ converges.

The series is said to converge *conditionally* when it converges but $\sum_{k=1}^{\infty} |a_k|$ *diverges*.

Thus,

$$\sum_{k=1}^{\infty} (-1)^{k+1} \frac{1}{k^2} \text{ converges absolutely,}$$

$$\sum_{k=1}^{\infty} (-1)^{k+1} \frac{1}{k} \text{ converges conditionally.}$$

In general,

> A convergent series with nonnegative terms converges absolutely.

Throughout this section, we have used series of the form $\sum_{k=1}^{\infty} (-1)^{k+1} a_k$, where $a_k > 0$. Because in such series the terms are alternately positive and negative, they are called *alternating series*. These series are particularly easy to analyze, as the following theorem shows.

THEOREM 4
If $a_1 \geq a_2 \geq a_3 \geq \cdots \geq a_k \geq \cdots \geq 0$ and $\lim_{k \to \infty} a_k = 0$, then the alternating series $\sum_{k=1}^{\infty} (-1)^{k+1} a_k$ converges.

Proof
From the equations

$$s_{2n} = (a_1 - a_2) + (a_3 - a_4) + \cdots + (a_{2n-1} - a_{2n})$$

and
$$s_{2n+2} = (a_1 - a_2) + \cdots + (a_{2n-1} - a_{2n}) + (a_{2n+1} - a_{2n+2})$$
and the fact that
$$a_{2n+1} - a_{2n+2} \geq 0$$
we see that the sequence
$$s_2, s_4, \ldots, s_{2n}, s_{2n+2}, \ldots$$
is increasing.

From the equations
$$s_{2n-1} = a_1 - (a_2 - a_3) - \cdots - (a_{2n-2} - a_{2n-1})$$
and
$$s_{2n+1} = a_1 - (a_2 - a_3) - \cdots - (a_{2n-2} - a_{2n-1}) - (a_{2n} - a_{2n+1})$$
we see that the sequence
$$s_1, s_3, \ldots, s_{2n-1}, s_{2n+1}, \ldots$$
is decreasing (see Figure 11.19).

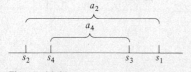
Figure 11.19

In addition, since
$$s_{2n} = s_{2n-1} - a_{2n} < s_{2n-1} < s_1$$
the even sums s_{2n} are less than the odd sums s_{2n-1} and the sequence s_{2n} is bounded. By Theorem 1 of Section 11.2, the sequence s_{2n} has a limit:
$$\lim_{n \to \infty} s_{2n} = A$$

Since $s_{2n-1} = s_{2n} + a_{2n}$, we have
$$\lim_{n \to \infty} s_{2n-1} = \lim_{n \to \infty} (s_{2n} + a_{2n}) = \lim_{n \to \infty} s_{2n} + \lim_{n \to \infty} a_{2n} = A + 0$$
and thus
$$\lim_{n \to \infty} s_{2n-1} = A$$

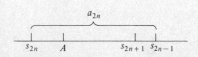
Figure 11.20

It is now clear that the limit A is less than the odd sums and greater than the even sums. It follows (see Figure 11.20) that
$$0 < A - s_{2n} < s_{2n+1} - s_{2n} = a_{2n+1}$$
and
$$0 < s_{2n-1} - A < s_{2n-1} - s_{2n} = a_{2n}$$
It follows from this that
$$|A - s_N| < a_{N+1} \quad \text{for all } N = 1, 2, 3, \ldots \tag{3}$$

and since
$$\lim_{n \to \infty} a_n = 0$$
we have
$$\lim_{n \to \infty} s_n = A$$

The inequality given in (3) is very easy to apply, as we shall see in the next example; we can restate it by saying that the difference between the sum A of the alternating series and the Nth partial sum does not exceed a_{N+1}.

Example 4

Show that the series
$$1 - \frac{1}{\sqrt{2}} + \frac{1}{\sqrt{3}} - \frac{1}{\sqrt{4}} + \frac{1}{\sqrt{5}} - \cdots$$
converges, and find the error when its sum is approximated by the partial sum s_{99}.

Solution
The series given in the problem is the alternating series
$$\sum_{k=1}^{\infty} (-1)^{k+1} \frac{1}{\sqrt{k}}$$
Since
$$1 > \frac{1}{\sqrt{2}} > \frac{1}{\sqrt{3}} > \frac{1}{\sqrt{4}} > \cdots > \frac{1}{\sqrt{k}} > \cdots > 0$$
and
$$\lim_{k \to \infty} \frac{1}{\sqrt{k}} = 0$$
we know from Theorem 4 that the series converges. Using now (3) with $N = 99$, we see that the error in question is $a_{N+1} = a_{100} = 1/\sqrt{100} = \frac{1}{10} = 0.1$. Thus, if we designate the sum of the series by A, then
$$|A - s_{99}| < 0.1$$

EXERCISES

Find the sums in Exercises 1–6.

1. $\sum_{k=0}^{\infty} \left(\frac{1}{2^k} + \frac{1}{k!} \right)$ (by definition, $0! = 1$)

2. $\sum_{k=0}^{\infty}\left(\dfrac{1}{k!}+\dfrac{1}{(k+1)!}\right)$ (by definition, $0!=1$)

3. $\sum_{k=0}^{\infty}\left(1-\dfrac{3}{k!}\right)\dfrac{1}{3^k}$

4. $\sum_{k=0}^{\infty}\left(\dfrac{1}{2^k}+\dfrac{1}{3^k}+\dfrac{1}{4^k}+\dfrac{1}{5^k}\right)$

5. $\sum_{k=0}^{\infty}\left(\dfrac{1}{(k+2)!}-\dfrac{1}{(k+1)!}\right)$

6. $\sum_{k=0}^{\infty}(x^k+x^{k+1}+x^{k+2}+x^{k+3})$ for $|x|<1$

In Exercises 7–12, decide which series *diverges* according to Theorem 2; when the theorem does not apply, state so.

7. $\sum_{k=0}^{\infty}\dfrac{k}{10k+1}$

8. $\sum_{k=0}^{\infty}\dfrac{k}{10k^2+1}$

9. $\sum_{k=0}^{\infty}\left(1-\dfrac{k}{k+3}\right)$

Hint: $k\sin\dfrac{1}{k}=\dfrac{\sin u}{u}$, where $u=\dfrac{1}{k}$.   10. $\sum_{k=0}^{\infty}k\sin\dfrac{1}{k}$

11. $\sum_{k=0}^{\infty}\left(\dfrac{1}{\sqrt{k}}-\dfrac{1}{k^2}\right)$

12. $1-\dfrac{1}{2^2}+\dfrac{1}{2^3}-\dfrac{1}{2^4}+\dfrac{1}{2^5}-\dfrac{1}{2^6}+\dfrac{1}{2^7}-\dfrac{1}{2^8}+\cdots$

In Exercises 13–16, write out the first five terms of the series, then find a value N for which the Nth partial sum gives an error not exceeding 0.0001.

13. $\sum_{k=1}^{\infty}(-1)^{k+1}\dfrac{1}{\sqrt[3]{k}}$ 14. $\sum_{k=1}^{\infty}(-1)^{k+1}\dfrac{1}{k^k}$

15. $\sum_{k=2}^{\infty}(-1)^k\dfrac{1}{\ln k}$ 16. $\sum_{k=2}^{\infty}(-1)^k\dfrac{1}{\ln(\ln k)}$

11.8 CONVERGENCE TESTS

In the preceding two sections we discussed some basic tests for deciding the convergence of infinite series. In this section we develop two more such tests. The reason for having several convergence tests at our disposal is that no single test can be used for all series.

The first test we discuss is a direct consequence of the boundedness theorem in Section 11.6:

COMPARISON TEST

Let $\sum_{k=0}^{\infty} a_k$ and $\sum_{k=0}^{\infty} b_k$ be series with nonnegative terms ($a_k \geq 0$ and $b_k \geq 0$) such that $a_k \leq b_k$ for all values of k. Then

1. if $\sum_{k=0}^{\infty} b_k$ converges, then $\sum_{k=0}^{\infty} a_k$ converges;

2. if $\sum_{k=0}^{\infty} a_k$ diverges, then $\sum_{k=0}^{\infty} b_k$ diverges.

Proof

Part 1

Let $\sum_{k=0}^{\infty} b_k$ converge to the sum B. Then, for each value of n,

$$\sum_{k=0}^{n} b_k \leq B$$

since the partial sums of the series form a nondecreasing sequence. From the condition $a_k \leq b_k$, we get the inequalities

$$\sum_{k=0}^{n} a_k \leq \sum_{k=0}^{n} b_k \leq B$$

and they show that the partial sums of $\sum_{k=0}^{\infty} a_k$ are bounded. We therefore conclude, from the boundedness theorem in Section 11.6, that the series converges.

Part 2

This statement follows at once from statement 1, since if $\sum_{k=0}^{\infty} b_k$ would converge, then so would $\sum_{k=0}^{\infty} a_k$. The student with a basic knowledge of logic will recognize, in fact, that statement 2 is the contrapositive of statement 1, and as such is automatically true.

Example 1

Does the series

$$\sum_{k=1}^{\infty} \frac{1}{5k^2 - 1}$$

converge?

Solution
If we can show that

$$\frac{1}{5k^2 - 1} \leq \frac{1}{k^2} \tag{1}$$

then it will follow from the comparison test that the series converges because $\sum_{k=1}^{\infty} 1/k^2$ converges. Using simple algebra, we get for (1)

$$k^2 \leq 5k^2 - 1$$

which is the same as $1 \leq 4k^2$. The last inequality is true for all values of k, and hence (1) is true.

Example 2

Do the series

$$\sum_{k=1}^{\infty} \frac{1}{7k - 2}$$

and

$$\sum_{k=1}^{\infty} \frac{1}{7k + 2}$$

converge?

Solution
Both series diverge. This is very easy to show for the first series, since

$$\frac{1}{7k - 2} \geq \frac{1}{7k}$$

for all values of k, and $\sum_{k=1}^{\infty} 1/7k$ diverges. To show that the second series diverges, we seek a constant $m \neq 0$ such that

$$\frac{1}{7k + 2} \geq \frac{1}{mk}$$

for all values of k, since

$$\sum_{k=1}^{\infty} \frac{1}{mk} = \frac{1}{m} \sum_{k=1}^{\infty} \frac{1}{k}$$

diverges no matter what value m has. Clearly, we must have $m > 7$, and a simple computation shows that $m = 8$ does not work for $k = 1$, but $m = 9$ works for all values of k.

The comparison test can be used to prove Theorem 3 of the last section. The theorem states that:

> If $\sum_{k=0}^{\infty} |a_k|$ converges, then $\sum_{k=0}^{\infty} a_k$ converges.

Proof
The inequalities

$$-|a_k| \leq a_k \leq |a_k|$$

which are always true, can be written in the form

$$0 \leq a_k + |a_k| \leq |a_k| + |a_k| = 2|a_k|$$

Since $\sum_{k=0}^{\infty} 2|a_k| = 2 \sum_{k=0}^{\infty} |a_k|$ converges by assumption, it follows from the comparison test that $\sum_{k=0}^{\infty} (a_k + |a_k|)$ converges.

By Theorem 1 of Section 11.7, therefore, the series

$$\sum_{k=0}^{\infty} a_k = \sum_{k=0}^{\infty} (a_k + |a_k|) - \sum_{k=0}^{\infty} |a_k|$$

converges and this completes the proof.

We now make the following observation: From any series $\sum_{k=0}^{\infty} a_k$, we can now form the series $\sum_{k=N}^{\infty} a_k$ by omitting the terms $a_0, a_1, a_2, \ldots, a_{N-1}$. It is easy to see that:

> If $\sum_{k=N}^{\infty} a_k$ converges for some integer N, then $\sum_{k=0}^{\infty} a_k$ converges.

(2)

To verify this fact, it suffices to write

$$\sum_{k=0}^{\infty} a_k = \sum_{k=0}^{N-1} a_k + \sum_{k=N}^{\infty} a_k$$

and note that $\sum_{k=0}^{N-1} a_k$, being a finite sum, is always defined. As a first application of the above fact, we shall use a generalization of the comparison test.

Example 3

Show that the series

$$\sum_{k=1}^{\infty} \sqrt{\frac{k+20}{2k^5}}$$

converges.

Solution
Write

$$\sqrt{\frac{k+20}{2k^5}} = \frac{1}{k^2}\sqrt{\frac{k+20}{2k}}$$

We see that

$$\sqrt{\frac{k+20}{2k}} \leq 1$$

when

$$\frac{k+20}{2k} \leq 1$$

that is, when $k \geq 20$. Hence,

$$\sqrt{\frac{k+20}{2k^5}} \leq \frac{1}{k^2} \quad \text{for } k \geq 20$$

Since the series $\sum_{k=20}^{\infty} (1/k^2)$ converges, we know that the series

$$\sum_{k=20}^{\infty} \sqrt{(k+20)/2k^5}$$ also converges, but this shows that the series in question converges.

In our next test we shall describe a class of infinite series that can be compared with a geometric series

$$\sum_{k=0}^{\infty} q^k = \frac{1}{1-q} \quad \text{for } 0 < q < 1$$

> **THE RATIO TEST**
> Consider an infinite series $\sum_{k=0}^{\infty} a_k$ such that
> $$\lim_{k \to \infty} \left| \frac{a_{k+1}}{a_k} \right| = r$$
> Then
> 1. the series *converges absolutely* when $r < 1$;
> 2. the series *diverges* when $r > 1$;
> 3. the test is inconclusive when $r = 1$.

The proof of this theorem will be found following the applications below.

Example 4

To see why the ratio test fails when $r = 1$, consider the two series

$$\sum_{k=1}^{\infty} a_k = \sum_{k=1}^{\infty} \frac{1}{k}$$

and

$$\sum_{k=1}^{\infty} a_k = \sum_{k=1}^{\infty} \frac{1}{k^2}$$

For the first series we have

$$\left| \frac{a_{k+1}}{a_k} \right| = \frac{1/(k+1)}{1/k} = \frac{k}{k+1} = \frac{1}{1 + 1/k}$$

and hence

$$\lim_{k \to \infty} \left| \frac{a_{k+1}}{a_k} \right| = \lim_{k \to \infty} \frac{1}{1 + 1/k} = 1$$

For the second series we have

$$\left| \frac{a_{k+1}}{a_k} \right| = \frac{1/(k+1)^2}{1/k^2} = \frac{k^2}{(k+1)^2} = \left(\frac{k}{k+1} \right)^2 = \frac{1}{(1 + 1/k)^2}$$

so that

$$\lim_{k \to \infty} \left| \frac{a_{k+1}}{a_k} \right| = \lim_{k \to \infty} \frac{1}{(1 + 1/k)^2} = 1$$

Thus, in both cases $r = 1$, yet the first series diverges, whereas the second series converges.

Example 5

Use the ratio test to decide if

$$\sum_{k=1}^{\infty} (-1)^k \frac{k}{2^k}$$

converges absolutely.

Solution
Taking $a_k = (-1)^k(k/2^k)$, $a_{k+1} = (-1)^{k+1}[(k+1)/2^{k+1}]$, we find after simplification that

$$\left|\frac{a_{k+1}}{a_k}\right| = \frac{2^k(k+1)}{2^{k+1}k} = \frac{k+1}{2k} = \frac{1}{2}\left(1 + \frac{1}{k}\right)$$

Hence,

$$\lim_{k\to\infty} \left|\frac{a_{k+1}}{a_k}\right| = \lim_{k\to\infty} \frac{1}{2}\left(1 + \frac{1}{k}\right) = \frac{1}{2}$$

and this tells us that the series converges absolutely.

Example 6

Decide if the series

$$\sum_{k=1}^{\infty} \frac{2^k}{k^{10}}$$

converges.

Solution
Putting $a_k = 2^k/k^{10}$, we apply the ratio test:

$$\left|\frac{a_{k+1}}{a_k}\right| = \frac{2^{k+1}/(k+1)^{10}}{2^k/k^{10}} = \frac{2^{k+1} \cdot k^{10}}{2^k(k+1)^{10}} = 2\frac{k^{10}}{(k+1)^{10}} = 2\frac{1}{(1+1/k)^{10}}$$

From this we conclude that

$$\lim_{k\to\infty} \left|\frac{a_{k+1}}{a_k}\right| = \lim_{k\to\infty} 2\frac{1}{(1+1/k)^{10}} = 2 > 1$$

and owing to the ratio test the series is seen to diverge.

We now prove the ratio test.

Proof of the Ratio Test

Part 1
Suppose $r < 1$; let a number $\epsilon > 0$ be chosen so small that

$$r + \epsilon = q < 1$$

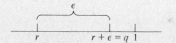

Figure 11.21

(see Figure 11.21). The statement

$$\lim_{k \to \infty} \left| \frac{a_{k+1}}{a_k} \right| = r \tag{3}$$

is the same as the statement that there is an integer N such that

$$\left| \left| \frac{a_{k+1}}{a_k} \right| - r \right| < \epsilon \quad \text{for all integers } k \geq N \tag{4}$$

(see Section 11.1). This fact implies that

$$\left| \frac{a_{k+1}}{a_k} \right| < r + \epsilon = q \quad \text{for all integers } k \geq N$$

which is the same as saying that

$$|a_{k+1}| < |a_k| q \quad \text{for } k = N, N+1, N+2, \ldots$$

Using an iteration gives the following inequalities:

$$|a_{N+1}| < |a_N| q$$
$$|a_{N+2}| < |a_{N+1}| q < |a_N| q^2$$
$$|a_{N+3}| < |a_{N+2}| q < |a_N| q^3$$

and, in general,

$$|a_{N+m}| < |a_N| q^m \quad \text{for } m = 1, 2, 3, 4, \ldots$$

Since the series

$$\sum_{m=1}^{\infty} |a_N| q^m = |a_N| \sum_{m=1}^{\infty} q^m \quad (0 < q < 1)$$

converges, it follows from the comparison test that the series $\sum_{k=N+1}^{\infty} a_k$

converges absolutely, and from (2) we conclude that $\sum_{k=0}^{\infty} a_k$ converges absolutely.

Part 2

Suppose $r > 1$; let a number $\epsilon > 0$ be chosen so small that

$$r - \epsilon = q > 1$$

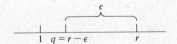

Figure 11.22

(see Figure 11.22). Then (3) is the same as (4) for the new choice of ϵ, but this time we use the fact that (4) implies that

$$\left| \frac{a_{k+1}}{a_k} \right| > r - \epsilon = q \quad \text{for all integers } k \geq N$$

that is,

$$|a_{k+1}| > |a_k| q \quad \text{for } k = N, N+1, N+2, \ldots$$

An iteration similar to the one performed above shows that
$$|a_{N+m}| > |a_N|q^m \quad \text{for } m = 1, 2, 3, \ldots$$
This tells us that
$$\lim_{m \to \infty} |a_{N+m}| \neq 0$$
and hence
$$\lim_{m \to \infty} a_{N+m} \neq 0$$
implying that
$$\lim_{k \to \infty} a_k \neq 0$$

By Theorem 2 in Section 11.7, therefore, the series diverges.

The truth of assertion 3 of the theorem was already demonstrated in Example 4.

EXERCISES

In Exercises 1–13, decide which series converges by using the tests of this section. When no test gives you the necessary information, say so.

1. $\sum_{k=0}^{\infty} \dfrac{1}{100k + 25}$

2. $\sum_{k=2}^{\infty} \dfrac{1}{k(k - 1)}$

3. $\sum_{k=1}^{\infty} \dfrac{3}{2(k^{3/2} + k^2)}$

4. $\sum_{k=0}^{\infty} \dfrac{1}{2^{k/20}}$

5. $\sum_{k=1}^{\infty} \dfrac{k^k}{k!}$

6. $\sum_{k=0}^{\infty} e^{-k}$

7. $\sum_{k=0}^{\infty} k\left(\dfrac{3}{4}\right)^k$

8. $\sum_{k=1}^{\infty} \sqrt{2k - 1}\left(\dfrac{2}{3}\right)^k$

9. $\sum_{k=1}^{\infty} \left(\dfrac{1}{k} - \dfrac{1}{k + 1}\right)$

10. $\sum_{k=0}^{\infty} \dfrac{2^k}{2^{2k} + 1}$

11. $\sum_{k=0}^{\infty} \dfrac{k^2}{k^3 + 1}$

12. $\sum_{k=1}^{\infty} \dfrac{1}{1 + \ln k}$

13. $\sum_{k=1}^{\infty} \dfrac{\cos k}{k^2}$

14. If $\lim_{k \to \infty} |a_{k+1}/a_k|$ does not exist, but $|a_{k+1}/a_k| < r < 1$ for $k = 0, 1, 2, 3, \ldots$, can you conclude that the series $\sum_{k=0}^{\infty} a_k$ converges absolutely? Justify your answer.

15. Show that if $\sqrt[k]{|a_k|} < r < 1$ for $k = 1, 2, 3, 4, \ldots$, then $\sum_{k=1}^{\infty} a_k$ converges absolutely.

In Exercises 16–24, use the tests at your disposal to decide which series converges absolutely or conditionally, and which series diverges. Give the test used in each case.

16. $\sum_{k=1}^{\infty} (-1)^{k+1} \dfrac{\cos k\pi}{k^2}$

17. $\sum_{k=1}^{\infty} (-1)^k \dfrac{k^2}{3k^3 + 1}$

18. $\sum_{k=0}^{\infty} \dfrac{\sqrt{k+1}}{k^2 + 1}$

19. $\sum_{k=1}^{\infty} (-1)^{k+1} \dfrac{k^2}{2^k}$

20. $\sum_{k=1}^{\infty} (-1)^{k+1} \dfrac{2^k}{k^2}$

21. $\sum_{k=1}^{\infty} (-1)^{k+1} \left(\dfrac{k-1}{k} \right)^k$

22. $\sum_{k=1}^{\infty} (-1)^{k+1} \dfrac{\ln k}{k}$

23. $\sum_{k=1}^{\infty} \dfrac{1 \cdot 3 \cdot 5 \cdot \cdots \cdot (2k-1)}{2 \cdot 4 \cdot 6 \cdot \cdots \cdot 2k}$

24. $\sum_{k=1}^{\infty} \dfrac{1 + 2 + 3 + \cdots + k}{1^2 + 2^2 + 3^2 + \cdots + k^2}$

11.9 POWER SERIES

In Section 11.5 we studied functions $f(x)$ having a Taylor series expansion

$$f(x) = f(b) + \sum_{k=1}^{\infty} \dfrac{f^{(k)}(b)}{k!}(x - b)^k$$

about a point b. When, instead of the special coefficients $f^{(k)}(b)/k!$, we take arbitrary numbers a_k, we obtain infinite series,

$$a_0 + \sum_{k=1}^{\infty} a_k(x - b)^k = \sum_{k=0}^{\infty} a_k(x - b)^k \qquad (1)$$

called a *power series*. For each value of x, the power series gives an infinite series. To discover for what values of x the power series converges, let us suppose that

$$\lim_{k \to \infty} \left| \dfrac{a_{k+1}}{a_k} \right| = r$$

where r may be infinite. Putting

$$u_k = a_k(x - b)^k$$

we apply the ratio test to the series

$$\sum_{k=0}^{\infty} u_k$$

finding that

$$\lim_{k\to\infty}\left|\frac{u_{k+1}}{u_k}\right| = \lim_{k\to\infty}\left|\frac{a_{k+1}(x-b)^{k+1}}{a_k(x-b)^k}\right| = \lim_{k\to\infty}\left|\frac{a_{k+1}}{a_k}\right||x-b| = r|x-b|$$

This tells us the following about the power series (1): If $r=0$, then the series converges absolutely for all values of x (since then $r|x-b| = 0 < 1$ for all values of x); if $r \neq 0$ and $r \neq \infty$, then the series converges absolutely for all values of x satisfying the inequality $|x-b| < 1/r$; if $r = \infty$, then the series converges only when $x = b$.

It is customary to put

$$R = \begin{cases} 1/r & \text{for } r \neq 0, r \neq \infty \\ 0 & \text{for } r = \infty \\ \infty & \text{for } r = 0 \end{cases} \qquad (2)$$

and call R the *radius of convergence* of the power series.

We summarize the above discussion as follows:

THEOREM 1

Let $\sum_{k=0}^{\infty} a_k(x-b)^k$ be a given power series and suppose that $\lim_{k\to\infty}|a_{k+1}/a_k| = r$. Then the power series converges absolutely for values of x satisfying the inequality $|x-b| < R$, and diverges for $|x-b| > R$, where R is as defined above.

REMARK

Observe that nothing was said about the convergence of a power series at the points $x = b - R$ and $x = b + R$ when $R \neq 0$ and $R \neq \infty$. As we shall see, a power series may converge or diverge at one or both of these points.

A better understanding of the convergence of power series will be gained through the following examples.

Example 1

Examine the power series

$$\sum_{k=1}^{\infty} a_k(x-b)^k = \sum_{k=1}^{\infty} k^k(x-1)^k$$

We have

$$\left|\frac{a_{k+1}}{a_k}\right| = \left|\frac{(k+1)^{k+1}}{k^k}\right| = \frac{(k+1)^k}{k^k}(k+1) = \left(1+\frac{1}{k}\right)^k(k+1) > k+1$$

Since

$$\lim_{k\to\infty}(k+1) = \infty$$

we have

$$\lim_{k\to\infty}\left|\frac{a_{k+1}}{a_k}\right| = r = \infty$$

Thus, the radius of convergence is $R=0$ and hence *the power series converges only at the point $x=1$.*

Example 2

Consider the power series

$$\sum_{k=0}^{\infty} a_k(x-b)^k = \sum_{k=0}^{\infty} k(x-1)^k$$

Here

$$\lim_{k\to\infty}\left|\frac{a_{k+1}}{a_k}\right| = \lim_{k\to\infty}\left|\frac{k+1}{k}\right| = 1$$

and, by (2), the radius of convergence is $R = 1$. According to Theorem 1, therefore, *the series converges for $|x-1|<1$, diverges for $|x-1|>1$.* Note that $|x-1|<1$ is the same as $0<x<2$, and it is easy to see that the series diverges at $x=0$ and at $x=2$.

Example 3

From Section 11.5 we know that the power series

$$\sum_{k=1}^{\infty} a_k(x-b)^k = \sum_{k=1}^{\infty} \frac{1}{k!}(x-1)^k$$

converges to $e^{x-1} - 1$. Hence, *this series converges absolutely for all values of x,* since $e^{x-1} - 1$ is defined for all values of x. Indeed,

$$\lim_{k\to\infty}\left|\frac{a_{k+1}}{a_k}\right| = \lim_{k\to\infty}\frac{1/(k+1)!}{1/k!} = \lim_{k\to\infty}\frac{k!}{(k+1)!} = \lim_{k\to\infty}\frac{1}{k+1} = 0 = r$$

and hence the radius of convergence is $R = \infty$.

Example 4

Discuss the convergence of the power series

$$\sum_{k=1}^{\infty} \frac{1}{k}x^k$$

Solution
The radius of convergence of this series is $R=1$, and hence it converges for all values $|x|<1$. For $x=1$ we obtain the harmonic series

$$\sum_{k=1}^{\infty} \frac{1}{k}$$

which is known to diverge (Example 7, Section 11.6); for $x=-1$ we have the alternating series

$$\sum_{k=1}^{\infty}(-1)^k \frac{1}{k}$$

which is known to converge (see Theorem 4, Section 11.7). Hence, *the above power series converges for the values $-1 \le x < 1$ and diverges for all other values of x.*

REMARK
Every power series has a radius of convergence R, even in cases in which the ratio test cannot be used. This fact is proved in more advanced texts.

We now have a closer look at the functions defined by means of power series; for convenience we take series with $b=0$. By way of motivation, consider the power series expansion of $1/(1+x)$ and $\ln(1+x)$ (see Section 11.5):

$$\frac{1}{1+x} = 1 - x + x^2 - x^3 + x^4 - x^5 + \ldots \tag{3}$$

$$\ln(1+x) = x - \tfrac{1}{2}x^2 + \tfrac{1}{3}x^3 - \tfrac{1}{4}x^4 + \tfrac{1}{5}x^5 - \ldots$$

A known fact is that

$$\int \frac{1}{x+1}\, dx = \ln(1+x)$$

and we observe that if the series expansion of $1/(1+x)$ is integrated term by term, we obtain the series for $\ln(1+x)$:

$$\int 1\, dx - \int x\, dx + \int x^2\, dx - \int x^3\, dx + \int x^4\, dx - \ldots$$
$$= x - \tfrac{1}{2}x^2 + \tfrac{1}{3}x^3 - \tfrac{1}{4}x^4 + \tfrac{1}{5}x^5 - \ldots$$

In general, the following result holds:

INTEGRATION OF POWER SERIES
Let R be a positive number and let $f(x)$ be defined on the interval $(-R, R)$ by the power series

$$f(x) = \sum_{k=0}^{\infty} a_k x^k$$

Then

$$\int_0^x f(t)\, dt = \sum_{k=0}^{\infty} \frac{a_k}{k+1} x^{k+1}$$

That is, a power series that converges on an interval $(-R, R)$ can be integrated term by term on each subinterval $[0, x]$. The proof of this fact is omitted here.

Example 5

Find the power series expansion of $\tan^{-1} x$.

Solution
The relation

$$\tan^{-1} x = \int_0^x \frac{1}{1+t^2}\, dt$$

[see formula (2) in Section 9.8], which holds for $|x| < 1$, suggests that we seek the power series expansion of $1/(1+t^2)$. This is easily accomplished by replacing x by t^2 in formula (3); we get

$$\frac{1}{1+t^2} = 1 - t^2 + t^4 - t^6 + t^8 - t^{10} + \ldots$$

and this series converges for $|t| < 1$. Integrating term by term gives

$$\int_0^x \frac{1}{1+t^2}\, dt = x - \tfrac{1}{3}x^3 + \tfrac{1}{5}x^5 - \tfrac{1}{7}x^7 + \tfrac{1}{9}x^9 - \tfrac{1}{11}x^{11} + \ldots$$

Hence,

$$\tan^{-1} x = \sum_{k=0}^{\infty} (-1)^k \frac{1}{2k+1} x^{2k+1} \quad \text{for } |x| < 1$$

Example 6

Let $f(x)$ be defined by the formula

$$f(x) = \int_0^x e^{-t^2}\, dt$$

Find the power series expansion of $f(x)$ and approximate the function for $|x| < 0.1$ with an error less than 3×10^{-9}.

Solution
We mentioned in Section 8.7 that there is no simple way for computing the integral defining $f(x)$. Power series offer an effective method for doing just that. We start with the power series expansion of e^x:

$$e^x = 1 + \sum_{k=1}^{\infty} \frac{1}{k!} x^k = 1 + x + \frac{1}{2!}x^2 + \frac{1}{3!}x^3 + \frac{1}{4!}x^4 + \ldots$$

This expansion holds for all values of x and we use the substitution $x = -t^2$. Thus,

$$e^{-t^2} = 1 - t^2 + \frac{1}{2!}t^4 - \frac{1}{3!}t^6 + \frac{1}{4!}t^8 - \cdots$$

By the above theorem we can integrate the right side term by term to give the relation

$$\int_0^x e^{-t^2}\,dt = x - \frac{1}{3}x^3 + \frac{1}{5 \cdot 2!}x^5 - \frac{1}{7 \cdot 3!}x^7 + \frac{1}{9 \cdot 4!}x^9 - \cdots$$

Using the summation notation, this is

$$\int_0^x e^{-t^2}\,dt = x + \sum_{k=1}^{\infty} (-1)^k \frac{1}{(2k+1)k!} x^{2k+1}$$

Since the series we obtained is alternating, we know from formula (3), Section 11.7, that the error in approximating the integral by the first three terms is less than the absolute value of the fourth term. Thus, for $|x| < 0.1$ the error is less than

$$\frac{1}{7 \cdot 3!} \cdot 0.1^7 = \tfrac{1}{42}\, 10^{-7} < 0.03 \times 10^{-7} = 3 \times 10^{-9}$$

and accordingly,

$$\left| \int_0^x e^{-t^2}\,dt - (x - \tfrac{1}{3}x^3 + \tfrac{1}{10}x^5) \right| < 3 \times 10^{-9} \qquad \text{for } |x| < 0.1$$

The possibility of term-by-term differentiation may have occurred already to the reader. We state without proof the following fact.

DIFFERENTIATION OF POWER SERIES
Let $f(x)$ be defined by the power series

$$f(x) = \sum_{k=0}^{\infty} a_k x^k$$

for $|x| < R$. Then $f'(x)$ exists for all values $|x| < R$ and, moreover,

$$f'(x) = \sum_{k=1}^{\infty} k a_k x^{k-1}$$

That is, a power series that converges on an interval $|x| < R$ can be differentiated term by term on that interval.

Example 7

Find the power series expansion for $1/(1-x)^2$ about the point $x = 0$.

Solution
Starting with the power series

$$\frac{1}{1-x} = 1 + x + x^2 + x^3 + x^4 + \ldots \quad \text{for } |x| < 1$$

we find with term-by-term differentiation that

$$\frac{1}{(1-x)^2} = 1 + 2x + 3x^2 + 4x^3 + \ldots$$

or

$$\frac{1}{(1-x)^2} = \sum_{k=1}^{\infty} kx^{k-1} \quad \text{for } |x| < 1$$

REMARK
We cannot conclude this section without mentioning the important fact that any power series can be written in the form of a Taylor series. To indicate that this is true, suppose that

$$f(x) = \sum_{k=0}^{\infty} a_k(x-b)^k$$
$$= a_0 + a_1(x-b) + a_2(x-b)^2 + a_3(x-b)^3 + a_4(x-b)^4 + \ldots$$

Taking successive derivatives gives

$$f^{(1)}(x) = a_1 + 2a_2(x-b) + 3a_3(x-b)^2 + 4a_4(x-b)^3 + \ldots$$
$$f^{(2)}(x) = 2!a_2 + 3!a_3(x-b) + 3 \cdot 4a_4(x-b)^2 + \ldots$$
$$f^{(3)}(x) = 3!a_3 + 4!a_4(x-b) + \ldots$$
$$\vdots$$
$$f^{(k)}(x) = k!a_k + (k+1)!a_{k+1}(x-b) + \ldots$$
$$\vdots$$

Putting $x = b$ and solving for $a_1, a_2, a_3, \ldots, a_k, \ldots$ gives

$$a_1 = \frac{f^{(1)}(b)}{1!}$$

$$a_2 = \frac{f^{(2)}(b)}{2!}$$

$$a_3 = \frac{f^{(3)}(b)}{3!}$$
$$\vdots$$
$$a_k = \frac{f^{(k)}(b)}{k!}$$
$$\vdots$$

and using the fact that $a_0 = f(b)$ we can write the above series in the form

$$f(x) = f(b) + \sum_{k=1}^{\infty} \frac{f^{(k)}(b)}{k!}(x-b)^k$$

Thus:

> Any power series expansion of a function $f(x)$ about b is its Taylor series expansion.

This uniqueness of power series is the basis for much advanced work in calculus.

EXERCISES

In Exercises 1–9, find the radius of convergence R of the given power series $\sum_{k=0}^{\infty} a_k(x-b)^k$. When R is finite and not zero, find out if the series converges at $x = b - R$ or $x = b + R$.

1. $\sum_{k=1}^{\infty} k^{k/2} x^k$

2. $\sum_{k=0}^{\infty} \frac{k}{k+1}(x-2)^k$

3. $\sum_{k=1}^{\infty} \frac{1}{k} x^k$

4. $\sum_{k=1}^{\infty} \frac{1}{k!}(x-3)^k$

5. $\sum_{k=1}^{\infty} k x^k$

6. $\sum_{k=1}^{\infty} k(2x)^k$

7. $\sum_{k=1}^{\infty} k\left(\frac{x}{2}\right)^k$

8. $\sum_{k=0}^{\infty} 2^k x^k$

9. $\sum_{k=1}^{\infty} (-1)^k \frac{x^{2k}}{(2k)!}$

In Exercises 10–16, find the power series expansion of $f(x)$ about the given point $x = b$.

Hint: Use the power series for $\frac{1}{1-t}$, substitute $t = x^2$, and differentiate.

10. $f(x) = \dfrac{x}{(1-x^2)^2}$ $b = 0$

11. $f(x) = \displaystyle\int_0^x \ln(1+t^2)\, dt$ $b = 0$

12. $f(x) = \sin^2 x$ $b = \dfrac{\pi}{2}$

13. $f(x) = \sin^2 x$ $b = \pi$

14. $f(x) = \dfrac{1}{(1-x)^3}$ $b = 0$

15. $f(x) = \sin x \cos x \qquad b = \dfrac{\pi}{2}$

16. $f(x) = \displaystyle\int_0^x t^2 e^{-t^2}\, dt \qquad b = 0$

Compute the integrals in Exercises 17–20 with an error less than 10^{-5} by using power series.

17. $\displaystyle\int_0^{0.5} \dfrac{1}{1-x^5}\, dx$

18. $\displaystyle\int_0^1 \sin x^2\, dx$

19. $\displaystyle\int_{0.01}^{0.02} \dfrac{1}{x} e^{-x}\, dx$

20. $\displaystyle\int_{0.1}^{0.2} \dfrac{1}{x} \sin x\, dx$

11.10 INDETERMINATE FORMS

Consider the limit

$$\lim_{x \to 0} \frac{\sin x}{x} = 1 \qquad (1)$$

which was the subject of Theorem 2 in Section 9.2. We had to resort to a complicated argument because both $\sin x$ and x vanish at the origin and no simplification was available to us. In this section we present a general method for evaluating limits of functions having an indeterminate form.

> The function $f(x)/g(x)$ is said to have the *indeterminate form* $0/0$ at a when $f(a) = 0$ and $g(a) = 0$.

Other indeterminate forms are described below. The method we alluded to above is known as *L'Hôpital's rule*.

> **L'HÔPITAL'S RULE FOR THE INDETERMINATE FORM 0/0**
> Let $f(x)$ and $g(x)$ be given functions such that
>
> 1. $f(a) = 0$ and $g(a) = 0$;
> 2. $f'(x)$ and $g'(x)$ exist and are continuous in an interval containing a.
>
> If
> $$\lim_{x \to a} \frac{f'(x)}{g'(x)} = A$$
> then
> $$\lim_{x \to a} \frac{f(x)}{g(x)} = A$$

Proof
We prove this rule only for the case $g'(a) \neq 0$, but this will give you a good insight into the mechanics of this theorem. The proof for

the case $g'(a) = 0$ is somewhat more involved, and it can be found in many advanced calculus texts. Thus, when $g'(a) \neq 0$, there is an interval containing a in which $g'(x) \neq 0$. Picking any x in this interval, we have, by the Mean Value Theorem (Section 4.6),

$$\frac{f(x)}{g(x)} = \frac{f(x) - f(a)}{g(x) - g(a)} = \frac{f'(c_1)}{g'(c_2)}$$

where c_1 and c_2 lie between x and a. This implies that $c_1 \to a$ and $c_2 \to a$ as $x \to a$, and hence

$$\lim_{x \to a} \frac{f(x)}{g(x)} = \lim_{x \to a} \frac{f'(c_1)}{g'(c_2)} = \frac{f'(a)}{g'(a)}$$

Since, by the continuity of $f'(x)$ and $g'(x)$,

$$\lim_{x \to a} \frac{f'(x)}{g'(x)} = \frac{f'(a)}{g'(a)}$$

we have shown that

$$\lim_{x \to a} \frac{f(x)}{g(x)} = \lim_{x \to a} \frac{f'(x)}{g'(x)}$$

Example 1

Evaluate

$$\lim_{x \to 0} \frac{\sin 5x}{\tan x}$$

Solution
Since $\sin 0 = 0$ and $\tan 0 = 0$, and the functions have continuous derivatives in an interval containing the origin, we can apply L'Hôpital's rule.
Putting $f(x) = \sin 5x$ and $g(x) = \tan x$, we have $f'(x) = 5 \cos 5x$ and $g'(x) = \sec^2 x$. Hence,

$$\lim_{x \to 0} \frac{f'(x)}{g'(x)} = \lim_{x \to 0} \frac{5 \cos 5x}{\sec^2 x} = 5$$

and consequently

$$\lim_{x \to 0} \frac{\sin 5x}{\tan x} = 5$$

Sometimes $f'(x)/g'(x)$ has itself the indeterminate form 0/0 at a. When this happens we apply L'Hôpital's rule also to this function. Such a situation is considered in the problem below.

Example 2

Evaluate

$$\lim_{x \to 0} \frac{e^x + e^{-x} - 2}{2(1 - \cos x)}$$

Solution
Putting $f(x) = e^x + e^{-x} - 2$ and $g(x) = 2(1 - \cos x)$ we see that $f(0) = 0$ and $g(0) = 0$, and since the derivatives exist, L'Hôpital's rule applies. Since

$$f'(x) = e^x - e^{-x} \quad \text{and} \quad g'(x) = 2 \sin x$$

we have

$$\lim_{x \to 0} \frac{f(x)}{g(x)} = \lim_{x \to 0} \frac{f'(x)}{g'(x)} = \lim_{x \to 0} \frac{e^x - e^{-x}}{2 \sin x}$$

However, $f'(0) = 0$ and $g'(0) = 0$, and the rule has to be applied again. Since

$$f''(x) = e^x + e^{-x} \quad \text{and} \quad g''(x) = 2 \cos x$$

we find that

$$\lim_{x \to 0} \frac{f''(x)}{g''(x)} = \lim_{x \to 0} \frac{e^x + e^{-x}}{2 \cos x} = 1$$

and hence

$$\lim_{x \to 0} \frac{e^x + e^{-x} - 2}{2(1 - \cos x)} = \lim_{x \to 0} \frac{e^x - e^{-x}}{2 \sin x} = \lim_{x \to 0} \frac{e^x + e^{-x}}{2 \cos x} = 1$$

We now consider other indeterminate forms. We say that:

The function $f(x)/g(x)$ has the *indeterminate form* $\pm\infty/\infty$ at a when

$$\lim_{x \to a} f(x) = \pm\infty \quad \text{and} \quad \lim_{x \to a} g(x) = \pm\infty$$

In this case L'Hôpital's rule also applies:

L'HÔPITAL'S RULE FOR THE INDETERMINATE FORM $\pm\infty/\infty$
Let $f(x)$ and $g(x)$ be given functions such that

1. $\lim_{x \to a} f(x) = \pm\infty$ and $\lim_{x \to a} g(x) = \pm\infty$;
2. $f'(x)$ and $g'(x)$ exist and are continuous in an interval containing a.

If

$$\lim_{x \to a} \frac{f'(x)}{g'(x)} = A$$

then

$$\lim_{x \to a} \frac{f(x)}{g(x)} = A$$

The proof of this fact is omitted here.

Example 3

Evaluate

$$\lim_{x \to 0+} \frac{\ln \sin^2 x}{\cot x}$$

Solution
Here we are asked to evaluate a right limit; for the definition, see Definition 1 in Section 4.4. Putting $f(x) = \ln \sin^2 x$ and $g(x) = \cot x$, we note that

$$\lim_{x \to 0+} f(x) = \lim_{x \to 0+} \ln \sin^2 x = -\infty$$

and

$$\lim_{x \to 0+} g(x) = \lim_{x \to 0+} \cot x = \infty$$

Since

$$f'(x) = \frac{1}{\sin^2 x} 2 \sin x \cos x = 2 \frac{\cos x}{\sin x}$$

and

$$g'(x) = -\csc^2 x = -\frac{1}{\sin^2 x}$$

we have

$$\lim_{x \to 0+} \frac{f'(x)}{g'(x)} = \lim_{x \to 0+} \frac{2(\cos x/\sin x)}{-(1/\sin^2 x)} = \lim_{x \to 0+} (-2 \sin x \cos x) = 0$$

Hence,

$$\lim_{x \to 0+} \frac{\ln \sin^2 x}{\cot x} = 0$$

Example 4

Evaluate

$$\lim_{x \to \infty} \frac{\ln(x + e^x)}{x}$$

Solution
This problem differs from the preceding ones, because here x does not tend to a finite limit. Since, however,

$$\lim_{x \to \infty} \ln(x + e^x) = \infty \quad \text{and} \quad \lim_{x \to \infty} x = \infty$$

we have the indeterminate form ∞/∞ and L'Hôpital's rule still holds.

With the notation
$$f(x) = \ln(x + e^x) \quad \text{and} \quad g(x) = x$$
we have
$$f'(x) = \frac{1 + e^x}{x + e^x} \quad \text{and} \quad g'(x) = 1$$
Hence, using the fact that
$$\lim_{x \to \infty} xe^{-x} = 0$$
gives
$$\lim_{x \to \infty} \frac{f'(x)}{g'(x)} = \lim_{x \to \infty} \frac{(1 + e^x)/(x + e^x)}{1} = \lim_{x \to \infty} \frac{1 + e^x}{x + e^x}$$
$$= \lim_{x \to \infty} \frac{e^{-x} + 1}{xe^{-x} + 1} = 1$$

By L'Hôpital's rule, therefore,
$$\lim_{x \to \infty} \frac{\ln(x + e^x)}{x} = 1$$

THE INDETERMINATE FORM $0 \cdot \infty$

Example 5

Evaluate
$$\lim_{x \to 0+} x \ln x$$

Solution
This function is of the form $f(x)g(x)$, where
$$\lim_{x \to 0+} f(x) = 0 \quad \text{and} \quad \lim_{x \to 0+} g(x) = -\infty$$
This is an example of a function of indeterminate form $0 \cdot \infty$. Writing
$$x \ln x = \frac{\ln x}{1/x}$$
however, gives the indeterminate form $-\infty/\infty$, since
$$\lim_{x \to 0+} \frac{1}{x} = \infty$$
Putting
$$f(x) = \ln x \quad \text{and} \quad g(x) = \frac{1}{x}$$
we have
$$f'(x) = \frac{1}{x} \quad \text{and} \quad g'(x) = -\frac{1}{x^2}$$

and hence

$$\lim_{x \to 0+} \frac{f'(x)}{g'(x)} = \lim_{x \to 0+} \frac{1/x}{-1/x^2} = \lim_{x \to 0+}(-x) = 0$$

By L'Hôpital's rule,

$$\lim_{x \to 0+} x \ln x = 0$$

THE INDETERMINATE FORM 1^∞

Example 6

Evaluate

$$\lim_{x \to \infty}\left(1 + \frac{2}{x} + \frac{3}{x^2}\right)^x$$

Solution
This function is of the form $[f(x)]^{g(x)}$, where

$$\lim_{x \to \infty} f(x) = 1 \quad \text{and} \quad \lim_{x \to \infty} g(x) = \infty$$

Such a function is said to have indeterminate form 1^∞. This is reduced to the indeterminate form $0/0$ by taking logarithms:

$$\ln\left(1 + \frac{2}{x} + \frac{3}{x^2}\right)^x = x \ln\left(1 + \frac{2}{x} + \frac{3}{x^2}\right) = \frac{\ln[1 + (2/x) + (3/x^2)]}{1/x}$$

and since

$$\lim_{x \to \infty} \ln\left(1 + \frac{2}{x} + \frac{3}{x^2}\right) = 0 \quad \text{and} \quad \lim_{x \to \infty} \frac{1}{x} = 0$$

we have the form $0/0$. Putting

$$f(x) = \ln\left(1 + \frac{2}{x} + \frac{3}{x^2}\right) \quad \text{and} \quad g(x) = \frac{1}{x}$$

we have

$$f'(x) = \frac{1}{1 + (2/x) + (3/x^2)}\left(-\frac{2}{x^2} - \frac{6}{x^3}\right) = \frac{-(1/x^2)(2 + (6/x))}{1 + (2/x) + (3/x^2)}$$

and $g'(x) = -1/x^2$. Hence

$$\lim_{x \to \infty} \frac{f'(x)}{g'(x)} = \lim_{x \to \infty} \frac{2 + 6/x}{1 + (2/x) + (3/x^2)} = 2$$

and consequently,

$$\lim_{x \to \infty}\left(1 + \frac{2}{x} + \frac{3}{x^2}\right)^x = e^2$$

THE INDETERMINATE FORM ∞^0

Example 7

Evaluate

$$\lim_{x \to \infty} x^{e^{-x}}$$

Solution

This function is of the form $[f(x)]^{g(x)}$ when

$$\lim_{x \to \infty} f(x) = \infty \quad \text{and} \quad \lim_{x \to \infty} g(x) = 0$$

Such a function is said to have indeterminate form ∞^0. Taking logarithms again, we have

$$\ln x^{e^{-x}} = e^{-x} \ln x = \frac{\ln x}{e^x}$$

and this function has indeterminate form ∞/∞. Hence,

$$\lim_{x \to \infty} \frac{\ln x}{e^x} = \lim_{x \to \infty} \frac{1/x}{e^x} = \lim_{x \to \infty} \frac{1}{xe^x} = 0$$

and consequently,

$$\lim_{x \to \infty} x^{e^{-x}} = e^0 = 1$$

EXERCISES

In Exercises 1–12, the functions have indeterminate form 0/0 or $\pm\infty/\infty$. Find the limits whenever they exist.

1. $\lim\limits_{x \to 0} \dfrac{x}{\tan x}$

2. $\lim\limits_{x \to 2} \dfrac{x^{10} - 2^{10}}{x - 2}$

3. $\lim\limits_{x \to 2} \dfrac{\sqrt{x} - \sqrt{2}}{x - 2}$

4. $\lim\limits_{x \to 0} \dfrac{\sin^2 x}{\sin x^2}$

5. $\lim\limits_{x \to \infty} \dfrac{(\ln x)^2}{x}$

6. $\lim\limits_{x \to 0+} \dfrac{\ln(e^x - 1)}{1/x}$

7. $\lim\limits_{x \to \pi/2-} \dfrac{\ln \tan x}{\tan x}$

8. $\lim\limits_{x \to 1} \dfrac{\ln x}{x - \sqrt{x}}$

9. $\lim\limits_{x \to 0} \dfrac{e^x - e^{-x}}{x^2}$

10. $\lim\limits_{x \to 1} \dfrac{(x - 1)^2}{(x^2 - 1)^2}$

11. $\lim\limits_{x \to \infty} \dfrac{5^x}{x^5}$

12. $\lim\limits_{x \to \pi} \dfrac{1 + \cos 2x}{1 - \sin x}$

Find the limits in Exercises 13–26.

13. $\lim_{x \to 0} x^x$

14. $\lim_{x \to 0+} x \ln^2 x$

15. $\lim_{x \to 0+} \left(\dfrac{1}{x} - \dfrac{1}{\sin x} \right)$

16. $\lim_{x \to 1} x^{1/(1-x)}$

17. $\lim_{x \to \infty} \left(1 + \dfrac{1}{x} \right)^{x^2}$

18. $\lim_{x \to \infty} \left(1 + \dfrac{1}{x^2} \right)^x$

19. $\lim_{x \to 0} (1 + x)^{1/x}$

20. $\lim_{x \to 0} (1 + x^2)^{1/x}$

21. $\lim_{x \to \infty} x \ln \dfrac{x+1}{x-1}$

22. $\lim_{x \to 0} x \csc x$

23. $\lim_{x \to 0+} \dfrac{x}{\sin x}$

24. $\lim_{x \to \infty} x^{1/x}$

25. $\lim_{x \to \infty} x(\sqrt{x^2 - 1} - x)$

26. $\lim_{x \to 0} \dfrac{5 + 6x}{3 - 4x}$

QUIZ 1

1. Find the following limits.

   (a) $\lim_{n \to \infty} \dfrac{\sqrt{n} - 1}{\sqrt{n} + 1}$

   (b) $\lim_{n \to \infty} \exp\left(\dfrac{n^2 + 2n}{n^3 + 2n} \right)$

2. Find the nth-degree Maclaurin approximation to

   (a) $\dfrac{1}{x+1}$

   (b) e^{2x}

3. Find the Taylor series of $f(x)$ about a.

   (a) $f(x) = 10^x \qquad a = 0$

   (b) $f(x) = \exp(2x) \qquad a = 1$

4. Decide if the following infinite series converge or diverge, and give your reason.

   (a) $\sum_{k=1}^{\infty} k e^{-3k}$

   (b) $\sum_{k=1}^{\infty} \dfrac{k}{(k^3 + 1)}$

   (c) $\sum_{k=0}^{\infty} \dfrac{k^2}{2^k}$

   (d) $\sum_{k=1}^{\infty} \dfrac{\sin k}{k^2}$

5. Evaluate the following series.

   (a) $\sum_{k=1}^{\infty} \left(\frac{10}{11}\right)^{k+3}$

   (b) $\sum_{k=1}^{\infty} \frac{1}{(k+1)!}$

6. Find the radius of convergence of each power series below.

   (a) $\sum_{k=1}^{\infty} k^2 x^k$

   (b) $\sum_{k=1}^{\infty} \frac{10^k}{k!}(x-1)^k$

   (c) $\sum_{k=1}^{\infty} k^k x^{2k}$

7. Evaluate the following limits when they exist.

   (a) $\lim_{x \to 0+} x^{1/x}$

   (b) $\lim_{x \to 1} \frac{\ln x}{x^2 - x}$

QUIZ 2

1. Find the following limits.

   (a) $\lim_{n \to \infty} \frac{\sqrt{n} - 2}{\sqrt{2n} - 2}$

   (b) $\lim_{n \to \infty} \sin \sqrt{\frac{n\pi^2}{n+1}}$

2. Find the nth-degree Maclaurin approximation to

   (a) $\ln(1 + x^2)$

   (b) $\frac{1}{1 - x^2}$

3. Find the Taylor series to $f(x)$ about a.

   (a) $f(x) = \sin(2x + 1) \qquad a = -\frac{1}{2}$

   (b) $f(x) = x \ln x \qquad a = 1$

4. Decide if the following series converge or diverge.

   (a) $\sum_{k=1}^{\infty} \frac{1}{k^2} \exp \frac{1}{k}$ (b) $\sum_{k=1}^{\infty} \frac{\sin k}{k^{3/2}}$

(c) $\sum_{k=0}^{\infty} \sqrt[k]{\frac{1}{k}}$

(d) $\sum_{k=1}^{\infty} \frac{k^{1/2}}{1 + k^{3/2}}$

5. Evaluate the following series.

    (a) $\sum_{k=1}^{\infty} (-1)^k 2^{-k}$

    (b) $\sum_{k=1}^{\infty} \frac{10}{3^k}$

6. Find the radius of convergence of each of the following power series.

    (a) $\sum_{k=1}^{\infty} \frac{k!}{10^k}(x-2)^k$

    (b) $\sum_{k=1}^{\infty} 3^k x^k$

    (c) $\sum_{k=1}^{\infty} (-1)^k \left(\frac{k}{k+1}\right) x^k$

7. Evaluate the following limits when they exist.

    (a) $\lim_{x \to 0+} \sqrt{x} \ln x$

    (b) $\lim_{x \to \infty} (1 + x)^{1/x}$

CHAPTER 12

PLANE GEOMETRY

12.1 CURVES IN PARAMETRIC FORM

Up to this point we have represented functions and curves by means of equations of the form $y = f(x)$. At times this is not advantageous, and we give here an alternative representation that involves a third variable, called a *parameter*. As an application for the use of parameters, we shall obtain a formula for finding the length of a curve. The following example will serve to introduce parametric representations of curves and to exhibit their superiority in certain situations.

Example 1

Consider a circle of radius r rolling on a straight line. A fixed point P on this circle will trace a curve, as illustrated in Figure 12.1. The

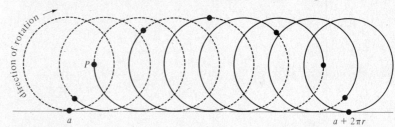

Figure 12.1
Following a fixed point P on a circle rolling on a straight line.

442 PLANE GEOMETRY

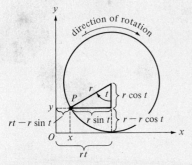

Figure 12.2
The coordinates of P after the circle has turned through an angle of t radians.

curve is called a *cycloid*, and we wish to find an equation for this curve.

Let us take the x axis for the straight line and let the circle start rolling with P at the origin of the coordinate system. Examining the position of P after the circle has rotated through an angle of t radians (see Figure 12.2), we see that its coordinates can be read off from the figure as follows: The distance the circle rolls on the x axis corresponds to the length of the arc determined by the angle of t radians. By Example 3 of Section 9.1, this length is rt. Hence, we see that

$$\left. \begin{array}{l} x = r(t - \sin t) \\ y = r(1 - \cos t) \end{array} \right\} \tag{1}$$

In these equations the variables x and y are expressed in terms of the variable t. This third variable is called a *parameter*, and equations (1) are called the *parametric representation* of the curve (see below). The xy coordinates of P can be computed from the parametric equations (1), as illustrated in Table 12.1 for the case $r = 1$. Figure 12.3 gives the graph of the cycloid for this case.

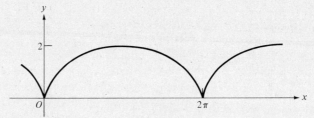

Figure 12.3
The cycloid generated with a circle of radius 1.

Table 12.1
Using the Parametric Equations (1) to Plot the Graph of the Cycloid

t	$\sin t$	$\cos t$	$t - \sin t$	$1 - \cos t$
0	0	1	0	0
$\pi/6 = 0.5236$	0.5000	0.8660	0.0236	0.1340
$2\pi/6 = 1.0472$	0.8660	0.5000	0.1812	0.5000
$3\pi/6 = 1.5708$	1.0000	0	0.5708	1.0000
$4\pi/6 = 2.0944$	0.8660	-0.5000	1.2284	1.5000
$5\pi/6 = 2.6180$	0.5000	-0.8660	2.1180	1.8660
$6\pi/6 = 3.1416$	0	-1.0000	3.1416	2.0000
$7\pi/6 = 3.6652$	-0.5000	-0.8660	4.1652	1.8660
$8\pi/6 = 4.1888$	-0.8660	-0.5000	5.0548	1.5000
$9\pi/6 = 4.7124$	-1.0000	0	5.7124	1.0000
$10\pi/6 = 5.2360$	-0.8660	0.5000	6.1020	0.5000
$11\pi/6 = 5.7596$	-0.5000	0.8660	6.2596	0.1340
$12\pi/6 = 6.2832$	0	1.0000	6.2832	0

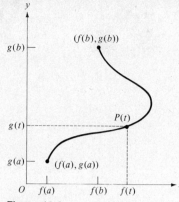

Figure 12.4
A parametric curve
$P(t) = (f(t), g(t))$.

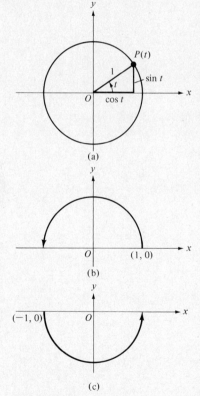

Figure 12.5
The unit circle. (a) The unit circle as traced by $P(t)$. (b) The arc $P(t) = (\cos t, \sin t)$ as t increases from 0 to π. (c) The arc $P(t) = (\cos t, \sin t)$ as t increases from π to 2π.

REMARK
The parameter t in (1) can be eliminated (see Exercise 19 at the end of this section), resulting in the following equation for the cycloid:

$$x = r \cos^{-1} \frac{r-y}{r} \pm \sqrt{2ry - y^2}$$

This equation is difficult to handle, and to express y in terms of x is virtually impossible.

We are now ready to say more generally what we mean by parametric representations of curves:

Let the functions

$$\left.\begin{array}{l} x = f(t) \\ y = g(t) \end{array}\right\} \tag{2}$$

be continuous on the common domain $a \leq t \leq b$, and put

$$P(t) = (f(t), g(t)) \tag{3}$$

Then as t increases from a to b, the point $P(t)$ traces a curve in the xy plane joining the points $(f(a), g(a))$ and $(f(b), g(b))$ (see Figure 12.4). The curve described by either (2) or (3) is said to be given in *parametric form*, and the variable t is called a *parameter*. The following examples and problems will make the concept of parametric representations clear.

Example 2

Consider the parametric equations

$$x = \cos t$$
$$y = \sin t$$

These equations are known from Section 9.1 to represent the unit circle $x^2 + y^2 = 1$ (see Figure 12.5a). The point

$$P(t) = (\cos t, \sin t)$$

is obtained from the point $(1, 0)$ as the point is rotated about the origin through an angle of t radians. Thus, as t increases from 0 to π, the point $P(t)$ traces the upper half of the circle (Figure 12.5b), and as t increases from π to 2π, the point $P(t)$ traces the lower half of the circle (Figure 12.5c). In general, if a is any number, then $P(t)$ traces the entire circle as t increases from a to $a + 2\pi$.

Example 3

Plot the parametric curve

$$\left.\begin{array}{l} x = t^2 \\ y = t \end{array}\right\} \quad -2 \leq t \leq 2 \tag{4}$$

t	$P(t)$
-2	$(4, -2)$
$-\frac{3}{2}$	$(\frac{9}{4}, -\frac{3}{2})$
-1	$(1, -1)$
$-\frac{1}{2}$	$(\frac{1}{4}, -\frac{1}{2})$
0	$(0, 0)$
$\frac{1}{2}$	$(\frac{1}{4}, \frac{1}{2})$
1	$(1, 1)$
$\frac{3}{2}$	$(\frac{9}{4}, \frac{3}{2})$
2	$(4, 2)$

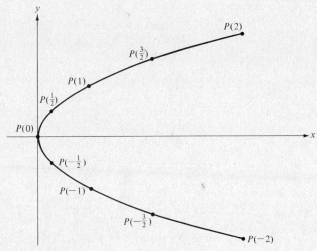

Figure 12.6
The curve
$$\left. \begin{array}{l} x = t^2 \\ y = t \end{array} \right\}$$

Solution
The procedure is simply to compile a list of points $P(t) = (t^2, t)$, plot them in the xy plane, and connect them with a smooth curve. This is done in Figure 12.6. We note that the parameter t can be eliminated in equations (4), giving the single equation $x = y^2$, $-2 \leq y \leq 2$.

Example 4

Consider the curve
$$\left. \begin{array}{l} x = \cos t \\ y = \cos 2t \end{array} \right\}$$

Show that this curve can be written in the form
$$y = 2x^2 - 1$$

and describe how $P(t) = (\cos t, \cos 2t)$ moves on the curve as t increases from 0.

Solution
In Exercise 17 of Section 9.1, we have $\cos 2t = 2 \cos^2 t - 1$, and this gives at once the equation $y = 2x^2 - 1$. The curve is graphed in Figure 12.7. The behavior of $P(t)$ as t increases from 0 to 2π can be found from the table accompanying Figure 12.8, in which the motion of $P(t)$ is presented graphically. As t increases from 2π to 4π, from 4π to 6π, and so on, the motion of $P(t)$ is simply repeated.

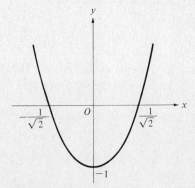

Figure 12.7
The curve
$$\left. \begin{array}{l} x = \cos t \\ y = \cos 2t \end{array} \right\}$$

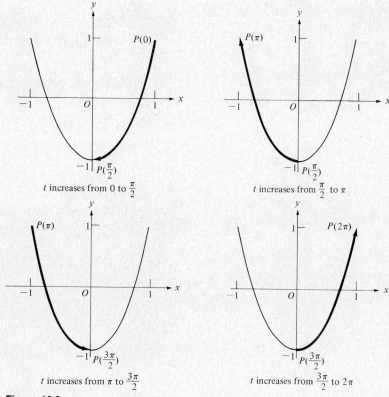

Figure 12.8
The motion of $P(t)$ as t increases from 0 to 2π.

t	$\cos t$	$\cos 2t$	$P(t)$
$0 \to \pi/2$	$1 \to 0$	$1 \to -1$	$(1, 1) \to (0, -1)$
$\pi/2 \to \pi$	$0 \to -1$	$-1 \to 1$	$(0, -1) \to (-1, 1)$
$\pi \to 3\pi/2$	$-1 \to 0$	$1 \to -1$	$(-1, 1) \to (0, -1)$
$3\pi/2 \to 2\pi$	$0 \to 1$	$-1 \to 1$	$(0, -1) \to (1, 1)$

We shall now obtain a formula for the slope of the tangent to a curve in parametric form.

THEOREM 1
Let the functions $f(t)$ and $g(t)$ be differentiable at $t = t_0$ and suppose that $f'(t_0) \neq 0$. Then the tangent to the curve

$$\left. \begin{array}{l} x = f(t) \\ y = g(t) \end{array} \right\}$$

at the point $P(t_0) = (f(t_0), g(t_0))$ has slope $g'(t_0)/f'(t_0)$.

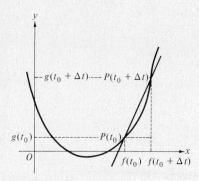

Figure 12.9

Proof

Let $P(t_0 + \Delta t) = (f(t_0 + \Delta t), g(t_0 + \Delta t))$ be another point on the curve (see Figure 12.9). Then the slope of the line joining it to $P(t_0)$ is

$$\frac{g(t_0 + \Delta t) - g(t_0)}{f(t_0 + \Delta t) - f(t_0)} = \frac{[g(t_0 + \Delta t) - g(t_0)]/\Delta t}{[f(t_0 + \Delta t) - f(t_0)]/\Delta t}$$

where $\Delta t \neq 0$. Since

$$\lim_{\Delta t \to 0} \frac{g(t_0 + \Delta t) - g(t_0)}{\Delta t} = g'(t_0)$$

$$\lim_{\Delta t \to 0} \frac{f(t_0 + \Delta t) - f(t_0)}{\Delta t} = f'(t_0)$$

it follows that

$$\text{slope} = \lim_{\Delta t \to 0} \frac{g(t_0 + \Delta t) - g(t_0)}{f(t_0 + \Delta t) - f(t_0)} = \frac{g'(t_0)}{f'(t_0)}$$

which is as asserted.

It is noteworthy that

$$\boxed{\frac{g'(t_0)}{f'(t_0)} = \frac{dy}{dt} \bigg/ \frac{dx}{dt}}$$

Example 5

Find the slope of the tangent to the curve

$$\left. \begin{array}{l} x = t^2 \\ y = t \end{array} \right\}$$

at the point $P(\frac{3}{2}) = (\frac{9}{4}, \frac{3}{2})$.

Solution

This curve was discussed in detail in Example 3. Since $dx/dt = 2t$ and $dy/dt = 1$, we have

$$\frac{dy}{dt} \bigg/ \frac{dx}{dt} = \frac{1}{2t} = \frac{1}{2 \cdot \frac{3}{2}} = \frac{1}{3} \quad \text{for } t = \frac{3}{2}$$

THE LENGTH OF A CURVE

We conclude this section by applying parametric representations to obtain a formula for the length of a curve; this length is sometimes referred to as *arc length*. Like area, this quantity will be given by means of an integral.

Consider differentiable functions $f(t)$ and $g(t)$ with domain $a \leq t \leq b$ and the curve $P(t) = (f(t), g(t))$. Let $s(t)$ give the length of the curve joining $P(a)$ and $P(t)$ (see Figure 12.10); we let $s(a) = 0$. The length of the curve joining two points $P(t)$ and $P(t + \Delta t)$ is

$$\Delta s = s(t + \Delta t) - s(t)$$

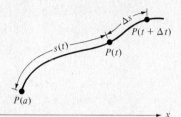

Figure 12.10
$s(t)$ represents the length of the curve joining $P(a)$ to $P(t)$.

(see Figure 12.10). If the increment Δt is sufficiently small, then we have the approximate equality

$$\Delta s \approx \sqrt{(\Delta x)^2 + (\Delta y)^2} \qquad (5)$$

where

$$\Delta x = f(t + \Delta t) - f(t)$$
$$\Delta y = g(t + \Delta t) - g(t)$$

(see Figure 12.11). We see intuitively that

$$\Delta s - \sqrt{(\Delta x)^2 + (\Delta y)^2} \to 0 \qquad \text{as } \Delta t \to 0$$

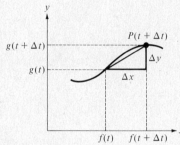

Figure 12.11
The length of the hypotenuse of the right triangle is
$$\sqrt{(\Delta x)^2 + (\Delta y)^2}$$

Dividing both sides in (5) by Δt gives

$$\frac{\Delta s}{\Delta t} \approx \sqrt{\left(\frac{\Delta x}{\Delta t}\right)^2 + \left(\frac{\Delta y}{\Delta t}\right)^2} \qquad (6)$$

and since $x = f(t)$ and $y = g(t)$ are differentiable functions, it follows that

$$\lim_{\Delta t \to 0} \frac{\Delta x}{\Delta t} = f'(t)$$

$$\lim_{\Delta t \to 0} \frac{\Delta y}{\Delta t} = g'(t)$$

Using this information with the observation made in (6) shows that

$$\frac{ds}{dt} = \lim_{\Delta t \to 0} \frac{\Delta s}{\Delta t} = \sqrt{[f'(t)]^2 + [g'(t)]^2}$$

Integrating ds/dt from a to b thus gives us the length of the curve joining $P(a)$ and $P(b)$:

$$s(b) - s(a) = \int_a^b \frac{ds}{dt}\, dt = \int_a^b \sqrt{[f'(t)]^2 + [g'(t)]^2}\, dt$$

Formally, we summarize what we have shown in the following theorem.

THEOREM 2. LENGTH OF A CURVE IN THE PLANE
Let $f(t)$ and $g(t)$ be differentiable functions with domain $a \leq t \leq b$. Then the length L of the curve $P(t) = (f(t), g(t))$ joining $P(a)$ and $P(b)$ is given by the formula

$$L = \int_a^b \sqrt{[f'(t)]^2 + [g'(t)]^2}\, dt$$

Another way of writing this formula is

$$L = \int_a^b \sqrt{\left(\frac{dx}{dt}\right)^2 + \left(\frac{dy}{dt}\right)^2}\, dt$$

For the curves we treat, the derivatives $f'(t)$ and $g'(t)$ are continuous; this guarantees that the integral of Theorem 2 is well defined.

Example 6

Find the length of the cycloid

$$\left. \begin{array}{l} x = 2(t - \sin t) \\ y = 2(1 - \cos t) \end{array} \right\} \quad 0 \leq t \leq 2\pi$$

(see Example 1).

Solution
We have

$$\frac{dx}{dt} = 2(1 - \cos t)$$

$$\frac{dy}{dt} = 2 \sin t$$

and hence

$$\begin{aligned}\left(\frac{dx}{dt}\right)^2 + \left(\frac{dy}{dt}\right)^2 &= 4(1 - \cos t)^2 + 4 \sin^2 t \\ &= 4(1 - 2 \cos t + \cos^2 t + \sin^2 t) \\ &= 8(1 - \cos t) = 8 \cdot 2 \sin^2 \frac{t}{2}\end{aligned}$$

(see Exercise 18, Section 9.1 for the last step). Using the formula obtained above gives

$$L = \int_0^{2\pi} \sqrt{\left(\frac{dx}{dt}\right)^2 + \left(\frac{dy}{dt}\right)^2}\, dt = \int_0^{2\pi} \sqrt{4^2 \sin^2 \frac{t}{2}}\, dt$$

$$= \int_0^{2\pi} 4 \sin \frac{t}{2} \, dt = -8 \cos \frac{t}{2} \Big|_0^{2\pi}$$

$$= -8 \cos \pi + 8 \cos 0 = 8 + 8 = 16$$

EXERCISES

In Exercises 1-8, write the equations of the curves in the form $y = f(x)$ or $x = g(y)$.

1. $\left. \begin{array}{l} x = t \\ y = \sqrt{t} \end{array} \right\} t \geq 0$

2. $\left. \begin{array}{l} x = \cos^2 t - \sin^2 t \\ y = \sin t \end{array} \right\}$

3. $\left. \begin{array}{l} x = e^t \\ y = t^2 \end{array} \right\} t \geq 0$

4. $\left. \begin{array}{l} x = a \sin t \\ y = b \cos t \end{array} \right\} 0 \leq t \leq \pi$

5. $\left. \begin{array}{l} x = \sin t \\ y = \cos t \end{array} \right\} -\frac{\pi}{2} \leq t \leq \frac{\pi}{2}$

6. $\left. \begin{array}{l} x = 1 - t \\ y = 1 + t \end{array} \right\}$

7. $\left. \begin{array}{l} x = \dfrac{1-t}{1+t} \\ y = t^2 \end{array} \right\}$

8. $\left. \begin{array}{l} x = \sin 2t \\ y = \sin t \end{array} \right\}$

In Exercises 9-12, find t_0 and then find the slope of the tangent to the given curve at $P(t_0)$.

9. $\left. \begin{array}{l} x = \dfrac{1-t}{1+t} \\ y = t^3 \end{array} \right\} P(t_0) = (0, 1)$

10. $\left. \begin{array}{l} x = \cos 2t \\ y = \cos t \end{array} \right\} P(t_0) = (-1, 0)$

11. $\left. \begin{array}{l} x = \ln t \\ y = t^2 - 1 \end{array} \right\} P(t_0) = (2, e^4 - 1)$

12. $\left. \begin{array}{l} x = \sqrt{t} \\ y = \dfrac{1}{\sqrt{t}} \end{array} \right\} P(t_0) = (2, \tfrac{1}{2})$

Plot the curves in Exercises 13-18.

13. $\left. \begin{array}{l} x = t \\ y = \sqrt{t} \end{array} \right\} 0 \leq t \leq 2$

14. $\left. \begin{array}{l} x = 1 - t \\ y = 1 + t \end{array} \right\}$

15. $\left. \begin{array}{l} x = 1 - t^2 \\ y = 1 + t^2 \end{array} \right\}$

16. $\left. \begin{array}{l} x = t - \sin t \\ y = t \end{array} \right\}$

17. $\left. \begin{array}{l} x = 2 \sin t \\ y = 4 \cos t \end{array} \right\}$

18. $\left. \begin{array}{l} x = 3 \sec t \\ y = 2 \tan t \end{array} \right\} -\frac{\pi}{2} < t < 0$

19. Derive the equation

$$x = r\cos^{-1}\frac{r-y}{r} \pm \sqrt{2ry - y^2}$$

from the parametric equations

$$x = r(t - \sin t)$$
$$y = r(1 - \cos t)$$

Procedure:

(a) Solve the first equation for $\sin t$, the second for $\cos t$.

(b) Square both sides of the two equations you get, and add them.

(c) From the second equation in step 1 you get

$$t = \cos^{-1}\left(\frac{r-y}{r}\right)$$

Substitute this value for t in step 2 and solve for x.

Find the length of the curves in Exercises 20–24.

20. $\left.\begin{array}{l} x = t^2 \\ y = t \end{array}\right\} 0 \le t \le 2$
21. $\left.\begin{array}{l} x = \cos t \\ y = \sin t \end{array}\right\} -\pi \le t \le \pi$

22. $\left.\begin{array}{l} x = t^3 \\ y = t^2 \end{array}\right\} 0 \le t \le 2$
23. $\left.\begin{array}{l} x = \ln \cos t \\ y = t \end{array}\right\} 0 \le t \le \frac{\pi}{3}$

24. $\left.\begin{array}{l} x = \cos 2t \\ y = 1 - 6\sin^2 t \end{array}\right\} 0 \le t \le \frac{\pi}{2}$

12.2 POLAR COORDINATES

Returning to Section 9.1, we recall that the coordinates of any point $P = (x, y)$ that is distinct from the origin O can be written in the form

$$x = r\cos\theta$$
$$y = r\sin\theta$$

where $r = \sqrt{x^2 + y^2}$ is the distance from O to P, and θ is the radian measure of the angle from the positive x axis to the line segment $\overline{OP}$ (see Figure 12.12). The numbers r and θ are called the *polar coordinates* of $P = (x, y)$; the point will be expressed as $[r, \theta]$, where the square brackets serve to distinguish polar coordinates from rectangular coordinates.

We now make two basic observations:

1. For any constant c, the equation $r = c$ gives a circle of radius c and center at the origin (Figure 12.13).

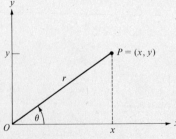

Figure 12.12

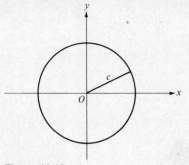

Figure 12.13
The circle $r = c$.

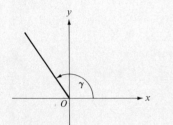

Figure 12.14
The ray $\theta = \gamma$.

2. For any constant γ the equation $\theta = \gamma$ gives a ray issuing from the origin and making an angle of γ radians with the positive x axis (Figure 12.14).

The point $[c, \gamma]$ is thus the point of intersection of the circle $r = c$ with the ray $\theta = \gamma$ (Figure 12.15).

To give a clearer idea of polar coordinates, examine Figure 12.16, where a number of points are given in polar coordinates.

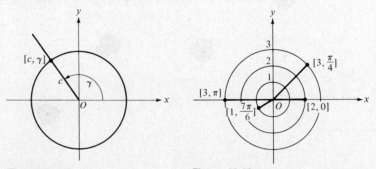

Figure 12.15
The point $[c, \gamma]$.

Figure 12.16
Points in polar coordinates.

The above discussion of polar coordinates was limited to positive values of r and values of θ in the interval $0 \leq \theta < 2\pi$. It is useful to lift these limitations, and we therefore give the following, more general definition of polar coordinates.

> **DEFINITION 1**
> $[r, \theta]$ are the polar coordinates of (x, y) whenever $x = r \cos \theta$ and $y = r \sin \theta$.

According to this definition, the polar coordinates of the origin $(0, 0)$ are $[0, \gamma]$ for any value of γ. The fact that polar coordinates in general are not unique comes out in the examples below.

Example 1

Find the rectangular coordinates of the points below and plot them:

$[2, \tfrac{5}{6}\pi]$ $[2, \tfrac{5}{6}\pi + \pi]$ $[2, \tfrac{5}{6}\pi + 2\pi]$ $[2, \tfrac{5}{6}\pi - 2\pi]$
$[-2, \tfrac{5}{6}\pi]$ $[2, -\tfrac{5}{6}\pi]$ $[2, -\tfrac{5}{6}\pi + \pi]$ $[-2, -\tfrac{5}{6}\pi]$

Solution
Using the definition of polar coordinates, we find the following:

$[2, \tfrac{5}{6}\pi] = (-\sqrt{3}, 1)$ $\qquad [-2, \tfrac{5}{6}\pi] = (\sqrt{3}, -1)$
$[2, \tfrac{5}{6}\pi + \pi] = (\sqrt{3}, -1)$ $\qquad [2, -\tfrac{5}{6}\pi] = (-\sqrt{3}, -1)$

$$[2, \tfrac{5}{6}\pi + 2\pi] = (-\sqrt{3}, 1) \qquad [2, -\tfrac{5}{6}\pi + \pi] = (\sqrt{3}, 1)$$
$$[2, \tfrac{5}{6}\pi - 2\pi] = (-\sqrt{3}, 1) \qquad [-2, -\tfrac{5}{6}\pi] \;\; = (\sqrt{3}, 1)$$

These points are plotted in Figure 12.17.

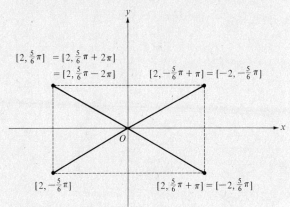

Figure 12.17
Plotting points in polar coordinates.

The relations given in Table 12.2, which are derived from Definition 1, will help you to clarify the relation between the different polar coordinate representations of a given point. These relations are illustrated in Figure 12.18.

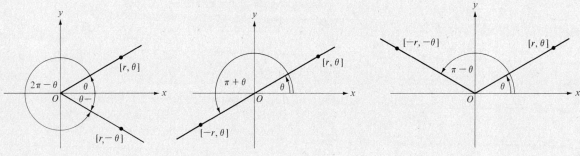

Figure 12.18
The relations between polar coordinate representations.

Table 12.2

Relation		Comment
$[r, -\theta] = [r, 2\pi - \theta]$		$[r, \theta]$ and $[r, -\theta]$ are symmetric with respect to the x axis
$[-r, \theta] = [r, \pi + \theta]$		$[r, \theta]$ and $[-r, \theta]$ are symmetric with respect to the origin
$[-r, -\theta] = [r, \pi - \theta]$		$[r, \theta]$ and $[-r, -\theta]$ are symmetric with respect to the y axis

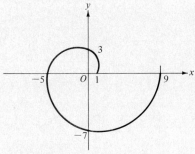

Figure 12.19
The graph of $r = 1 + (4/\pi)\theta$ for $0 \leq \theta \leq 2\pi$.

θ	r
0	1
$\dfrac{\pi}{4}$	2
$\dfrac{\pi}{2}$	3
$\dfrac{3\pi}{4}$	4
π	5
$\dfrac{5\pi}{4}$	6
$\dfrac{3\pi}{2}$	7
$\dfrac{7\pi}{4}$	8
2π	9

CURVES IN POLAR COORDINATES
A relation

$$r = f(\theta)$$

which specifies r in terms of θ, gives a curve that consists of all points $[r, \theta] = [f(\theta), \theta]$. The curve is said to be given in *polar form*. A good idea of the techniques involved in graphing curves given in polar coordinates can be obtained by studying Examples 2–4 below.

Example 2

Plot the curve

$$r = 1 + \frac{4}{\pi}\theta \quad \text{for } 0 \leq \theta \leq 6\pi$$

Solution
To begin with, compile a table of values $[r, \theta]$ for $0 \leq \theta \leq 2\pi$, plot the points, and connect them with a smooth curve. This is done in Figure 12.19. The graph we obtain is that of a spiral. It is clear that for $2\pi \leq \theta \leq 4\pi$ we obtain another spiral, which winds around the first one, and so on. With a set of values $[r, \theta]$ for $0 \leq \theta \leq 6\pi$, we obtain the graph in Figure 12.20.

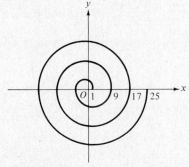

Figure 12.20
The graph of $r = 1 + (4/\pi)\theta$ for $0 \leq \theta \leq 6\pi$.

Example 3

Plot the curve

$$r = p \cos \theta \quad \text{for } 0 \leq \theta \leq \pi$$

where p is a positive constant.

Solution
As before, we compile a table of values $[r, \theta]$, plot the points, and connect them with a smooth curve (Figure 12.21). We seem to get

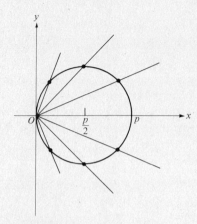

Figure 12.21
The graph of $r = p \cos \theta$ for $0 \leq \theta \leq \pi$.

θ	r
0	p
$\pi/6$	$\frac{\sqrt{3}}{2}p$
$\pi/4$	$\frac{\sqrt{2}}{2}p$
$\pi/3$	$\frac{1}{2}p$
$\pi/2$	0
$2\pi/3$	$-\frac{1}{2}p$
$3\pi/4$	$-\frac{\sqrt{2}}{2}p$
$5\pi/6$	$-\frac{\sqrt{3}}{2}p$
π	$-p$

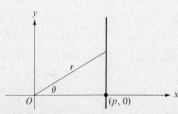

Figure 12.22
The line $r = p/\cos \theta$.

a circle with center at $(p/2, 0)$ and radius $p/2$. That this is, indeed, correct, is easily proved algebraically, as follows: With the fact that $\cos \theta = x/r$, the equation $r = p \cos \theta$ can be written

$$r = p\frac{x}{r} \quad \text{or} \quad r^2 = px$$

Since $r^2 = x^2 + y^2$, we get

$$x^2 + y^2 = px$$

Completing the square, this equation can be written as

$$\left(x - \frac{p}{2}\right)^2 + y^2 = \left(\frac{p}{2}\right)^2$$

which is the equation of a circle with center $(p/2, 0)$ and radius $p/2$.

We mention in passing that the circle is traversed again as θ increases from π to 2π, and in general the equation $r = p \cos \theta$ specifies the same circle when $\alpha \leq \theta \leq \alpha + \pi$, α being an arbitrary number.

Example 4

Plot the curve

$$r = \frac{p}{\cos \theta} \quad \text{for } 0 \leq \theta \leq 2\pi \quad \text{and} \quad \theta \neq \frac{\pi}{2}, \frac{3\pi}{2}$$

where p is any nonzero constant.

Solution
Since $x = r \cos \theta$, we see at once that the equation $r = p/\cos \theta$ is the same as $x = p$, which is the equation of the vertical line through $(p, 0)$ (see Figure 12.22). Very revealing is the relation between r and θ as tabulated in Table 12.3.

Table 12.3

θ	$\cos \theta$	$r = p/\cos \theta$
increases from 0 to $\pi/2$	decreases from 1 to 0	increases from p to ∞
increases from $\pi/2$ to π	decreases from 0 to -1	increases from $-\infty$ to $-p$
increases from π to $3\pi/2$	increases from -1 to 0	decreases from $-p$ to $-\infty$
increases from $3\pi/2$ to 2π	increases from 0 to 1	decreases from ∞ to p

REMARK
The fact that $r = p/\cos \theta$ is undefined for $\theta = \pi/2, 3\pi/2$ has a simple geometric interpretation. Namely, it tells us that the rays $\theta = \pi/2$ and $\theta = 3\pi/2$ are parallel to the line $x = p$, and thus have no point in common with it.

EXERCISES

1. Give *three* polar coordinate representations for each of the given points.

   (a) $(1, 1)$ (b) $(0, -1)$

   (c) $(-1, -1)$ (d) $(5/2, 5\sqrt{3}/2)$

   (e) $(a, a) \quad a > 0$ (f) $(\sqrt{3}, 1)$

2. Give the rectangular coordinates of each of the given points.

   (a) $[2, \pi/3]$ (b) $[-1, 3\pi]$

   (c) $[-2, \pi/6]$ (d) $[3, \pi/6]$

   (e) $[1/2, \pi/3]$ (f) $[1, -3\pi]$

   (g) $[\pi/4, \pi/4]$ (h) $[-1/2, -\pi/3]$

3. A rectangle with sides parallel to the coordinate axes is inscribed in the circle $x^2 + y^2 = 1$ (see Figure 12.23). Find the remaining three vertices in polar coordinates if one vertex is $[r, \pi/8]$.

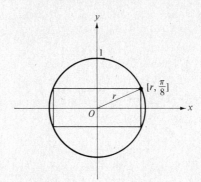

Figure 12.23

4. Give polar coordinates for the points $(3, 0)$, $(3, 4)$, and $(3, -4)$.

5. Give polar coordinates for the points $(-3, 4)$, $(0, 4)$, and $(3, 4)$.

Plot the curves given in Exercises 6–14.

6. $r = \dfrac{3}{\sin \theta} \quad -\pi < \theta < \pi, \ \theta \neq 0$

7. $r = 1 + \dfrac{1}{\pi}\theta \quad$ for $0 \leq \theta \leq 4\pi$

8. $r = \dfrac{4}{\sin(\theta + \pi/2)} \quad$ for $-\pi < \theta < \pi, \ \theta \neq -\dfrac{\pi}{2}, \dfrac{\pi}{2}$

9. $r = 5 \cos\left(\theta - \dfrac{\pi}{2}\right)$

10. $r = p \sin \theta \quad p > 0$

11. $r = 2 \sin\left(\theta + \dfrac{\pi}{4}\right)$

12. $r = \dfrac{1}{2} + \dfrac{1}{\pi} \cos \theta$

13. $r = e^{a\theta} \quad (a \neq 0)$

14. $r = 2 \cos \theta + 3 \sin \theta$

Give the polar form of the lines in Exercises 15–18.

15. $y = -x$ 16. $y = 5$

17. $x = 7$ 18. $y = \sqrt{3}x$

12.3 SYMMETRIES AND GRAPHING IN POLAR COORDINATES

We continue in this section the discussion of curves given in polar form. In graphing, it is very useful to know if the curve in question is symmetric with respect to the x or y coordinate axes or the origin. For this we have the following general rule.

> **SYMMETRIES OF CURVES IN POLAR FORM**
> Let $r = f(\theta)$ be a given curve in polar form.
>
> 1. If $f(-\theta) = f(\theta)$, then the curve is symmetric with respect to the x axis.
> 2. If $f(\pi + \theta) = f(\theta)$, then the curve is symmetric with respect to the *origin*.
> 3. If $f(\pi - \theta) = f(\theta)$, then the curve is symmetric with respect to the y axis.

The proof of these symmetries is contained in Figure 12.24.

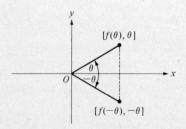
Symmetry with respect to the x axis

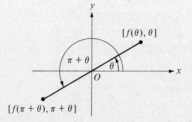
Symmetry with respect to the origin

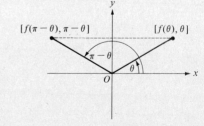
Symmetry with respect to the y axis

Figure 12.24
Symmetries of curves in polar form.

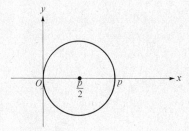

Figure 12.25

Example 1

(a) The curve $f(\theta) = p \cos \theta$ $(p > 0)$ is symmetric with respect to the x axis, since

$$f(-\theta) = p \cos(-\theta) = p \cos \theta = f(\theta)$$

(see Example 3, Section 12.2, and Figure 12.25).

(b) Consider the curve $f(\theta) = \cos 2\theta$. Then

$$f(-\theta) = \cos 2(-\theta) = \cos 2\theta = f(\theta)$$
$$f(\pi + \theta) = \cos 2(\pi + \theta) = \cos(2\pi + 2\theta) = \cos 2\theta = f(\theta)$$
$$f(\pi - \theta) = \cos(2\pi - 2\theta) = \cos 2\theta = f(\theta)$$

Hence $f(\theta)$ is symmetric with respect to the x axis, the y axis, and the origin. The graph of this curve is obtained in Example 3 that follows.

(c) The curve $f(\theta) = p \sin \theta$ $(p > 0)$ is symmetric with respect to the y axis, since

$$f(\pi - \theta) = p \sin(\pi - \theta) = p(\sin \pi \cos \theta - \sin \theta \cos \pi)$$
$$= p \sin \theta = f(\theta)$$

From Exercise 10 of the last section, we know in fact that this curve is a circle with center at $(0, p/2)$ and radius $p/2$ (see Figure 12.26).

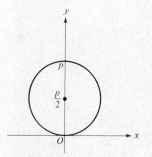

Figure 12.26
The circle $r = p \sin \theta$.

The following fact is invaluable in graphing many curves in polar form.

THEOREM 1
Let $r = f(\theta)$ be a given curve such that

$$f(\theta_0) = 0$$

Then the tangent to the curve at the origin makes an angle θ_0 with the positive x axis.

This theorem is illustrated in Figure 12.27; the proof will follow Example 2.

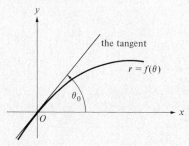

Figure 12.27
The curve $r = f(\theta)$ with $f(\theta_0) = 0$.

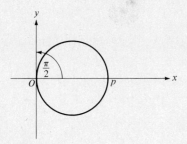

Figure 12.28

Example 2

Consider the circle $r = p\cos\theta$ (Example 3, Section 12.2, and Figure 12.28). Since $r = 0$ for $\theta = \pi/2$, it follows that the tangent to the circle at the origin is perpendicular to the x axis, as we already know.

Proof of Theorem 1
Substituting $r = f(\theta)$ in the equations $x = r\cos\theta$ and $y = r\sin\theta$ gives

$$\left.\begin{array}{l} x = f(\theta)\cos\theta \\ y = f(\theta)\sin\theta \end{array}\right\}$$

These equations can be taken as parametric equations for the curve $r = f(\theta)$, where θ is the parameter. According to Section 12.1, the slope of the tangent to the curve at $\theta = \theta_0$ can be obtained from the formula $(dy/d\theta)/(dx/d\theta)$. But

$$\frac{dx}{d\theta} = f'(\theta)\cos\theta - f(\theta)\sin\theta$$

$$\frac{dy}{d\theta} = f'(\theta)\sin\theta + f(\theta)\cos\theta$$

and since $f(\theta_0) = 0$, we get for the slope at the origin

$$\frac{dy/d\theta}{dx/d\theta} = \frac{f'(\theta_0)\sin\theta_0 + f(\theta_0)\cos\theta_0}{f'(\theta_0)\cos\theta_0 - f(\theta_0)\sin\theta_0} = \frac{\sin\theta_0}{\cos\theta_0} = \tan\theta_0$$

Hence, the tangent to the curve at the origin makes an angle of θ_0 radians with the positive x axis. Another way of thinking about this result is given in Figure 12.29. Namely, when θ is close to θ_0, then the line $\overline{OP}$ is close to the tangent at O. The line $\overline{OP}$ makes an angle θ with the positive x axis; thus, as P approaches O along the curve, the line $\overline{OP}$ approaches the tangent and θ approaches θ_0.

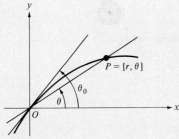

Figure 12.29
The curve $r = f(\theta)$ with the line $\overline{OP}$ approaching the tangent at O.

Example 3

Plot the curve

$$r = \cos 2\theta \qquad \text{for } 0 \leq \theta \leq 2\pi$$

Solution
From Example 1(b), we know this curve to be symmetric with respect to the x axis, the y axis, and the origin. It thus suffices to plot that portion of the curve for $0 \leq \theta \leq \pi/2$, and then we can take mirror images of it.
  Since, for $\theta_0 = \pi/4$,

$$r = \cos 2\theta_0 = \cos 2\left(\frac{\pi}{4}\right) = \cos\frac{\pi}{2} = 0$$

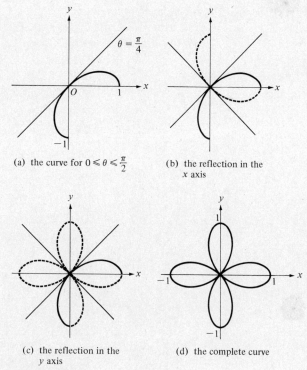

(a) the curve for $0 \leq \theta \leq \frac{\pi}{2}$

(b) the reflection in the x axis

(c) the reflection in the y axis

(d) the complete curve

Figure 12.30
The construction of the curve $r = \cos 2\theta$.

we see that the tangent to the curve at the origin is the line $\theta = \pi/4$. Next, we observe that as θ increases from 0 to $\pi/4$, $\cos 2\theta$ decreases from 1 to 0; as θ increases from $\pi/4$ to $\pi/2$, $\cos 2\theta$ decreases from 0 to -1. With an appropriate table of values we construct the graph given in Figure 12.30a. Reflecting this curve in the x axis and y axis, as in Figure 12.30b and 12.30c, gives the complete graph in Figure 12.30d.

Example 4

Plot the curve

$$r = a(1 - \cos \theta)$$

Cardioid means heart-shaped.

where a is a positive constant. This curve is called a *cardioid*.

Solution
The fact that $\cos(\theta) = \cos(-\theta)$ tells us that this curve is symmetric with respect to the x axis. Since $\cos 0 = 1$, we have

$$r = a(1 - \cos 0) = a(1 - 1) = 0$$

and hence the x axis is tangent to the curve at the origin. With the tables of values for $0 \leq \theta \leq \pi$ and a reflection in the x axis, we obtain the graph in Figure 12.31.

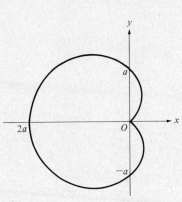

θ	$a(1 - \cos \theta)$
0	0
$\dfrac{\pi}{6}$	$0.13a$
$\dfrac{\pi}{4}$	$0.29a$
$\dfrac{\pi}{3}$	$0.50a$
$\dfrac{\pi}{2}$	a
$\dfrac{2\pi}{3}$	$1.50a$
$\dfrac{3\pi}{4}$	$1.71a$
$\dfrac{5\pi}{6}$	$1.87a$
π	$2a$

Figure 12.31
The cardioid $r = a(1 - \cos \theta)$.

EXERCISES

Graph the curves in Exercises 1–11.

1. $r = \theta$

2. $r = \dfrac{1}{\theta} \quad (\theta > 0)$

3. $r = a(1 + \cos \theta) \quad (a > 0)$

4. $r = a(1 - \sin \theta) \quad (a > 0)$

5. $r = a(1 + \sin \theta) \quad (a > 0)$

6. $r = \dfrac{1}{1 - \cos \theta}$

7. $r^2 = 4a^2 \cos \theta$

Hint: Plot the curves $r = 2a\sqrt{\cos \theta}$ and $r = -2a\sqrt{\cos \theta}$.

8. $r^2 = a^2 \sin 2\theta$

Hint: Plot the curves $r = a\sqrt{\sin 2\theta}$ and $r = -a\sqrt{\sin 2\theta}$.

9. $r = a \cos 3\theta$

10. $r = a \sin 3\theta$

11. $r = 4 \sin 2\theta$

12.4 AREA AND LENGTH IN POLAR COORDINATES

Two fundamental problems of calculus, finding the area enclosed by a curve and the length of a curve, are discussed here for curves given in polar form.

AREA

Consider a curve $r = f(\theta)$ and suppose that $\alpha \le \theta \le \beta$. As the point P moves on this curve from $[\alpha, f(\alpha)]$ to $[\beta, f(\beta)]$, the line segment $\overline{OP}$ sweeps out an area, A, given by the formula

$$A = \tfrac{1}{2} \int_\alpha^\beta [f(\theta)]^2 \, d\theta \qquad (1)$$

(see Figure 12.32). This formula is also written

$$A = \tfrac{1}{2} \int_\alpha^\beta r^2 \, d\theta \qquad (1')$$

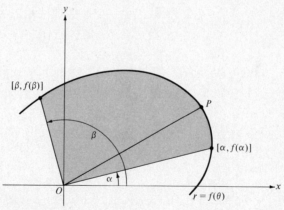

Figure 12.32
The area swept out by the line segment $\overline{OP}$.

In verifying this formula, we make use of the following fact from Section 9.1 (see Figure 12.33):

The area of a sector of a circle of radius r and central angle θ radians is $\tfrac{1}{2} r^2 \theta$.

Figure 12.33
A sector of a circle of radius r and central angle of θ radians.

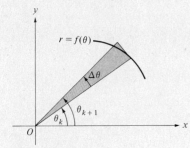

Figure 12.34
An approximating sector.

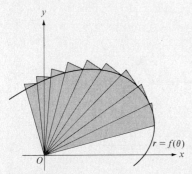

Figure 12.35
The approximate area swept out by the line segment $\overline{OP}$.

Let us divide the interval $[\alpha, \beta]$ into n equal subintervals, each of length $\Delta\theta = (\beta - \alpha)/n$, and put

$$\theta_k = \alpha + k\,\Delta\theta \qquad \text{for } k = 0, 1, 2, \ldots, n-1$$

Then the area of the sector swept out by the line segment $\overline{OP}$ as P moves from $[\theta_k, f(\theta_k)]$ to $[\theta_{k+1}, f(\theta_{k+1})]$ is approximately

$$\tfrac{1}{2}[f(\theta_k)]^2\,\Delta\theta$$

which is the area of a sector of a circle of radius $f(\theta_k)$ and central angle $\Delta\theta$ (see Figure 12.34). Summing the areas of the sectors gives the approximate formula (see Figure 12.35)

$$A \approx \sum_{k=0}^{n-1} \tfrac{1}{2}[f(\theta_k)]^2\,\Delta\theta$$

Taking the limit as $n \to \infty$, we obtain formula (1).

Example 1

Let us test our formula in two known cases.

(a) Let $f(\theta) = c$, $c > 0$. This is the equation of a circle of radius c, and we have

$$\tfrac{1}{2}\int_\alpha^\beta c^2\,d\theta = \tfrac{1}{2}c^2\theta\Big|_\alpha^\beta = \tfrac{1}{2}c^2(\beta - \alpha)$$

This agrees with the formula previously obtained for the area of a sector of a circle of radius c and central angle $\beta - \alpha$ (see Figure 12.36).

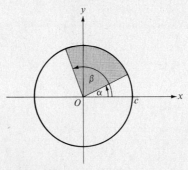

Figure 12.36
The sector of a circle with central angle of $\beta - \alpha$ radians.

(b) Let $r = p\cos\theta$ for $p > 0$ and $0 \leq \theta \leq \pi/2$. This equation gives the upper half of a circle of radius $p/2$ (see Figure 12.37), and, indeed,

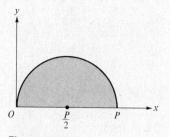

Figure 12.37
The area bounded by the upper half of the circle $r = p\cos\theta$ and the x axis.

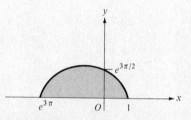

Figure 12.38
The area bounded by the curve $r = e^{3\theta}$ for $0 \le \theta \le \pi$ and the x axis.

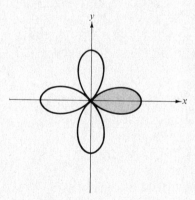

Figure 12.39
One petal of the curve $r = \cos 2\theta$.

$$\frac{1}{2}\int_0^{\pi/2} p^2 \cos^2\theta \, d\theta = \frac{1}{2}p^2 \int_0^{\pi/2} \cos^2\theta \, d\theta$$

$$= \frac{1}{2}p^2 \left(\frac{1}{2}\theta + \frac{1}{4}\sin 2\theta\right)\Big|_0^{\pi/2}$$

$$= \frac{1}{2}p^2 \frac{\pi}{4} = \frac{1}{2}\pi\left(\frac{p}{2}\right)^2$$

Example 2

Find the area swept out by the ray $r = e^{3\theta}$ for $0 \le \theta \le \pi$ (see Figure 12.38).

Solution

$$A = \frac{1}{2}\int_0^\pi (e^{3\theta})^2 \, d\theta = \frac{1}{2}\int_0^\pi e^{6\theta} \, d\theta = \frac{1}{2}\frac{e^{6\theta}}{6}\Big|_0^\pi = \frac{1}{12}(e^{6\pi} - 1)$$

Example 3

Find the area bounded by one petal of the curve $r = \cos 2\theta$ (see Example 3 of the last section).

Solution
One petal is swept out as θ increases from $-\pi/4$ to $\pi/4$ (see Figure 12.39), and thus

$$A = \frac{1}{2}\int_{-\pi/4}^{\pi/4} \cos^2 2\theta \, d\theta$$

The substitution $t = 2\theta$ gives, by Section 9.7,

$$A = \frac{1}{2}\int_{-\pi/2}^{\pi/2} \cos^2 t \cdot \frac{1}{2} \, dt = \frac{1}{4}\left(\frac{1}{2}t + \frac{1}{4}\sin 2t\right)\Big|_{-\pi/2}^{\pi/2} = \frac{\pi}{8}$$

LENGTH OF A CURVE

Let $r = f(\theta)$ be a given curve in polar form, and suppose that $\alpha \le \theta \le \beta$. Then the length L of the curve is given by the integral

$$L = \int_\alpha^\beta \sqrt{[f(\theta)]^2 + [f'(\theta)]^2} \, d\theta \qquad (2)$$

This formula is also written

$$L = \int_\alpha^\beta \sqrt{r^2 + \left(\frac{dr}{d\theta}\right)^2} \, d\theta \qquad (2')$$

In deriving this formula, we use the argument used in the proof of Theorem 1 of the last section. Namely, since $r = f(\theta)$, we can write the equations $x = r \cos \theta$ and $y = r \sin \theta$ as a pair of parametric equations

$$\left. \begin{array}{l} x = f(\theta) \cos \theta \\ y = f(\theta) \sin \theta \end{array} \right\} \quad \alpha \leq \theta \leq \beta \tag{3}$$

which represent the curve $r = f(\theta)$. Differentiating these equations gives

$$\frac{dx}{d\theta} = f'(\theta) \cos \theta - f(\theta) \sin \theta$$

$$\frac{dy}{d\theta} = f'(\theta) \sin \theta + f(\theta) \cos \theta$$

and using the identity $\sin^2 \theta + \cos^2 \theta = 1$ gives

$$\left(\frac{dx}{d\theta}\right)^2 + \left(\frac{dy}{d\theta}\right)^2 = [f(\theta)]^2 + [f'(\theta)]^2 = r^2 + \left(\frac{dr}{d\theta}\right)^2$$

This is now substituted in the formula

$$L = \int_\alpha^\beta \sqrt{\left(\frac{dx}{d\theta}\right)^2 + \left(\frac{dy}{d\theta}\right)^2} \, d\theta$$

which was obtained in Section 12.1, to obtain formulas (2) and (2'). We test this formula in two known cases.

Example 4

(a) Consider the circle $f(\theta) = c$, $\alpha = 0$ and $\beta = 2\pi$. Since c is a constant, $f'(\theta) = 0$ and hence

$$L = \int_0^{2\pi} \sqrt{c^2 + 0^2} \, d\theta = \int_0^{2\pi} c \, d\theta = 2\pi c$$

This is, of course, the correct answer for the circumference of a circle of radius c.

(b) Let us now take the semicircle $r = p \cos \theta$, where $p > 0$ and $0 \leq \theta \leq \pi/2$ [see Example 1(b) above]. Here $dr/d\theta = -p \sin \theta$ and, consequently,

$$L = \int_0^{\pi/2} \sqrt{(p \cos \theta)^2 + (-p \sin \theta)^2} \, d\theta$$

$$= \int_0^{\pi/2} \sqrt{p^2(\cos^2 \theta + \sin^2 \theta)} \, d\theta = \int_0^{\pi/2} p \, d\theta = \pi \cdot \frac{p}{2}$$

This is again the correct answer.

Example 5

Find the length of the curve $r = e^{3\theta}$ for $0 \le \theta \le 4\pi$.

Solution

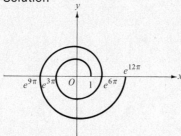

Figure 12.40
The logarithmic spiral $r = e^{3\theta}$.

The curve, which is called a *logarithmic spiral*, is graphed in Figure 12.40. Since $dr/d\theta = 3e^{3\theta}$, we have

$$L = \int_0^{4\pi} \sqrt{(e^{3\theta})^2 + (3e^{3\theta})^2}\, d\theta = \int_0^{4\pi} \sqrt{e^{6\theta} + 9e^{6\theta}}\, d\theta$$

$$= \int_0^{4\pi} \sqrt{10}\, e^{3\theta}\, d\theta = \sqrt{10}\, \frac{e^{3\theta}}{3}\bigg|_0^{4\pi} = \frac{\sqrt{10}}{3}(e^{12\pi} - 1)$$

EXERCISES

In Exercises 1–8, find the area bounded by the given curve.

1. $r = 2a \sin\theta$
2. $r = a(2 - \cos\theta)$
3. $r = a(1 + \cos\theta)$
4. $r = 2 + 2\cos 2\theta$
5. $r^2 = a^2 \cos 2\theta$
6. $r = a \sin 2\theta$
7. $r = a \sin 3\theta$
8. $r = a \cos 3\theta$

9. Find the area swept out by the line segments from the origin to the points $[e^{\theta/3}, \theta]$ for $\pi \le \theta \le 2\pi$, and give a diagram of the area.

In Exercises 10–13, find the length of the given curve.

10. $r = 2a \sin\theta$
11. $r = a(1 - \cos\theta)$
12. $r = a \cos\theta + b \sin\theta$
13. $r = \theta^2$ for $0 \le \theta \le 2\pi$

12.5 PARABOLAS

In this section we discuss the first of three types of curves that play a special role in geometry and certain applications. The curves to which we allude are parabolas, ellipses, and hyperbolas. These curves are called *conic sections*, because they can be obtained by intersecting a cone with appropriate planes (see Figure 12.41). Rather than use this approach, we introduce the curves by means of defining equations in the *xy* plane.

 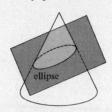 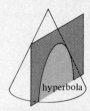

The plane giving the parabola is parallel to a side of the cone.

The plane giving the ellipse is parallel neither to a side nor to the axis of the cone.

The plane giving the hyperbola is parallel to the axis of the cone.

Figure 12.41
The parabola, ellipse, and hyperbola as conic sections.

Parabolas are not new to us; they are curves given by quadratic equations $y = mx^2 + px + q$, where $m \neq 0$. Here we wish to study certain properties of these curves, and this is best begun with the case $p = q = 0$.

> The curve
> $$y = mx^2 \quad \text{with} \quad m > 0$$
> is called a *parabola* in *standard position*.

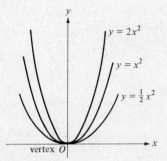

Figure 12.42
Parabolas in standard position.

For a parabola in standard position, the origin 0 is called its *vertex*, and the *y* axis its *axis* (see Figure 12.42).

Parabolas are endowed with a special property that is often used to define them. This property is described in the following theorem.

> **THEOREM 1**
> Any point $P = (x, y)$ on the parabola $y = mx^2$ is equidistant from the point $(0, 1/4m)$ and the line $y = -1/4m$.

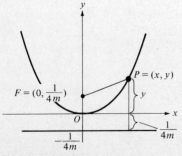

Figure 12.43
Illustrating Theorem 1.

Proof
The situation is described in Figure 12.43. Note that the distance from *P* to the line $y = -1/4m$ is the length of the perpendicular line segment from *P* to the line. The length of this segment is $y + 1/4m$. To compute the distance from *P* to $F = (0, 1/4m)$, we begin by writing $y = mx^2$ in the form

$$x^2 = \frac{1}{m}y = 4 \cdot \frac{1}{4m}y$$

12.5 PARABOLAS

With this we see that

$$\text{dist}(P, F) = \sqrt{(x-0)^2 + \left(y - \frac{1}{4m}\right)^2} = \sqrt{x^2 + y^2 - 2\frac{1}{4m}y + \left(\frac{1}{4m}\right)^2}$$

$$= \sqrt{4 \cdot \frac{1}{4m}y + y^2 - 2\frac{1}{4m}y + \left(\frac{1}{4m}\right)^2}$$

$$= \sqrt{y^2 + 2\frac{1}{4m}y + \left(\frac{1}{4m}\right)^2}$$

$$= \sqrt{\left(y + \frac{1}{4m}\right)^2} = y + \frac{1}{4m}$$

and this completes the proof.

The following terminology is standard:

> Let $y = mx^2$ and put $p = 1/4m$. Then the point $F = (0, p)$ is called the *focus* of the parabola $y = mx^2$, and the line $y = -p$ is called the *directrix* of the parabola $y = mx^2$.

The focus and directrix of a parabola in standard position are identified in Figure 12.44.

Getting away now from the narrow concept of parabola in standard position, we introduce the following, more general definition based on Theorem 1.

> A *parabola* is the locus of points that are equidistant from a fixed point, called its *focus,* and a fixed line, called its *directrix.*

Some simple parabolas are given in Figure 12.45.

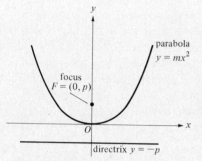

Figure 12.44
The focus and directrix of a parabola in standard position.

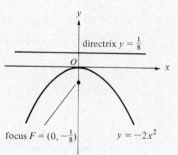
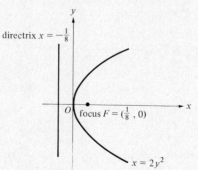
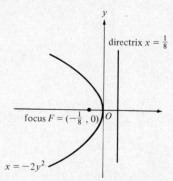

Figure 12.45
Parabolas not in standard position.

In discussing general parabolas, we make use of the notion of translation of graphs.

468 PLANE GEOMETRY

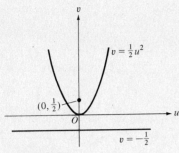

Figure 12.46
The parabola $v = \frac{1}{2}u^2$ with focus $(0, \frac{1}{2})$ and directrix $v = -\frac{1}{2}$.

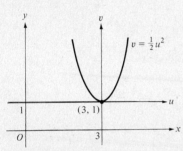

Figure 12.47
uv coordinate axes superimposed in an xy coordinate system.

Example 1

Let us describe the curve

$$y - 1 = \tfrac{1}{2}(x - 3)^2 \tag{1}$$

With the change of variables

$$u = x - 3 \quad \text{and} \quad v = y - 1 \tag{2}$$

our equation becomes

$$v = \tfrac{1}{2}u^2$$

In a uv coordinate system, this is the equation of a parabola in standard position with focus $F = (0, \frac{1}{2})$ and directrix $v = -\frac{1}{2}$ (see Figure 12.46). This same interpretation would remain valid if we were to draw a horizontal and a vertical line through the point $(x, y) = (3, 1)$ in the xy plane, calling these the u axis and v axis, respectively (see Figure 12.47). To establish a connection between equation (1) and Figure 12.47, we make the following observation concerning the two superimposed coordinate systems: If P has coordinates (u, v) in the uv coordinate system, then the coordinates (x, y) of P in the xy coordinate system are found from the equations

$$x = u + 3 \quad \text{and} \quad y = v + 1 \tag{3}$$

which are obtained from (2) (see Figure 12.48).

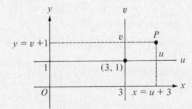

Figure 12.48
A point P in the uv and xy coordinate systems.

Using (3), we see that $y - 1 = \frac{1}{2}(x - 3)^2$ is the equation of a parabola with vertex $(3, 1)$, axis $x = 3$, focus $F = (3, 1 + \frac{1}{2}) = (3, \frac{3}{2})$ and directrix $y = 1 - \frac{1}{2} = \frac{1}{2}$ (see Figure 12.49).

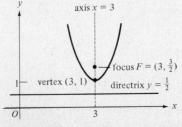

Figure 12.49
The parabola $y - 1 = \frac{1}{2}(x - 3)^2$.

Let
$$\frac{x^2}{a^2} + \frac{y^2}{b^2} = 1$$
be a given ellipse with $a > b > 0$. The points
$$F^+ = (ea, 0) \quad \text{and} \quad F^- = (-ea, 0)$$
are called the *foci* of the ellipse. The lines
$$l^+ \colon x = \frac{a}{e} \quad \text{and} \quad l^- \colon x = -\frac{a}{e}$$
are called the *directrices* of the ellipse.

Since $0 < e < 1$, we see at once that
$$-\frac{a}{e} < -a < -ea \quad \text{and} \quad ea < a < \frac{a}{e}$$
and from these inequalities we find the relative position of the foci, directrices, and the vertices $(-a, 0)$ and $(a, 0)$ (see Figure 12.56).

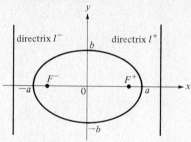

Figure 12.56
The foci and directrices of an ellipse in standard position.

Example 1

Plot the ellipse
$$\frac{x^2}{5^2} + \frac{y^2}{3^2} = 1$$
with its foci and directrices.

Solution
Here $a = 5$, $b = 3$, and hence the eccentricity is
$$e = \sqrt{1 - (\tfrac{3}{5})^2} = \sqrt{\tfrac{16}{25}} = \tfrac{4}{5}$$

Thus we have
$$F^+ = (\tfrac{4}{5} \cdot 5, 0) = (4, 0)$$
$$F^- = (-\tfrac{4}{5} \cdot 5, 0) = (-4, 0)$$

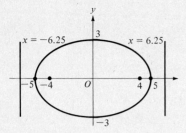

Figure 12.57
The ellipse $\dfrac{x^2}{5^2} + \dfrac{y^2}{3^2} = 1$.

$$l^+: x = \frac{5}{\frac{4}{5}} = \frac{25}{4} = 6.25$$

$$l^-: x = -\frac{5}{\frac{4}{5}} = -6.25$$

The graph in Figure 12.57 is obtained with an appropriate table of values.

We now prove two theorems dealing with the geometric properties of the foci and directrices alluded to above.

> **THEOREM 1**
> If $P = (x, y)$ is any point on the ellipse
> $$\frac{x^2}{a^2} + \frac{y^2}{b^2} = 1$$
> then
> $$\text{dist}(P, F^+) = e \,\text{dist}(P \text{ to } l^+)$$
> $$\text{dist}(P, F^-) = e \,\text{dist}(P \text{ to } l^-)$$

Proof

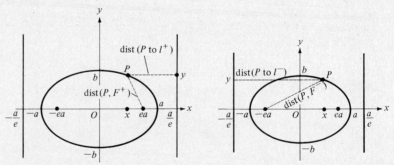

Figure 12.58
Illustrating Theorem 1.

Let us verify the first equation in the statement of the theorem. Consulting Figure 12.58, we see that

$$\text{dist}(P \text{ to } l^+) = \frac{a}{e} - x = \frac{1}{e}(a - ex) \tag{1}$$

and we shall show that

$$\text{dist}(P, F^+) = a - ex$$

To show this, we use the relation

$$y^2 = b^2 - \left(\frac{b}{a}\right)^2 x^2 \tag{2}$$

which is obtained from the equation

$$\frac{x^2}{a^2} + \frac{y^2}{b^2} = 1$$

and the relation

$$a^2 = (ea)^2 + b^2 \qquad (3)$$

which is obtained from $e = \sqrt{1 - (b/a)^2}$. This gives

$$\text{dist}(P, F^+) = \sqrt{(x - ea)^2 + (y - 0)^2} = \sqrt{x^2 - 2eax + (ea)^2 + y^2}$$

$$= \sqrt{x^2 - 2eax + (ea)^2 + b^2 - \left(\frac{b}{a}\right)^2 x^2} \qquad \text{by (2)}$$

$$= \sqrt{\left[1 - \left(\frac{b}{a}\right)^2\right] x^2 - 2eax + [(ea)^2 + b^2]}$$

$$= \sqrt{e^2 x^2 - 2eax + a^2} \qquad \text{by (3)}$$

$$= \sqrt{(a - ex)^2} = a - ex \qquad (\text{since } a - ex > 0)$$

Thus, $\text{dist}(P, F^+) = a - ex$, and comparing this with (1) gives the equation

$$\text{dist}(P, F^+) = e \, \text{dist}(P \text{ to } l^+)$$

The second equation in the theorem is proved in the same way, except that we use the relations

$$\text{dist}(P \text{ to } l^-) = \frac{a}{e} + x = \frac{1}{e}(a + ex) \qquad (4)$$

and

$$\text{dist}(P, F^-) = \sqrt{(x + ea)^2 + (y - 0)^2}$$

The latter relation can be simplified to

$$\text{dist}(P, F^-) = a + ex \qquad (5)$$

and the desired result follows from (4) and (5).

THEOREM 2
Let

$$\frac{x^2}{a^2} + \frac{y^2}{b^2} = 1$$

be a given ellipse with $a > b > 0$. Then for any point P on the ellipse

$$\text{dist}(P, F^+) + \text{dist}(P, F^-) = 2a$$

Stated verbally, this theorem says that the sum of the distances

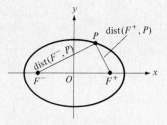

Figure 12.59
Illustrating Theorem 2.

from any point P on the ellipse to its foci F^+ and F^- is constant (see Figure 12.59).

Proof
From the proof of Theorem 1, we have

$$\text{dist}(P, F^+) + \text{dist}(P, F^-) = (a - ex) + (a + ex) = 2a$$

Getting away from ellipses in standard position, we use Theorem 2 to give the following general definition.

> An *ellipse* is the locus of points P the sum of whose distances from two fixed points, F^+ and F^-, is constant.

This definition is explained in Figure 12.60. We begin the discussion of general ellipses with an example of an ellipse whose foci lie on the y axis.

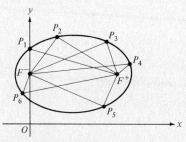

Figure 12.60
The lengths of the line segments joining P_j to F^+ and F^- have the same sum for all P_j.

Example 2

Let us consider the curve

$$\frac{x^2}{3^2} + \frac{y^2}{5^2} = 1$$

We observe that this is the curve of Example 1 with x and y interchanged. This tells us that the curve is an ellipse with the same eccentricity,

$$e = \tfrac{4}{5}$$

foci,

$$F^+ = (0, 4) \quad \text{and} \quad F^- = (0, -4)$$

and directrices,

$$l^+: y = 6.25 \quad \text{and} \quad l^-: y = -6.25$$

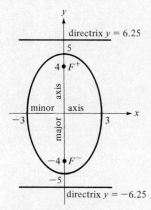

Figure 12.61
The ellipse $\dfrac{x^2}{3^2} + \dfrac{y^2}{5^2} = 1$.

(see Figure 12.61). The major axis is now the line segment joining $(0, 5)$ and $(0, -5)$, and the minor axis is the line segment joining $(-3, 0)$ to $(3, 0)$.

In Table 12.4 we compare the two types of ellipses discussed above.

Table 12.4

	Ellipse $\dfrac{x^2}{a^2} + \dfrac{y^2}{b^2} = 1$	
	$a > b > 0$	$b > a > 0$
center	origin	origin
major axis	$(-a, 0)$ to $(a, 0)$	$(0, b)$ to $(0, -b)$
minor axis	$(0, b)$ to $(0, -b)$	$(a, 0)$ to $(-a, 0)$
eccentricity	$e = \sqrt{1 - (b/a)^2}$	$e = \sqrt{1 - (a/b)^2}$
foci	$F^+ = (ea, 0)$	$F^+ = (0, eb)$
	$F^- = (-ea, 0)$	$F^- = (0, -eb)$
directrices	$l^+: x = a/e$	$l^+: y = b/e$
	$l^-: x = -a/e$	$l^-: y = -b/e$

In discussing general ellipses, we again make use of translations (see Example 1 of Section 12.5). Thus, consider the curve

$$\frac{(x - \alpha)^2}{a^2} + \frac{(y - \beta)^2}{b^2} = 1 \tag{6}$$

Putting $u = x - \alpha$ and $v = y - \beta$, we get

$$\frac{u^2}{a^2} + \frac{v^2}{b^2} = 1$$

which is the equation of an ellipse in a uv coordinate system. Hence, if we superimpose these coordinates on the xy coordinate system so that the origin of the uv coordinates coincides with the point $(x, y) = (\alpha, \beta)$, then we see that the curve (6) is an ellipse that is situated relative to the uv coordinate system with origin at (α, β) as the ellipse

$$\frac{x^2}{a^2} + \frac{y^2}{b^2} = 1$$

is situated relative to the xy coordinate system (see Figure 12.62).

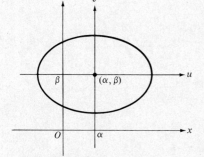

Figure 12.62
The ellipse
$\dfrac{(x - \alpha)^2}{a^2} + \dfrac{(y - \beta)^2}{b^2} = 1$.

Example 3

Describe the ellipse

$$\frac{(x + 5)^2}{5^2} + \frac{(y - 2)^2}{4^2} = 1$$

Solution

Here $\alpha = -5$ and $\beta = 2$, so that the ellipse is situated relative to axes going through the point $(x, y) = (-5, 2)$ in exactly the same way that the ellipse

$$\frac{x^2}{5^2} + \frac{y^2}{4^2} = 1$$

is situated in the xy coordinate system. We can thus write down the information shown in Table 12.5. The two ellipses are given in Figure 12.63.

Table 12.5

Ellipse	$\dfrac{x^2}{5^2} + \dfrac{y^2}{4^2} = 1$	$\dfrac{(x+5)^2}{5^2} + \dfrac{(y-2)^2}{4^2} = 1$
center	$(0, 0)$	$(-5, 2)$
major axis	$(-5, 0)$ to $(5, 0)$	$(-10, 2)$ to $(0, 2)$
minor axis	$(0, 4)$ to $(0, -4)$	$(-5, 6)$ to $(-5, -2)$
eccentricity	$e = \sqrt{1 - (\frac{4}{5})^2} = \frac{3}{5}$	$e = \sqrt{1 - (\frac{4}{5})^2} = \frac{3}{5}$
foci	$F^+ = (3, 0)$	$F^+ = (3-5, 0+2) = (-2, 2)$
	$F^- = (-3, 0)$	$F^- = (-3-5, 0+2) = (-8, 2)$
directrices	$l^+: x = \frac{25}{3}$	$l^+: x = \frac{25}{3} - 5 = \frac{10}{3}$
	$l^-: x = -\frac{25}{3}$	$l^-: x = -\frac{25}{3} - 5 = -\frac{40}{3}$

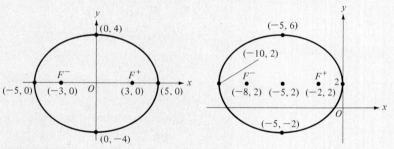

Figure 12.63
The ellipses $(x^2/5^2) + (y^2/4^2) = 1$ and $[(x+5)^2/5^2] + [(y-2)^2/4^2] = 1$.

GENERAL QUADRATIC EQUATIONS
Consider the equation

$$Ax^2 + By^2 + Cx + Dy + E = 0 \tag{7}$$

By completing the square and simplifying, this equation can often be written in the form (6):

$$\frac{(x-\alpha)^2}{a^2} + \frac{(y-\beta)^2}{b^2} = 1$$

When this is the case, all our previous work applies and we can describe the ellipse. We shall have to say more about general equations of the form (7) at the end of the next section.

Example 4

Decide if
$$9x^2 + 4y^2 + 36x - 8y + 4 = 0$$
is the equation of an ellipse. If it is, describe the ellipse.

Solution
Completing the square, we write
$$9(x^2 + 4x + 4) + 4(y^2 - 2y + 1) = -4 + 36 + 4 = 36$$
Hence
$$9(x + 2)^2 + 4(y - 1)^2 = 36$$
and this can be written as
$$\frac{(x + 2)^2}{2^2} + \frac{(y - 1)^2}{3^2} = 1$$

Putting $\alpha = -2$, $\beta = 1$, $a = 2$, and $b = 3$, we can write down the following information:

center	$(-2, 1)$
major axis	$(-2, -2)$ to $(-2, 4)$
minor axis	$(-4, 1)$ to $(0, 1)$
eccentricity	$e = \sqrt{1 - (2/3)^2} = \sqrt{5}/3$
foci	$F^+ = (-2, 1 + \sqrt{5})$
	$F^- = (-2, 1 - \sqrt{5})$
directrices	$l^+: y = 1 + 9\sqrt{5}/5$
	$l^-: y = 1 - 9\sqrt{5}/5$

The graph of the ellipse is given in Figure 12.64.

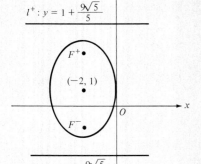

Figure 12.64
The ellipse
$9x^2 + 4y^2 + 36x - 8y + 4 = 0$.

REMARK ON PLANETARY ORBITS
Astronomy provided one of the greatest moving forces in the development of calculus, and some of its most important and profound applications are in this field. We mention here only one application, which is directly related to the material of this section. This is Kepler's first law, which states that planets move in planes that contain the sun, and their orbits are ellipses with the sun at one of their foci.

Table 12.6 shows the eccentricities of the planetary orbits and the lengths of their major axes; the latter are expressed in astronomical units. Note that except for Mercury and Pluto, the eccentricities are very small and hence the orbits of these planets are nearly circular.

Table 12.6
The Eccentricities of Planetary Orbits

Planet	Eccentricity	Length of major axis (astronomical units)[a]
Mercury	0.205	0.774
Venus	0.006	1.446
Earth	0.016	2.000
Mars	0.093	3.046
Jupiter	0.048	10.404
Saturn	0.055	19.108
Uranus	0.046	38.436
Neptune	0.008	60.218
Pluto	0.246	79.200

[a] An *astronomical unit* is defined as the mean distance from earth to sun, about 93 million miles.

EXERCISES

In Exercises 1–10, find the center, major and minor axes, foci, and directrices of the given ellipses and plot their graphs.

1. $x^2 + \dfrac{y^2}{2^2} = 1$

2. $\dfrac{(x-1)^2}{2^2} + y^2 = 1$

3. $\dfrac{(x-1)^2}{3} + \dfrac{(y+1)^2}{4} = 1$

4. $\dfrac{(x-\frac{1}{2})^2}{3} + \dfrac{(y-\frac{1}{4})^2}{4} = 1$

5. $3x^2 + 4y^2 = 1$

6. $x^2 + \frac{1}{4}y^2 - 2x - y + 1 = 0$

7. $x^2 + 2y^2 - 4y + 1 = 0$

8. $x^2 + 3y^2 + 6x - 30y + 81 = 0$

9. $16x^2 + 9y^2 - 32x + 18y - 119 = 0$

10. $x^2 + 4y^2 + x + \frac{8}{3}y - \frac{1}{4} = 0$

11. Find the equation of the tangent to the ellipse $x^2 + y^2/4 = 1$ at $P = (\frac{1}{2}, \sqrt{3})$.

12. Find the two tangents to the ellipse

    $$\dfrac{x^2}{3^2} + \dfrac{y^2}{4^2} = 1$$

    having slope $m = 1$.

13. Find the ellipse having foci $F^+ = (-2, -2)$ and $F^- = (-2, -10)$, and a directrix $y = 0$.

14. An ellipse has eccentricity $e = \frac{1}{2}$ and directrices $l^+: x = -3$, $l^-: x = 3$. Find it.

15. Find two ellipses with foci $F^+ = (3, 0)$ and $F^- = (-3, 0)$ and vertices $(4, 0)$ and $(-4, 0)$.

16. Find an ellipse with vertices $(0, 1)$ and $(0, -1)$, and foci $F^+ = (1, 0)$ and $F^- = (-1, 0)$. Is there more than one ellipse satisfying these conditions?

17. Find the ellipse with one focus at the origin, one directrix $y = 10$, and eccentricity $e = 0.25$.

18. Show that $x = a \cos t$, $y = b \sin t$ $(0 \leq t \leq 2\pi)$ is a parametric representation of the ellipse

$$\frac{x^2}{a^2} + \frac{y^2}{b^2} = 1$$

12.7 HYPERBOLAS

The third of the conic sections is the hyperbola.

> The curve
> $$\frac{x^2}{a^2} - \frac{y^2}{b^2} = 1 \text{ with } a > 0 \text{ and } b > 0 \quad (1)$$
> is called a *hyperbola* in *standard position*.

From this equation we see that the curve, which consists of two branches, is symmetric with respect to both the x axis and the y axis, and hence also with respect to the origin (see Figure 12.65). The origin is called the *center* of the hyperbola in standard position, and the x axis is called its *axis*. The points $(-a, 0)$ and $(a, 0)$ are the *vertices* of the hyperbola.

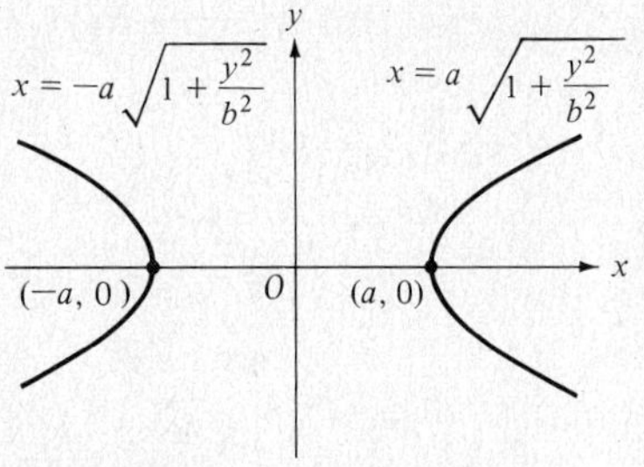

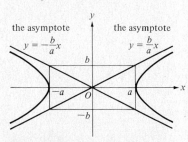

Figure 12.65
A hyperbola $(x^2/a^2) - (y^2/b^2) = 1$ in standard position.

Figure 12.66
The asymptotes to a hyperbola $(x^2/a^2) - (y^2/b^2) = 1$ in standard position.

ASYMPTOTES

In Section 4.3 we discussed asymptotes. We shall show below that the lines

$$y = \frac{b}{a}x \quad \text{and} \quad y = -\frac{b}{a}x$$

are approached by the hyperbola (1) as $x \to \infty$ and $x \to -\infty$ (see Figure 12.66). These lines are called the *asymptotes* of the hyperbola. As shown in Figure 12.66, they form the diagonals of the rectangle $|x| \leq a$, $|y| \leq b$.

The claims just made will be verified for the first quadrant, where the hyperbola is given by the equation

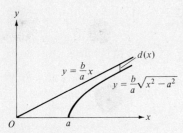

Figure 12.67
The distance between the hyperbola and its asymptote in the first quadrant.

$$y = b\sqrt{\frac{x^2}{a^2} - 1} = \frac{b}{a}\sqrt{x^2 - a^2}$$

Putting

$$d(x) = \frac{b}{a}x - \frac{b}{a}\sqrt{x^2 - a^2} = \frac{b}{a}(x - \sqrt{x^2 - a^2})$$

(see Figure 12.67), we see first that $d(x) > 0$ since $x - \sqrt{x^2 - a^2} > 0$. Hence, the curve lies under the asymptote $y = (b/a)x$. Next, the identity

$$x - \sqrt{x^2 - a^2} = \frac{a^2}{x + \sqrt{x^2 - a^2}}$$

which is verified by multiplying both sides by $x + \sqrt{x^2 - a^2}$ and simplifying, shows that

$$\lim_{x \to \infty} d(x) = \lim_{x \to \infty} \frac{b}{a}[x - \sqrt{x^2 - a^2}]$$

$$= \frac{b}{a} \lim_{x \to \infty} \frac{a^2}{x + \sqrt{x^2 - a^2}} = 0$$

This shows that the line $y = (b/a)x$ is an asymptote as claimed.

As for the ellipse, we define for hyperbola (1) its *eccentricity*, which is given by the number

$$e = \sqrt{1 + \left(\frac{b}{a}\right)^2}$$

Clearly $e > 1$, and from the fact that the hyperbola lies between its asymptotes, it follows that the closer b/a is to zero, the closer e is to 1, and the more elongated is the hyperbola (see Figure 12.68).

We define foci and directrices by formulas that are the same as for the ellipse.

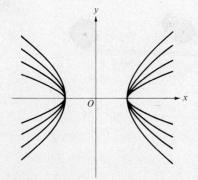

Figure 12.68
Hyperbolas with decreasing eccentricity.

> Let
> $$\frac{x^2}{a^2} - \frac{y^2}{b^2} = 1$$
> be a given hyperbola with $a > 0$ and $b > 0$. The points
> $$F^+ = (ea, 0) \quad \text{and} \quad F^- = (-ea, 0)$$
> are called the *foci* of the hyperbola. The lines
> $$l^+: x = \frac{a}{e} \quad \text{and} \quad l^-: x = -\frac{a}{e}$$
> are called the *directrices* of the hyperbola.

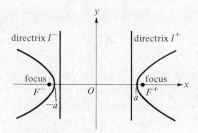

Figure 12.69
The foci and directrices of a hyperbola in standard position.

Since $e > 1$, we see that

$$-ea < -a < -\frac{a}{e} \quad \text{and} \quad \frac{a}{e} < a < ea$$

and these inequalities give the relative positions of the foci, directrices, and vertices (see Figure 12.69).

Example 1
Plot the hyperbola

$$\frac{x^2}{4^2} - \frac{y^2}{3^2} = 1$$

with its foci, directrices, and asymptotes.

Solution
Here $a = 4$, $b = 3$, and hence the asymptotes are

$$y = \tfrac{3}{4}x \quad \text{and} \quad y = -\tfrac{3}{4}x$$

The eccentricity being

$$e = \sqrt{1 + \left(\frac{3}{4}\right)^2} = \sqrt{\frac{4^2 + 3^2}{4^2}} = \frac{5}{4}$$

we find that the foci are

$$F^+ = (\tfrac{5}{4} \cdot 4, 0) = (5, 0) \quad \text{and} \quad F^- = (-\tfrac{5}{4} \cdot 4, 0) = (-5, 0)$$

and the directrices are

$$l^+: x = \frac{4}{\frac{5}{4}} = \frac{16}{5} \quad \text{and} \quad l^-: x = -\frac{4}{\frac{5}{4}} = -\frac{16}{5}$$

The graph in Figure 12.70 is obtained with an appropriate table of values.

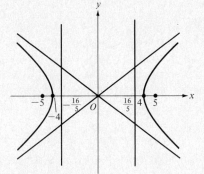

Figure 12.70
The hyperbola $(x^2/4^2) - (y^2/3^2) = 1$.

Theorem 1 of Section 12.6 is also true for hyperbolas:

THEOREM 1
If $P = (x, y)$ is any point on the hyperbola

$$\frac{x^2}{a^2} - \frac{y^2}{b^2}$$

then

$$\text{dist}(P, F^+) = e \,\text{dist}(P \text{ to } l^+)$$
$$\text{dist}(P, F^-) = e \,\text{dist}(P \text{ to } l^-)$$

The proof of this theorem is almost identical to the proof with ellipses; it is therefore omitted. We just remark that in proving the theorem, you obtain the relations

$$\text{dist}(P, F^+) = \pm(ex - a) \quad \text{and} \quad \text{dist}(P, F^-) = \pm(ex + a) \qquad (2)$$

where the $+$ sign is taken on the right-hand branch $x \geq a$, and the $-$ sign is taken on the left-hand branch $x \leq -a$.

The analogue of Theorem 2 of Section 12.6 is the following theorem.

> **THEOREM 2**
> Let
> $$\frac{x^2}{a^2} - \frac{y^2}{b^2} = 1$$
> be a given hyperbola with $a > 0$ and $b > 0$. If P is any point on the right-hand branch, then
> $$\text{dist}(P, F^-) - \text{dist}(P, F^+) = 2a$$
> If P is on the left-hand branch, then
> $$\text{dist}(P, F^+) - \text{dist}(P, F^-) = 2a$$

The proof follows at once from equation (2).
Theorem 2 is used to define a hyperbola in general.

> A *hyperbola* is the locus of points P the difference of whose distances from two fixed points, F^+ and F^-, is constant.

This definition is explained in Figure 12.71.

Example 2

Plot the graph of the hyperbola

$$-\frac{x^2}{4^2} + \frac{y^2}{3^2} = 1$$

with its foci, directrices, and asymptotes.

Solution
If we put

$$y = u \quad \text{and} \quad x = v$$

then the equation of the hyperbola becomes

$$\frac{u^2}{3^2} - \frac{v^2}{4^2} = 1$$

This is the equation of a hyperbola in standard position relative to

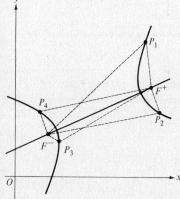

Figure 12.71
The lengths of the line segments joining P_j to F^+ and F^- have the same difference for all points P_j.

12.7 HYPERBOLAS   485

a uv coordinate system. We can thus write down the data as in Table 12.7. We observe that the asymptotes of our hyperbola are the same as those in Example 1. The graph of the hyperbola

$$-\frac{x^2}{4^2} + \frac{y^2}{3^2} = 1$$

is given in Figure 12.72, and in Figure 12.73 this hyperbola is compared with the hyperbola of Example 1.

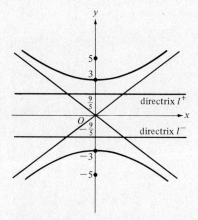

Figure 12.72
The hyperbola $-\frac{x^2}{4^2} + \frac{y^2}{3^2} = 1$.

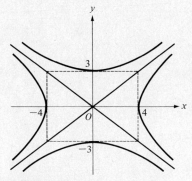

Figure 12.73
The hyperbolas $\frac{x^2}{4^2} - \frac{y^2}{3^2} = 1$ and $-\frac{x^2}{4^2} + \frac{y^2}{3^2} = 1$.

Table 12.7

Hyperbola	$\frac{u^2}{3^2} - \frac{v^2}{4^2} = 1$	$-\frac{x^2}{4^2} + \frac{y^2}{3^2} = 1$
axis	$v = 0$	$x = 0$
vertices	$(3, 0)$ and $(-3, 0)$	$(0, 3)$ and $(0, -3)$
asymptotes	$v = \frac{4}{3}u$	$x = \frac{4}{3}y$ or $y = \frac{3}{4}x$
	$v = -\frac{4}{3}u$	$x = -\frac{4}{3}y$ or $y = -\frac{3}{4}x$
eccentricity	$e = \sqrt{1 + (\frac{4}{3})^2} = \frac{5}{3}$	$e = \sqrt{1 + (\frac{4}{3})^2} = \frac{5}{3}$
foci	$F^+ = (5, 0)$	$F^+ = (0, 5)$
	$F^- = (-5, 0)$	$F^- = (0, -5)$
directrices	$l^+: u = \frac{3}{\frac{5}{3}} = \frac{9}{5}$	$l^+: y = \frac{9}{5}$
	$l^-: u = -\frac{3}{\frac{5}{3}} = -\frac{9}{5}$	$l^-: y = -\frac{9}{5}$

Generalizing the above example, we have the comparison given in Table 12.8.

Table 12.8

Hyperbola	$\frac{x^2}{a^2} - \frac{y^2}{b^2} = 1$	$-\frac{x^2}{a^2} + \frac{y^2}{b^2} = 1$
axis	$y = 0$	$x = 0$
vertices	$(a, 0)$ and $(-a, 0)$	$(0, b)$ and $(0, -b)$
asymptotes	$y = \frac{b}{a}x$	same
	$y = -\frac{b}{a}x$	
eccentricity	$e = \sqrt{1 + \left(\frac{b}{a}\right)^2}$	$e = \sqrt{1 + \left(\frac{a}{b}\right)^2}$
foci	$F^+ = (ea, 0)$	$F^+ = (0, eb)$
	$F^- = (-ea, 0)$	$F^- = (0, -eb)$
directrices	$l^+: x = \frac{a}{e}$	$l^+: y = \frac{b}{e}$
	$l^-: x = -\frac{a}{e}$	$l^-: y = -\frac{b}{e}$

We now consider general hyperbolas

$$\frac{(x-\alpha)^2}{a^2} - \frac{(y-\beta)^2}{b^2} = 1$$

$$-\frac{(x-\alpha)^2}{a^2} + \frac{(y-\beta)^2}{b^2} = 1 \qquad (3)$$

As in the preceding two sections, we make use of translation of coordinates and put $u = x - \alpha$, $v = y - \beta$. This gives the equations

$$\frac{u^2}{a^2} - \frac{v^2}{b^2} = 1 \quad \text{and} \quad -\frac{u^2}{a^2} + \frac{v^2}{b^2} = 1$$

These hyperbolas are the same as those compared in Table 12.7, with u and v replacing x and y. Thus, let us superimpose the uv coordinates on the xy coordinates so that the origin of the former coincides with the point $(x, y) = (\alpha, \beta)$ (see Figure 12.74). Then we see that the hyperbolas (3) are situated relative to the uv coordinate system in the same way that the hyperbolas

$$\frac{x^2}{a^2} - \frac{y^2}{b^2} = 1 \quad \text{and} \quad -\frac{x^2}{a^2} + \frac{y^2}{b^2} = 1$$

are situated relative to the xy coordinate system.

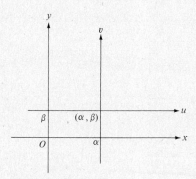

Figure 12.74
The superposition of uv and xy coordinate systems.

Example 3

Describe the hyperbola

$$-\frac{(x-5)^2}{3^2} + \frac{(y+2)^2}{4^2} = 1$$

Solution
Without further ado we can write down the data in Table 12.9. The hyperbola is graphed in Figure 12.75.

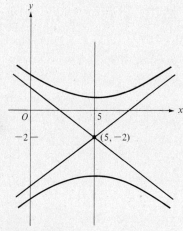

Figure 12.75
The hyperbola
$-[(x-5)^2/3^2] + [(y+2)^2/4^2] = 1$.

Table 12.9

Hyperbola	$-\dfrac{x^2}{4^2} + \dfrac{y^2}{3^2} = 1$	$-\dfrac{(x-5)^2}{4^2} + \dfrac{(y+2)^2}{3^2} = 1$
center	$(0, 0)$	$(5, -2)$
axis	$x = 0$	$x = 5$
vertices	$(0, 3)$ and $(0, -3)$	$(5, 1)$ and $(5, -5)$
asymptotes	$y = \frac{3}{4}x$	$y + 2 = \frac{3}{4}(x - 5)$
	$y = -\frac{3}{4}x$	$y + 2 = -\frac{3}{4}(x - 5)$
eccentricity	$e = \sqrt{1 + (\frac{4}{3})^2} = \frac{5}{3}$	$e = \frac{5}{3}$
foci	$F^+ = (0, 5)$	$F^+ = (0 + 5, 5 - 2) = (5, 3)$
	$F^- = (0, -5)$	$F^- = (0 + 5, -5 - 2) = (5, -7)$
directrices	$l^+: y = \frac{9}{5}$	$l^+: y = -2 + \frac{9}{5} = -\frac{1}{5}$
	$l^-: y = -\frac{9}{5}$	$l^-: y = -2 - \frac{9}{5} = -\frac{19}{5}$

GENERAL QUADRATIC EQUATIONS

We indicated in Section 12.6 that the equation

$$Ax^2 + By^2 + Cx + Dy + E = 0 \qquad (4)$$

may be the equation of an ellipse. Actually, this equation may give any one of the three conic sections, but there are exceptions. These exceptions, which are called *degenerate conics,* are listed in Table 12.10 with some examples.

> **THEOREM 3**
>
> The quadratic equation
>
> $$Ax^2 + By^2 + Cx + Dy + E = 0$$
>
> where not both A and B are zero, represents either a conic or a degenerate conic. When it represents a conic, it gives
>
> 1. a parabola when $AB = 0$,
> 2. an ellipse when $AB > 0$,
> 3. a hyperbola when $AB < 0$.

Condition 1 implies that $A = 0$ or $B = 0$.
Condition 2 implies that A and B have the same sign.
Condition 3 implies that A and B have opposite signs.

This theorem is proved by completing the square and examining the resulting equation (see, in this connection, Example 2 in Section 12.5, and Example 4 in Section 12.6).

Table 12.10
Examples of Degenerate Conics

Description of the degenerate conic	Value of the constants in equation (4)	Equation	Locus
1. Two intersecting lines	$\begin{cases} A = 1, B = -1, \\ C = D = E = 0 \end{cases}$	$x^2 - y^2 = 0$	$y = x$ and $y = -x$
2. Two parallel lines	$\begin{cases} A = 1, B = C = D = 0, \\ E = -1 \end{cases}$	$x^2 - 1 = 0$	$x = 1$ and $x = -1$
3. One line	$\begin{cases} A = 1 \\ B = C = D = E = 0 \end{cases}$	$x^2 = 0$	$x = 0$
4. One point	$\begin{cases} A = B = 1 \\ C = D = E = 0 \end{cases}$	$x^2 + y^2 = 0$	$(0, 0)$
5. No points	$\begin{cases} A = B = E = 1 \\ C = D = 0 \end{cases}$	$x^2 + y^2 + 1 = 0$	no points

Example 4

Decide if the equation

$$9x^2 - 4y^2 + 36x + 8y + 32 = 0$$

represents a conic; if it does, find the conic.

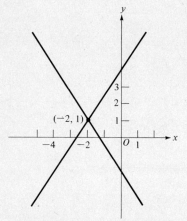

Figure 12.76
The degenerate conic
$9x^2 - 4y^2 + 36x + 8y + 32 = 0$.

Solution
Completing the square reduces this equation to the form
$$9(x+2)^2 - 4(y-1)^2 = 0$$
Hence the equation gives a degenerate conic. Writing
$$(y-1)^2 = \tfrac{9}{4}(x+2)^2$$
and taking square roots gives the two lines
$$y - 1 = \tfrac{3}{2}(x+2) \quad \text{and} \quad y - 1 = -\tfrac{3}{2}(x+2)$$
(see Figure 12.76).

Example 5

Decide if the equation
$$9x^2 - 4y^2 + 36x + 8y + 4 = 0$$
represents a conic; if it does, find the conic.

Solution
Completing the square gives
$$9(x+2)^2 - 4(y-1)^2 = 28$$
or
$$\frac{(x+2)^2}{\frac{28}{9}} - \frac{(y-1)^2}{7} = 1$$

This is the equation of a hyperbola, and we leave it to the reader to plot its graph.

EXERCISES

In Exercises 1–8, find the center, axis, vertices, asymptotes, eccentricity, foci, and directrices of the given hyperbolas.

1. $x^2 - \dfrac{y^2}{2} = 1$

2. $-\dfrac{(x-5)^2}{3^2} + \dfrac{(y+2)^2}{4^2} = 1$

3. $(x-1)^2 - \dfrac{(y+1)^2}{4} = 1$

4. $-\dfrac{(x-1)^2}{4} + \dfrac{(y-1)^2}{25} = 1$

5. $2(x+1)^2 - 4(y-1)^2 = 1$

6. $2x^2 - y^2 + 2y - 3 = 0$

7. $4x^2 - 16y^2 - 4x + 8y - 1 = 0$

8. $-\dfrac{(x-\frac{1}{2})^2}{\frac{1}{4}} + \dfrac{(y+\frac{1}{4})^2}{\frac{1}{16}} = 1$

In Exercises 9–20, decide which equations represent conics. Identify the conics and the degenerate conics and plot their graphs.

9. $y^2 - 2y - 1 = 0$
10. $2x^2 - 3y^2 = 0$
11. $x^2 + y^2 + 2x - 2y + 2 = 0$
12. $-x^2 + 3y^2 + 2x + 12y + 11 = 0$
13. $x^2 + y^2 - x - \frac{1}{2}y + 1 = 0$
14. $3y^2 + \frac{1}{2}x + \frac{1}{2}y - 1 = 0$
15. $-x^2 - \frac{1}{2}y^2 + \frac{1}{4}y - 5 = 0$
16. $\frac{1}{2}x^2 + \frac{1}{2}y^2 - 3x + 6y + 1 = 0$
17. $x^2 - y^2 + x + 2 = 0$
18. $x^2 + y^2 - 2x + 2y + 4 = 0$
19. $4x^2 - x - y + 2 = 0$
20. $2x^2 - 4x + 2 = 0$

21. Find all tangents to the hyperbola
$$\frac{x^2}{3^2} - \frac{y^2}{4^2} = 1$$
that pass through the point $(0, 1)$.

22. Find the two tangents to the hyperbola
$$-\frac{x^2}{3^2} + \frac{y^2}{5^2} = 1$$
having slope -1.

23. Find the hyperbola having asymptotes $y = 3x$ and $y = -3x$, and one focus at $(10, 0)$.

24. Find the hyperbola with center at $(0, 0)$, vertex at $(0, 1)$, and focus at $(0, 2)$.

25. The definition of the hyperbolic sine and cosine is
$$\sinh t = \tfrac{1}{2}(e^t - e^{-t})$$
$$\cosh t = \tfrac{1}{2}(e^t + e^{-t})$$
The equations
$$x = a \cosh t \quad \text{and} \quad y = b \sinh t \qquad -\infty < t < \infty$$
give a parametric equation of one branch of a hyperbola: find its equation in terms of x and y.

12.8 ROTATION OF COORDINATE AXES

In several examples above we have seen how an equation of a curve can be simplified through a translation of the coordinate system. We recall that a translation is a parallel displacement, that is, a displacement resulting in new coordinate axes that are parallel to the old ones. In this section we shall see how equations of curves change when the coordinates are rotated about the origin.

By way of motivation, consider the curve
$$xy = 1 \tag{1}$$

490 PLANE GEOMETRY

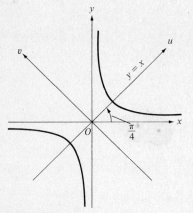

Figure 12.77
The hyperbola $xy = 1$.

(see Figure 12.77). We shall show in Example 2 that this curve, relative to the uv coordinate system in Figure 12.77, has the equation

$$\frac{u^2}{2} - \frac{v^2}{2} = 1 \tag{2}$$

In this situation the u axis coincides with the line $y = x$, and the v axis coincides with the line $y = -x$. We observe that the uv coordinate axes can be obtained from the xy coordinates by a rotation through an angle of $\pi/4$ radians in the counterclockwise direction.

By Section 12.5, equation (2) is a hyperbola with the line $y = x$ as axis. Hence, equation (1) is a hyperbola with axis $y = x$.

In general, consider two rectangular coordinate systems with a common origin and an angle of α radians between the x axis and the u axis (see Figure 12.78). In this case we view the uv coordinates as having been obtained from the xy coordinates by rotating the former through an angle of α radians, the rotation being about the origin. Formulas for the coordinates of a point P relative to the two coordinate systems are given in Theorem 1.

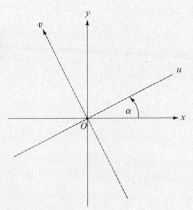

Figure 12.78
xy coordinates and uv coordinates.

THEOREM 1
Let a uv coordinate system be obtained from an xy coordinate system by a counterclockwise rotation about the common origin through an angle of α radians. Then the respective coordinates (x, y) and (u, v) of any point P in the plane are related through the following equations:

$$\left. \begin{array}{l} u = x \cos \alpha + y \sin \alpha \\ v = -x \sin \alpha + y \cos \alpha \end{array} \right\} \tag{3}$$

$$\left. \begin{array}{l} x = u \cos \alpha - v \sin \alpha \\ y = u \sin \alpha + v \cos \alpha \end{array} \right\} \tag{4}$$

12.8 ROTATION OF COORDINATE AXES

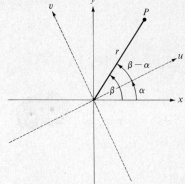

Figure 12.79   **Figure 12.80**

Proof

Consult Figure 12.79. Let P have coordinates (x, y) and (u, v) in the respective coordinate systems; designate the distance from O to P by r. Finally, let α and β be as designated in Figure 12.80. Then

$$u = r\cos(\beta - \alpha)$$
$$v = r\sin(\beta - \alpha)$$
$$x = r\cos\beta$$
$$y = r\sin\beta$$

We have

$$r\cos(\beta - \alpha) = r(\cos\beta\cos\alpha + \sin\beta\sin\alpha)$$
$$= (r\cos\beta)\cos\alpha + (r\sin\beta)\sin\alpha = x\cos\alpha + y\sin\alpha$$
$$r\sin(\beta - \alpha) = r(\sin\beta\cos\alpha - \cos\beta\sin\alpha)$$
$$= (r\sin\beta)\cos\alpha - (r\cos\beta)\sin\alpha = y\cos\alpha - x\sin\alpha$$

Hence we get the equations in (3). Also,

$$r\cos\beta = r\cos[(\beta - \alpha) + \alpha]$$
$$= r[\cos(\beta - \alpha)\cos\alpha - \sin(\beta - \alpha)\sin\alpha]$$
$$= [r\cos(\beta - \alpha)]\cos\alpha - [r\sin(\beta - \alpha)]\sin\alpha$$
$$= u\cos\alpha - v\sin\alpha$$

$$r\sin\beta = r\sin[(\beta - \alpha) + \alpha]$$
$$= r[\sin(\beta - \alpha)\cos\alpha + \cos(\beta - \alpha)\sin\alpha]$$
$$= [r\sin(\beta - \alpha)]\cos\alpha + [r\cos(\beta - \alpha)]\sin\alpha$$
$$= v\cos\alpha + u\sin\alpha$$

These relations gives us the equations in (4).

Example 1

When $\alpha = \pi/4$, then $\cos\alpha = \sin\alpha = \sqrt{2}/2$. The uv coordinates of a point $P = (x, y)$ are therefore given by the equations

$$u = \frac{\sqrt{2}}{2}(x+y)$$
$$v = \frac{\sqrt{2}}{2}(-x+y)$$
(5)

The xy coordinates, expressed in terms of u and v, are given by the equations

$$x = \frac{\sqrt{2}}{2}(u-v)$$
$$y = \frac{\sqrt{2}}{2}(u+v)$$
(6)

Example 2

Find the equation of the hyperbola $xy = 1$ in a uv coordinate system obtained through a counterclockwise rotation through an angle of $\alpha = \pi/4$ radians.

Solution
Using the equations in (6) gives

$$xy = \frac{\sqrt{2}}{2}(u-v)\frac{\sqrt{2}}{2}(u+v) = \frac{1}{2}(u^2 - v^2) = \frac{u^2}{2} - \frac{v^2}{2}$$

Hence, as claimed at the beginning of the section, the equation $xy = 1$ transforms into the equation

$$\frac{u^2}{2} - \frac{v^2}{2} = 1$$

Rotations of coordinates play an important part in studying the general quadratic equation

$$Ax^2 + Bxy + Cy^2 + Dx + Ey + F = 0 \qquad (7)$$

In Section 12.7 we have learned how to handle such equations when $B = 0$. With an appropriate rotation of coordinates, equation (7) can be transformed to an equation of the form

$$\bar{A}u^2 + \bar{C}v^2 + \bar{D}u + \bar{E}v + \bar{F} = 0 \qquad (8)$$

and we know how to determine the conic given by (8). When $A = C$ in (7), then the angle of rotation can be taken to be $\pi/4$, as already illustrated in Example 2 (see also Example 3). The case $A \neq C$ involves complicated calculations, and it will not be taken up here; the angle of rotation α can be determined from the equation

$$\cot 2\alpha = \frac{A-C}{B}$$

Example 3

What curve is given by the equation

$$x^2 + xy + y^2 = 3$$

Solution
We again use equations (6). Since

$$x^2 = \tfrac{1}{2}(u^2 - 2uv + v^2)$$
$$y^2 = \tfrac{1}{2}(u^2 + 2uv + v^2)$$
$$xy = \tfrac{1}{2}(u^2 - v^2)$$

we find that

$$x^2 + xy + y^2 = \tfrac{1}{2}(3u^2 + v^2)$$

Since $x^2 + xy + y^2 = 3$, we get the equation $3u^2 + v^2 = 6$, or

$$\frac{u^2}{2} + \frac{v^2}{6} = 1$$

This, as we know, is the equation of an ellipse (see Figure 12.81).

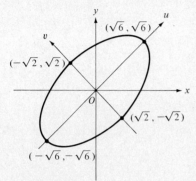

Figure 12.81
The ellipse $x^2 + xy + y^2 = 3$.

Example 4

Find which conic is given by the equation

$$-\tfrac{1}{2}x^2 + xy - \tfrac{1}{2}y^2 + \sqrt{2}x + \sqrt{2}y - 1 = 0$$

Solution
Once more we use equations (6). With these, our equation becomes

$$-\tfrac{1}{2}[\tfrac{1}{2}(u^2 - 2uv + v^2)] + \tfrac{1}{2}(u^2 - v^2) - \tfrac{1}{2}[\tfrac{1}{2}(u^2 + 2uv + v^2)]$$
$$+ (u - v) + (u + v) - 1 = 0$$

When simplified, this equation becomes

$$-v^2 + 2u - 1 = 0$$

This, as we know, is the equation of a parabola (see Figure 12.82).

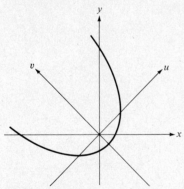

Figure 12.82
The parabola $-v^2 + 2u - 1 = 0$.

EXERCISES

In the exercises below, identify the conics and plot their graphs.

1. $x^2 + 2xy + y^2 + 2\sqrt{2}x - 2\sqrt{2}y + 1 = 0$
2. $34x^2 + 32xy + 34y^2 - 450 = 0$
3. $\dfrac{1}{2}x^2 - xy + \dfrac{1}{2}y^2 + \dfrac{\sqrt{2}}{2}x + \dfrac{\sqrt{2}}{2}y + 1 = 0$
4. $7x^2 + 2xy + 7y^2 - 24 = 0$

5. $7x^2 - 50xy + 7y^2 - 288 = 0$

6. $\frac{1}{2}x^2 - xy + \frac{1}{2}y^2 + \sqrt{2}x - \sqrt{2}y - 1 = 0$

7. $3x^2 + 2xy + 3y^2 + \frac{\sqrt{2}}{2}x - \frac{\sqrt{2}}{2}y - 20 = 0$

8. $-7x^2 + 50xy - 7y^2 - 288 = 0$

9. $x^2 - 2xy + y^2 + 3x - 5y + 5\sqrt{2} = 0$

10. $2x^2 + 4xy + 2y^2 - \sqrt{2}y + 2 = 0$

11. $x^2 + y^2 - 2\sqrt{2}x - 1 = 0$

12. $x^2 + 6xy + y^2 - 4 = 0$

13. $3x^2 - 10xy + 3y^2 = 0$

14. $x^2 + 2xy + y^2 + 2x - y = 0$

15. $x^2 + 4xy = 0$

16. Determine the conic given by the equation $17x^2 - 12xy + 8y^2 - 80 = 0$ by rotating the coordinate axes through an angle α such that $\cos \alpha = \sqrt{5}/5$ and $\sin \alpha = 2\sqrt{5}/5$.

QUIZ 1

1. Below is a curve in parametric form. Write it in the form $y = f(x)$ or $x = g(y)$.

    $x = t$
    $y = t^{3/2}$

2. Plot the curve

    $x = \frac{3}{4} + \frac{1}{4} \cos 2t$
    $y = \sin t$

3. Find the length of the curve

    $x = \sin t$
    $y = \sin^{3/2} t$ $\quad 0 \le t \le \frac{\pi}{2}$

4. Give three polar coordinates representing the points $(1, -1)$ and $(\sqrt{2}, \sqrt{2})$.

5. Plot the curve $r = 4 \sin 6\theta$ given in polar form.

6. Find the area bounded by the curve $r = 4 \sin 6\theta$.

7. Find the axis, vertex, focus, and directrix of the parabola $x = y^2 - 4y + 4$ and plot its graph.

8. Find and graph the parabola having focus at $(-2, -4)$ and directrix $y = 1$.

9. Find the center, major and minor axes, foci, and directrices of the ellipse

$$\frac{(x+1)^2}{2} + \frac{(y-2)^2}{4} = 1$$

10. Find the ellipse having foci $F^+ = (4, -4)$ and $F^- = (-4, -4)$ and a directrix $x = 5$.

11. Find the center, axis, vertices, asymptotes, eccentricity, foci, and directrices of the hyperbola

$$3x^2 - 2y^2 - 6x - 4y - 5 = 0$$

and plot its graph.

12. Identify the following conics and plot their graphs.

  (a) $7x^2 + 2xy + 7y^2 - 24 = 0$

  (b) $x^2 + y^2 - 2x + 6y + 1 = 0$

QUIZ 2

1. Below is a curve in parametric form. Write it in the form $y = f(x)$ or $x = g(y)$.

   $x = \sin 2t$
   $y = \cos 2t$

2. Plot the curve

   $x = 2 \sin t$
   $y = \sin 2t$

3. Find the length of the curve

   $\begin{array}{l} x = (1-t)^3 \\ y = (1+t)^3 \end{array} \quad -1 \leq t \leq 1$

4. Give three polar coordinates representing the points $(1, \sqrt{3})$ and $(-1, 0)$.

5. Plot the curve $r = e^{2\theta}$ given in polar form.

6. Plot the curve $r = 2(1 - \cos \theta)$ given in polar form.

7. Find the area bounded by the curve $r = 2(1 - \cos \theta)$.

8. Find the axis, vertex, focus, and directrix of the parabola $y = 4x^2 - 4x + 2$ and plot its graph.

9. Find the two parabolas that intersect in the points $(0, -4)$ and $(0, 4)$ and have a common focus at $F = (0, 0)$.

10. Find the center, major and minor axes, foci, directrices, and eccentricity of the ellipse

$$\frac{(x+2)^2}{9} + (y-4)^2 = 1$$

and plot its graph.

11. Find the ellipse with one focus at $(4, 4)$, one directrix $y = 16$, and eccentricity $e = 0.16$.

12. Find the center, axis, vertices, asymptotes, eccentricity, foci, and directrices of the hyperbola

$$-\frac{(x+5)^2}{25} + (y+5)^2 = 1$$

and plot its graph.

13. Identify the conics and degenerate conics below and plot their graphs.

   (a) $x^2 - 4x - 1 = 0$

   (b) $x^2 + y^2 - 9x + 9y + 1 = 0$

   (c) $4x^2 - 2xy + 4y^2 - 100 = 0$

CHAPTER 13
FUNCTIONS OF SEVERAL VARIABLES

13.1 COORDINATES IN THREE DIMENSIONS

We recall that the graph of a function $y = f(x)$ of a single independent variable was obtained by plotting points (x, y) in the plane. For this purpose we introduced a coordinate system. In this chapter we shall deal with functions of two or more independent variables. The graph of a function $z = f(x, y)$ of the two independent variables x and y is obtained by plotting points (x, y, z) in three-dimensional space, and for this we need a three-dimensional coordinate system. Functions of more than two independent variables cannot be graphed.

The coordinate system constructed for the plane is augmented by erecting through the origin O a line that is perpendicular to the plane containing the x and y axes (see Figure 13.1). This new axis, usually designated the z axis, is thus perpendicular to both the x axis and the y axis. A unit of length is marked off on the z axis as shown in Figure 13.1. The xyz coordinates are always read in the order

$$x \to y \to z$$

that is, we read the coordinates from left to right and then up. With this scheme every point P in three-dimensional space is uniquely identified by an ordered triple (x, y, z). The information in Figure 13.2 will be found useful in locating points and in visualizing graphs in three-dimensional space.

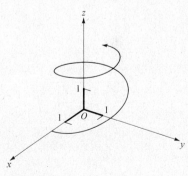

Figure 13.1
The xyz coordinate system.

498 FUNCTIONS OF SEVERAL VARIABLES

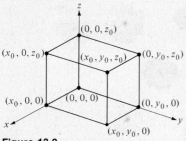

Figure 13.2
Location of points in space.

Point	Location of point
$(0, 0, 0)$	origin of coordinate system
$(x_0, 0, 0)$	x axis
$(0, y_0, 0)$	y axis
$(0, 0, z_0)$	z axis
$(x_0, y_0, 0)$	xy plane
$(x_0, 0, z_0)$	xz plane
$(0, y_0, z_0)$	yz plane

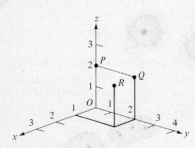

Figure 13.3
The coordinate planes.

The xy plane, xz plane, and yz plane referred to in Figure 13.2 are called coordinate planes. Thus, a *coordinate plane* is a plane which contains two coordinate axes. These planes are mutually perpendicular because so are the coordinate axes. Observe that any point in the xy plane, for example, is identified by a tuple $(x, y, 0)$. In this 3-tuple, x and y are unrestricted and hence the condition $z = 0$ identifies the xy plane. Similarly, the condition $y = 0$ identifies the xz plane, and $x = 0$ identifies the yz plane.

Example 1

Plot the points $P = (0, 0, 2)$, $Q = (0, 2, 2)$, and $R = (1, 2, 2)$.

Solution
Consult Figure 13.4. The presence of the two zeros in $(0, 0, 2)$ tells us that P lies on the z axis. Since $2 > 0$, we know that P lies 2 units above the xy plane. The zero in $(0, 2, 2)$ tells us that Q lies in the yz plane. It lies 2 units above the point $(0, 2, 0)$.

To plot R, first locate the point $(1, 2, 0)$ in the xy plane, then plot R on the perpendicular through this point 2 units above the plane.

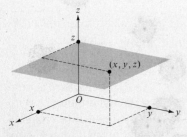

Figure 13.4

Example 2

Find the equation of the plane that is parallel to the xy plane and is 2 units above it in the direction of the positive z axis.

Solution
For a point (x, y, z) to lie 2 units above the xy plane, we must fix $z = 2$, but x and y are not restricted. Hence, the plane we are considering consists of all points $(x, y, 2)$, and it is therefore identified by the condition $z = 2$ (see Figure 13.5).

Figure 13.5
The first quadrant of the plane $z = 2$.

In general, the planes that are parallel to one of the coordinate planes can be described as in Table 13.1. These planes are depicted in Figure 13.6. Notice, however, that only a portion of each plane is drawn.

Table 13.1

Equation	Plane
$x = x_0$	parallel to the yz plane and containing the point $(x_0, 0, 0)$
$y = y_0$	parallel to the xz plane and containing the point $(0, y_0, 0)$
$z = z_0$	parallel to the xy plane and containing the point $(0, 0, z_0)$

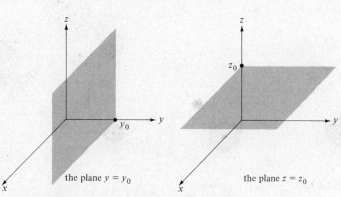

Figure 13.6
Planes parallel to the coordinate planes.

THE DISTANCE FORMULA

In analogy to the distance formula in Section 1.6, we derive here a formula for the distance between two points in three-dimensional space.

Consider points $P = (x_1, y_1, z_1)$ and $Q = (x_2, y_2, z_2)$ as in Figure 13.7. The points P, Q, and R are the vertices of a right triangle, and the theorem of Pythagoras tells us that

$$\text{dist}(P, Q) = \sqrt{[\text{dist}(P, R)]^2 + [\text{dist}(R, Q)]^2} \tag{1}$$

We note in Figure 13.7 that

$$\text{dist}(P, R) = |z_2 - z_1|$$

so that

$$[\text{dist}(P, R)]^2 = (z_2 - z_1)^2 \tag{2}$$

In addition,

$$\text{dist}(R, Q) = \text{dist}(R', Q')$$

and the theorem of Pythagoras is used again to get

$$[\text{dist}(R', Q')]^2 = (x_2 - x_1)^2 + (y_2 - y_1)^2 \tag{3}$$

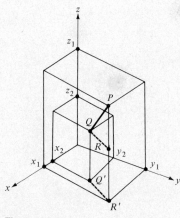

Figure 13.7
The distance between P and Q.

Substituting (2) and (3) into (1) gives the distance formula

$$\text{dist}(P, Q) = \sqrt{(x_2 - x_1)^2 + (y_2 - y_1)^2 + (z_2 - z_1)^2} \qquad (4)$$

Example 3

Find the distance between $P = (1, -1, 3)$ and $Q = (-1, -1, 2)$.

Solution

$$\text{dist}(P, Q) = \sqrt{[1 - (-1)]^2 + [-1 - (-1)]^2 + [3 - 2]^2}$$
$$= \sqrt{4 + 0 + 1} = \sqrt{5}$$

SPHERES

A sphere in three-dimensional space is the locus of all points that are equidistant from a fixed point, called its *center*. This suggests that the distance formula can be used to obtain the formula for any sphere.

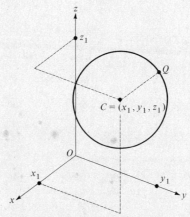

Figure 13.8
The sphere with center C and radius r.

Given a point $C = (x_1, y_1, z_1)$ and a number $r > 0$, then the sphere of radius r and center C is the set of all points $Q = (x, y, z)$ for which $\text{dist}(Q, C) = r$ (see Figure 13.8). Replacing (x_2, y_2, z_2) by (x, y, z) in (4) and squaring both sides gives

$$(x - x_1)^2 + (y - y_1)^2 + (z - z_1)^2 = r^2 \qquad (5)$$

and this is the desired equation of the sphere.

Example 4

Find the center and radius of the sphere $x^2 + y^2 + z^2 = 4$.

Solution

Our equation can be written in the form

$$(x-0)^2 + (y-0)^2 + (z-0)^2 = 2^2$$

Comparing this with formula (5) shows that the sphere has center $C = (0, 0, 0)$ and radius $r = 2$.

Example 5

Find the equation of the sphere having center at $C = (0, 2, -1)$ and radius 5.

Solution

Substituting $(x_1, y_1, z_1) = (0, 2, -1)$ and $r = 5$ into formula (5) gives

$$(x-0)^2 + (y-2)^2 + [z-(-1)]^2 = 5^2$$

or

$$x^2 + (y-2)^2 + (z+1)^2 = 5^2$$

EXERCISES

In Exercises 1–4, find the distance between P and Q.

1. $P = (0, 0, 1)$; $Q = (1, 0, 1)$
2. $P = \left(\dfrac{1}{2}, \dfrac{1}{6}, \dfrac{\sqrt{34}}{12}\right)$; $Q = \left(\dfrac{1}{3}, 0, \dfrac{\sqrt{34}}{12}\right)$
3. $P = (-1, -1, -1)$; $Q = (1, 1, 1)$
4. $P = \left(\dfrac{1}{5}, \dfrac{2}{5}, \dfrac{1}{2}\right)$; $Q = \left(\dfrac{1}{5}, \dfrac{1}{5}, \dfrac{1}{2}\right)$
5. Find the equation of the sphere with center at $(-1, 2, -3)$ and radius 1.
6. Find the center and radius of the sphere
$(x-2)^2 + y^2 + (z+2)^2 = 100$.
7. Does the sphere $(x - \frac{1}{2})^2 + y^2 + z^2 = \frac{1}{4}$ lie *inside* the sphere $x^2 + y^2 + z^2 = 1$? Justify your answer.
8. Does the sphere $(x-1)^2 + y^2 + z^2 = \frac{1}{16}$ lie *outside* the sphere $x^2 + y^2 + z^2 = 1$? Justify your answer.
9. The order in which the entries x_0, y_0, and z_0 are listed in the formula $P(x_0, y_0, z_0)$ is important. To see why this is so, plot the points

$$(0, 1, 2) \quad (0, 2, 1) \quad (1, 0, 2) \quad (2, 0, 1) \quad (1, 2, 0) \quad (2, 1, 0)$$

on Figure 13.9.

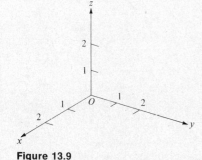

Figure 13.9

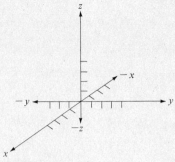

Figure 13.10

10. Plot the following points on Figure 13.10.

$(1, 1, 0)$	$(-2, 0, 0)$	$(-1, -2, 0)$
$(2, 0, 2)$	$(0, -2, 0)$	$(0, 0, -3)$
$(0, -1, -1)$	$(-1, 1, 0)$	$(3, 0, 3)$

13.2 FUNCTIONS OF TWO VARIABLES

In Section 2.1 a function was thought of as a relation that converts one variable quantity into another; we had one independent variable and one dependent variable. We now consider functions that depend on two independent variable quantities. For example, the distance d from a point (x, y) in the plane to the origin is given by the formula

$$d = \sqrt{x^2 + y^2}$$

The equation determines a function of the two independent variables x and y; d is the dependent variable.

A somewhat more formal approach is to think of a function of two variables as a relation that assigns a unique number to each point (x, y) in the plane. Often a function is defined only for points (x, y) belonging to a particular set D in the plane, called the *domain* of the function. All this is explained in the following examples. Functions of three or more variables will be discussed in Section 13.6.

Example 1

Consider a rectangle with sides x and y (see Figure 13.11). Its area is given by the formula

$$A(x, y) = xy$$

which specifies a function of two independent variables. It is defined only for values $x \geq 0$ and $y \geq 0$, because x and y represent length. Hence, the *domain* of the function A is the first quadrant

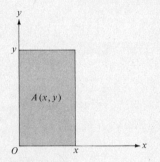

Figure 13.11

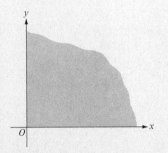

Figure 13.12
The domain of the function $A(x, y)$.

Figure 13.13

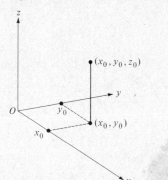

Figure 13.14

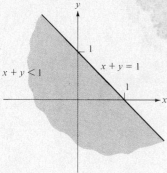

Figure 13.15
The domain $x + y \leq 1$.

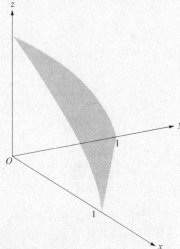

Figure 13.16
The graph of $f(x, y) = \sqrt{1 - x - y}$ for $x \geq 0$ and $y \geq 0$.

(see Figure 13.12). Each point in the domain determines the unique number $z = A(x, y)$. Thus,

$$A(0, 1) = 0 \cdot 1 = 0$$
$$A(2, 3) = 2 \cdot 3 = 6$$

and so on. The diagram in Figure 13.13 can be used to convey the idea that the function A converts the point (x, y) into the number $z = A(x, y)$. This is also expressed symbolically as

$$(x, y) \to A(x, y)$$

If (x_0, y_0) is in the domain of A, then the number $z_0 = A(x_0, y_0)$, which is nonnegative in this example, can be placed z_0 units above the xy plane on the line through (x_0, y_0) that is perpendicular to the plane (see Figure 13.14). The graph of $z = A(x, y)$ is the locus of points $(x, y, z) = (x, y, A(x, y))$ for which (x, y) lies in the domain of A. This graph is given in Figure 13.31 accompanying Example 1 of Section 13.3.

Example 2

Consider the relation

$$z = \sqrt{1 - x - y} \tag{1}$$

To determine the domain of this relation, we observe that the square root of a negative number is not a real number. Hence, equation (1) is defined only when $1 - x - y \geq 0$, or $x + y \leq 1$ (see Figure 13.15). To each point (x_0, y_0) in the domain there corresponds exactly one number $z_0 = \sqrt{1 - x_0 - y_0}$. The relation (1) is, therefore, a function of the variables x and y; a portion of the graph is given in Figure 13.16. The functional notation in this case is

$$f(x, y) = \sqrt{1 - x - y}$$

Thus,

$$f(0, 0) = \sqrt{1 - 0 - 0} = 1$$
$$f(1, 0) = \sqrt{1 - 1 - 0} = 0$$
$$f(\tfrac{1}{2}, \tfrac{1}{4}) = \sqrt{1 - \tfrac{1}{2} - \tfrac{1}{4}} = \tfrac{1}{2}$$

and so on.

Example 3

Consider the function

$$f(x, y) = \sqrt{y - x}$$

Its domain is the set of points (x, y) for which $y - x \geq 0$ (see Figure 13.17). This function is graphed in Figure 13.18.

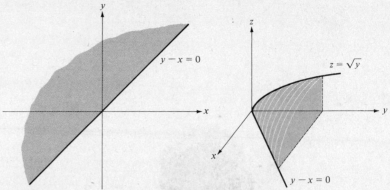

Figure 13.17
The domain $y - x \geq 0$.

Figure 13.18
The graph of $f(x, y) = \sqrt{y - x}$ for $y \geq 0$.

The set of values $f(x, y)$ of a given function f is called its *range*. For example, the range of each of the functions in Examples 1–3 is seen to be interval $[0, \infty)$.

LEVEL CURVES

From the preceding examples it is apparent that drawing three-dimensional surfaces in perspective is rather difficult. A two-dimensional representation of functions of two variables that is quite simple involves the concept of level curves:

> **DEFINITION 1**
> A *level curve* of $z = f(x, y)$ is a set of points (x, y) in the xy plane for which $f(x, y) = $ constant.

In maps, level curves are used to mark points of equal elevation above sea level.

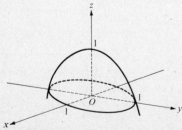

Figure 13.19
The graph of $f(x, y) = 1 - x^2 - y^2$.

Example 4

Consider the function

$$f(x, y) = 1 - x^2 - y^2$$

(see Figure 13.19). The level curve $f(x, y) = \frac{3}{4}$ is the curve

$$1 - x^2 - y^2 = \tfrac{3}{4}$$

or

$$x^2 + y^2 = \tfrac{1}{4}$$

13.2 FUNCTIONS OF TWO VARIABLES

Thus, the level curve is a circle with center at the origin and radius $\frac{1}{2}$. This curve is the projection, or "shadow," of the intersection of the surface $z = f(x, y)$ with the plane $z = \frac{3}{4}$ (see Figure 13.20). Such intersections are called *contour curves*. Different level curves and the corresponding contour curves are given in Figure 13.21.

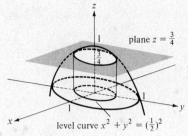

Figure 13.20
The intersection of $f(x, y) = 1 - x^2 - y^2$ with the plane $z = \frac{3}{4}$ and the corresponding level curve.

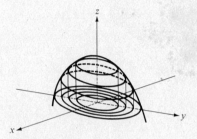

Figure 13.21
Contour curves and level curves of $f(x, y) = 1 - x^2 - y^2$.

Example 5

Let
$$f(x, y) = \frac{1 - x^2}{1 + y} \qquad y \geq 0$$

What are the level curves $f(x, y) = c$ for the values
$$c = 0, \tfrac{1}{5}, \tfrac{2}{5}, \tfrac{3}{5}, \tfrac{4}{5}, 1$$

Solution
$$f(x, y) = \frac{1 - x^2}{1 + y} = 0$$

only when $1 - x^2 = 0$. This is true for $x = -1$ and $x = 1$. Hence, the level curve $f(x, y) = 0$ consists of two parallel lines (see Figure 13.22). For $c \neq 0$, we have $1 - x^2 = c(1 + y)$, or
$$y = \frac{1 - c}{c} - \frac{1}{c} x^2$$

These curves, which are parabolas, are given in Figure 13.23 for the given values of c. Note that the positions of the x and y axes are not the customary ones. The axes have been so positioned that the level curves can easily be compared with the corresponding contour lines on the graph in Figure 13.24.

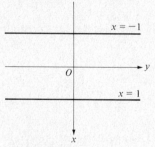

Figure 13.22
The level curve $(1 - x^2)/(1 + y) = 0$.

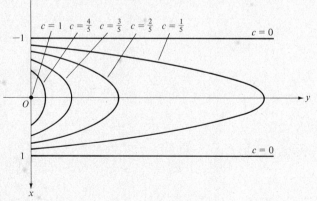

Figure 13.23
Level curves $(1 - x^2)/(1 + y) = c$ for $y \geq 0$.

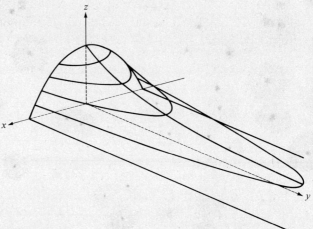

Figure 13.24
The graph of $f(x, y) = (1 - x^2)/(1 + y)$ for $y \geq 0$ with contour lines.

Example 6

Let
$$f(x, y) = x^2 - 4y^2$$

Find the level curves $f(x, y) = c$ for $c = 0, \pm 1, \pm 2, \pm 3$.

Solution
For $c = 0$, we have
$$x^2 - 4y^2 = (x - 2y)(x + 2y) = 0$$

This gives the two straight lines
$$y = \tfrac{1}{2}x \quad \text{and} \quad y = -\tfrac{1}{2}x$$

(see Figure 13.25). For $c \neq 0$, we have
$$x^2 - 4y^2 = c$$

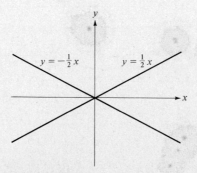

Figure 13.25
The level curve $x^2 - 4y^2 = 0$.

and from Section 12.7 we know that this equation specifies a hyperbola for each value of c. The level curves are given in Figure 13.26. The graph of the function, known as a *hyperbolic paraboloid*, is given in Figure 13.27.

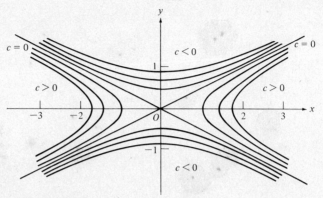

Figure 13.26
The level curves $x^2 - 4y^2 = c$.

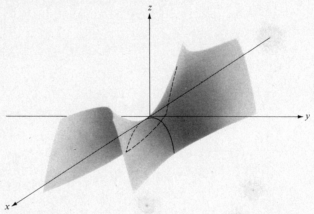

Figure 13.27
The graph of $f(x, y) = x^2 - 4y^2$.

EXERCISES

Find the domain and range of each of the functions in Exercises 1–11, and plot the level curves $f(x, y) = c$ for the given values of c.

1. $f(x, y) = xy$; $c = \pm 1$
2. $f(x, y) = x + y$; $c = 0, \pm 1$
3. $f(x, y) = \dfrac{x}{y}$; $c = 0, \pm \dfrac{1}{2}$
4. $f(x, y) = x^2 y$; $c = \pm \dfrac{1}{4}$
5. $f(x, y) = x^2 + y^2$; $c = 0, 1, 4$
6. $f(x, y) = \dfrac{x^2}{4} + \dfrac{y^2}{9}$; $c = 0, 1$

7. $f(x,y) = -\dfrac{x^2}{4} + \dfrac{y^2}{9}$; $c = \pm\dfrac{1}{9}, \pm 1$ 8. $f(x,y) = 1 - x - y$; $c = 0, \pm 2$

9. $f(x,y) = \dfrac{1-x}{1+y^2}$; $c = 0, \pm 2$ 10. $f(x,y) = \dfrac{1-yx}{1+xy}$; $c = 1, 2$

11. $f(x,y) = 1 - x^2$; $c = 0, \pm 1$

13.3 CROSS SECTIONS AND SURFACES

The level curves discussed in the preceding section were obtained from the intersection of a surface with horizontal planes. A good idea of the shape of a surface is often obtained from intersections with vertical planes.

Consider a surface $z = f(x, y)$. We realize that if $y = y_0$ is held constant, then $f(x, y_0)$ becomes a function of x alone. The graph of $z = f(x, y_0)$ is a curve in the plane $y = y_0$ that is the intersection of the surface $z = f(x, y)$ with that plane (see Figure 13.28). Such an intersection is called a *cross section* of $z = f(x, y)$, and a good idea of the shape of the graph is obtained by drawing several cross sections, $z = f(x, y_0)$, $z = f(x, y_1)$, and so on. The same can be accomplished with cross sections $z = f(x_0, y)$ obtained when x is held constant (see Figure 13.29). The curve $z = f(x_0, y)$ lies in the plane $x = x_0$, and it can be thought of as the intersection of the surface $z = f(x, y)$ with that plane. These ideas are now illustrated with some examples.

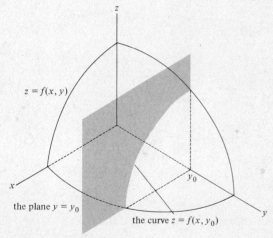
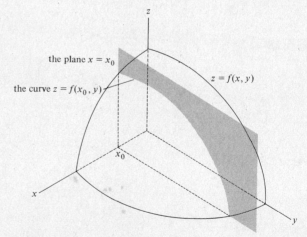

Figure 13.28
The cross section $z = f(x, y_0)$.

Figure 13.29
The cross section $z = f(x_0, y)$.

Example 1

Let us attempt to get an idea of the shape of the surface $z = xy$ for $x > 0$ and $y > 0$ (see Example 1 of Section 13.2). The cross sec-

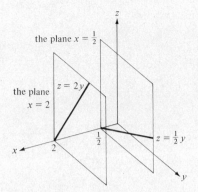

Figure 13.30
Two cross sections of $z = x \cdot y$.

tions of this surface are straight lines, $z = x_0 y$ and $z = x y_0$, with slopes x_0 and y_0, respectively (see Figure 13.30). Numerous cross sections have been drawn in Figure 13.31. The planes of intersection $x = x_0$ have not been drawn in order to give a better picture of the emerging graph.

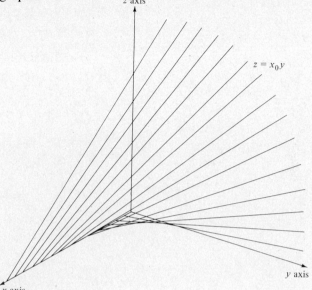

Figure 13.31
Cross section of the surface $z = xy$.

Example 2

Consider the function $z = \sqrt{1 - x^2}$. The absence of y from the formula tells us that the intersection of this graph with each plane $y = y_0$ is the same (see Figure 13.32). The cross section with each plane $x = x_0$ is the horizontal line $z = \sqrt{1 - x_0^2} = $ constant (see Figure 13.33).

Observe that the function $z = \sqrt{1 - x^2}$ gives the upper half of a cylinder. It can be generated by taking lines parallel to the y axis and passing through the semicircle $z = \sqrt{1 - x^2}$ in the plane $y = 0$ (the xz plane).

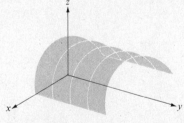

Figure 13.32
Cross sections $z = \sqrt{1 - x^2}$, $y = y_0$.

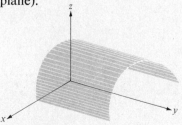

Figure 13.33
Cross sections $z = \sqrt{1 - x_0^2}$.

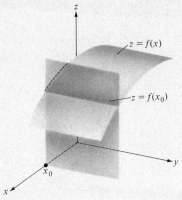

Figure 13.34
A surface $z = f(x)$.

In general, the equation $z = f(x)$ describes a surface whose cross sections with planes $x = x_0$ are the lines given by $z = f(x_0)$ parallel to the y axis as in Example 2; the surface $z = f(x)$ can be thought of as generated by moving lines parallel to the y axis along the curve $z = f(x)$ in the xz plane (see Figure 13.34). Likewise, the surface $z = f(y)$ is generated by lines parallel to the x axis, and the surface $y = f(x)$ is generated by lines parallel to the z axis. A surface generated by parallel lines is called a *cylindrical surface*.

Example 3

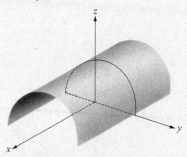

Figure 13.35
Cross sections of the surface $z = \sqrt{1 - y^2}$.

The cross section of the surface $z = \sqrt{1 - y^2}$ in the plane $x = 0$ is a semicircle. The surface (see Figure 13.35) is generated by lines parallel to the x axis and passing through this semicircle. For comparison, the surface $y = \sqrt{1 - x^2}$ (see Figure 13.36) is generated by lines parallel to the z axis and passing through the semicircle in the plane $z = 0$.

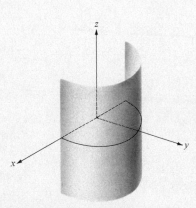

Figure 13.36
Cross sections of the surface $y = \sqrt{1 - x^2}$.

Example 4

The equation $z = -2x + 4$ generates a surface with lines parallel to the y axis and passing through the line $z = -2x + 4$ in the plane $y = 0$.

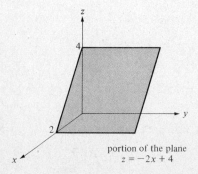
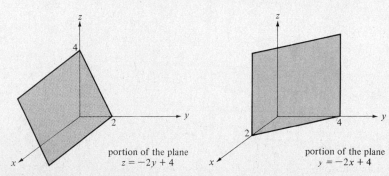

Figure 13.37

portion of the plane $z = -2x + 4$

portion of the plane $z = -2y + 4$

portion of the plane $y = -2x + 4$

13.3 CROSS SECTIONS AND SURFACES

This surface is a plane (see Figure 13.37). The planes $z = -2y + 4$ and $y = -2x + 4$ are given for comparison.

SURFACES OF REVOLUTION

In Example 4 of Section 13.2 we considered the surface

$$z = 1 - x^2 - y^2$$

whose level curves are circles with center at the origin. Such a surface is a surface of revolution about the z axis (surfaces of revolution were discussed in Section 8.3 in connection with integration). Thus, the surface in question has the parabola $z = 1 - x^2$, as its cross section with the plane $y = 0$ (the xz plane) and the surface is obtained by revolving this parabola about the z axis (see Figure 13.38).

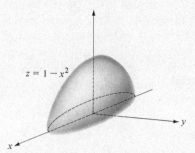

Figure 13.38
The surface $z = 1 - x^2 - y^2$.

Example 5

The cross section of the surface $z = 2\sqrt{x^2 + y^2}$ in the plane $y = 0$ is the line $z = 2x$ for $x \geq 0$. The surface, which is the cone depicted in Figure 13.39, is generated by revolving this line about the z axis. Note that for each constant $z_0 > 0$, the level curve $z_0 = 2\sqrt{x^2 + y^2}$ is a circle with center at the origin and radius $z_0/2$.

The remainder of this section is devoted to surfaces like the sphere $x^2 + y^2 + z^2 = 1$, which can be described by an equation of the form $F(x, y, z) = c$, where c is some constant. It should be kept in mind that such an equation may not define a function, as is the case when $F(x, y, z) = x^2 + y^2 + z^2$.

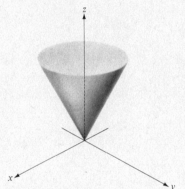

Figure 13.39
The surface $z = 2\sqrt{x^2 + y^2}$.

Example 6

The surface specified by the equation

$$\frac{x^2}{3^2} + \frac{y^2}{4^2} + \frac{z^2}{5^2} = 1$$

is an example of an *ellipsoid* (see Figure 13.40). It intersects the coordinate axes in the six points $(\pm 3, 0, 0)$, $(0, \pm 4, 0)$, and $(0, 0, \pm 5)$, and the cross sections cut out by the coordinate planes are ellipses. We also mention that this ellipsoid is symmetric with respect to each coordinate plane.

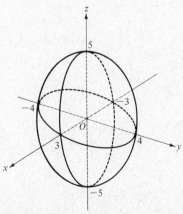

Figure 13.40
The ellipsoid $\frac{x^2}{3^2} + \frac{y^2}{4^2} + \frac{z^2}{5^2} = 1$.

Example 7

The surface specified by the equation

$$\frac{x^2}{3^2} + \frac{y^2}{4^2} - \frac{z^2}{5^2} = 1$$

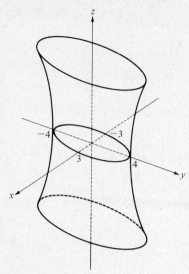

Figure 13.41
Hyperboloid of one sheet
$\frac{x^2}{3^2} + \frac{y^2}{4^2} - \frac{z^2}{5^2} = 1.$

is called a *hyperboloid* of *one sheet* (see Figure 13.41). It intersects the x and y axes in the points $(\pm 3, 0, 0)$ and $(0, \pm 4, 0)$, respectively. It does not intersect the z axis, since the equation $-z^2/5^2 = 1$ has no real solution. The surface is again symmetric with respect to each coordinate plane. We observe that the cross sections cut out by the coordinate planes $x = 0$ and $y = 0$ are hyperbolas, whereas the cross section cut out by the coordinate plane $z = 0$ is an ellipse. The surface in Figure 13.42 is an example of a *hyperboloid* of *two sheets;* its equation is

$$-\frac{x^2}{3^2} - \frac{y^2}{4^2} + \frac{z^2}{5^2} = 1$$

Its vertical cross sections are still hyperbolas, but the surface, which consists of two symmetric parts, is not intersected by the plane $z = 0$.

Example 8

The surfaces in Figure 13.43 are examples of *cylinders*. Observe that in each defining equation one of the variables is missing. As explained in Example 2, these cylinders can be generated with straight lines.

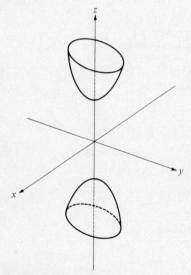

Figure 13.42
Hyperboloid of two sheets
$-\frac{x^2}{3^2} - \frac{y^2}{4^2} + \frac{z^2}{5^2} = 1.$

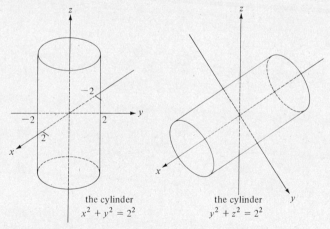

the cylinder $x^2 + y^2 = 2^2$

the cylinder $y^2 + z^2 = 2^2$

Figure 13.43

EXERCISES

In Exercises 1–8, describe the cross sections of the given surfaces cut by the coordinate planes. The names of the surfaces defined by these equations are given in parentheses.

1. $z^2 - 4(x^2 + y^2) = 0$ (cone)

2. $x^2 - 4(y^2 + z^2) = 0$ (cone)

3. $\dfrac{x^2}{3^2} + \dfrac{y^2}{4^2} - z = 0$ (elliptic paraboloid)

4. $\dfrac{x^2}{3^2} - \dfrac{y^2}{4^2} - z = 0$ (hyperbolic paraboloid)

5. $y - x^2 = 0$ (parabolic cylinder)

6. $x - z^2 = 0$ (parabolic cylinder)

7. $y^2 + 4z^2 = 1$ (elliptic cylinder)

8. $x^2 + 16z^2 = 1$ (elliptic cylinder)

13.4 PARTIAL DERIVATIVES

In this section we apply the notion of derivative to functions of two variables. For the time being we discuss the derivative only in the case in which one of the variables is held constant. This means, of course, that we are applying the notion of derivative to cross sections.

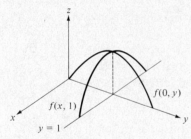

Figure 13.44
Cross sections of $f(x, y) = 1 - x^2 - (y - 1)^2$.

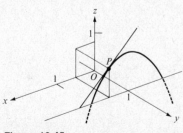

Figure 13.45
The derivative $f_x(\tfrac{1}{2}, 1) = -1$ is the slope of the tangent to the curve $f(x, 1) = 1 - x^2$ at $P = (\tfrac{1}{2}, 1, \tfrac{3}{4})$.

Example 1

Consider the function

$$f(x, y) = 1 - x^2 - (y - 1)^2$$

and two of its cross sections,

$$f(x, 1) = 1 - x^2$$
$$f(0, y) = 1 - (y - 1)^2$$

(see Figure 13.44). The functions $f(x, 1)$ and $f(0, y)$ are functions of one variable each, and they have derivatives

$$\frac{d}{dx}[f(x, 1)] = -2x$$

$$\frac{d}{dy}[f(0, y)] = -2(y - 1)$$

These derivatives, called *partial derivatives* of $f(x, y)$, are often denoted

$$f_x(x, 1) = -2x$$
$$f_y(0, y) = -2(y - 1)$$

A partial derivative, like an ordinary derivative, can be interpreted as the slope of a tangent to a curve. This is shown in Figures 13.45 and 13.46.

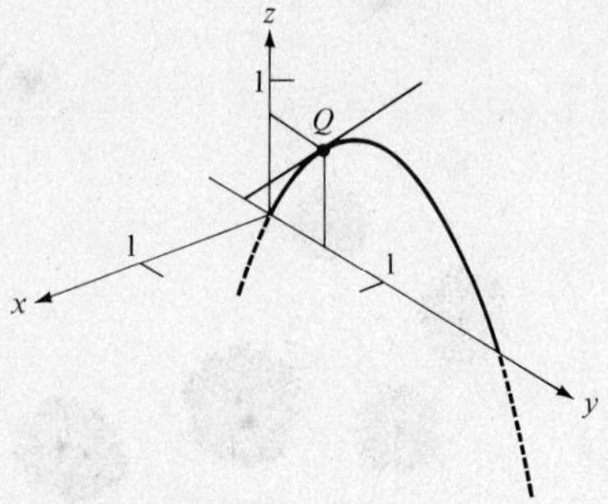

514 FUNCTIONS OF SEVERAL VARIABLES

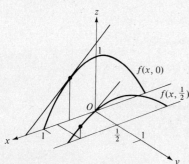

Figure 13.46
The derivative $f_y(0, \frac{1}{2}) = 1$ is the slope of the tangent to the curve $f(0, y) = 1 - (y - 1)^2$ at $Q = (0, \frac{1}{2}, \frac{3}{4})$.

Figure 13.47
Tangents to cross sections of $f(x, y) = (1 - x^2)(1 - y)$.

Example 2

Let
$$f(x, y) = (1 - x^2)(1 - y)$$
Find $f_x(x_0, y_0)$ when $(x_0, y_0) = (\frac{1}{2}, 0)$ and $(x_0, y_0) = (\frac{3}{4}, \frac{1}{2})$.

Solution
Holding y constant, $y = y_0$, gives
$$f_x(x, y_0) = \frac{d}{dx}[(1 - x^2)(1 - y_0)]$$
$$= (1 - y_0)\frac{d}{dx}(1 - x^2) = (1 - y_0)(-2x)$$

Hence,
$$f_x(\tfrac{1}{2}, 0) = -1$$
$$f_x(\tfrac{3}{4}, \tfrac{1}{2}) = -\tfrac{3}{4}$$

(see Figure 13.47).

To summarize, the partial derivative of $f(x, y)$ with respect to x is found by holding y constant, $y = y_0$, and differentiating $f(x, y_0)$ according to the rules established for functions of one variable. This leads us to the following definition.

DEFINITION 1
The partial derivative of $f(x, y)$ at (x_0, y_0) with respect to $\dot{x}$ is
$$f_x(x_0, y_0) = \lim_{\Delta x \to 0} \frac{f(x_0 + \Delta x, x_0) - f(x_0, y_0)}{\Delta x}$$
whenever the limit exists; the partial derivative of $f(x, y)$ at (x_0, y_0) with respect to y is
$$f_y(x_0, y_0) = \lim_{\Delta y \to 0} \frac{f(x_0, y_0 + \Delta y) - f(x_0, y_0)}{\Delta y}$$
whenever the limit exists.

NOTATION
The following symbols will be used interchangeably:
$$f_x(x, y) = f_x = \frac{\partial f}{\partial x}$$

$$f_y(x, y) = f_y = \frac{\partial f}{\partial y}$$

> $f_x(x, y)$ is found by considering y a constant and differentiating the function with respect to x; $f_y(x, y)$ is found by considering x a constant and differentiating the function with respect to y.

Example 3

If $f(x, y) = x^4 y^5$, then

$$f_x(x, y) = 4x^3 y^5$$

and

$$f_y(x, y) = 5x^4 y^4$$

SECOND-ORDER PARTIAL DERIVATIVES

As in the one-variable case, a partial derivative is a function that may itself be differentiated. The following symbols for second-order derivatives are conventional:

$$\frac{\partial}{\partial x}\left(\frac{\partial f}{\partial x}\right) = \frac{\partial^2 f}{\partial x^2} = f_{xx}$$

$$\frac{\partial}{\partial y}\left(\frac{\partial f}{\partial x}\right) = \frac{\partial^2 f}{\partial y\, \partial x} = f_{yx}$$

$$\frac{\partial}{\partial x}\left(\frac{\partial f}{\partial y}\right) = \frac{\partial^2 f}{\partial x\, \partial y} = f_{xy}$$

$$\frac{\partial}{\partial y}\left(\frac{\partial f}{\partial y}\right) = \frac{\partial^2 f}{\partial y^2} = f_{yy}$$

Example 4

Find $f_{xx}, f_{xy}, f_{yx},$ and f_{yy} when $f(x, y) = x^3 y^2 - \frac{x}{y}$.

Solution

$$f_x = 3x^2 y^2 - \frac{1}{y} \qquad f_{xx} = 6xy^2 \qquad f_{xy} = 6x^2 y + \frac{1}{y^2}$$

$$f_y = 2x^3 y + \frac{x}{y^2} \qquad f_{yy} = 2x^3 - 2\frac{x}{y^3} \qquad f_{yx} = 6x^2 + \frac{1}{y^2}$$

Notice that $f_{xy} = f_{yx}$.

Example 5

If $u = x^2 - y^2$, show that

$$\frac{\partial^2 u}{\partial x^2} + \frac{\partial^2 u}{\partial y^2} = 0$$

Solution

$$\frac{\partial u}{\partial x} = 2x \qquad \frac{\partial^2 u}{\partial x^2} = 2$$

$$\frac{\partial u}{\partial y} = -2y \qquad \frac{\partial^2 u}{\partial y^2} = -2$$

Hence,

$$\frac{\partial^2 u}{\partial x^2} + \frac{\partial^2 u}{\partial y^2} = 2 - 2 = 0$$

Higher-order partial derivatives will be treated in greater detail in later sections.

EXERCISES

Find $\partial u/\partial x$ and $\partial u/\partial y$ in Exercises 1–14.

1. $u = xy$
2. $u = \dfrac{x}{y} + \dfrac{y}{x}$
3. $u = \dfrac{x-y}{x+y}$
4. $u = \dfrac{x}{\sqrt{x^2+y^2}} + \dfrac{y}{\sqrt{x^2+y^2}}$
5. $u = x^3 y^2$
6. $u = (x+y)^5$
7. $u = x$
8. $u = y$

Ans. $\dfrac{\partial u}{\partial x} = f'(xy)y,\ \dfrac{\partial u}{\partial y} = f'(xy)x.$

9. $u = f(xy)$
10. $u = f(x)g(y)$
11. $u = f\left(\dfrac{x}{y}\right)$
12. $u = xf(y)$
13. $u = f(x+y)$
14. $u = \dfrac{\partial f}{\partial x} - \dfrac{\partial f}{\partial y}$

15. Let $u = x^3 - 3xy^2$ and $v = 3x^2 y - y^3$. Show that
    (a) $u_x = v_y$ and $u_y = -v_x$
    (b) $u_{xx} + u_{yy} = 0$ and $v_{xx} + v_{yy} = 0$

16. Let $f(x, y) = \sqrt{1 - x^2 - y^2}$. Find the tangent to the curve $z = f(x, 1/2)$ at $P = (1/2, 1/2, \sqrt{2}/2)$.

17. Show that if $u = \exp(x + cy) + \exp(x - cy)$, then $u_{yy} = c^2 u_{xx}$.

13.5 LIMITS AND CONTINUITY

The method used to define the concepts of limit and continuity for functions of one variable (Chapters 2 and 4) can be adapted to define these concepts for functions of several variables. In analogy to the statement

$$\lim_{x \to x_0} f(x) = A$$

we must give meaning to a statement of the form

$$\lim_{\substack{x \to x_0 \\ y \to y_0}} f(x, y) = A$$

In analogy to the statement of continuity

$$\lim_{x \to x_0} f(x) = f(x_0)$$

we shall consider the statement

$$\lim_{\substack{x \to x_0 \\ y \to y_0}} f(x, y) = f(x_0, y_0)$$

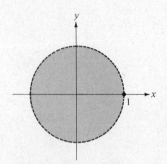

Figure 13.48
The domain $x^2 + y^2 < 1$.

Let us first look at some examples.

Example 1

Consider the function

$$f(x, y) = \frac{1}{\sqrt{1 - x^2 - y^2}}$$

Its domain is the disk $x^2 + y^2 < 1$, that is, the *region* contained inside the circle $x^2 + y^2 = 1$ (see Figure 13.48). We see that when we consider only points (x, y) lying in a small disk $x^2 + y^2 < \delta^2$ (Figure 13.49), then $f(x, y)$ is close to 1; in fact, the smaller δ is, the closer is $f(x, y)$ to 1, and we express this by writing

$$\lim_{\substack{x \to 0 \\ y \to 0}} f(x, y) = 1$$

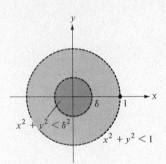

Figure 13.49

We say that *the limit of $f(x, y)$ as (x, y) approaches $(0, 0)$ is 1*. Observe that there are infinitely many paths along which the variable point (x, y) can approach the point $(0, 0)$. Along each of these paths the function must, of course, be defined.

Using the terminology introduced for intervals (see Section 1.3) we call the disk $(x - x_0)^2 + (y - y_0)^2 < c^2$ an *open disk* and the disk $(x - x_0)^2 + (y - y_0)^2 \leq c^2$ a *closed disk* (see Figure 13.50). Thus, the open disk excludes its bounding circle (boundary), whereas the closed disk includes it.

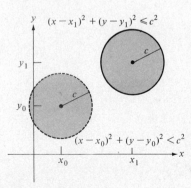

Figure 13.50
Open and closed disks.

In Example 1 we noted that to speak of a limit at $(0, 0)$ the func-

518 FUNCTIONS OF SEVERAL VARIABLES

tion had to be defined in an open disk with center at $(0, 0)$. A different situation is described in the next example.

Example 2

The function
$$f(x, y) = \sqrt{x^2 + y^2 - 1}$$
is defined for all points (x, y) for which $x^2 + y^2 \geq 1$. The domain here consists, therefore, of all points *outside* the disk $x^2 + y^2 < 1$ (see Figure 13.51). In particular, this function is defined at the point $(1, 0)$, but *every* disk $(x - 1)^2 + (y - 0)^2 < \delta^2$, no matter how small $\delta > 0$ is, contains points not in the domain; that is, each disk with center at $(1, 0)$ contains points at which $f(x, y)$ is undefined (see Figure 13.52). We say, therefore, that

$$\lim_{\substack{x \to 1 \\ y \to 0}} f(x, y)$$

is undefined in this case. On the other hand,

$$\lim_{\substack{x \to 1 \\ y \to 2}} f(x, y) = 2$$

since the point $(1, 2)$ has an open disk in the domain of f, and as (x, y) approaches this point, $f(x, y)$ approaches the value 2 (Figure 13.53).

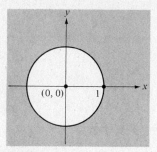

Figure 13.51
The domain $x^2 + y^2 \geq 1$.

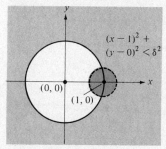

Figure 13.52

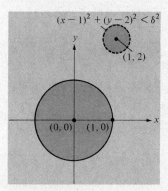

Figure 13.53

The definition of a limit in the case of two variables is quite similar to the one-variable case. A formal definition is as follows:

DEFINITION 1
Let f be defined in a truncated disk $0 < (x - x_0)^2 + (y - y_0)^2 < d^2$. Then $f(x, y)$ has the limit A at (x_0, y_0), written

$$\lim_{\substack{x \to x_0 \\ y \to y_0}} f(x, y) = A$$

if for each error $\epsilon > 0$, no matter how small, there is a number $\delta > 0$, such that

$$|f(x, y) - A| < \epsilon \quad \text{whenever} \quad 0 < (x - x_0)^2 + (y - y_0)^2 < \delta^2$$

The idea behind Definition 1 is described in Figure 13.54. A *truncated disk* is a disk with its center deleted.

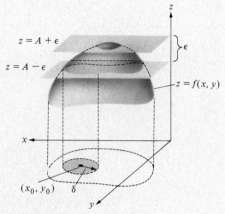

Figure 13.54
Illustrating Definition 1: If $(x - x_0)^2 + (y - y_0)^2 < \delta^2$, then the values $z = f(x, y)$ lie between the planes $z = A + \epsilon$ and $z = A - \epsilon$.

In the following examples we illustrate some of the reasons for which a limit may fail to exist.

Example 3

The function

$$f(x, y) = \frac{1}{x^2 + y^2}$$

is defined at all points except $(0, 0)$. Since $1/(x^2 + y^2)$ increases beyond all bounds as (x, y) approaches $(0, 0)$, we say that f has no limit at $(0, 0)$.

Example 4

A function need not be defined at a point to have a limit there. Let

$$f(x, y) = \frac{\sin(x^2 + y^2)}{x^2 + y^2}$$

Then $f(0, 0)$ is undefined, but from Section 9.2 we know that

$$\lim_{t \to 0} \frac{\sin t}{t} = 1$$

Putting $t = x^2 + y^2$, we thus see that

$$\lim_{\substack{x \to 0 \\ y \to 0}} \frac{\sin(x^2 + y^2)}{x^2 + y^2} = 1$$

Corresponding to the definition of continuity in Section 4.5, we state the following definition.

DEFINITION 2

A function $f(x, y)$ is *continuous* at (x_0, y_0) if

$$\lim_{\substack{x \to x_0 \\ y \to y_0}} f(x, y) = f(x_0, y_0)$$

Implicit in this definition is the assumption that f is defined at (x_0, y_0) as well as in some disk $(x - x_0)^2 + (y - y_0)^2 < \delta^2$. When f is not continuous at a point, it is said to be *discontinuous* there.

Example 5

The function $f(x, y) = \sin(x^2 + y^2)/(x^2 + y^2)$ (see Example 4) is not continuous at $(0, 0)$ since $f(0, 0)$ is undefined. This function, however, is continuous at all other points, and by Example 4,

$$\lim_{\substack{x \to 0 \\ y \to 0}} \frac{\sin(x^2 + y^2)}{x^2 + y^2} = 1$$

Hence, the function

$$g(x, y) = \begin{cases} \dfrac{\sin(x^2 + y^2)}{x^2 + y^2} & \text{for } (x, y) \neq (0, 0) \\ 1 & \text{for } (x, y) = (0, 0) \end{cases}$$

is continuous at $(0, 0)$.

Example 6

Consider the function

$$f(x, y) = \frac{xy}{x^2 + y^2}$$

What happens to the values $f(x, y)$ as (x, y) approaches $(0, 0)$ along straight lines $y = mx$? A substitution gives

$$f(x, mx) = \frac{x \cdot mx}{x^2 + (mx)^2} = \frac{m}{1 + m^2}$$

Hence, for different values of m, the limit at $(0, 0)$ has different values. For instance, as $(x, y) \to (0, 0)$ along the line $y = x$, we have $f(x, x) = \frac{1}{2}$; for the line $y = 3x$, we have $f(x, 3x) = \frac{3}{10}$, and so on. It follows that $f(x, y)$ has no limit at $(0, 0)$.

Many functions, like that in Example 1, are continuous at each point of their domain. Since continuity at a point involves a limit that, in turn, requires the function to be defined in a small open disk centered at that point, we introduce the notion of open region:

> A region is said to be *open* if each of its points lies in a small open disk that lies entirely in the region.

With this we can state the following definition.

DEFINITION 3
If f is continuous at each point of an open region R in the plane, then f is said to be *continuous on R*.

Example 7
(a) The function $f(x, y) = 1/\sqrt{1 - x^2 - y^2}$, discussed in Example 1, is continuous in the region $x^2 + y^2 < 1$.
(b) The function $f(x, y) = \sin(x^2 + y^2)/(x^2 + y^2)$ (see Examples 4 and 5) is continuous in the plane with the origin $(0, 0)$ omitted.

The following simple fact will be found very useful in the sections to follow.

LEMMA 1
Let $f(x, y)$ be continuous on an open region R. If the line segment joining the points (a, y_0) and (b, y_0) lies entirely in R, then the function $f(x, y_0)$ is continuous for $a \leq x \leq b$.

The lemma is illustrated in Figure 13.55. It is worthwhile pointing out that $f(x, y_0)$ is a function of the single variable x, since y is held constant.

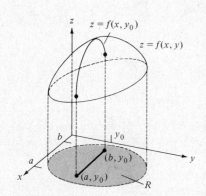

Figure 13.55

Example 8

If, in Example 1, we put $y = \tfrac{1}{2}$, then we get

$$f(x, \tfrac{1}{2}) = \frac{1}{\sqrt{\tfrac{3}{4} - x^2}}$$

This function is continuous for $-\tfrac{\sqrt{3}}{2} < x < \tfrac{\sqrt{3}}{2}$.

EXERCISES

Find the limits in Exercises 1–4.

1. $\lim\limits_{\substack{x \to 0 \\ y \to 0}} \dfrac{\sin(x+y)}{x+y}$

2. $\lim\limits_{\substack{x \to 1 \\ y \to 1}} \dfrac{e^x - e^y}{e^x + e^y}$

3. $\lim\limits_{\substack{x \to 1 \\ y \to 0}} (x+y) \ln(x-y)$

4. $\lim\limits_{\substack{x \to 1 \\ y \to 2}} \dfrac{(x-1)(x+1)}{y}$

In Exercises 5–10, determine the region of continuity of the given functions.

5. $f(x, y) = \tan \dfrac{x}{y}$

6. $f(x, y) = \ln \dfrac{x}{y}$

7. $f(x, y) = e^{xy}$

8. $f(x, y) = \dfrac{x+y}{x-y}$

9. $f(x, y) = \tan^{-1} \dfrac{y}{x}$

10. $f(x, y) = \begin{cases} \dfrac{x^2 - y^2}{x - y} & \text{when } x \neq y \\ x + y & \text{when } x = y \end{cases}$

In Exercises 11–14, decide if f is continuous at (x_0, y_0). Justify your answer.

11. $f(x, y) = \begin{cases} \dfrac{\sin xy}{x} & \text{when } x \neq 0 \\ y & \text{when } x = 0 \end{cases}$ $\quad (x_0, y_0) = (0, 0)$

12. $f(x, y) = \dfrac{x - y}{x + y} \quad (x_0, y_0) = (1, -1)$

13. $f(x, y) = \begin{cases} \dfrac{x+y}{x^2 + y^2} & \text{when } (x, y) \neq (0, 0) \\ 0 & \text{when } (x, y) = (0, 0) \end{cases}$ $\quad (x_0, y_0) = (0, 0)$

14. $f(x, y) = e^{-1/(x^2 + y^2)} \quad (x_0, y_0) = (0, 0)$

13.6 FUNCTIONS OF THREE OR MORE VARIABLES

The concept of function can be extended to three or more independent variables. In analogy to the formula

$$d = \sqrt{x^2 + y^2}$$

13.6 FUNCTIONS OF THREE OR MORE VARIABLES

considered in Section 13.2, consider the formula

$$d = \sqrt{x^2 + y^2 + z^2}$$

For each point (x_0, y_0, z_0) this formula determines the unique number $d_0 = \sqrt{x_0^2 + y_0^2 + z_0^2}$. The formula defines, therefore, a function of *three* independent variables. Likewise, the formula

$$d = \sqrt{v^2 + x^2 + y^2 + z^2}$$

defines a function of *four* independent variables, and so on. As in Section 13.2, the set of points for which a function is defined is its *domain*.

Example 1

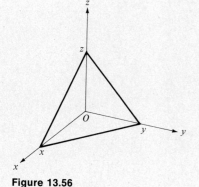

Figure 13.56

Consider the pyramid in Figure 13.56. Its volume is given by the formula

$$V = \tfrac{1}{6} xyz$$

Since x, y, and z may be assigned any nonnegative values, V is a function of three independent variables, a fact expressed by writing

$$f(x, y, z) = \tfrac{1}{6} xyz$$

Thus,

$$f(0, 1, 1) = 0$$
$$f(1, \tfrac{1}{4}, \tfrac{1}{8}) = \tfrac{1}{192}$$

and so on. The domain of this function is the set of points (x, y, z) such that $x \geq 0$, $y \geq 0$, and $z \geq 0$. Observe that no graph can be drawn here, since this would call for $vxyz$ coordinates in four-dimensional space.

Example 2

The relation

$$f(x, y, z) = \sqrt{1 - x^2 - y^2 - z^2}$$

defines a function of three variables. Since the square root of a negative number is undefined (in the context of real numbers), we see that the domain of this function is the solid ball

$$x^2 + y^2 + z^2 \leq 1$$

Example 3

The relation

$$f(x, y, z, u, v) = x + y + z + u + v$$

defines a function of five independent variables. It is defined for all points (x, y, z, u, v).

We mention in passing that the concepts of limit and continuity can be defined for functions of three or more variables by following a procedure similar to that in Section 13.5. These developments, which are very technical in nature and not very illuminating, are only of peripheral interest in the coming chapters, and they are therefore omitted.

PARTIAL DERIVATIVES

The definition and rules of partial derivatives for functions of three or more variables are similar to the case of two variables (see Section 13.4).

Example 4

Let
$$f(x, y, z) = xy^2 + xz^4 + x^2yz$$
Find $\partial f/\partial x$, $\partial f/\partial y$, $\partial f/\partial z$, and $\partial^2 f/\partial z\, \partial x$.

Solution

$$\frac{\partial f}{\partial x} = y^2 + z^4 + 2xyz$$

$$\frac{\partial f}{\partial y} = 2xy + x^2z$$

$$\frac{\partial f}{\partial z} = 4xz^3 + x^2y$$

$$\frac{\partial^2 f}{\partial z\, \partial x} = \frac{\partial}{\partial z}\left(\frac{\partial f}{\partial x}\right) = \frac{\partial}{\partial z}(y^2 + z^4 + 2xyz) = 4z^3 + 2xy$$

Observe that

$$\frac{\partial^2 f}{\partial z\, \partial x} = \frac{\partial^2 f}{\partial x\, \partial z}$$

Example 5

If $u = t^2 - x^2 - y^2 - z^2$, find a constant k such that

$$\frac{\partial^2 u}{\partial t^2} + k\left(\frac{\partial^2 u}{\partial x^2} + \frac{\partial^2 u}{\partial y^2} + \frac{\partial^2 u}{\partial z^2}\right) = 0 \tag{1}$$

Solution

$$\frac{\partial u}{\partial t} = 2t, \qquad \frac{\partial^2 u}{\partial t^2} = 2$$
$$\frac{\partial u}{\partial x} = -2x, \qquad \frac{\partial^2 u}{\partial x^2} = -2$$

and similarly,

$$\frac{\partial^2 u}{\partial y^2} = \frac{\partial^2 u}{\partial z^2} = -2$$

Hence, equation (1) becomes

$$2 + k[(-2) + (-2) + (-2)] = 2 - 6k = 0$$

and taking $k = \frac{1}{3}$ gives the desired result.

CYLINDRICAL COORDINATES

In dealing with curves in the plane, we found it useful to introduce polar coordinates $x = r\cos\theta$ and $y = r\sin\theta$ (see Figure 13.57). Cylindrical coordinates are an extension of polar coordinates to three dimensions: The *cylindrical coordinates* of a point (x, y, z) are (r, θ, z), where

$$\left. \begin{array}{l} x = r\cos\theta \\ y = r\sin\theta \\ z = z \end{array} \right\} \qquad \text{cylindrical coordinates} \qquad (2)$$

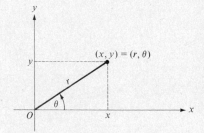

Figure 13.57
Polar coordinates.

(see Figure 13.58). From the fact that $x^2 + y^2 = r^2$, we see that the locus of points (r, θ, z) for fixed r is a cylinder (see Figure 13.59) and hence the name cylindrical coordinates. These coordinates and the spherical coordinates discussed below are very useful in the

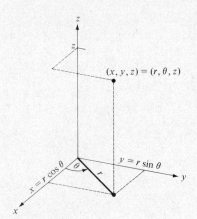

Figure 13.58
Cylindrical coordinates.

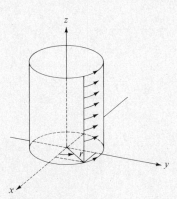

Figure 13.59
Portion of the cylinder generated by points (r, θ, z) for fixed r.

Example 6

Express

$$u = \frac{zx}{\sqrt{x^2 + y^2}}$$

in terms of cylindrical coordinates, then show that $u_{\theta\theta} + u = 0$.

Solution
From (2) we have

$$u = \frac{zr\cos\theta}{\sqrt{r^2}} = z\cos\theta$$

Hence

$$u_{\theta\theta} + u = z(-\cos\theta) + z\cos\theta = 0$$

SPHERICAL COORDINATES

The *spherical coordinates* of a point $Q = (x, y, z)$ are a triplet (ρ, θ, ϕ), where $\rho = \text{dist}(O, Q)$, θ is the radian measure of the angle between the projection $\overline{OP}$ (shadow) in the xy plane of the line segment $\overline{OQ}$ and the x axis, and ϕ is the radian measure of the angle between $\overline{OQ}$ and the z axis (see Figure 13.60). To relate the rectangular coordinates (x, y, z) with the spherical coordinates (ρ, θ, ϕ), consult Figures 13.60 and 13.61. With $p = \text{dist}(O, P)$ we obtain the following relations:

$x = p\cos\theta$
$y = p\sin\theta$
$z = \rho\cos\phi$
$p = \rho\sin\phi$

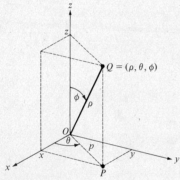

Figure 13.60
Spherical coordinates.

Substituting for p gives the equations

$$\left.\begin{array}{l} x = \rho\sin\phi\cos\theta \\ y = \rho\sin\phi\sin\theta \\ z = \rho\cos\phi \end{array}\right\} \quad \text{spherical coordinates} \qquad (3)$$

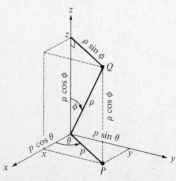

Figure 13.61

Squaring each equation in (3) and adding them gives

$$\begin{aligned} x^2 + y^2 + z^2 &= \rho^2\sin^2\phi\cos^2\theta + \rho^2\sin^2\phi\sin^2\theta + \rho^2\cos^2\phi \\ &= \rho^2[\sin^2\phi(\cos^2\theta + \sin^2\theta) + \cos^2\phi] \\ &= \rho^2[\sin^2\phi + \cos^2\phi] = \rho^2 \end{aligned}$$

13.6 FUNCTIONS OF THREE OR MORE VARIABLES

For fixed ρ this is the equation of a sphere of radius ρ and hence the designation spherical coordinates.

Example 7

Consult Figure 13.62. Holding ϕ constant and revolving the line $\overline{OQ}$ about the z axis gives a circular cone. In particular, the inequalities

$$0 \le \phi \le \frac{\pi}{4}$$

$$0 \le \rho \le 2$$

give a solid resembling an ice cream cone as in Figure 13.62. Note that no restriction is put on θ.

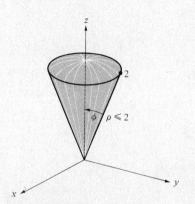

Figure 13.62
The solid determined by $0 \le \phi \le \pi$, $0 \le \rho \le 2$.

Example 8

The function $u = \rho(1 - \cos \phi)$ is given in spherical coordinates. Write it in terms of rectangular coordinates, and then show that

$$(u_x)^2 + (u_y)^2 + (u_z + 1)^2 = 1$$

Solution
By the above,

$$u = \rho - \rho \cos \phi = \sqrt{x^2 + y^2 + z^2} - z$$

Now,

$$u_x = \frac{\partial}{\partial x}[(x^2 + y^2 + z^2)^{1/2} - z] = \frac{1}{2}(x^2 + y^2 + z^2)^{-1/2} \cdot 2x$$

$$= \frac{x}{\sqrt{x^2 + y^2 + z^2}}$$

and hence

$$(u_x)^2 = \frac{x^2}{x^2 + y^2 + z^2}$$

Similarly,

$$(u_y)^2 = \frac{y^2}{x^2 + y^2 + z^2}$$

and

$$(u_z + 1)^2 = \frac{z^2}{x^2 + y^2 + z^2}$$

Hence,

$$(u_x)^2 + (u_y)^2 + (u_z + 1)^2 = 1$$

EXERCISES

Find the domains of the functions in Exercises 1–7.

1. $f(x, y, z) = x + y + z$ **2.** $f(x, y, z) = \dfrac{1}{\sqrt{x^2 + y^2 + z^2}}$

3. $f(x, y, z) = xy \dfrac{\sin z}{z}$ **4.** $f(x, y, z) = \exp\left(-\dfrac{1}{x^2 + y^2 + z^2}\right)$

5. $f(x, y, z) = \tan(x^2 + y^2)$ **6.** $f(x, y, z) = \dfrac{1}{x}$

7. $f(u, x, y, z) = \dfrac{|u| + |x|}{|u| - |x|} + \dfrac{|y| + |z|}{|y| - |z|}$

8. If $u = \sin(x + y) + \sin(x + z) + \sin(y + z)$, show that $u_{xx} + u_{yy} + u_{zz} - 2u = 0$.

9. Find u_x and u_{xy} if
$$u = \exp\left(-\dfrac{1}{x^2 + y^2 + z^2}\right)$$

10. Show that if $u = f(x + cy) + g(x - cy) + z$, then $u_{yy} = c^2 u_{xx}$.

11. Express the functions in Exercises 1–6 in spherical and cylindrical coordinates.

13.7 LINEAR APPROXIMATIONS

In Section 5.5 we discovered the approximation formula
$$f(x) \approx f(x_0) + f'(x_0)(x - x_0) \quad \text{when } x \approx x_0$$
for a differentiable function f of a single variable. With the notation $x = x_0 + \Delta x$, this formula becomes
$$f(x_0 + \Delta x) \approx f(x_0) + f'(x_0)\Delta x \quad \text{when } \Delta x \approx 0$$
We now derive the analogous formula for a function of two variables having partial derivatives (the formula for three variables is discussed later in this section). The formula is obtained from the following theorem.

THEOREM 1
Let a function f have partial derivatives f_x and f_y which, together with f, are continuous in an open disk with center at (x_0, y_0). Then there is an error ϵ such that
$$f(x_0 + \Delta x, y_0 + \Delta y) = f(x_0, y_0) + f_x(x_0, y_0)\,\Delta x + f_y(x_0, y_0)\,\Delta y + \epsilon \tag{1}$$
where $\epsilon = \epsilon_1 \Delta x + \epsilon_2 \Delta y$.

and

$$\lim_{\substack{\Delta x \to 0 \\ \Delta y \to 0}} \epsilon_1 = \lim_{\substack{\Delta x \to 0 \\ \Delta y \to 0}} \epsilon_2 = 0$$

This theorem is proved later in this section. It is implicit in its statement that ϵ depends on Δx and Δy, and using approximate equality we obtain the *linear approximation formula*

$$f(x_0 + \Delta x, y_0 + \Delta y) \approx f(x_0, y_0) + f_x(x_0, y_0) \Delta x + f_y(x_0, y_0) \Delta y \quad \text{when } \Delta x \approx 0 \text{ and } \Delta y \approx 0 \qquad (2)$$

With the notation $\Delta x = x - x_0$ and $\Delta y = y - y_0$, this formula becomes

$$f(x, y) \approx f(x_0, y_0) + f_x(x_0, y_0)(x - x_0) + f_y(x_0, y_0)(x - x_0) \quad \text{when } x \approx x_0 \text{ and } y \approx y_0 \qquad (3)$$

Example 1

Consider the area of a rectangle of sides x and y:

$$A(x, y) = xy$$

For any pair of values (x_0, y_0) we have

$$A_x(x_0, y_0) = y_0$$
$$A_y(x_0, y_0) = x_0$$

and hence formula (2) gives

$$(x_0 + \Delta x)(y_0 + \Delta y) \approx x_0 y_0 + y_0 \Delta x + x_0 \Delta y$$

Expanding the left side shows that actually

$$(x_0 + \Delta x)(y_0 + \Delta y) = x_0 y_0 + y_0 \Delta x + x_0 \Delta y + \Delta x \Delta y$$

and hence the error of the approximation here is

$$\epsilon = \Delta x \Delta y$$

Example 2

Find an approximation value for

$$\Delta f = f(2 + \Delta x, -3 + \Delta y) - f(2, -3)$$

when

$$f(x, y) = \sqrt{x^2 + y^2 - 1}$$

Solution

We use formula (2) with $(x_0, y_0) = (2, -3)$. Now,

$$f_x(x_0, y_0) = \frac{x_0}{\sqrt{x_0^2 + y_0^2 - 1}}$$

$$f_x(2, -3) = \frac{2}{\sqrt{12}} = \frac{\sqrt{3}}{3}$$

$$f_y(x_0, y_0) = \frac{y_0}{\sqrt{x_0^2 + y_0^2 - 1}}$$

$$f_y(2, -3) = \frac{-3}{\sqrt{12}} = -\frac{\sqrt{3}}{2}$$

Formula (2) thus gives

$$\Delta f \approx \frac{\sqrt{3}}{3} \Delta x - \frac{\sqrt{3}}{2} \Delta y = \sqrt{3}\left(\frac{1}{3}\Delta x - \frac{1}{2}\Delta y\right)$$

A theorem entirely analogous to Theorem 1 can be formulated for functions of more than two variables, and formulas like (2) can then be derived. In the case of three variables, the linear approximation formula is

$$f(x_0 + \Delta x, y_0 + \Delta y, z_0 + \Delta z) \approx f(x_0, y_0, z_0) + f_x(x_0, y_0, z_0)\Delta x$$
$$+ f_y(x_0, y_0, z_0)\Delta y + f_z(x_0, y_0, z_0)\Delta z$$
$$\text{when } \Delta x \approx 0, \Delta y \approx 0, \text{ and } \Delta z \approx 0 \quad (4)$$

With the notation $\Delta x = x - x_0$, $\Delta y = y - y_0$, and $\Delta z = z - z_0$, this formula becomes

$$f(x, y, z) \approx f(x_0, y_0, z_0) + f_x(x_0, y_0, z_0)(x - x_0)$$
$$+ f_y(x_0, y_0, z_0)(y - y_0) + f_z(x_0, y_0, z_0)(z - z_0)$$
$$\text{when } x \approx x_0, y \approx y_0, \text{ and } z \approx z_0 \quad (5)$$

Example 3

Find the linear approximation to

$$f(x, y, z) = \sqrt{3x^2 - 10y^2 + 2z^2}$$

when

$$(x_0, y_0, z_0) = (2, -1, 1)$$

Solution
We use here formula (5). Direct differentiation gives

$$f_x(x_0, y_0, z_0) = \frac{3x_0}{\sqrt{3x_0^2 - 10y_0^2 + 2z_0^2}}$$

$$f_y(x_0, y_0, z_0) = \frac{-10y_0}{\sqrt{3x_0^2 - 10y_0^2 + 2z_0^2}}$$

$$f_z(x_0, y_0, z_0) = \frac{2z_0}{\sqrt{3x_0^2 - 10y_0^2 + 2z_0^2}}$$

Hence

$$f_x(2, -1, 1) = \frac{3 \cdot 2}{2} = 3$$

$$f_y(2, -1, 1) = \frac{(-10) \cdot (-1)}{2} = 5$$

$$f_z(2, -1, 1) = \frac{2 \cdot 1}{2} = 1$$

and formula (5) gives

$$\sqrt{3x^2 - 10y^2 + 2z^2} \approx 2 + 3(x - 2) + 5(y + 1) + (z - 1)$$
$$\text{when } x \approx 2,\ y \approx -1, \text{ and } z \approx 1 \qquad (6)$$

We now turn to the proof of Theorem 1.

Proof
Our starting point is the following equation, obtained by adding and subtracting the quantity $f(x_0, y_0 + \Delta y)$:

$$\begin{aligned} f(x_0 + \Delta x, y_0 + \Delta y) - f(x_0, y_0) \\ = [f(x_0 + \Delta x, y_0 + \Delta y) - f(x_0, y_0 + \Delta y)] \\ + [f(x_0, y_0 + \Delta y) - f(x_0, y_0)] \end{aligned} \qquad (7)$$

For simplicity, it will be assumed that $\Delta x > 0$ and $\Delta y > 0$. Our first objective is to rewrite each of the bracketed expressions in a more advantageous form.

Consider the function

$$G(x) = f(x, y_0 + \Delta y)$$

By Lemma 1 of Section 13.6, this function is continuous on the interval $x_0 \leq x \leq x_0 + \Delta x$. We note that

$$G(x_0 + \Delta x) - G(x_0) = f(x_0 + \Delta x, y_0 + \Delta y) - f(x_0, y_0 + \Delta y) \qquad (8)$$

and

$$G'(x) = f_x(x, y_0 + \Delta y) \qquad (9)$$

By the Mean Value Theorem (see Section 4.6), there is a point x_0^*, $x_0 < x_0^* < x_0 + \Delta x$, such that

$$G(x_0 + \Delta x) - G(x_0) = G'(x_0^*)\Delta x$$

By (9),

$$G'(x_0^*) = f_x(x_0^*, y_0 + \Delta y)$$

and hence

$$G(x_0 + \Delta x) - G(x_0) = f_x(x_0^*, y_0 + \Delta y)\Delta x \qquad (10)$$

Combining (8) and (10) now gives

$$f(x_0 + \Delta x, y_0 + \Delta y) - f(x_0, y_0 + \Delta y) = f_x(x_0^*, y_0 + \Delta y)\, \Delta x \quad (11)$$

To rewrite the second bracketed term in (7), consider the function

$$H(y) = f(x_0, y)$$

which is continuous on the interval $y_0 \leq y \leq y_0 + \Delta y$. We observe that

$$H(y_0 + \Delta y) - H(y_0) = f(x_0, y_0 + \Delta y) - f(x_0, y_0) \quad (12)$$

and

$$H'(y) = f_y(x_0, y) \quad (13)$$

Again by the mean value theorem,

$$H(y_0 + \Delta y) - H(y_0) = H'(y_0^*)\, \Delta y \quad (14)$$

for some point $y_0 < y_0^* < y_0 + \Delta y$. By (13),

$$H'(y_0^*) = f_y(x_0, y_0^*)$$

and by (12) and (14), therefore,

$$f(x_0, y_0 + \Delta y) - f(x_0, y_0) = f_y(x_0, y_0^*)\, \Delta y \quad (15)$$

Substituting (11) and (15) into (7) thus gives

$$\begin{aligned}f(x_0 + \Delta x, y_0 + \Delta y) - f(x_0, y_0) &= f_x(x_0^*, y_0 + \Delta y)\, \Delta x \\ &\quad + f_y(x_0, y_0^*)\, \Delta y\end{aligned} \quad (16)$$

We now define

$$\begin{aligned}\epsilon_1 &= f_x(x_0^*, y_0 + \Delta y) - f_x(x_0, y_0) \\ \epsilon_2 &= f_y(x_0, y_0^*) - f_y(x_0, y_0)\end{aligned} \quad (17)$$

and then

$$\begin{aligned}f_x(x_0^*, y_0 + \Delta y) &= f_x(x_0, y_0) + \epsilon_1 \\ f_y(x_0, y_0^*) &= f_y(x_0, y_0) + \epsilon_2\end{aligned}$$

In particular, equation (16) can be written in the form

$$\begin{aligned}f(x_0 + \Delta x, y_0 + \Delta y) - f(x_0, y_0) &= f_x(x_0, y_0)\, \Delta x + f_y(x_0, y_0)\, \Delta y \\ &\quad + [\epsilon_1\, \Delta x + \epsilon_2\, \Delta y]\end{aligned}$$

Putting now

$$\epsilon = \epsilon_1\, \Delta x + \epsilon_2\, \Delta y$$

we obtain equation (1) appearing in the statement of the theorem. From the continuity of the partial derivatives f_x and f_y in an open disk with center at (x_0, y_0), it follows from (17) that ϵ_1 and ϵ_2 approach zero as $\Delta x \to 0$ and $\Delta y \to 0$, and this completes the proof of Theorem 1.

EXERCISES

In Exercises 1–10, find the linear approximation to f near the given point.

1. $f(x, y) = \ln(2x + y)$ $(x_0, y_0) = (0.1, 0.1)$
2. $f(x, y) = e^{x-y}$ $(x_0, y_0) = (0, 1)$
3. $f(x, y) = \dfrac{1}{\sqrt{x^2 + y}}$ $(x_0, y_0) = (2, 1)$
4. $f(x, y) = \sin(xy)$ $(x_0, y_0) = \left(\dfrac{\pi}{3}, \dfrac{\pi}{3}\right)$
5. $f(x, y) = \sqrt{\dfrac{1 + x^2}{1 + y^2}}$ $(x_0, y_0) = (2, 3)$
6. $f(x, y) = \dfrac{xy - 1}{xy + 1}$ $(x_0, y_0) = (\tfrac{1}{2}, \tfrac{1}{4})$
7. $f(x, y, z) = \sqrt{z^2 + xy}$ $(x_0, y_0, z_0) = (2, \tfrac{1}{2}, \tfrac{1}{4})$
8. $f(x, y, z) = xe^y + ye^z$ $(x_0, y_0, z_0) = (0.1, -0.1, 1)$
9. $f(x, y, z) = \sqrt{1 - x^2 - y^2 - z^2}$ $(x_0, y_0, z_0) = (\tfrac{1}{2}, \tfrac{1}{2}, \tfrac{1}{2})$
10. $f(x, y, z) = \dfrac{1}{\sqrt{z^2 - x^2 - y^2}}$ $(x_0, y_0, z_0) = (2, 3, 4)$

11. Find an approximate value for $f(2.3, -0.1, 0.1)$ when $f(x, y, z) = xe^y + ye^z$.

13.8 THE CHAIN RULE

Let $F(x, y)$ be a function of two variables, and suppose that x and y are functions of t, say, $x = g(t)$ and $y = h(t)$. Then a substitution gives $F[g(t), h(t)]$, and F is seen to be a function of the single variable t. The chain rule gives us a formula for differentiating such a function (in this connection, see Section 3.6). Functions of more than two variables are discussed later in this section.

THEOREM 1
Let $F(x, y)$ be a given function and suppose that $x = g(t)$ and $y = h(t)$. Then

$$\frac{dF}{dt} = \frac{\partial F}{\partial x}\frac{dx}{dt} + \frac{\partial F}{\partial y}\frac{dy}{dt} \tag{1}$$

provided the derivatives on the right side exist.

The proof of this theorem is given following an example.

Example 1

Find dF/dt if $F(x, y) = x^2y^3$, $x = e^t$, and $y = e^{-t}$.

Solution
Direct differentiation gives

$$\frac{\partial F}{\partial x} = 2xy^3 \qquad \frac{\partial F}{\partial y} = 3x^2y^2$$

$$\frac{dx}{dt} = e^t \qquad \frac{dy}{dt} = -e^{-t}$$

Hence

$$\frac{dF}{dt} = 2xy^3 e^t + 3x^2y^2(-e^{-t}) = 2xy^3 e^t - 3x^2y^2 e^{-t}$$

Substituting for x and y gives

$$\frac{dF}{dt} = 2e^t e^{-3t} e^t - 3e^{2t} e^{-2t} e^{-t} = 2e^{-t} - 3e^{-t} = -e^{-t}$$

Proof of Theorem 1
By Theorem 1 of Section 13.7, we have the formula

$$F(x + \Delta x, y + \Delta y) - F(x, y) = \frac{\partial F}{\partial x} \Delta x + \frac{\partial F}{\partial y} \Delta y + \epsilon_1 \Delta x + \epsilon_2 \Delta y \qquad (2)$$

where $\epsilon_1 \to 0$ and $\epsilon_2 \to 0$ as $\Delta x \to 0$ and $\Delta y \to 0$. With the substitutions

$$x = g(t)$$
$$y = h(t)$$

and

$$\Delta x = g(t + \Delta t) - g(t)$$
$$\Delta y = h(t + \Delta t) - h(t)$$

we obtain from (2) the formula

$$F[g(t + \Delta t), h(t + \Delta t)] - F[g(t), h(t)]$$

$$= \frac{\partial F}{\partial x}[g(t + \Delta t) - g(t)] + \frac{\partial F}{\partial y}[h(t + \Delta t) - h(t)]$$

$$+ \epsilon_1[g(t + \Delta t) - g(t)] + \epsilon_2[h(t + \Delta t) - h(t)]$$

Designating the left side by ΔF and dividing both sides by Δt gives

$$\frac{\Delta F}{\Delta t} = \frac{\partial F}{\partial x} \frac{g(t + \Delta t) - g(t)}{\Delta t} + \frac{\partial F}{\partial y} \frac{h(t + \Delta t) - h(t)}{\Delta t}$$

$$+ \epsilon_1 \frac{g(t + \Delta t) - g(t)}{\Delta t} + \epsilon_2 \frac{h(t + \Delta t) - h(t)}{\Delta t} \qquad (3)$$

Now,

$$\lim_{\Delta t \to 0} \frac{g(t + \Delta t) - g(t)}{\Delta t} = \frac{dx}{dt}$$

$$\lim_{\Delta t \to 0} \frac{h(t + \Delta t) - h(t)}{\Delta t} = \frac{dy}{dt}$$

Also, $\Delta x \to 0$ and $\Delta y \to 0$ as $\Delta t \to 0$, and hence also $\epsilon_1 \to 0$ and $\epsilon_2 \to 0$ as $\Delta t \to 0$. When this is applied to (3) we obtain formula (1), and this completes the proof.

IMPLICIT DIFFERENTIATION

In Section 13.2 we considered the level curves $F(x, y) = c$ of functions of two variables. The equation $F(x, y) = c$ defines a relation between x and y, and in many cases this relation is a function $y = h(x)$. We shall now establish a formula for finding $dy/dx = h'(x)$ in terms of the partial derivatives of F.

COROLLARY 1
If $F(x, y) = c$, then

$$\frac{dy}{dx} = -\frac{\partial F}{\partial x} \bigg/ \frac{\partial F}{\partial y} \qquad (4)$$

provided the partial derivatives exist and $\partial F/\partial y \neq 0$.

Proof
If we designate by $h(x)$ the function determined by the relation $F(x, y) = c$, then a substitution gives

$$F(x, h(x)) = c$$

Putting

$$x = t$$
$$y = h(t)$$

we can apply Theorem 1 to give

$$\frac{dF}{dx} = \frac{\partial F}{\partial x}\frac{dx}{dx} + \frac{\partial F}{\partial y}\frac{dy}{dx} = \frac{\partial F}{\partial x} + \frac{\partial F}{\partial y}\frac{dy}{dx}$$

Since F is the constant c we have $dF/dx = 0$, hence

$$\frac{\partial F}{\partial x} + \frac{\partial F}{\partial y}\frac{dy}{dx} = 0$$

and formula (4) follows.

Example 2

Putting $F(x, y) = x^2 + y^2$ gives

$$\frac{\partial F}{\partial x} = 2x \qquad \frac{\partial F}{\partial y} = 2y$$

and hence, if $x^2 + y^2 = 1$, then

$$\frac{dy}{dx} = -\frac{2x}{2y} = -\frac{x}{y}$$

A more sophisticated version of the chain rule concerns the case when x and y are functions of two variables.

THEOREM 2

If $F(x, y)$ is a given function of x and y, and

$$\left. \begin{array}{l} x = g(u, v) \\ y = h(u, v) \end{array} \right\}$$

then

$$\boxed{\begin{aligned} \frac{\partial F}{\partial u} &= \frac{\partial F}{\partial x}\frac{\partial x}{\partial u} + \frac{\partial F}{\partial y}\frac{\partial y}{\partial u} \\ \frac{\partial F}{\partial v} &= \frac{\partial F}{\partial x}\frac{\partial x}{\partial v} + \frac{\partial F}{\partial y}\frac{\partial y}{\partial v} \end{aligned}} \qquad \begin{aligned}(5)\\(6)\end{aligned}$$

whenever the derivatives exist.

This theorem is proved by observing that these formulas are obtained from formula (1) by holding, in turn, u and v constant.

Example 3

Consider the function

$$F(x, y) = xe^y$$

in polar coordinates

$$\begin{aligned} x &= r \cos \theta \\ y &= r \sin \theta \end{aligned}$$

Find $\partial F / \partial r$ and $\partial F / \partial \theta$.

Solution

$$\frac{\partial F}{\partial x} = e^y \qquad \frac{\partial F}{\partial y} = xe^y$$

Also,

$$\left.\begin{array}{ll} \dfrac{\partial x}{\partial r} = \cos\theta & \dfrac{\partial y}{\partial r} = \sin\theta \\[2mm] \dfrac{\partial x}{\partial \theta} = -r\sin\theta & \dfrac{\partial y}{\partial \theta} = r\cos\theta \end{array}\right\} \quad (7)$$

Putting $u = r$ and $v = \theta$ in Theorem 2 gives

$$\frac{\partial F}{\partial r} = e^y \cos\theta + xe^y \sin\theta = (1 + r\sin\theta)\cos\theta\, e^{r\sin\theta}$$

$$\frac{\partial F}{\partial \theta} = e^y(-r\sin\theta) + xe^y r\cos\theta = (r\cos^2\theta - \sin\theta)re^{r\sin\theta}$$

Example 4

If F is a function of x and y, show that, in polar coordinates,

$$\boxed{\left(\frac{\partial F}{\partial x}\right)^2 + \left(\frac{\partial F}{\partial y}\right)^2 = \left(\frac{\partial F}{\partial r}\right)^2 + \frac{1}{r^2}\left(\frac{\partial F}{\partial \theta}\right)^2} \quad (8)$$

Solution
By Theorem 2 and the formulas in (7) we have

$$\left.\begin{array}{l} \dfrac{\partial F}{\partial r} = \dfrac{\partial F}{\partial x}\cos\theta + \dfrac{\partial F}{\partial y}\sin\theta \\[2mm] \dfrac{1}{r}\dfrac{\partial F}{\partial \theta} = -\dfrac{\partial F}{\partial x}\sin\theta + \dfrac{\partial F}{\partial y}\cos\theta \end{array}\right\} \quad (9)$$

Hence,

$$\left(\frac{\partial F}{\partial r}\right)^2 = \left(\frac{\partial F}{\partial x}\right)^2 \cos^2\theta + \left(\frac{\partial F}{\partial y}\right)^2 \sin^2\theta + 2\frac{\partial F}{\partial x}\frac{\partial F}{\partial y}\cos\theta\sin\theta$$

$$\frac{1}{r^2}\left(\frac{\partial F}{\partial \theta}\right)^2 = \left(\frac{\partial F}{\partial x}\right)^2 \sin^2\theta + \left(\frac{\partial F}{\partial y}\right)^2 \cos^2\theta - 2\frac{\partial F}{\partial x}\frac{\partial F}{\partial y}\cos\theta\sin\theta$$

and adding these equations gives the desired formula.

Using simple algebra we can solve the equations in (9) for $\partial F/\partial x$ and $\partial F/\partial y$ to get

$$\boxed{\begin{array}{l} \dfrac{\partial F}{\partial x} = \dfrac{\partial F}{\partial r}\cos\theta - \dfrac{\partial F}{\partial \theta}\dfrac{\sin\theta}{r} \\[3mm] \dfrac{\partial F}{\partial y} = \dfrac{\partial F}{\partial r}\sin\theta + \dfrac{\partial F}{\partial \theta}\dfrac{\cos\theta}{r} \end{array}} \quad (10)$$

There are similar differentiation rules for functions of three and more variables. Without proof we state the following theorem.

THEOREM 3
If $F(x, y, z)$ is a given function of x, y, and z, and

$$x = f(t)$$
$$y = g(t)$$
$$z = h(t)$$

then

$$\frac{dF}{dt} = \frac{\partial F}{\partial x}\frac{dx}{dt} + \frac{\partial F}{\partial y}\frac{dy}{dt} + \frac{\partial F}{\partial z}\frac{dz}{dt}$$

whenever all derivatives exist.

Example 5

Let $w = z^2 - x^2 - y^2 - xy$, and

$$x = \cos t$$
$$y = \sin t$$
$$z = t$$

Find dw/dt.

Solution
Putting $F(x, y, z) = z^2 - x^2 - y^2 - xy$, we have

$$\frac{\partial F}{\partial x} = -2x - y \qquad \frac{dx}{dt} = -\sin t$$

$$\frac{\partial F}{\partial y} = -2y - x \qquad \frac{dy}{dt} = \cos t$$

$$\frac{\partial F}{\partial z} = 2z \qquad \frac{dz}{dt} = 1$$

and hence

$$\begin{aligned}\frac{dw}{dt} &= -(2x + y)(-\sin t) - (2y + x)\cos t + 2z \\ &= (2x + y)\sin t - (2y + x)\cos t + 2z \\ &= (2\cos t + \sin t)\sin t - (2\sin t + \cos t)\cos t + 2t \\ &= \sin^2 t - \cos^2 t + 2t = -\cos 2t + 2t\end{aligned}$$

A general statement of the chain rule is as follows:

THEOREM 4
If
$$w = F(x, y, z, \ldots, v)$$

and

$$x = f(r, s, t, \ldots, u)$$
$$y = g(r, s, t, \ldots, u)$$
$$z = h(r, s, t, \ldots, u)$$
$$\vdots$$
$$v = k(r, s, t, \ldots, u)$$

then

$$\frac{\partial F}{\partial r} = \frac{\partial F}{\partial x}\frac{\partial x}{\partial r} + \frac{\partial F}{\partial y}\frac{\partial y}{\partial r} + \frac{\partial F}{\partial z}\frac{\partial z}{\partial r} + \cdots + \frac{\partial F}{\partial v}\frac{\partial v}{\partial r}$$

$$\frac{\partial F}{\partial s} = \frac{\partial F}{\partial x}\frac{\partial x}{\partial s} + \frac{\partial F}{\partial y}\frac{\partial y}{\partial s} + \frac{\partial F}{\partial z}\frac{\partial z}{\partial s} + \cdots + \frac{\partial F}{\partial v}\frac{\partial v}{\partial s}$$

$$\frac{\partial F}{\partial t} = \frac{\partial F}{\partial x}\frac{\partial x}{\partial t} + \frac{\partial F}{\partial y}\frac{\partial y}{\partial t} + \frac{\partial F}{\partial z}\frac{\partial z}{\partial t} + \cdots + \frac{\partial F}{\partial v}\frac{\partial v}{\partial t}$$
$$\vdots$$
$$\frac{\partial F}{\partial u} = \frac{\partial F}{\partial x}\frac{\partial x}{\partial u} + \frac{\partial F}{\partial y}\frac{\partial y}{\partial u} + \frac{\partial F}{\partial z}\frac{\partial z}{\partial u} + \cdots + \frac{\partial F}{\partial u}\frac{\partial v}{\partial u}$$

provided the partial derivatives exist.

EXERCISES

Find dw/dt in Exercises 1–6.

1. $w = \dfrac{x-y}{x+y}$, $x = \ln t$, $y = t$

2. $w = e^x - e^y$, $x = e^{t^2}$, $y = e^{-t^2}$

3. $w = x^2 - y^2$, $x = e^t \cos t$, $y = e^t \sin t$

4. $w = \sin xy$, $x = \cos t$, $y = \sin t$

5. $w = \tan \dfrac{x}{y}$, $x = \tan t$, $y = 1 + \tan t$

6. $w = \sum_{k=0}^{n} x^k y^{n-k}$, $x = \dfrac{t}{t+1}$, $y = 1 - t$

Find $\partial w/\partial r$ and $\partial w/\partial \theta$ in Exercises 7–12.

7. $w = \dfrac{x-y}{x^2+y^2}$, $x = r \cos \theta$, $y = r \sin \theta$

8. $w = \sin x \sin y$, $x = r \cos \theta$, $y = r \sin \theta$

9. $w = \exp\left(-\dfrac{1}{x^2+y^2}\right)$, $x = re^\theta$, $y = re^{-\theta}$

10. $w = \tan^{-1} \dfrac{y}{x}$, $x = re^\theta - 1$, $y = re^\theta + 1$

11. $w = x^2 - 10y^2$, $x = r \sec\theta$, $y = r \tan\theta$

12. $w = e^{xy}$, $x = \ln(r^2 + \theta^2)$, $y = \tan^{-1}\dfrac{r}{\theta}$

Find dy/dx in Exercises 13–18.

13. $\tan^{-1}\dfrac{y}{x} + \dfrac{x}{y} = 0$

14. $x^2 y^3 + y \sin x = 0$

15. $x^3 + y^3 - 5xy = 0$

16. $\exp(x^2 - y^2) = \ln(x + y)$

17. $\sin xy + \cos(x + y) = 1$

18. $\sin(e^x + e^{-x}) = \tfrac{1}{2}$

13.9 HIGHER-ORDER PARTIAL DERIVATIVES

In this section we shall be concerned with applications of the chain rule discussed in the last section to higher-order derivatives. Basic to our computations will be the following fact, proved at the end of the section.

THEOREM 1
Let $F(x, y)$ be a given function. If $\partial F/\partial x$, $\partial F/\partial y$, $\partial^2 F/\partial x\, \partial y$, and $\partial^2 F/\partial y\, \partial x$ exist and are continuous in an open disk about (x_0, y_0), then

$$\frac{\partial^2 F}{\partial x\, \partial y} = \frac{\partial^2 F}{\partial y\, \partial x}$$

at (x_0, y_0).

Thus, under the conditions of this theorem, the order in which successive mixed partial derivatives is taken is unimportant. Some basis formulas will now be derived using this fact.

In Section 13.8 we derived the formula

$$\frac{dF}{dt} = \frac{\partial F}{\partial x}\frac{dx}{dt} + \frac{\partial F}{\partial y}\frac{dy}{dt} \tag{1}$$

Differentiating this formula with respect to t gives

$$\frac{d^2 F}{dt^2} = \frac{d}{dt}\left(\frac{\partial F}{\partial x}\frac{dx}{dt} + \frac{\partial F}{\partial y}\frac{dy}{dt}\right) = \frac{d}{dt}\left(\frac{\partial F}{\partial x}\frac{dx}{dt}\right) + \frac{d}{dt}\left(\frac{\partial F}{\partial y}\frac{dy}{\partial t}\right) \tag{2}$$

By the product rule for derivatives,

$$\frac{d}{dt}\left(\frac{\partial F}{\partial x}\frac{dx}{dt}\right) = \frac{d}{dt}\left(\frac{\partial F}{\partial x}\right)\frac{dx}{dt} + \frac{\partial F}{\partial x}\frac{d^2 x}{dt^2}$$

$$\frac{d}{dt}\left(\frac{\partial F}{\partial y}\frac{dy}{dt}\right) = \frac{d}{dt}\left(\frac{\partial F}{\partial y}\right)\frac{dy}{dt} + \frac{\partial F}{\partial y}\frac{d^2 y}{dt^2}$$

13.9 HIGHER-ORDER PARTIAL DERIVATIVES

If we apply, in turn, formula (1) to $\partial F/\partial x$ and $\partial F/\partial y$, we get

$$\frac{d}{dt}\left(\frac{\partial F}{\partial x}\right) = \frac{\partial}{\partial x}\left(\frac{\partial F}{\partial x}\right)\frac{dx}{dt} + \frac{\partial}{\partial y}\left(\frac{\partial F}{\partial x}\right)\frac{dy}{dt} = \frac{\partial^2 F}{\partial x^2}\frac{dx}{dt} + \frac{\partial^2 F}{\partial y\,\partial x}\frac{dy}{dt}$$

$$\frac{d}{dt}\left(\frac{\partial F}{\partial y}\right) = \frac{\partial}{\partial x}\left(\frac{\partial F}{\partial y}\right)\frac{dx}{dt} + \frac{\partial}{\partial y}\left(\frac{\partial F}{\partial y}\right)\frac{dy}{dt} = \frac{\partial^2 F}{\partial x\,\partial y}\frac{dx}{dt} + \frac{\partial^2 F}{\partial y^2}\frac{dy}{dt}$$

Hence

$$\frac{d}{dt}\left(\frac{\partial F}{\partial x}\frac{dx}{dt}\right) = \frac{\partial^2 F}{\partial x^2}\left(\frac{dx}{dt}\right)^2 + \frac{\partial^2 F}{\partial y\,\partial x}\frac{dx}{dt}\frac{dy}{dt} + \frac{\partial F}{\partial x}\frac{d^2 x}{dt^2}$$

$$\frac{d}{dt}\left(\frac{\partial F}{\partial y}\frac{dy}{dt}\right) = \frac{\partial^2 F}{\partial x\,\partial y}\frac{dx}{dt}\frac{dy}{dt} + \frac{\partial^2 F}{\partial y^2}\left(\frac{dy}{dt}\right)^2 + \frac{\partial F}{\partial y}\frac{d^2 y}{dt^2}$$

Substituting the latter relations into (2) and simplifying gives the formula

$$\boxed{\frac{d^2 F}{dt^2} = \frac{\partial^2 F}{\partial x^2}\left(\frac{dx}{dt}\right)^2 + 2\frac{\partial^2 F}{\partial x\,\partial y}\frac{dx}{dt}\frac{dy}{dt} + \frac{\partial^2 F}{\partial y^2}\left(\frac{dy}{dt}\right)^2 + \frac{\partial F}{\partial x}\frac{d^2 x}{dt^2} + \frac{\partial F}{\partial y}\frac{d^2 y}{dt^2}} \quad (3)$$

Example 1

Find d^2F/dt^2 when $F(x, y) = x^2 y^2$, $x = e^{2t}$, and $y = e^{3t}$.

Solution
By direct differentiation,

$$\frac{\partial F}{\partial x} = 2xy^2 \qquad \frac{\partial^2 F}{\partial x^2} = 2y^2$$

$$\frac{\partial F}{\partial y} = 2x^2 y \qquad \frac{\partial^2 F}{\partial y^2} = 2x^2$$

$$\frac{\partial^2 F}{\partial x\,\partial y} = \frac{\partial^2 F}{\partial y\,\partial x} = 4xy$$

$$\frac{dx}{dt} = 2e^{2t} \qquad \frac{d^2 x}{dt^2} = 4e^{2t}$$

$$\frac{dy}{dt} = 3e^{3t} \qquad \frac{d^2 y}{dt^2} = 9e^{3t}$$

Hence,

$$\frac{d^2 F}{dt^2} = 2y^2(2e^{2t})^2 + 2\cdot 4xy \cdot 2e^{2t} \cdot 3e^{3t} + 2x^2(3e^{3t})^2$$
$$+ 2xy^2 \cdot 4e^{2t} + 2x^2 y \cdot 9e^{3t}$$

Simplifying and rearranging terms gives

$$\frac{d^2 F}{dt^2} = 2e^{2t}(4xy^2 + 9x^2 y e^t + 4y^2 e^{2t} + 24xy e^{3t} + 9x^2 e^{4t})$$
$$= 100 e^{10t}$$

Next, let us examine the case in which x and y are functions of two independent variables.

Example 2

Let $F(x, y)$ be a given function and suppose that

$$x = u + v \qquad y = u - v$$

Find $\partial^2 F / \partial u\, \partial v$.

Solution
By Section 13.8,

$$\frac{\partial F}{\partial u} = \frac{\partial F}{\partial x}\frac{\partial x}{\partial u} + \frac{\partial F}{\partial y}\frac{\partial y}{\partial u} = \frac{\partial F}{\partial x} + \frac{\partial F}{\partial y}$$

Hence,

$$\frac{\partial^2 F}{\partial u\, \partial v} = \frac{\partial}{\partial v}\left(\frac{\partial F}{\partial x} + \frac{\partial F}{\partial y}\right) = \frac{\partial}{\partial v}\left(\frac{\partial F}{\partial x}\right) + \frac{\partial}{\partial v}\left(\frac{\partial F}{\partial y}\right)$$

$$= \left(\frac{\partial^2 F}{\partial x^2}\frac{\partial x}{\partial v} + \frac{\partial^2 F}{\partial x\, \partial y}\frac{\partial y}{\partial v}\right) + \left(\frac{\partial^2 F}{\partial x\, \partial y}\frac{\partial x}{\partial v} + \frac{\partial^2 F}{\partial y^2}\frac{\partial y}{\partial v}\right)$$

$$= \frac{\partial^2 F}{\partial x^2} - \frac{\partial^2 F}{\partial x\, \partial y} + \frac{\partial^2 F}{\partial x\, \partial y} - \frac{\partial^2 F}{\partial y^2}$$

Thus,

$$\frac{\partial^2 F}{\partial u\, \partial v} = \frac{\partial^2 F}{\partial x^2} - \frac{\partial^2 F}{\partial y^2}$$

General formulas for second-order partial derivatives in which x and y depend on two variables are given in the following theorem (see Theorem 2 of Section 13.8).

THEOREM 2
If $F(x, y)$ is a given function of x and y, and

$$x = g(u, v) \qquad y = h(u, v)$$

then

$$\frac{\partial^2 F}{\partial u^2} = \frac{\partial^2 F}{\partial x^2}\left(\frac{\partial x}{\partial u}\right)^2 + 2\frac{\partial^2 F}{\partial x\, \partial y}\frac{\partial x}{\partial u}\frac{\partial y}{\partial u} + \frac{\partial^2 F}{\partial y^2}\left(\frac{\partial y}{\partial u}\right)^2 + \frac{\partial F}{\partial x}\frac{\partial^2 x}{\partial u^2} + \frac{\partial F}{\partial y}\frac{\partial^2 y}{\partial u^2}$$

$$\frac{\partial^2 F}{\partial v^2} = \frac{\partial^2 F}{\partial x^2}\left(\frac{\partial x}{\partial v}\right)^2 + 2\frac{\partial^2 F}{\partial x\, \partial y}\frac{\partial x}{\partial v}\frac{\partial y}{\partial v} + \frac{\partial^2 F}{\partial y^2}\left(\frac{\partial y}{\partial v}\right)^2 + \frac{\partial F}{\partial x}\frac{\partial^2 x}{\partial v^2} + \frac{\partial F}{\partial y}\frac{\partial^2 y}{\partial v^2}$$

$$\frac{\partial^2 F}{\partial u\, \partial v} = \frac{\partial^2 F}{\partial x^2}\frac{\partial x}{\partial u}\frac{\partial x}{\partial v} + \frac{\partial^2 F}{\partial x\, \partial y}\left(\frac{\partial x}{\partial u}\frac{\partial y}{\partial v} + \frac{\partial x}{\partial v}\frac{\partial y}{\partial u}\right) + \frac{\partial^2 F}{\partial y^2}\frac{\partial y}{\partial u}\frac{\partial y}{\partial v}$$

$$+ \frac{\partial F}{\partial x}\frac{\partial^2 x}{\partial u\, \partial v} + \frac{\partial F}{\partial y}\frac{\partial^2 y}{\partial u\, \partial v}$$

THE ERROR IN THE LINEAR APPROXIMATION
In Theorem 1 of Section 13.7 we established the formula

$$f(x_0 + h, y_0 + k) = f(x_0, y_0) + f_x(x_0, y_0)h + f_y(x_0, y_0)k + \epsilon$$

where

$$\epsilon = \epsilon_1 h + \epsilon_2 k$$

and

$$\lim_{\substack{h \to 0 \\ k \to 0}} \epsilon_1 = \lim_{\substack{h \to 0 \\ k \to 0}} \epsilon_2 = 0$$

(in Section 13.7 we used the notation $h = \Delta x$ and $k = \Delta y$). We shall now derive a formula for the error ϵ, and we begin with the following example.

Example 3

Given the function $F(x, y)$, find formulas for dF/dt and d^2F/dt^2 when $x = x_0 + ht$ and $y = y_0 + kt$, where h and k are constants.

Solution
Since

$$\frac{dx}{dt} = h \quad \text{and} \quad \frac{dy}{dt} = k$$

we have, by formula (1) of Section 13.8,

$$\frac{dF}{dt} = F_x(x, y)h + F_y(x, y)k \tag{4}$$

since

$$\frac{d^2x}{dt^2} = 0 \quad \text{and} \quad \frac{d^2y}{dt^2} = 0$$

we have, by formula (3),

$$\frac{d^2F}{dt^2} = F_{xx}(x, y)h^2 + 2F_{xy}(x, y)hk + F_{yy}(x, y)k^2 \tag{5}$$

Let us now look at the formula

$$f(t) = f(0) + tf'(0) + \frac{t^2}{2!}f''(\alpha t) \qquad 0 < \alpha < 1 \tag{6}$$

(see Section 11.3). Setting

$$f(t) = F(x_0 + ht, y_0 + kt)$$

we have, using formulas (4) and (5),

$$F(x_0 + ht, y_0 + kt) = f(x_0, y_0) + t[F_x(x_0, y_0)h + F_y(x_0, y_0)k]$$
$$+ \frac{t^2}{2}[F_{xx}(x_0^*, y_0^*)h^2 + 2F_{xy}(x_0^*, y_0^*)hk$$
$$+ F_{yy}(x_0^*, y_0^*)k^2] \tag{7}$$

for some points

$$x_0 < x_0^* < x_0 + ht \quad \text{and} \quad y_0 < y_0^* < y_0 + kt$$

Letting $t = 1$, we get the second-order *Taylor formula*

$$F(x_0 + h, y_0 + k) = F(x_0, y_0) + [F_x(x_0, y_0)h + F_y(x_0, y_0)k]$$
$$+ \tfrac{1}{2}[F_{xx}(x_0^*, y_0^*)h^2 + 2F_{xy}(x_0^*, y_0^*)hk$$
$$+ F_{yy}(x_0^*, y_0^*)k^2] \tag{8}$$

With the notation $x = x_0 + h$, $y = y_0 + k$, this formula can also be written as

$$F(x, y) = F(x_0, y_0) + [F_x(x_0, y_0)(x - x_0) + F_y(x_0, y_0)(y - y_0)]$$
$$+ \tfrac{1}{2}[F_{xx}(x_0^*, y_0^*)(x - x_0)^2$$
$$+ 2F_{xy}(x_0^*, y_0^*)(x - x_0)(y - y_0)$$
$$+ F_{yy}(x_0^*, y_0^*)(y - y_0)^2] \tag{9}$$

Formula (9) will be needed in the proof of the theorem in the next section.

This section is concluded with a proof of Theorem 1.

Proof of Theorem 1

The proof of this theorem, like many others, is quite technical and depends on a trick: We start with the expression

$$U = F(x_0 + h, y_0 + k) - F(x_0 + h, y_0) - F(x_0, y_0 + k) + F(x_0, y_0)$$

The increments h and k are chosen to be positive so as to simplify the proof, but the other cases are very similar. Also, h and k are chosen sufficiently small to assume that $(x_0 + h, y_0 + k)$ lies in the disk specified in the statement of the theorem. Putting now

$$G(y) = F(x_0 + h, y) - F(x_0, y)$$

we have

$$G'(y) = F_y(x_0 + h, y) - F_y(x_0, y) \tag{10}$$

On the other hand, by the mean value theorem (see Section 4.6), we have

$$G(y_0 + k) - G(y_0) = kG'(y_0^*)$$

for some point $y_0 < y_0^* < y_0 + k$; from (10) we have

$$G'(y_0{}^*) = F_y(x_0 + h, y_0{}^*) - F_y(x_0, y_0{}^*)$$

Hence,
$$G(y_0 + k) - G(y_0) = k[F_y(x_0 + h, y_0{}^*) - F_y(x_0, y_0{}^*)]$$

A simple calculation shows that
$$U = G(y_0 + k) - G(y_0)$$

and thus
$$U = k[F_y(x_0 + h, y_0{}^*) - F_y(x_0, y_0{}^*)]$$

Now, using the mean value theorem once more, we have
$$F_y(x_0 + h, y_0{}^*) - F_y(x_0, y_0{}^*) = hF_{yx}(x_0{}^*, y_0{}^*)$$

for some point $x_0 < x_0{}^* < x_0 + h$, and thus
$$U = hkF_{yx}(x_0{}^*, y_0{}^*) \tag{11}$$

Repeating a similar argument with the function
$$H(x) = F(x, y_0 + k) - F(x, y_0)$$

instead of $G(y)$, we arrive at the conclusion that
$$U = hkF_{xy}(x_0{}^{**}, y_0{}^{**}) \tag{12}$$

for some points $x_0 < x_0{}^{**} < x_0 + h$, $y_0 < y_0{}^{**} < y_0 + k$. Comparing (11) with (12) shows that
$$F_{xy}(x_0{}^{**}, y_0{}^{**}) = F_{yx}(x_0{}^*, y_0{}^*) \tag{13}$$

where
$$x_0 < x_0{}^* < x_0 + h \qquad y_0 < y_0{}^* < y_0 + k$$

and
$$x_0 < x_0{}^{**} < x_0 + h \qquad y_0 < y_0{}^{**} < y_0 + k$$

By the continuity of F_{xy} and F_{yx} at (x_0, y_0), we have
$$\lim_{\substack{h \to 0 \\ k \to 0}} F_{yx}(x_0^*, y_0^*) = F_{yx}(x_0, y_0)$$

$$\lim_{\substack{h \to 0 \\ k \to 0}} F_{xy}(x_0{}^{**}, y_0{}^{**}) = F_{xy}(x_0, y_0)$$

By (13),
$$\lim_{\substack{h \to 0 \\ k \to 0}} F_{yx}(x_0^*, y_0^*) = \lim_{\substack{h \to 0 \\ k \to 0}} F_{xy}(x_0{}^{**}, y_0{}^{**})$$

and this completes the proof.

EXERCISES

1. Let $u = e^y \sin x$, $v = e^y \cos x$. Show that

   (a) $\dfrac{\partial u}{\partial x} = \dfrac{\partial v}{\partial y}$

   $\dfrac{\partial u}{\partial y} = -\dfrac{\partial v}{\partial x}$

   (b) $\dfrac{\partial^2 u}{\partial x^2} + \dfrac{\partial^2 u}{\partial y^2} = 0$ and $\dfrac{\partial^2 v}{\partial x^2} + \dfrac{\partial^2 v}{\partial y^2} = 0$

   REMARK
   The equations in (a) are called *Cauchy-Riemann equations;* the equations in (b) are called *Laplace equations.*

2. Let $F(x, y) = \ln(x^2 + y^2) + \tan^{-1}(y/x)$. Show that

   $$\dfrac{\partial^2 F}{\partial x^2} + \dfrac{\partial^2 F}{\partial y^2} = 0$$

3. If $F(x, y)$ is a given function, show that in polar coordinates $x = r \cos \theta$ and $y = r \sin \theta$ we have

   $$\dfrac{\partial^2 F}{\partial x^2} + \dfrac{\partial^2 F}{\partial y^2} = \dfrac{\partial^2 F}{\partial r^2} + \dfrac{1}{r}\dfrac{\partial F}{\partial r} + \dfrac{1}{r^2}\dfrac{\partial^2 F}{\partial \theta^2}$$

4. If $F(x, y)$ is a given function, $x = \tfrac{1}{2}u(e^v + e^{-v})$ and $y = \tfrac{1}{2}u(e^v - e^{-v})$, show that

   $$\dfrac{\partial^2 F}{\partial x^2} - \dfrac{\partial^2 F}{\partial y^2} = \dfrac{\partial^2 F}{\partial u^2} + \dfrac{1}{u}\dfrac{\partial F}{\partial u} - \dfrac{1}{u^2}\dfrac{\partial^2 F}{\partial v^2}$$

5. Let $F(x, y, z) = 1/\sqrt{x^2 + y^2 + z^2}$. Show that

   $$\dfrac{\partial^2 F}{\partial x^2} + \dfrac{\partial^2 F}{\partial y^2} + \dfrac{\partial^2 F}{\partial z^2} = 0$$

6. Let $F(x, t) = f(x + ct) + g(x - ct)$. Show that

   $$\dfrac{\partial^2 F}{\partial t^2} = c^2 \dfrac{\partial^2 F}{\partial x^2}$$

7. Let $F(x, y)$ be given, and suppose that $x = e^r \cos \theta$ and $y = e^r \sin \theta$. Show that

   $$\dfrac{\partial^2 F}{\partial x^2} + \dfrac{\partial^2 F}{\partial y^2} = e^{-2r}\left(\dfrac{\partial^2 F}{\partial r^2} + \dfrac{\partial^2 F}{\partial \theta^2}\right)$$

13.10 MAXIMA AND MINIMA

Consider the surface $z = F(x, y)$ in Figure 13.63. At the maximum point $P = (x_0, y_0, z_0)$ both of the cross sections $z = F(x, y_0)$ and

13.10 MAXIMA AND MINIMA

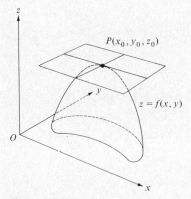

Figure 13.63
A surface with a maximum point at P.

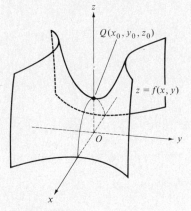

Figure 13.64
A surface with a saddle point at Q.

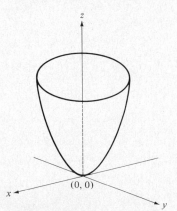

Figure 13.65
The minimum of the surface $F(x, y) = x^2 + y^2$.

$z = F(x_0, y)$ have a maximum point. Assuming f to have partial derivatives, we can conclude from our knowledge of functions of one variable that

$$F_x(x_0, y_0) = 0$$
$$F_y(x_0, y_0) = 0 \qquad (1)$$

Since the cross sections are curves, we already know that the vanishing of the first derivative is not sufficient to guarantee either a maximum or a minimum.

An interesting situation is described in Figure 13.64, where the surface has a *saddle point* at $Q = (x_0, y_0, z_0)$: The point Q is a maximum point of the cross section $z = F(x, y_0)$, and a minimum point of the cross section $z = F(x_0, y)$. Here equations (1) hold, yet the surface has neither a maximum nor a minimum at Q.

In analogy to the definition of local maximum and minimum of a function of a single variable (see Section 4.1), we state the following definition:

DEFINITION 1
The function $F(x, y)$ has a *local maximum* at (x_0, y_0) if

$$F(x, y) < F(x_0, y_0)$$

for all points $(x, y) \ne (x_0, y_0)$ in a disk $(x-x_0)^2 + (y-y_0)^2 < \delta^2$; the function has a *local minimum* at (x_0, y_0) if

$$F(x, y) > F(x_0, y_0)$$

for all points $(x, y) \ne (x_0, y_0)$ in a disk $(x-x_0)^2 + (y-y_0)^2 < \delta$.

There is a corresponding definition of absolute maximum and absolute minimum. A function $F(x, y)$ defined on a domain D has an *absolute maximum* at (x_0, y_0) if $F(x, y) \le F(x_0, y_0)$ for all points (x, y) in D, an *absolute minimum* at (x_0, y_0) if $F(x, y) \ge F(x_0, y_0)$ for all (x, y) in D. In general, maximum or minimum points are referred to as *extremum points*.

Example 1

The function $F(x, y) = x^2 + y^2$, described in Figure 13.65, has a local minimum at $(0, 0)$. This follows from the fact that $x^2 + y^2 > 0$ when $(x, y) \ne (0, 0)$. Note also that $F_x(x, y) = 2x$ and $F_y(x, y) = 2y$, so that $F_x(0, 0) = 0$ and $F_y(0, 0) = 0$.

This local minimum is also an absolute minimum.

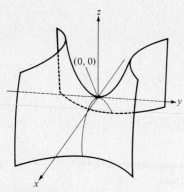

Figure 13.66
The surface $F(x, y) = x^2 - y^2$.

Example 2

The function $F(x, y) = x^2 - y^2$, described in Figure 13.66, has a saddle point at $(0, 0)$ since $F(x, 0) = x^2$ has a minimum at $x = 0$ and $F(0, y) = -y^2$ has a maximum at $y = 0$.

What follows next is a test to ensure the existence of a local maximum or minimum; the proof is delayed until the end of the section.

THEOREM 1
Let $F(x, y)$ and its first- and second-order partial derivatives be continuous on some open disk $(x - x_0)^2 + (y - y_0)^2 < \delta^2$. Suppose also that

$$F_x(x_0, y_0) = 0$$
$$F_y(x_0, y_0) = 0$$
$$[F_{xy}(x_0, y_0)]^2 - F_{xx}(x_0, y_0) F_{yy}(x_0, y_0) < 0$$

Then F has a *local maximum* at (x_0, y_0) if

$$F_{xx}(x_0, y_0) < 0$$

and F has a *local minimum* at (x_0, y_0) if

$$F_{xx}(x_0, y_0) > 0$$

We remark that F has neither a local maximum nor a local minimum at (x_0, y_0) when

$$[F_{xy}(x_0, y_0)]^2 - F_{xx}(x_0, y_0) F_{yy}(x_0, y_0) > 0$$

No conclusion can be reached when

$$[F_{xy}(x_0, y_0)]^2 - F_{xx}(x_0, y_0) F_{yy}(x_0, y_0) = 0$$

Example 3

Discuss the maxima and minima of the function

$$F(x, y) = x^2 + xy + y^2.$$

Solution
We begin by locating all possible candidates (x, y) for maxima and minima. According to our test, this is done by finding the zeros of F_x and F_y. The equations

$$F_x = 2x + y = 0$$
$$F_y = x + 2y = 0$$

have the single solution $(x, y) = (0, 0)$. Since

$$F_{xx} = 2 \qquad F_{yy} = 2 \qquad F_{xy} = 1$$

we find that, at $(0, 0)$,

$$(F_{xy})^2 - F_{xx}F_{yy} = 1 - 2 \cdot 2 = -3 < 0$$

Since

$$F_{xx} = 2 > 0$$

we conclude that f has a minimum at $(0, 0)$.

We note in passing that this problem can be solved without Theorem 1 once we observe that

$$x^2 + xy + y^2 = \left(x + \frac{y}{2}\right)^2 + \tfrac{3}{4}y^2$$

This identity tells us that $F(0, 0) = 0$, whereas $F(x, y) > 0$ for all points $(x, y) \neq (0, 0)$. Hence, f has a local minimum at $(0, 0)$ that is also an absolute minimum.

Example 4

Discuss the maxima and minima of

$$F(x, y) = x^3 - 3x - y^2$$

Solution
To find the candidates for extremum points, we solve the equations

$$F_x = 3x^2 - 3 = 0$$
$$F_y = -2y = 0$$

The solutions are $(-1, 0)$ and $(1, 0)$. Now,

$$F_{xx} = 6x$$
$$F_{yy} = -2$$
$$F_{xy} = 0$$

We thus have

$$[F_{xy}(-1, 0)]^2 - F_{xx}(-1, 0)F_{yy}(-1, 0)$$
$$= 0 - (-6)(-2) = -12 < 0$$
$$[F_{xy}(1, 0)]^2 - F_{xx}(1, 0)F_{yy}(1, 0)$$
$$= 0 - 6(-2) = 12 > 0$$

and hence there is no extremum point at $(1, 0)$. Since

$$F_{xx}(-1, 0) = -6 < 0$$

we conclude that F has a local maximum at $(-1, 0)$. Note that this local maximum is not an absolute maximum, since $F(-1, 0) = 2$ whereas, for example, $F(3, 0) = 18$. The given function has no absolute maximum since

$$\lim_{x \to \infty} F(x, 0) = \lim_{x \to \infty} (x^3 - 3x) = \infty$$

Example 5

Discuss the maxima and minima of

$$f(x, y) = A(x - x_0)^2 + 2B(x - x_0)(y - y_0) + C(y - y_0)^2 \qquad (2)$$

Solution

We apply Theorem 1 to find conditions on the coefficients A, B, and C, under which f has a local maximum or a local minimum at (x_0, y_0). Direct differentiation gives

$$\begin{array}{ll} f_x(x, y) = 2A(x - x_0) + 2B(y - y_0) & f_{xx}(x, y) = 2A \\ f_y(x, y) = 2B(x - x_0) + 2C(y - y_0) & f_{yy}(x, y) = 2C \\ f_{xy}(x, y) = 2B & \end{array}$$

and we see that the only candidate for zeros of f_x and f_y is (x_0, y_0). Hence,

$$f_x(x_0, y_0) = 0$$
$$f_y(x_0, y_0) = 0$$
$$[f_{xy}(x_0, y_0)]^2 - f_{xx}(x_0, y_0)f_{yy}(x_0, y_0) = 4(B^2 - AC)$$

and by Theorem 1,

if $B^2 - AC < 0$ and $A < 0$, then f has a local maximum at (x_0, y_0);
if $B^2 - AC < 0$ and $A > 0$, then f has a local minimum at (x_0, y_0).

The solution to Example 5 can be obtained algebraically as follows: When $A \neq 0$, then $f(x, y)$ can be written in the form

$$f(x, y) = \frac{1}{A}\{[A(x - x_0) + B(y - y_0)]^2 - (B^2 - AC)(y - y_0)^2\}$$

Thus, if $B^2 - AC < 0$, then $-(B^2 - AC) > 0$ and it is seen that the expression in brackets $\{\ \}$ is always positive. It follows that the sign of $f(x, y)$ coincides with the sign of A. Since $f(x_0, y_0) = 0$, we can conclude that f has a maximum at (x_0, y_0) if $A < 0$, a minimum at (x_0, y_0) if $A > 0$.

We now turn to the proof of Theorem 1.

Proof

Refer to formula (9) of Section 13.3. Suppose that $F_x(x_0, y_0) = 0$ and $F_y(x_0, y_0) = 0$, then this formula becomes

$$\begin{aligned} F(x, y) = F(x_0, y_0) &+ \tfrac{1}{2}[F_{xx}(x_0^*, y_0^*)(x - x_0)^2 \\ &+ 2F_{xy}(x_0^*, y_0^*)(x - x_0)(y - y_0) \\ &+ F_{yy}(x_0^*, y_0^*)(y - y_0)^2] \end{aligned} \qquad (3)$$

Put

$$A = F_{xx}(x_0^*, y_0^*)$$
$$B = F_{xy}(x_0^*, y_0^*)$$
$$C = F_{yy}(x_0^*, y_0^*).$$

Then with the notation in (2) we can write (3) as

$$F(x, y) = F(x_0, y_0) + \tfrac{1}{2} f(x, y)$$

If now

$$[F_{xy}(x_0, y_0)]^2 - F_{xx}(x_0, y_0) F_{yy}(x_0, y_0) < 0$$

then by the continuity of F_{xx}, F_{xy}, and F_{yy} in an open disk about (x_0, y_0) we have also

$$[F_{xy}(x_0^*, y_0^*)]^2 - F_{xx}(x_0^*, y_0^*) F_{yy}(x_0^*, y_0^*) = B^2 - AC < 0$$

if (x_0^*, y_0^*) is sufficiently close to (x_0, y_0). From the remarks following Example 5, we know that $f(x, y)$ has the same sign as $A = F_{xx}(x_0^*, y_0^*)$. Thus, if (x_0^*, y_0^*) is sufficiently close to (x_0, y_0), then $F_{xx}(x_0^*, y_0^*) < 0$ when $F_{xx}(x_0, y_0) < 0$, and $F_{xx}(x_0^*, y_0^*) > 0$ when $F_{xx}(x_0, y_0) > 0$. Hence, we see that

if $F_{xx}(x_0, y_0) < 0$, then $F(x, y) - F(x_0, y_0) < 0$, (4)
if $F_{xx}(x_0, y_0) > 0$, then $F(x, y) - F(x_0, y_0) > 0$. (5)

The statement in (4) tells us that F has a local maximum at (x_0, y_0), and (5) tells us that F has a local minimum at (x_0, y_0). This concludes the proof.

EXERCISES

Find the local and absolute maximum and minimum points of the functions in Exercises 1–11; when none exist, state so.

1. $F(x, y) = (x - 2)^2 + (y + 2)^2$
2. $F(x, y) = x^2 y$
3. $F(x, y) = x^2 y^2$
4. $F(x, y) = (x + 3)^3 - 3(x + 3) - y^2$
5. $F(x, y) = x^3 - 2x + y^2$
6. $F(x, y) = x^4 + y^4 - 2(x - y)^2$
7. $F(x, y) = x^2 + xy + y^3$
8. $F(x, y) = \dfrac{1}{x^2 + y^2}$
9. $F(x, y) = xy^2(x + 2y - \tfrac{2}{3})$
10. $F(x, y) = \sin(x + y)$
11. $F(x, y) = x^3 + y^3$
12. A fuel reservoir at D (see Figure 13.67) is to service plants located at A, B, and C, as shown. The cost in thousands of

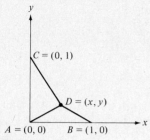

Figure 13.67

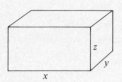

Figure 13.68

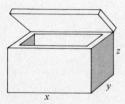

Figure 13.69

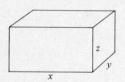

Figure 13.70

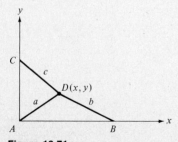

Figure 13.71

dollars connecting the plants to D is determined by the formula
$$C = 6x^2 + 6y^2 - 4x - 6y + 5$$
Find the location of D that will minimize this cost.

13. A metal detector is used to locate an underground pipe. When several meter readings of the detector are compared, it is found that the reading at an arbitrary point (x, y) is given by the formula
$$M = y(x - x^2) - y^2 \text{ volts} \qquad (x \geq 0, y \geq 0)$$
Find the point (x, y) where the reading is largest.

14. A closed rectangular box has surface area p (see Figure 13.68). What are the lengths of the sides that yield maximum volume?

15. A rectangular picnic cooler (see Figure 13.69) is to be constructed of a 1-in.-thick insulating material except for the lid, which is 2 in. thick. If capacity is 12,000 in.³, what inside dimensions will require the least volume $V = xyz$?

16. A steel manufacturer produces two grades of steel, x tons of grade A, and y tons of grade B. His cost c and revenue r are given in dollars by the formulas
$$c = \tfrac{1}{20}x^2 + 700x + y^2 - 150y - \tfrac{1}{2}xy$$
$$r = 2700x - \tfrac{3}{20}x^2 + 1000y - y^2 + \tfrac{1}{2}xy + 10,000$$
If
$$P = \text{profit} = r - c$$
find the production in tons of grades A and B of steel that maximizes the manufacturer's profit.

17. A toy manufacturer buys shipments of x tons of plastic which is used at the rate of y tons per week. It is found that waste due to storage and interplant distribution amounts to
$$W = \tfrac{1}{100}[\tfrac{1}{20}x^2 + 25y^2 - x(y+4)] \text{ tons}$$
For what values of x and y is the waste minimum?

18. The cost of material for the sides of a rectangular shipping container is a cents/ft²; the cost of the top and bottom material is $\tfrac{3}{2}a$ cents/ft². If the volume is to be $\tfrac{3}{2}$ ft³, what dimensions of the container in Figure 13.70 will minimize its cost? Find the location of the point $D(x, y)$ for which the sum of the squares of the distances a, b, and c is minimized (see Figure 13.71).

QUIZ 1

1. Find the center and radius of the sphere $x^2 + y^2 + z^2 - 2x + 2y + 4z + 5 = 0$.

2. Plot the level curves $f(x, y) = c$ when $f(x, y) = xy^2$ and when $c = -1, 1, 2$.

3. Find $\partial u/\partial x$, $\partial u/\partial y$, and $\partial^2 u/\partial x \partial y$ when

    (a) $u = \dfrac{x - y}{x + y}$

    (b) $u = \sin^{-1}\left(\dfrac{x}{y}\right)$

    (c) $u = \sqrt{x^2 + y^2}$

4. Find $\lim\limits_{\substack{x \to \infty \\ y \to \infty}} (x + y) \exp(-x^2 - y^2)$.

5. Find the domain of the function $F(x, y, z) = \ln(x^2 + y^2 - z)$.

6. Find the linear approximation to $f(x, y) = x \ln y + y \ln x$ at $(\tfrac{1}{2}, \tfrac{1}{4})$.

7. Find dw/dt when $w = \ln(x^2 + y^2)$, $x = \ln t$, and $y = t$.

8. Find $\partial w/\partial r$ and $\partial w/\partial \theta$ when $w = \sin x \sin^{-1} y$, $x = r \cos \theta$, and $y = r \sin \theta$.

9. Find dy/dx when

$$\frac{y}{x} \cos^{-1}\left(\frac{x}{y}\right) = 1$$

10. Given that $F(x, y) = \sin(x + 2y) + \cos(x - 2y)$. Show that $F_{yy} - 4F_{xx} = 0$.

11. Find the absolute and local maximum and minimum points of $F(x, y)$ when

    (a) $F(x, y) = x \sin y$

    (b) $F(x, y) = x^2 + y^3 - 3y$

    (c) $F(x, y) = \tfrac{1}{2}x^2 + 2y^2 - x(y + 1)$

QUIZ 2

1. Find the equation of the sphere with center $(\tfrac{1}{2}, -1, -\tfrac{1}{3})$ and radius 4.

2. Describe the cross sections of the surface $x - y^2 = 1$ cut by the coordinate planes.

3. Find $\partial u/\partial x$, $\partial u/\partial y$, and $\partial^2 u/\partial x \partial y$ when
   (a) $u = x \sin^{-1} y$
   (b) $u = \exp(x^2 + y^2)$
   (c) $u = \ln[x \ln y]$

4. Is the function
$$f(x, y) = \begin{cases} (x+y) \ln(x^2 + y^2) & \text{for } (x, y) \neq (0, 0) \\ 0 & \text{for } (x, y) = (0, 0) \end{cases}$$
continuous at $(0, 0)$? Justify your answer.

5. Find the domain of the function $F(x, y, z) = x/y$.

6. Find the linear approximation to $f(x, y) = \tan^{-1}(x/y)$ at $(\frac{1}{10}, \frac{1}{3})$.

7. Find dw/dt when
$$w = \exp\left(\frac{x-y}{x+y}\right), \quad x = \ln t, \quad \text{and} \quad y = t^2$$

8. Find $\partial w/\partial r$ and $\partial w/\partial \theta$ when
$$w = \tan^{-1}\left(\frac{x-y}{x+y}\right), \quad x = re^\theta, \quad \text{and} \quad y = re^{-\theta}$$

9. Find dy/dx when $\ln(x^2 + y^2) = \tan^{-1}(x + y)$.

10. Let $F(x, y)$ be a given function, and let $x = r \cos t$, $y = r \sin t$. Find $\partial^2 F/\partial r \partial t$.

11. Find the absolute and local maximum and minimum points of $F(x, y) = 8x^2 - xy + y^2$.

12. Of all triangles with sides x, y, and z, such that $x + y + z = c$, find the one having maximum area.

CHAPTER 14
MULTIPLE INTEGRALS

14.1 INTRODUCTION TO DOUBLE INTEGRALS

Consider a surface $z = f(x, y)$ and its "shadow" A in the xy plane as pictured in Figure 14.1. Connecting the surface and A with vertical line segments produces a solid whose volume we are seeking.

Our attempts in Chapter 6 to find the area under a curve led to the development of the *definite integral*. The problem of finding volumes will lead us in a similar way to consider *double integrals*, designated with a symbol

$$\iint_A f(x, y) \, dx \, dy \tag{1}$$

This symbol will be used here to designate volume. In the next section we shall define the double integral using approximations similar to those used in defining the definite integral. In this case we shall use rectangular solids in place of the rectangular areas used earlier.

To fix ideas, consult Table 14.1, in which areas and volumes are compared.

Example 1

Consider the equation of a sphere with center at the origin and radius 3:

$$x^2 + y^2 + z^2 = 3^2$$

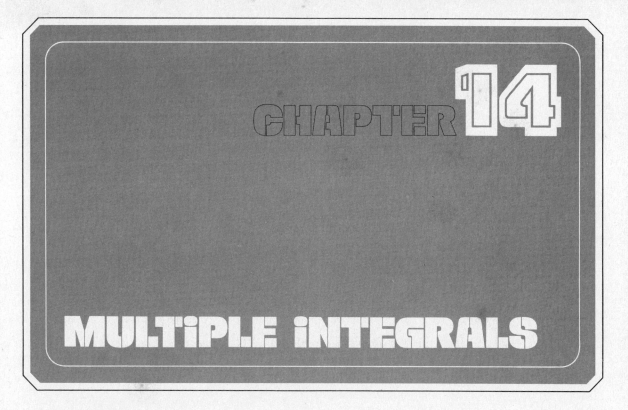

Figure 14.1
A solid under a surface $z = f(x, y)$.

556 MULTIPLE INTEGRALS

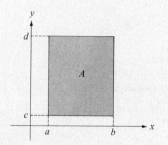

Figure 14.2
The hemisphere $z = \sqrt{9 - x^2 - y^2}$.

With restriction $z \geq 0$, we can write
$$z = \sqrt{9 - x^2 - y^2}$$

This is the equation of the upper hemisphere intersecting the xy plane in the circle $x^2 + y^2 = 3^2$ and lying above it (see Figure 14.2). From Section 8.3 we know this volume to be $\frac{1}{2}(\frac{4}{3}\pi 3^3) = 18\pi$. Using integral notation, we represent this volume with the symbol (1), where $f(x, y) = \sqrt{9 - x^2 - y^2}$ and A is the disk $x^2 + y^2 \leq 3$. Thus,

$$\iint_A \sqrt{9 - x^2 - y^2} \, dx \, dy = 18\pi$$

Table 14.1

	Area	Volume
bounded between	curve $y = f(x)$ and the x axis	surface $z = f(x, y)$ and the xy plane
domain of integration	an interval $[a, b]$ on the x axis	a set A in the xy plane
integral symbol	$\int_a^b f(x) \, dx$	$\iint_A f(x, y) \, dx \, dy$

Figure 14.3

The next example illustrates a general method that can be used when the region A is a rectangle of the form
$$a \leq x \leq b \quad \text{and} \quad c \leq y \leq d$$
(see Figure 14.3). The method used is an application of volumes by slicing given in Section 8.4. After this example with its two solutions, we shall state the general principle as a theorem.

Example 2

Find the volume of the solid with base $0 \leq x \leq 1$, $1 \leq y \leq 3$, lying between the xy plane and the surface $z = \frac{1}{3}(x^2 - x + 1)y^2$ (see Figure 14.4).

First Solution
We slice the solid by planes $x = x_0$ (see Figure 14.5). For each value x_0, where $0 \leq x_0 \leq 1$, we get the curve $z = \frac{1}{3}(x_0^2 - x_0 + 1)y^2$ in the plane $x = x_0$. The area, $A(x_0)$, under this curve is

$$A(x_0) = \int_1^3 \frac{1}{3}(x_0^2 - x_0 + 1)y^2 \, dy \tag{2}$$

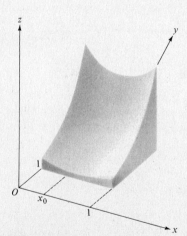

Figure 14.4
The solid under the surface $z = \frac{1}{3}(x^2 - x + 1)y^2$.

It is the area of the cross section determined by that plane. Evaluating the integral gives

$$A(x_0) = \tfrac{1}{3}(x_0^2 - x_0 + 1) \int_1^3 y^2 \, dy$$

$$= \tfrac{1}{3}(x_0^2 - x_0 + 1)\tfrac{1}{3}y^3 \Big|_1^3 = \tfrac{26}{9}(x_0^2 - x_0 + 1)$$

Replacing x_0 by x gives for each x in the interval $[0, 1]$ the formula

$$A(x) = \tfrac{26}{9}(x^2 - x + 1)$$

for the area of the cross section. The volume of the solid is

$$V = \int_0^1 A(x) \, dx = \int_0^1 \tfrac{26}{9}(x^2 - x + 1) \, dx = \tfrac{26}{9}(\tfrac{1}{3}x^3 - \tfrac{1}{2}x^2 + x)\Big|_0^1$$

$$= \tfrac{26}{9}(\tfrac{1}{3} - \tfrac{1}{2} + 1) = \tfrac{26}{9} \cdot \tfrac{5}{6} = \tfrac{65}{27}$$

We have thus completed the problem. Let us examine, however, how we arrive at the answer. Replacing x_0 by x in (2) gives the formula

$$A(x) = \int_1^3 \tfrac{1}{3}(x^2 - x + 1)y^2 \, dy \tag{3}$$

Observe that y is only a variable of integration (a dummy variable) and, as indicated by the notation $A(x)$, the expression is a function of x alone. Substituting (3) into the formula

$$V = \int_0^1 A(x) \, dx$$

gives the volume of the solid as

$$V = \int_0^1 \left[\int_1^3 \tfrac{1}{3}(x^2 - x + 1)y^2 \, dy \right] dx \tag{4}$$

This is an example of an *iterated integral*.

Second Solution

In the above solution we computed the volume by slicing the solid with cross sections $x = x_0$ that are parallel to the yx plane. We now take cross sections $y = y_0$ parallel to the xz plane (see Figure 14.6). In the plane $y = y_0$ we get the curve

$$z = \tfrac{1}{3}(x^2 - x + 1)y_0^2$$

The area under this curve is

$$A(y_0) = \int_0^1 \tfrac{1}{3}(x^2 - x + 1)y_0^2 \, dx = \tfrac{1}{3}y_0^2 \int_0^1 (x^2 - x + 1) \, dx$$

$$= \tfrac{1}{3}y_0^2(\tfrac{1}{3}x^3 - \tfrac{1}{2}x^2 + x)\Big|_0^1 = \tfrac{5}{18}y_0^2$$

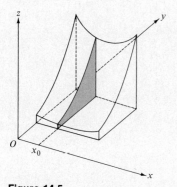

Figure 14.5
The area under the curve $z = \tfrac{1}{3}(x_0^2 - x_0 + 1)y^2$ is the cross section determined by the plane $x = x_0$.

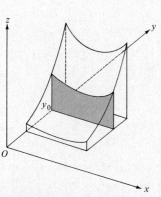

Figure 14.6
The area under the curve $z = \tfrac{1}{3}(x^2 - x + 1)y_0^2$ is the cross section determined by the plane $y = y_0$.

For each value of y in the interval $1 \leq y \leq 3$ we have the formula

$$A(y) = \tfrac{5}{18} y^2$$

and the volume of the solid is therefore

$$V = \int_1^3 A(y)\, dy = \int_1^3 \tfrac{5}{18} y^2\, dy = \tfrac{5}{18} \cdot \tfrac{1}{3} y^3 \Big|_1^3 = \tfrac{5}{18} \cdot \tfrac{26}{9} = \tfrac{65}{27}$$

Corresponding to (4) we can write this time

$$V = \int_1^3 \left[\int_0^1 \tfrac{1}{3}(x^2 - x + 1) y^2\, dx \right] dy \tag{5}$$

and comparing (4) with (5) we find that the order in which the integration is carried out is irrelevant, and hence

$$\int_0^1 \left[\int_1^3 \tfrac{1}{3}(x^2 - x + 1) y^2\, dy \right] dx = \int_1^3 \left[\int_3^1 \tfrac{1}{3}(x^2 - x + 1) y^2\, dx \right] dy$$

The preceding example leads to the following general statement.

THEOREM 1. VOLUME UNDER A SURFACE
If $f(x, y) \geq 0$ is continuous, then the solid bounded by surface $z = f(x, y)$ and the xy plane with $a \leq x \leq b$ and $c \leq y \leq d$ has volume given by either of the two formulas

$$V = \int_a^b \left[\int_c^d f(x, y)\, dy \right] dx \tag{6}$$

or

$$V = \int_c^d \left[\int_a^b f(x, y)\, dx \right] dy \tag{7}$$

With the notation (1) for volume we have

$$\iint_R f(x, y)\, dx\, dy = \int_a^b \left[\int_c^d f(x, y)\, dy \right] dx = \int_c^d \left[\int_a^b f(x, y)\, dx \right] dy$$

where R represents the rectangle $a \leq x \leq b$, $c \leq y \leq d$.

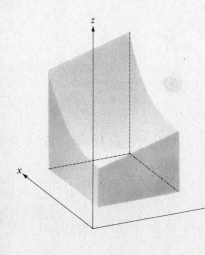

Example 3

Find the volume V of the solid contained between the xy plane and the surface $z = x^2 + y$ and having base R: $1 \leq x \leq 2$, $1 \leq y \leq 2$ (see Figure 14.7).

Figure 14.7
The solid under the surface $z = x^2 + y$.

Solution
Using formula (6), we can write for the volume

$$V = \iint_R (x^2 + y) \, dx \, dy = \int_1^2 \left[\int_1^2 (x^2 + y) \, dx \right] dy$$

Since

$$\int_1^2 (x^2 + y) \, dx = \tfrac{1}{3}x^3 + xy \Big|_{x=1}^{x=2} = \tfrac{7}{3} + y \tag{8}$$

we have

$$V = \int_1^2 (\tfrac{7}{3} + y) \, dy = \tfrac{7}{3}y - \tfrac{1}{2}y^2 \Big|_1^2 = \tfrac{23}{6}$$

Notice that in (8) we have written

$$\tfrac{1}{3}x^3 + xy \Big|_{x=1}^{x=2} \quad \text{instead of} \quad \tfrac{1}{3}x^2 + xy \Big|_1^2$$

to indicate the fact that y is treated as a constant. Calculating the volume with formula (7) gives

$$V = \int_1^2 \left[\int_1^2 (x^2 + y) \, dy \right] dx$$

and since

$$\int_1^2 (x^2 + y) \, dy = x^2 y + \tfrac{1}{2}y^2 \Big|_{y=1}^{y=2} = x^2 + \tfrac{3}{2}$$

we find once more that

$$V = \int_1^2 (x^2 + \tfrac{3}{2}) \, dx = \tfrac{1}{3}x^3 + \tfrac{3}{2}x \Big|_1^2 = \tfrac{23}{6}$$

When a function $f(x, y)$ of two variables can be written in the form

$$f(x, y) = g(x)h(y)$$

we say that the variables are *separated*. For such functions there is a very useful special case of Theorem 1.

THEOREM 2
If $f(x, y) = g(x)h(y)$ is continuous and R represents the rectangle $a \le x \le b$, $c \le y \le d$, then

$$\iint_R g(x)h(y) \, dx \, dy = \int_a^b g(x) \, dx \cdot \int_c^d h(y) \, dy$$

Proof
By Theorem 1

$$\iint_R g(x)h(y)\ dx\ dy = \int_a^b \left[\int_c^d g(x)h(y)\ dy \right] dx$$

Now,

$$\int_c^d g(x)h(y)\ dy = g(x) \int_c^d h(y)\ dy$$

since x is held constant when integrating with respect to y. Putting

$$A = \int_c^d h(y)\ dy$$

therefore gives

$$\int_a^b \left[\int_c^d g(x)h(y)\ dy \right] dx = \int_a^b [g(x) \cdot A]\ dx = \left(\int_a^b g(x)\ dx \right) A$$

$$= \int_a^b g(x)\ dx \cdot \int_c^d h(y)\ dy$$

and the theorem is seen to be true.

Example 4
Evaluate the integral $\iint_R \frac{1}{3}(x^2 - x + 1)y^2\ dx\ dy$ when R represents the rectangle $0 \le x \le 4$, $1 \le y \le 3$.

Solution
By Theorem 2,

$$\iint_R \frac{1}{3}(x^2 - x + 1)y^2\ dx\ dy = \int_0^4 \frac{1}{3}(x^2 - x + 1)\ dx \cdot \int_1^3 y^2\ dy$$

$$= \left[\frac{1}{3}(\frac{1}{3}x^3 - \frac{1}{2}x^2 + x) \Big|_0^4 \right] \cdot \left[\frac{1}{3}y^3 \Big|_1^3 \right]$$

$$= \frac{52}{9} \cdot \frac{26}{3} = \frac{1352}{27}$$

EXERCISES

Evaluate $\iint_A f(x, y)\ dx\ dy$ for the functions given in Exercises 1–16.

1. $f(x, y) = 3x - 1$ $A: -1 \le x \le 2, 0 \le y \le 5$
2. $f(x, y) = 3y - 1$ $A: -1 \le x \le 2, 0 \le y \le 5$
3. $f(x, y) = xye^{x^2}$ $A: 0 \le x \le 1, 2 \le y \le 3$

4. $f(x, y) = xy^2 e^{x^2}$ $A: 0 \leq x \leq 1, 2 \leq y \leq 3$

5. $f(x, y) = (x - y)^2$ $A: -1 \leq x \leq 1, -1 \leq y \leq 1$

6. $f(x, y) = (x - y)^4$ $A: -1 \leq x \leq 1, -1 \leq y \leq 1$

7. $f(x, y) = xy \ln x$ $A: 1 \leq x \leq 3, 2 \leq y \leq 4$

8. $f(x, y) = x \ln(xy)$ $A: 1 \leq x \leq 3, 2 \leq y \leq 4$

9. $f(x, y) = y \cos(xy)$ $A: 0 \leq x \leq 2, 0 \leq y \leq \pi$

10. $f(x, y) = y^2 \sin(xy)$ $A: 0 \leq x \leq 2\pi, 0 \leq y \leq 1$

11. $f(x, y) = \sin(x + y)$ $A: -\pi \leq x \leq \pi, 0 \leq y \leq \pi$

12. $f(x, y) = \cos(x - y)$ $A: 0 \leq x \leq \pi, -\frac{\pi}{2} \leq x \leq \frac{\pi}{2}$

13. $f(x, y) = \dfrac{xy}{|x|}$ $A: -1 \leq x \leq 1, 0 \leq y \leq 1$

14. $f(x, y) = 2x \sin(x^2 + y)$ $A: 0 \leq x \leq 2\pi, 0 \leq y \leq \pi$

15. $f(x, y) = 2^{x+y}$ $A: 0 \leq x \leq 1, -1 \leq y \leq 1$

16. $f(x, y) = x^y$ $A: 0 \leq x \leq 1, 1 \leq y \leq 2$

14.2 DOUBLE INTEGRALS AS LIMITS

In the previous section we gave a tentative interpretation of the double integral as volume. We have shown how double integrals can be evaluated as iterated integrals when the region of integration is rectangular. This section shows how to define the double integral as a limit, without relying on the notion of volume. The strategy for this is similar to the one used in defining the definite integral.

To have a concrete picture in mind, we shall define the double integral of $f(x, y)$ over a circular region A, but the procedure we use is a general one that applies to other shaped regions as well.

Let $R: a \leq x \leq b, c \leq y \leq d$ be a rectangle containing A (see Figure 14.8). This rectangle is divided into n^2 equal subrectangles as follows:

1. Divide the interval $[a, b]$ into n subintervals of lengths $\Delta x = (b - a)/n$ by means of the points

$$x_0 = a \quad x_1 = a + \Delta x \quad x_2 = a + 2\Delta x \ldots x_n = a + n\Delta x = b$$

A general formula for these points is

$$x_i = x_0 + i\Delta x \quad \text{for } i = 1, 2, \ldots, n$$

(This notation was first introduced in Chapter 6.)

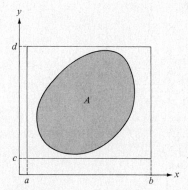

Figure 14.8

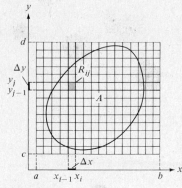

Figure 14.9

2. Divide the interval $[c, d]$ into n subintervals of lengths $\Delta y = (d - c)/n$ by means of the points

$$y_0 = c \quad y_1 = c + \Delta y \quad y_2 = c + 2\Delta y \ldots y_n = c + n\Delta y = d$$

A general formula for these points is

$$y_j = y_0 + j\Delta y \quad \text{for } j = 1, 2, \ldots, n$$

3. For each value of i and j, form the rectangle

$$R_{ij}: x_{i-1} \leq x \leq x_i, \quad y_{j-1} \leq y \leq y_j$$

(see Figure 14.9). Each rectangle R_{ij} has area $\Delta x \Delta y$.

Example 1

Before proceeding, we consider the special case with $R: 1 \leq x \leq 2$, $0 \leq y \leq 2$, and $n = 4$. Here

$$\Delta x = \frac{2-1}{4} = \frac{1}{4}$$

$$x_0 = 1, \, x_1 = \frac{5}{4}, \, x_2 = \frac{6}{4}, \, x_3 = \frac{7}{4}, \, x_4 = \frac{8}{4} = 2$$

$$x_i = 1 + \frac{1}{4}i$$

$$\Delta y = \frac{2-0}{4} = \frac{1}{2}$$

$$y_0 = 0, \, y_1 = \frac{1}{2}, \, y_2 = 1, \, y_3 = \frac{3}{2}, \, y_4 = \frac{4}{2} = 2$$

$$y_j = 0 + \frac{1}{2}j = \frac{1}{2}j$$

The rectangles R_{ij} we get are

$$R_{ij}: 1 + \tfrac{1}{4}(i-1) \leq x \leq 1 + \tfrac{1}{4}i, \quad \tfrac{1}{2}(j-1) \leq y \leq \tfrac{1}{2}j$$
for $i = 1, 2, 3, 4$ and $j = 1, 2, 3, 4$

This situation is described in Figure 14.10.

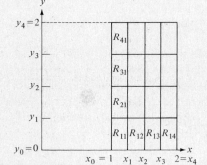

Figure 14.10

We now continue the general discussion. Consider the rectangular solid with base R_{ij} and height $f(x_i, y_j)$ (see Figure 14.11a). Intuitively, its volume, $f(x_i, y_j)\Delta x \Delta y$, is close to the volume under the surface $z = f(x, y)$ with the same base R_{ij}, if n is large (see Figure 14.11b). This is so because $f(x, y)$ does not change much over a small region R_{ij}. To approximate the double integral, we add up all such quantities and form the nth approximation S_n. For notational convenience we set $f(x_i, y_j) = 0$ when the point (x_i, y_j) is not in A. This has the effect of disregarding all rectangles whose upper right vertex (x_i, y_j) is not in A.

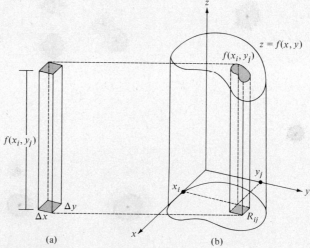

Figure 14.11

For each value of n let

$$S_n = \sum_{i,j=1}^{n} f(x_i, y_j) \, \Delta x \Delta y$$

where

$$\Delta x = \frac{b-a}{n} \qquad x_i = a + i \, \Delta x$$

$$\Delta y = \frac{d-c}{n} \qquad y_j = b + j \, \Delta y$$

The double integral of $f(x, y)$ over A is defined as the limit

$$\iint_A f(x, y) \, dx \, dy = \lim_{n \to \infty} S_n$$

provided this limit exists.

Example 2

Consider the function $f(x, y) = x + y$ and double integral

$$\iint_E (x + y) \, dx \, dy$$

taken over the unit square $E: 0 \leq x \leq 1, 0 \leq y \leq 1$. Let us approximate this integral with S_n for $n = 4$. Here our scheme looks as follows:

$$\Delta x = \frac{1-0}{4} = \frac{1}{4} \qquad x_i = 0 + i \cdot \frac{1}{4} = \frac{1}{4}i$$

$$\Delta y = \frac{1-0}{4} = \frac{1}{4} \qquad y_j = 0 + j \cdot \frac{1}{4} = \frac{1}{4}j$$

$$f(x_i, y_j) = x_i + y_j = \frac{1}{4}i + \frac{1}{4}j = \frac{1}{4}(i+j)$$

$$S_4 = \sum_{i,j=1}^{4} f(x_i, y_j) \Delta x \Delta y = \sum_{i,j=1}^{4} \frac{1}{4}(i+j)\frac{1}{4} \cdot \frac{1}{4} = \frac{1}{4^3} \sum_{i,j=1}^{4} (i+j)$$

$$= \frac{1}{4^3} \sum_{j=1}^{4} [(1+j) + (2+j) + (3+j) + (4+j)]$$

$$= \frac{1}{4^3}[(2+3+4+5) + (3+4+5+6) + (4+5+6+7)$$

$$+ (5+6+7+8)] = \tfrac{80}{64} = 1.25$$

Hence

$$\iint_E (x+y) \, dx \, dy \approx 1.25$$

From Section 14.1 we know that

$$\iint_E (x+y) \, dx \, dy = \int_0^1 \left[\int_0^1 (x+y) \, dx \right] dy$$

Since

$$\int_0^1 (x+y) \, dx = \tfrac{1}{2}x^2 + xy \Big|_{x=0}^{x=1} = \tfrac{1}{2} + y$$

we have

$$\int_0^1 \left[\int_0^1 (x+y) \, dx \right] dy = \int_0^1 (\tfrac{1}{2} + y) \, dy = \tfrac{1}{2}y + \tfrac{1}{2}y^2 \Big|_0^1 = 1$$

and thus

$$\iint_E (x+y) \, dx \, dy = 1$$

Considering that we used $n = 4$, the error in the approximation is reasonable. For the purpose of approximate integration, however, better methods are available. For instance, the analogue of Simpson's rule (Section 8.7) can be developed for double integrals.

14.2 DOUBLE INTEGRALS AS LIMITS

With Definition 1, it is possible to prove the existence of the double integral for a large class of functions and regions A. One can also derive the following properties, stated here without proof.

THEOREM 1. PROPERTIES OF DOUBLE INTEGRALS
The following properties hold whenever the integrals exist.

1. $\iint_A [f(x, y) + g(x, y)] \, dx \, dy$
$$= \iint_A f(x, y) \, dx \, dy + \iint_A g(x, y) \, dx \, dy;$$

2. $\iint_A [f(x, y) - g(x, y)] \, dx \, dy$
$$= \iint_A f(x, y) \, dx \, dy - \iint_A g(x, y) \, dx \, dy;$$

3. $\iint_A cf(x, y) \, dx \, dy = c \iint_A f(x, y) \, dx \, dy$, for any constant c.

4. If A is made up of two regions B and C having only part of their boundary in common (see Figure 14.12), then
$$\iint_A f(x, y) \, dx \, dy = \iint_B f(x, y) \, dx \, dy + \iint_C f(x, y) \, dx \, dy$$

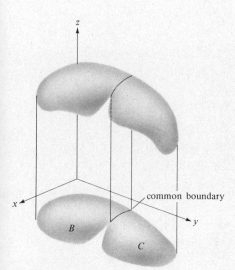

Figure 14.12

These properties should be compared with the corresponding properties of definite integrals (Section 6.3).

EXERCISES

Find the approximations S_n to the double integrals in Exercises 1–4 for $n = 4$.

1. $\iint_A xy \, dx \, dy \qquad A: -1 \leq x \leq 1, -1 \leq y \leq 1$

2. $\iint_A |xy| \, dx \, dy \qquad A: -1 \leq x \leq 1, -1 \leq y \leq 1$

3. $\iint_A x \, dx \, dy \qquad A: 0 \leq x \leq 1, 2 \leq y \leq 3$

4. $\iint_A \ln(x + y) \, dx \, dy \qquad A: 1 \leq x \leq 2, 0 \leq y \leq 1$

Evaluate the double integrals in Exercises 5–8.

5. $\iint_A (e^x + e^y)\, dx\, dy \quad A: -1 \leq x \leq 1, -1 \leq y \leq 1$

6. $\iint_A (\sin x - \sin y)\, dx\, dy \quad A: 0 \leq x \leq \frac{\pi}{2}, 0 \leq y \leq \frac{\pi}{2}$

7. $\iint_A (2x^2 + 4xy + y^2)\, dx\, dy \quad A: 0 \leq x \leq 1, -1 \leq y \leq 0$

8. $\iint_A 1\, dx\, dy \quad A: -1 \leq x \leq 1, -1 \leq y \leq 1$

9. Express the volume of a sphere of radius r as a double integral.

10. Express the volume of the ellipsoid

$$\frac{x^2}{2^2} + \frac{y^2}{3^2} + \frac{z^2}{4^2} = 1$$

as a double integral.

11. State the property of definite integrals having Property 4 of Theorem 1 as its analogue.

14.3 INTEGRATION OVER GENERAL REGIONS

In Section 14.2 we defined the double integral

$$\iint_A f(x, y)\, dx\, dy$$

for a general region A, but we offered no method for evaluating such integrals. The method of Section 14.1, we recall, applied only to rectangular regions A; in this section we develop methods that apply to other regions. We motivate the discussion with an example.

Example 1

Evaluate the integral

$$\iint_A (5 - x^2 - y^2)\, dx\, dy$$

when A is a triangular region bounded by the lines

$y = x, \quad y = 2x, \quad \text{and} \quad x = 1$

(see Figure 14.13).

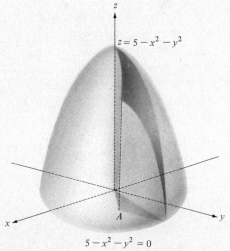

Figure 14.13
The surface $z = 5 - x^2 - y^2$ and the solid with triangular base A.

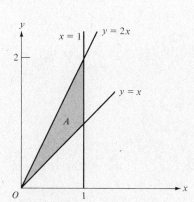

Figure 14.14
The region A of integration.

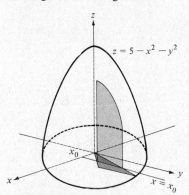

Figure 14.15
A cross section of the solid bounded by the surface $z = 5 - x^2 - y^2$ and above A.

Solution
We shall show here how to use the method of finding volume by slicing (Section 8.4) to evaluate double integrals. The first and most important step is to draw a good sketch of the region of integration. This is done here in Figure 14.14. In Figure 14.15 we identify the cross section lying in the plane $x = x_0$. We see that the area $A(x_0)$ of this cross section is given by the integral

$$A(x_0) = \int_{x_0}^{2x_0} (5 - x_0^2 - y^2)dy = (5 - x_0^2)y - \tfrac{1}{3}y^3 \Big|_{x_0}^{2x_0} = 5x_0 - \tfrac{10}{3}x_0^3$$

Thus the area of any cross section is given by the formula

$$A(x) = 5x - \tfrac{10}{3}x^3$$

for a value x between 0 and 1. Using the method of finding volume by slicing (Section 8.4), we find the volume to be

$$V = \int_0^1 A(x)\,dx = \int_0^1 (5x - \tfrac{10}{3}x^3)\,dx = \tfrac{5}{3}$$

Generalizing the procedure used in Example 1, consider a region A in the xy plane bounded by the curves $y = \alpha(x)$ and $y = \beta(x)$, and the lines $x = a$ and $x = b$ (see Figure 14.16). It is assumed that $\alpha(x) \leq \beta(x)$ for $a \leq x \leq b$.

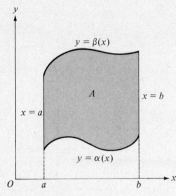

Figure 14.16

THEOREM 1
Let $f(x, y)$ be continuous over the region A bounded by the lines $x = a$ and $x = b$ and the curves $y = \alpha(x)$ and $y = \beta(x)$, where $\alpha(x) \leq \beta(x)$ for $a \leq x \leq b$. Then

$$\iint_A f(x, y) \, dx \, dy = \int_a^b \left[\int_{\alpha(x)}^{\beta(x)} f(x, y) \, dy \right] dx \tag{1}$$

Proof
Looking at Figure 14.17, we see that for the cross section lying in the plane $x = x_0$ we have

$$A(x_0) = \int_{\alpha(x_0)}^{\beta(x_0)} f(x_0, y) \, dy$$

Being true for each point x_0 in the interval $[a, b]$, we have

$$A(x) = \int_{\alpha(x)}^{\beta(x)} f(x, y) \, dx \quad \text{for } a \leq x \leq b$$

and hence by the method of slicing,

$$\iint_A f(x, y) \, dx \, dy = \int_a^b A(x) \, dx = \int_a^b \left[\int_{\alpha(x)}^{\beta(x)} f(x, y) \, dy \right] dx$$

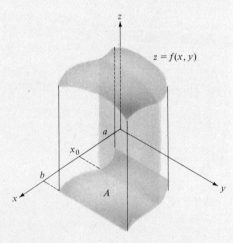

Figure 14.17
A cross section of a solid under $z = f(x, y)$.

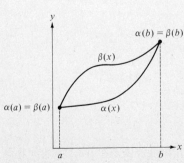

Figure 14.18

REMARK
When $\alpha(a) = \beta(a)$ and $\alpha(b) = \beta(b)$ (see Figure 14.18), then there is no need to mention the boundaries $x = a$ and $x = b$.

The roles of x and y can be interchanged in Theorem 1 to yield the following fact.

14.3 INTEGRATION OVER GENERAL REGIONS   569

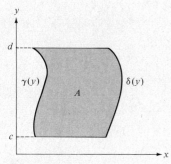

Figure 14.19

THEOREM 2
Let $f(x, y)$ be continuous over the region A bounded by the lines $y = c$ and $y = d$, and the curves $x = \gamma(y)$ and $y = \delta(y)$, where $\gamma(y) \leq \delta(y)$ for $c \leq y \leq d$ (see Figure 14.19). Then

$$\iint_A f(x, y) \, dx \, dy = \int_c^d \left[\int_{\gamma(y)}^{\delta(y)} f(x, y) \, dx \right] dy \qquad (2)$$

Example 2

Find the volume under the surface $z = xy^2$ with base A bounded by the curves $x = y^2$ and $x = y$.

Solution 1
Consult Figure 14.20. The curves $x = y^2$ and $x = y$ intersect when $y^2 = y$. This is true when $y = 0$ and $y = 1$, and the points of intersection are found to be $(0, 0)$ and $(1, 1)$. Taking $c = 0$, $d = 1$, $\gamma(y) = y^2$, and $\delta(y) = y$ in formula (2), we have

$$\iint_A xy^2 \, dx \, dy = \int_0^1 \left[\int_{y^2}^y xy^2 \, dx \right] dy$$

Since

$$\int_{y^2}^y xy^2 \, dx = y^2 \int_{y^2}^y x \, dx = \tfrac{1}{2} y^2 x^2 \Big|_{x=y^2}^{x=y} = \tfrac{1}{2}(y^4 - y^6)$$

and

$$\int_0^1 \tfrac{1}{2}(y^4 - y^6) \, dy = \tfrac{1}{2}(\tfrac{1}{5} y^5 - \tfrac{1}{7} y^7) \Big|_0^1 = \tfrac{1}{35}$$

we find that the volume is

$$\iint_A xy^2 \, dx \, dy = \tfrac{1}{35} \qquad (3)$$

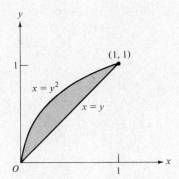

Figure 14.20

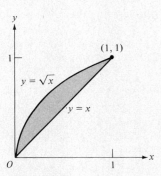

Figure 14.21

Solution 2
Let us find the volume using formula (1) instead of (2). To do this we replace the equations

$$x = y \quad \text{and} \quad x = y^2 \qquad 0 \leq x \leq 1$$

by the equivalent equations

$$y = x \quad \text{and} \quad y = \sqrt{x} \qquad 0 \leq x \leq 1$$

(see Figure 14.21). With $a = 0$, $b = 1$, $\alpha(x) = x$, and $\beta(x) = \sqrt{x}$, formula (1) becomes

$$\iint_A xy^2 \, dx \, dy = \int_0^1 \left[\int_x^{\sqrt{x}} xy^2 \, dy \right] dx$$

Since

$$\int_x^{\sqrt{x}} xy^2 \, dy = \tfrac{1}{3} xy^3 \Big|_{y=x}^{y=\sqrt{x}} = \tfrac{1}{3}(x^{5/2} - x^4)$$

and

$$\int_0^1 \tfrac{1}{3}(x^{5/2} - x^4) \, dx = \tfrac{1}{35}$$

We once more obtain (3).

Example 3

Evaluate

$$\iint_A (x^2 + y) \, dx \, dy$$

when A is the square having vertices

$$(1, 0), \quad (2, 1), \quad (1, 2), \quad \text{and} \quad (0, 1)$$

Solution

In evaluating this integral we make use of Property (4) of Theorem 1 in Section 14.2. In Figure 14.22 A is represented as the union of triangles B and C having a line segment in common. We thus have

$$\iint_A (x^2 + y) \, dx \, dy = \iint_B (x^2 + y) \, dx \, dy + \iint_C (x^2 + y) \, dx \, dy \qquad (4)$$

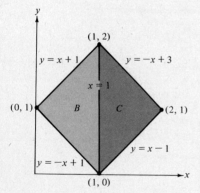

Figure 14.22

The integrals over B and C will now be evaluated using Theorem 1. Now, B is bounded by the lines $\alpha(x) = -x + 1$ and $\beta(x) = x + 1$, $(0 \le x \le 1)$, and $x = 1$. Hence

$$\iint_B (x^2 + y) \, dx \, dy = \int_0^1 \left[\int_{-x+1}^{x+1} (x^2 + y) \, dy \right] dx$$

Since

$$\int_{-x+1}^{x+1} (x^2 + y) \, dy = x^2 y + \tfrac{1}{2} y^2 \Big|_{y=-x+1}^{y=x+1} = 2(x^3 + x)$$

and

$$\int_0^1 2(x^3 + x) \, dx = \tfrac{3}{2}$$

we have

$$\iint_B (x^2 + y)\ dx\ dy = \tfrac{3}{2} \tag{5}$$

Similarly, C is bounded by the lines $\alpha(x) = x - 1$ and $\beta(x) = -x + 3$ for $1 \leq x \leq 2$, and $x = 1$ (see Figure 14.22). Hence

$$\iint_C (x^2 + y)\ dx\ dy = \int_1^2 \left[\int_{x-1}^{-x+3} (x^2 + y)\ dy \right] dx$$

$$= \int_1^2 (-2x^3 + 2x^2 - 2x + 4)\ dx = \tfrac{115}{24} \tag{6}$$

From (4), (5), and (6), we now find that

$$\iint_A (x^2 + y)\ dx\ dy = \tfrac{3}{2} + \tfrac{115}{24} = \tfrac{151}{24}$$

Example 4

Evaluate

$$\iint_A xy\ dx\ dy$$

when A is the triangle with vertices $(0, 0)$, $(1, 0)$, and $(1, 1)$.

Using Theorems 1 and 2 we shall evaluate this integral in two different ways.

Solution 1

Referring to Figure 14.23, we see that A is bounded by the lines $\alpha(x) = 0$ and $\beta(x) = x$ ($0 \leq x \leq 1$) and the line $x = 1$. By Theorem 1, therefore,

$$\iint_A xy\ dx\ dy = \int_0^1 \left[\int_0^x xy\ dy \right] dx = \int_0^1 \tfrac{1}{2} x^3\ dx = \tfrac{1}{8}$$

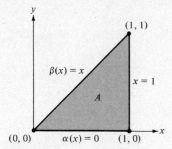

Figure 14.23

Solution 2

Now consult Figure 14.24. Using Theorem 2 with $\gamma(y) = y$ and $\delta(y) = 1$ gives

$$\iint_A xy\ dx\ dy = \int_0^1 \left[\int_y^1 xy\ dx \right] dy = \int_0^1 \tfrac{1}{2}(y - y^3)\ dy = \tfrac{1}{8}$$

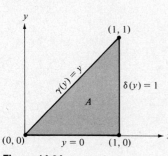

Figure 14.24

EXERCISES

In Exercises 1–4, find the volumes under $z = f(x, y)$ having base A bounded by $y = x^2 + 1$, $y = 0$, $x = 0$, and $x = 1$.

1. $z = 1$ **2.** $z = x$

3. $z = y$ **4.** $z = x + y$

In Exercises 5–8, evaluate the double integrals when A is the region bounded by the curves $y = \sqrt{x}$ and $y = x^3$.

5. $\displaystyle\iint_A x^2 \, dx \, dy$ **6.** $\displaystyle\iint_A y^2 \, dx \, dy$

7. $\displaystyle\iint_A x^2 y^2 \, dx \, dy$ **8.** $\displaystyle\iint_A (x + y)^2 \, dx \, dy$

9. Find α and β in the equation

$$\iint_A f(x, y) \, dx \, dy = \int_0^1 \left[\int_\alpha^\beta f(x, y) \, dx \right] dy$$

if A is the region bounded by the curves $y = \sqrt{x}$ and $y = x^3$.

10. Find α and β in the equation

$$\iint_A f(x, y) \, dx \, dy = \int_0^1 \left[\int_0^1 f(x, y) \, dx \right] dy + \int_1^2 \left[\int_\alpha^\beta f(x, y) \, dx \right] dy$$

if the region A is bounded by $y = x^2 + 1$, $y = 0$, $x = 0$, and $x = 1$.

Hint: See Example 4.

11. Evaluate $\displaystyle\int_0^1 \left[\int_y^1 e^{x^2} \, dx \right] dy$ by reversing the order of integration.

12. Evaluate

$$\iint_A x \, dx \, dy$$

When A is the region bounded by the curves $x^2 + y^2 = 1$ and $x^2 + (y - 1)^2 = 1$, and $x = 0$.

14.4 MASS, DENSITY, AND CENTER OF GRAVITY

Suppose that a mass is distributed over a region A in the xy plane. The *average density* of the material is defined to be

$$\text{average density} = \frac{\text{mass in } A}{\text{area of } A}$$

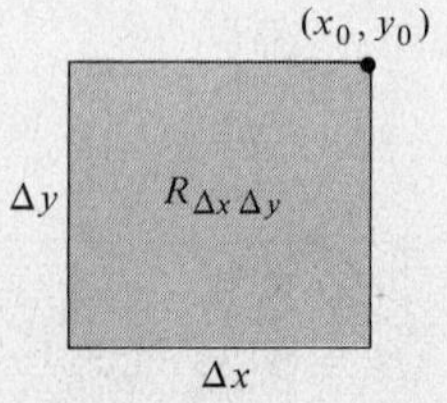

Now consider a point (x_0, y_0) in A and a rectangle $R_{\Delta x \, \Delta y}$ of sides Δx and Δy and upper right vertex (x_0, y_0) (see Figure 14.25). If the mass in the rectangle is designated $m(R_{\Delta x \, \Delta y})$, then over this rectangle we have

$$\text{average density} = \frac{m(R_{\Delta x \, \Delta y})}{\Delta x \, \Delta y}$$

Figure 14.25

At the point (x_0, y_0), density is defined as follows:

DEFINITION 1
The *density* $\rho(x_0, y_0)$ of a mass distribution at the point (x_0, y_0) is given by the limit

$$\rho(x_0, y_0) = \lim_{\substack{\Delta x \to 0 \\ \Delta y \to 0}} \frac{m(R_{\Delta x\, \Delta y})}{\Delta x\, \Delta y}$$

provided the limit exists. The function $\rho(x, y)$, which gives the density at each point of A, is called a *density function*.

We observe that for small values of Δx and Δy we have

$$m(R_{\Delta x\, \Delta y}) \approx \rho(x_0, y_0)\, \Delta x\, \Delta y \tag{1}$$

This is a convenient approximate formula for the mass over a small region with a known mass density. Theorem 1 tells us that the mass can be found by integrating the density function.

THEOREM 1
If a mass is distributed over a region A in the xy plane with density function $\rho(x, y)$, then the mass $m(A)$ over A is given by the formula

$$m(A) = \iint_A \rho(x, y)\, dx\, dy \tag{2}$$

To give an indication of the proof, recall from Section 14.2 how the double integral is defined. The region A is divided into rectangles R_{ij} of sides of lengths Δx and Δy, and selecting the upper right-hand vertex (x_i, y_j) of each rectangle R_{ij}, we have

$$\iint_A \rho(x, y)\, dx\, dy = \lim_{n \to \infty} \sum_{i,j=1}^{n} \rho(x_i, y_j)\, \Delta x\, \Delta y$$

Hence

$$\iint_A \rho(x, y)\, dx\, dy \approx \sum_{i,j=1}^{n} \rho(x_i, y_j)\, \Delta x\, \Delta y$$

and by relation (1),

$$m(R_{ij}) \approx \rho(x_i, y_j)\, \Delta x\, \Delta y$$

We now use the intuitive notion that mass is additive:

$$m(A) = \sum_{i,j=1}^{n} m(R_{ij})$$

From this we see that

$$\sum_{i,j=1}^{n} \rho(x_i, y_j)\, \Delta x\, \Delta y \approx \sum_{i,j=1}^{n} m(R_{ij}) = m(A)$$

and hence

$$\iint_A \rho(x, y)\, dx\, dy \approx m(A)$$

A rigorous procedure along the lines suggested above with a limit as $n \to \infty$ shows that we get actual equality.

Example 1

A mass is distributed over the unit square $E: 0 \leq x \leq 1, 0 \leq y \leq 1$ with density $\rho(x, y) = x^2 y$.

(a) Find the mass $m(E)$.
(b) Find the mass $m(A)$, where A is the triangle with vertices $(0, 0)$, $(1, 0)$, and $(1, \tfrac{1}{2})$.

Solution
(a) The mass, according to formula (2), is

$$m(E) = \iint_E x^2 y\, dx\, dy = \int_0^1 x^2\, dx \cdot \int_0^1 y\, dy = \tfrac{1}{3} \cdot \tfrac{1}{2} = \tfrac{1}{6}$$

(b) The boundaries of A (see Figure 14.26) are found to be the lines $y = 0$, $y = \tfrac{1}{2}x$, $x = 0$, and $x = 1$. Hence,

$$m(A) = \iint_A x^2 y\, dx\, dy = \int_0^1 \left[\int_0^{x/2} x^2 y\, dy \right] dx = \int_0^1 \tfrac{1}{8} x^4\, dx = \tfrac{1}{40}$$

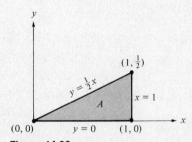

Figure 14.26

We observe that if the density function equals a constant c, $\rho(x, y) = c$, then mass and area are proportional, since in this case mass/area $= c$ and so mass $= c \times$ area. The following special case of Theorem 1 can therefore be stated.

COROLLARY 1
For any constant c,

$$\iint_A c\, dx\, dy = c \times \text{area of } A$$

Example 2

Consider a region A bounded by $y = \alpha(x)$, $y = \beta(x)$, $x = a$, and $x = b$

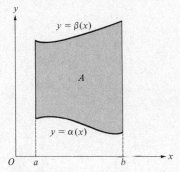

Figure 14.27

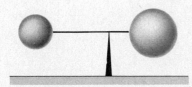

Figure 14.28

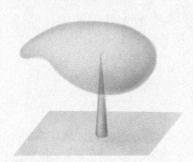

Figure 14.29

(see Figure 14.27). Using Corollary 1 with $c = 1$ and Theorem 1 of Section 14.3, we find the area to be as follows:

$$\text{area of } A = \iint_A 1 \, dx \, dy = \int_a^b \left[\int_{\alpha(x)}^{\beta(x)} 1 \, dy \right] dx = \int_a^b [\beta(x) - \alpha(x)] \, dx$$

This result agrees with Theorem 2 of Section 6.3.

CENTER OF GRAVITY

Center of gravity, also called *center of mass,* can be intuitively understood by considering two balls connected with a line (see Figure 14.28). The mass center is that point on the line about which the balls stay in equilibrium, that is, the point about which the "seesaw" remains horizontal. For a mass distribution over an area A, the center of gravity can be thought of as the point of support that keeps A horizontal (see Figure 14.29). Rather than derive the formulas for center of mass, we have given them without proof and show how they are applied (see also Section 14.5).

DEFINITION 2

The center of gravity of a mass $m(A)$ distributed over a region A with density $\rho(x, y)$ is the point $(\bar{x}, \bar{y})$ given by the equations

$$\bar{x} = \frac{1}{m(A)} \iint_A x\rho(x, y) \, dx \, dy$$

$$\bar{y} = \frac{1}{m(A)} \iint_A y\rho(x, y) \, dx \, dy$$

center of gravity

Example 3

Referring to Example 1, find the center of gravity of the mass distributed over the square $E: 0 \le x \le 1, 0 \le y \le 1$, with density $\rho(x, y) = x^2 y$.

Solution
We have

$$\iint_E x\rho(x, y) \, dx \, dy = \iint_E x^3 y \, dx \, dy = \int_0^1 x^3 \, dx \cdot \int_0^1 y \, dy = \tfrac{1}{4} \cdot \tfrac{1}{2} = \tfrac{1}{8}$$

$$\iint_E y\rho(x, y) \, dx \, dy = \iint_E x^2 y^2 \, dx \, dy = \int_0^1 x^2 \, dx \cdot \int_0^1 y^2 \, dy = \tfrac{1}{3} \cdot \tfrac{1}{3} = \tfrac{1}{9}$$

By Example 1, $m(E) = \tfrac{1}{6}$, and hence

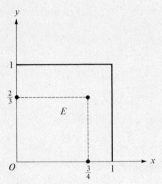

Figure 14.30

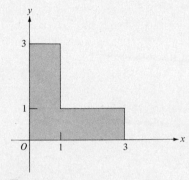

Figure 14.31

$$\bar{x} = \frac{1}{\frac{1}{6}} \cdot \frac{1}{8} = \frac{6}{8} = \frac{3}{4}$$

$$\bar{y} = \frac{1}{\frac{1}{6}} \cdot \frac{1}{9} = \frac{6}{9} = \frac{2}{3}$$

and the center of gravity is $(\bar{x}, \bar{y}) = (\frac{3}{4}, \frac{2}{3})$ (see Figure 14.30).

EXERCISES

In Exercises 1 through 4, find the mass and center of gravity for a mass distribution with density function $\rho(x, y)$ over the rectangle $A: -1 \leq x \leq 1, -1 \leq y \leq 1$.

1. $\rho(x, y) = 1$
2. $\rho(x, y) = (x + y)^2$
3. $\rho(x, y) = x^2 + y^2$
4. $\rho(x, y) = x^2$

In Exercises 5–10, find the mass and center of gravity for a mass distribution with density function $\rho(x, y)$ over the domain A.

5. $\rho(x, y) = x$ A is bounded by $y = x^2$ and $y = x$.
6. $\rho(x, y) = x^2$ A is bounded by $y = x^2$ and $y = x$.
7. $\rho(x, y) = x^2 + y^2$ A is the triangle with vertices $(0, 0)$, $(0, 1)$, and $(1, 0)$.
8. $\rho(x, y) = 1$ A is the triangle with vertices $(1, 0)$, $(-1, 0)$, and $(0, \sqrt{3})$.
9. $\rho(x, y) = 1$ A is as given in Figure 14.31.
10. $\rho(x, y) = 1$ A is as given in Figure 14.32.

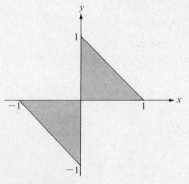

Figure 14.32

14.5 DOUBLE INTEGRALS IN POLAR COORDINATES

The evaluation of double integrals is often facilitated through the use of polar coordinates. We recall that rectangular coordinates

14.5 DOUBLE INTEGRALS IN POLAR COORDINATES

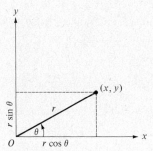

Figure 14.33
Polar coordinates.

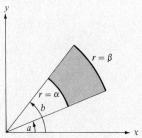

Figure 14.34

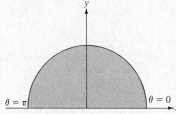

Figure 14.35

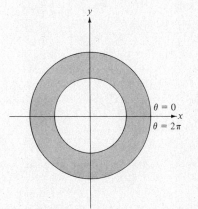

Figure 14.36

(x, y) and polar coordinates $[r, \theta]$ are related through the equations

$$\left. \begin{array}{l} x = r \cos \theta \\ y = r \sin \theta \end{array} \right\} \tag{1}$$

(see Figure 14.33).

As in the case of double integrals in rectangular coordinates, we start here with regions of a special shape. The region considered in Theorem 1 below is an area bounded by curves $r = \alpha$, $r = \beta$, and lines $\theta = a$ and $\theta = b$ (see Figure 14.34). Such a region will be referred to as a *wedge*. Regions such as disks or half-disks (Figure 14.35) or disks with a hole (Figure 14.36) are also considered as wedges.

THEOREM 1
Let A be a wedge bounded by the curves

$$r = \alpha, \quad r = \beta, \quad \theta = a, \quad \text{and} \quad \theta = b$$

Then

$$\iint_A f(x, y) \, dx \, dy = \int_a^b \left[\int_\alpha^\beta f(r \cos \theta, r \sin \theta) r \, dr \right] d\theta$$

This theorem will not be proved here, but the presence of r under the integral sign should be explained. It comes from the formula for the area of a wedge. In rectangular coordinates, the double integral was approximated with solids having rectangular base $\Delta x \Delta y$. In polar coordinates, we use instead solids having a wedge-shaped base.

By Section 9.1, the length of the lower arc of the shaded wedge in Figure 14.37 is $r \Delta \theta$. When Δr and $\Delta \theta$ are small compared with r, the wedge is nearly rectangular and its area is approximately base × height, or

$$A \approx (r \Delta \theta) \Delta r = r \Delta r \Delta \theta$$

The integral is thus approximated with summands of the form

$$f(r \cos \theta, r \sin \theta) r \Delta r \Delta \theta$$

and this gives rise to the integral in the right side.

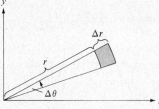

Figure 14.37

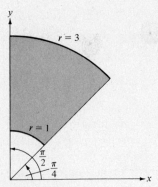

Figure 14.38

Example 1

Evaluate

$$\iint_A xy \, dx \, dy$$

when A is the wedge bounded by $r = 1$, $r = 3$, $\theta = \pi/4$, and $\theta = \pi/2$ (see Figure 14.38).

Solution
By Theorem 1,

$$\iint_A xy \, dx \, dy = \int_{\pi/4}^{\pi/2} \left[\int_1^3 r \cos \theta \cdot r \sin \theta \, r \, dr \right] d\theta \qquad (2)$$

Since

$$\int_1^3 r \cos \theta \cdot r \sin \theta \, r \, dr = \cos \theta \sin \theta \int_1^3 r^3 \, dr$$

$$= \cos \theta \sin \theta \, (\tfrac{1}{4} r^4) \Big|_1^3 = 20 \cos \theta \sin \theta$$

and

$$\int_{\pi/4}^{\pi/2} 20 \cos \theta \sin \theta \, d\theta = 10 \sin^2 \theta \Big|_{\pi/4}^{\pi/2} = 5$$

we have

$$\iint_A xy \, dx \, dy = 5$$

Observe that by Theorem 1 of Section 14.1 we could have written the right side of (2) as a product of integrals

$$\int_{\pi/4}^{\pi/2} \left[\int_1^3 r \cos \theta \, r \sin \theta \, r \, dr \right] dr = \left(\int_{\pi/4}^{\pi/2} \cos \theta \sin \theta \, d\theta \right)\left(\int_1^3 r^3 \, dr \right)$$

The result would, of course, be the same.

Example 2

Find the center of gravity of a mass with density $\rho(x, y) = 1$ distributed on the wedge A bounded by $r = 1$, $r = 2$, $\theta = 0$, and $\theta = \pi$ (see Figure 14.39).

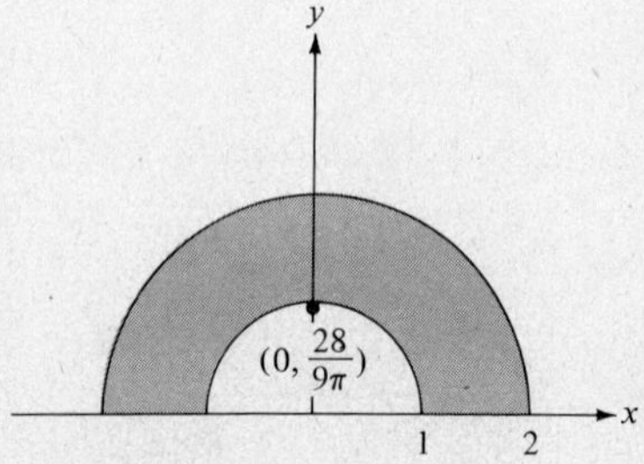

Solution
By Theorem 1 of Section 14.4, the mass $m(A)$ of the wedge A is

$$m(A) = \iint_A \rho(x, y) \, dx \, dy = \iint_A 1 \, dx \, dy$$

Figure 14.39

By Theorem 1 we have

$$\iint_A 1 \, dx \, dy = \int_0^\pi \left[\int_1^2 1 \cdot r \, dr \right] d\theta = \int_0^\pi \tfrac{3}{2} \, d\theta = \tfrac{3}{2}\pi$$

and hence

$$m(A) = \tfrac{3}{2}\pi$$

Observe that this answer can also be obtained directly, since the area of A is the difference of two half-disks.

To find the center of gravity $(\bar{x}, \bar{y})$ we use Definition 2 of Section 14.4, according to which

$$\bar{x} = \frac{1}{m(A)} \iint_A x\rho(x, y) \, dx \, dy = \frac{1}{3\pi/2} \iint_A x \cdot 1 \, dx \, dy$$

$$\bar{y} = \frac{1}{m(A)} \iint_A y\rho(x, y) \, dx \, dy = \frac{1}{3\pi/2} \iint_A y \cdot 1 \, dx \, dy$$

Again by Theorem 1,

$$\iint_A x \, dx \, dy = \int_0^\pi \left[\int_1^2 r \cos\theta \, r \, dr \right] d\theta = \int_0^\pi \tfrac{7}{3} \cos\theta \, d\theta = \tfrac{7}{3} \sin\theta \Big|_0^\pi = 0$$

$$\iint_A y \, dx \, dy = \int_0^\pi \left[\int_1^2 r \sin\theta \, r \, dr \right] d\theta = \int_0^\pi \tfrac{7}{3} \sin\theta \, d\theta = \tfrac{14}{3}$$

Hence

$$\bar{x} = 0 \quad \text{and} \quad \bar{y} = \frac{2}{3\pi} \cdot \frac{14}{3} = \frac{28}{9\pi}$$

and the center of gravity is

$$(\bar{x}, \bar{y}) = \left(0, \frac{28}{9\pi} \right)$$

(see Figure 14.39). Note that the center of gravity is *outside* A.

A more general form of Theorem 1, also given without proof, is the following result (see Figure 14.40).

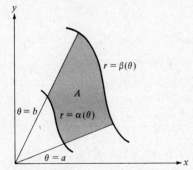

Figure 14.40

THEOREM 2
Let the region A be bounded by the curves

$$r = \alpha(\theta), \quad r = \beta(\theta), \quad \theta = a, \quad \text{and} \quad \theta = b$$

given in polar coordinates. Then

$$\iint_A f(x, y) \, dx \, dy = \int_a^b \left[\int_{\alpha(\theta)}^{\beta(\theta)} f(r\cos\theta, r\sin\theta) \, r \, dr \right] d\theta$$

580 MULTIPLE INTEGRALS

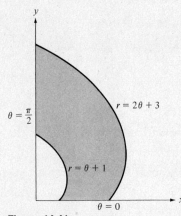

Figure 14.41

Example 3

Evaluate

$$\iint_A \sqrt{x^2 + y^2}\, dx\, dy$$

when A is bounded by

$$r = \theta + 1, \quad r = 2\theta + 3, \quad \theta = 0, \quad \text{and} \quad \theta = \frac{\pi}{2}$$

(see Figure 14.41).

Solution

Since $x = r \cos \theta$ and $y = r \sin \theta$, we have

$$\sqrt{x^2 + y^2} = r$$

and hence, using Theorem 2 with $\alpha(\theta) = \theta + 1$ and $\beta(\theta) = 2\theta + 3$ gives

$$\iint_A \sqrt{x^2 + y^2}\, dx\, dy = \int_0^{\pi/2} \left[\int_{\theta+1}^{2\theta+3} r \cdot r\, dr \right] d\theta$$

$$= \int_0^{\pi/2} \tfrac{1}{3}[(2\theta + 3)^3 - (\theta + 1)^3]\, d\theta$$

$$= \tfrac{1}{3}\left[\tfrac{1}{8}(2\theta + 3)^4 - \tfrac{1}{4}(\theta + 1)^4\right]\Big|_0^{\pi/2}$$

$$= \tfrac{1}{24}\left[(\pi + 3)^4 - 2\left(\frac{\pi}{2} + 1\right)^4 - 79\right]$$

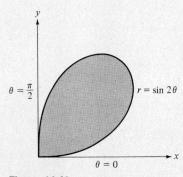

Figure 14.42

Example 4

Evaluate

$$\iint_A \frac{y}{x}\, dx\, dy$$

when A is bounded by

$$r = \sin 2\theta \quad \text{for } 0 \le \theta \le \frac{\pi}{2}$$

(see Figure 14.42).

Solution

Using Theorem 2 with $\alpha(\theta) = 0$ and $\beta(\theta) = \sin 2\theta$ gives

$$\iint_A \frac{y}{x}\, dx\, dy = \int_0^{\pi/2} \left[\int_0^{\sin 2\theta} \tan \theta \cdot r\, dr \right] d\theta$$

Now,

$$\int_0^{\sin 2\theta} \tan\theta \, r \, dr = \tan\theta \int_0^{\sin 2\theta} r \, dr = \tan\theta \left(\frac{1}{2}\sin^2 2\theta\right) = \frac{1}{2}\tan\theta \sin^2 2\theta$$

$$= \frac{1}{2}\frac{\sin\theta}{\cos\theta}(2\sin\theta\cos\theta)^2 = 2\sin^3\theta \cos\theta$$

Since

$$\int_0^{\pi/2} 2\sin^3\theta \cos\theta \, d\theta = \frac{1}{2}\sin^4\theta \Big|_0^{\pi/2} = \frac{1}{2}$$

we have

$$\iint_A \frac{y}{x} \, dx \, dy = \frac{1}{2}$$

EXERCISES

Evaluate the double integrals in Exercises 1–5.

1. $\iint_A (x^2 + y^2) \, dx \, dy$ A is bounded by $r = 1$, $\theta = 0$, and $\theta = \frac{\pi}{3}$.

2. $\iint_A \sqrt{x^2 + y^2} \, dx \, dy$ A is bounded by $r = 2$, $r = 1$, $\theta = 0$, and $\theta = \frac{2\pi}{3}$.

3. $\iint_A x^2 \, dx \, dy$ A is the disk $x^2 + y^2 \leq 2$.

4. $\iint_A x^2 y^2 \, dx \, dy$ A is the disk $x^2 + y^2 \leq 2$.

5. $\iint_A x^3 \, dx \, dy$ A is the disk $x^2 + y^2 \leq 1$.

Evaluate the double integrals in Exercises 6–8 using polar coordinates.

6. $\int_{-a}^{a} \left[\int_{-\sqrt{a^2-x^2}}^{\sqrt{a^2-x^2}} e^{x^2+y^2} \, dy \right] dx$

7. $\int_0^a \left[\int_0^{\sqrt{a^2-y^2}} \sqrt{x^2 + y^2} \, dx \right] dy$

8. $\int_{-a/\sqrt{2}}^{0} \left[\int_{-\sqrt{a^2-y^2}}^{-y} x \, dx \right] dy$

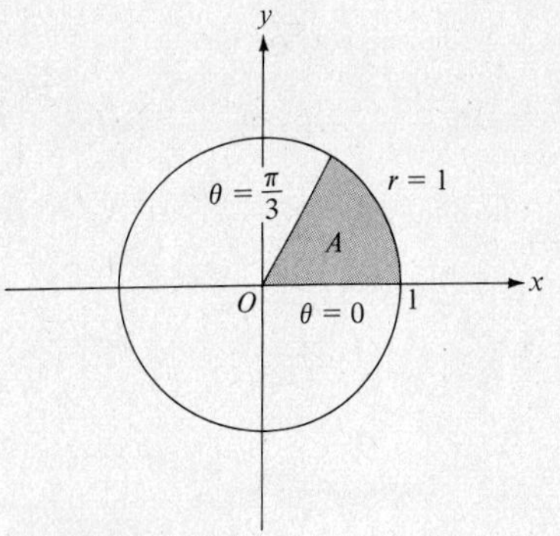

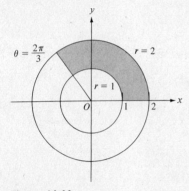

Figure 14.43

Figure 14.44

In Exercises 9–12, find the mass and center of gravity for a mass distribution with density function $\rho(x, y)$ over the top half of the disk $A: x^2 + y^2 \leq 1$.

9. $\rho(x, y) = 1$
10. $\rho(x, y) = x^2 + y^2$
11. $\rho(x, y) = y^2$
12. $\rho(x, y) = x^2$

13. Find the volume under the section of the cone $z = \sqrt{x^2 + y^2}$ having base A as in Figure 14.43.

14. Find the volume under the section of the paraboloid $z = x^2 + y^2$ having base A as in Figure 14.44.

15. If A is the disk $x^2 + y^2 \leq a^2$, then there are constants c and k such that
$$\iint_A x^8 y^{16} \, dx \, dy = ca^k$$
Find k.

16. Let A be the disk $x^2 + y^2 \leq 1$ and suppose that
$$\int f(x) \, dx = F(x) + C$$
Show that
$$\iint_A f(x^2 + y^2) \, dx \, dy = \pi[F(1) - F(0)]$$

14.6 TRIPLE INTEGRALS

To specify a point in a solid, we need three coordinates (x, y, z). In particular, three coordinates $(\bar{x}, \bar{y}, \bar{z})$ are needed to specify the center of gravity of a solid. The double integral, used in Section 14.4 to define center of gravity $(\bar{x}, \bar{y})$ of a mass distributed over a domain A in the plane, must be replaced by a triple integral,
$$\iiint_S f(x, y, z) \, dx \, dy \, dz$$
as explained below.

A region S in this section will refer to a three-dimensional solid, such as a pyramid, a cube, a ball, an ellipsoid, and so on. The definition of the triple integral is entirely analogous to the definition of the double integral.

Consider a region S and a rectangular solid R containing it (see Figure 14.45), where R is given by the inequalities $a \leq x \leq b$, $c \leq y \leq d$, and $e \leq z \leq f$. Each of the intervals $[a, b]$, $[c, d]$, and $[e, f]$ is divided into n subintervals by means of the points

$$x_i = a + i\,\Delta x \quad i = 1, 2, \ldots, n \quad \Delta x = \frac{b-a}{n}$$

$$y_j = c + j\,\Delta y \quad j = 1, 2, \ldots, n \quad \Delta y = \frac{d-c}{n} \quad (1)$$

$$z_k = e + k\,\Delta z \quad k = 1, 2, \ldots, n \quad \Delta z = \frac{f-e}{n}$$

For each set of integers i, j, k we consider the number

$$f(x_i, y_j, z_k)\,\Delta x\,\Delta y\,\Delta z$$

where, as in Section 14.2, we set $f(x_i, y_j, z_k) = 0$ when the point (x_i, y_i, z_k) is not in S.

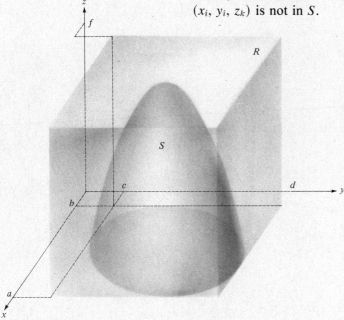

Figure 14.45

DEFINITION 1
For each value of n let

$$S_n = \sum_{i,j,k=1}^{n} f(x_i, y_j, z_k)\,\Delta x\,\Delta y\,\Delta z \quad (2)$$

[see (1)]. If $f(x, y, z)$ is defined on S, then the *triple integral* of $f(x, y, z)$ over S is defined as

$$\iiint_S f(x, y, z)\,dx\,dy\,dz = \lim_{n \to \infty} S_n$$

provided the limit exists.

As in the case of integrals in general, the definition does not tell us how to evaluate them in practice. The triple integral, like the double integral, can be written as an iterated integral.

For the following theorem consult Figure 14.46.

THEOREM 1
Let $f(x, y, z)$ be continuous on a region S in three-dimensional space bounded by surfaces $z = \alpha(x, y)$ and $z = \beta(x, y)$ where (x, y) is restricted to some region A in the xy plane. If $\alpha(x, y) \leq \beta(x, y)$ for all points (x, y) in A, then

$$\iiint_S f(x, y, z) \, dx \, dy \, dz = \iint_A \left[\int_{\alpha(x,y)}^{\beta(x,y)} f(x, y, z) \, dz \right] dx \, dy \quad (3)$$

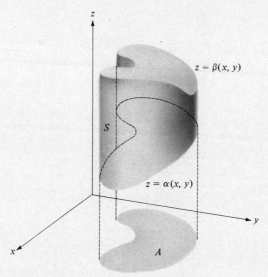

Figure 14.46

Formula (3) is similar to formula (1) in Section 14.3. An indication of the proof is given at the end of this section. Observe that evaluating the inner integral reduces the triple integral to a double integral.

Example 1

Evaluate the triple integral

$$\iiint_S x^3 y z^3 \, dx \, dy \, dz$$

taken over the region S bounded by the surfaces $z = 0$ and $z = 1/x^2$

with (x, y) restricted to the region A bounded by $x = 1$, $x = 2$, $y = 0$, and $y = x^3$ (see Figure 14.47).

Solution
In using formula (3), we first evaluate the inner integral. With $\alpha(x, y) = 0$ and $\beta(x, y) = 1/x^2$, we have

$$\iiint_S x^3 y z^3 \, dx \, dy \, dz = \iint_A \left(\int_0^{1/x^2} x^3 y z^3 \, dz \right) dx \, dy$$

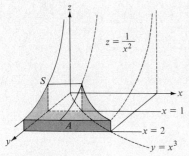

Figure 14.47

Since

$$\int_0^{1/x^2} x^3 y z^3 \, dx = x^3 y \int_0^{1/x^2} z^3 \, dz = x^3 y \left(\frac{1}{4} z^4 \right) \Big|_{z=0}^{z=1/x^2} = \frac{1}{4} \frac{y}{x^5}$$

we now have

$$\iiint_S x^3 y z^3 \, dx \, dy \, dz = \iint_A \frac{1}{4} \cdot \frac{y}{x^5} \, dx \, dy$$

To evaluate the double integral over A, begin with a sketch of the region A (see Figure 14.48). By Theorem 1 of Section 14.3,

$$\iint_A \frac{1}{4} \frac{y}{x^5} \, dx \, dy = \int_1^2 \left(\int_0^{x^3} \frac{1}{4} \frac{y}{x^5} \, dy \right) dx = \int_1^2 \frac{1}{8} x \, dx = \frac{3}{8}$$

and hence

$$\iiint_S x^3 y z^3 \, dx \, dy \, dz = \frac{3}{8}$$

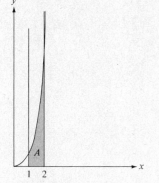

Figure 14.48

Note that the method in this case can be summarized as

$$\iiint_S x^3 y z^3 \, dx \, dy \, dz = \int_1^2 \left[\int_0^{x^3} \left(\int_0^{1/x^2} x^3 y z^3 \, dz \right) dy \right] dx$$

Example 2

Evaluate

$$\iiint_S 2z \, dx \, dy \, dz$$

taken over the region S bounded by $z = x - y$ and $z = x + y$, and restricted to the unit square E: $0 \leq x \leq 1$, $0 \leq y \leq 1$.

Solution
Using formula (3) with $\alpha(x, y) = x - y$ and $\beta(x, y) = x + y$ gives

$$\iiint_S 2z \, dx \, dy \, dz = \iint_E \left(\int_{x-y}^{x+y} 2z \, dz \right) dx \, dy$$

Since
$$\int_{x-y}^{x+y} 2z\, dz = z^2 \Big|_{x-y}^{x+y} = (x+y)^2 - (x-y)^2 = 4xy$$

we have
$$\iiint_S 2z\, dx\, dy\, dz = \iint_A 4xy\, dx\, dy$$
$$= \int_0^1 \left(\int_0^1 4xy\, dx\right) dy = 4 \int_0^1 x\, dx \cdot \int_0^1 y\, dy$$
$$= 4 \cdot \tfrac{1}{2} \cdot \tfrac{1}{2} = 1$$

Example 3

Evaluate
$$\iiint_S xy^2z^3\, dx\, dy\, dz$$

taken over the cube S: $0 \le x \le 1$, $0 \le y \le 2$, $0 \le z \le 3$.

Solution
By formula (3), with $\alpha(x, y) = 0$ and $\beta(x, y) = 3$ we have
$$\iiint_S xy^2z^3\, dx\, dy\, dz = \iint_A \left(\int_0^3 xy^2z^3\, dz\right) dx\, dy$$

where the double integral is taken over the square A: $0 \le x \le 1$, $0 \le y \le 2$. Since
$$\int_0^3 xy^2z^3\, dz = xy^2 \int_0^3 z^3\, dz = \tfrac{81}{4} xy^2$$

we have, by Theorem 1 of Section 14.1,
$$\iint_A \left(\int_0^3 xy^2z^3\, dz\right) dx\, dy = \iint_A \tfrac{81}{4} xy^2\, dx\, dy = \int_0^1 \left(\int_0^2 \tfrac{81}{4} xy^2\, dy\right) dx$$
$$= \tfrac{81}{4} \int_0^1 x\, dx \cdot \int_0^2 y^2\, dy = \tfrac{81}{4} \cdot \tfrac{1}{2} \cdot \tfrac{8}{3} = 27$$

This procedure can be summarized as
$$\iiint_S xy^2z^3\, dx\, dy\, dz = \int_0^1 x\, dx \cdot \int_0^2 y^2\, dy \cdot \int_0^3 z^3\, dz$$

In concluding this section we outline a proof of Theorem 1. Suppose that

$$\iiint_S f(x, y, z)\, dx\, dy\, dz = \lim_{n \to \infty} S_n$$

where

$$S_n = \sum_{i,j,k=1}^{n} f(x_i, y_j, z_k)\, \Delta x\, \Delta y\, \Delta z$$

(see Definition 1). The order of summation is unimportant, and we consider first summation over k:

$$\sum_{k=1}^{n} f(x_i, y_j, z_k)\, \Delta x\, \Delta y\, \Delta z$$

In the definition of S_n, we agreed to put $f(x_i, y_j, z_k) = 0$ when the point (x_i, y_j, z_k) is not in S. Thus, $f(x_i, y_j, z_k) = 0$ when $z_k < \alpha(x_i, y_j)$ or $z_k > \beta(x_i, y_j)$.

From the definition of the definite integral we have that for each point (x_i, y_j),

$$\sum_{k=1}^{n} f(x_i, y_j, z_k)\, \Delta x \approx \int_{\alpha(x_i,y_j)}^{\beta(x_i,y_j)} f(x_i, y_j, z)\, dz$$

and multiplying both sides by $\Delta x\, \Delta y$ gives

$$\sum_{k=1}^{n} f(x_i, y_j, z_k)\, \Delta z\, \Delta x\, \Delta y \approx \int_{\alpha(x_i,y_j)}^{\beta(x_i,y_j)} f(x_i, y_j, z)\, dz\, \Delta x\, \Delta y$$

Summing now both sides over i and j gives [see formula (2)]

$$S_n \approx \sum_{i,j=1}^{n} \left[\int_{\alpha(x_i,y_j)}^{\beta(x_i,y_j)} f(x_i, y_j, z)\, dz \right] \Delta x\, \Delta y$$

From the definition of the double integral (Section 14.2), we know that the right side is an approximation to

$$\iint_A \left[\int_{\alpha(x,y)}^{\beta(x,y)} f(x, y, z)\, dz \right] dx\, dy$$

Hence

$$S_n \approx \iint_A \left[\int_{\alpha(x,y)}^{\beta(x,y)} f(x, y, z)\, dz \right] dx\, dy$$

and by Definition 1, therefore,

$$\iiint_S f(x, y, z)\, dx\, dy\, dz \approx \iint_A \left[\int_{\alpha(x,y)}^{\beta(x,y)} f(x, y, z)\, dz \right] dx\, dy$$

Appropriate technical detail and a passage to the limit as $n \to \infty$ gives an exact equality.

EXERCISES

Evaluate the following triple integrals.

1. $\iiint_S x \, dx \, dy \, dz \qquad S: 0 \leq x \leq 1, -1 \leq y \leq 1, -1 \leq z \leq 1$

2. $\iiint_S (x+y+z) \, dx \, dy \, dz \qquad S$ is bounded by $z = -x - y$ and $z = x + y$ for $0 \leq x \leq 1$ and $0 \leq y \leq 1$

3. $\iiint_S y \, dx \, dy \, dz \qquad S$ is bounded by $z = 2x - y$ and $z = 4x + y$ for (x, y) in the region A bounded by $y = x$, $y = 1 - x$, and $y = 0$

4. $\iiint_S z e^{x+y} \, dx \, dy \, dz \qquad S$ is bounded by $z = x - y$ and $z = x + y$ for (x, y) in the region bounded by $y = x$, $y = 0$, and $x = 1$

5. $\iiint_S (x+y+z) \, dx \, dy \, dz \qquad S: -1 \leq x \leq 0, 1 \leq y \leq 2, 2 \leq z \leq 3$

6. $\iiint_S (-5) \, dx \, dy \, dz \qquad S$ is the pyramid with vertices $(0, 0, 0)$, $(2, 0, 0)$, $(0, 3, 0)$, and $(0, 0, 4)$

7. $\iiint_S z e^{x+y} \, dx \, dy \, dz \qquad S: 0 \leq x \leq 1, 0 \leq y \leq 2, 0 \leq z \leq 3$

8. $\iiint_S (x+y+z) \, dx \, dy \, dz \qquad S$ is bounded by $\alpha(x, y) = 0$ and $\beta(x, y) = 1$ for $0 \leq x + y \leq 1$ and $0 \leq x - y \leq 1$

14.7 MASS, DENSITY, AND CENTER OF GRAVITY OF SOLIDS

Continuing the discussion in Section 14.4, consider a solid S. The average density of a mass distributed in S is

$$\text{average density} = \frac{\text{mass in } S}{\text{volume of } S}$$

If (x_0, y_0, z_0) is a point in S and $R_{(\Delta x \, \Delta y \, \Delta z)}$ a rectangular box with sides Δx, Δy, and Δz containing it (see Figure 14.49), then over the box

$$\text{average density} = \frac{m(R_{\Delta x \, \Delta y \, \Delta z})}{\Delta x \Delta y \Delta z}$$

where $m(R_{\Delta x \, \Delta y \, \Delta z})$ designates the mass in the box. In analogy to Definition 1 of Section 14.4, we have the following definition.

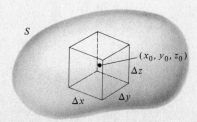

Figure 14.49

14.7 MASS, DENSITY, AND CENTER OF GRAVITY OF SOLIDS

DEFINITION 1
The density $\delta(x_0, y_0, z_0)$ of a mass distribution at the point (x_0, y_0, z_0) is given by the limit

$$\delta(x_0, y_0, z_0) = \lim_{\substack{\Delta x \to 0 \\ \Delta y \to 0 \\ \Delta z \to 0}} \frac{m(R_{\Delta x \, \Delta y \, \Delta z})}{\Delta x \, \Delta y \, \Delta z}$$

provided the limit exists. The function $\delta(x, y, z)$, which gives the density at each point of A, is called a *density function*.

Again corresponding to the two-dimensional case, we have the following theorem.

THEOREM 1
If $\delta(x, y, z)$ is the density function of a mass distribution in a solid S, then the mass, $m(S)$, is given by the formula

$$m(S) = \iiint_S \delta(x, y, z) \, dx \, dy \, dz \tag{1}$$

An argument similar to the one used to justify formula (2) in Section 14.4 can be used to justify formula (1).

Example 1

A mass is distributed in the cube $C: 0 \leq x \leq 2, 0 \leq y \leq 2, 0 \leq z \leq 2$ with density $\delta(x, y, z) = 1$.

(a) Find the mass $m(C)$.
(b) Find the mass $m(S)$ when S is the cone $z = 1 - \sqrt{x^2 + y^2}$ with base $x^2 + y^2 \leq 1$ (see Figure 14.50).

Solution
(a) The mass, according to formula (1), is

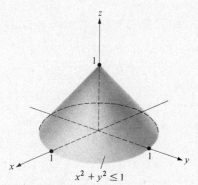

Figure 14.50
The cone $z = 1 - \sqrt{x^2 + y^2}$ with base $x^2 + y^2 \leq 1$.

$$m(C) = \iiint_C 1 \, dx \, dy \, dz = \int_0^2 1 \, dx \cdot \int_0^2 1 \, dy \cdot \int_0^2 1 \, dz = 2 \cdot 2 \cdot 2 = 8$$

Observe that numerically this gives us the volume of the cube C.

(b) We use now Theorem 1 of Section 14.6. The solid S is bounded by the surfaces $\alpha(x, y) = 0$ and $\beta(x, y) = 1 - \sqrt{x^2 + y^2}$. Hence

$$m(S) = \iiint_S 1 \, dx \, dy \, dz = \iint_A \left(\int_0^{1-\sqrt{x^2+y^2}} 1 \, dz \right) dx \, dy$$

where A is the disk $x^2 + y^2 \leq 1$. Since

$$\int_0^{1-\sqrt{x^2+y^2}} 1\ dz = 1 - \sqrt{x^2+y^2}$$

we have

$$m(S) = \iint_A (1 - \sqrt{x^2+y^2})\ dx\ dy$$

With polar coordinates (Section 14.5), the disk A is described by the equations $r = 1$, $\theta = 0$, and $\theta = 2\pi$, and since $x^2 + y^2 = r^2$ we get $\sqrt{x^2+y^2} = \sqrt{r^2} = r$ and hence

$$m(S) = \int_0^{2\pi} \left[\int_0^1 (1-r)r\ dr \right] d\theta = \int_0^{2\pi} \frac{1}{6}\ d\theta = \frac{\pi}{3}$$

We observe that this is the volume of the given cone.

Generalizing the observations made in the above example, we have the following:

The volume $V(S)$ of a solid S is given by the formula

$$V(S) = \iiint_S 1\ dx\ dy\ dz \qquad (2)$$

This is analogous to the formula for area given in Section 14.4. Extending Definition 2 of Section 14.4 to solids, we have the following definition.

DEFINITION 2

The *center of gravity* of a mass $m(S)$ distributed in a solid S with density $\delta(x, y, z)$ is the point $(\bar{x}, \bar{y}, \bar{z})$ given by the equations

---

$$\bar{x} = \frac{1}{m(S)} \iiint_S x\delta(x, y, z)\ dx\ dy\ dz$$

$$\bar{y} = \frac{1}{m(S)} \iiint_S y\delta(x, y, z)\ dx\ dy\ dz \qquad \text{center of gravity}$$

$$\bar{z} = \frac{1}{m(S)} \iiint_S z\delta(x, y, z)\ dx\ dy\ dz$$

Example 2

A uniform mass distribution on the tetrahedron T with vertices $(0, 0, 0)$, $(1, 0, 0)$, $(0, 1, 0)$, and $(0, 0, 1)$ (see Figure 14.51), has density $\delta(x, y, z) = 1$.

14.7 MASS, DENSITY, AND CENTER OF GRAVITY OF SOLIDS

(a) Find the volume and mass of T.
(b) Find the center of gravity of T.

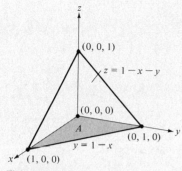

Figure 14.51
The tetrahedron with vertices $(0, 0, 0)$, $(1, 0, 0)$, $(0, 1, 0)$, and $(0, 0, 1)$.

Solution
(a) In this case volume and mass are numerically equal. By formula (2) we have

$$m(T) = V(T) = \iiint_T 1 \, dx \, dy \, dz$$

From Figure 14.51 we see that T is bounded by the planes $z = 0$ and $z = 1 - x - y$ and it has base A bounded by the lines $y = 1 - x$, $x = 0$, and $y = 0$. Hence,

$$V(T) = \iint_A \left(\int_0^{1-x-y} 1 \, dz \right) dx \, dy = \iint_A (1 - x - y) \, dx \, dy$$

$$= \int_0^1 \left[\int_0^{1-x} (1 - x - y) \, dy \right] dx$$

We find that

$$\int_0^{1-x} (1 - x - y) \, dy = y - xy - \tfrac{1}{2} y^2 \bigg|_{y=0}^{y=1-x} = \tfrac{1}{2} x^2 - x + \tfrac{1}{2}$$

and

$$\int_0^1 (\tfrac{1}{2} x^2 - x + \tfrac{1}{2}) \, dx = \tfrac{1}{6}$$

so that

$$m(T) = V(T) = \tfrac{1}{6}$$

(b) To find the center of gravity, let us begin with $\bar{x}$. We find first that

$$\iiint_T x \cdot 1 \, dx \, dy \, dz = \iint_A \left(\int_0^{1-x-y} x \, dz \right) dx \, dy$$

$$= \iint_A x(1 - x - y) \, dx \, dy$$

$$= \int_0^1 \left[\int_0^{1-x} x(1 - x - y) \, dy \right] dx$$

$$= \int_0^1 x \left[\int_0^{1-x} (1 - x - y) \, dy \right] dx$$

$$= \int_0^1 x(\tfrac{1}{2} x^2 - x + \tfrac{1}{2}) \, dx = \tfrac{1}{24}$$

Hence

$$\bar{x} = \frac{1}{m(T)} \iiint_T x \cdot 1 \, dx \, dy \, dz = \frac{1}{\frac{1}{6}} \cdot \frac{1}{24} = \frac{1}{4}$$

In Exercise 1 you are asked to check that $\bar{y} = \bar{z} = \frac{1}{4}$; and this gives for the center of gravity

$$(x, y, z) = (\tfrac{1}{4}, \tfrac{1}{4}, \tfrac{1}{4})$$

EXERCISES

1. Referring to Example 2, show that $\bar{y} = \bar{z} = \frac{1}{4}$.

2. Find the mass and center of gravity of a solid S: $0 \le x \le 1$, $0 \le y \le 2$, $0 \le z \le 3$ with density $\delta(x, y, z) = x^3 y^2 z$.

3. Find the volume of the tetrahedron having vertices $(0, 0, 0)$, $(a, 0, 0)$, $(0, b, 0)$, and $(0, 0, c)$.

Hint: The equation of the plane that passes through the vertices $(a, 0, 0)$, $(0, b, 0)$, and $(0, 0, c)$ is

$$\frac{x}{a} + \frac{y}{b} + \frac{z}{c} = 1$$

The tetrahedron has base A bounded by

$$\frac{x}{a} + \frac{y}{b} = 1, \quad x = 0, \quad \text{and} \quad y = 0$$

4. Find the center of gravity of the tetrahedron of Exercise 3 when $\delta(x, y, z) = 1$.

Find the mass for the following solids S and given densities.

5. $\delta(x, y, z) = x^2 + y^2$, S is the cone of Example 1.

6. $\delta(x, y, z) = x^2 y^2$, S is the cone of Example 1.

7. $\delta(x, y, z) = x^2$, S is the tetrahedron with vertices $(0, 0, 0)$, $(1, 0, 0)$, $(0, 2, 0)$, and $(0, 0, 3)$.

8. $\delta(x, y, z) = y^2$, S is the tetrahedron of Exercise 7.

14.8 CYLINDRICAL AND SPHERICAL COORDINATES

In Section 14.5 we saw how polar coordinates can be used to evaluate double integrals. We shall now see how the cylindrical and spherical coordinates introduced in Section 13.6 are used to evaluate triple integrals.

Cylindrical coordinates of a point (x, y, z) are (r, θ, z), where

$$\begin{aligned} x &= r \cos \theta \\ y &= r \sin \theta \\ z &= z \end{aligned} \tag{1}$$

(see Figure 14.52). Observe that cylindrical coordinates are the natural extension of polar coordinates to three-dimensional space. An application to integration was already considered in Example 1 of Section 14.7.

14.8 CYLINDRICAL AND SPHERICAL COORDINATES

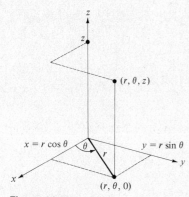

Figure 14.52

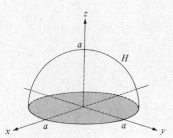

Figure 14.53
The solid hemisphere
$x^2 + y^2 + z^2 \leq a^2$.

Example 1

Consider the solid hemisphere $H: x^2 + y^2 + z^2 \leq a^2$ for $z \geq 0$ (see Figure 14.53) with uniform mass distribution of density $\delta(x, y, z) = 1$.

(a) Find the mass $m(H)$.
(b) Find the center of gravity $(\bar{x}, \bar{y}, \bar{z})$.

Solution

(a) If we let $V(H)$ stand for the volume of H, then from Section 14.7 we know that

$$m(H) = V(H) = \iiint_H 1 \, dx \, dy \, dz$$

The volume of a solid sphere of radius a is $\frac{4}{3}\pi a^3$ and hence $V(H) = \frac{1}{2}(\frac{4}{3}\pi a^3) = \frac{2}{3}\pi a^3$. In particular,

$$m(H) = \iiint_H 1 \, dx \, dy \, dz = \frac{2}{3}\pi a^3$$

(b) The solid H is bounded by the surfaces $z = 0$ and $z = \sqrt{a^2 - x^2 - y^2}$ for $x^2 + y^2 \leq a^2$. By Theorem 1 of Section 14.6, therefore,

$$\iiint_H x \cdot 1 \, dx \, dy \, dz = \iint_A \left(\int_0^{\sqrt{a^2-x^2-y^2}} x \, dz \right) dx \, dy$$

$$= \iint_A x\sqrt{a^2 - x^2 - y^2} \, dx \, dy$$

Using polar coordinates (see Example 1 in Section 14.7), we see that

$$\iint_A x\sqrt{a^2 - x^2 - y^2} \, dx \, dy = \int_0^{2\pi} \left(\int_0^a r \cos \theta \sqrt{a^2 - r^2} \, r \, dr \right) d\theta$$

$$= \int_0^{2\pi} \cos \theta \, d\theta \cdot \int_0^a r^2 \sqrt{a^2 - r^2} \, dr$$

But

$$\int_0^{2\pi} \cos \theta \, d\theta = \sin \theta \Big|_0^{2\pi} = 0$$

and consequently we do not even have to evaluate the second integral. Hence

$$\bar{x} = \frac{1}{m(H)} \iint_A x\sqrt{a^2 - x^2 - y^2} \, dx \, dy = \frac{1}{m(H)} \cdot 0 = 0$$

A similar calculation shows that $\bar{y} = 0$. To calculate $\bar{z}$, we evaluate

$$\iiint\limits_H z\,dx\,dy\,dz = \iint\limits_A \left(\int_0^{\sqrt{a^2-x^2-y^2}} z\,dz \right) dx\,dy$$

$$= \iint\limits_A \tfrac{1}{2}(a^2 - x^2 - y^2)\,dx\,dy$$

Using polar coordinates,

$$\iint\limits_A \tfrac{1}{2}(a^2 - x^2 - y^2)\,dx\,dy = \int_0^{2\pi} \left[\int_0^a \tfrac{1}{2}(a^2 - r^2)r\,dr \right] d\theta$$

Now,

$$\int_0^a \tfrac{1}{2}(a^2 - r^2)r\,dr = \tfrac{1}{2}(\tfrac{1}{2}a^2 r^2 - \tfrac{1}{4}r^4)\Big|_0^a = \tfrac{1}{8}a^4$$

and

$$\int_0^{2\pi} \tfrac{1}{8}a^4\,d\theta = \tfrac{1}{8}a^4 \cdot 2\pi = \tfrac{1}{4}\pi a^4$$

Hence

$$\bar{z} = \frac{1}{m(H)} \iiint\limits_H z\,dx\,dy\,dz = \frac{1}{\tfrac{2}{3}\pi a^3} \cdot \tfrac{1}{4}\pi a^4 = \tfrac{3}{8}a$$

The center of gravity of H is $(\bar{x}, \bar{y}, \bar{z}) = (0, 0, \tfrac{3}{8}a)$.

SPHERICAL COORDINATES

Recall from Section 13.6 that the spherical coordinate of a point (x, y, z) are (ρ, θ, ϕ), where

$$\begin{aligned} x &= \rho \sin \phi \cos \theta \\ y &= \rho \sin \phi \sin \theta \\ z &= \rho \cos \phi \end{aligned} \tag{2}$$

(see Figure 14.54). We already observed that holding ϕ and ρ constant and revolving the line $\overline{OP}$ about the z axis gives a circular cone (see Figure 14.55). In particular, the inequalities

$$0 \le \phi \le \frac{\pi}{4}$$

$$0 \le \rho \le 2 \tag{3}$$

give a cone with a circular top (see Figure 14.56).

In what follows we shall use a dual notation. A solid that can be specified by inequalities as in (3) will be designated C in rectangular coordinates and C_s in spherical coordinates.

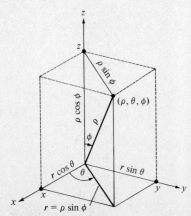

Figure 14.54
Spherical coordinates.

14.8 CYLINDRICAL AND SPHERICAL COORDINATES

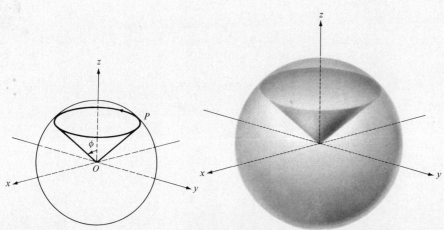

Figure 14.55

Figure 14.56
The solid C: $0 \leq \phi \leq \pi/4$, $0 \leq \rho \leq 2$.

THEOREM 1
If $f(x, y, z)$ is continuous on the solid C, then

$$\iiint_C f(x, y, z) \, dx \, dy \, dz$$

$$= \iiint_{C_s} f(\rho \sin \phi \cos \theta, \rho \sin \phi \sin \theta, \rho \cos \phi) \rho^2 \sin \phi \, d\rho \, d\phi \, d\theta$$

Although the right side of the above equation looks more complicated than the left side, in practice the integration is actually simplified.

Example 2

Find the volume of the solid in Figure 14.56.

Solution
From Section 14.7 we know that the volume $V(C)$ is

$$V(C) = \iiint_C 1 \, dx \, dy \, dz$$

The cone C is described by the inequalities

$$0 \leq \rho \leq 2$$

$$0 \leq \phi \leq \frac{\pi}{4}$$

596 MULTIPLE INTEGRALS

$$0 \leq \theta \leq 2\pi$$

These inequalities, in $\rho\phi\theta$ coordinates, specify a rectangular box. Hence, using Theorem 1 with $f(x, y, z) = 1$ gives

$$V(C) = \iiint_{C_s} \rho^2 \sin \phi \, d\rho \, d\phi \, d\theta$$

$$= \int_0^{2\pi} \left[\int_0^{\pi/4} \left(\int_0^2 \rho^2 \sin \phi \, d\rho \right) d\phi \right] d\theta$$

$$= \int_0^{2\pi} d\theta \cdot \int_0^{\pi/4} \sin \phi \, d\phi \cdot \int_0^2 \rho^2 \, d\rho = 2\pi \cdot \left(1 - \cos \frac{\pi}{4}\right) \cdot \frac{8}{3}$$

$$= \frac{16\pi}{3}\left(1 - \cos \frac{\pi}{4}\right) = \frac{16\pi}{3}\left(1 - \frac{\sqrt{2}}{2}\right) = \frac{8\pi(2 - \sqrt{2})}{3}$$

Example 3

A mass distribution in the solid C of Figure 14.56 has density $\delta(x, y, z) = x^2 + y^2 + z^2$.

(a) Find the mass $m(C)$.
(b) Find the center of gravity $(\bar{x}, \bar{y}, \bar{z})$.

Solution
(a) The mass is given by

$$m(C) = \iiint_C \delta(x, y, z) \, dx \, dy \, dz = \iiint_C (x^2 + y^2 + z^2) \, dx \, dy \, dz$$

Using spherical coordinates, $x^2 + y^2 + z^2 = \rho^2$ and so, by Theorem 1,

$$m(C) = \iiint_{C_s} \rho^2 \cdot \rho^2 \sin \phi \, d\rho \, d\phi \, d\theta$$

$$= \int_0^{2\pi} \left[\int_0^{\pi/4} \left(\int_0^2 \rho^4 \sin \phi \, d\rho \right) d\phi \right] d\theta \quad \text{(by Example 2)}$$

$$= \int_0^{2\pi} d\theta \cdot \int_0^{\pi/4} \sin \phi \, d\phi \cdot \int_0^2 \rho^4 \, d\rho$$

$$= 2\pi \left(1 - \cos \frac{\pi}{4}\right) \frac{32}{5} = \frac{32\pi(2 - \sqrt{2})}{5}$$

(b) Once more we use spherical coordinates and Example 2 to get

$$\iiint_C x\delta(x, y, z) \, dx \, dy \, dz = \iiint_{C_s} \rho \sin \phi \cos \theta \cdot \rho^2 \cdot \rho^2 \sin \phi \, d\rho \, d\phi \, d\theta$$

$$= \int_0^{2\pi} \cos \theta \, d\theta \cdot \int_0^{\pi/4} \sin^2 \phi \, d\phi \cdot \int_0^2 \rho^5 \, d\rho$$

Since, however,
$$\int_0^{2\pi} \cos\theta \, d\theta = 0$$
we have $\bar{x} = 0$, and the integral for $\bar{y}$ shows that also $\bar{y} = 0$. This should be expected from the symmetry about the z axis of both the solid C and the density function $\delta(x, y, z) = x^2 + y^2 + z^2$. To calculate $\bar{z}$ we evaluate

$$\iiint_C z\delta(x, y, z) \, dx \, dy \, dz = \iiint_{C_S} \rho \cos\phi \cdot \rho^2 \cdot \rho^2 \sin\phi \, d\rho \, d\phi \, d\theta$$

$$= \int_0^{2\pi} d\theta \cdot \int_0^{\pi/4} \cos\phi \sin\phi \, d\phi \cdot \int_0^2 \rho^5 \, d\rho$$

$$= 2\pi \cdot \frac{1}{2}\left(\sin^2\frac{\pi}{4} - \sin^2 0\right) \cdot \frac{64}{4}$$

$$= \frac{32\pi}{3} \sin^2\frac{\pi}{4} = \frac{32\pi}{3}\left(\frac{1}{\sqrt{2}}\right)^2 = \frac{8\pi}{3}$$

Hence
$$\bar{z} = \frac{1}{32\pi(2-\sqrt{2})/5} \cdot \frac{8\pi}{3} = \frac{5}{12(2-\sqrt{2})} = \frac{5(2+\sqrt{2})}{12(2-\sqrt{2})(2+\sqrt{2})}$$
$$= \frac{5(2+\sqrt{2})}{24}$$

and accordingly,
$$(\bar{x}, \bar{y}, \bar{z}) = \left(0, 0, \frac{5(2+\sqrt{2})}{24}\right)$$

COMMENTS ON THEOREM 1
To explain the presence of the factor $\rho^2 \sin\phi$ in the integrand, we examine the solid S bounded by

$\rho = \rho_0 \quad\quad \rho = \rho_0 + \Delta\rho$
$\phi = \phi_0 \quad\quad \phi = \phi_0 + \Delta\phi$
$\theta = \theta_0 \quad\quad \theta = \theta_0 + \Delta\theta$

(see Figure 14.57). We claim that the volume of S is approximately $\rho_0^2 \sin\phi_0 \, \Delta\rho \, \Delta\phi \, \Delta\theta$ when $\Delta\rho$, $\Delta\phi$, and $\Delta\theta$ are small.

If we hold θ constant for the moment, then the corresponding cross section has approximately the area $\rho_0 \, \Delta\rho \, \Delta\phi$ (see Section 14.5). Now, as θ varies from θ_0 to $\theta_0 + \Delta\theta$, the point $P = (\rho_0, \phi_0, \theta_0)$ swings along a circular arc of radius $\rho_0 \sin\phi_0$ through an angle $\Delta\theta$. The length of this arc is $\rho_0 \sin\phi_0 \, \Delta\theta$ (see Section 9.1). For small values of $\Delta\theta$ this arc is approximated by a straight line segment that is perpendicular to the plane $\theta = \theta_0$. Hence, the volume swept out in the process is approximately

$$(\rho_0 \, \Delta\rho \, \Delta\phi)(\rho_0 \sin\phi_0 \, \Delta\theta) = \rho_0^2 \sin\phi_0 \, \Delta\rho \, \Delta\phi \, \Delta\theta$$

With the appropriate limit defining the integral, this gives rise to the notation $\rho^2 \sin\phi \, d\rho \, d\phi \, d\theta$.

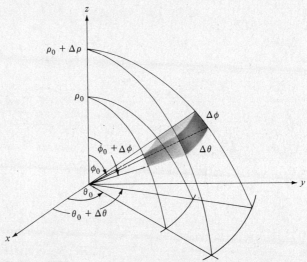

Figure 14.57

EXERCISES

In Exercises 1–4, find the mass and center of gravity of a mass distributed with density $\delta(x, y, z)$ in the cylinder $C: x^2 + y^2 \leq 2^2$, $0 \leq z \leq 1$.

1. $\delta(x, y, z) = z$
2. $\delta(x, y, z) = z^2$
3. $\delta(x, y, z) = 4 + x$
4. $\delta(x, y, z) = 5 + y$

In Exercises 5–8, find the mass and center of gravity of a mass distributed with density $\delta(x, y, z)$ in the section of a cone given by

$$0 \leq \phi \leq \frac{\pi}{3}$$

$$0 \leq \rho \leq 1$$

$$0 \leq \theta \leq \frac{3\pi}{2}$$

5. $\delta(x, y, z) = x^2 + y^2 + z^2$
6. $\delta(x, y, z) = 1$
7. $\delta(x, y, z) = z$
8. $\delta(x, y, z) = (x^2 + y^2 + z^2)^2$

9. Find the volume cut out of the solid sphere $x^2 + y^2 + z^2 \leq 9$ by the cylinder $x^2 + y^2 \leq 4$ (see Figure 14.58).

Hint: Find the curve in which the surfaces intersect.

10. Find the volume bounded by the surfaces $z = x^2 + y^2$ and $z = 18 - x^2 - y^2$.

11. Find the volume bounded by the paraboloid $z = x^2 + y^2$ and the plane $z = 4x$.

Hint: Find the curve in which the paraboloid intersects the plane.

QUIZ 1

1. Mass is distributed over the region bounded by the curves $y = x$ and $y = x^2$ and the density is given by $\rho(x, y) = (x - y)^2$. Find the total mass.

2. Evaluate the following double integral using polar coordinates.
$$\int_0^a \left[\int_0^{\sqrt{a^2-y^2}} (x^2 + y^2) \, dx \right] dy$$

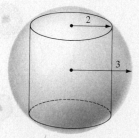

Figure 14.58

3. Evaluate the triple integral
$$\iiint_S (x + yz) \, dx \, dy \, dz$$
where S is given by $-1 \leq x \leq 1$, $0 \leq y \leq 1$, and $-1 \leq z \leq 0$.

4. Find the mass and center of gravity of the solid under the cone $z = 1 - \sqrt{x^2 - y^2}$ and above the xy plane. The density is given by $\delta(x, y, z) = x^2 + y^2$.

QUIZ 2

1. Evaluate
$$\iint_A (x - y)^3 \, dx \, dy$$
where A is the region bounded by $y = x^2$ and $y = \sqrt{x}$.

2. Use polar coordinates to find
$$\iint_A x^2 \, dx \, dy$$
where A is bounded by $r = 1$, $r = 2$, $\theta = 0$ and $\theta = \pi/2$.

3. Find the mass and the center of gravity of the hemisphere $x^2 + y^2 + z^2 \leq 1$ with $z \geq 0$ and density given by $\delta(x, y, z) = \sqrt{x^2 + y^2 + z^2}$. Use spherical coordinates.

CHAPTER 15
VECTORS

15.1 INTRODUCTION TO VECTORS

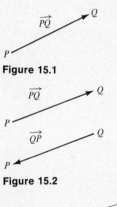

Figure 15.1

Figure 15.2

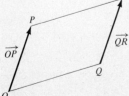

Figure 15.3

Vectors are useful tools in mathematics and its applications. The classical description of a vector is that it has direction and magnitude.

In physics, force, velocity, electric and magnetic fields, and gravitational fields are represented by vectors; in mathematics, displacement and rotation. We mention in passing that vector notation has had a great impact on algebra by enabling a single (vector) equation to take the place of n ordinary equations.

Vectors can be considered in spaces of different dimensions. In this text vectors are restricted to two- and three-dimensional spaces. Before defining what a vector is, we give some intuitive ideas about this concept.

Intuitively, a vector is often thought of as a directed line segment from a point P to a point Q; an arrow as in Figure 15.1 is used to convey this thought. The vector is referred to with the symbol $\overrightarrow{PQ}$, P being its *initial point* and Q its *terminal point*. The vector $\overrightarrow{QP}$ has the same magnitude as $\overrightarrow{PQ}$, but its direction is reversed (see Figure 15.2). Two vectors are considered *equivalent* if they have the same direction and magnitude. Referring to Figure 15.3, we see that in the parallelogram $OPQR$ the vectors $\overrightarrow{OP}$ and $\overrightarrow{QR}$ are equivalent.

Whether in the plane or in three-dimensional space, for any

vector $\overrightarrow{QR}$ there is just one vector with initial point O that is equivalent to it: This is the vector $\overrightarrow{OP}$ in Figure 15.3. The key reason for this is the following basic axiom in plane geometry: Given a line and a point not on it, then there is exactly one line through the point that is parallel to the given line. It is intuitively clear that there is only one way to match magnitude and direction.

In Definition 1 below, vectors are thought of as starting from a fixed origin O. This permits the identification of any vector with a point. With this reasoning there is no difference between vectors and points. A distinction is made, however, because different things are done with vectors than with points. For instance, later in this section we shall introduce addition, subtraction, and multiplication of vectors.

NOTATION

From now on, a point in the xy plane will be designated with a single letter and subscripts, for instance, (a_1, a_2), (b_1, b_2), (x_1, x_2), and so on. Likewise, points in three-dimensional space will be designated (a_1, a_2, a_3), (b_1, b_2, b_3), (x_1, x_2, x_3), and so on. Vectors will be designated with boldface letters, such as **a**, **b**, **x**, **R**, and so on.

DEFINITION 1

A *vector* **a** in the plane is a point

$$\mathbf{a} = (a_1, a_2)$$

A vector **a** in three-dimensional space is a point

$$\mathbf{a} = (a_1, a_2, a_3)$$

The coordinates a_1, a_2, and a_3, are called *components*. This definition is illustrated in Figures 15.4 and 15.5.

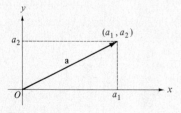

Figure 15.4

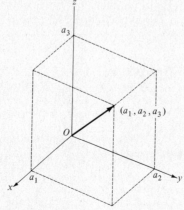

Figure 15.5

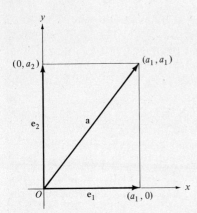

Figure 15.6

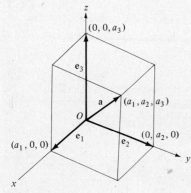

Figure 15.7

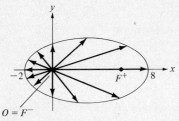

Figure 15.8

To be consistent, we designate the origin in the plane

$$\mathbf{0} = (0, 0)$$

and in three-dimensional space

$$\mathbf{0} = (0, 0, 0)$$

and call $\mathbf{0}$ the *zero vector*.

Example 1

The vectors $\mathbf{a} = (a_1, a_2)$, $\mathbf{e}_1 = (a_1, 0)$, $\mathbf{e}_2 = (0, a_2)$, and $\mathbf{0} = (0, 0)$ give the vertices of a rectangle (see Figure 15.6). The vectors $\mathbf{a} = (a_1, a_2, a_3)$, $\mathbf{e}_1 = (a_1, 0, 0)$, $\mathbf{e}_2 = (0, a_2, 0)$, $\mathbf{e}_3 = (0, 0, a_3)$, and $\mathbf{0} = (0, 0, 0)$, give five of the vertices of a rectangular box (see Figure 15.7). In Exercise 1 you are asked to find the vectors giving the remaining three vertices.

Example 2

Consider the vectors $\mathbf{x} = (x_1, x_2)$ whose components are related through the equation

$$\frac{(x_1 - 3)^2}{25} + \frac{x_2^2}{16} = 1$$

This is the equation of an ellipse with the foci $F^- = (0, 0)$ and $F^+ = (6, 0)$, and hence the vectors are as shown in Figure 15.8.

We now introduce three operations on vectors. These are addition, multiplication by a scalar (number), and subtraction. These operations will define for any two vectors $\mathbf{a}$ and $\mathbf{b}$ the vectors $\mathbf{a} + \mathbf{b}$, $k\mathbf{a}$, and $\mathbf{a} - \mathbf{b}$. In this section the operations are treated in a formal algebraic manner, but in Section 15.2 they are given a geometric interpretation in terms of directed line segments. Both the algebraic and the geometric approaches are important.

DEFINITION 2
If $\mathbf{a} = (a_1, a_2)$ and $\mathbf{b} = (b_1, b_2)$, then

$$\mathbf{a} + \mathbf{b} = (a_1 + b_1, a_2 + b_2)$$

If $\mathbf{a} = (a_1, a_2, a_3)$ and $\mathbf{b} = (b_1, b_2, b_3)$, then

$$\mathbf{a} + \mathbf{b} = (a_1 + b_1, a_2 + b_2, a_3 + b_3)$$

The vector $\mathbf{a} + \mathbf{b}$ is called the *sum* of $\mathbf{a}$ and $\mathbf{b}$. Observe that the sum is defined in terms of the sums of corresponding components.

Example 3

(a) $(1, 2) + (3, -1) = (4, 1)$
(b) $(1, 2) + (2, 1) = (3, 3)$
(c) $(1, 2) + (-1, -2) = (0, 0)$
(d) $(1, 2) + (0, 0) = (1, 2)$

Observe in Figure 15.9 the parallelograms formed in cases (a) and (b). We shall use this fact in Section 15.2 when we discuss the addition of vectors geometrically.

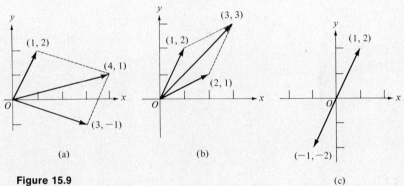

Figure 15.9

Example 4

(a) $(\frac{3}{2}, -2, 3) + (-2, 3, 1) = (-\frac{1}{2}, 1, 4)$
(b) $(\frac{3}{2}, -2, 3) + (-\frac{3}{2}, -2, -3) = (0, -4, 0)$

DEFINITION 3
Let k be any constant. If

$\mathbf{a} = (a_1, a_2)$

then

$k\mathbf{a} = (ka_1, ka_2)$

If

$\mathbf{a} = (a_1, a_2, a_3)$

then

$k\mathbf{a} = (ka_1, ka_2, ka_3)$

As in Definition 2, the operation on a vector is defined through an operation on its components.

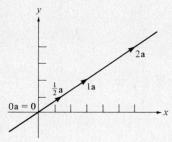

Figure 15.10

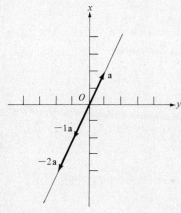

Figure 15.11

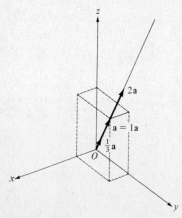

Figure 15.12

Example 5

Let $\mathbf{a} = (3, 2)$. Then

$0\mathbf{a} = (0, 0)$
$\tfrac{1}{2}\mathbf{a} = (\tfrac{3}{2}, 1)$
$1\mathbf{a} = (3, 2)$
$2\mathbf{a} = (6, 4)$

In Figure 15.10 you will observe that all these vectors lie along the same line. Also,

$$1 \cdot \mathbf{a} = \mathbf{a}$$

and in all other cases the magnitude changes but not the direction.

Example 6

Let $\mathbf{a} = (1, 2)$. Then

$-1\mathbf{a} = (-1, -2)$
$-2\mathbf{a} = (-2, -4)$

Observe in Figure 15.11 that here also the vectors lie along the same line, but multiplication by a negative number reverses the direction.

Example 7

Let $\mathbf{a} = (1, 2, 3)$. Then

$\tfrac{1}{3}\mathbf{a} = (\tfrac{1}{3}, \tfrac{2}{3}, 1)$
$1\mathbf{a} = (1, 2, 3)$
$2\mathbf{a} = (2, 4, 6)$

As in the plane, these vectors lie along the same line (see Figure 15.12).

DEFINITION 4

For any vectors $\mathbf{a}$ and $\mathbf{b}$, we define

$$-\mathbf{a} = -1\mathbf{a}$$
$$\mathbf{b} - \mathbf{a} = \mathbf{b} + (-\mathbf{a})$$

Thus, if

$$\mathbf{a} = (a_1, a_2) \quad \text{and} \quad \mathbf{b} = (b_1, b_2)$$

then

$$-\mathbf{a} = (-a_1, -a_2)$$

and

$$\mathbf{b} - \mathbf{a} = (b_1 - a_1, b_2 - a_2)$$

If
$$\mathbf{a} = (a_1, a_2, a_3) \quad \text{and} \quad \mathbf{b} = (b_1, b_2, b_3)$$
then
$$-\mathbf{a} = (-a_1, -a_2, -a_3)$$
and
$$\mathbf{b} - \mathbf{a} = (b_1 - a_1, b_2 - a_2, b_3 - a_3)$$

From the foregoing definitions and examples it is evident that the same properties and procedures apply to both vectors in the plane and vectors in three-dimensional space. This remains true also in spaces of other dimensions, and for this reason one often studies vectors abstractly, without reference to dimension. Notice that we have done so in Definition 4.

Collected below are some basic consequences of our definitions. Again, they are stated without reference to dimension and they are all easy to verify.

THEOREM 1. PROPERTIES OF VECTORS
Let $\mathbf{a}$, $\mathbf{b}$, and $\mathbf{c}$ be arbitrary vectors. Then

$$\mathbf{a} + \mathbf{b} = \mathbf{b} + \mathbf{a} \tag{1}$$
$$\mathbf{a} + (\mathbf{b} + \mathbf{c}) = (\mathbf{a} + \mathbf{b}) + \mathbf{c} \tag{2}$$
$$\mathbf{a} + \mathbf{0} = \mathbf{a} \tag{3}$$
$$\mathbf{a} - \mathbf{a} = \mathbf{0} \tag{4}$$
$$1\mathbf{a} = \mathbf{a} \tag{5}$$
$$0\mathbf{a} = \mathbf{0} \tag{6}$$
$$(k + h)\mathbf{a} = k\mathbf{a} + h\mathbf{a} \tag{7}$$
$$k(\mathbf{a} + \mathbf{b}) = k\mathbf{a} + k\mathbf{b} \tag{8}$$

EXERCISES

1. Find the vectors giving the missing vertices in Example 1.

In Exercises 2–14, find $\mathbf{a} + \mathbf{b}$ and $\mathbf{a} - \mathbf{b}$, and in Exercises 2–8 draw in the same diagram $\mathbf{a}$, $\mathbf{b}$, $\mathbf{a} + \mathbf{b}$, and $\mathbf{a} - \mathbf{b}$.

2. $\mathbf{a} = (0, 1)$, $\mathbf{b} = (1, 0)$

3. $\mathbf{a} = (-1, -1)$, $\mathbf{b} = (0, 1)$

4. $\mathbf{a} = (1, 3)$, $\mathbf{b} = (2, 6)$

5. $\mathbf{a} = (1, 2)$, $\mathbf{b} = (1, -2)$

6. $\mathbf{a} = (1, 2)$, $\mathbf{b} = (2, 1)$

7. $\mathbf{a} = (-1, -2)$, $\mathbf{b} = (-2, -1)$

8. $\mathbf{a} = (-2, 3)$, $\mathbf{b} = (2, 3)$

9. $\mathbf{a} = (1, 1, 0)$, $\mathbf{b} = (1, 0, 1)$

10. $\mathbf{a} = (-3, -1, 2)$, $\mathbf{b} = (2, -1, -2)$

11. $\mathbf{a} = (\frac{1}{2}, \frac{1}{3}, \frac{1}{2})$, $\mathbf{b} = (-\frac{1}{3}, \frac{1}{2}, \frac{1}{3})$

12. $\mathbf{a} = (-2, 3)$, $\mathbf{b} = \frac{1}{2}\mathbf{a}$

13. $\mathbf{a} = (-2, 4)$, $\mathbf{b} = -\mathbf{a}$

14. $\mathbf{a} = (2, -3, 3)$, $\mathbf{b} = \frac{1}{3}\mathbf{a}$

15.2 GEOMETRIC PROPERTIES OF VECTORS

To begin the geometric investigation of vectors, we define the length of a vector, called its magnitude.

> **DEFINITION 1**
> The *magnitude* of a vector $\mathbf{a} = (a_1, a_2)$ is the number
> $$|\mathbf{a}| = \sqrt{a_1^2 + a_2^2}$$
> the *magnitude* of a vector $\mathbf{a} = (a_1, a_2, a_3)$ is the number
> $$|\mathbf{a}| = \sqrt{a_1^2 + a_2^2 + a_3^2}$$

We observe that the magnitude of a vector $\mathbf{a}$ is the distance of $\mathbf{a}$ to the origin (see Figures 15.13 and 15.14). In general, the following theorem is true.

> **THEOREM 1**
> For any two vectors $\mathbf{a}$ and $\mathbf{b}$, the number $|\mathbf{b} - \mathbf{a}|$ is the distance from $\mathbf{a}$ to $\mathbf{b}$.

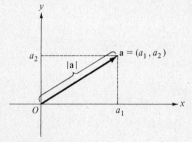

Figure 15.13

Proof
Let $\mathbf{a} = (a_1, a_2)$, $\mathbf{b} = (b_1, b_2)$, and put
$$\mathbf{b} - \mathbf{a} = (b_1 - a_1, b_2 - a_2) = (c_1, c_2) = \mathbf{c}$$
Then
$$|\mathbf{c}| = \sqrt{c_1^2 + c_2^2} = \sqrt{(b_1 - a_1)^2 + (b_2 - a_2)^2}$$
and since
$$|\mathbf{c}| = |\mathbf{b} - \mathbf{a}|$$
we arrive at the formula
$$|\mathbf{b} - \mathbf{a}| = \sqrt{(b_1 - a_1)^2 + (b_2 - a_2)^2}$$
(see Figure 15.15). In a like manner, we show that if $\mathbf{a} = (a_1, a_2, a_3)$ and $\mathbf{b} = (b_1, b_2, b_3)$, then

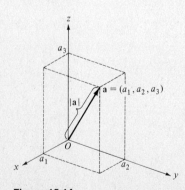

Figure 15.14

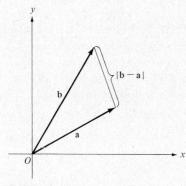

Figure 15.15

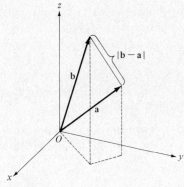

Figure 15.16

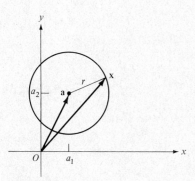

Figure 15.17

$$|\mathbf{b} - \mathbf{a}| = \sqrt{(b_1 - a_1)^2 + (b_2 - a_2)^2 + (b_3 - a_3)^2}$$

(see Figure 15.16).

In particular, we note that $|\mathbf{a}|$, $|\mathbf{b}|$, and $|\mathbf{b} - \mathbf{a}|$ are the sides of a triangle.

Example 1

If $\mathbf{a} = (2, -3)$ and $\mathbf{b} = (-3, 2)$, then

$$|\mathbf{a}| = \sqrt{2^2 + (-3)^2} = \sqrt{13}$$

and

$$|\mathbf{b} - \mathbf{a}| = \sqrt{(-3 - 2)^2 + [2 - (-3)]^2} = \sqrt{5^2 + 5^2} = 5\sqrt{2}$$

If $\mathbf{a} = (3, 0, 4)$ and $\mathbf{b} = (-1, 2, 0)$, then

$$|\mathbf{b} - \mathbf{a}| = \sqrt{(-1 - 3)^2 + (2 - 0)^2 + (0 - 4)^2}$$
$$= \sqrt{4^2 + 2^2 + 4^2} = 6$$

Example 2

Using vector notation, the equation of a circle in the plane with center $\mathbf{a} = (a_1, a_2)$ and radius r becomes

$$|\mathbf{x} - \mathbf{a}| = r \tag{1}$$

where $\mathbf{x} = (x_1, x_2)$ (see Figure 15.17). It should be noted that if $\mathbf{a} = (a_1, a_2, a_3)$ and $\mathbf{x} = (x_1, x_2, x_3)$, then (1) is the equation of a sphere in three-dimensional space with center $\mathbf{a}$ and radius r.

We now return to the geometric relation between the vectors $\mathbf{a}$, $\mathbf{b}$, $\mathbf{a} + \mathbf{b}$, and $\mathbf{a} - \mathbf{b}$ alluded to in the preceding section.

THEOREM 2

The points $\mathbf{0}$, $\mathbf{a}$, $\mathbf{a} + \mathbf{b}$, and $\mathbf{b}$ form the vertices of a parallelogram.

Proof

If $\mathbf{b} = k\mathbf{a}$ for some constant k, then by Section 15.1 all the points lie on a straight line. In this case we say that the parallelogram is degenerate. For the case when the points are not collinear, consult Figure 15.18: We show that opposite sides of the figure have equal lengths, and this will show that the figure is a parallelogram. By Theorem 1,

$$L_3 = |(\mathbf{a} + \mathbf{b}) - \mathbf{b}| = |\mathbf{a}| = L_1$$

and

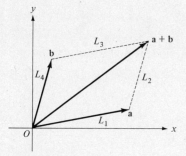

Figure 15.18

$$L_2 = |(a + b) - a| = |b| = L_4$$

This theorem suggests a geometric scheme for adding vectors **a** and **b** with a common origin **0**: Complete the parallelogram whose three vertices are **0**, **a**, and **b**; **a** + **b** is then the fourth vertex. Before elaborating on this further, we prove the following fact.

THEOREM 3
The points **0**, **a**, **b**, and **b** − **a** form the vertices of a parallelogram.

Proof
Consult Figure 15.19. Writing **b** = **a** + (**b** − **a**), we see that this theorem is true by virtue of Theorem 2.

Intuitively, Theorem 3 says that the displacement from **a** to **b** is equivalent to the displacement from **0** to **b** − **a**. This idea is conveyed in Figure 15.20, where the vector **b** − **a** has been repositioned to lead from **a** to **b**, thereby forming a triangle. Similarly, Theorem 2 tells us that the displacement from **a** to **a** + **b** is equivalent to the displacement from **0** to **b** (see Figure 15.21), and that the displacement from **b** to **a** + **b** is equivalent to the displacement from **0** to **a** (see Figure 15.22).

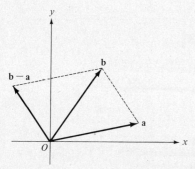

Figure 15.19

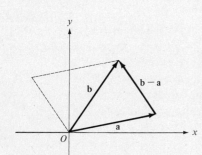

Figure 15.20

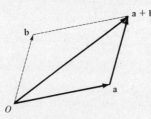

Figure 15.21

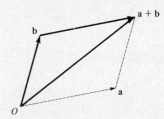

Figure 15.22

Example 3

Express the vector **a** = (2, 3) in terms of the vectors **b** = (5, 0) and **c** = (0, 1).

Solution
Consult Figure 15.23. We have to find constants k_1 and k_2 such that

$$a = k_1 b + k_2 c$$

Writing the vectors in terms of their components gives

$$(2, 3) = k_1(5, 0) + k_2(0, 1)$$
$$= (5k_1, 0) + (0, k_2) = (5k_1, k_2)$$

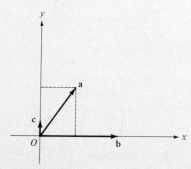

Figure 15.23

and hence we must have
$$5k_1 = 2 \quad \text{and} \quad k_2 = 3$$
We see that
$$\mathbf{a} = \tfrac{2}{5}\mathbf{b} + 3\mathbf{c}$$

Example 4

Express the vectors $\mathbf{a} = (\tfrac{1}{2}, 2)$ and $\mathbf{b} = (3, -1)$ in terms of the vectors $\mathbf{i} = (1, 0)$ and $\mathbf{j} = (0, 1)$.

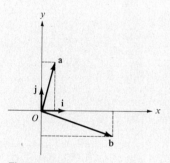

Figure 15.24

Solution
Consult Figure 15.24. As in Example 3, we seek constants k_1, k_2, m_1, and m_2, such that
$$\mathbf{a} = k_1\mathbf{i} + k_2\mathbf{j}$$
$$\mathbf{b} = m_1\mathbf{i} + m_2\mathbf{j}$$

In terms of components, this gives
$$(\tfrac{1}{2}, 2) = k_1(1, 0) + k_2(0, 1) = (k_1, k_2)$$
$$(3, -1) = m_1(1, 0) + m_2(0, 1) = (m_1, m_2)$$

Hence,
$$k_1 = \tfrac{1}{2}, \quad k_2 = 2, \quad m_1 = 3, \quad m_2 = -1$$
and we find that
$$\mathbf{a} = \tfrac{1}{2}\mathbf{i} + 2\mathbf{j}$$
$$\mathbf{b} = 3\mathbf{i} - \mathbf{j}$$

In general, the following is seen to be true.

THEOREM 4

If $\mathbf{i} = (1, 0)$ and $\mathbf{j} = (0, 1)$, then for any vector $\mathbf{a} = (a_1, a_2)$ we have the representation
$$\mathbf{a} = a_1\mathbf{i} + a_2\mathbf{j}$$

If $\mathbf{i} = (1, 0, 0)$, $\mathbf{j} = (0, 1, 0)$, and $\mathbf{k} = (0, 0, 1)$, then for any vector $\mathbf{a} = (a_1, a_2, a_3)$ we have the representation
$$\mathbf{a} = a_1\mathbf{i} + a_2\mathbf{j} + a_3\mathbf{k}$$

Thus, all vectors can be represented as linear combinations of fixed unit vectors:

DEFINITION 2

$\mathbf{u}$ is a *unit vector* if $|\mathbf{u}| = 1$.

While we are on the subject of unit vectors, we should mention a fact that will be used later: If $\mathbf{a} \neq \mathbf{0}$, then the vector

$$\mathbf{u} = \frac{1}{|\mathbf{a}|}\mathbf{a}$$

is a unit vector, since by Theorem 5 below,

$$|\mathbf{u}| = \left|\frac{1}{|\mathbf{a}|}\mathbf{a}\right| = \frac{1}{|\mathbf{a}|} \cdot |\mathbf{a}| = 1$$

> **DEFINITION 3**
> The vector
> $$\mathbf{u} = \frac{1}{|\mathbf{a}|}\mathbf{a}$$
> is called the *normalized form* of **a**.

Example 5

Find the normalized form of $\mathbf{a} = (2, -3)$.

Solution
Since

$$|\mathbf{a}| = \sqrt{2^2 + (-3)^2} = \sqrt{13}$$

the normalized form of **a** is

$$\mathbf{u} = \frac{1}{\sqrt{13}}(2, -3) = \left(\frac{2}{\sqrt{13}}, \frac{-3}{\sqrt{13}}\right)$$

In the remainder of this section we examine vectors $k\mathbf{a}$ for a fixed vector $\mathbf{a} \neq \mathbf{0}$. We begin with the following consequence of Definition 1.

> **THEOREM 5**
> If $k > 0$, then $|k\mathbf{a}| = k|\mathbf{a}|$.

Proof
Let $\mathbf{a} = (a_1, a_2)$. Then $k\mathbf{a} = (ka_1, ka_2)$, and with the fact that $\sqrt{k^2} = k$ for $k > 0$, we see that

$$|k\mathbf{a}| = \sqrt{(ka_1)^2 + (ka_2)^2} = \sqrt{k^2(a_1^2 + a_2^2)}$$
$$= k\sqrt{a_1^2 + a_2^2} = k|\mathbf{a}|$$

The proof when $\mathbf{a} = (a_1, a_2, a_3)$ is similar.

With this we can establish the following theorem.

15.2 GEOMETRIC PROPERTIES OF VECTORS

THEOREM 6
For any constant k, $k\mathbf{a}$ lies on the line passing through the points **0** and **a**.

Proof
The proof relies on the fact that the shortest distance between two points is a straight line. The theorem is proved here only for the case $0 \leq k \leq 1$. The general case is much easier to prove with the help of *direction cosines*, and this is done in Section 15.3.

Consult Figure 15.25. If $L_2 + L_3 = L_1$, then $k\mathbf{a}$ lies on the line in question. Since

$$L_1 = |\mathbf{a}|$$
$$L_2 = |k\mathbf{a}| = k|\mathbf{a}|$$
$$L_3 = |\mathbf{a} - k\mathbf{a}| = |(1 - k)\mathbf{a}| = (1 - k)|\mathbf{a}|$$

it follows that

$$L_2 + L_3 = k|\mathbf{a}| + (1 - k)|\mathbf{a}| = |\mathbf{a}| = L_1$$

Note that we have actually proved that if $0 \leq k \leq 1$, then $k\mathbf{a}$ lies on the line segment joining **0** and **a**. In particular, the *midpoint* of this line segment is $\frac{1}{2}\mathbf{a}$. The relation between **a** and $k\mathbf{a}$ for some values of k is pictured in Figure 15.26.

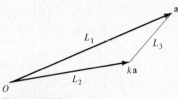
Figure 15.25

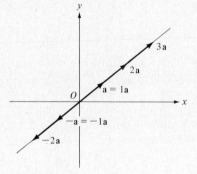
Figure 15.26

EXERCISES

Find the magnitude and normalized form of the vectors in Exercises 1–9, and draw the two-dimensional vectors.

1. $\mathbf{a} = (-\frac{1}{3}, -\frac{1}{4})$
2. $\mathbf{b} = (-1, 1)$
3. $\mathbf{c} = (-\sqrt{2}, -\sqrt{2})$
4. $\mathbf{d} = (-\frac{2}{3}, \frac{2}{5})$
5. $\mathbf{e} = (-1, 0, 1)$
6. $\mathbf{a} = (3, 4, 5)$
7. $\mathbf{b} = (1 - k, 1 + k, \sqrt{2k^2 - 2})$
8. $\mathbf{c} = \left(\frac{\sqrt{3}}{5}, \frac{\sqrt{3}}{5}, \frac{\sqrt{3}}{5}\right)$
9. $\mathbf{d} = (x, 2x, 4x) \ (x > 0)$
10. Express the circle $(x - 1)^2 + (y - 2)^2 = 3$ in vector notation.
11. Express the vertices of the triangle formed by $\mathbf{a} = (2, 0)$, $\mathbf{b} = (1, 4)$, and $\mathbf{c} = (0, 2)$ in terms of $\mathbf{v} = (-1, 0)$ and $\mathbf{w} = (-3, -3)$.
12. Find the vector $\mathbf{x} = (x_1, x_2)$ of minimum magnitude when $x_2 = x_1(1 - x_1)$ and $0 \leq x_1 \leq 1$.
13. Find the vector $\mathbf{x} = (x_1, x_2)$ of minimum magnitude when $x_2 = \ln(x_1 + e)$ and $x_1 + e \geq 1$.

15.3 DIRECTION COSINES AND THE SCALAR PRODUCT

In the preceding section we discussed the magnitude of a vector, and now we take up the question of direction. To this end, consider a vector $\mathbf{a} = (a_1, a_2, a_3)$, $\mathbf{a} \neq \mathbf{0}$; let α_1, α_2, and α_3 be the radian measures of the respective angles between $\mathbf{a}$ and the coordinate axes (see Figure 15.27). The α_i are chosen in the interval $0 \leq \alpha_i \leq \pi$, and it is seen that

$$\frac{a_1}{|\mathbf{a}|} = \cos \alpha_1$$

$$\frac{a_2}{|\mathbf{a}|} = \cos \alpha_2 \qquad (1)$$

$$\frac{a_3}{|\mathbf{a}|} = \cos \alpha_3$$

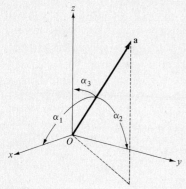

Figure 15.27

We observe that

$$\frac{1}{|\mathbf{a}|}\mathbf{a} = (\cos \alpha_1, \cos \alpha_2, \cos \alpha_3) \qquad (2)$$

and, since $(1/|\mathbf{a}|)\mathbf{a}$ is a unit vector,

$$\cos^2 \alpha_1 + \cos^2 \alpha_2 + \cos^2 \alpha_3 = 1 \qquad (3)$$

Thus, the numbers $\cos \alpha_1$, $\cos \alpha_2$, and $\cos \alpha_3$ are the coordinates of the normalized form of the vector $\mathbf{a}$; they are called the *direction cosines* of $\mathbf{a}$. As a rule, direction cosines are not used in two dimensions.

Example 1

Find the direction cosines of $\mathbf{a} = (3, -4, 12)$.

Solution
The magnitude of $\mathbf{a}$ is

$$|\mathbf{a}| = \sqrt{3^2 + (-4)^2 + 12^2} = 13$$

and hence

$$\frac{1}{|\mathbf{a}|}\mathbf{a} = (\tfrac{3}{13}, -\tfrac{4}{13}, \tfrac{12}{13})$$

The direction cosines are

$$\cos \alpha_1 = \tfrac{3}{13} \qquad \cos \alpha_2 = -\tfrac{4}{13} \qquad \cos \alpha_3 = \tfrac{12}{13}$$

REMARK
The actual angles α_i can be found by using a table of cosines. The fact that $\cos \alpha_2 < 0$ indicates that $\pi/2 < \alpha_2 < \pi$.

We now define the first of two ways in which vectors are multiplied: The first of these results in a number; the second (see Section 15.10) results in a vector.

> **DEFINITION 1**
> The *scalar product* of the vectors **a** and **b**, designated $\mathbf{a} \cdot \mathbf{b}$, is defined as follows: If $\mathbf{a} = (a_1, a_2)$ and $\mathbf{b} = (b_1, b_2)$, then
> $$\mathbf{a} \cdot \mathbf{b} = a_1 b_1 + a_2 b_2$$
> If $\mathbf{a} = (a_1, a_2, a_3)$ and $\mathbf{b} = (b_1, b_2, b_3)$, then
> $$\mathbf{a} \cdot \mathbf{b} = a_1 b_1 + a_2 b_2 + a_3 b_3$$

Other names for the scalar product are *inner product* and *dot product*.

Example 2

If $\mathbf{a} = (1, 2)$ and $\mathbf{b} = (2, -3)$, then
$$\mathbf{a} \cdot \mathbf{b} = 1 \cdot 2 + 2 \cdot (-3) = -4$$
If $\mathbf{a} = (3, -1)$ and $\mathbf{b} = (1, 3)$, then
$$\mathbf{a} \cdot \mathbf{b} = 3 \cdot 1 + (-1) \cdot 3 = 0$$
If $\mathbf{a} = (2, 1, -3)$ and $\mathbf{b} = (1, -4, -2)$, then
$$\mathbf{a} \cdot \mathbf{b} = 2 \cdot 1 + 1 \cdot (-4) + (-3) \cdot (-2) = 4$$

Example 3

For the unit vectors
$$\mathbf{i} = (1, 0, 0) \qquad \mathbf{j} = (0, 1, 0) \qquad \mathbf{k} = (0, 0, 1)$$
we have
$$\begin{array}{ll} \mathbf{i} \cdot \mathbf{i} = 1 & \mathbf{i} \cdot \mathbf{j} = \mathbf{j} \cdot \mathbf{i} = 0 \\ \mathbf{j} \cdot \mathbf{j} = 1 & \mathbf{j} \cdot \mathbf{k} = \mathbf{k} \cdot \mathbf{j} = 0 \\ \mathbf{k} \cdot \mathbf{k} = 1 & \mathbf{k} \cdot \mathbf{i} = \mathbf{i} \cdot \mathbf{k} = 0 \end{array}$$

The following theorem shows how the scalar product is related to magnitude.

> **THEOREM 1**
> For any vector **a** we have
> $$|\mathbf{a}| = \sqrt{\mathbf{a} \cdot \mathbf{a}}$$

Proof

If $\mathbf{a} = (a_1, a_2)$, then

$$\sqrt{\mathbf{a} \cdot \mathbf{a}} = \sqrt{a_1^2 + a_2^2} = |\mathbf{a}|$$

If $\mathbf{a} = (a_1, a_2, a_3)$, then

$$\sqrt{\mathbf{a} \cdot \mathbf{a}} = \sqrt{a_1^2 + a_2^2 + a_3^2} = |\mathbf{a}|$$

We recall that $|\mathbf{a}|$, $|\mathbf{b}|$, and $|\mathbf{b} - \mathbf{a}|$ are the lengths of the sides of a triangle (see Figure 15.28). If, as in Figure 15.28, it is a right triangle, then the theorem of Pythagoras says that

$$|\mathbf{b} - \mathbf{a}|^2 = |\mathbf{a}|^2 + |\mathbf{b}|^2 \tag{4}$$

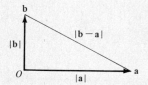

Figure 15.28

For an arbitrary triangle, the following theorem is true.

THEOREM 2
For any vectors $\mathbf{a}$ and $\mathbf{b}$,

$$|\mathbf{b} - \mathbf{a}|^2 = |\mathbf{a}|^2 + |\mathbf{b}|^2 - 2\,\mathbf{a} \cdot \mathbf{b} \tag{5}$$

REMARK
Formula (4), it is noted, is true when and only when we have a right triangle, that is, when and only when the vectors $\mathbf{a}$ and $\mathbf{b}$ are mutually perpendicular. Comparing formulas (4) and (5) leads us to conjecture that $\mathbf{a} \cdot \mathbf{b} = 0$ whenever $\mathbf{a}$ and $\mathbf{b}$ are mutually perpendicular. It will follow from Theorem 3 that this is, indeed, the case.

Proof
The proof is carried out for vectors in three-dimensional space; in the case of the plane the proof is similar.
If $\mathbf{a} = (a_1, a_2, a_3)$ and $\mathbf{b} = (b_1, b_2, b_3)$, then

$$\begin{aligned}
|\mathbf{b} - \mathbf{a}|^2 &= (b_1 - a_1)^2 + (b_2 - a_2)^2 + (b_3 - a_3)^2 \\
&= (b_1^2 - 2a_1b_1 + a_1^2) + (b_2^2 - 2a_2b_2 + a_2^2) \\
&\quad + (b_3^2 - 2a_3b_3 + b_3^2) \\
&= (a_1^2 + a_2^2 + a_3^2) + (b_1^2 + b_2^2 + b_3^2) \\
&\quad - 2(a_1b_1 + a_2b_2 + a_3b_3) \\
&= |\mathbf{a}|^2 + |\mathbf{b}|^2 - 2\,\mathbf{a} \cdot \mathbf{b}
\end{aligned}$$

Hence, formula (5) is correct.

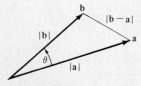

Figure 15.29

Examine Figure 15.29. If θ is the angle between the vectors $\mathbf{a}$ and $\mathbf{b}$, then the *law of cosines* says that

$$|\mathbf{b} - \mathbf{a}|^2 = |\mathbf{a}|^2 + |\mathbf{b}|^2 - 2|\mathbf{a}||\mathbf{b}|\cos\theta \tag{6}$$

This formula, when compared with formula (5), shows that

$$\mathbf{a} \cdot \mathbf{b} = |\mathbf{a}||\mathbf{b}|\cos\theta \tag{7}$$

and from this formula we can calculate the cosine of the angle between the vectors **a** and **b**. Finally, we state this result as follows.

> **THEOREM 3**
> Consider vectors $\mathbf{a} \neq \mathbf{0}$ and $\mathbf{b} \neq \mathbf{0}$ with an angle θ between them. Then
> $$\cos \theta = \frac{\mathbf{a} \cdot \mathbf{b}}{|\mathbf{a}||\mathbf{b}|} \qquad (8)$$

Example 4

Find the angle between the vectors $\mathbf{a} = (2, 1, 1)$ and $\mathbf{b} = (1, -3, 5)$.

Solution
$$|\mathbf{a}| = \sqrt{2^2 + 1^2 + 1^2} = \sqrt{6}$$
$$|\mathbf{b}| = \sqrt{1^2 + (-3)^2 + 5^2} = \sqrt{35}$$
$$\mathbf{a} \cdot \mathbf{b} = 2 \cdot 1 + 1 \cdot (-3) + 1 \cdot 5 = 4$$

Hence
$$\cos \theta = \frac{4}{\sqrt{6} \cdot \sqrt{35}} = \frac{4}{\sqrt{210}} \approx \frac{4}{14.49} \approx 0.2684$$

and solving for θ gives
$$\theta = \cos^{-1} \frac{4}{\sqrt{210}} \approx \cos^{-1} 0.2684 \approx 1.3 \text{ radians}$$

Example 5

Find the angle between the vectors $\mathbf{a}$ and $\mathbf{b} - \mathbf{a}$ when $\mathbf{a} = (1, 2)$ and $\mathbf{b} = (-1, 4)$.

Solution
Here we use formula (8) above with $\mathbf{b} - \mathbf{a}$ instead of $\mathbf{b}$. We have $\mathbf{b} - \mathbf{a} = (-2, 2)$ and hence
$$|\mathbf{a}| = \sqrt{5}$$
$$|\mathbf{b} - \mathbf{a}| = \sqrt{8}$$
$$\mathbf{a} \cdot (\mathbf{b} - \mathbf{a}) = 1 \cdot (-2) + 2 \cdot 2 = 2$$

and
$$\cos \theta = \frac{2}{\sqrt{5} \cdot \sqrt{8}} = \frac{1}{\sqrt{10}}$$

Solving for θ gives

$$\theta = \cos^{-1}\frac{1}{\sqrt{10}} \approx \cos^{-1}\frac{1}{3.1623} \approx \cos^{-1} 0.3162 \approx 1.2490 \text{ radians}$$

Using formula (8) we can easily prove Theorem 6 of Section 15.2, repeated here for convenience:

For any constant k, $k\mathbf{a}$ lies on the line passing through $\mathbf{0}$ and $\mathbf{a}$.

This result is clear when $k = 0$ or $\mathbf{a} = \mathbf{0}$. Otherwise, $|k\mathbf{a}| = \pm k|\mathbf{a}|$, and formula (8), with $\mathbf{b} = k\mathbf{a}$, gives

$$\cos\theta = \frac{\mathbf{a}\cdot\mathbf{b}}{|\mathbf{a}||k\mathbf{a}|} = \frac{k(\mathbf{a}\cdot\mathbf{a})}{\pm k|\mathbf{a}|^2} = \frac{k|\mathbf{a}|^2}{\pm k|\mathbf{a}|^2} = \pm 1$$

Hence $\theta = \cos^{-1}(\pm 1)$, telling us that $\theta = 0$ when $k > 0$ and $\theta = \pi$ when $k < 0$. In either case, however, the theorem is seen to be true.

Returning to the remark made following Theorem 2, we introduce the following concept.

DEFINITION 2
If $\mathbf{a}\cdot\mathbf{b} = 0$, then the vectors $\mathbf{a}$ and $\mathbf{b}$ are said to be *orthogonal*.

Thus, if $\mathbf{a}$ and $\mathbf{b}$ are orthogonal and $\mathbf{a} \neq \mathbf{0}$ and $\mathbf{b} \neq \mathbf{0}$, then the angle between the vectors is $\theta = \pi/2$. Since $\mathbf{0}\cdot\mathbf{b} = 0$ for any vector $\mathbf{b}$, we say for convenience that the vector $\mathbf{0}$ is orthogonal to any vector.

Example 5

The vectors $\mathbf{i} = (1, 0, 0)$, $\mathbf{j} = (0, 1, 0)$, and $\mathbf{k} = (0, 0, 1)$ are mutually orthogonal since $\mathbf{i}\cdot\mathbf{j} = 0$, $\mathbf{j}\cdot\mathbf{k} = 0$, and $\mathbf{i}\cdot\mathbf{k} = 0$.

Example 6

Find the unit vectors $\mathbf{u} = (u_1, u_2, u_3)$ that are orthogonal to

$$\mathbf{a} = (2, 1, 1) \quad \text{and} \quad \mathbf{b} = (1, 2, 1).$$

Solution
To satisfy the orthogonality requirements we must solve the equations $\mathbf{u}\cdot\mathbf{a} = 0$ and $\mathbf{u}\cdot\mathbf{b} = 0$. In terms of components we have the equations

$$2u_1 + u_2 + u_3 = 0$$
$$u_1 + 2u_2 + u_3 = 0$$

Solving these gives

$$u_2 = u_1 \qquad u_3 = -3u_1$$

and hence the vector **u** can be written as

$$\mathbf{u} = (u_1, u_1, -3u_1) = u_1(1, 1, -3)$$

This vector will have unit length if

$$|\mathbf{u}|^2 = u_1^2[1^2 + 1^2 + (-3)^2] = 11u_1^2 = 1$$

Hence

$$u_1 = \pm \frac{1}{\sqrt{11}}$$

and the unit vectors we are seeking are

$$\mathbf{u} = \left(\frac{1}{\sqrt{11}}, \frac{1}{\sqrt{11}}, -\frac{3}{\sqrt{11}}\right)$$

and

$$\mathbf{u} = \left(-\frac{1}{\sqrt{11}}, -\frac{1}{\sqrt{11}}, \frac{3}{\sqrt{11}}\right)$$

In concluding this section, we state without proof the following basic facts about the dot product (see Exercise 17).

THEOREM 4. PROPERTIES OF THE SCALAR PRODUCT
For all vectors **a**, **b**, and **c**, the following relations hold:

$$\mathbf{a} \cdot \mathbf{b} = \mathbf{b} \cdot \mathbf{a} \tag{9}$$
$$k\mathbf{a} \cdot \mathbf{b} = k(\mathbf{a} \cdot \mathbf{b}) \quad \text{for any constant } k \tag{10}$$
$$\mathbf{a} \cdot (\mathbf{b} + \mathbf{c}) = \mathbf{a} \cdot \mathbf{b} + \mathbf{a} \cdot \mathbf{c} \tag{11}$$
$$\mathbf{a} \cdot \mathbf{a} > 0 \quad \text{unless } \mathbf{a} = \mathbf{0} \tag{12}$$

EXERCISES

Find the dot product of the vectors in Exercises 1–6.

1. $\mathbf{a} = (-6, 3, 1)$, $\mathbf{b} = (2, 0, -3)$
2. $\mathbf{x} = (1, 1, 0)$, $\mathbf{y} = (0, 1, 1)$
3. $\mathbf{a} = (1, 0, -3)$, $\mathbf{b} = (2, 0, -3) + (-6, 3, 1)$
4. $\mathbf{u} = \mathbf{i} + \mathbf{j} + \mathbf{k}$, $\mathbf{v} = -\mathbf{i} + \mathbf{j} + \mathbf{k}$
5. $\mathbf{a} = -6\mathbf{i} + 3\mathbf{j} + \mathbf{k}$, $\mathbf{b} = 2\mathbf{i} - \mathbf{j} + 2\mathbf{k}$
6. $\mathbf{a} = 3\mathbf{i} + 4\mathbf{j} - 2\mathbf{k}$, $\mathbf{b} = 4\mathbf{i} - 2\mathbf{j} + 2\mathbf{k}$

Find the direction cosines of the vectors in Exercises 7–10.

7. $\mathbf{x} = (-1, 4, 5)$
8. $\mathbf{y} = (1, 0, 1)$
9. $\mathbf{z} = (1, 0, -1)$
10. $\mathbf{e} = (1, 1, 1)$

Find the cosine of the angles between the vectors in Exercises 11–16.

11. $\mathbf{a} = (1, 1, 1)$, $\mathbf{i} = (1, 0, 0)$
12. $\mathbf{a} = (1, 1, 1)$, $\mathbf{j} + \mathbf{k} = (0, 1, 1)$
13. $\mathbf{a} = (\tfrac{1}{3}, 3)$, $\mathbf{b} = (-3, -1)$
14. $\mathbf{a} = (1, 1, 1)$, $\mathbf{b} = (-1, -1, 1)$
15. $\mathbf{u} = (-1, -1, -1)$, $\mathbf{v} = (1, 1, 1)$
16. $\mathbf{x} = \mathbf{i} - \mathbf{j} - \mathbf{k}$, $\mathbf{y} = 2\mathbf{i} - \mathbf{j} - \mathbf{k}$

17. Find the cosines of the interior angles of the triangle having vertices $\mathbf{0}$, $\mathbf{a} = (1, 1)$, and $\mathbf{b} = (\tfrac{1}{2}, 2)$.

18. Find the cosines of the interior angles of the triangle having vertices $\mathbf{0}$, $\mathbf{a} = (1, 1, 1)$, $\mathbf{b} = (\tfrac{1}{3}, 1, 3)$.

19. Show that $\mathbf{a} \cdot (\mathbf{b} + \mathbf{c}) = \mathbf{a} \cdot \mathbf{b} + \mathbf{a} \cdot \mathbf{c}$ for all vectors $\mathbf{a}$, $\mathbf{b}$, and $\mathbf{c}$.

20. Let the vector $\mathbf{a}$ have direction cosines, $\cos \alpha_1$, $\cos \alpha_2$, $\cos \alpha_3$, and let the vector $\mathbf{b}$ have direction cosines $\cos \beta_1$, $\cos \beta_2$, $\cos \beta_3$. What does the equation $\cos \alpha_1 \cos \beta_1 + \cos \alpha_2 \cos \beta_2 + \cos \alpha_3 \cos \beta_3 = 0$ tell us about the relation between $\mathbf{a}$ and $\mathbf{b}$?

21. Consider vectors $\mathbf{a} = (a_1, a_2, a_3)$ and $\mathbf{b} = (b_1, b_2, b_3)$. Explain the meaning of the expression

$$\frac{a_1 b_1 + a_2 b_2 + a_3 b_3}{\sqrt{a_1^2 + a_2^2 + a_3^2} \; \sqrt{b_1^2 + b_2^2 + b_3^2}}$$

15.4 VECTOR FUNCTIONS AND THEIR DERIVATIVES

We recall from Section 12.1 that a curve in the plane is given in parametric form by two equations such as

$$\left. \begin{array}{l} x = f_1(t) \\ y = f_2(t) \end{array} \right\} \quad a \leq t \leq b \tag{1}$$

The parameter t gives each point (x, y) on the curve in the form

$$(x, y) = (f_1(t), f_2(t))$$

as shown in Figure 15.30 (compare this with Figure 12.4 of Section 12.1). Taking advantage of vector notation, we put

$$\mathbf{R} = (x, y)$$
$$\mathbf{f}(t) = (f_1(t), f_2(t))$$

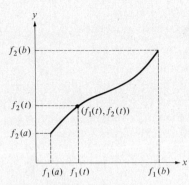

Figure 15.30
A curve in parametric form.

and replace (1) by the single equation

$$\mathbf{R} = \mathbf{f}(t) \qquad a \le t \le b \qquad (2)$$

The geometric interpretation of this equation is that, for each value of t, $\mathbf{R}$ is a vector with terminal point on the curve (1). The function $\mathbf{f}(t)$ is called a *vector function*. Similarly, a curve in three-dimensional space is given by three equations such as

$$\left.\begin{array}{l} x = f_1(t) \\ y = f_2(t) \\ z = f_3(t) \end{array}\right\} a \le t \le b \qquad (3)$$

These equations are also replaced by the single equation (2), this time putting

$$\mathbf{R} = (x, y, z)$$
$$\mathbf{f}(t) = (f_1(t), f_2(t), f_3(t))$$

We observe that there is no conceptual difference between vector functions in the plane and in three-dimensional space. This is, of course, consistent with our observation in the preceding sections of this chapter.

Example 1

A vector notation for the semicircle of radius 2,

$$\left.\begin{array}{l} x = 2\cos t \\ y = 2\sin t \end{array}\right\} 0 \le t \le \pi$$

is

$$\mathbf{R} = (2\cos t, 2\sin t) \qquad 0 \le t \le \pi$$

With the unit vectors $\mathbf{i}$ and $\mathbf{j}$, this equation can also be written as

$$\mathbf{R} = 2\cos t\,\mathbf{i} + 2\sin t\,\mathbf{j} \qquad 0 \le t \le \pi$$

Example 2

Consider the curve

$$\left.\begin{array}{l} x = 2\cos t \\ y = 2\sin t \\ z = 3t \end{array}\right\} 0 \le t \le 4\pi$$

In vector notation this curve is given as

$$\mathbf{R} = (2\cos t, 2\sin t, 3t) \qquad 0 \le t \le 4\pi$$

or

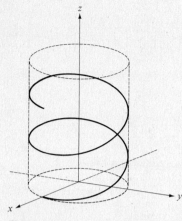

Figure 15.31
The helix $\mathbf{R} = (2\cos t, 2\sin t, 3t)$.

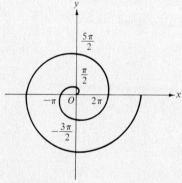

Figure 15.32
The spiral
$\begin{cases} x = t\cos t \\ y = t\sin t \end{cases}$

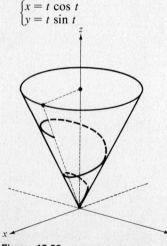

Figure 15.33
The curve $\mathbf{R} = (t\cos t, t\sin t, t)$.

$$\mathbf{R} = 2\cos t\,\mathbf{i} + 2\sin t\,\mathbf{j} + 3t\,\mathbf{k} \qquad 0 \le t \le 4\pi$$

To get an idea of the shape of this curve, write

$$\mathbf{R} - 3t\,\mathbf{k} = 2\cos t\,\mathbf{i} + 2\sin t\,\mathbf{j}$$

By Example 1, the right-hand side of this equation is the equation of a circle of radius 2. As t varies from 0 to 4π, this circle is traced exactly twice, but at the same time the curve moves up a cylinder of radius 2, as indicated in Figure 15.31. This curve is a *helix*.

Example 3

Consider the curve

$$\mathbf{R} = t\cos t\,\mathbf{i} + t\sin t\,\mathbf{j} + t\,\mathbf{k} \qquad 0 \le t \le 4\pi$$

If we write it in the form

$$\mathbf{R} - t\,\mathbf{k} = t\cos t\,\mathbf{i} + t\sin t\,\mathbf{j} \qquad 0 \le t \le 4\pi$$

then we see from Section 11.1 that the right side describes a spiral (see Figure 15.32). Putting

$$x = t\cos t$$
$$y = t\sin t$$
$$z = t$$

we have $x^2 + y^2 = t^2$ and hence

$$z = \sqrt{x^2 + y^2}$$

The latter is the equation of a cone, and hence the spiral travels up the cone as t increases from 0 to 4π (see Figure 15.33).

Example 4

Find the curve $\mathbf{R} = \mathbf{a} + t(\mathbf{b} - \mathbf{a})$, where we have $\mathbf{a} = (a_1, a_2, a_3)$ and $\mathbf{b} = (b_1, b_2, b_3)$ as fixed vectors, and t varies in the interval $0 \le t \le 1$.

Solution

From Theorem 6 of Section 15.2, we see that $t(\mathbf{b} - \mathbf{a})$ gives for $0 \le t \le 1$ the line segment joining $\mathbf{0}$ and $\mathbf{b} - \mathbf{a}$. Hence $\mathbf{a} + t(\mathbf{b} - \mathbf{a})$ gives the line segment joining $\mathbf{a} + \mathbf{0} = \mathbf{a}$ and $\mathbf{a} + (\mathbf{b} - \mathbf{a}) = \mathbf{b}$ (see Figure 15.34).

Example 5

Find the curve $\mathbf{R} = \mathbf{a} + t\mathbf{b}$, where $\mathbf{a}$, $\mathbf{b}$, and t are as in Example 4.

15.4 VECTOR FUNCTIONS AND THEIR DERIVATIVES

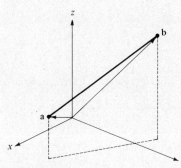

Figure 15.34
The curve $R = a + t(b - a)$ for $0 \leq t \leq 1$.

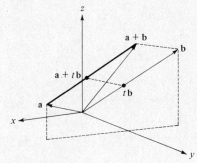

Figure 15.35
The curve $R = a + tb$.

Solution
Again by Theorem 6 of Section 15.2, $t\mathbf{b}$ gives, for $0 \leq t \leq 1$, the line segment joining $\mathbf{0}$ and $\mathbf{b}$. Hence $\mathbf{a} + t\mathbf{b}$ gives the line segment joining $\mathbf{a}$ to $\mathbf{a} + \mathbf{b}$ (see Figure 15.35). The line $\mathbf{R} = \mathbf{a} + t\mathbf{b}$ with unrestricted parameter t is called *the line through* $\mathbf{a}$ *in the direction* $\mathbf{b}$.

DERIVATIVES

The concept of derivative of a vector function is the same as that for ordinary functions (see Section 2.1). For a vector function $\mathbf{f}(t) = (f_1(t), f_2(t))$, we consider the difference quotient

$$\frac{1}{\Delta t}[\mathbf{f}(t + \Delta t) - \mathbf{f}(t)] = \left(\frac{f_1(t + \Delta t) - f_1(t)}{\Delta t}, \frac{f_2(t + \Delta t) - f_2(t)}{\Delta t}\right) \quad (4)$$

If $f_1(t)$ and $f_2(t)$ are differentiable, then

$$\lim_{\Delta t \to 0} \frac{f_1(t + \Delta t) - f_1(t)}{\Delta t} = f_1'(t)$$

$$\lim_{\Delta t \to 0} \frac{f_2(t + \Delta t) - f_2(t)}{\Delta t} = f_2'(t)$$

and we express this fact by writing

$$\lim_{\Delta t \to 0} \frac{1}{\Delta t}[\mathbf{f}(t + \Delta t) - \mathbf{f}(t)] = \mathbf{f}'(t)$$

A similar argument applies to vector functions

$$\mathbf{f}(t) = (f_1(t), f_2(t), f_3(t))$$

DEFINITION 1
The derivative $\mathbf{f}'(t)$ of the vector function $\mathbf{f}(t)$ is defined as follows: If $\mathbf{f}(t) = (f_1(t), f_2(t))$, then

$$\mathbf{f}'(t) = (f_1'(t), f_2'(t))$$

If $\mathbf{f}(t) = (f_1(t), f_2(t), f_3(t))$, then

$$\mathbf{f}'(t) = (f_1'(t), f_2'(t), f_3'(t))$$

provided f_1', f_2', and f_3' exist. We say that in this case $\mathbf{f}(t)$ is *differentiable*.

In the next section we shall see that the derivative of a vector function also has an interpretation in terms of a tangent. It should also be noted that once more an operation on a vector was defined through an operation on its components. Thus, a vector function is differentiable if and only if its component functions are differentiable. The second derivative of $\mathbf{f}(t) = (f_1(t), f_2(t))$ is defined as $\mathbf{f}''(t) = (f_1''(t), f_2''(t))$, and so on.

Example 6

Find $\mathbf{f}'(t)$ when

(a) $\mathbf{f}(t) = (t \cos t, t \sin t)$;
(b) $\mathbf{f}(t) = (t \cos t, t \sin t, t^2)$.

Solution

(a) $\mathbf{f}'(t) = (\cos t - t \sin t, \sin t + t \cos t)$;
(b) $\mathbf{f}'(t) = (\cos t - t \sin t, \sin t + t \cos t, 2t)$.

Example 7

Find $\mathbf{f}'(t)$ and $\mathbf{f}''(t)$ when $\mathbf{f}(t) = e^t \mathbf{i} + e^{-t} \mathbf{j} - t^{-2} \mathbf{k}$.

Solution

$\mathbf{f}'(t) = e^t \mathbf{i} - e^{-t} \mathbf{j} + 2t^{-3} \mathbf{k}$
$\mathbf{f}''(t) = e^t \mathbf{i} + e^{-t} \mathbf{j} - 6t^{-4} \mathbf{k}$

We now derive differentiation formulas for products $h(t)\mathbf{g}(t)$ and $\mathbf{f}(t) \cdot \mathbf{g}(t)$.

THEOREM 1

Let $h(t)$, $\mathbf{f}(t)$, and $\mathbf{g}(t)$ be differentiable. Then

1. $\dfrac{d}{dt}[h(t)\mathbf{g}(t)] = h'(t)\mathbf{g}(t) + h(t)\mathbf{g}'(t)$;

2. $\dfrac{d}{dt}[\mathbf{f}(t) \cdot \mathbf{g}(t)] = \mathbf{f}'(t) \cdot \mathbf{g}(t) + \mathbf{f}(t) \cdot \mathbf{g}'(t)$.

Proof

The proof is carried out for functions $\mathbf{f}(t) = (f_1(t), f_2(t))$ and $\mathbf{g}(t) = (g_1(t), g_2(t))$. In the case of vector functions in three-dimensional space the procedure is similar.

1. We have

$$h(t)\mathbf{g}(t) = (h(t)g_1(t), h(t)g_2(t))$$

and by Definition 1,

$$\frac{d}{dt}[h(t)\mathbf{g}(t)] = \left(\frac{d}{dt}[h(t)g_1(t)], \frac{d}{dt}[h(t)g_2(t)]\right)$$

$$= (h'(t)g_1(t) + h(t)g_1'(t), h'(t)g_2(t) + h(t)g_2'(t))$$
$$= (h'(t)g_1(t), h'(t)g_2(t)) + (h(t)g_1'(t), h(t)g_2'(t))$$

$$= h'(t)(g_1(t), g_2(t)) + h(t)(g_1'(t), g_2'(t))$$
$$= h'(t)\mathbf{g}(t) + h(t)\mathbf{g}'(t)$$

2. We have
$$\mathbf{f}(t) \cdot \mathbf{g}(t) = f_1(t)g_1(t) + f_2(t)g_2(t)$$

and hence
$$\frac{d}{dt}[\mathbf{f}(t) \cdot \mathbf{g}(t)] = [f_1'(t)g_1(t) + f_1(t)g_1'(t)]$$
$$+ [f_2'(t)g_2(t) + f_2(t)g_2'(t)]$$
$$= [f_1'(t)g_1(t) + f_2'(t)g_2(t)]$$
$$+ [f_1(t)g_1'(t) + f_2(t)g_2'(t)]$$
$$= \mathbf{f}'(t) \cdot \mathbf{g}(t) + \mathbf{f}(t) \cdot \mathbf{g}'(t)$$

Example 8

Let
$$h(t) = t$$
$$\mathbf{f}(t) = (1, t)$$
$$\mathbf{g}(t) = (e^t, e^{-t})$$

Find
$$\frac{d}{dt}[h(t)\mathbf{g}(t)] \quad \text{and} \quad \frac{d}{dt}[\mathbf{f}(t) \cdot \mathbf{g}(t)]$$

Solution
$$\frac{d}{dt}[h(t)\mathbf{g}(t)] = 1(e^t, e^{-t}) + t(e^t, -e^{-t})$$
$$= (e^t, e^{-t}) + (te^t, -te^{-t})$$
$$= ((1+t)e^t, (1-t)e^{-t})$$
$$\frac{d}{dt}[\mathbf{f}(t) \cdot \mathbf{g}(t)] = (0, 1) \cdot (e^t, e^{-t}) + (1, t)(e^t, -e^{-t})$$
$$= (0e^t + 1e^{-t}) + (1e^t - te^{-t})$$
$$= e^t + (1-t)e^{-t}$$

From Theorem 1 we derive the following result.

COROLLARY 1

If $\mathbf{f}(t)$ is a differentiable function such that $|\mathbf{f}(t)| = c$ for all values of t, then $\mathbf{f}(t) \cdot \mathbf{f}'(t) = 0$ for all values of t.

Proof
Since $|\mathbf{f}(t)| = \sqrt{\mathbf{f}(t) \cdot \mathbf{f}(t)} = c$, we must have $\mathbf{f}(t) \cdot \mathbf{f}(t) = c^2$, and hence

$$\frac{d}{dt}[\mathbf{f}(t) \cdot \mathbf{f}(t)] = \mathbf{f}'(t) \cdot \mathbf{f}(t) + \mathbf{f}(t) \cdot \mathbf{f}'(t) = 2\mathbf{f}(t) \cdot \mathbf{f}'(t) = 0$$

Thus also

$$\mathbf{f}(t) \cdot \mathbf{f}'(t) = 0$$

Restated, Corollary 1 says that if $\mathbf{f}(t)$ is differentiable and $|\mathbf{f}(t)| = c$, then $\mathbf{f}(t)$ and $\mathbf{f}'(t)$ are orthogonal. The implication of this fact will be discussed in Section 15.5.

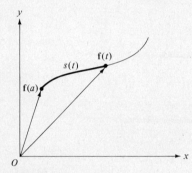

Figure 15.36

ARC LENGTH
Consider a vector function $\mathbf{f}(t)$ and suppose that $\mathbf{f}'(t)$ is continuous on some interval $a \leq t \leq b$. If $s(t)$ represents the length of the curve $\mathbf{R} = \mathbf{f}(t)$ from $\mathbf{f}(a)$ to $\mathbf{f}(t)$ (see Figure 15.36), then by Section 12.1,

$$\frac{ds}{dt} = \sqrt{[f_1'(t)]^2 + [f_2'(t)]^2}$$

when $\mathbf{f}(t) = (f_1(t), f_2(t))$, and

$$\frac{ds}{dt} = \sqrt{[f_1'(t)]^2 + [f_2'(t)]^2 + [f_3'(t)]^2}$$

when $\mathbf{f}(t) = (f_1(t), f_2(t), f_3(t))$. In vector notation either of these equations can be expressed as

$$\frac{ds}{dt} = |\mathbf{f}'(t)|$$

and the length L of the curve from $\mathbf{f}(a)$ to $\mathbf{f}(b)$ is given by the formula

$$L = \int_a^b |\mathbf{f}'(t)|\, dt$$

Example 9
Find the length of the curve $\mathbf{R} = (t^2, 1)$ for $0 \leq t \leq 2$.

Solution
Putting $\mathbf{f}(t) = (t^2, 1)$ gives $\mathbf{f}'(t) = (2t, 0)$ and $|\mathbf{f}'(t)| = \sqrt{4t^2 + 0} = 2t$. Hence

$$L = \int_0^2 2t\, dt = 4$$

Example 10

Find the length of the curve $\mathbf{f}(t) = (\cos t, \sin t, t)$ for $0 \le t \le 4\pi$.

Solution
We have $|\mathbf{f}'(t)| = \sqrt{2}$ and hence

$$L = \int_0^{4\pi} \sqrt{2}\, dt = 4\sqrt{2}\pi$$

Example 11

Find the length of the curve $\mathbf{f}(t) = (t \cos t, t \sin t, t)$ for $0 \le t \le 4\pi$.

Solution
Since

$$\mathbf{f}'(t) = (\cos t - t \sin t, \sin t + t \cos t, 1)$$

we have

$$|\mathbf{f}'(t)| = \sqrt{2 + t^2}$$

and hence

$$L = \int_0^{4\pi} \sqrt{2 + t^2}\, dt$$

Using the integration formula,

$$\int \sqrt{a^2 + x^2}\, dx = \frac{1}{2}x\sqrt{a^2 + x^2} + \frac{1}{2}a^2 \ln|x + \sqrt{a^2 + x^2}| + C$$

gives

$$L = \left(\tfrac{1}{2} t \sqrt{2 + t^2} + \tfrac{1}{2} \cdot 2 \cdot \ln|t + \sqrt{2 + t^2}|\right)\Big|_0^{4\pi}$$

$$= 2\pi\sqrt{2 + 16\pi^2} + \ln|4\pi + \sqrt{2 + 16\pi^2}| - \ln\sqrt{2}$$

$$= 2\pi\sqrt{2 + 16\pi^2} + \ln\frac{4\pi + \sqrt{2 + 16\pi^2}}{\sqrt{2}}$$

EXERCISES

Find $\mathbf{f}'(t)$ and $\mathbf{f}''(t)$ in Exercises 1–10.

1. $\mathbf{f}(t) = \left(te^t, \dfrac{1}{t}e^{-t}\right)$

2. $\mathbf{f}(t) = (t^2 \cos t, t^2 \sin t)$

3. $\mathbf{f}(t) = (2 \cos t, \cos 2t)$

4. $\mathbf{f}(t) = \dfrac{t}{t+1}\mathbf{i} + \dfrac{1}{t+1}\mathbf{j}$

5. $\mathbf{f}(t) = \dfrac{\sqrt{t}}{\sqrt{t}+1}\mathbf{i} - \dfrac{1}{\sqrt{t}+1}\mathbf{j}$

6. $\mathbf{f}(t) = (e^t, e^{2t}, e^{3t})$

7. $\mathbf{f}(t) = (t\cos t, t\sin t, t\cos t + t\sin t)$

8. $\mathbf{f}(t) = \left(\tan t, \tan\left(t+\dfrac{\pi}{2}\right), \tan\left(t-\dfrac{\pi}{2}\right)\right)$

9. $\mathbf{f}(t) = \mathbf{i} + \mathbf{j} + \mathbf{k}$

10. $\mathbf{f}(t) = e^{t^2}\mathbf{i} - e^{-t^2}\mathbf{j} + t\mathbf{k}$

11. Let

$$\mathbf{f}(t) = \left(\dfrac{1}{\sqrt{1+t^2}}, \dfrac{t}{\sqrt{1+t^2}}\right)$$

Show that $\mathbf{f}(t)$ and $\mathbf{f}'(t)$ are orthogonal.

12. Let

$$\mathbf{f}(t) = \dfrac{1}{\sqrt{2}}\cos t\,\mathbf{i} + \dfrac{1}{\sqrt{2}}\sin t\,\mathbf{j} + \dfrac{1}{\sqrt{2}}\mathbf{k}$$

Show that $\mathbf{f}(t)$ and $\mathbf{f}'(t)$ are orthogonal.

Find formulas for the differentiation indicated in Exercises 13–18.

13. $\dfrac{d^2}{dt^2}[h(t)\mathbf{g}(t)]$

14. $\dfrac{d^2}{dt^2}[\mathbf{f}(t)\cdot\mathbf{g}(t)]$

15. $\dfrac{d}{dt}\left[\mathbf{f}\left(\dfrac{t-1}{t+1}\right)\right]$

16. $\dfrac{d}{dt}\{\mathbf{f}[h(t)]\}$

17. $\dfrac{d}{dt}[|\mathbf{f}(t)|^2]$

18. $\dfrac{d}{dt}[h(t)\mathbf{f}(t)\cdot\mathbf{g}(t)]$

Find the length L of the curves in Exercises 19–22.

19. $\mathbf{f}(t) = (e^t\cos t, e^t\sin t) \qquad 0 \le t \le 3\pi$

20. $\mathbf{f}(t) = 2(t - \sin t, 1 - \cos t) \qquad 0 \le t \le 2\pi$

21. $\mathbf{f}(t) = t^2\mathbf{i} - t^2\mathbf{j} + \mathbf{k} \qquad 0 \le t \le 1$

22. $\mathbf{f}(t) = (\cos 2t, \sin 2t, t^2) \qquad -1 \le t \le 1$

15.5 TANGENTS AND NORMALS TO CURVES

Let us derive the formula for the tangent to a curve $\mathbf{R} = \mathbf{f}(t)$ at a point $\mathbf{f}(t_0)$. We shall see that the analogous procedure to the one used in Sections 2.5 and 2.6 (see also Section 11.1) for finding tangents yields the vector $\mathbf{f}'(t_0)$. We use a curve in the plane, but

15.5 TANGENTS AND NORMALS TO CURVES

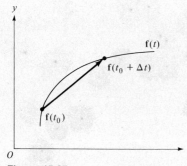

Figure 15.37

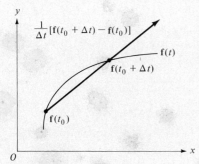

Figure 15.38

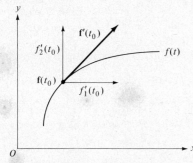

Figure 15.39

the procedure is the same for curves in three-dimensional space. Consider the curve $\mathbf{f}(t) = (f_1(t), f_2(t))$. To find the tangent at $\mathbf{f}(t_0)$, we begin with the difference,

$$\mathbf{f}(t_0 + \Delta t) - \mathbf{f}(t_0) = (f_1(t_0 + \Delta t) - f_1(t_0), f_2(t_0 + \Delta t) - f_2(t_0))$$

which gives the displacement from $\mathbf{f}(t_0)$ to $\mathbf{f}(t_0 + \Delta t)$ (see Figure 15.37). The expression

$$\frac{1}{\Delta t}[\mathbf{f}(t_0 + \Delta t) - \mathbf{f}(t_0)] \tag{1}$$

for $\Delta t > 0$ gives a vector whose direction is that of the displacement vector $\mathbf{f}(t_0 + \Delta t) - \mathbf{f}(t_0)$. If t is thought of as time, then the magnitude of the vector (1) gives the average velocity of the moving point $\mathbf{f}(t)$ over the time interval Δt (see Figure 15.38). In terms of components, formula (1) becomes formula (4) of Section 15.4, and taking limits gives the formula $\mathbf{f}'(t_0) = (f_1'(t_0), f_2'(t_0))$ (see Figure 15.39).

> **DEFINITION 1**
> The *tangent vector* to the curve $\mathbf{R} = \mathbf{f}(t)$ at the point $\mathbf{f}(t_0)$ is the vector $\mathbf{f}'(t_0)$. When t is time, then $\mathbf{f}'(t)$ is called the *velocity vector*.

It should be noted that this definition is consistent with the discussion in Section 12.1, according to which the slope of the tangent line is $f_2'(t_0)/f_1'(t_0)$.

Example 1

Find the tangent vector to the circle $\mathbf{f}(t) = (\cos t, \sin t)$ at $\mathbf{f}(\pi/6)$.

Solution
We have

$$\mathbf{f}'(t) = (-\sin t, \cos t)$$

and hence the tangent vector is

$$\mathbf{f}'\left(\frac{\pi}{6}\right) = \left(-\sin \frac{\pi}{6}, \cos \frac{\pi}{6}\right) = \left(-\frac{1}{2}, \frac{\sqrt{3}}{2}\right)$$

(see Figure 15.40).

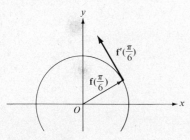

Figure 15.40
Tangent vector to
$\mathbf{f}(t) = (\cos t, \sin t)$.

Example 2

Find the tangent vector to the cycloid $\mathbf{f}(t) = (t - \sin t)\mathbf{i} + (1 - \cos t)\mathbf{j}$ at $\mathbf{f}(\pi/3)$.

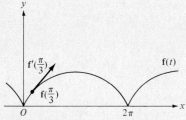

Figure 15.41
The tangent vector to the cycloid
$f(t) = (t \sin t)i + (1 - \cos t)j$
at $f(\pi/3)$.

Solution
This cycloid was discussed in Section 12.1. We have

$$f'(t) = (1 - \cos t)i + \sin t\, j$$

and hence

$$f'\left(\frac{\pi}{3}\right) = \left(1 - \cos \frac{\pi}{3}\right)i + \sin \frac{\pi}{3} j = \left(1 - \frac{1}{2}\right)i + \frac{\sqrt{3}}{2} j$$

$$= \frac{1}{2}i + \frac{\sqrt{3}}{2} j$$

(see Figure 15.41).

In many cases we are interested only in the direction of the tangent vector and not its magnitude. We thus introduce the following concept.

DEFINITION 2
The vector

$$T(t_0) = \frac{1}{|f'(t_0)|} f'(t_0) \qquad (f'(t_0) \neq 0)$$

is called the *unit tangent vector* to the curve $R = f(t)$ at $f(t_0)$.

Example 3
Find the unit tangent vector to the curve $f(t) = (t \cos t, t \sin t)$ at $f(2\pi)$.

Solution
We have

$$f'(t) = (\cos t - t \sin t, \sin t + t \cos t)$$

Hence

$$f'(2\pi) = (\cos 2\pi - 2\pi \sin 2\pi, \sin 2\pi + 2\pi \cos 2\pi)$$

$$= (1, 2\pi)$$

$$|f'(2\pi)| = \sqrt{1 + 4\pi^2}$$

and the unit tangent vector is

$$T(2\pi) = \frac{1}{\sqrt{1 + 4\pi^2}}(1, 2) = \left(\frac{1}{\sqrt{1 + 4\pi^2}}, \frac{2\pi}{\sqrt{1 + 4\pi^2}}\right)$$

(see Figure 15.42).

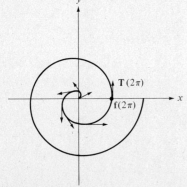

Figure 15.42
The unit tangent to
$f(t) = (t \cos t, t \sin t)$
at $f(2\pi)$.

Example 4

Find $\mathbf{f}'(0)$ and $\mathbf{T}(0)$ when

$$\mathbf{f}(t) = \cos t\, \mathbf{i} + \sin t\, \mathbf{j} + t\, \mathbf{k}$$

Solution

$$\mathbf{f}'(t) = -\sin t\, \mathbf{i} + \cos t\, \mathbf{j} + \mathbf{k}$$
$$|\mathbf{f}'(t)| = \sqrt{\sin^2 t + \cos^2 t + 1} = \sqrt{2}$$

and hence

$$\mathbf{T}(t) = -\frac{1}{\sqrt{2}} \sin t\, \mathbf{i} + \frac{1}{\sqrt{2}} \cos t\, \mathbf{j} + \frac{1}{\sqrt{2}} \mathbf{k}$$

At the point $\mathbf{f}(0)$ we have

$$\mathbf{f}'(0) = \mathbf{j} + \mathbf{k}$$
$$\mathbf{T}(0) = \frac{1}{\sqrt{2}} \mathbf{j} + \frac{1}{\sqrt{2}} \mathbf{k}$$

NORMALS

If, in Corollary 1 of Section 15.4, we replace $\mathbf{f}(t)$ by $\mathbf{T}(t)$, then we find that

$$\mathbf{T}(t) \cdot \mathbf{T}'(t) = 0$$

that is, the vectors $\mathbf{T}(t)$ and $\mathbf{T}'(t)$ are orthogonal (see Figure 15.43). The vector $\mathbf{T}'(t)$ is of importance in many applications, and it is given a special name.

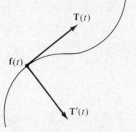

Figure 15.43
The vectors $\mathbf{T}(t)$ and $\mathbf{T}'(t)$.

DEFINITION 3
If $\mathbf{T}(t)$ is the unit tangent vector to the curve $\mathbf{R} = \mathbf{f}(t)$, then $\mathbf{T}'(t)$ is called the *normal* to the curve. The vector

$$\mathbf{N}(t) = \frac{1}{|\mathbf{T}'(t)|} \mathbf{T}'(t) \qquad (\mathbf{T}'(t) \neq \mathbf{0})$$

is called the *unit normal* or *principal normal*.

Example 5

Find the unit tangent vector and the unit normal to the circle

$$\mathbf{f}(t) = (2 \cos 3t,\ 2 \sin 3t)$$

at the point $\mathbf{f}(\pi/12)$.

Solution
Since $\mathbf{f}'(t) = (-6 \sin 3t,\ 6 \cos 3t)$ and

$$|\mathbf{f}'(t)| = 6$$

we have

$$\mathbf{T}(t) = (-\sin 3t, \cos 3t)$$

since

$$\mathbf{T}'(t) = (-3\cos 3t, -3\sin 3t)$$

and

$$|\mathbf{T}'(t)| = 3$$

we have

$$\mathbf{N}(t) = (-\cos 3t, -\sin 3t)$$

For $t = \pi/12$, we have

$$\mathbf{f}\left(\frac{\pi}{12}\right) = (\sqrt{2}, \sqrt{2})$$

$$\mathbf{T}\left(\frac{\pi}{12}\right) = \left(-\frac{\sqrt{2}}{2}, \frac{\sqrt{2}}{2}\right)$$

$$\mathbf{N}\left(\frac{\pi}{12}\right) = \left(-\frac{\sqrt{2}}{2}, -\frac{\sqrt{2}}{2}\right)$$

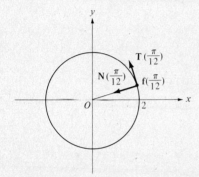

Figure 15.44
Unit tangent and unit normal to a circle.

(see Figure 15.44).

Example 6

Find the unit tangent vector and principal normal to the helix

$$\mathbf{f}(t) = (2\cos 3t, 2\sin 3t, 4t)$$

Solution
A similar helix was discussed in Example 2 of Section 15.4. We have

$$\mathbf{f}'(t) = (-6\sin 3t, 6\cos 3t, 4)$$
$$|\mathbf{f}'(t)| = \sqrt{52}$$

and hence

$$\mathbf{T}(t) = \left(-\frac{6}{\sqrt{52}}\sin 3t, \frac{6}{\sqrt{52}}\cos 3t, \frac{4}{\sqrt{52}}\right)$$

Since

$$\mathbf{T}'(t) = \left(-\frac{18}{\sqrt{52}}\cos 3t, -\frac{18}{\sqrt{52}}\sin 3t, 0\right)$$

and

15.6 CURVATURE, VELOCITY, AND ACCELERATION   631

$$|\mathbf{T}'(t)| = \frac{18}{\sqrt{52}}$$

we have

$$\mathbf{N}(t) = (-\cos 3t, -\sin 3t, 0)$$

(see Figure 15.45).

REMARKS ABOUT NORMALS

In the plane there are two orthogonal unit vectors to the tangent at a point, because if $\mathbf{U}(t) \cdot \mathbf{T}(t) = 0$, then also $-\mathbf{U}(t) \cdot \mathbf{T}(t) = 0$ (see Figure 15.45). In our definition the normal is that vector which lies on the same side of the tangent as the curve. In three-dimensional space our choice is one out of infinitely many (see Figure 15.46).

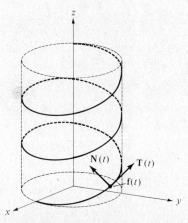

Figure 15.45
Unit tangent and principal normal to a helix.

EXERCISES

Find the unit tangents $\mathbf{T}(t)$ and principal normals $\mathbf{N}(t)$ to the curves in Exercises 1–10.

1. $\mathbf{f}(t) = (e^t \cos t, e^t \sin t)$
2. $\mathbf{f}(t) = (t, 4t)$
3. $\mathbf{f}(t) = \left(\dfrac{1}{1+t^2}, -\dfrac{t}{1+t^2}\right)$
4. $\mathbf{f}(t) = t^3 \mathbf{i} + t^2 \mathbf{j} \quad (t > 0)$
5. $\mathbf{f}(t) = 2(t - \sin t)\mathbf{i} + 2(1 - \cos t)\mathbf{j}$
6. $\mathbf{f}(t) = (\cos 2t, \sin 2t, t^2)$
7. $\mathbf{f}(t) = (t, t, 3t)$
8. $\mathbf{f}(t) = e^t \mathbf{i} + e^{-t} \mathbf{j} + \sqrt{2} t \mathbf{k}$
9. $\mathbf{f}(t) = t^2 \mathbf{i} - t^2 \mathbf{j} + \mathbf{k} \quad (t < 0)$
10. $\mathbf{f}(t) = \mathbf{i} + t\mathbf{j} + 6t^2 \mathbf{k}$

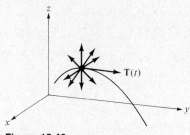

Figure 15.46
Normal vectors to $\mathbf{T}(t)$.

15.6 CURVATURE, VELOCITY, AND ACCELERATION

We now wish to associate with each point on a curve a number that will measure its curvature: A straight line is assigned curvature zero at each point, and curvature at a point for any smooth curve can be thought of as its deviation from the tangent line at that point. Thus, a circle has constant curvature, since its behavior in the neighborhood of each point is the same; the parabola $\mathbf{R} = (t, t^2)$ has its greatest curvature at the point $\mathbf{0}$.

To obtain a function that measures curvature, let us recall that for a curve $\mathbf{R} = \mathbf{f}(t)$, the number $|\mathbf{T}'(t)|$ measures the rate of change of the unit tangent vector to the curve. This gives us the information we want about the curvature of $\mathbf{R} = \mathbf{f}(t)$, but to standardize this measurement for the purpose of comparing the curvature of different curves we consider instead the number $|\mathbf{T}'(t)|/|\mathbf{f}'(t)|$.

DEFINITION 1

The *curvature* $\kappa(t_0)$ of the curve $\mathbf{R} = \mathbf{f}(t)$ at the point $\mathbf{f}(t_0)$ is the number

$$\kappa(t_0) = \frac{|\mathbf{T}'(t_0)|}{|\mathbf{f}'(t_0)|} \tag{1}$$

provided $\mathbf{f}'(t_0) \neq \mathbf{0}$.

Observe that curvature is, by definition, nonnegative.

Example 1

Find the curvature of the circle $\mathbf{f}(t) = (r \cos \omega t, r \sin \omega t)$, $r > 0$, $\omega > 0$.

Solution
Since $\mathbf{f}'(t) = (-r\omega \sin \omega t, r\omega \cos \omega t)$, we have

$$|\mathbf{f}'(t)| = r\omega$$
$$\mathbf{T}(t) = (-\sin \omega t, \cos \omega t)$$
$$\mathbf{T}'(t) = (-\omega \cos t, -\sin \omega t)$$

and hence

$$|\mathbf{T}'(t)| = \omega$$

By formula (1),

$$\kappa(t) = \frac{\omega}{r\omega} = \frac{1}{r}$$

We notice from this that a large circle has small curvature, whereas a small circle has large curvature and is independent of ω, which is called *angular velocity*.

A particularly simple formula for curvature can be derived for smooth curves that can be written in the form $\mathbf{f}(t) = (t, F(t))$, which is the parametric form of the curve $y = F(x)$.

THEOREM 1
If $\mathbf{f}(t) = (t, F(t))$ and $F''(t)$ exists, then

$$\kappa(t) = \frac{|F''(t)|}{\{1 + [F'(t)]^2\}^{3/2}}$$

Proof
We have

$$\mathbf{f}'(t) = (1, F'(t))$$
$$|\mathbf{f}'(t)| = (1 + [F'(t)]^2)^{1/2}$$
$$\mathbf{T}(t) = \frac{\mathbf{f}'(t)}{|\mathbf{f}'(t)|} = \left(\frac{1}{(1 + [F'(t)]^2)^{1/2}}, \frac{F'(t)}{(1 + [F'(t)]^2)^{1/2}}\right)$$

Leaving out intermediate calculations,

$$\mathbf{T}'(t) = \left(\frac{-F'(t)F''(t)}{(1 + [F'(t)]^2)^{3/2}}, \frac{F''(t)}{(1 + [F'(t)]^2)^{3/2}}\right)$$

and hence

$$|\mathbf{T}'(t)| = \left\{\frac{[F'(t)]^2[F''(t)]^2}{(1 + [F'(t)]^2)^3} + \frac{[F''(t)]^2}{(1 + [F'(t)]^2)^3}\right\}^{1/2}$$
$$= \left\{\frac{([F'(t)]^2 + 1)[F''(t)]^2}{(1 + [F'(t)]^2)^3}\right\}^{1/2} = \left|\frac{F''(t)}{1 + [F'(t)]^2}\right|$$

By formula (1), therefore,

$$\kappa(t) = \frac{|\mathbf{T}'(t)|}{|\mathbf{f}'(t)|} = \frac{|F''(t)|}{(1 + [F'(t)]^2)^{3/2}}$$

Example 2

Find the curvature of the curve $y = x^3$.

Solution
This curve has the parametric representation

$$\left.\begin{array}{l} x = t \\ y = t^3 \end{array}\right\}$$

Putting $F(t) = t^3$, we have $F'(t) = 3t^2$, $F''(t) = 6t$, and hence

$$\kappa(t) = \frac{|6t|}{(1 + 9t^4)^{3/2}}$$

An easy formula for computing the curvature of a curve in parametric form in the plane or in three-dimensional space is given in Section 15.11.

VELOCITY AND ACCELERATION
We recall from Section 3.5 that if $x = f(t)$ describes the motion of a particle along a straight line, then velocity and acceleration are defined as follows:

velocity: $v = f'(t)$
acceleration: $a = f''(t)$

For motion in space we now define a velocity vector and an acceleration vector.

DEFINITION 2
If the motion of a particle is described by the equation $\mathbf{R} = \mathbf{f}(t)$, then the *velocity vector* is

$$\mathbf{v}(t) = \mathbf{f}'(t) \qquad \text{velocity vector}$$

and the *acceleration vector* is

$$\mathbf{a}(t) = \mathbf{f}''(t) \qquad \text{acceleration vector}$$

It is convenient to consider the scalar quantity

$$v(t) = |\mathbf{v}(t)|$$

which is called the *speed* of the particle. The unit vector

$$\mathbf{T}(t) = \frac{1}{|\mathbf{f}'(t)|}\mathbf{f}'(t)$$

can now be expressed as

$$\mathbf{T}(t) = \frac{1}{v(t)}\mathbf{v}(t)$$

and we thus get for the velocity vector

$$\mathbf{v}(t) = v(t)\mathbf{T}(t) \tag{2}$$

To obtain an alternative representation of the acceleration vector, let us differentiate both sides of (2):

$$\mathbf{a}(t) = \mathbf{v}'(t) = v'(t)\mathbf{T}(t) + v(t)\mathbf{T}'(t) \tag{3}$$

Using the unit normal vector $\mathbf{N}(t)$ (see Definition 3 of Section 15.5), we can write

$$\mathbf{T}'(t) = |\mathbf{T}'(t)|\mathbf{N}(t) \tag{4}$$

and using formula (1) gives

$$|\mathbf{T}'(t)| = |\mathbf{f}'(t)|\kappa(t) = v(t)\kappa(t)$$

Substituting for $|\mathbf{T}'(t)|$ in equation (4) thus gives

$$\mathbf{T}'(t) = v(t)\kappa(t)\mathbf{N}(t)$$

and a substitution in (3) gives the acceleration vector as

$$\mathbf{a}(t) = v'(t)\mathbf{T}(t) + v^2(t)\kappa(t)\mathbf{N}(t) \tag{5}$$

The scalar functions $v'(t)$ and $v^2(t)\kappa(t)$ are the components of the acceleration vector along the tangent vector and the unit normal vector, respectively. The acceleration $v^2(t)\kappa(t)$ in the direction of the normal is called the *centripetal acceleration*.

Example 3

Find the velocity vector (2) and the acceleration vector (5) for the circle $\mathbf{f}(t) = (r \cos \omega t, r \sin \omega t)$, $r > 0$ and $\omega > 0$.

Solution
For the velocity $\mathbf{v}(t)$, we have

$$\mathbf{v}(t) = \mathbf{f}'(t) = (-r\omega \sin \omega t, r\omega \cos \omega t)$$

The speed $v(t)$ is

$$v(t) = [(-r\omega \sin \omega t)^2 + (r\omega \cos \omega t)^2]^{1/2}$$
$$= [(r\omega)^2(\sin^2 \omega t + \cos^2 \omega t)]^{1/2} = r\omega$$

and hence

$$\mathbf{v}(t) = r\omega \mathbf{T}(t)$$

Now, speed is constant and hence $v'(t) = 0$. This shows that the acceleration vector has no component in the direction of the unit tangent. From Example 1 we know that $\kappa(t) = 1/r$, and so

$$v^2(t)\kappa(t) = (r\omega)^2 \cdot \frac{1}{r} = r\omega^2$$

We thus have

$$\mathbf{a}(t) = r\omega^2 \mathbf{N}(t)$$

EXERCISES

In Exercises 1–12, find

(a) the vectors $\mathbf{T}(t)$ and $\mathbf{N}(t)$, and the curvature $\kappa(t)$;

(b) the components of the acceleration $\mathbf{a}$ in the directions $\mathbf{T}(t)$ and $\mathbf{N}(t)$.

1. $\mathbf{f}(t) = (t, t^3)$ $(t > 0)$
2. $\mathbf{f}(t) = (t, t^4)$ $(t > 0)$
3. $\mathbf{f}(t) = t^3\mathbf{i} + t^2\mathbf{j}$ $(t > 0)$
4. $\mathbf{f}(t) = (t, 4t)$
5. $\mathbf{f}(t) = (e^t, e^{-t})$
6. $\mathbf{f}(t) = 2(t - \sin t)\mathbf{i} + 2(1 - \cos t)\mathbf{j}$
7. $\mathbf{f}(t) = (\cos t, \sin t, t)$

8. $\mathbf{f}(t) = (t, t, 3t)$

9. $\mathbf{f}(t) = (e^t \sin t, e^t \cos t, 3e^t)$

10. $\mathbf{f}(t) = \mathbf{i} + \mathbf{j} + t\mathbf{k}$

11. $\mathbf{f}(t) = t^2\mathbf{i} + t^2\mathbf{j} + \mathbf{k}$ $\quad (t > 0)$

12. $\mathbf{f}(t) = (t^2 \cos t, t^2 \sin t)$ $\quad (t > 0)$

15.7 PLANES

A discussion of planes is particularly simple with the use of vector notation and the concept of orthogonality, which allows us to view a plane as the collection of vectors that are perpendicular to a fixed (normal) vector.

We begin our discussion with the following definition.

DEFINITION 1

A *plane* is the locus of points (x, y, z) determined by an equation

$$a_1 x + a_2 y + a_3 z = c \tag{1}$$

With the vector notation $\mathbf{a} = (a_1, a_2, a_3)$ and $\mathbf{R} = (x, y, z)$, we can write (1) as the scalar product

$$\mathbf{a} \cdot \mathbf{R} = c \tag{2}$$

Throughout this section, $\mathbf{R}$ will stand for the variable point (x, y, z) and $\mathbf{a}$ will be assumed to be fixed.

Example 1

Describe the plane $\mathbf{a} \cdot \mathbf{R} = c$ when $\mathbf{a} = (0, 1, 0)$.

Solution

Here equation (1) reduces to $y = c$. This is the equation of a plane parallel to the xz plane (see Figure 15.47).

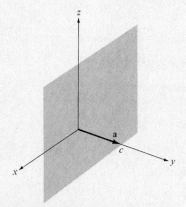

Figure 15.47
Portion of the plane $\mathbf{a} \cdot \mathbf{R} = c$.

Example 2

Find the equation of the plane containing the points $\mathbf{A} = (2, 0, 0)$, $\mathbf{B} = (0, 3, 0)$, and $\mathbf{C} = (0, 0, 4)$.

Solution

If these points are to lie in the plane $\mathbf{a} \cdot \mathbf{R} = c$, then the following equations must be satisfied:

$\mathbf{a} \cdot \mathbf{A} = c$
$\mathbf{a} \cdot \mathbf{B} = c$
$\mathbf{a} \cdot \mathbf{C} = c$

15.7 PLANES

These equations give
$$a_1 = \tfrac{1}{2}c \qquad a_2 = \tfrac{1}{3}c \qquad a_3 = \tfrac{1}{4}c$$
and hence
$$\mathbf{a} = (a_1, a_2, a_3) = (\tfrac{1}{2}c, \tfrac{1}{3}c, \tfrac{1}{4}c) = c(\tfrac{1}{2}, \tfrac{1}{3}, \tfrac{1}{4})$$

The equation of our plane is, therefore $c(\tfrac{1}{2}, \tfrac{1}{3}, \tfrac{1}{4}) \cdot \mathbf{R} = c$, or
$$(\tfrac{1}{2}, \tfrac{1}{3}, \tfrac{1}{4}) \cdot \mathbf{R} = 1$$
(see Figure 15.48). An equivalent equation for the plane,
$$\tfrac{1}{2}x + \tfrac{1}{3}y + \tfrac{1}{4}z = 1$$

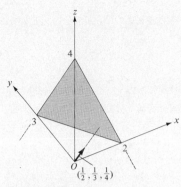

Figure 15.48
Portion of the plane $(\tfrac{1}{2}, \tfrac{1}{3}, \tfrac{1}{4}) \cdot \mathbf{R} = 1$.

Consider now the special equation
$$\mathbf{a} \cdot \mathbf{R} = 0 \tag{3}$$

We observe that $\mathbf{0}$ satisfies equation (3), since $\mathbf{a} \cdot \mathbf{0} = 0$; and recalling that $\mathbf{a} \cdot \mathbf{R} = 0$ means that the vectors $\mathbf{a}$ and $\mathbf{R}$ are orthogonal (see Section 15.3), we can state the following result.

THEOREM 1
Let $\mathbf{a} = (a_1, a_2, a_3)$ be a nonzero vector. The equation $\mathbf{a} \cdot \mathbf{R} = 0$ is the equation of a plane containing the origin $\mathbf{0}$ and consisting of all vectors $\mathbf{R} = (x, y, z)$ orthogonal to $\mathbf{a}$.

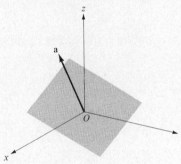

Figure 15.49
Portion of a plane $\mathbf{a} \cdot \mathbf{R} = 0$.

The vector $\mathbf{a}$ is said to be *normal* to the plane (see Figure 15.49).

Example 3
Find the normal to the plane $2x - y + 3z = 0$.

Solution
Putting $\mathbf{a} = (2, -1, 3)$ and $\mathbf{R} = (x, y, z)$, our plane can be written as $\mathbf{a} \cdot \mathbf{R} = 0$ and hence the normal is $\mathbf{a} = (2, -1, 3)$ (see Figure 15.50).

Example 4
Describe the plane $\mathbf{a} \cdot \mathbf{R} = 0$ when $\mathbf{a} = (1, -1, 0)$.

Solution
The equation $\mathbf{a} \cdot \mathbf{R} = 0$ reduces here to $x - y = 0$. This equation does not involve z explicitly. It determines the plane that is orthogonal to the xy plane and contains the line $x = y$.

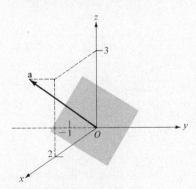

Figure 15.50
Portion of the plane $(2, -1, 3) \cdot \mathbf{R} = 0$.

Theorem 1 is readily generalized to include planes not containing the origin (see Figure 15.51).

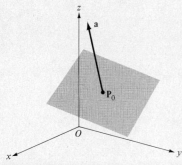

Figure 15.51
Portion of a plane
$\mathbf{a} \cdot (\mathbf{R} - \mathbf{P}_0) = 0$.

THEOREM 2
Let $\mathbf{a} = (a_1, a_2, a_3)$ be a nonzero vector and let $\mathbf{P}_0 = (x_0, y_0, z_0)$ be any point. Then the equation

$$\mathbf{a} \cdot (\mathbf{R} - \mathbf{P}_0) = 0 \qquad (4)$$

specifies the plane that passes through $\mathbf{P}_0$ and consists of all vectors $\mathbf{R} = (x, y, z)$ for which $\mathbf{R} - \mathbf{P}_0$ is orthogonal to $\mathbf{a}$.

Now, what about the general linear equation (1)? Since

$$\mathbf{a} \cdot (\mathbf{R} - \mathbf{P}_0) = \mathbf{a} \cdot \mathbf{R} - \mathbf{a} \cdot \mathbf{P}_0 = 0$$

we can put $c = \mathbf{a} \cdot \mathbf{P}_0$ and write equation (4) in the form

$$\mathbf{a} \cdot \mathbf{R} = c \qquad (5)$$

Hence we see that:

> Every plane $\mathbf{a} \cdot \mathbf{R} = c$ with $\mathbf{a} \neq \mathbf{0}$ can be written in the form $\mathbf{a} \cdot (\mathbf{R} - \mathbf{P}_0) = 0$ where $\mathbf{P}_0$ is any point on the plane and thus satisfies $\mathbf{a} \cdot \mathbf{P}_0 = c$.

We mention that equation (4) can also be written as

$$a_1(x - x_0) + a_2(y - y_0) + a_3(z - z_0) = 0 \qquad (6)$$

Example 5

Find the equation of the plane through $\mathbf{P}_0 = (2, 0, -5)$ and having normal $\mathbf{a} = (-2, 3, -4)$.

Solution
Using equation (6) gives

$$-2(x - 2) + 3(y - 0) - 4(z + 5) = 0$$

Notice that this equation is equivalent to

$$-2x + 3y - 4z = 16$$

Example 6

Find the plane passing through the point $\mathbf{P}_0 = (1, 0, 0)$ and parallel to the plane $2x - y + z = 0$.

Solution
With the notation $\mathbf{a} = (2, -1, 1)$, we can write the given plane as $\mathbf{a} \cdot \mathbf{R} = 0$. Let the plane we are seeking be given by the formula $\mathbf{b} \cdot (\mathbf{R} - \mathbf{P}_0) = 0$. If these two planes are parallel, then their normals

must be parallel and thus may be taken equal. Hence our plane is $\mathbf{a} \cdot (\mathbf{R} - \mathbf{P}_0) = 0$; that is,

$$2(x - 1) - (y - 0) + (z - 0) = 0$$

or

$$2x - y + z = 2$$

EXERCISES

In Exercises 1–8, find an equation for the plane passing through $\mathbf{P}_0$ and having normal $\mathbf{a}$ in terms of x, y, and z.

1. $\mathbf{P}_0 = \mathbf{0}$, $\mathbf{a} = (1, 2, 0)$
2. $\mathbf{P}_0 = \mathbf{0}$, $\mathbf{a} = (-2, 1, 0)$
3. $\mathbf{P}_0 = \mathbf{0}$, $\mathbf{a} = (0, 3, 0)$
4. $\mathbf{P}_0 = (-1, 2, -3)$, $\mathbf{a} = (-1, 2, -3)$
5. $\mathbf{P}_0 = (-1, 2, -3)$, $\mathbf{a} = (-1, 1, 1)$
6. $\mathbf{P}_0 = (-1, 2, -3)$, $\mathbf{a} = (0, 3, 0)$
7. $\mathbf{P}_0 = (-1, -1, -1)$, $\mathbf{a} = (-1, -1, -1)$
8. $\mathbf{P}_0 = (1, 0, 0)$, $\mathbf{a} = (0, 0, 1)$

In Exercises 9–12, find the planes passing through the given points.

9. $\mathbf{A} = \mathbf{0}$, $\mathbf{B} = (0, 0, 1)$, $\mathbf{C} = (0, 1, 1)$
10. $\mathbf{A} = (1, 1, 1)$, $\mathbf{B} = (-1, -1, -1)$, $\mathbf{C} = (2, 0, 0)$
11. $\mathbf{A} = (-1, 0, 0)$, $\mathbf{B} = (0, -1, 0)$, $\mathbf{C} = (0, 0, -1)$
12. $\mathbf{A} = (\frac{1}{2}, \frac{1}{3}, \frac{1}{4})$, $\mathbf{B} = (1, 0, 1)$, $\mathbf{C} = (-\frac{1}{2}, -\frac{1}{3}, -\frac{1}{4})$
13. Find the plane passing through $\mathbf{P}_0 = (1, 1, 1)$ and parallel to $x + 2y + 3z = 7$.
14. Find the plane passing through $\mathbf{P} = (1, 0, 0)$ and $\mathbf{Q} = (0, 1, 1)$ and perpendicular to the plane $y - z = 0$.
15. Find the plane passing through $\mathbf{P} = (1, 0, 0)$ and $\mathbf{Q} = (0, 1, 1)$ and perpendicular to $3x + 2y + z = 0$.
16. Find two planes passing through $\mathbf{P} = (1, 0, 0)$ and perpendicular to $3x + 2y + z = 0$.

15.8 TANGENT PLANES AND THE GRADIENT

We now consider the problem of finding formulas for the normal and tangent plane to a surface

$$F(x, y, z) = c \tag{1}$$

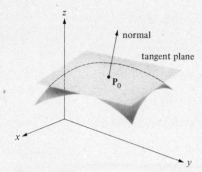

Figure 15.52
Normal and tangent plane to a surface at P_0.

at a point $\mathbf{P}_0 = (x_0, y_0, z_0)$ on it (see Figure 15.52). Surfaces of this form are discussed in Section 13.3. We assume in the ensuing discussion that the functions occurring are differentiable as required.

Let the curve

$$x = f_1(t)$$
$$y = f_2(t)$$
$$z = f_3(t)$$

which in vector notation can be written as

$$\mathbf{f}(t) = (f_1(t), f_2(t), f_3(t)) \qquad (2)$$

lie on the surface. Then

$$F(f_1(t), f_2(t), f_3(t)) = c \qquad (3)$$

and, using the chain rule, we differentiate both sides of (3) with respect to t. Evaluated at t_0, this gives

$$\frac{\partial F}{\partial x} f_1'(t_0) + \frac{\partial F}{\partial y} f_2'(t_0) + \frac{\partial F}{\partial z} f_3'(t_0) = 0$$

and in vector notation we have

$$\left(\frac{\partial F}{\partial x}, \frac{\partial F}{\partial y}, \frac{\partial F}{\partial z}\right) \cdot \mathbf{f}'(t_0) = 0$$

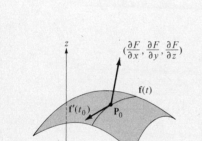

Figure 15.53
Illustrating the orthogonality of $\mathbf{f}'(t)$ and $(\partial F/\partial x, \partial F/\partial y, \partial F/\partial z)$.

This tells us that the vectors $(\partial F/\partial x, \partial F/\partial y, \partial F/\partial z)$ and $\mathbf{f}'(t_0)$ are always orthogonal: Observe, however, that the partial derivatives $\partial F/\partial x$, $\partial F/\partial y$, and $\partial F/\partial z$ are evaluated at the point $\mathbf{f}(t_0)$ (see Figure 15.53). Since $\mathbf{f}'(t)$ is the tangent vector to the curve $\mathbf{f}(t)$, we see that $(\partial F/\partial x, \partial F/\partial y, \partial F/\partial z)$ is orthogonal to the tangent vector at the point $\mathbf{f}(t_0)$.

Looking now at all curves $\mathbf{f}(t)$ on the surface (1) passing through a given point $\mathbf{P}_0$ on (1), we see that the vector $(\partial F/\partial x, \partial F/\partial y, \partial F/\partial z)$ is orthogonal to the tangent vectors to *all* these curves at $\mathbf{P}_0$. These vectors determine a plane, the *tangent plane* to the surface at $\mathbf{f}(t_0)$.

We thus introduce the following terminology.

DEFINITION 1
The vector

$$\left(\frac{\partial F}{\partial x}, \frac{\partial F}{\partial y}, \frac{\partial F}{\partial z}\right)$$

evaluated at $\mathbf{P}_0 = (x_0, y_0, z_0)$ on the surface $F(x, y, z) = c$ is called the *normal* to the surface at $\mathbf{P}_0$.

The tangent plane to the surface at a point is now defined using the normal.

15.8 TANGENT PLANES AND THE GRADIENT

DEFINITION 2
The plane

$$\frac{\partial F}{\partial x}(x - x_0) + \frac{\partial F}{\partial y}(y - y_0) + \frac{\partial F}{\partial z}(z - z_0) = 0 \qquad (4)$$

is the *tangent plane* to the surface $F(x, y, z) = c$ at $\mathbf{P}_0 = (x_0, y_0, z_0)$.

As above, $\partial F/\partial x$, $\partial F/\partial y$, and $\partial F/\partial z$ are evaluated at $\mathbf{P}_0$. In vector notation, (4) becomes

$$\left(\frac{\partial F}{\partial x}, \frac{\partial F}{\partial y}, \frac{\partial F}{\partial z}\right) \cdot (\mathbf{R} - \mathbf{P}_0) = 0 \qquad (5)$$

where $\mathbf{R} = (x, y, z)$.

Example 1

Find the normal and tangent plane to the surface

$$x^2 + xyz + y^2 = -1$$

at $\mathbf{P}_0 = (-1, 2, 3)$.

Solution
Putting

$$F(x, y, z) = x^2 + xyz + y^2$$

gives

$$\frac{\partial F}{\partial x} = 2x + yz$$

$$\frac{\partial F}{\partial y} = xz + 2y$$

$$\frac{\partial F}{\partial z} = xy$$

Evaluated at $\mathbf{P}_0$, these partial derivatives are

$$\frac{\partial F}{\partial x} = 4$$

$$\frac{\partial F}{\partial y} = 1$$

$$\frac{\partial F}{\partial z} = -2$$

The normal at $\mathbf{P}_0$ is, therefore,

$$\left(\frac{\partial F}{\partial x}, \frac{\partial F}{\partial y}, \frac{\partial F}{\partial z}\right) = (4, 1, -2)$$

and the tangent plane at P_0 is

$$4(x+1) + (y-2) - 2(z-3) = 0$$

Example 2

Find the tangent plane to the sphere

$$x^2 + y^2 + z^2 = 4$$

at $P_0 = (0, 2, 0)$.

Solution
Taking

$$F(x, y, z) = x^2 + y^2 + z^2$$

gives

$$\left.\frac{\partial F}{\partial x}\right|_{(0,2,0)} = 2x\Big|_{x=0} = 0$$

$$\left.\frac{\partial F}{\partial y}\right|_{(0,2,0)} = 2y\Big|_{y=2} = 4$$

$$\left.\frac{\partial F}{\partial z}\right|_{(0,2,0)} = 2z\Big|_{z=0} = 0$$

The tangent plane is, therefore, $4(y-2) = 0$, or $y = 2$ (see Figure 15.54).

For surfaces given in the form $z = f(x, y)$, the following theorem is true.

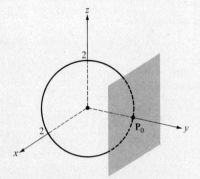

Figure 15.54
The tangent plane to the sphere $x^2 + y^2 + z^2 = 4$ at $P_0 = (0, 2, 0)$.

THEOREM 1
For a surface

$$z = f(x, y) \tag{6}$$

the normal at $P_0 = (x_0, y_0, z_0)$ is given by

$$\left(\frac{\partial f}{\partial x}, \frac{\partial f}{\partial x}, -1\right)$$

and the tangent plane at P_0 is given by

$$\frac{\partial f}{\partial x}(x - x_0) + \frac{\partial f}{\partial y}(y - y_0) - (z - z_0) = 0$$

where $\partial f/\partial x$ and $\partial f/\partial y$ are evaluated at (x_0, y_0) and $z_0 = f(x_0, y_0)$.

Proof
Let
$$F(x, y, z) = f(x, y) - z$$
Then
$$F(x, y, z) = 0$$
is an equation of the surface and hence (6) is a special case of (1), and we have

$$\frac{\partial F}{\partial x} = \frac{\partial f}{\partial x} \qquad \frac{\partial F}{\partial y} = \frac{\partial f}{\partial y} \qquad \frac{\partial F}{\partial z} = -1$$

Example 3

Find the normal and tangent plane to the surface
$$z = x^3 + y^2 - 2xy$$
when $x_0 = 1$ and $y_0 = -2$.

Solution
Letting
$$f(x, y) = x^3 + y^2 - 2xy$$
gives
$$\left.\frac{\partial f}{\partial x}\right|_{(1,-2)} = 3x^2 - 2y \bigg|_{(1,-2)} = 7$$
$$\left.\frac{\partial f}{\partial y}\right|_{(1,-2)} = 2y - 2x \bigg|_{(1,-2)} = -6$$
and hence the normal is
$$\left(\frac{\partial f}{\partial x}, \frac{\partial f}{\partial y}, -1\right) = (7, -6, -1)$$
Now, $(1, -2, z_0)$ is on the surface if
$$z_0 = x_0^3 + y_0^2 - 2x_0 y_0 = 1 + (-2)^2 - 2 \cdot 1 \cdot (-2) = 9$$
and hence the tangent plane is
$$7(x - 1) - 6(y + 2) - (z - 9) = 0$$

CURVES
Results analogous to the above can be obtained for curves in the plane given implicitly. These are summarized in Theorem 2.

THEOREM 2
For the curve
$$F(x, y) = c$$
the normal at $\mathbf{P}_0 = (x_0, y_0)$ is given by
$$\left(\frac{\partial f}{\partial x}, \frac{\partial f}{\partial y}\right)$$
and the tangent at $\mathbf{P}_0$ is given by
$$\frac{\partial f}{\partial x}(x - x_0) + \frac{\partial f}{\partial y}(y - y_0) = 0 \tag{7}$$
where $\partial f/\partial x$ and $\partial f/\partial y$ are evaluated at $\mathbf{P}_0$.

Proof
The relation $F(x, y) = c$ is assumed to determine a function $y = g(x)$. Hence, differentiating $F(x, y) = c$ implicitly (see Section 3.6) gives
$$\frac{\partial f}{\partial x} + \frac{\partial f}{\partial y} \cdot \frac{dy}{dx} = 0$$
Thus
$$\frac{dy}{dx} = -\frac{\partial f}{\partial x} \Big/ \frac{\partial f}{\partial y}$$
and the tangent is seen to be
$$\frac{y - y_0}{x - x_0} = -\frac{\partial f}{\partial x} \Big/ \frac{\partial f}{\partial y}$$

This equation, however, is equivalent to (7).

GRADIENT
The normal vectors $(\partial F/\partial x, \partial F/\partial y, \partial F/\partial z)$ and $(\partial F/\partial x, \partial F/\partial y)$, which were the key to our method of finding tangents, are of great importance in many applications. They are given the special name *gradient* and designation **grad F**.

DEFINITION 3
The *gradient* of the function $F(x, y)$ is
$$\mathbf{grad\ F} = \left(\frac{\partial F}{\partial x}, \frac{\partial F}{\partial y}\right)$$
and the *gradient* of the function $F(x, y, z)$ is
$$\mathbf{grad\ F} = \left(\frac{\partial F}{\partial x}, \frac{\partial F}{\partial y}, \frac{\partial F}{\partial z}\right)$$
provided the partial derivatives exist.

Example 4

Find the gradient of $x^2 - xy + y^2$ at $(2, 3)$.

Solution
Putting
$$F(x, y) = x^2 - xy + y^2$$
gives
$$\left.\frac{\partial F}{\partial x}\right|_{(2,3)} = 2x - y \bigg|_{(2,3)} = 1$$
$$\left.\frac{\partial F}{\partial y}\right|_{(2,3)} = -x + 2y \bigg|_{(2,3)} = 4$$
Hence
$$\text{grad } \mathbf{F} = (1, 4)$$

Example 5

Find the gradient of xy^2z^3.

Solution
Putting
$$F(x, y, z) = xy^2z^3$$
gives
$$\text{grad } \mathbf{F} = (y^2z^3, 2xyz^3, 3xy^2z^2)$$

With the gradient notation, equation (5) of the tangent plane assumes the simpler form
$$\text{grad } \mathbf{F} \cdot (\mathbf{R} - \mathbf{P}_0) = 0$$

The gradient is discussed further in Section 15.10.

EXERCISES

Find the normal and tangent plane to the surfaces in Exercises 1–6 at the given point $\mathbf{P}_0$.

1. $x^3y^2 - 3yz^4 = -81$, $\mathbf{P}_0 = (-2, 3, 1)$
2. $(x + y)^2 - z = 25$, $\mathbf{P}_0 = (0, 0, -25)$ and $\mathbf{P}_0 = (4, 2, 9)$
3. $z - xe^y = 0$, $\mathbf{P}_0 = (-3, 0, -3)$
4. $xy = 1$, $\mathbf{P}_0 = (2, \frac{1}{2}, 4)$ and $\mathbf{P}_0 = (2, \frac{1}{2}, 0)$

5. $x^2 + y^2 = 2^2$, $P_0 = (\sqrt{3}, 1, 1)$

6. $z - \sin x \cos y = 0$, $P_0 = (\pi/2, -\pi, -1)$

Find the normal and tangent line to the curves in Exercises 7–10 at the given point P_0.

7. $x^2 y^{-3} = -\frac{1}{2}$, $P_0 = (2, -2)$

8. $xe^y = 1$, $P_0 = (1, 0)$

9. $\cos(x + y) = 1/\sqrt{2}$, $P_0 = (-\pi/4, \pi/2)$

10. $x^4 - xy + y^4 = 15$, $P_0 = (1, 2)$

Find **grad F** in Exercises 11–15.

11. $F(x, y, z) = x^3 y^2 - 3yz^4$
12. $F(x, y, z) = e^{x+y} - e^{x-y}$
13. $F(x, y, z) = xe^y + ye^z$
14. $F(x, y) = x^2 y^{-3}$
15. $F(x, y) = e^{-x} y$

In the theory of electromagnetism, the *electric field* **E** can be found from a function $V(x, y, z)$ by the formula $\mathbf{E} = -\mathbf{grad}\ V$. Find **E** in Exercises 16–19 at the given point.

16. $V = x^2 - 2z^2 + y^2$ at $(1, -1, 1)$
17. $V = (x^2 + y^2)z$ at $(-1, 2, 1)$
18. $V = \cos 2x \cdot e^{-(y+z)}$ at $(\pi/4, 0, 0)$
19. $V = \cos 2x \cdot e^{-y}$ at $(\pi/4, 0, 0)$

15.9 DIRECTIONAL DERIVATIVES

To introduce the concept of directional derivative, consider a function $F(x, y)$ and a line

$$\left. \begin{array}{l} x = x_0 + u_1 t \\ y = y_0 + u_2 t \end{array} \right\} \tag{1}$$

expressed in terms of a parameter t. We can ask the following question: How fast is the function $F(x, y)$ changing as we move along this line? We note that the parameter t completely determines our position as well as the observed values of $F(x, y)$, which can also be expressed in terms of t, say,

$$G(t) = F(x_0 + u_1 t, y_0 + u_2 t)$$

Since rate of change is the derivative, we use the chain rule to get

$$G'(t) = \frac{\partial F}{\partial x} \cdot \frac{dx}{dt} + \frac{\partial F}{\partial y} \cdot \frac{dy}{dt} = \frac{\partial F}{\partial x} u_1 + \frac{\partial F}{\partial y} u_2$$

We now know how fast G is changing with respect to t.

To answer the above question, we look once more at (1). We find that

$$(x - x_0)^2 + (y - y_0)^2 = (u_1 t)^2 + (u_2 t)^2 = (u_1^2 + u_2^2) t^2$$

Hence, if $u_1^2 + u_2^2 = 1$, then we get

$$|t| = \sqrt{(x - x_0)^2 + (y - y_0)^2}$$

The parameter t is thus seen to measure distance along the line (1), and this leads us to the following conclusion:

> If $\mathbf{u} = (u_1, u_2)$ is a unit vector, then
>
> $$G'(t) = \frac{\partial F}{\partial x} u_1 + \frac{\partial F}{\partial x} u_2$$
>
> measures the rate of change of $F(x, y)$ with respect to the distance in the direction $\mathbf{u}$.

DEFINITION 1
Let $F(x, y)$ have partial derivatives $\partial F/\partial x$ and $\partial F/\partial y$ and let $\mathbf{u} = (u_1, u_2)$ be a unit vector. The *directional derivative* of $F(x, y)$ at $\mathbf{P}_0 = (x_0, y_0)$ in the direction $\mathbf{u}$ is

$$D_\mathbf{u} F(x_0, y_0) = \frac{\partial F}{\partial x} u_1 + \frac{\partial F}{\partial y} u_2 \qquad (2)$$

The reader is reminded once more that $\partial F/\partial x$ and $\partial F/\partial y$ are evaluated at (x_0, y_0).

The above discussion can be repeated for a function $F(x, y, z)$ of three variables. In this case we would consider a line

$$\left. \begin{array}{l} x = x_0 + u_1 t \\ y = y_0 + u_2 t \\ z = z_0 + u_3 t \end{array} \right\} \qquad (3)$$

where $\mathbf{u} = (u_1, u_2, u_3)$ is a unit vector. Putting

$$G(t) = F(x_0 + u_1 t, y_0 + u_2 t, z_0 + u_3 t)$$

we then get

$$G'(t) = \frac{\partial F}{\partial x} u_1 + \frac{\partial F}{\partial y} u_2 + \frac{\partial F}{\partial z} u_3$$

DEFINITION 2
Let $F(x, y, z)$ have partial derivatives $\partial F/\partial x$, $\partial F/\partial y$, and $\partial F/\partial z$. If $\mathbf{u} = (u_1, u_2, u_3)$ is a unit vector, then the *directional derivative* of $F(x, y, z)$ at $\mathbf{P}_0 = (x_0, y_0, z_0)$ in the direction $\mathbf{u}$ is given by

$$D_{\mathbf{u}} F(x_0, y_0, z_0) = \frac{\partial F}{\partial x} u_1 + \frac{\partial F}{\partial y} u_2 + \frac{\partial F}{\partial z} u_3 \tag{4}$$

Example 1

Find the directional derivative of $F(x, y) = x^2 y^3$ at $(-1, 2)$ in the direction $\mathbf{u} = (-\sqrt{2}/2, \sqrt{2}/2)$.

Solution

$$\left.\frac{\partial F}{\partial x}\right|_{(-1,2)} = 2xy^3 \bigg|_{(-1,2)} = -16$$

$$\left.\frac{\partial F}{\partial y}\right|_{(-1,2)} = 3x^2 y^2 \bigg|_{(-1,2)} = 12$$

Hence

$$D_{\mathbf{u}} F(-1, 2) = -16\left(-\frac{\sqrt{2}}{2}\right) + 12\frac{\sqrt{2}}{2} = 14\sqrt{2}$$

Example 2

Find the directional derivative $D_{\mathbf{u}} F(1, 2, -1)$ in the direction $\mathbf{u} = (-\sqrt{3}/3, \sqrt{3}/3, -\sqrt{3}/3)$ when $F(x, y, z) = xy - y^2 z^2$.

Solution

$$\left.\frac{\partial F}{\partial x}\right|_{(1,2,-1)} = y \bigg|_{(1,2,-1)} = 2$$

$$\left.\frac{\partial F}{\partial y}\right|_{(1,2,-1)} = (x - 2yz^2) \bigg|_{(1,2,-1)} = -3$$

$$\left.\frac{\partial F}{\partial z}\right|_{(1,2,-1)} = -2y^2 z \bigg|_{(1,2,-1)} = 8$$

Hence

$$D_{\mathbf{u}} F(1, 2, -1) = 2\left(-\frac{\sqrt{3}}{3}\right) + (-3)\left(\frac{\sqrt{3}}{3}\right) + 8\left(-\frac{\sqrt{3}}{3}\right) = -\frac{13\sqrt{3}}{3}$$

Using now vector notation, we observe that both equations (2) and (4) can be put in the form

$$D_{\mathbf{u}} F(\mathbf{P}_0) = \text{grad } \mathbf{F}(\mathbf{P}_0) \cdot \mathbf{u} \tag{5}$$

THEOREM 1

$$D_{\mathbf{u}} F(\mathbf{P}_0) = |\text{grad } \mathbf{F}(\mathbf{P}_0)| \cos \theta \tag{6}$$

where θ is the angle between the vector $\mathbf{u}$ and grad $\mathbf{F}(\mathbf{P}_0)$.

Proof
We prove this theorem using the formula for the angle between two vectors. By Section 15.3,

grad $F(P_0) \cdot \mathbf{u} = |\mathbf{grad}\ F(P_0)||\mathbf{u}|\cos\theta$

and since $\mathbf{u}$ is a unit vector we have $|\mathbf{u}| = 1$ and formula (6) follows.

Formula (6) gives us valuable information about the directional derivative. It tells us that its magnitude depends on the direction we pick. It follows that $D_\mathbf{u} F(P_0)$ can be maximized at any point P_0 by taking $\theta = 0$, in which case $\mathbf{u}$ and **grad** $F(P_0)$ have the same direction. Hence we see that

> **grad** F always points in the direction of largest rate of increase of F.

Since $-1 \leq \cos\theta \leq 1$, we see from formula (6) that

$$-|\mathbf{grad}\ F(P_0)| \leq D_\mathbf{u} F(P_0) \leq |\mathbf{grad}\ F(P_0)| \tag{7}$$

for all unit vectors $\mathbf{u}$, and the following theorem is seen to be true.

THEOREM 2
If **grad** $F(P_0) \neq \mathbf{0}$ and $\mathbf{u}$ is a unit vector, then $D_\mathbf{u} F(P_0)$ assumes the maximum value $|\mathbf{grad}\ F(P_0)|$ when

$$\mathbf{u} = \frac{1}{|\mathbf{grad}\ F(P_0)|}\ \mathbf{grad}\ F(P_0)$$

and it assumes the minimum value $-|\mathbf{grad}\ F(P_0)|$ when

$$\mathbf{u} = -\frac{1}{|\mathbf{grad}\ F(P_0)|}\ \mathbf{grad}\ F(P_0)$$

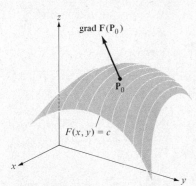

Figure 15.55
The gradient of $F(x, y)$ at P_0.

The previous discussion applies to functions of two and three variables, but it is worthwhile to explain these ideas geometrically in the case of two variables. Take a function $F(x, y)$ and curves $F(x, y) = c$ for different values of c (see Figure 15.55). If P_0 is a point on such a curve, then **grad** $F(P_0)$ is perpendicular to that level curve at P_0 (see Section 15.8), and by the above discussion it points in the direction of greatest rate of increase.

EXERCISES

1. Show that for $F(x, y, z)$,

$$|\mathbf{grad}\ F| = \sqrt{\left(\frac{\partial F}{\partial x}\right)^2 + \left(\frac{\partial F}{\partial y}\right)^2 + \left(\frac{\partial F}{\partial z}\right)^2}$$

In Exercises 2–9, find the directional derivatives of F at P_0 in the given direction.

2. $F(x, y) = xy^2 + \cos xy$, $P_0 = (1, \pi/2)$, direction of the vector making a 30° angle with the positive x axis.

3. $F(x, y) = e^{x+y} + e^{-(x+y)}$, $P_0 = (1, -1)$, direction $\mathbf{u} = (\sqrt{2}/2, \sqrt{2}/2)$.

4. $F(x, y) = \sqrt{x^2 + y^2}$, $P_0 = (3, 4)$, direction $\mathbf{u} = (1, 0)$.

5. $F(x, y) = x/y$, $P_0 = (1, 1/2)$, direction $\mathbf{u} = (3/5, 4/5)$.

6. $F(x, y, z) = z - x^2 - y^2$, $P_0 = (3, 4, 0)$, direction $\mathbf{u} = (-\sqrt{2}/2, -\sqrt{2}/2, 0)$.

7. $F(x, y, z) = x^3 y + z e^{x-1}$, $P_0 = (1, 2, -3)$, direction $\mathbf{u} = (-1/2, -1/2, \sqrt{2}/2)$.

8. $F(x, y, z) = z \ln[(x + z)/(x - y)]$, $P_0 = (2, 1, 2)$, direction $\mathbf{u} = (\sqrt{3}/3, -\sqrt{3}/3, \sqrt{3}/3)$.

9. $F(x, y, z) = x^2 y^3 z$, $P_0 = (1, -1, 1)$, direction $\mathbf{u} = (\cos \theta, \sin \theta, 0)$.

15.10 THE VECTOR PRODUCT

Given two vectors $\mathbf{a}$ and $\mathbf{b}$ in three-dimensional space, we define a new vector, designated $\mathbf{a} \times \mathbf{b}$ and called the vector product of $\mathbf{a}$ and $\mathbf{b}$. The vector product is very important in applications to physics and other areas of mathematics. We shall develop vector products just enough to derive a formula for the curvature of a curve in three-dimensional space (see Section 15.6).

We begin with a formal definition.

DEFINITION 1

The *vector product* of $\mathbf{a} = (a_1, a_2, a_3)$ and $\mathbf{b} = (b_1, b_2, b_3)$, designated $\mathbf{a} \times \mathbf{b}$, is given by the formula

$$\mathbf{a} \times \mathbf{b} = (a_2 b_3 - a_3 b_2, a_3 b_1 - a_1 b_3, a_1 b_2 - a_2 b_1) \qquad (1)$$

This strange definition will be explained as we go along. Formula (1) is often expressed in the form

$$\mathbf{a} \times \mathbf{b} = (a_2 b_3 - a_3 b_2)\mathbf{i} + (a_3 b_1 - a_1 b_3)\mathbf{j} + (a_1 b_2 - a_2 b_1)\mathbf{k} \qquad (2)$$

Example 1

Find $\mathbf{a} \times \mathbf{b}$ when $\mathbf{a} = (3, 2, -1)$ and $\mathbf{b} = (2, -3, 4)$.

Solution

$$\mathbf{a} \times \mathbf{b} = (2 \cdot 4 - (-1)(-3), (-1) \cdot 2 - 3 \cdot 4, 3 \cdot 2 - 2 \cdot 2)$$
$$= (5, -14, 2)$$

An alternative representation for the vector product is

$$\mathbf{a} \times \mathbf{b} = 5\mathbf{i} - 14\mathbf{j} + 2\mathbf{k}$$

Example 2

Find the vector product of the unit vectors

$$\mathbf{i} = (1, 0, 0), \quad \mathbf{j} = (0, 1, 0), \quad \text{and} \quad \mathbf{k} = (0, 0, 1)$$

Solution

$$\begin{array}{ll} \mathbf{i} \times \mathbf{j} = \mathbf{k} & \mathbf{j} \times \mathbf{i} = -\mathbf{k} \\ \mathbf{j} \times \mathbf{k} = \mathbf{i} & \mathbf{k} \times \mathbf{j} = -\mathbf{i} \\ \mathbf{k} \times \mathbf{i} = \mathbf{j} & \mathbf{i} \times \mathbf{k} = -\mathbf{j} \end{array}$$

A notational device that is very useful here is the *determinant*, studied in linear algebra.

DEFINITION 2

A second-order determinant is defined by

$$\begin{vmatrix} \alpha & \beta \\ \gamma & \delta \end{vmatrix} = \alpha\delta - \beta\gamma$$

Example 3

$$\begin{vmatrix} 1 & 0 \\ 0 & 1 \end{vmatrix} = 1 \cdot 1 - 0 \cdot 0 = 1$$

$$\begin{vmatrix} 2 & 1 \\ 1 & -3 \end{vmatrix} = 2 \cdot (-3) - 1 \cdot 1 = -7$$

$$\begin{vmatrix} 2 & 4 \\ -3 & \frac{1}{2} \end{vmatrix} = 2 \cdot \frac{1}{2} - 4 \cdot (-3) = 13$$

Using determinants, equation (2) can be written as

$$\mathbf{a} \times \mathbf{b} = \begin{vmatrix} a_2 & a_3 \\ b_2 & b_3 \end{vmatrix} \mathbf{i} + \begin{vmatrix} a_3 & a_1 \\ b_3 & b_1 \end{vmatrix} \mathbf{j} + \begin{vmatrix} a_1 & a_2 \\ b_1 & b_2 \end{vmatrix} \mathbf{k} \qquad (3)$$

REMARK

We remark in passing that this formula is conveniently expressed in terms of a third-order determinant

$$\mathbf{a} \times \mathbf{b} = \begin{vmatrix} \mathbf{i} & \mathbf{j} & \mathbf{k} \\ a_1 & a_2 & a_3 \\ b_1 & b_2 & b_3 \end{vmatrix} \quad (4)$$

The right side of (3) can be used as a formal definition of the determinant in (4).

Basic properties of the vector product are given in Theorem 1.

THEOREM 1. PROPERTIES OF THE VECTOR PRODUCT
Let $\mathbf{a}$, $\mathbf{b}$, and $\mathbf{c}$ be vectors in three-dimensional space. Then

1. $\mathbf{a} \times \mathbf{a} = \mathbf{0}$.
2. $\mathbf{a} \times \mathbf{0} = \mathbf{0}$.
3. $\mathbf{a} \times \mathbf{b} = -(\mathbf{b} \times \mathbf{a})$.
4. $\mathbf{a} \times (\mathbf{b} + \mathbf{c}) = \mathbf{a} \times \mathbf{b} + \mathbf{a} \times \mathbf{c}$.
5. $(k\mathbf{a}) \times (m\mathbf{b}) = km(\mathbf{a} \times \mathbf{b})$ for any constants k and m.
6. $|\mathbf{a} \times \mathbf{b}|^2 = |\mathbf{a}|^2 |\mathbf{b}|^2 - (\mathbf{a} \cdot \mathbf{b})^2$.

The properties listed in this theorem are proved by direct calculation, using the definition. To illustrate the general method of proof, we establish the last property.

Proof of Property 6
Let $\mathbf{a} = (a_1, a_2, a_3)$ and $\mathbf{b} = (b_1, b_2, b_3)$. Then

$$\begin{aligned}
|\mathbf{a} \times \mathbf{b}|^2 &= (a_2 b_3 - a_3 b_2)^2 + (a_3 b_1 - a_1 b_3)^2 + (a_1 b_2 - a_2 b_1)^2 \\
&= a_2^2 b_3^2 + a_3^2 b_2^2 - 2 a_2 a_3 b_2 b_3 \\
&\quad + a_3^2 b_1^2 + a_1^2 b_3^2 - 2 a_1 a_3 b_1 b_3 \\
&\quad + a_1^2 b_2^2 + a_2^2 b_1^2 - 2 a_1 a_2 b_1 b_2
\end{aligned}$$

On the other hand,

$$\begin{aligned}
|\mathbf{a}|^2 |\mathbf{b}|^2 - (\mathbf{a} \cdot \mathbf{b})^2 &= (a_1^2 + a_2^2 + a_3^2)(b_1^2 + b_2^2 + b_3^2) \\
&\quad - (a_1 b_1 + a_2 b_2 + a_3 b_3)^2 \\
&= a_1^2 b_1^2 + a_1^2 b_2^2 + a_1^2 b_3^2 \\
&\quad + a_2^2 b_1^2 + a_2^2 b_2^2 + a_2^2 b_3^2 \\
&\quad + a_3^2 b_1^2 + a_3^2 b_2^2 + a_3^2 b_3^2 \\
&\quad - a_1^2 b_1^2 - a_2^2 b_2^2 - a_3^2 b_3^2 \\
&\quad - 2 a_1 b_1 a_2 b_2 - 2 a_1 b_1 a_3 b_3 - 2 a_2 b_2 a_3 b_3
\end{aligned}$$

After cancellation, a comparison with the preceding equation shows that Property (6) is true.

Example 4

Show that if $|\mathbf{a}| = 1$, $|\mathbf{b}| = 1$, and $\mathbf{a} \cdot \mathbf{b} = 0$, then $|\mathbf{a} \times \mathbf{b}| = 1$.

Solution
Using Property (6), we have

$$|\mathbf{a} \times \mathbf{b}|^2 = |\mathbf{a}|^2|\mathbf{b}|^2 - (\mathbf{a} \cdot \mathbf{b})^2 = 1 \cdot 1 - 0 = 1$$

Since $|\mathbf{a} \times \mathbf{b}|$ is positive and $|\mathbf{a} \times \mathbf{b}|^2 = 1$, we must have $|\mathbf{a} \times \mathbf{b}| = 1$.

As an application of Property (6), we prove the following fact.

THEOREM 2
Let $\mathbf{a}$ and $\mathbf{b}$ be nonzero vectors in three-dimensional space, and let θ be the radian measure of the angle between them, $0 \leq \theta \leq \pi$. Then

$$|\mathbf{a} \times \mathbf{b}| = |\mathbf{a}||\mathbf{b}| \sin \theta \tag{5}$$

Proof
By Property (6) of Theorem 1,

$$|\mathbf{a} \times \mathbf{b}|^2 = |\mathbf{a}|^2|\mathbf{b}|^2 - (\mathbf{a} \cdot \mathbf{b})^2$$

and by formula (7) of Section 15.3,

$$(\mathbf{a} \cdot \mathbf{b})^2 = |\mathbf{a}|^2|\mathbf{b}|^2 \cos^2 \theta$$

Hence

$$\begin{aligned} |\mathbf{a} \times \mathbf{b}|^2 &= |\mathbf{a}|^2|\mathbf{b}|^2 - |\mathbf{a}|^2|\mathbf{b}|^2 \cos^2 \theta \\ &= |\mathbf{a}|^2|\mathbf{b}|^2(1 - \cos^2 \theta) \\ &= |\mathbf{a}|^2|\mathbf{b}|^2 \sin^2 \theta \end{aligned}$$

and formula (5) follows.

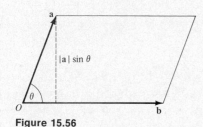

Figure 15.56

Formula (5) has the following geometric interpretation: Consider nonzero vectors $\mathbf{a}$ and $\mathbf{b}$. We note in Figure 15.56 that $|\mathbf{a}| \sin \theta$ is the height of the parallelogram determined by $\mathbf{a}$ and $\mathbf{b}$. Hence, $|\mathbf{a} \times \mathbf{b}|$ gives the area of the parallelogram determined by $\mathbf{a}$ and $\mathbf{b}$.

We recall from Section 15.3 (Definition 2) that nonzero vectors $\mathbf{a}$ and $\mathbf{b}$ are orthogonal if and only if $\mathbf{a} \cdot \mathbf{b} = 0$. With this we can prove the following fact (see Figure 15.57).

THEOREM 3
For any nonzero vectors $\mathbf{a}$ and $\mathbf{b}$ in three-dimensional space, $\mathbf{a} \times \mathbf{b}$ is orthogonal to both $\mathbf{a}$ and $\mathbf{b}$.

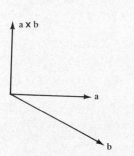

Figure 15.57

Proof
Let $\mathbf{a} = (a_1, a_2, a_3)$ and $\mathbf{b} = (b_1, b_2, b_3)$. Then

$$\mathbf{a} \cdot (\mathbf{a} \times \mathbf{b})$$
$$= a_1(a_2b_3 - a_3b_2) + a_2(a_3b_1 - a_1b_3) + a_3(a_1b_2 - a_2b_1)$$
$$= a_1a_2a_3 - a_1a_3b_2 + a_2a_3b_1 - a_1a_2b_3 + a_1a_3b_2 - a_2a_3b_1$$
$$= (a_1a_2b_3 - a_1a_2b_3) + (a_2a_3b_1 - a_2a_3b_1) + (a_1a_3b_2 - a_1a_3b_2) = 0$$

A similar calculation shows that $\mathbf{b} \cdot (\mathbf{a} \times \mathbf{b}) = 0$.

Example 5

Let $\mathbf{a} = (1, 4, 0)$ and $\mathbf{b} = (2, 3, 5)$. Show by direct calculation that $\mathbf{a}$ and $\mathbf{a} \times \mathbf{b}$ are orthogonal.

Solution
$$\mathbf{a} \times \mathbf{b} = (4 \cdot 5 - 0 \cdot 3, \, 0 \cdot 2 - 1 \cdot 5, \, 1 \cdot 3 - 4 \cdot 2) = (20, -5, 5)$$

Hence
$$\mathbf{a} \cdot (\mathbf{a} \times \mathbf{b}) = (1, 4, 0) \cdot (20, -5, 5) = 1 \cdot 20 + 4 \cdot (-5) + 0 \cdot 5 = 0$$

EXERCISES

Evaluate the vector products $\mathbf{a} \times \mathbf{b}$ in Exercises 1–6.

1. $\mathbf{a} = (1, 0, 0)$, $\mathbf{b} = (0, 1, 0)$
2. $\mathbf{a} = (2, 2, 3)$, $\mathbf{b} = (1, 1, \frac{3}{2})$
3. $\mathbf{a} = (2, -2, 4)$, $\mathbf{b} = (-2, 2, 0)$
4. $\mathbf{a} = (\frac{1}{2}, \frac{1}{3}, \frac{1}{5})$, $\mathbf{b} = (\frac{1}{2}, -\frac{1}{3}, \frac{1}{5})$
5. $\mathbf{a} = (1, 0, 1)$, $\mathbf{b} = (0, 1, 1)$
6. $\mathbf{a} = (1/\sqrt{3}, -1/\sqrt{3}, 1/\sqrt{3})$, $\mathbf{b} = (2/\sqrt{5}, 1/\sqrt{5}, 0)$

Evaluate the determinants in Exercises 7–12.

7. $\begin{vmatrix} 1 & 2 \\ 2 & -1 \end{vmatrix}$

8. $\begin{vmatrix} -4 & 3 \\ 1 & 1 \end{vmatrix}$

9. $\begin{vmatrix} 3 & 5 \\ 1 & \frac{5}{3} \end{vmatrix}$

10. $\begin{vmatrix} \mathbf{i} & \mathbf{j} & \mathbf{k} \\ 1 & 1 & 1 \\ 1 & \frac{1}{2} & 1 \end{vmatrix}$

11. $\begin{vmatrix} \mathbf{i} & \mathbf{j} & \mathbf{k} \\ 2 & 7 & -5 \\ 0 & 1 & -7 \end{vmatrix}$

12. $\begin{vmatrix} \mathbf{i} & \mathbf{j} & \mathbf{k} \\ 1 & -1 & 1 \\ 2 & 4 & 8 \end{vmatrix}$

Verify the properties in Exercises 13–17.

13. $\mathbf{a} \times \mathbf{a} = \mathbf{0}$
14. $\mathbf{a} \times \mathbf{b} + \mathbf{b} \times \mathbf{a} = \mathbf{0}$
15. $\mathbf{a} \times (\mathbf{b} + \mathbf{c}) = \mathbf{a} \times \mathbf{b} + \mathbf{a} \times \mathbf{c}$
16. $\mathbf{a} \cdot (\mathbf{b} \times \mathbf{c}) = \mathbf{b} \cdot (\mathbf{c} \times \mathbf{a}) = \mathbf{c} \cdot (\mathbf{a} \times \mathbf{b})$
17. $(\mathbf{a} \times \mathbf{b}) \times \mathbf{c} = (\mathbf{a} \cdot \mathbf{c})\mathbf{b} - (\mathbf{b} \cdot \mathbf{c})\mathbf{a}$

15.11 AN APPLICATION OF VECTOR PRODUCTS TO CURVATURE

In Section 15.6 we introduced the concept of *curvature* of a curve $\mathbf{R} = \mathbf{f}(t)$; our definition was

$$\kappa(t) = \frac{|\mathbf{T}'(t)|}{|\mathbf{f}'(t)|} \qquad (\mathbf{f}'(t) \neq \mathbf{0})$$

where $\mathbf{T}(t)$ was the unit tangent to the curve at $\mathbf{f}(t)$. An alternative formula was derived in Theorem 1 for curves that can be represented as $\mathbf{f}(t) = (t, F(t))$. Still another formula can be expressed in terms of the vector product. For this we recall the notation (also introduced in Section 15.6)

$$\mathbf{v}(t) = \mathbf{f}'(t) \qquad \text{(velocity vector)}$$
$$\mathbf{a}(t) = \mathbf{f}''(t) \qquad \text{(acceleration vector)}$$

THEOREM 1

Let $\mathbf{R} = \mathbf{f}(t)$ be a given curve. The curvature $\kappa(t)$ is given by the formula

$$\kappa(t) = \frac{|\mathbf{a}(t) \times \mathbf{v}(t)|}{|\mathbf{v}(t)|^3} \qquad (1)$$

whenever $\mathbf{v}$ and $\mathbf{a}$ exist and $\mathbf{v} \neq \mathbf{0}$.

Proof

In Section 15.6 we derived the formula

$$\mathbf{a}(t) = v'(t)\mathbf{T}(t) + v^2(t)\kappa(t)\mathbf{N}(t)$$

where $v(t) = |\mathbf{v}(t)|$ was speed and $\mathbf{N}(t)$ the unit normal to the curve (see Definition 3, Section 15.5). Hence

$$\mathbf{a}(t) \times \mathbf{v}(t) = [v'(t)\mathbf{T}(t) + v^2(t)\kappa(t)\mathbf{N}(t)] \times [v(t)\mathbf{T}(t)]$$
$$= v'(t)v(t)[\mathbf{T}(t) \times \mathbf{T}(t)] + v^3(t)\kappa(t)[\mathbf{N}(t) \times \mathbf{T}(t)]$$

Since $\mathbf{T}(t) \times \mathbf{T}(t) = \mathbf{0}$, we have

$$\mathbf{a}(t) \times \mathbf{v}(t) = v^3(t)\kappa(t)[\mathbf{N}(t) \times \mathbf{T}(t)] \qquad (2)$$

Now, $\mathbf{N}(t)$ and $\mathbf{T}(t)$ are orthogonal unit vectors, and so by Example 4 of Section 15.5

$$|\mathbf{N}(t) \times \mathbf{T}(t)| = 1 \tag{3}$$

Furthermore, $\kappa(t) > 0$ and $v(t) > 0$, and hence

$$|\mathbf{a}(t) \times \mathbf{v}(t)| = v^3(t)\kappa(t) = |\mathbf{v}(t)|^3 \kappa(t)$$

Then $\mathbf{v}(t) \neq \mathbf{0}$. We can solve for $\kappa(t)$ to obtain formula (1).

Omitting reference to t, formula (1) can be written as

$$\kappa = \frac{|\mathbf{a} \times \mathbf{v}|}{|\mathbf{v}|^3}$$

By Property 6, Theorem 1 of Section 15.10, formula (1) can also be written as

$$\kappa = \frac{\sqrt{|\mathbf{a}|^2 |\mathbf{v}|^2 - (\mathbf{a} \cdot \mathbf{v})^2}}{|\mathbf{v}|^3} \tag{4}$$

This form of the formula does not involve the vector product, and it also gives the curvature of curves in the plane. This remark is justified by the fact that a curve $(f_1(t), f_2(t))$ in the plane has the same curvature as the curve $(f_1(t), f_2(t), 0)$ in three dimensions.

The vector

$$\mathbf{B} = \mathbf{T} \times \mathbf{N} \tag{5}$$

which appeared in the proof of Theorem 1 is called the *binormal* to the curve. As we know, this vector is orthogonal to both $\mathbf{N}$ and $\mathbf{T}$, and by (3) it is a unit vector.

Example 1

Find the curvature of $\mathbf{f}(t) = (t \cos t, t \sin t, t)$ when $t = \pi/2$.

Solution
We shall use formula (4). We have

$$\mathbf{v}(t) = (\cos t - t \sin t, \sin t + t \cos t, 1)$$
$$\mathbf{a}(t) = (-2 \sin t - t \cos t, 2 \cos t - t \sin t, 0)$$

and hence, at $t = \pi/4$,

$$\mathbf{v} = \left(-\frac{\pi}{2}, 1, 1\right)$$

$$\mathbf{a} = \left(-2, -\frac{\pi}{2}, 0\right)$$

Accordingly,

$$|\mathbf{v}|^2 = \frac{\pi^2}{4} + 2$$

$$|\mathbf{a}|^2 = 4 + \frac{\pi^2}{4}$$

$$\mathbf{a} \cdot \mathbf{v} = \pi - \frac{\pi}{2} = \frac{\pi}{2}$$

and we have

$$\kappa = \frac{\sqrt{(4 + \pi^2/4)^2 (2 + \pi^2/4)^2 - (\pi/2)^2}}{(2 + \pi^2/4)^{3/2}}$$

A simplification gives

$$\kappa = \frac{2\sqrt{(16 + \pi^2)^2 (8 + \pi^2)^2 - 4\pi^2}}{(8 + \pi^2)^{3/2}}$$

Example 2

Find the curvature of the ellipse

$x = 3 \cos t$
$y = 4 \sin t$

when $t = 0$.

Solution
With the notation $\mathbf{f}(t) = (3 \cos t, 4 \sin t)$, we have

$\mathbf{v}(t) = (-3 \sin t, 4 \cos t)$
$\mathbf{a}(t) = (-3 \cos t, -4 \sin t)$

When $t = 0$,

$\mathbf{v} = (0, 4)$
$\mathbf{a} = (-3, 0)$

so that

$|\mathbf{v}|^2 = 16$
$|\mathbf{a}|^2 = 9$
$\mathbf{a} \cdot \mathbf{v} = 0$

By formula (4),

$$\kappa = \frac{\sqrt{4^2 \cdot 3^2 - 0}}{4^3} = \frac{4 \cdot 3}{4^3} = \frac{3}{16}$$

EXERCISES

Find the curvature of the curves in Exercises 1–8.

1. $\mathbf{f}(t) = (t \cos t, t \sin t)$ when $t = -\pi$

2. $\mathbf{f}(t) = (t, t^2)$ when $t = 1$
3. $\mathbf{f}(t) = (e^t, e^{-t})$ when $t = 0$
4. $y = x^3$ at $(-2, -8)$
5. $\mathbf{f}(t) = (t, t^2, 1)$ when $t = 1$
6. $\mathbf{f}(t) = (te^t, te^{-t}, 1)$ when $t = 1$
7. $\left.\begin{array}{l} x = t - \cos t \\ y = t - \sin t \\ z = t \end{array}\right\}$ when $t = 0$
8. $\left.\begin{array}{l} x = t^2 \\ y = t^2 \\ z = t \end{array}\right\}$ when $t = 2$
9. Show that the binormal vector $\mathbf{B}$ can be represented as

$$\mathbf{B} = \frac{1}{|\mathbf{v}|^3 \kappa}(\mathbf{v} \times \mathbf{a})$$

QUIZ 1

1. Find the normalized form of the vector $c(\mathbf{a} - \mathbf{b})$ when we have $c = (1, -2, -1) \cdot (2, 3, 1)$, $\mathbf{a} = (1, -2, -3)$, and $\mathbf{b} = (2, 3, -1)$.

2. Find the length of the curve $\mathbf{R} = (t, e^t \cos t, e^t \sin 2t)$, $0 \leq t \leq 2$. Give the answer in the form of a definite integral; it is not required to find the numerical value.

3. Find the curvature of the curve $\mathbf{R} = (e^t, -e^{-2t})$ at the point $(1, -1)$.

4. Find the gradient of the function $\sin x + x^2 y + z^2 x$ at the point $(\pi, 1, 2)$. What is the equation of the plane tangent to the surface $\sin x + x^2 y + z^2 x = \pi^2 + 4\pi$ at the point $(\pi, 1, 2)$?

5. Find the vector $\mathbf{A} = (1, 2, -1) \times (2, 2, 3)$. The plane $\mathbf{A} \cdot \mathbf{R} = 0$ contains the origin; find two other points on this plane (using a minimum of calculations).

QUIZ 2

1. Find the number x so that the vectors $(-1, 2, x)$ and $(2, 3, -1)$ are orthogonal.

2. Find the unit tangent and the unit normal to the curve $\mathbf{R} = (t, t^2)$ at the point $(-1, 1)$.

3. Find the tangent plane to the surface $x^3 + y^2 + 2xyz = 0$ at the point $(1, -1, 1)$.

4. In what direction **u** (a unit vector) is the function x^2y^3z decreasing most rapidly at the point $(1, 2, -1)$?

5. Use the formula

$$\kappa = \frac{|\mathbf{a} \times \mathbf{v}|}{|\mathbf{v}|^3}$$

to find the curvature of the curve $\mathbf{R} = (t + \cos t, t - 2 \sin t, t)$ when $t = 0$.

Table 1
Powers and Roots

n	n^2	$\sqrt{n}$	n^3	$\sqrt[3]{n}$	n	n^2	$\sqrt{n}$	n^3	$\sqrt[3]{n}$
1	1	1.000	1	1.000	51	2,601	7.141	132,651	3.708
2	4	1.414	8	1.260	52	2,704	7.211	140,608	3.732
3	9	1.732	27	1.442	53	2,809	7.280	148,877	3.756
4	16	2.000	64	1.587	54	2,916	7.348	157,464	3.780
5	25	2.236	125	1.710	55	3,025	7.416	166,375	3.803
6	36	2.449	216	1.817	56	3,136	7.483	175,616	3.826
7	49	2.646	343	1.913	57	3,249	7.550	185,193	3.848
8	64	2.828	512	2.000	58	3,364	7.616	195,112	3.871
9	81	3.000	729	2.080	59	3,481	7.681	205,379	3.893
10	100	3.162	1,000	2.154	60	3,600	7.746	216,000	3.915
11	121	3.317	1,331	2.224	61	3,721	7.810	226,981	3.936
12	144	3.464	1,728	2.289	62	3,844	7.874	238,328	3.958
13	169	3.606	2,197	2.351	63	3,969	7.937	250,047	3.979
14	196	3.742	2,744	2.410	64	4,096	8.000	262,144	4.000
15	225	3.873	3,375	2.466	65	4,225	8.062	274,625	4.021
16	256	4.000	4,096	2.520	66	4,356	8.124	287,496	4.041
17	289	4.123	4,913	2.571	67	4,489	8.185	300,763	4.062
18	324	4.243	5,832	2.621	68	4,624	8.246	314,432	4.082
19	361	4.359	6,859	2.668	69	4,761	8.307	328,509	4.102
20	400	4.472	8,000	2.714	70	4,900	8.367	343,000	4.121
21	441	4.583	9,261	2.759	71	5,041	8.426	357,911	4.141
22	484	4.690	10,648	2.802	72	5,184	8.485	373,248	4.160
23	529	4.796	12,167	2.844	73	5,329	8.544	389,017	4.179
24	576	4.899	13,824	2.884	74	5,476	8.602	405,224	4.198
25	625	5.000	15,625	2.924	75	5,625	8.660	421,875	4.217
26	676	5.099	17,576	2.962	76	5,776	8.718	438,976	4.236
27	729	5.196	19,683	3.000	77	5,929	8.775	456,533	4.254
28	784	5.291	21,952	3.037	78	6,084	8.832	474,552	4.273
29	841	5.385	24,389	3.072	79	6,241	8.888	493,039	4.291
30	900	5.477	27,000	3.107	80	6,400	8.944	512,000	4.309
31	961	5.568	29,791	3.141	81	6,561	9.000	531,441	4.327
32	1,024	5.657	32,768	3.175	82	6,724	9.055	551,368	4.344
33	1,089	5.745	35,937	3.208	83	6,889	9.110	571,787	4.362
34	1,156	5.831	39,304	3.240	84	7,056	9.165	592,704	4.380
35	1,225	5.916	42,875	3.271	85	7,225	9.220	614,124	4.397
36	1,296	6.000	46,656	3.302	86	7,396	9.274	636,056	4.414
37	1,369	6.083	50,653	3.332	87	7,569	9.327	658,503	4.431
38	1,444	6.164	54,872	3.362	88	7,744	9.381	681,472	4.448
39	1,521	6.245	59,319	3.391	89	7,921	9.434	704,969	4.465
40	1,600	6.325	64,000	3.420	90	8,100	9.487	729,000	4.481
41	1,681	6.403	68,921	3.448	91	8,281	9.539	753,571	4.498
42	1,764	6.481	74,088	3.476	92	8,464	9.592	778,688	4.514
43	1,849	6.557	79,507	3.503	93	8,649	9.643	804,357	4.531
44	1,936	6.633	85,184	3.530	94	8,836	9.695	830,584	4.547
45	2,025	6.708	91,125	3.557	95	9,025	9.747	857,375	4.563
46	2,116	6.782	97,336	3.583	96	9,216	9.798	884,736	4.579
47	2,209	6.856	103,823	3.609	97	9,409	9.849	912,673	4.595
48	2,304	6.928	110,592	3.634	98	9,604	9.899	941,192	4.610
49	2,401	7.000	117,649	3.659	99	9,801	9.950	970,299	4.626
50	2,500	7.071	125,000	3.684	100	10,000	10.000	1,000,000	4.642

Table 2
Natural Logarithms

N	0	1	2	3	4	5	6	7	8	9
1.0	0000	0100	0198	0296	0392	0488	0583	0677	0770	0862
1.1	0953	1044	1133	1222	1310	1398	1484	1570	1655	1740
1.2	1823	1906	1989	2070	2151	2231	2311	2390	2469	2546
1.3	2624	2700	2776	2852	2927	3001	3075	3148	3221	3293
1.4	3365	3436	3507	3577	3646	3716	3784	3853	3920	3988
1.5	4055	4121	4187	4253	4318	4383	4447	4511	4574	4637
1.6	4700	4762	4824	4886	4947	5008	5068	5128	5188	5247
1.7	5306	5365	5423	5481	5539	5596	5653	5710	5766	5822
1.8	5878	5933	5988	6043	6098	6152	6206	6259	6313	6366
1.9	6419	6471	6523	6575	6627	6678	6729	6780	6831	6881
2.0	6931	6981	7031	7080	7129	7178	7227	7275	7324	7372
2.1	7419	7467	7514	7561	7608	7655	7701	7747	7793	7839
2.2	7885	7930	7975	8020	8065	8109	8154	8198	8242	8286
2.3	8329	8372	8416	8459	8502	8544	8587	8629	8671	8713
2.4	8755	8796	8838	8879	8920	8961	9002	9042	9083	9123
2.5	9163	9203	9243	9282	9322	9361	9400	9439	9478	9517
2.6	9555	9594	9632	9670	9708	9746	9783	9821	9858	9895
2.7	9933	9969	1.0006	1.0043	1.0080	1.0116	1.0152	1.0188	1.0225	1.0260
2.8	1.0296	0332	0367	0403	0438	0473	0508	0543	0578	0613
2.9	0647	0682	0716	0750	0784	0818	0852	0886	0919	0953
3.0	1.0986	1019	1053	1086	1119	1151	1184	1217	1249	1282
3.1	1314	1346	1378	1410	1442	1474	1506	1537	1569	1600
3.2	1632	1663	1694	1725	1756	1787	1817	1848	1878	1909
3.3	1939	1969	2000	2030	2060	2090	2119	2149	2179	2208
3.4	2238	2267	2296	2326	2355	2384	2413	2442	2470	2499
3.5	1.2528	2556	2585	2613	2641	2669	2698	2726	2754	2782
3.6	2809	2837	2865	2892	2920	2947	2975	3002	3029	3056
3.7	3083	3110	3137	3164	3191	3218	3244	3271	3297	3324
3.8	3350	3376	3403	3429	3455	3481	3507	3533	3558	3584
3.9	3610	3635	3661	3686	3712	3737	3762	3788	3813	3838
4.0	1.3863	3888	3913	3938	3962	3987	4012	4036	4061	4085
4.1	4110	4134	4159	4183	4207	4231	4255	4279	4303	4327
4.2	4351	4375	4398	4422	4446	4469	4493	4516	4540	4563
4.3	4586	4609	4633	4656	4679	4702	4725	4748	4770	4793
4.4	4816	4839	4861	4884	4907	4929	4951	4974	4996	5019
4.5	1.5041	5063	5085	5107	5129	5151	5173	5195	5217	5239
4.6	5261	5282	5304	5326	5347	5369	5390	5412	5433	5454
4.7	5476	5497	5518	5539	5560	5581	5602	5623	5644	5665
4.8	5686	5707	5728	5748	5769	5790	5810	5831	5851	5872
4.9	5892	5913	5933	5953	5974	5994	6014	6034	6054	6074
5.0	1.6094	6114	6134	6154	6174	6194	6214	6233	6253	6273
5.1	6292	6312	6332	6351	6371	6390	6409	6429	6448	6467
5.2	6487	6506	6525	6544	6563	6582	6601	6620	6639	6658
5.3	6677	6696	6715	6734	6752	6771	6790	6808	6827	6845
5.4	6864	6882	6901	6919	6938	6956	6974	6993	7011	7029
5.5	1.7047	7066	7084	7102	7120	7138	7156	7174	7192	7210
5.6	7228	7246	7263	7281	7299	7317	7334	7352	7370	7387
5.7	7405	7422	7440	7457	7475	7492	7509	7527	7544	7561
5.8	7579	7596	7613	7630	7647	7664	7681	7699	7716	7733
5.9	7750	7766	7783	7800	7817	7834	7851	7867	7884	7901

Table 2 (*continued*)

N	0	1	2	3	4	5	6	7	8	9
6.0	1.7918	7934	7951	7967	7984	8001	8017	8034	8050	8066
6.1	8083	8099	8116	8132	8148	8165	8181	8197	8213	8229
6.2	8245	8262	8278	8294	8310	8326	8342	8358	8374	8390
6.3	8405	8421	8437	8453	8469	8485	8500	8516	8532	8547
6.4	8563	8579	8594	8610	8625	8641	8656	8672	8687	8703
6.5	1.8718	8733	8749	8764	8779	8795	8810	8825	8840	8856
6.6	8871	8886	8901	8916	8931	8946	8961	8976	8991	9006
6.7	9021	9036	9051	9066	9081	9095	9110	9125	9140	9155
6.8	9169	9184	9199	9213	9228	9242	9257	9272	9286	9301
6.9	9315	9330	9344	9359	9373	9387	9402	9416	9430	9445
7.0	1.9459	9473	9488	9502	9516	9530	9544	9559	9573	9587
7.1	9601	9615	9629	9643	9657	9671	9685	9699	9713	9727
7.2	9741	9755	9769	9782	9796	9810	9824	9838	9851	9865
7.3	9879	9892	9906	9920	9933	9947	9961	9974	9988	2.0001
7.4	2.0015	0028	0042	0055	0069	0082	0096	0109	0122	0136
7.5	2.0149	0162	0176	0189	0202	0215	0229	0242	0255	0268
7.6	0281	0295	0308	0321	0334	0347	0360	0373	0386	0399
7.7	0412	0425	0438	0451	0464	0477	0490	0503	0516	0528
7.8	0541	0554	0567	0580	0592	0605	0618	0630	0643	0656
7.9	0669	0681	0694	0707	0719	0732	0744	0757	0769	0782
8.0	2.0794	0807	0819	0832	0844	0857	0869	0882	0894	0906
8.1	0919	0931	0943	0956	0968	0980	0992	1005	1017	1029
8.2	1041	1054	1066	1078	1090	1102	1114	1126	1138	1150
8.3	1163	1175	1187	1199	1211	1223	1235	1247	1258	1270
8.4	1282	1294	1306	1318	1330	1342	1353	1365	1377	1389
8.5	2.1401	1412	1424	1436	1448	1459	1471	1483	1494	1506
8.6	1518	1529	1541	1552	1564	1576	1587	1599	1610	1622
8.7	1633	1645	1656	1668	1679	1691	1702	1713	1725	1736
8.8	1748	1759	1770	1782	1793	1804	1815	1827	1838	1849
8.9	1861	1872	1883	1894	1905	1917	1928	1939	1950	1961
9.0	2.1972	1983	1994	2006	2017	2028	2039	2050	2061	2072
9.1	2083	2094	2105	2116	2127	2138	2148	2159	2170	2181
9.2	2192	2203	2214	2225	2235	2246	2257	2268	2279	2289
9.3	2300	2311	2322	2332	2343	2354	2364	2375	2386	2396
9.4	2407	2418	2428	2439	2450	2460	2471	2481	2492	2502
9.5	2.2513	2523	2534	2544	2555	2565	2576	2586	2597	2607
9.6	2618	2628	2638	2649	2659	2670	2680	2690	2701	2711
9.7	2721	2732	2742	2752	2762	2773	2783	2793	2803	2814
9.8	2824	2834	2844	2854	2865	2875	2885	2895	2905	2915
9.9	2925	2935	2946	2956	2966	2976	2986	2996	3006	3016

NOTE: To find the logarithm of a number $N > 10$, use scientific notation to write $N = A \times 10^n$, where $0 < A \leq 1$, and compute $\ln N$ from the formula

$$\ln N = \ln A + n \times \ln 10$$

A table of values for $n \times \ln 10$ is included for your convenience.

n	$n \times \ln 10$	n	$n \times \ln 10$
1	2.30259	6	13.81551
2	4.60517	7	16.11810
3	6.90776	8	18.42068
4	9.21034	9	20.72327
5	11.51293	10	23.02585

Table 3
Common Logarithms

N	0	1	2	3	4	5	6	7	8	9
10	0000	0043	0086	0128	0170	0212	0253	0294	0334	0374
11	0414	0453	0492	0531	0569	0607	0645	0682	0719	0755
12	0792	0828	0864	0899	0934	0969	1004	1038	1072	1106
13	1139	1173	1206	1239	1271	1303	1335	1367	1399	1430
14	1461	1492	1523	1553	1584	1614	1644	1673	1703	1732
15	1761	1790	1818	1847	1875	1903	1931	1959	1987	2014
16	2041	2068	2095	2122	2148	2175	2201	2227	2253	2279
17	2304	2330	2355	2380	2405	2430	2455	2480	2504	2529
18	2553	2577	2601	2625	2648	2672	2695	2718	2742	2765
19	2788	2810	2833	2856	2878	2900	2923	2945	2967	2989
20	3010	3032	3054	3075	3096	3118	3139	3160	3181	3201
21	3222	3243	3263	3284	3304	3324	3345	3365	3385	3404
22	3424	3444	3464	3483	3502	3522	3541	3560	3579	3598
23	3617	3636	3655	3674	3692	3711	3729	3747	3766	3784
24	3802	3820	3838	3856	3874	3892	3909	3927	3945	3962
25	3979	3997	4014	4031	4048	4065	4082	4099	4116	4133
26	4150	4166	4183	4200	4216	4232	4249	4265	4281	4298
27	4314	4330	4346	4362	4378	4393	4409	4425	4440	4456
28	4472	4487	4502	4518	4533	4548	4564	4579	4594	4609
29	4624	4639	4654	4669	4683	4698	4713	4728	4742	4757
30	4771	4786	4800	4814	4829	4843	4857	4871	4886	4900
31	4914	4928	4942	4955	4969	4983	4997	5011	5024	5038
32	5051	5065	5079	5092	5105	5119	5132	5145	5159	5172
33	5185	5198	5211	5224	5237	5250	5263	5276	5289	5302
34	5315	5328	5340	5353	5366	5378	5391	5403	5416	5428
35	5441	5453	5465	5478	5490	5502	5514	5527	5539	5551
36	5563	5575	5587	5599	5611	5623	5635	5647	5658	5670
37	5682	5694	5705	5717	5729	5740	5752	5763	5775	5786
38	5798	5809	5821	5832	5843	5855	5866	5877	5888	5899
39	5911	5922	5933	5944	5955	5966	5977	5988	5999	6010
40	6021	6031	6042	6053	6064	6075	6085	6096	6107	6117
41	6128	6138	6149	6160	6170	6180	6191	6201	6212	6222
42	6232	6243	6253	6263	6274	6284	6294	6304	6314	6325
43	6335	6345	6355	6365	6375	6385	6395	6405	6415	6425
44	6435	6444	6454	6464	6474	6484	6493	6503	6513	6522
45	6532	6542	6551	6561	6571	6580	6590	6599	6609	6618
46	6628	6637	6646	6656	6665	6675	6684	6693	6702	6712
47	6721	6730	6739	6749	6758	6767	6776	6785	6794	6803
48	6812	6821	6830	6839	6848	6857	6866	6875	6884	6893
49	6902	6911	6920	6928	6937	6946	6955	6964	6972	6981
50	6990	6998	7007	7016	7024	7033	7042	7050	7059	7067
51	7076	7084	7093	7101	7110	7118	7126	7135	7143	7152
52	7160	7168	7177	7185	7193	7202	7210	7218	7226	7235
53	7243	7251	7259	7267	7275	7284	7292	7300	7308	7316
54	7324	7332	7340	7348	7356	7364	7372	7380	7388	7396

Table 3 (*continued*)

N	0	1	2	3	4	5	6	7	8	9
55	7404	7412	7419	7427	7435	7443	7451	7459	7466	7474
56	7482	7490	7497	7505	7513	7520	7528	7536	7543	7551
57	7559	7566	7574	7582	7589	7597	7604	7612	7619	7627
58	7634	7642	7649	7657	7664	7672	7679	7686	7694	7701
59	7709	7716	7723	7731	7738	7745	7752	7760	7767	7774
60	7782	7789	7796	7803	7810	7818	7825	7832	7839	7846
61	7853	7860	7868	7875	7882	7889	7896	7903	7910	7917
62	7924	7931	7938	7945	7952	7959	7966	7973	7980	7987
63	7993	8000	8007	8014	8021	8028	8035	8041	8048	8055
64	8062	8069	8075	8082	8089	8096	8102	8109	8116	8122
65	8129	8136	8142	8149	8156	8162	8169	8176	8182	8189
66	8195	8202	8209	8215	8222	8228	8235	8241	8248	8254
67	8261	8267	8274	8280	8287	8293	8299	8306	8312	8319
68	8325	8331	8338	8344	8351	8357	8363	8370	8376	8382
69	8388	8395	8401	8407	8414	8420	8426	8432	8439	8445
70	8451	8457	8463	8470	8476	8482	8488	8494	8500	8506
71	8513	8519	8525	8531	8537	8543	8549	8555	8561	8567
72	8573	8579	8585	8591	8597	8603	8609	8615	8621	8627
73	8633	8639	8645	8651	8657	8663	8669	8675	8681	8686
74	8692	8698	8704	8710	8716	8722	8727	8733	8739	8745
75	8751	8756	8762	8768	8774	8779	8785	8791	8797	8802
76	8808	8814	8820	8825	8831	8837	8842	8848	8854	8859
77	8865	8871	8876	8882	8887	8893	8899	8904	8910	8915
78	8921	8927	8932	8938	8943	8949	8954	8960	8965	8971
79	8976	8982	8987	8993	8998	9004	9009	9015	9020	9025
80	9031	9036	9042	9047	9053	9058	9063	9069	9074	9079
81	9085	9090	9096	9101	9106	9112	9117	9122	9128	9133
82	9138	9143	9149	9154	9159	9165	9170	9175	9180	9186
83	9191	9196	9201	9206	9212	9217	9222	9227	9232	9238
84	9243	9248	9253	9258	9263	9269	9274	9279	9284	9289
85	9294	9299	9304	9309	9315	9320	9325	9330	9335	9340
86	9345	9350	9355	9360	9365	9370	9375	9380	9385	9390
87	9395	9400	9405	9410	9415	9420	9425	9430	9435	9440
88	9445	9450	9455	9460	9465	9469	9474	9479	9484	9489
89	9494	9499	9504	9509	9513	9518	9523	9528	9533	9538
90	9542	9547	9552	9557	9562	9566	9571	9576	9581	9586
91	9590	9595	9600	9605	9609	9614	9619	9624	9628	9633
92	9638	9643	9647	9652	9657	9661	9666	9671	9675	9680
93	9685	9689	9694	9699	9703	9708	9713	9717	9722	9727
94	9731	9736	9741	9745	9750	9754	9759	9763	9768	9773
95	9777	9782	9786	9791	9795	9800	9805	9809	9814	9818
96	9823	9827	9832	9836	9841	9845	9850	9854	9859	9863
97	9868	9872	9877	9881	9886	9890	9894	9899	9903	9908
98	9912	9917	9921	9926	9930	9934	9939	9943	9948	9952
99	9956	9961	9965	9969	9974	9978	9983	9987	9991	9996

Table 4
Exponential Functions

x	e^x	e^{-x}	x	e^x	e^{-x}
0.00	1.0000	1.000000	**0.50**	1.6487	0.606531
0.01	1.0101	0.990050	0.51	1.6653	.600496
0.02	1.0202	.980199	0.52	1.6820	.594521
0.03	1.0305	.970446	0.53	1.6989	.588605
0.04	1.0408	.960789	0.54	1.7160	.582748
0.05	1.0513	0.951229	**0.55**	1.7333	0.576950
0.06	1.0618	.941765	0.56	1.7507	.571209
0.07	1.0725	.932394	0.57	1.7683	.565525
0.08	1.0833	.923116	0.58	1.7860	.559898
0.09	1.0942	.913931	0.59	1.8040	.554327
0.10	1.1052	0.904837	**0.60**	1.8221	0.548812
0.11	1.1163	.895834	0.61	1.8404	.543351
0.12	1.1275	.886920	0.62	1.8589	.537944
0.13	1.1388	.878095	0.63	1.8776	.532592
0.14	1.1503	.869358	0.64	1.8965	.527292
0.15	1.1618	0.860708	**0.65**	1.9155	0.522046
0.16	1.1735	.852144	0.66	1.9348	.516851
0.17	1.1853	.843665	0.67	1.9542	.511709
0.18	1.1972	.835270	0.68	1.9739	.506617
0.19	1.2092	.826959	0.69	1.9937	.501576
0.20	1.2214	0.818731	**0.70**	2.0138	0.496585
0.21	1.2337	.810584	0.71	2.0340	.491644
0.22	1.2461	.802519	0.72	2.0544	.486752
0.23	1.2586	.794534	0.73	2.0751	.481909
0.24	1.2712	.786628	0.74	2.0959	.477114
0.25	1.2840	0.778801	**0.75**	2.1170	0.472367
0.26	1.2969	.771052	0.76	2.1383	.467666
0.27	1.3100	.763379	0.77	2.1598	.463013
0.28	1.3231	.755784	0.78	2.1815	.458406
0.29	1.3364	.748264	0.79	2.2034	.453845
0.30	1.3499	0.740818	**0.80**	2.2255	0.449329
0.31	1.3634	.733447	0.81	2.2479	.444858
0.32	1.3771	.726149	0.82	2.2705	.440432
0.33	1.3910	.718924	0.83	2.2933	.436049
0.34	1.4049	.711770	0.84	2.3164	.431711
0.35	1.4191	0.704688	**0.85**	2.3396	0.427415
0.36	1.4333	.697676	0.86	2.3632	.423162
0.37	1.4477	.690734	0.87	2.3869	.418952
0.38	1.4623	.683861	0.88	2.4109	.414783
0.39	1.4770	.677057	0.89	2.4351	.410656
0.40	1.4918	0.670320	**0.90**	2.4596	0.406570
0.41	1.5068	.663650	0.91	2.4843	.402524
0.42	1.5220	.657047	0.92	2.5093	.398519
0.43	1.5373	.650509	0.93	2.5345	.394554
0.44	1.5527	.644036	0.94	2.5600	.390628
0.45	1.5683	0.637628	**0.95**	2.5857	0.386741
0.46	1.5841	.631284	0.96	2.6117	.382893
0.47	1.6000	.625002	0.97	2.6379	.379083
0.48	1.6161	.618783	0.98	2.6645	.375311
0.49	1.6323	.612626	0.99	2.6912	.371577

Table 4 (*continued*)

x	e^x	e^{-x}	x	e^x	e^{-x}
1.00	2.7183	0.367879	**3.50**	33.115	0.030197
1.05	2.8577	.349938	3.55	34.813	.028725
1.10	3.0042	.332871	3.60	36.598	.027324
1.15	3.1582	.316637	3.65	38.475	.025991
1.20	3.3201	.301194	3.70	40.447	.024724
1.25	3.4903	0.286505	**3.75**	42.521	0.023518
1.30	3.6693	.272532	3.80	44.701	.022371
1.35	3.8574	.259240	3.85	46.993	.021280
1.40	4.0552	.246597	3.90	49.402	.020242
1.45	4.2631	.234570	3.95	51.935	.019255
1.50	4.4817	0.223130	**4.00**	54.598	0.018316
1.55	4.7115	.212248	4.05	57.397	.017422
1.60	4.9530	.201897	4.10	60.340	.016573
1.65	5.2070	.192050	4.15	63.434	.015764
1.70	5.4739	.182684	4.20	66.686	.014996
1.75	5.7546	0.173774	**4.25**	70.105	0.014264
1.80	6.0496	.165299	4.30	73.700	.013569
1.85	6.3598	.157237	4.35	77.478	.012907
1.90	6.6859	.149569	4.40	81.451	.012277
1.95	7.0287	.142274	4.45	85.627	.011679
2.00	7.3891	0.135335	**4.50**	90.017	0.011109
2.05	7.7679	.128735	4.55	94.632	.010567
2.10	8.1662	.122456	4.60	99.484	.010052
2.15	8.5849	.116484	4.65	104.58	.009562
2.20	9.0250	.110803	4.70	109.95	.009095
2.25	9.4877	0.105399	**4.75**	115.58	0.008652
2.30	9.9742	.100259	4.80	121.51	.008230
2.35	10.486	.095369	4.85	127.74	.007828
2.40	11.023	.090718	4.90	134.29	.007477
2.45	11.588	.086294	4.95	141.17	.007083
2.50	12.182	0.082085	**5.00**	148.41	0.006738
2.55	12.807	.078082	5.05	156.02	.006409
2.60	13.464	.074274	5.10	164.02	.006097
2.65	14.154	.070651	5.15	172.43	.005799
2.70	14.880	.067206	5.20	181.27	.005517
2.75	15.643	0.063928	**5.25**	190.57	0.005248
2.80	16.445	.060810	5.30	200.34	.004992
2.85	17.288	.057844	5.35	210.61	.004748
2.90	18.174	.055023	5.40	221.41	.004517
2.95	19.106	.052340	5.45	232.76	.004296
3.00	20.086	0.049787	**5.50**	244.69	0.0040868
3.05	21.115	.047359	5.55	257.24	.0038875
3.10	22.198	.045049	5.60	270.43	.0036979
3.15	23.336	.042852	5.65	284.29	.0035175
3.20	24.533	.040764	5.70	298.87	.0033460
3.25	25.790	0.038774	**5.75**	314.19	0.0031828
3.30	27.113	.036883	5.80	330.30	.0030276
3.35	28.503	.035084	5.85	347.23	.0028799
3.40	29.964	.033373	5.90	365.04	.0027394
3.45	31.500	.031746	5.95	383.75	.0026058

Table 4 (*continued*)

x	e^x	e^{-x}	x	e^x	e^{-x}
6.00	403.43	0.0024788	**8.00**	2981.0	0.0003355
6.05	424.11	.0023579	8.05	3133.8	.0003191
6.10	445.86	.0022429	8.10	3294.5	.0003035
6.15	468.72	.0021335	8.15	3463.4	.0002887
6.20	492.75	.0020294	8.20	3641.0	.0002747
6.25	518.01	0.0019305	**8.25**	3827.6	0.0002613
6.30	544.57	.0018363	8.30	4023.9	.0002485
6.35	572.49	.0017467	8.35	4230.2	.0002364
6.40	601.85	.0016616	8.40	4447.1	.0002249
6.45	632.70	.0015805	8.45	4675.1	.0002139
6.50	665.14	0.0015034	**8.50**	4914.8	0.0002036
6.55	699.24	.0014301	8.55	5166.8	.0001935
6.60	735.10	.0013604	8.60	5431.7	.0001841
6.65	772.78	.0012940	8.65	5710.0	.0001751
6.70	812.41	.0012309	8.70	6002.9	.0001666
6.75	854.06	0.0011709	**8.75**	6310.7	0.0001585
6.80	897.85	.0011138	8.80	6634.2	.0001507
6.85	943.88	.0010595	8.85	6974.4	.0001434
6.90	992.27	.0010078	8.90	7332.0	.0001364
6.95	1043.1	.0009586	8.95	7707.9	.0001297
7.00	1096.6	0.0009119	**9.00**	8103.1	0.0001234
7.05	1152.9	.0008674	9.05	8518.5	.0001174
7.10	1212.0	.0008251	9.10	8955.3	.0001117
7.15	1274.1	.0007849	9.15	9414.4	.0001062
7.20	1339.4	.0007466	9.20	9897.1	.0001010
7.25	1408.1	0.0007102	**9.25**	10405	0.0000961
7.30	1480.3	.0006755	9.30	10938	.0000914
7.35	1556.2	.0006426	9.35	11499	.0000870
7.40	1636.0	.0006113	9.40	12088	.0000827
7.45	1719.9	.0005814	9.45	12708	.0000787
7.50	1808.0	0.0005531	**9.50**	13360	0.0000749
7.55	1900.7	.0005261	9.55	14045	.0000712
7.60	1998.2	.0005005	9.60	14765	.0000677
7.65	2100.6	.0004760	9.65	15522	.0000644
7.70	2208.3	.0004528	9.70	16318	.0000613
7.75	2321.6	0.0004307	**9.75**	17154	0.0000583
7.80	2440.6	.0004097	9.80	18034	.0000555
7.85	2565.7	.0003898	9.85	18958	.0000527
7.90	2697.3	.0003707	9.90	19930	.0000502
7.95	2835.6	.0003527	9.95	20952	.0000477
			10.00	22026	0.0000454

Table 5
Trigonometric Functions of Certain Angles

Degrees	0°	30°	45°	60°	90°	180°	270°	360°
Radians	0	$\dfrac{\pi}{6}$	$\dfrac{\pi}{4}$	$\dfrac{\pi}{3}$	$\dfrac{\pi}{2}$	π	$\dfrac{3\pi}{2}$	2π
Sine	0	$\dfrac{1}{2}$	$\dfrac{\sqrt{2}}{2}$	$\dfrac{\sqrt{3}}{2}$	1	0	-1	0
Cosine	1	$\dfrac{\sqrt{3}}{2}$	$\dfrac{\sqrt{2}}{2}$	$\dfrac{1}{2}$	0	-1	0	1

Table 6
Trigonometric Functions

Angle		Sine	Cosine	Tangent	Angle		Sine	Cosine	Tangent
Degree	Radian				Degree	Radian			
0°	0.000	0.000	1.000	0.000					
1°	0.017	0.017	1.000	0.017	46°	0.803	0.719	0.695	1.036
2°	0.035	0.035	0.999	0.035	47°	0.820	0.731	0.682	1.072
3°	0.052	0.052	0.999	0.052	48°	0.838	0.743	0.669	1.111
4°	0.070	0.070	0.998	0.070	49°	0.855	0.755	0.656	1.150
5°	0.087	0.087	0.996	0.087	50°	0.873	0.766	0.643	1.192
6°	0.105	0.105	0.995	0.105	51°	0.890	0.777	0.629	1.235
7°	0.122	0.122	0.993	0.123	52°	0.908	0.788	0.616	1.280
8°	0.140	0.139	0.990	0.141	53°	0.925	0.799	0.602	1.327
9°	0.157	0.156	0.988	0.158	54°	0.942	0.809	0.588	1.376
10°	0.175	0.174	0.985	0.176	55°	0.960	0.819	0.574	1.428
11°	0.192	0.191	0.982	0.194	56°	0.977	0.829	0.559	1.483
12°	0.209	0.208	0.978	0.213	57°	0.995	0.839	0.545	1.540
13°	0.227	0.225	0.974	0.231	58°	1.012	0.848	0.530	1.600
14°	0.244	0.242	0.970	0.249	59°	1.030	0.857	0.515	1.664
15°	0.262	0.259	0.966	0.268	60°	1.047	0.866	0.500	1.732
16°	0.279	0.276	0.961	0.287	61°	1.065	0.875	0.485	1.804
17°	0.297	0.292	0.956	0.306	62°	1.082	0.883	0.469	1.881
18°	0.314	0.309	0.951	0.325	63°	1.100	0.891	0.454	1.963
19°	0.332	0.326	0.946	0.344	64°	1.117	0.899	0.438	2.050
20°	0.349	0.342	0.940	0.364	65°	1.134	0.906	0.423	2.145
21°	0.367	0.358	0.934	0.384	66°	1.152	0.914	0.407	2.246
22°	0.384	0.375	0.927	0.404	67°	1.169	0.921	0.391	2.356
23°	0.401	0.391	0.921	0.424	68°	1.187	0.927	0.375	2.475
24°	0.419	0.407	0.914	0.445	69°	1.204	0.934	0.358	2.605
25°	0.436	0.423	0.906	0.466	70°	1.222	0.940	0.342	2.748
26°	0.454	0.438	0.899	0.488	71°	1.239	0.946	0.326	2.904
27°	0.471	0.454	0.891	0.510	72°	1.257	0.951	0.309	3.078
28°	0.489	0.469	0.883	0.532	73°	1.274	0.956	0.292	3.271
29°	0.506	0.485	0.875	0.554	74°	1.292	0.961	0.276	3.487
30°	0.524	0.500	0.866	0.577	75°	1.309	0.966	0.259	3.732
31°	0.541	0.515	0.857	0.601	76°	1.326	0.970	0.242	4.011
32°	0.559	0.530	0.848	0.625	77°	1.344	0.974	0.225	4.332
33°	0.576	0.545	0.839	0.649	78°	1.361	0.978	0.208	4.705
34°	0.593	0.559	0.829	0.675	79°	1.379	0.982	0.191	5.145
35°	0.611	0.574	0.819	0.700	80°	1.396	0.985	0.174	5.671
36°	0.628	0.588	0.809	0.727	81°	1.414	0.988	0.156	6.314
37°	0.646	0.602	0.799	0.754	82°	1.431	0.990	0.139	7.115
38°	0.663	0.616	0.788	0.781	83°	1.449	0.993	0.122	8.144
39°	0.681	0.629	0.777	0.810	84°	1.466	0.995	0.105	9.514
40°	0.698	0.643	0.766	0.839	85°	1.484	0.996	0.087	11.43
41°	0.716	0.656	0.755	0.869	86°	1.501	0.998	0.070	14.30
42°	0.733	0.669	0.743	0.900	87°	1.518	0.999	0.052	19.08
43°	0.750	0.682	0.731	0.933	88°	1.536	0.999	0.035	28.64
44°	0.768	0.695	0.719	0.966	89°	1.553	1.000	0.017	57.29
45°	0.785	0.707	0.707	1.000	90°	1.571	1.000	0.000	

Table 7
Differentiation Formulas

In the following formulas u, v, and w represent functions of x; a, c, and n represent real numbers.

1. $\dfrac{d}{dx}(cu) = c\dfrac{du}{dx}$

2. $\dfrac{d}{dx}(u+v-w) = \dfrac{du}{dx} + \dfrac{dv}{dx} - \dfrac{dw}{dx}$

3. $\dfrac{d}{dx}(uv) = u\dfrac{dv}{dx} + v\dfrac{du}{dx}$

4. $\dfrac{d}{dx}\left(\dfrac{u}{v}\right) = \dfrac{v\dfrac{du}{dx} - u\dfrac{dv}{dx}}{v^2}$

5. $\dfrac{d}{dx}[f(u)] = \dfrac{d}{du}[f(u)]\dfrac{du}{dx}$

6. $\dfrac{d}{dx}(u^n) = nu^{n-1}\dfrac{du}{dx}$ $\qquad \dfrac{d}{dx}(x^n) = nx^{n-1}$

7. $\dfrac{d}{dx}(\ln u) = \dfrac{1}{u}\dfrac{du}{dx}$ $\qquad \dfrac{d}{dx}(\ln x) = \dfrac{1}{x}$

8. $\dfrac{d}{dx}(\log_a u) = \dfrac{1}{u \ln a}\dfrac{du}{dx}$ $\qquad \dfrac{d}{dx}(\log_a x) = \dfrac{1}{x \ln a}$

9. $\dfrac{d}{dx}(e^u) = e^u\dfrac{du}{dx}$ $\qquad \dfrac{d}{dx}(e^x) = e^x$

10. $\dfrac{d}{dx}(a^u) = a^u \ln a \dfrac{du}{dx}$ $\qquad \dfrac{d}{dx}(a^x) = a^x \ln a$

11. $\dfrac{d}{dx}(u^v) = u^v \ln u \dfrac{dv}{dx} + vu^{v-1}\dfrac{du}{dx}$

12. $\dfrac{d}{dx}\displaystyle\int_a^x f(t)\,dt = f(x)$

13. $\dfrac{d}{dx}(\sin u) = \cos u \dfrac{du}{dx}$

14. $\dfrac{d}{dx}(\cos u) = -\sin u \dfrac{du}{dx}$

15. $\dfrac{d}{dx}(\tan u) = \sec^2 u \dfrac{du}{dx}$

16. $\dfrac{d}{dx}(\cot u) = -\csc^2 u \dfrac{du}{dx}$

17. $\dfrac{d}{dx}(\sec u) = \sec u \tan u \dfrac{du}{dx}$

18. $\dfrac{d}{dx}(\csc u) = -\csc u \cot u \dfrac{du}{dx}$

19. $\dfrac{d}{dx}(\sin^{-1} u) = \dfrac{1}{\sqrt{1-u^2}}\dfrac{du}{dx}, \ \left(-\dfrac{\pi}{2} \leq \sin^{-1} u \leq \dfrac{\pi}{2}\right)$

20. $\dfrac{d}{dx}(\cos^{-1} u) = -\dfrac{1}{\sqrt{1-u^2}}\dfrac{du}{dx}, \ (0 \leq \cos^{-1} u \leq \pi)$

Table 7 (continued)

21. $\dfrac{d}{dx}(\tan^{-1} u) = \dfrac{1}{1+u^2} \dfrac{du}{dx}, \left(-\dfrac{\pi}{2} < \tan^{-1} u < \dfrac{\pi}{2}\right)$

22. $\dfrac{d}{dx}(\cot^{-1} u) = -\dfrac{1}{1+u^2} \dfrac{du}{dx}, \ (0 \le \cot^{-1} u \le \pi)$

23. $\dfrac{d}{dx}(\sec^{-1} u) = \dfrac{1}{u\sqrt{u^2-1}} \dfrac{du}{dx}, \left(0 \le \sec^{-1} u < \dfrac{\pi}{2}, -\pi \le \sec^{-1} u < -\dfrac{\pi}{2}\right)$

24. $\dfrac{d}{dx}(\csc^{-1} u) = -\dfrac{1}{u\sqrt{u^2-1}} \dfrac{du}{dx}, \left(0 < \csc^{-1} u \le \dfrac{\pi}{2}, -\pi < \csc^{-1} u \le -\dfrac{\pi}{2}\right)$

Table 8
Table of Integrals

1. $\displaystyle\int u\, dv = uv - \int v\, du$

2. $\displaystyle\int a^x\, dx = \dfrac{a^x}{\ln a} + C, \ a \ne 1, \ a > 0$

3. $\displaystyle\int \cos x\, dx = \sin x + C$

4. $\displaystyle\int \sin x\, dx = -\cos x + C$

5. $\displaystyle\int (ax+b)^n\, dx = \dfrac{(ax+b)^{n+1}}{a(n+1)} + C, \ n \ne -1$

6. $\displaystyle\int (ax+b)^{-1}\, dx = \dfrac{1}{a}\ln|ax+b| + C$

7. $\displaystyle\int x(ax+b)^n\, dx = \dfrac{(ax+b)^{n+1}}{a^2}\left[\dfrac{ax+b}{n+2} - \dfrac{b}{n+1}\right] + C, \ n \ne -1, -2$

8. $\displaystyle\int x(ax+b)^{-1}\, dx = \dfrac{x}{a} - \dfrac{b}{a^2}\ln|ax+b| + C$

9. $\displaystyle\int x(ax+b)^{-2}\, dx = \dfrac{1}{a^2}\left[\ln|ax+b| + \dfrac{b}{ax+b}\right] + C$

10. $\displaystyle\int \dfrac{dx}{x(ax+b)} = \dfrac{1}{b}\ln\left|\dfrac{x}{ax+b}\right| + C$

11. $\displaystyle\int (\sqrt{ax+b})^n\, dx = \dfrac{2}{a}\dfrac{(\sqrt{ax+b})^{n+2}}{n+2} + C, \ n \ne -2$

12. $\displaystyle\int \dfrac{\sqrt{ax+b}}{x}\, dx = 2\sqrt{ax+b} + b\int \dfrac{dx}{x\sqrt{ax+b}}$

13. (a) $\displaystyle\int \dfrac{dx}{x\sqrt{ax+b}} = \dfrac{2}{\sqrt{-b}}\tan^{-1}\sqrt{\dfrac{ax+b}{-b}} + C, \ \text{if } b < 0$

    (b) $\displaystyle\int \dfrac{dx}{x\sqrt{ax+b}} = \dfrac{1}{\sqrt{b}}\ln\left|\dfrac{\sqrt{ax+b} - \sqrt{b}}{\sqrt{ax+b} + \sqrt{b}}\right| + C, \ \text{if } b > 0$

Table 8 (*continued*)

14. $\int \dfrac{\sqrt{ax+b}}{x^2}\,dx = -\dfrac{\sqrt{ax+b}}{x} + \dfrac{a}{2}\int \dfrac{dx}{x\sqrt{ax+b}} + C$

15. $\int \dfrac{dx}{x^2\sqrt{ax+b}} = -\dfrac{\sqrt{ax+b}}{bx} - \dfrac{a}{2b}\int \dfrac{dx}{x\sqrt{ax+b}} + C$

16. $\int \dfrac{dx}{a^2+x^2} = \dfrac{1}{a}\tan^{-1}\dfrac{x}{a} + C$

17. $\int \dfrac{dx}{(a^2+x^2)^2} = \dfrac{x}{2a^2(a^2+x^2)} + \dfrac{1}{2a^3}\tan^{-1}\dfrac{x}{a} + C$

18. $\int \dfrac{dx}{a^2-x^2} = \dfrac{1}{2a}\ln\left|\dfrac{x+a}{x-a}\right| + C$

19. $\int \dfrac{dx}{(a^2-x^2)^2} = \dfrac{x}{2a^2(a^2-x^2)} + \dfrac{1}{2a^2}\int \dfrac{dx}{a^2-x^2}$

20. $\int \dfrac{dx}{\sqrt{a^2+x^2}} = \ln|x+\sqrt{a^2+x^2}| + C$

21. $\int \sqrt{a^2+x^2}\,dx = \dfrac{x}{2}\sqrt{a^2+x^2} + \dfrac{a^2}{2}\ln|x+\sqrt{a^2+x^2}| + C$

22. $\int x^2\sqrt{a^2+x^2}\,dx = \dfrac{x(a^2+2x^2)\sqrt{a^2+x^2}}{8} - \dfrac{a^4}{8}\ln|x+\sqrt{a^2+x^2}| + C$

23. $\int \dfrac{\sqrt{a^2+x^2}}{x}\,dx = \sqrt{a^2+x^2} - a\sinh^{-1}\left|\dfrac{a}{x}\right| + C$

24. $\int \dfrac{\sqrt{a^2+x^2}}{x^2}\,dx = \ln|x+\sqrt{a^2+x^2}| - \dfrac{\sqrt{a^2+x^2}}{x} + C$

25. $\int \dfrac{x^2}{\sqrt{a^2+x^2}}\,dx = -\dfrac{a^2}{2}\ln|x+\sqrt{a^2+x^2}| + \dfrac{x\sqrt{a^2+x^2}}{2} + C$

26. $\int \dfrac{dx}{x\sqrt{a^2+x^2}} = -\dfrac{1}{a}\ln\left|\dfrac{a+\sqrt{a^2+x^2}}{x}\right| + C$

27. $\int \dfrac{dx}{x^2\sqrt{a^2+x^2}} = -\dfrac{\sqrt{a^2+x^2}}{a^2 x} + C$

28. $\int \dfrac{dx}{\sqrt{a^2-x^2}} = \sin^{-1}\dfrac{x}{a} + C$

29. $\int \sqrt{a^2-x^2}\,dx = \dfrac{x}{2}\sqrt{a^2-x^2} + \dfrac{a^2}{2}\sin^{-1}\dfrac{x}{a} + C$

30. $\int x^2\sqrt{a^2-x^2}\,dx = \dfrac{a^4}{8}\sin^{-1}\dfrac{x}{a} - \dfrac{1}{8}x\sqrt{a^2-x^2}\,(a^2-2x^2) + C$

31. $\int \dfrac{\sqrt{a^2-x^2}}{x}\,dx = \sqrt{a^2-x^2} - a\ln\left|\dfrac{a+\sqrt{a^2-x^2}}{x}\right| + C$

32. $\int \dfrac{\sqrt{a^2-x^2}}{x^2}\,dx = -\sin^{-1}\dfrac{x}{a} - \dfrac{\sqrt{a^2-x^2}}{x} + C$

33. $\int \dfrac{x^2}{\sqrt{a^2-x^2}}\,dx = \dfrac{a^2}{2}\sin^{-1}\dfrac{x}{a} - \dfrac{1}{2}x\sqrt{a^2-x^2} + C$

34. $\int \dfrac{dx}{x\sqrt{a^2-x^2}} = -\dfrac{1}{a}\ln\left|\dfrac{a+\sqrt{a^2-x^2}}{x}\right| + C$

Table 8 (*continued*)

35. $\int \dfrac{dx}{x^2\sqrt{a^2-x^2}} = -\dfrac{\sqrt{a^2-x^2}}{a^2 x} + C$

36. $\int \dfrac{dx}{\sqrt{x^2-a^2}} = \ln\left|x+\sqrt{x^2-a^2}\right| + C$

37. $\int \sqrt{x^2-a^2}\, dx = \dfrac{x}{2}\sqrt{x^2-a^2} - \dfrac{a^2}{2}\ln\left|x+\sqrt{x^2-a^2}\right| + C$

38. $\int (\sqrt{x^2-a^2})^n\, dx = \dfrac{x(\sqrt{x^2-a^2})^n}{n+1} - \dfrac{na^2}{n+1}\int (\sqrt{x^2-a^2})^{n-2}\, dx, \quad n \neq -1$

39. $\int \dfrac{dx}{(\sqrt{x^2-a^2})^n} = \dfrac{x(\sqrt{x^2-a^2})^{2-n}}{(2-n)a^2} - \dfrac{n-3}{(n-2)a^2}\int \dfrac{dx}{(\sqrt{x^2-a^2})^{n-2}}, \quad n \neq 2$

40. $\int x(\sqrt{x^2-a^2})^n\, dx = \dfrac{(\sqrt{x^2-a^2})^{n+2}}{n+2} + C, \quad n \neq -2$

41. $\int x^2\sqrt{x^2-a^2}\, dx = \dfrac{x}{8}(2x^2-a^2)\sqrt{x^2-a^2} - \dfrac{a^4}{8}\ln\left|x+\sqrt{x^2-a^2}\right| + C$

42. $\int \dfrac{\sqrt{x^2-a^2}}{x}\, dx = \sqrt{x^2-a^2} - a\sec^{-1}\left|\dfrac{x}{a}\right| + C$

43. $\int \dfrac{\sqrt{x^2-a^2}}{x^2}\, dx = \ln\left|x+\sqrt{x^2-a^2}\right| - \dfrac{\sqrt{x^2-a^2}}{x} + C$

44. $\int \dfrac{x^2}{\sqrt{x^2-a^2}}\, dx = \dfrac{a^2}{2}\ln\left|x+\sqrt{x^2-a^2}\right| + \dfrac{x}{2}\sqrt{x^2-a^2} + C$

45. $\int \dfrac{dx}{x\sqrt{x^2-a^2}} = \dfrac{1}{a}\sec^{-1}\left|\dfrac{x}{a}\right| + C = \dfrac{1}{a}\cos^{-1}\left|\dfrac{a}{x}\right| + C$

46. $\int \dfrac{dx}{x^2\sqrt{x^2-a^2}} = \dfrac{\sqrt{x^2-a^2}}{a^2 x} + C$

47. $\int \dfrac{dx}{\sqrt{2ax-x^2}} = \sin^{-1}\left(\dfrac{x-a}{a}\right) + C$

48. $\int \sqrt{2ax-x^2}\, dx = \dfrac{x-a}{2}\sqrt{2ax-x^2} + \dfrac{a^2}{2}\sin^{-1}\left(\dfrac{x-a}{a}\right) + C$

49. $\int (\sqrt{2ax-x^2})^n\, dx = \dfrac{(x-a)(\sqrt{2ax-x^2})^n}{n+1} + \dfrac{na^2}{n+1}\int (\sqrt{2ax-x^2})^{n-2}\, dx$

50. $\int \dfrac{dx}{(\sqrt{2ax-x^2})^n} = \dfrac{(x-a)(\sqrt{2ax-x^2})^{2-n}}{(n-2)a^2} + \dfrac{n-3}{(n-2)a^2}\int \dfrac{dx}{(\sqrt{2ax-x^2})^{n-2}}$

51. $\int x\sqrt{2ax-x^2}\, dx = \dfrac{(x+a)(2x-3a)\sqrt{2ax-x^2}}{6} + \dfrac{a^3}{2}\sin^{-1}\dfrac{x-a}{a} + C$

52. $\int \dfrac{\sqrt{2ax-x^2}}{x}\, dx = \sqrt{2ax-x^2} + a\sin^{-1}\dfrac{x-a}{a} + C$

53. $\int \dfrac{\sqrt{2ax-x^2}}{x^2}\, dx = -2\sqrt{\dfrac{2a-x}{x}} - \sin^{-1}\left(\dfrac{x-a}{a}\right) + C$

54. $\int \dfrac{x\, dx}{\sqrt{2ax-x^2}} = a\sin^{-1}\dfrac{x-a}{a} - \sqrt{2ax-x^2} + C$

55. $\int \dfrac{dx}{x\sqrt{2ax-x^2}} = -\dfrac{1}{a}\sqrt{\dfrac{2a-x}{x}} + C$

Table 8 (*continued*)

56. $\int \sin ax \, dx = -\frac{1}{a} \cos ax + C$

57. $\int \cos ax \, dx = \frac{1}{a} \sin ax + C$

58. $\int \sin^2 ax \, dx = \frac{x}{2} - \frac{\sin 2ax}{4a} + C$

59. $\int \cos^2 ax \, dx = \frac{x}{2} + \frac{\sin 2ax}{4a} + C$

60. $\int \sin^n ax \, dx = \frac{-\sin^{n-1} ax \cos ax}{na} + \frac{n-1}{n} \int \sin^{n-2} ax \, dx$

61. $\int \cos^n ax \, dx = \frac{\cos^{n-1} ax \sin ax}{na} + \frac{n-1}{n} \int \cos^{n-2} ax \, dx$

62. (a) $\int \sin ax \cos bx \, dx = -\frac{\cos (a+b)x}{2(a+b)} - \frac{\cos (a-b)x}{2(a-b)} + C, \quad a^2 \neq b^2$

    (b) $\int \sin ax \sin bx \, dx = \frac{\sin (a-b)x}{2(a-b)} - \frac{\sin (a+b)x}{2(a+b)}, \quad a^2 \neq b^2$

    (c) $\int \cos ax \cos bx \, dx = \frac{\sin (a-b)x}{2(a-b)} + \frac{\sin (a+b)x}{2(a+b)}, \quad a^2 \neq b^2$

63. $\int \sin ax \cos ax \, dx = -\frac{\cos 2ax}{4a} + C$

64. $\int \sin^n ax \cos ax \, dx = \frac{\sin^{n+1} ax}{(n+1)a} + C, \quad n \neq -1$

65. $\int \frac{\cos ax}{\sin ax} dx = \frac{1}{a} \ln |\sin ax| + C$

66. $\int \cos^n ax \sin ax \, dx = -\frac{\cos^{n+1} ax}{(n+1)a} + C, \quad n \neq -1$

67. $\int \frac{\sin ax}{\cos ax} dx = -\frac{1}{a} \ln |\cos ax| + C$

68. $\int \sin^n ax \cos^m ax \, dx = -\frac{\sin^{n-1} ax \cos^{m+1} ax}{a(m+n)} + \frac{n-1}{m+n} \int \sin^{n-2} ax \cos^m ax \, dx,$
    $n \neq -m$ (If $n = -m$, use no. 86.)

69. $\int \sin^n ax \cos^m ax \, dx = \frac{\sin^{n+1} ax \cos^{m-1} ax}{a(m+n)} + \frac{m-1}{m+n} \int \sin^n ax \cos^{m-2} ax \, dx,$
    $m \neq -n$ (If $m = -n$, use no. 87.)

70. $\int \frac{dx}{b+c \sin ax} = \frac{-2}{a\sqrt{b^2-c^2}} \tan^{-1} \left[\sqrt{\frac{b-c}{b+c}} \tan \left(\frac{\pi}{4} - \frac{ax}{2} \right) \right] + C, \quad b^2 > c^2$

71. $\int \frac{dx}{b+c \sin ax} = \frac{-1}{a\sqrt{c^2-b^2}} \ln \left| \frac{c+b \sin ax + \sqrt{c^2-b^2} \cos ax}{b+c \sin ax} \right| + C, \quad b^2 < c^2$

72. $\int \frac{dx}{1+\sin ax} = -\frac{1}{a} \tan \left(\frac{\pi}{4} - \frac{ax}{2} \right) + C$

73. $\int \frac{dx}{1-\sin ax} = \frac{1}{a} \tan \left(\frac{\pi}{4} + \frac{ax}{2} \right) + C$

74. $\int \frac{dx}{b+c \cos ax} = \frac{2}{a\sqrt{b^2-c^2}} \tan^{-1} \left[\sqrt{\frac{b-c}{b+c}} \tan \frac{ax}{2} \right] + C, \quad b^2 > c^2$

Table 8 (*continued*)

75. $\int \dfrac{dx}{b + c \cos ax} = \dfrac{1}{a\sqrt{c^2 - b^2}} \ln \left| \dfrac{c + b \cos ax + \sqrt{c^2 - b^2} \sin ax}{b + c \cos ax} \right| + C, \quad b^2 < c^2$

76. $\int \dfrac{dx}{1 + \cos ax} = \dfrac{1}{a} \tan \dfrac{ax}{2} + C$

77. $\int \dfrac{dx}{1 - \cos ax} = -\dfrac{1}{a} \cot \dfrac{ax}{2} + C$

78. $\int x \sin ax \, dx = \dfrac{1}{a^2} \sin ax - \dfrac{x}{a} \cos ax + C$

79. $\int x \cos ax \, dx = \dfrac{1}{a^2} \cos ax + \dfrac{x}{a} \sin ax + C$

80. $\int x^n \sin ax \, dx = -\dfrac{x^n}{a} \cos ax + \dfrac{n}{a} \int x^{n-1} \cos ax \, dx$

81. $\int x^n \cos ax \, dx = \dfrac{x^n}{a} \sin ax - \dfrac{n}{a} \int x^{n-1} \sin ax \, dx$

82. $\int \tan ax \, dx = -\dfrac{1}{a} \ln |\cos ax| + C$

83. $\int \cot ax \, dx = \dfrac{1}{a} \ln |\sin ax| + C$

84. $\int \tan^2 ax \, dx = \dfrac{1}{a} \tan ax - x + C$

85. $\int \cot^2 ax \, dx = -\dfrac{1}{a} \cot ax - x + C$

86. $\int \tan^n ax \, dx = \dfrac{\tan^{n-1} ax}{a(n-1)} - \int \tan^{n-2} ax \, dx, \quad n \neq 1$

87. $\int \cot^n ax \, dx = -\dfrac{\cot^{n-1} ax}{a(n-1)} - \int \cot^{n-2} ax \, dx, \quad n \neq 1$

88. $\int \sec ax \, dx = \dfrac{1}{a} \ln |\sec ax + \tan ax| + C$

89. $\int \csc ax \, dx = -\dfrac{1}{a} \ln |\csc ax + \cot ax| + C$

90. $\int \sec^2 ax \, dx = \dfrac{1}{a} \tan ax + C$

91. $\int \csc^2 ax \, dx = -\dfrac{1}{a} \cot ax + C$

92. $\int \sec^n ax \, dx = \dfrac{\sec^{n-2} ax \tan ax}{a(n-1)} + \dfrac{n-2}{n-1} \int \sec^{n-2} ax \, dx, \quad n \neq 1$

93. $\int \csc^n ax \, dx = -\dfrac{\csc^{n-2} ax \cot ax}{a(n-1)} + \dfrac{n-2}{n-1} \int \csc^{n-2} ax \, dx, \quad n \neq 1$

94. $\int \sec^n ax \tan ax \, dx = \dfrac{\sec^n ax}{na} + C, \quad n \neq 0$

95. $\int \csc^n ax \cot ax \, dx = -\dfrac{\csc^n ax}{na} + C, \quad n \neq 0$

96. $\int \sin^{-1} ax \, dx = x \sin^{-1} ax + \dfrac{1}{a} \sqrt{1 - a^2 x^2} + C$

Table 8 (*continued*)

97. $\int \cos^{-1} ax \, dx = x \cos^{-1} ax - \frac{1}{a}\sqrt{1-a^2x^2} + C$

98. $\int \tan^{-1} ax \, dx = x \tan^{-1} ax - \frac{1}{2a} \ln(1+a^2x^2) + C$

99. $\int x^n \sin^{-1} ax \, dx = \frac{x^{n+1}}{n+1} \sin^{-1} ax - \frac{a}{n+1} \int \frac{x^{n+1} \, dx}{\sqrt{1-a^2x^2}}, \quad n \neq -1$

100. $\int x^n \cos^{-1} ax \, dx = \frac{x^{n+1}}{n+1} \cos^{-1} ax + \frac{a}{n+1} \int \frac{x^{n+1} \, dx}{\sqrt{1-a^2x^2}}, \quad n \neq -1$

101. $\int x^n \tan^{-1} ax \, dx = \frac{x^{n+1}}{n+1} \tan^{-1} ax - \frac{a}{n+1} \int \frac{x^{n+1} \, dx}{1+a^2x^2}, \quad n \neq -1$

102. $\int e^{ax} \, dx = \frac{1}{a} e^{ax} + C$

103. $\int b^{ax} \, dx = \frac{1}{a \ln b} b^{ax} + C, \quad b > 0, \, b \neq 1$

104. $\int xe^{ax} \, dx = \frac{e^{ax}}{a^2}(ax-1) + C$

105. $\int x^n e^{ax} \, dx = \frac{1}{a} x^n e^{ax} - \frac{n}{a} \int x^{n-1} e^{ax} \, dx$

106. $\int x^n b^{ax} \, dx = \frac{x^n b^{ax}}{a \ln b} - \frac{n}{a \ln b} \int x^{n-1} b^{ax} \, dx, \quad b > 0, \, b \neq 1$

107. $\int e^{ax} \sin bx \, dx = \frac{e^{ax}}{a^2 + b^2}(a \sin bx - b \cos bx) + C$

108. $\int e^{ax} \cos bx \, dx = \frac{e^{ax}}{a^2 + b^2}(a \cos bx + b \sin bx) + C$

109. $\int \ln ax \, dx = x \ln ax - x + C$

110. $\int x^n \ln ax \, dx = \frac{x^{n+1}}{n+1} \ln ax - \frac{x^{n+1}}{(n+1)^2} + C, \quad n \neq -1$

111. $\int x^{-1} \ln ax \, dx = \frac{1}{2}(\ln ax)^2 + C$

112. $\int \frac{dx}{x \ln ax} = \ln|\ln ax| + C$

Formulas 113–138 involve *hyperbolic functions* whose definition is as follows:

$\sinh x = \frac{1}{2}(e^x - e^{-x})$

$\cosh x = \frac{1}{2}(e^x + e^{-x})$

$\tanh x = \frac{\sinh x}{\cosh x} = \frac{e^x - e^{-x}}{e^x + e^{-x}}$

$\coth x = \frac{\cosh x}{\sinh x} = \frac{e^x + e^{-x}}{e^x - e^{-x}}$

Table 8 (*continued*)

$$\operatorname{sech} x = \frac{1}{\cosh x} = \frac{2}{e^x + e^{-x}}$$

$$\operatorname{csch} x = \frac{1}{\sinh x} = \frac{2}{e^x - e^{-x}}$$

113. $\displaystyle\int \sinh ax\, dx = \frac{1}{a} \cosh ax + C$

114. $\displaystyle\int \cosh ax\, dx = \frac{1}{a} \sinh ax + C$

115. $\displaystyle\int \sinh^2 ax\, dx = \frac{\sinh 2ax}{4a} - \frac{x}{2} + C$

116. $\displaystyle\int \cosh^2 ax\, dx = \frac{\sinh 2ax}{4a} + \frac{x}{2} + C$

117. $\displaystyle\int \sinh^n ax\, dx = \frac{\sinh^{n-1} ax \cosh ax}{na} - \frac{n-1}{n} \int \sinh^{n-2} ax\, dx, \quad n \neq 0$

118. $\displaystyle\int \cosh^n ax\, dx = \frac{\cosh^{n-1} ax \sinh ax}{na} + \frac{n-1}{n} \int \cosh^{n-2} ax\, dx, \quad n \neq 0$

119. $\displaystyle\int x \sinh ax\, dx = \frac{x}{a} \cosh ax - \frac{1}{a^2} \sinh ax + C$

120. $\displaystyle\int x \cosh ax\, dx = \frac{x}{a} \sinh ax - \frac{1}{a^2} \cosh ax + C$

121. $\displaystyle\int x^n \sinh ax\, dx = \frac{x^n}{a} \cosh ax - \frac{n}{a} \int x^{n-1} \cosh ax\, dx$

122. $\displaystyle\int x^n \cosh ax\, dx = \frac{x^n}{a} \sinh ax - \frac{n}{a} \int x^{n-1} \sinh ax\, dx$

123. $\displaystyle\int \tanh ax\, dx = \frac{1}{a} \ln (\cosh ax) + C$

124. $\displaystyle\int \coth ax\, dx = \frac{1}{a} \ln |\sinh ax| + C$

125. $\displaystyle\int \tanh^2 ax\, dx = x - \frac{1}{a} \tanh ax + C$

126. $\displaystyle\int \coth^2 ax\, dx = x - \frac{1}{a} \coth ax + C$

127. $\displaystyle\int \tanh^n ax\, dx = -\frac{\tanh^{n-1} ax}{(n-1)a} + \int \tanh^{n-2} ax\, dx, \quad n \neq 1$

128. $\displaystyle\int \coth^n ax\, dx = -\frac{\coth^{n-1} ax}{(n-1)a} + \int \coth^{n-2} ax\, dx, \quad n \neq 1$

129. $\displaystyle\int \operatorname{sech} ax\, dx = \frac{1}{a} \sin^{-1}(\tanh ax) + C$

130. $\displaystyle\int \operatorname{csch} ax\, dx = \frac{1}{a} \ln \left|\tanh \frac{ax}{2}\right| + C$

131. $\displaystyle\int \operatorname{sech}^2 ax\, dx = \frac{1}{a} \tanh ax + C$

Table 8 (*continued*)

132. $\int \text{csch}^2\, ax\, dx = -\dfrac{1}{a} \coth ax + C$

133. $\int \text{sech}^n\, ax\, dx = \dfrac{\text{sech}^{n-2} ax \tanh ax}{(n-1)a} + \dfrac{n-2}{n-1} \int \text{sech}^{n-2} ax\, dx, \quad n \neq 1$

134. $\int \text{csch}^n\, ax\, dx = -\dfrac{\text{csch}^{n-2} ax \coth ax}{(n-1)a} - \dfrac{n-2}{n-1} \int \text{csch}^{n-2} ax\, dx, \quad n \neq 1$

135. $\int \text{sech}^n ax \tanh ax\, dx = -\dfrac{\text{sech}^n ax}{na} + C, \quad n \neq 0$

136. $\int \text{csch}^n ax \coth ax\, dx = -\dfrac{\text{csch}^n ax}{na} + C, \quad n \neq 0$

137. $\int e^{ax} \sinh bx\, dx = \dfrac{e^{ax}}{2}\left[\dfrac{e^{bx}}{a+b} - \dfrac{e^{-bx}}{a-b}\right] + C, \quad a^2 \neq b^2$

138. $\int e^{ax} \cosh bx\, dx = \dfrac{e^{ax}}{2}\left[\dfrac{e^{bx}}{a+b} + \dfrac{e^{-bx}}{a-b}\right] + C, \quad a^2 \neq b^2$

ANSWERS TO ODD-NUMBERED EXERCISES

SECTION 1.1 **1.** (a) $=$ (b) $=$ (c) $\neq$ (d) $=$ (e) $\neq$ (f) $=$ (g) $=$ (h) $=$ (i) $\neq$ (j) $\neq$ (k) $=$ (l) $\neq$ (m) $\neq$ (n) $\neq$ (o) $=$
3. (a) 1×10^{-1} (b) 1×10^{-2} (c) 9.09×10^{-2} (d) 9.09×10^{-2} (e) 1.2×10^{-6} (f) 1.275×10^3 (g) -1.27525×10^3 (h) 1×10^8 (i) -5.00001×10^{-1} (j) 1.0000001×10 (k) 1.2×10^{-2} (l) 1.1×10^5 (m) 1.275×10^{11} (n) 2.5×10^{-1} (o) 4×10^{-5} (p) 7.5×10^{-4} (q) 8×10^7

SECTION 1.2 **1.** (a) T (b) T (c) T (d) T (e) F: take $a = 1$ (f) T (g) T (h) T (i) F: take $a = 0$ and $b = -1$ (j) F: take $a = 2$ and $b = -1$ (k) T (l) T (m) T (n) T (o) F: take $a = 1$ and $b = 2$ (p) F: take $a = b = 0$ (q) T (r) F: take $a = -1$ (s) F: take $t = 4$ (t) T (u) T (v) F: take $a = 2$ and $b = -1$
3. (a) $\geq$ (b) $\geq$ (c) $\geq$ (d) $=$ (e) $=$ (f) $=$ (g) $\geq$ (h) $\geq$ (i) $\leq$ (j) $=$ (k) $\leq$ (l) $\leq$

SECTION 1.3 **1.** (b)

(c)

(d)  **(g)** ●————●
                                    -1   1

(e) ←————————→ **(h)** ○————○
                             -3   3

(f) ○————●
        0   1

3. (a) $(-\infty, 0)$ (b) $[0, \infty)$ (c) $(-20, \infty)$
5. If b is any point in (a, ∞), it is not the largest since $b + 1$ is also in (a, ∞).
7. $(-1, 2)$
9. (a) F: take $x = -8$ (b) T (c) T (d) T (e) T
   (f) F: take $t = u = 3$

SECTION 1.4 1. (a) ○————○——— (d) ←————————○
                        $\frac{1}{2}$ $\frac{3}{2}$ 0

(b) ○————○ (e) ○————○
    $-\frac{1}{4}$ $\frac{1}{4}$ 1   3

(c) ●————● (f)
    -1   2 $\frac{3}{2}$ $\frac{5}{2}$

3. (a) $(-\infty, 0)$ (b) $(-\infty, \infty)$ (c) $(-\infty, 5]$ (d) $[0, \infty)$ (e) $(-5, \infty)$
5. (a) F: take $t = 0$ (b) F: take $t = 5.001$ (c) T (d) T
   (e) F: take $t = 7 + a$ (f) T (g) T (h) T (i) F: take $t = 4$

SECTION 1.5 1. $A = (-1, -1)$ is in the third quadrant. $B = (1, 0)$ is on the positive x axis. $C = (0, 1)$ is on the positive y axis. $D = (1, -1)$ is in the fourth quadrant. $E = (0, 0)$ is the origin, the intersection of the x and y axes.
3. $(2, 3)$ and $(1, 4)$
5. $(1, 2)$ is one choice; another choice is $(2, 3)$.
7. (a) $x = -3$ (b) $x = -3$ (c) $y = -x - 7$

SECTION 1.6 1. (a) $3\sqrt{2}$ (b) $6\sqrt{2}$ (c) $\sqrt{2}$ (d) $\sqrt{5}$ (e) $\sqrt{5}$ (f) $|a + b|\sqrt{2}$
   (g) $4\sqrt{x_0^2 + y_0^2}$
3. (a) $(x - 4)^2 + (y + 3)^2 = 98$
   (b) $(x + 1)^2 + (y - 1)^2 = 8$
   (c) $x^2 + (y - 9)^2 = 81$
5. The square in the middle has sides of length $(a - b)$ and thus has area $(a - b)^2$. The four triangles have total area $2ab$. Thus the sum of the areas of these pieces is

$$(a-b)^2 + 2ab = a^2 + b^2$$

But the area of the whole square is c^2. Thus $c^2 = a^2 + b^2$.

QUIZ 1.1 1. = 2. ≠ 3. ≠ 4. ≠ 5. = 6. ≠ 7. = 8. =

9. ←——○———————○——→
      $\frac{-\sqrt{5}-1}{2}$ $\frac{\sqrt{5}-1}{2}$

10. ———●———————●———
        0   1

11. $(x-2)^2 + (y-3)^2 = 5^2$
12. There are two such squares, one with vertices $(-3, 0)$ and $(0, -3)$, and the other with vertices $(6, 3)$ and $(3, 6)$.

SECTION 2.1

1. For $f(x) = 1/x$, $f(\frac{1}{2}) = 2$, $f(-\frac{1}{2}) = -2$, $f(2) = \frac{1}{2}$, and $f(-2) = -\frac{1}{2}$. The domain and range consists of all nonzero numbers.
3. For $g(t) = t - |t|$, $g(0) = 0$, $g(1) = 0$, $g(-1) = -2$, and $g(-\frac{1}{2}) = -1$. The domain is $(-\infty, \infty)$ and the range is $(-\infty, 0]$.
5. For $f(x) = |x-1|$, $f(a) = |a-1|$, $f(a+1) = |a|$, $f(1) = 0$, and $f(-1) = 2$. The domain is $(-\infty, \infty)$ and the range is $[0, \infty)$.
7. For $m(x) = x^3$, $m(2) - m(-2) = 16$, $m(2) + m(-2) = 0$, and $m(3^{1/2}) = 3^{3/2}$. The domain and range are both $(-\infty, \infty)$.
9. For $f(x) = x^3 - x$, $f(1) = 0$, $f(-1) = 0$, and $f(a) = a^{3/2} - a^{1/2}$. The domain and range are both $(-\infty, \infty)$.
11. $v = f(x) = x^3$. $f(1) = 1$ and $f(2) = 8$.
13. $s = f(t) = 10t^2$. $f(1) = 10$ and $f(\frac{1}{2}) = \frac{5}{2}$.
15. The function $f(x)$ is not defined for $x = 1$ and the function $g(x)$ is defined there.
17. $-\frac{5}{13} \cdot 32$ degrees Fahrenheit corresponds to $-\frac{10}{13} \cdot 32$ degrees Celsius.

SECTION 2.2

1.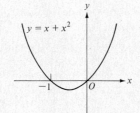

3.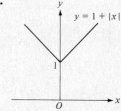

5.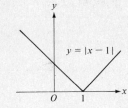

7.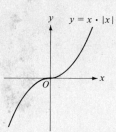

9. $y = \begin{cases} -1 & \text{for } -1 \leq x \leq 0 \\ 1 & \text{for } 0 < x \leq 1 \end{cases}$
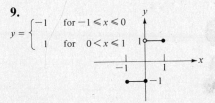

11.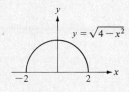

13. Even
15. Neither
17. Odd
19. Even
21. Odd
23. Yes: $L(x) = b$, where b is any constant.
25. For any even integer n.
27. Yes, because if $f(-x) = f(x)$, then $5f(-x) = 5f(x)$.
29. No, the function $|h|$ is even; for example, take $h(x) = x$. On the other hand, if h is even, then $|h|$ is also even.
31. $b = d = 0$, but a and c can be arbitrary numbers.

SECTION 2.3

1. (a) -1 (b) -1 (c) 1 (d) m is undefined (e) 0 (f) $b/(b-a)$, m is undefined when $b = a$. (g) m is undefined (h) $-\frac{1}{2}$ (i) -1 (j) m (k) m (l) m

3. For $(x_0, y_0) = (-1, 0)$ and $m = \frac{1}{2}$, we have $y - 0 = \frac{1}{2}(x + 1)$ or $y = \frac{1}{2}x + \frac{1}{2}$ or $x - 2y + 1 = 0$. For $(x_0, y_0) = (0, 0)$ and $m = 1$, we have $y - 0 = 1(x - 0)$ or $y = x$ or $x - y = 0$. For $(x_0, y_0) = (0, 4)$ and $m = 10$, we have $y - 4 = 10(x - 0)$ or $y = 10x + 4$ or $10x - y + 4 = 0$. For $(x_0, y_0) = (5, 0)$ and $m = -\frac{1}{2}$, we have $y - 0 = -\frac{1}{2}(x - 5)$ or $y = -\frac{1}{2}x + \frac{5}{2}$ or $x + 2y - 5 = 0$.

5. (a) $(-1, -1)$ (b) $(1, 1)$ (c) $(0, -1)$ (d) $(-3, 4)$ (e) $(\frac{4}{7}, \frac{23}{7})$ (f) $(1, 1)$

7. Part 4. To find the line $y = mx$ perpendicular to a given line $y = Mx$, refer to Figure 2.30. $[\text{dist}(O, P)]^2 = 1 + M^2$, $[\text{dist}(O, Q)]^2 = 1 + m^2$, $[\text{dist}(P, Q)]^2 = (M - m)^2$. The equation $(1 + M^2) + (1 + m^2) = (M - m)^2$ simplifies to $Mm = -1$ and thus $m = -1/M$.

Part 5. If M and m are slopes of perpendicular lines, then $Mm = -1$.

SECTION 2.4
1. $y - \frac{1}{4} = -(x + \frac{1}{2})$
3. $y - \frac{1}{4} = (x - \frac{1}{2})$
5. No
7. $y - 1 = -2(x + 1)$ and $y - 25 = 10(x - 5)$
9. No

SECTION 2.5
1. $\frac{\Delta y}{\Delta x} = \frac{(x_0 + \Delta x)^3 - x_0^3}{\Delta x} = 3x_0^2 + 3x_0 \Delta x + (\Delta x)^2$. Thus,
$$\lim_{\Delta x \to 0} \frac{\Delta y}{\Delta x} = 3x_0^2.$$
3. It is the line $y = 0$. It crosses the curve.
5. No
7. $\frac{\Delta y}{\Delta x} = \frac{[1/(x_0 + \Delta x)] - (1/x_0)}{\Delta x} = -\frac{1}{(x_0 + \Delta x)x_0}$. Thus,
$$\lim_{\Delta x \to 0} \frac{\Delta y}{\Delta x} = -\frac{1}{x_0^2}.$$
9. $(\sqrt{2}, 1/\sqrt{2})$ and $(-\sqrt{2}, -1/\sqrt{2})$
11. No
13.

Δx	$\Delta y / \Delta x$
0.1	1
0.01	1
-0.01	-1
-0.1	-1

15.

Δx	$\Delta y / \Delta x$
0.1	1
0.01	1
-0.01	0
-0.1	0

17. $\frac{\Delta y}{\Delta x} = 5$ and thus $\lim_{\Delta x \to 0} \frac{\Delta y}{\Delta x} = 5$. The tangent is $y = 5x + 3$.

SECTION 2.6
1. 1
3. 5
5. 0
7. -2
9. 1
11. 2
13. $\frac{1}{6}$
15. $-\frac{1}{2}$
17. 0
19. 1
21. 1
23. 0
25. 1
27. -1

29. 0
31. 1
33. −1

QUIZ 2.1

1. The domain consists of all points except −1 and 1; the range consists of all points except 0.
2.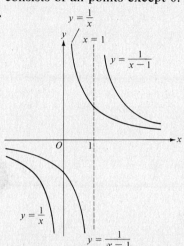
3. (a) even (b) odd (c) neither
4. $(-\frac{1}{3}, \frac{4}{3})$
5. $y = 5x - 4$
6. $y - 4 = -4(x + 2)$
7. (a) 32 (b) $\frac{1}{6}$ (c) $\frac{2}{3}$

SECTION 3.1

1. $10x^9$
3. $\frac{1}{3}x^{-2/3}$
5. $-5z^{-6}$
7. $1.1u^{0.1}$
9. $-100x^{-101}$
11. $f'(5) = 0$ $f'(-5) = 0$
13. $g'(1) = \frac{1}{4}$, $g'(5) = \frac{1}{4} \times 5^{-3/4}$
15. $f'(0) = 0$, $f'\left(-\frac{1}{2}\right) = \frac{11}{2^{10}}$
17. The tangent to the curve $y = c$ is just $y = c$, which has slope 0.
19. $y - \sqrt{2} = \frac{\sqrt{2}}{4}(x - 2)$

SECTION 3.2

1. 0
3. $4x^3 + 2$
5. $\dfrac{1}{a+b} - \dfrac{2x}{a-b}$
7. $2ax + \dfrac{1}{a}$

9. $\dfrac{1}{\sqrt{2t}}$

11. $1 - \dfrac{1}{u^2}$

13. $3ax^2 + 2bx + c$

15. $\dfrac{1}{x^2} - \dfrac{2}{x^3} + \dfrac{3}{x^4} - \dfrac{4}{x^5}$

17. $y = 2x$ and $y = -2x$

19. $(-1, \tfrac{17}{3})$ and $(1, \tfrac{13}{3})$

SECTION 3.3

1. $1 - \dfrac{2}{x^2}$
3. $2(2x^3 - x - 1)$
5. $3x^2 - 6x + 2$
7. $\dfrac{ad - bc}{(cz + d)^2}$
9. $\dfrac{-2}{(1 + x)^2}$
11. $2x + 1$
13. $\tfrac{1}{3}(10x^{7/3} + 8x^{5/3} - x^{-2/3} + x^{-4/3})$
15. $\dfrac{x^{-4/3}}{3(1 + x^{-1/3})^2} = \tfrac{1}{3}x^{-2/3}(x^{1/3} + 1)^{-2}$
17. $\dfrac{g(x) - xg'(x)}{[g(x)]^2}$
19. $vwz\dfrac{du}{dx} + uwz\dfrac{dv}{dx} + uvz\dfrac{dw}{dx} + uvw\dfrac{dz}{dx}$
21. $y - \tfrac{1}{3} = \tfrac{2}{9}(x - 2)$

SECTION 3.4

1. $-\tfrac{1}{2}x^{-3/2}$
3. $4x(x^2 - 1)^2(x^2 + 1)^4(4x^2 - 1)$
5. $3[x(x - 1)(x - 2)]^2(3x^2 - 6x + 2)$
7. $x^4(x^2 - 5)^6(19x^2 - 25)$
9. $\tfrac{1}{3}\left(x + \tfrac{1}{x}\right)^{-2/3}\left(1 - \tfrac{1}{x^2}\right) = \tfrac{1}{3}\left(\dfrac{x}{x^2 + 1}\right)^{2/3}\dfrac{x^2 - 1}{x^2}$
11. $\dfrac{1 - 4x^3}{(x^4 - x + 1)^2}$
13. $3(x + 1)^2$
15. $3[f(x)g(x)]^2[f(x)g'(x) + g(x)f'(x)]$
17. $\tfrac{1}{2}\dfrac{g(x)f'(x) - f(x)g'(x)}{[f(x)]^{1/2}[g(x)]^{3/2}}$
19. $2\sqrt{2}(2x + 1)^{\sqrt{2}-1}$
21. $-\sqrt{2}(x + 1)^{-(\sqrt{2}+1)}$
23. $-\dfrac{2(x - 1)}{(x^2 - 2x + 3)^2}$
25. $-\dfrac{2a}{(a + x)^2}$
27. $\dfrac{1 - x^2}{(x^2 + 1)^2}$

29. $2ax^3 \dfrac{2c + bx^2}{(c + bx^2)^2} = 2ax^3\left(1 + \dfrac{c}{(c + bx^2)^2}\right)$

31. $\dfrac{2x^3(2a^2 - x^2)}{(a^2 - x^2)^2}$

33. $\tfrac{1}{3}(x^{-2/3} - x^{-4/3})$

35. $2x + \tfrac{1}{2}x^{-1/2} - \tfrac{1}{2}x^{-3/2}$

37. $\dfrac{1}{2\sqrt{x}(\sqrt{x} + 1)^2}$

39. $-\tfrac{1}{2}x^{-3/2}(1 + x)^{-1/2}$

41. $\tfrac{7}{8}(x^2 + x + 1)^{-1/8}(2x + 1)$

43. $\dfrac{1 - x^2}{(x^2 + 1)^2}$

45. $\dfrac{x}{(x^2 - 1)^{1/2}}$

47. $\dfrac{1}{2\sqrt{x + \sqrt{x + \sqrt{x}}}}\left[1 + \dfrac{1}{2\sqrt{x + \sqrt{x}}}\left(1 + \dfrac{1}{2\sqrt{x}}\right)\right]$

49. $-\dfrac{2(x - 1)}{(x - 2)^3}$

51. $\dfrac{1}{(1 - x^2)^{3/2}}$

53. $(1, 2)$

SECTION 3.5

1. $y' = 2x,\ y'' = 2$
3. $y' = x^4(x + 1)^5(11x + 5),\ y'' = 10x^3(x + 1)^4[11x^2 + 10x + 2]$
5. $y' = 10x^9 - 7x^6 + 3x^2,\ y'' = 90x^8 - 42x^5 + 6x$
7. $y' = 5,\ y'' = 0$
9. $y' = 2f(x)f'(x),\ y'' = 2\{f(x)f''(x) + [f'(x)]^2\}$
11. $y' = xf'(x) + f(x),\ y'' = 2f'(x) + xf''(x)$
13. $v = \dfrac{ds}{dt} = 2(t - 1),\ a = \dfrac{d^2s}{dt^2} = 2$
15. $v = \dfrac{ds}{dt} = 2gt + b,\ a = \dfrac{d^2s}{dt^2} = 2g$
17. (a) $v = 3t^2 - 2t,\ a = 6t - 2$ (b) $a = 0$ when $t = \tfrac{1}{3}$
   (c) The velocity is minimum when $t = \tfrac{1}{3}$. The velocity and acceleration are maximum when $t = 10$.

SECTION 3.6

1. $-\dfrac{1}{(x + 1)^{1/2}(x - 1)^{3/2}}$

3. $-\dfrac{2x}{(2 - x^2)^2}$

5. $\dfrac{2x}{3}(x^2 + 1)^{-2/3}[1 - (x^2 + 1)^{-2/3}]$

7. $\dfrac{1}{2x^2\sqrt{x^2-1}}$

9. $\tfrac{1}{9}(1+x^{1/3})^{-2/3}x^{-2/3}$

11. $\sqrt{p}$

13. 1

15. -1

17. $-\dfrac{x_0 b^2}{y_0 a^2}$

19. $\dfrac{y_0 - 1}{1 - x_0 + 2y_0}$

21. $\dfrac{1 - (1/x_0^2)}{2[y_0 - (1/y_0^3)]}$

23. $\dfrac{A}{4y_0 - A}$, where $A = \dfrac{1}{\sqrt{x_0 + y_0}}$

25. $\dfrac{1 - x_0 + y_0}{(x_0 - y_0)^3}$

27. $\dfrac{dy}{dx} = 0$ at $\left(\dfrac{2}{\sqrt{3}}, -\dfrac{8}{\sqrt{3}}\right)$ and $\left(-\dfrac{2}{\sqrt{3}}, \dfrac{8}{\sqrt{3}}\right)$; $\dfrac{dy}{dx}$ is undefined at $\left(\dfrac{4}{\sqrt{3}}, -\dfrac{4}{\sqrt{3}}\right)$ and $\left(-\dfrac{4}{\sqrt{3}}, \dfrac{4}{\sqrt{3}}\right)$.

QUIZ 3.1

1. (a) $y' = (8x - 1)(x - 1)^6$

   (b) $y' = -(x^4 + x^{-4})^{-2}(4x^3 - 4x^{-5}) = -\dfrac{4x^3(x^8 - 1)}{(x^8 + 1)^2}$

   (c) $y = 1 - 2(\sqrt{x} + 1)^{-1}$ and hence $y' = \dfrac{1}{\sqrt{x}(\sqrt{x} + 1)^2}$

   (d) $y' = \dfrac{2\sqrt{x} + 1}{4\sqrt{x}\sqrt{x + \sqrt{x}}}$

   (e) $y' = (x - 1)^4(x + 5)^7(13x + 17)$

   (f) $y' = \dfrac{1}{(x + 1)^2}\sqrt{\dfrac{x + 1}{x - 1}} = \dfrac{1}{\sqrt{(x + 1)^3(x - 1)}}$

   (g) $y' = x^{-1/3}(3x^2 + 1)^{-2/3}(4x^2 + \tfrac{2}{3})$

2. $y = \tfrac{1}{2}(x - 1)$

3. $f'(4) = \tfrac{17}{16}$, $f'''(4) = 6 \cdot 4^{-4} = \tfrac{3}{128}$

4. (a) $3y^2 y' + y + xy' + 4x^3 = 0$ and hence $y' = -\dfrac{4x^3 + y}{x + 3y^2}$

   (b) $y' = \dfrac{\sqrt{x + y} - \sqrt{x}}{\sqrt{x}(1 - 2\sqrt{x + y})}$

5. $f[g(x)] = \left(\dfrac{x - 2}{x + 2}\right)^2 + 4$, $g[f(x)] = \dfrac{x^2 + 2}{x^2 + 6} = 1 - \dfrac{4}{x^2 + 6}$

SECTION 4.1

1. $f'(x) = 2x - 1$ (a) $[\tfrac{1}{2}, \infty)$ (b) $(-\infty, \tfrac{1}{2}]$
   (c) absolute minimum at $x = \tfrac{1}{2}$

3. $f'(x) = 3x^2$ (a) $(-\infty, \infty)$ (b) none (c) none
5. $f'(x) = x^2 + x - 6$ (a) $(-\infty, -3]$ and $[2, \infty)$ (b) $[-3, 2]$
 (c) local minimum at $x = 2$; local maximum at $x = -3$
7. $f'(x) = 1 + x^{-2}$ (b) Increasing over its entire domain
9. $f'(x) = \begin{cases} -2 & \text{for } -2 < x < 0 \\ 0 & \text{for } 0 < x < 3 \end{cases}$ (b) $[-2, 0]$
 (c) absolute minimum at $x = -2$; absolute maximum at each point $0 \le x \le 3$
11. $f'(x) = \begin{cases} 1 - x^{-2} & \text{for } x < -1 \text{ or } x > 1 \\ x^{-2} - 1 & \text{for } -1 < x < 1, x \ne 0 \end{cases}$
 (a) $[1, \infty)$ and $(-\infty, -1]$ (b) $[-1, 0)$ and $(0, 1]$
 (c) absolute minimum points at $x = -1$ and $x = 1$
13. $v = h' = -t + 3$; $a = v' = h'' = -1$; maximum height is $\frac{9}{2}$ ft
15. $f(x) = \begin{cases} 0 & \text{for } -\infty < x \le 0 \\ x/2 & \text{for } 0 \le x < \infty \end{cases}$

SECTION 4.2

1.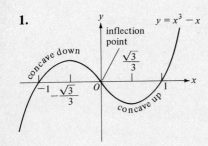

3.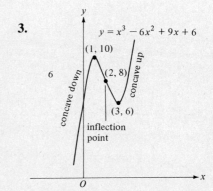

5.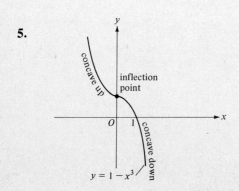

7.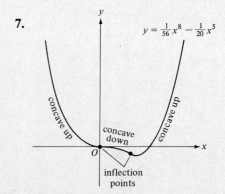

9.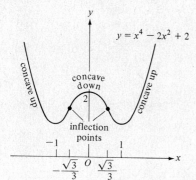

11.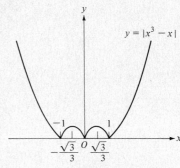

13. Same as Exercise 9.
17. There are no local maxima or minima, since $f'(x) \geq 0$ for all x.
19. There is a local minimum at $x = 0$, a local maximum at $x = -\frac{1}{2}$.
21. The only restriction is $a > 0$.
23. (a) all real numbers $m > 0$ (b) $m = 0$ (c) all real numbers $m < 0$
25. (a) all numbers m having the same sign as a (b) $m = 0$
(c) all numbers m having the opposite sign of a

SECTION 4.3

1. $y' = -\dfrac{2a}{(a+x)^2}$

3. 0

5. $\dfrac{2x^3(2c^2 - x^2)}{(c^2 - x^2)^2}$

7. $\dfrac{2}{(v+1)^2}\dfrac{dv}{dx}$

9. $2\dfrac{u(dv/dx) - v(du/dx)}{(v+u)^2}$

11. $x = 0$ and $x = 1$; $y = 1$

13. $x = 4^{1/3}$; $y = 0$
15. no vertical asymptotes; $y = 1$
17. (a) $x = 0$, $x = 3$, and $x = 5$; $y = 0$ (b) $x = 3$ and $x = 5$; $y = 0$
(c) $x = 3$ and $x = 5$; $y = 0$ (d) $x = 3$ and $x = 5$; $y = 1$
(e) $x = 3$ and $x = 5$; no horizontal asymptote

19.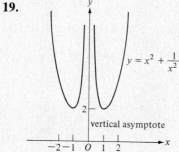

21.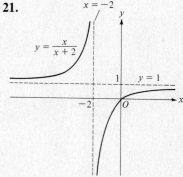

23.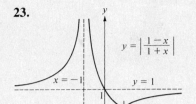
25.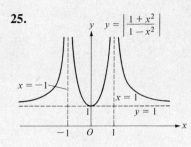

SECTION 4.4
1. $\lim_{x \to 0+} f(x) = -1$; $\lim_{x \to 0-} f(x) = 1$
3. $\lim_{x \to 0+} g(x)$ does not exist; $\lim_{x \to 0-} g(x) = 0$
5. $\lim_{x \to 0+} k(x)$ does not exist; $\lim_{x \to 0-} k(x) = 0$
7. $\lim_{x \to 0+} v(x) = \lim_{x \to 0-} v(x) = 0$
9. $f(x)$ has no limit at $x = 0$, because $\lim_{x \to 0+} f(x) = 1$ whereas $\lim_{x \to 0-} f(x) = -1$.
11. $f(x)$ has a limit at each point.

SECTION 4.5
1. $f(x)$ is discontinuous at $x = 1$, because $f(1)$ is undefined.
3. $h(x)$ is discontinuous at $x = 1$, because $h(1)$ is undefined.
5. $k(x)$ is discontinuous at $x = 0$, because $k(0)$ is undefined.
7. $f(x)$ is continuous for all x.
9. $h(x)$ is continuous for all x.
11. $u(x)$ is continuous for all x.
13. $M = 2$
15. $M = 27$
17. $M = 2$
19. The number 2 serves as upper and lower bound.
21. A lower bound is 0; there is no upper bound.
23. An upper bound is 0; there is no lower bound.
25. An upper bound is 0; a lower bound is -16.
27. This function is unbounded above and below.

SECTION 4.6
1. Theorem does not apply, because $f(x)$ has no derivative at $x = 0$.
3. Theorem does not apply, because $f(x)$ is not continuous in $[0, 1]$.
5. c can be taken to be any point in $[1, 2]$.
7. Theorem does not apply, because $f(x)$ has no derivative at $x = 0$.

9. $c = 2$

11. Theorem does not apply, because $f(x)$ is not continuous in $[-2, 1]$.

QUIZ 4.1

1. (a) Local maximum at $x = (9 - \sqrt{27})/6$; local minimum at $x = (9 + \sqrt{27})/6$. **(b)** Inflection point at $x = \frac{3}{2}$.
(c) Concave up on $(\frac{3}{2}, \infty)$; concave down on $(-\infty, \frac{3}{2})$.

2. (a) Local maximum at $x = 0$; local minimum at $x = 2$; inflection point at $x = 1$. **(b)** Concave up on $(1, \infty)$; concave down on $(-\infty, 1)$.

3. (a) The asymptotes are $x = -1$ and $y = 1$.

4. $\lim_{x \to 1+} f(x) = \lim_{x \to 1-} f(x) = 1$

5. (a) $f(x) = \begin{cases} x & \text{for } -1 < x < 1 \\ 0 & \text{for } x = -1 \text{ and } x = 1 \end{cases}$

6. $x = 2^{1/3}$

7. A lower bound is 0; an upper bound is 1.

SECTION 5.1

1. (a) When $q = 100$, $dc/dq = 300$ dollars/gallon; when $q = 10{,}000$, $dc/dq = 3{,}000$ dollars/gallon. **(b)** $f(101) - f(100) = 300.8$
(c) When $q = 100$, $\bar{c} = 203$; when $q = 10{,}000$, $\bar{c} = 2000.03$.

3. (a) Total revenue $= 100q - 0.01q^2$; marginal revenue $= 100 - 0.02q$; marginal cost $= 50$. **(b)** $q = 2500$

SECTION 5.2

1. $(\frac{10}{3})^{3/2}$ ft^3

3. $r = 1/\sqrt{6\pi}$; $h = 2/\sqrt{6\pi}$

5. $x = y = 2\sqrt{30}$ in.

7. $A = B = 5$

9. There is no maximum.

11. $x = \dfrac{\sqrt{3\pi}}{9 + \sqrt{3\pi}} L$; $L - x = \dfrac{9}{9 + \sqrt{3\pi}} L$

13. $12\sqrt{2}$

15. $\sqrt{b/a}$

17. $r = \dfrac{3}{2}$ in.; $h = \dfrac{12}{\pi}$ in.

19. 5000 tons of steel

SECTION 5.3

1. 10 ft^3/sec

3. 5 ft/min

5. $20\sqrt{202}$ mph

7. $\dfrac{\pi}{40}$ in.^{3}/sec

9. $\dfrac{4m_0}{3\sqrt{3}c}$

SECTION 5.4

1. $x_{k+1} = \frac{1}{3}\left(2x_k + \frac{b}{x_k^2}\right)$; $x_1 = 1.5$; $x_2 = 1.296$; $x_3 = 1.26093$
3. For $p(x) = x^3 - 2x^2 + 1$, $p(0) = 1$ and $p(-1) = -2$. Hence there is a root in the interval $(-1, 0)$. Since $p'(x) < 0$ for $-1 < x < 0$, this is the only root. $x_1 = -0.6363636$; $x_2 = -0.6183816$; $x_3 = -0.6180341$
5. Five more iterations ($k = 6$).

SECTION 5.5

1. $1 + \frac{3}{4}x$
3. $0.5 + (0.24 - 0.25) = 0.49$
5. $1 + \frac{2}{5} \times 0.1 = 1.04$
7. $4 + \frac{1}{8} \times 0.056 = 4.007$
9. (a) 1.01×10^{-4} (b) $\delta = 0.0951 \approx 0.1$

SECTION 5.6

1. $dy = -x^{-2}\, dx$
3. $dy = -dx$
5. $dv = 4\pi r^2\, dr$
7. $dy = (x+1)^9(11x+1)\, dx$
9. $dp = \dfrac{z}{\sqrt{z^2+1}}\, dz$
11. $dy = 2\left(x + \dfrac{1}{x}\right)\left(1 - \dfrac{1}{x^2}\right) dx$
13. $dy = 8x(x^2+1)^3\, dx$
15. $dy = (3x^2+1)\, dx$

QUIZ 5.1

1. $\dfrac{1}{4\pi}\left(\dfrac{2\pi}{15}\right)^{2/3}$
2. $\dfrac{1}{8}\left(\dfrac{2\pi}{8}\right)^{2/3}$
3. Square of sides $10\sqrt{2}$
4. $x_1 = \frac{3}{2}$; $x_2 = \frac{43}{31} \approx 1.385$
5. (a) $\dfrac{dc}{dq} = \dfrac{3}{10,000}q^2 + \dfrac{2}{5}q$; (b) for $q = 1000$, $\dfrac{dc}{dq} = 700$ dollars/ton.
6. $y \approx \frac{1}{4}(x-2)$
7. $dy = 2(x+2)^{-3/2}(x-2)^{-1/2}$

SECTION 6.1

1. $\frac{8}{3}$
3. $\frac{16}{3}$
5. $\frac{32}{3}$
7. $\frac{26}{3}$
9. 2

11. 0; the areas cancel out.
13. 0; the areas cancel out.
15. Not zero; entire area lies below the x axis.

SECTION 6.2
1. $\sum_{i=1}^{6} y_i$
3. $\sum_{i=1}^{6} \frac{1}{i}$
5. $\sum_{i=1}^{7} k^i$
7. $\sum_{i=1}^{6} x^i$
9. $\sum_{i=1}^{4} \frac{1}{x_i + 1} \Delta x$
11. $x^7 + x^8 + x^9 + x^{10}$
13. $(2 \times 1 + 1) + (2 \times 2 + 1) + (2 \times 3 + 1) + (2 \times 4 + 1) + (2 \times 5 + 1) = 35$
15. $0 + 1 + 2 + 3 + 4 + 5$
17. $x_2 - x_3 + x_4 - x_5 + x_6 - x_7 + x_8$

SECTION 6.3
1. $\frac{2}{3}$
3. 2
5. $\frac{9}{2}$
7. $-\frac{8}{3}$
9. x
11. $\frac{1}{3}(b^3 - a^3)$
13. $\frac{8\sqrt{2}}{3}$
15. Follow the hint.

SECTION 6.4
1. $\frac{14}{3}$
3. $\frac{1}{101} b^{101}$
5. $\frac{1}{4}(b^4 - a^4) - \frac{1}{2}(b^2 - a^2) = \frac{1}{4}(b^2 - a^2)(b^2 + a^2 - 2)$
7. $a_2\left(1 - \frac{1}{x}\right) + \frac{1}{2} a_3 \left(1 - \frac{1}{x^2}\right) + \frac{1}{3} a_4 \left(1 - \frac{1}{x^3}\right)$
9. 0
11. $2(\sqrt{x} - 1)$

SECTION 6.5
1. 0
3. $\frac{1}{a} - \frac{1}{b}$
5. $\frac{8}{9}$
7. 50
9. $\frac{15}{8}$
11. $f(b) - f(a)$
13. $\frac{1}{6}\{[f(b)]^6 - [f(a)]^6\}$

SECTION 6.6
1. $a_0 x + a_1 \frac{x^2}{2} + a_2 \frac{x^3}{3} + \cdots + a_k \frac{x^{k+1}}{k+1} + \cdots + a_n \frac{x^{n+1}}{n+1} + C$
3. $-\frac{x^{-4}}{4} + C$
5. $-\frac{(x+2)^{-6}}{6} + C$

7. $-\dfrac{1}{t} + C$

9. $\dfrac{(2x+1)^{11}}{22} + C$

11. $\dfrac{x^3}{3} - x^2 + x - 1$

13. $\left(1 + \dfrac{1}{x}\right)^3 - 8$

15. $\dfrac{15}{64}$

SECTION 6.7

1. $\dfrac{\sqrt{2}}{2}(x^2 + 1)^2 + C$
3. $\dfrac{1}{3}(x^4 - x^2 + 1)^{3/2} + C$
5. $\dfrac{3}{4}(t^2 + 1)^{2/3} + C$
7. $-\dfrac{3}{10}(4 - x^2)^{5/3} + C$
9. $\dfrac{(x^2 + 2x + 4)^{51}}{102} + C$
11. $\dfrac{(x^{10} + x)^{11}}{110} + C$
13. $-\dfrac{1}{(t^3 - t + 2)} + C$
15. $-\dfrac{1}{3(x+1)^3} + \dfrac{1}{4(x+1)^4} + C$
17. $-\dfrac{1}{5(x-1)^5} + C$
19. $-\dfrac{1}{3}$
21. Take $g(x) = x - A$ in Theorem 2.
23. Take $B = -1$ in Exercise 22.
25. $\displaystyle\int_{-a}^{0} f(x)\,dx = \int_{-a}^{0} f(-x)\,dx = \int_{0}^{a} f(x)\,dx$

SECTION 6.8

1. 0
3. $\dfrac{8}{3}$
5. 2
7. 4
9. 0
11. $\dfrac{1}{2}$

QUIZ 6.1

1. (a) $\sqrt{2x+5} + C$ (b) $\dfrac{2}{27}(3x^3 + 1)^{3/2} + C$ (c) $\dfrac{3}{8}[(x-3)^2 - 3]^{4/3} + C$
   (d) $-\dfrac{1}{3(t^4 - t)^3} + C$
2. (a) 0 (b) $\dfrac{1}{28}$ (c) 16 (d) 0
3. (a) $du = \dfrac{4x}{(x^2+1)^2}\,dx$ (b) $du = \dfrac{1}{2}\left(1 + \dfrac{1}{2\sqrt{x}}\right)\dfrac{1}{\sqrt{x + \sqrt{x}}}\,dx$
4. $\dfrac{4}{3}$
5. (a) = (b) ≠ (c) ≠ (d) = (e) =

SECTION 7.1

1. -7.6009
3. -2.9958
5. 5.4381
7. -16.1181

9. 17.7275
11. $2b$
13. $\frac{1}{3}(a+b)$
15. $4b$
17. $-6a - 3b$
19. $-2b$

SECTION 7.2
1. $1 + \frac{1}{x}$
3. $(1 - \ln x)x^{-2}$
5. $\frac{2\ln x}{x}$
7. $\frac{1}{2x\sqrt{1+\ln x}}$
9. $\frac{2}{t^2 - 1}$
11. $\frac{a}{ax+b}$
13. $-[(x^2+1)^{-1/2} + (x^2-1)^{-1/2}]$
15. $\frac{n}{v}\frac{dv}{dx}$
17. $(1 + \ln u)\frac{du}{dx}$
19. $-\ln 2$
21. $(\frac{5}{4}\ln 5, \frac{5}{4}\ln 5 - 1)$
23. No, because the tangents to $y = \ln x$ always have positive slope.
25. $y = x - 1$
27. The graph of $y = \ln(x - 2)$ is obtained by translating the graph of $y = \ln x$ two units to the right. $y = x - 3$ and $y - \ln 5 = \frac{1}{5}(x - 7)$.

SECTION 7.4
1. (a) $\frac{5}{2}$ (b) 2 (c) 3 (d) $-x^2$ (e) $x^{-1}y^2$ (f) x^2 (g) 2
3. (a) ≈ 2.66 (b) ≈ 1.63 (c) ≈ 15.15 (d) ≈ 0.37 (e) ≈ 2.56
5. (a) T (b) F (c) T (d) T (e) T
7. (a) 2 (b) 2 (c) $\frac{1}{2}$ (d) 1
9. (a) 0 (b) $\frac{\ln 3 + \ln 2}{\ln 3 - \ln 2}$ (c) $\sqrt{2}$ (d) $\sqrt{2}$
11. No. The curve $y = e^x$ lies above its tangent $y = x + 1$ at $(0, 1)$, and hence it does not meet $y = x$.
13. Yes. $x = 1$ is the only solution.

SECTION 7.5
1. $(x+1)^2 e^x$
3. $-2xe^{-x^2}$
5. $e^x \exp(e^x + 1)$
7. $\frac{2e^x}{(e^x+1)^2}$
9. $\ln 10$
11. $\frac{10^x \ln 10}{2\sqrt{1+10^x}}$
13. $\frac{e^x(x+1)(\ln x + x - 1)}{(\ln x + x)^2}$
15. $2^{2x+1} \ln 2$
17. $(x+1)^{(x+1)}[1 + \ln(x+1)]$
19. $2x^{2x}(1 + \ln x)$
21. $\frac{a^{\sqrt{x}} \ln x}{4\sqrt{x}}$
23. $5^x \ln 5 + 5x^4$
25. 0.5
27. $10 \ln 10$
29. $\frac{\ln 21}{21}$
31. $375 \ln 5$
33. (a) $y - e^{-10} = e^{-10}(x + 10)$ (b) $y - e = e(x - 1)$
35. $y - \frac{2}{3} = 2[x - \frac{1}{3}\ln(\frac{2}{3})]$
37. No, because e^x has a derivative for each x.
39. $y - \frac{1}{3}\ln 5 = \ln 5[x - \frac{1}{3}\ln(\frac{1}{3}\ln 5)]$

41. $y = xe^x$

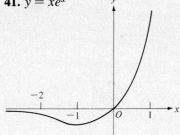

43. $y = e^x - e^{-x}$

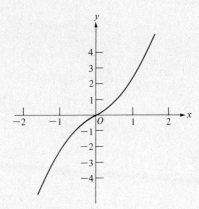

45. $y = e^{1/x}$

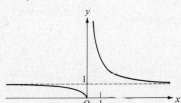

47. $y = x \ln x$

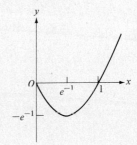

49. $-e^{-x}$

51. $\dfrac{1}{n} e^{x^n}$

53. $2(e^a - e^{-a})$

55. $\tfrac{1}{2} e + \tfrac{3}{2} e^{-1}$

57. $e^{b^2} - e^{a^2}$

SECTION 7.6

1. $(x^3 - 3x^2 + 6x - 6) e^x + C$
3. $\dfrac{x(x+5)^{21}}{21} - \dfrac{(x+5)^{22}}{21 \cdot 22} + C$
5. $x(\ln^2 x - 2 \ln x + 2) + C$
7. $-\tfrac{2}{15}(3x + 2)(1 - x)^{3/2} + C$
9. $\tfrac{13}{42}$
11. $4e^{-1}$
13. $\int \ln^n x \, dx = x \ln^n x - n \int \ln^{n-1} x \, dx$
15. Take the derivative of the right-hand side and use the fact that for any polynomial the higher derivatives all vanish from some point on.

17. $\int x^{2n}\sqrt{1-x^2}\,dx = -\dfrac{1}{2(n+1)}x^{2n-1}(1-x^2)^{3/2}$
$+\dfrac{2n-1}{2(n+1)}\int x^{2(n-1)}\sqrt{1-x^2}\,dx$

QUIZ 7.1 1. (a) $\dfrac{2}{x^2-1}$ (b) $\dfrac{1}{\sqrt{x^2+4}}$ (c) $\dfrac{1}{x^2}(1-\ln x)$ (d) $4xe^{2x^2}$ (e) $\dfrac{e^x}{e^x+1}$
(f) $4^x \ln 4 + 4x^3$ (g) $(1+xe^x)e^{e^x}$
2. (a) x^2 (b) $x+1$ (c) $2ex$
3. (a) 1.2 (b) 7.1 (c) 0.0704
4. (a) $\tfrac{1}{2}e^{4x^2+1}+C$ (b) $-\tfrac{1}{5}e^{-5x}(x+\tfrac{1}{5})+C$ (c) $\dfrac{1}{3}x^3 \ln 4x - \dfrac{1}{9}x^3 + C$
5. $\{[F(b)+1][\ln(F(b)+1)-1]\} - \{[F(a)+1][\ln(F(a)+1)-1]\}$

SECTION 8.1 1. 650 miles
3. $6(3/\pi)^{1/3}$
5. $\tfrac{10}{3}$ ft
7. 2 miles/min²
9. (a) 325 ft (b) $\tfrac{5}{4}(3+\sqrt{13})$ sec
11. $q(t) = 3t^2 + 0.02t + 1$
$Q(t) = t^3 + 0.01t^2 + t + 5$
13. $28/(q+1)$ dollars/ton
15. $m(q) = \dfrac{45}{2}q^2 - 5q - 7.5 \cdot 10^6$
$\dfrac{m(q)}{q} = \dfrac{45}{2}q - 5 - \dfrac{7.5 \cdot 10^6}{q}$

SECTION 8.2 1. 34.7 years
3. (a) All terms of the sequence equal $e^{\alpha T}$
(b) It is a geometric sequence, the nth term equals $k(e^{\alpha T})^n$, for a suitable constant k.
5. (a) $t = 0$ and $t = 4$ years (b) 7.6 years
7. Doubles in 11.55 years and trebles in 18.31 years.

SECTION 8.3 1. $0.21\pi h$ 9. $\tfrac{3}{4}\pi$
3. $\dfrac{5^3}{8}\pi$ 11. $\pi/6$
   13. 5π
5. $\tfrac{16}{3}\pi$ 15. 2.5π
7. $(\tfrac{7}{6}e^2 - \tfrac{1}{2})\pi$

SECTION 8.4 1. $\pi h^3/12$ 5. 6
3. $\tfrac{1}{5}h^{5/2}$ 7. $\tfrac{16}{15}h$

9. $\dfrac{25}{2}\pi$

11. $16(\ln 2)^2 - 16\ln 2 + 6 \approx 2.596$

SECTION 8.5
1. $\tfrac{1}{2}$
3. $\ln 2$
5. $\ln 2$
7. 1
9. -6
11. Diverges, because $\lim\limits_{x \to -\infty} xe^{-x} = -\infty$
13. 2
15. We use induction. The case $n = 1$ was verified in Exercise 2. The principle computation follows.

$$\int_0^b x^{n+1}e^{-x}\,dx = -x^{n+1}e^{-x}\Big|_0^b + (n+1)\int_0^b x^n e^{-x}\,dx$$

Letting $b \to \infty$, we have

$$\int_0^\infty x^{n+1}e^{-x}\,dx = (n+1)\int_0^\infty x^n e^{-x}\,dx$$

17. Letting $u = x^{2n-1}$ and $dv = xe^{-x^2}\,dx$, we have

$$\int_0^b x^{2n}e^{-x^2}\,dx = -\tfrac{1}{2}x^{2n-1}e^{-x^2}\Big|_0^b + \dfrac{2n-1}{2}\int_0^b x^{2n-2}e^{-x^2}\,dx$$

The result follows by letting $b \to \infty$.

SECTION 8.6
1. $\tfrac{3}{2}$
3. Diverges
5. 8
7. 3
9. 0
11. 2

SECTION 8.7
1. 8
3. 230.48
5. 157.64
7. 0.7825
9. 146.42
11. 9.3004

QUIZ 8.1
1. (a) $P(t) = 10^4 t^3 - 4t^5 + 10^6$ (b) 38.7 days (c) 50 days
2. price $= \dfrac{1}{q}\left[\dfrac{7}{4} - \dfrac{7}{2(q+2)}\right] = \dfrac{7}{4(q+2)}$
3. (a) $\tfrac{3}{4}\pi$ (b) $\tfrac{8}{21}\pi$
4. 10.517
5. (a) $\tfrac{1}{4}$ (b) Diverges (c) $2(\ln 2)^{1/2}$ (d) $\tfrac{2}{3}10^{1/2}$
6. 9.3741

SECTION 9.1

1.

degrees	270	30	295	5	π	400	$\dfrac{1}{60}$	$\dfrac{90}{\pi^2}$	60	900	$\dfrac{900}{\pi}$	$\dfrac{72{,}000}{\pi}$
radians	$\dfrac{3\pi}{2}$	$\dfrac{\pi}{6}$	$\dfrac{59\pi}{36}$	$\dfrac{\pi}{36}$	$\dfrac{\pi^2}{180}$	$\dfrac{20\pi}{9}$	$\dfrac{\pi}{10{,}800}$	$\dfrac{1}{2\pi}$	$\dfrac{\pi}{3}$	5π	5	400

3. neither; period 2π.
5. even; period π.
7. odd; period 2π.

11. 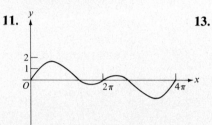   13.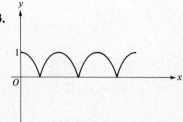

17. $\cos 2\theta = \cos^2 \theta - \sin^2 \theta = \cos^2 \theta - (1 - \cos^2 \theta) = 2\cos^2 \theta - 1$
19. Replace θ by $\theta/2$ in Exercise 17.
25. $\sin 3\theta = \sin(2\theta + \theta) = \sin 2\theta \cos \theta + \sin \theta \cos 2\theta$
    $= 2 \sin \theta \cos^2 \theta + \sin \theta (\cos^2 \theta - \sin^2 \theta)$
    $= 3 \sin \theta \cos^2 \theta - \sin^3 \theta$
27. $\cos 5x = -1$ for $x = \dfrac{\pi}{5} + \dfrac{2n\pi}{5}$; $\cos 5x = 1$ for $x = \dfrac{2n\pi}{5}$;
    $\cos 5x = 0$ for $x = \dfrac{\pi}{10} + \dfrac{n\pi}{5}$
29. Nine times

SECTION 9.2

1. $e^x \cos e^x$
3. $-\sin^{-2} x$
5. $\dfrac{\sin x}{\cos^2 x}$
7. $\cos 2x$
9. $4x^3 \cos(x^4)$
11. $-5 \sin 5x$
13. $(\sin x)^x \ln(\sin x) + x(\sin x)^{x-1} \cos x$
15. $-\dfrac{1}{x} \cos \dfrac{1}{x} + \sin \dfrac{1}{x}$
17. $\dfrac{\sin x}{\cos y}$
19. $\dfrac{1}{\cos y}$
21. $\dfrac{\sin 2x}{\sin 2y}$
23. 0
25. 3
27. $\tfrac{1}{2}$
29. $\tfrac{3}{2}$
33. Graph looks like that of $\cos x$, but with twice the amplitude.

700 ANSWERS

35. Graph looks like that of sin x, but with period $2\pi^2$.
37. $\cos x$
39. 0
41. (b) $x = \dfrac{\pi}{2} + 2k\pi$ (c) $x = -\dfrac{\pi}{2} + 2k\pi$
    (d) $(2k+1)\pi < x < (2k+2)\pi$ (e) $2k\pi < x < (2k+1)\pi$
    (f) $(2k-\tfrac{1}{2})\pi < x < (2k+\tfrac{1}{2})\pi$ (g) $(2k+\tfrac{1}{2})\pi < x < (2k+\tfrac{3}{2})\pi$

SECTION 9.3 1. $\cot x = \dfrac{\cos x}{\sin x} = \dfrac{-\cos(x+\pi)}{-\sin(x+\pi)} = \cot(x+\pi)$
3. Use the definition of the respective functions.
5. $-\dfrac{1}{x^2}\sec^2\left(\dfrac{1}{x}\right)$
7. $\cot x - x\csc^2 x$
9. $\cot x \sec^2 x$
11. $-\dfrac{x\csc^2\sqrt{x^2-1}}{\sqrt{x^2-1}}$
13. $x^{\sec x - 1}\sec x(x \ln x \tan x + 1)$
15. $-\sin^2(x+y) - 1$
17. $-\sec^2 x \sec^{-2} y$
19. $\tan(\theta_1 + \theta_2) = \dfrac{\sin(\theta_1+\theta_2)}{\cos(\theta_1+\theta_2)} = \dfrac{\sin\theta_1\cos\theta_2 + \cos\theta_1\sin\theta_2}{\cos\theta_1\cos\theta_2 - \sin\theta_1\sin\theta_2}$;
divide numerator and denominator by $\cos\theta_1\cos\theta_2$ and simplify.
21. $\tan 2\theta = \dfrac{\sin 2\theta}{\cos 2\theta} = \dfrac{2\sin\theta\cos\theta}{\cos^2\theta - \sin^2\theta}$; divide numerator and denominator by $\cos^2\theta$ and simplify.
23. $\dfrac{1-\cos\theta}{\sin\theta} = \dfrac{1-(\cos^2\theta/2 - \sin^2\theta/2)}{2\sin\theta/2\cos\theta/2} = \dfrac{2\sin^2\theta/2}{2\sin\theta/2\cos\theta/2} = \tan\dfrac{\theta}{2}$

27.
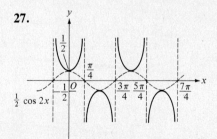

SECTION 9.4 1. $x = \dfrac{1+y}{1-y}$ $(y \neq 1)$
3. $x = \dfrac{1+y^2}{1-y^2}$ $(0 \leq y \leq 1)$
5. $y = \left(\dfrac{x+1}{x-1}\right)^2$ $(x > 1)$

7. $\dfrac{dx}{dy} = -\dfrac{1}{\sin x}$

9. No inverse exists.

11. $\dfrac{dx}{dy} = \dfrac{1}{\cos x - x \sin x}$

13. $\dfrac{dx}{dy} = \dfrac{x}{x+1}$

15. (a) $g'(\tfrac{6}{5}) = \tfrac{1}{2}$ (b) $g''(y) = -4x^3(x^4+1)^{-3}$, $g''(\tfrac{6}{5}) = -4 \cdot 1.2^3 \cdot (1 + 1.2^4)^{-3}$

SECTION 9.5

1. (a) $\pi/4$ (b) $\pi/4$ (c) π (d) $\pi/4$

3. $-\dfrac{a}{a^2 + x^2}$

5. 0

7. $-\dfrac{2}{(1+x^2)\sqrt{x^2+2}}$ $(x > 0)$

9. $-\dfrac{\cos x}{|\cos x|} = \begin{cases} -1 & \text{when } \cos x > 0 \\ 1 & \text{when } \cos x < 0 \end{cases}$

11. $-\dfrac{x}{\sqrt{1-x^2}}$

13. $-\dfrac{\csc^2 x}{\sqrt{1-(1+\cot x)^2}}$

15. $\dfrac{1}{a + b \sin x}$

17.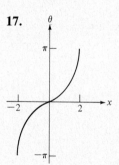

19.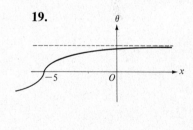

SECTION 9.6

1. -3.2×10^{-3} radians/min

3. If x is the distance along the ground between the searchlight and the vertical projection of the plane, then $x = 3000/\tan \theta$. Hence

$$\dfrac{dx}{dt} = -\dfrac{3000}{\sin^2 \theta} \dfrac{d\theta}{dt} = 2.64 \times 10^6 \text{ ft/hr}$$

When the distance from the light to the plane is 5000 ft,

$\sin \theta = 3000/5000$, and $d\theta/dt = -316.8$ radians/hr.
5. (a) $L(0) = 3$; $L(\frac{1}{2}) = 1$; $L(1) = 3$; $L(\frac{3}{2}) = 1$; $L(\frac{5}{8}) = 2 - \sqrt{2}/2$.
   (b) Velocity is $L'(t) = -2\pi \sin 2\pi t$, and hence $L'(\frac{1}{4}) = -2\pi$.
   (c) Acceleration is $L''(t) = -4\pi^2 \cos 2\pi t$, and hence $L''(\frac{1}{4}) = 0$.
   (d) $k - \frac{1}{2} < t < k$, where $k = 1, 2, 3, \ldots$.
7. We have $\theta = \tan^{-1} \dfrac{5x}{x^2 + 36}$ and $\dfrac{d\theta}{dx} = 0$ when $x = 6$, and the answer is 6 ft.
9. $\tan \theta = \sqrt[3]{1.5} \approx 1.1447$; $\theta \approx 48.85°$; $L \approx 14.03$ ft
11. $D = \text{dist}(A, P) + \text{dist}(P, B) = \sqrt{x^2 + 1} + \sqrt{x^2 - 10x + 29}$. Differentiating and setting equal to zero gives $3x^2 + 10x - 25 = 0$. The minimum is attained for $x = \frac{5}{3}$, and $\theta = \phi \approx 0.54$ radians.

SECTION 9.7

1. $-\frac{1}{7} \cos 7x + C$
3. $\frac{1}{2} \sin(2x + 1) + C$
5. $x - \frac{1}{2} \cos 2x + C$
7. $\frac{1}{4} \tan(2x^2 + 1) + C$
9. $2 \ln|\sin \sqrt{x + 1}| + C$
11. $-\frac{1}{2} \sin 2x + C$
13. $\frac{4}{3} \cos^2 x \sin x - \frac{1}{3} \sin x + C$
15. Write $\sin^n x = \sin^{n-1} x \sin x$ and use integration by parts with $u = \sin^{n-1} x$, $v' = \sin x$. This gives

$$\int \sin^n x \, dx = -\sin^{n-1} x \cos x + (n-1) \int \sin^{n-2} x \cos^2 x \, dx$$

$$= -\sin^{n-1} x \cos x + (n-1) \int \sin^{n-2} x (1 - \sin^2 x) \, dx$$

$$= -\sin^{n-1} x \cos x + (n-1) \int \sin^{n-2} x \, dx$$

$$- (n-1) \int \sin^n x \, dx$$

Solving for $\int \sin^n x \, dx$ gives the desired formula.

17. $\int \cot^n x \, dx = -\dfrac{1}{n-1} \cot^{n-1} x - \int \cot^{n-2} x \, dx$
19. $x \sin x + \cos x + C$
21. $x^2 \sin x + 2x \cos x - 2 \sin x + C$
23. $x \tan x + \ln|\cos x| + C$

SECTION 9.8

1. $\dfrac{\sqrt{2}}{2} \sin^{-1} \sqrt{2}x + C$
3. $\dfrac{\sqrt{14}}{14} \tan^{-1} \sqrt{\dfrac{2}{7}} x + C$
5. $\frac{1}{2} \ln|x^2 - 2x + 6| + \dfrac{1}{\sqrt{5}} \tan^{-1} \dfrac{x-1}{\sqrt{5}} + C$
7. $\dfrac{1}{10} \ln \left| \dfrac{x-5}{x+5} \right| + C$

9. $\frac{1}{6} \ln \left| \frac{x}{x+6} \right| + C$ 11. $\frac{1}{2} \tan^{-1}(x+1) + C$

13. $\sin^{-1} e^x + C$ (use the substitution $u = e^x$)
15. $\frac{1}{3} \ln |x^3 + 1| + C$ (use the substitution $u = x^3$)
17. $2 \sin^{-1} \sqrt{x} + C$ (use the substitution $u = \sqrt{x}$)
19. $\tan^{-1} x - \frac{1}{2} \ln \left| \frac{x-1}{x+1} \right| + C$
21. Use the substitution $u = x^2 + a$.

QUIZ 9.1

1.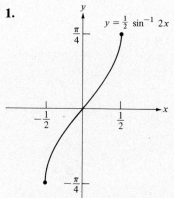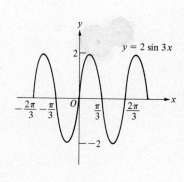

2. If $\tan(x + \beta) \sec 4(x + \beta) = \tan x \sec 4x$. When $x = 0$ we have $\tan \beta \sec 4\beta = 0$. Since $\sec 4\beta \neq 0$, we must have $\tan \beta = 0$. The smallest positive β satisfying this is $\beta = \pi$. Hence, the period is π.

3. (a) $x = (4k + 2)\pi$ (b) $x = \frac{(2k+1)\pi}{8}$

4. See Section 9.4, Exercise 19.

5. (a) $4 \cos 4x \cos \frac{x}{4} - \frac{1}{4} \sin 4x \sin \frac{x}{4}$ (b) $2 \cot x$

   (c) $x^{-2} \tan x (2x \sec^2 x - \tan x)$ (d) $\frac{1}{\sqrt{4^2 - x^2}}$ (e) $\frac{1}{1 + x^2}$

   (f) $\frac{12 \sin x}{\sqrt{1 - 144 \cos^2 x}}$ (g) $\frac{\cos x - 1}{\sin y}$ (h) $\frac{1 + y \csc^2 xy}{1 - x \csc^2 xy}$

6. (a) $\frac{1}{2} \ln |\sin(2x + 3)| + C$ (b) $\frac{1}{2}x - \frac{5}{4} \sin \frac{2}{5}x + C$
   (c) $-\ln |\cos(\ln x)| + C$
   (d) $(-\frac{1}{2}x^3 + \frac{3}{4}x) \cos 2x + (\frac{3}{4}x^2 - \frac{3}{8}) \sin 2x + C$
   (e) $\frac{1}{\sqrt{2}} \sin^{-1} \sqrt{\frac{2}{5}} x + C$ (f) $\tan^{-1} \frac{x}{2} + C$
   (g) $\frac{1}{2} \ln |x^2 - x + 1| + \frac{1}{\sqrt{3}} \tan^{-1} \frac{2}{\sqrt{3}} \left(x - \frac{1}{2} \right) + C$
   (h) $\frac{1}{2} \sin^{-1} 2x + C$

SECTION 10.1

1. $\frac{2}{5}\left(x+\frac{1}{x}\right)^{5/2} + C$

3. $\sqrt{x^2+4} + C$

5. $\frac{1}{8}e^{4x^2+2} + C$

7. $\frac{5}{6}(s^4+1)^{3/2} + C$

9. $\frac{1}{2\sqrt{3}} \ln\left|\frac{x-1-\sqrt{3}}{x-1+\sqrt{3}}\right| + C$

11. $-\frac{1}{3}\left(\frac{1}{x^2}-1\right)^{3/2} + C$

13. $\frac{1}{16}\left(\tan^{-1}\frac{x}{2} + \frac{2x}{x^2+4}\right) + C$

SECTION 10.2

1. $-\cos^{3/2} x \left(\frac{2}{3} - \frac{4}{7}\cos^2 x + \frac{2}{11}\cos^4 x\right) + C$

3. $\frac{1}{2}\sin^6 x \left(\frac{1}{3} - \frac{1}{2}\sin^2 x + \frac{1}{5}\sin^4 x\right) + C$

5. $\frac{1}{4}\sin^4 x + C$

7. $\sin^3 x \left(\frac{1}{3} - \frac{1}{5}\sin^2 x\right) + C$

9. $\frac{1}{5}\tan x \left(\frac{8}{3} + \frac{4}{3}\sec^2 x + \sec^4 x\right) + C$

11. $\frac{1}{3}\tan^3 x + C$

13. $\frac{1}{\sqrt{2}+2}\sec^{\sqrt{2}+2} x - \frac{1}{\sqrt{2}}\sec^{\sqrt{2}} x + C$

15. Follow hint.

SECTION 10.3

1. $\ln\left|\frac{1-\sqrt{1-x^2}}{x}\right| + C$

3. $-\frac{1}{3}\sqrt{9-x^2}(x^2+18) + C$

5. $-\frac{\sqrt{1-x^2}}{x} + C$

7. $\frac{\sqrt{9+x^2}}{243x}\left(2 - \frac{9}{x^2}\right) + C$

9. $-\frac{1}{2}\left[\frac{\sqrt{1-(x+2)^2}}{(x+2)^2} + \ln\left|\frac{1-\sqrt{1-(x+2)^2}}{x+2}\right|\right] + C$

11. $-\frac{1}{48}\sqrt{9-16x^2}(x^2+\frac{9}{8}) + C$

13. $\sec t = \frac{x}{a}$, $\sin t = \frac{\sqrt{x^2-a^2}}{x}$, $\cos t = \frac{a}{x}$, $\tan t = \frac{\sqrt{x^2-a^2}}{a}$, $\csc t = \frac{x}{\sqrt{x^2-a^2}}$, $\cot t = \frac{a}{\sqrt{x^2-a^2}}$

15. $\sec^{-1} x + C$

17. $\frac{1}{2}[x\sqrt{x^2-9} - 9\ln|x + \sqrt{x^2-9}|] + C$

SECTION 10.4

1. $\frac{7}{16}\ln\left|\frac{x-1}{x+3}\right| - \frac{1}{4(x+3)} + C$

3. $\ln(x^2+1) + \tan^{-1} x + \frac{1}{1-x} - 2\ln|x-1| + C$

5. $\frac{1}{4}\ln\left|\frac{x-1}{x+1}\right| + \frac{1}{2}\tan^{-1} x + C$

7. $x - \frac{4}{3}\ln|x+2| + \frac{1}{3}\ln|x-1| + C$

9. $\ln|x| - \dfrac{2}{\sqrt{3}} \tan^{-1} \dfrac{2x+1}{\sqrt{3}} + C$

11. Let $t = x + p/2$. Then $x^2 + px + q = t^2 + (q - p^2/4)$ and $Ax + B = At + (B - Ap/2)$. Let $\alpha = q - p^2/4$. Thus

$$\int \dfrac{Ax + B}{x^2 + px + q} \, dx = A \int \dfrac{t}{t^2 + \alpha} \, dt + \left(B - \dfrac{Ap}{2} \right) \int \dfrac{1}{t^2 + \alpha} \, dt$$

The result follows when $\alpha > 0$.

13. (a) $\dfrac{2x+1}{3(x^2+x+1)} + \dfrac{4}{3\sqrt{3}} \tan^{-1} \dfrac{2x+1}{\sqrt{3}} + C$

(b) $\dfrac{2x+1}{6(x^2+x+1)^2} + \dfrac{2x+1}{3(x^2+x+1)} + \dfrac{4}{3\sqrt{3}} \tan^{-1} \dfrac{2x+1}{\sqrt{3}} + C$

15. $-\dfrac{x+2}{6(x^2+x+1)^2} - \dfrac{2x+1}{6(x^2+x+1)} - \dfrac{2}{3\sqrt{3}} \tan^{-1} \dfrac{2x+1}{\sqrt{3}} + C$

SECTION 10.5

1. $\ln \left| \dfrac{1 + \tan x/2}{1 - \tan x/2} \right| + C = \ln \left| \tan \left(\dfrac{\pi}{4} + \dfrac{x}{2} \right) \right| + C$

3. $\dfrac{1}{8} \left(t^2 - \dfrac{1}{t^2} + 4 \ln|t| \right) + C$, where $t = \tan \dfrac{x}{2}$

5. $\ln \dfrac{1 + t^2}{(1 - t)^2} + C$, where $t = \tan \dfrac{x}{2}$

7. $\dfrac{2 + x^2}{\sqrt{1 + x^2}} + C$

9. $3 \tan^{-1} x^{1/3} + C$

11. $\dfrac{1}{\sqrt{2}} \tan^{-1} \left(\dfrac{\tan x}{\sqrt{2}} \right) + C$

QUIZ 10.1

1. $\dfrac{1}{9} \cos^{9/2} 2x - \dfrac{1}{5} \cos^{5/2} 2x + C$
2. $\dfrac{8}{45} \sin^{9/8} 5x + C$
3. $\dfrac{1}{4} \left(\dfrac{3}{2} x - \sin 2x + \dfrac{1}{8} \sin 4x \right) + C$
4. $\sec^3 x \left(\dfrac{1}{3} - \dfrac{2}{5} \sec^2 x + \dfrac{1}{7} \sec^4 x \right) + C$
5. $\dfrac{1}{2} \ln \left| \dfrac{2 - \sqrt{4 - x^2}}{x} \right| + C$
6. $\ln \left| \sqrt{16 + x^2} + x \right| + C$
7. $\dfrac{\sqrt{1 + x^2}(2x^2 - 1)}{3x^3} + C$
8. $\ln \left| \dfrac{x(x+4)}{x+2} \right| + C$
9. $\ln \left| \dfrac{1+x}{\sqrt{1+x^2}} \right| + \tan^{-1} x + C$
10. $\ln \left| \dfrac{x-1}{x} \right| - \dfrac{1}{x-1} + C$

SECTION 11.1

1. $\frac{1}{6}, \frac{1}{7}, \frac{1}{8}, \frac{1}{9}$
3. $-1, 2, -3, 4$
5. $-1, 2, 7, 14$
7. $\frac{2}{3}, \frac{4}{5}, \frac{6}{7}, \frac{8}{9}$
9. $1, 2, 6, 24, 120$
11. $\left\{\dfrac{1}{n+3}\right\}$
13. $\left\{\dfrac{1}{5n}\right\}$
15. $\{(-1)^{n+1}2\}$
21. 1
23. 1
25. 0

SECTION 11.2

1. Increasing, unbounded
3. Increasing, bounded
5. Increasing, unbounded
7. Neither increasing nor decreasing, unbounded
9. Neither increasing nor decreasing, bounded
11. Increasing, unbounded
13. 0
15. $\dfrac{a^2}{1-a}$
17. 1
19. $\frac{5}{3}$
21. 1

SECTION 11.3

1. The error in $\exp(10^{-6}) \approx 1 + (10)^{-6}$ is less than $(0.83)(10)^{-12}$. The error in $\exp(-10^{-6}) \approx 1 - (10)^{-6}$ is less than $(0.5)(10)^{-12}$.
3. For x in $(-0.1, 0.1)$, $\sin^{-1} x \approx x$ with error less than $\dfrac{0.05}{(0.99)^{3/2}}x^2$.

   For x in $(0.45, 0.55)$, $\sin^{-1} x \approx \pi/6 + (2/\sqrt{3})(x - \frac{1}{2})$ with error less than $\dfrac{1}{2}\dfrac{0.55}{(0.6975)^{3/2}}\left(x - \dfrac{1}{2}\right)^2$
5. $x^{10} \approx 1 + 10(x-1)$ with error less than $45(1.05)^8(x-1)^2$ for x in $(0.98, 1.05)$.
7. $\displaystyle\int_0^x e^{-t^2}\,dt \approx x$ with an error less than $0.2e^{-0.04}x^2$ for x in $(-0.2, 0.2)$.

SECTION 11.4

1. Taylor approximation: $\displaystyle\sum_{k=0}^{n} \dfrac{f^{(k)}(a)}{k!}(x-a)^k$

   Maclaurin approximation: $\displaystyle\sum_{k=0}^{n} \dfrac{f^{(k)}(0)}{k!}x^k$

3. $\dfrac{1}{1+x} \approx 1 - x + x^2 - x^3 + x^4$

5. $e^{-x} \approx 1 - x + \dfrac{x^2}{2} - \dfrac{x^3}{6} + \dfrac{x^4}{24}$

7. $5^x \approx 1 + (\ln 5)x + \dfrac{(\ln 5)^2}{2} x^2 + \dfrac{(\ln 5)^3}{6} x^3 + \dfrac{(\ln 5)^4}{24} x^4$

9. $(1+x)^{-1/2} \approx 1 - \tfrac{1}{2}x + \tfrac{3}{8}x^2 - \tfrac{15}{48}x^3 + \tfrac{105}{384}x^4$

11. $\dfrac{2^n e^{2a}}{n!}(x-a)^n$

13. $(-1)^{n-1}\dfrac{1}{n}\left(\dfrac{2}{2a+1}\right)^n (x-a)^n$

15. Differentiating the formula for $T_n(x)$, we have
$$T'_n(x) = f'(a) + f^{(2)}(a)(x-a) + \cdots + \dfrac{f^{(k)}(a)}{(k-1)!}(x-a)^{k-1} + \cdots$$

Setting $x = a$, we see that $T'_n(a) = f'(a)$. Successive differentiations gives the result.

17. The $(n-1)$st approximation is $A_0 + A_1 x + \cdots + A_{n-1} x^{n-1}$ with an error of $A_n x^n$. The nth and $(n+1)$st approximations are both $p(x)$ with 0 error.

SECTION 11.5

1. The derivatives of $\cos x$ are $\pm \sin x$ or $\pm \cos x$. Thus, $|f^{(k)}(x)| \le 1$ for all k and x. Thus,
$$|E_n(x)| \le \left|\dfrac{f^{(n+1)}(c)}{(n+1)!}(x-a)^{n+1}\right| \le \dfrac{|x-a|^{n+1}}{(n+1)!} \to 0 \quad \text{as } n \to \infty$$

3. $\sin x = 1 - \dfrac{1}{2}\left(x - \dfrac{\pi}{2}\right)^2 + \dfrac{1}{4!}\left(x - \dfrac{\pi}{2}\right)^4 + \cdots + \dfrac{(-1)^n}{(2n)!}\left(x - \dfrac{\pi}{2}\right)^{2n} + \cdots$

5. $\dfrac{1}{2}(e^x + e^{-x}) = 1 + \dfrac{x^2}{2!} + \dfrac{x^4}{4!} + \cdots + \dfrac{x^{2n}}{(2n)!} + \cdots$

7. $b^x = b^a + b^a \ln b(x-a) + \cdots + \dfrac{b^a [\ln b(x-a)]^k}{k!} + \cdots$

9. $\ln x = \ln a + \dfrac{x-a}{a} - \dfrac{(x-a)^2}{2a^2} + \dfrac{(x-a)^3}{3a^3} + \cdots$
$+ (-1)^{n+1} \dfrac{(x-a)^n}{na^n} + \cdots$

11. We use $|E_n(x)| \le \dfrac{\tfrac{1}{2}(e^{0.5} + (-1)^{n+1} e^{-0.5})}{(n+1)!}\left(\dfrac{1}{2}\right)^{n+1}$ for all x in $[-0.5, 0.5]$. This gives a bound of 2.4×10^{-5} for $n = 5$ and 8.1×10^{-7} for $n = 6$. We must take n to be at least 6.

SECTION 11.6

1. Diverges, because $\displaystyle\int_1^\infty (1/\sqrt{x})\,dx$ diverges.

3. Converges, because $\displaystyle\int_1^\infty (100/x^2)\,dx$ converges.

5. Converges, because $\displaystyle\int_1^\infty x^2 e^{-x}\,dx$ converges. Use

$$\int x^2 e^{-x}\, dx = -(x^2 + 2x + 2)e^{-x} + C.$$

7. Converges. Since $1/(k^3 + 1) < 1/k^3$, $\sum_{k=1}^{n} 1/(k^3 + 1) < \sum_{k=1}^{n} (1/k^3)$ for every n. Also, $\sum_{k=1}^{\infty} (1/k^3)$ converges by the integral test and thus $\sum_{k=1}^{n} (1/k^3) < M$. Thus $\sum_{k=1}^{n} 1/(k^3 + 1) < M$ for all n. Convergence follows from the boundedness theorem.

9. Diverges, because $\int_{2}^{\infty} \dfrac{1}{x \ln x}\, dx$ diverges. Use $\int \dfrac{1}{(x \ln x)}\, dx = \ln(\ln x) + C.$

11. Converges, because the Maclaurin series for $\cos x$ evaluated for $x = \pi$ converges.

13. $\sum_{k=10^6}^{\infty} (1/k)$ diverges because $\int_{10^6}^{\infty} (1/x)\, dx$ diverges. We can use the same reasoning as used in the proof of the integral test when the lower limit is not 1.

SECTION 11.7

1. $2 + e$
3. $3(\tfrac{1}{2} - e^{1/3})$
5. -1
7. Diverges, because
$$\lim_{k \to \infty} \dfrac{k}{10k + 1} = \dfrac{1}{10}.$$
9. Theorem 2 does not apply, because
$$\lim_{k \to \infty}\left(1 - \dfrac{k}{k+3}\right) = 1 - 1 = 0.$$
11. Theorem 2 does not apply, because
$$\lim_{k \to \infty}\left(\dfrac{1}{\sqrt{k}} - \dfrac{1}{k^2}\right) = 0 - 0 = 0.$$
13. $1 - \dfrac{1}{\sqrt[3]{2}} + \dfrac{1}{\sqrt[3]{3}} - \dfrac{1}{\sqrt[3]{4}} + \dfrac{1}{\sqrt[3]{5}}$. Take $N = 10^{12} - 1$.

15. $\dfrac{1}{\ln 2} - \dfrac{1}{\ln 3} + \dfrac{1}{\ln 4} - \dfrac{1}{\ln 5} + \dfrac{1}{\ln 6}$. Take $N = 10^{4343}.$

SECTION 11.8

1. Diverges by comparison with $1/126k$.
3. Converges by comparison with $3/4k^{3/2}$.
5. Diverges, because $a_{k+1}/a_k = (1 + 1/k)^k \to e > 1$ as $k \to \infty$.
7. Converges, because $a_{k+1}/a_k = (1 + 1/k)\tfrac{3}{4} \to \tfrac{3}{4} < 1$ as $k \to \infty$.
9. Converges, because
$$a_k = \left(\dfrac{1}{k} - \dfrac{1}{k+1}\right) = \dfrac{1}{k^2 + k} < \dfrac{1}{k^2}.$$

11. Diverges, because
$$\frac{k^2}{k^3+1} > \frac{k^2}{2k^3} = \frac{1}{2k}$$

13. Converges, because
$$\left|\frac{\cos k}{k^2}\right| \le \frac{1}{k^2}$$

15. If $\sqrt[k]{|a_k|} < r < 1$, then $|a_k| < r^k$ and thus $\sum_{k=1}^{\infty} |a_k|$ converges by comparison with $\sum_{k=1}^{\infty} r^k$.

17. Converges conditionally. Converges by the alternating series test. The series of absolute values diverges by comparison with $1/4k$.

19. Converges absolutely. We have a ratio of absolute values equal to $[(k+1)/k]^2 \frac{1}{2} \to \frac{1}{2} < 1$ as $k \to \infty$.

21. Diverges, because the kth term does not approach zero. Note that
$$\lim_{k\to\infty}\left(\frac{k-1}{k}\right)^k = \lim_{k\to\infty}\left(1-\frac{1}{k}\right)^k = e^{-1}$$

23. Diverges, because
$$\frac{1\cdot 3\cdot 5\cdots(2k-1)}{2\cdot 4\cdot 6\cdots 2k} > \frac{1\cdot 3\cdots(2k-1)}{3\cdot 5\cdots(2k-1)\cdot(2k+1)} = \frac{1}{2k+1}$$

SECTION 11.9

1. $R = 0$
3. $R = 1$; converges for $x = -1$, diverges for $x = 1$.
5. $R = 1$; diverges for $x = 1$ and $x = -1$.
7. $R = 2$; diverges for $x = 2$ and $x = -2$.
9. $R = \infty$
11. $\dfrac{x^3}{3} - \dfrac{1}{2}\dfrac{x^5}{5} + \dfrac{1}{3}\dfrac{x^7}{7} + \cdots + (-1)^{k+1}\dfrac{1}{k}\dfrac{x^{2k+1}}{2k+1} + \cdots$
13. $(x-\pi)^2 - \dfrac{8}{4!}(x-\pi)^4 + \dfrac{2^5}{6!}(x-\pi)^6 + \cdots$
$\quad + (-1)^{k+1}\dfrac{2^{2k-1}}{(2k)!}(x-\pi)^{2k} + \cdots$
15. $-\left(x - \dfrac{\pi}{2}\right) + \dfrac{2^2}{3!}\left(x - \dfrac{\pi}{2}\right)^3 - \dfrac{2^4}{5!}\left(x - \dfrac{\pi}{2}\right)^5 + \cdots$
$\quad + (-1)^{k+1}\dfrac{2^{2k}}{(2k+1)!}\left(x - \dfrac{\pi}{2}\right)^{2k+1} + \cdots$
17. Use $\displaystyle\int_0^{0.5}(1 + x^5 + x^{10})\,dx \approx 0.502648$.
19. Use $\displaystyle\int_{0.01}^{0.02}\left(\dfrac{1}{x} - 1 + \dfrac{x}{2}\right)dx \approx 0.68322$.

710 ANSWERS

SECTION 11.10
1. 1
3. $1/2\sqrt{2}$
5. 0
7. 0
9. ∞
11. ∞
13. 1
15. 0
17. ∞
19. e
21. 2
23. 1
25. $-\frac{1}{2}$

QUIZ 11.1
1. (a) 1 (b) 1
2. (a) $1 - x + x^2 - x^3 + \cdots + (-1)^n x^n$ (b) $1 + 2x + 2x^2 + \cdots + \frac{2^n}{n!}x^n$
3. (a) $1 + (\ln 10)x + \frac{(\ln 10)^2}{2}x^2 + \frac{(\ln 10)^3}{6}x^3 + \cdots$
   (b) $e^2(1 + 2(x-1) + \frac{2^2}{2!}(x-1)^2 + \frac{2^3}{3!}(x-1)^3 + \cdots)$
4. (a) Converges by the ratio test. (b) Converges by comparison with $1/k^2$. (c) Converges by the ratio test. (d) Converges by comparison with $1/k^2$.
5. (a) $11(\frac{10}{11})^4$ (b) $e - 2$
6. (a) 1 (b) ∞ (c) 0
7. (a) ∞ (b) 1

SECTION 12.1
1. $y = \sqrt{x}$
3. $y = \ln^2 x$ or $x = e^{\sqrt{y}}$
5. $y = \sqrt{1 - x^2}$
7. $y = \left(\frac{1-x}{1+x}\right)^2$
9. $t_0 = 1$, slope $= -6$
11. $t_0 = e^2$, slope $= 2e^4$
13.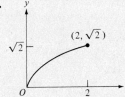
15.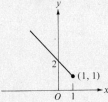
21. 2π
23. $\int_0^{\pi/3} \sqrt{1 + \tan^2 t}\, dt = \int_0^{\pi/3} \sec t\, dt = \ln|\sec t + \tan t|\Big|_0^{\pi/3}$
    $= \ln|2 + \sqrt{3}|$

SECTION 12.2
1. (a) $\left[\sqrt{2}, \frac{\pi}{4}\right], \left[-\sqrt{2}, -\frac{3\pi}{4}\right], \left[\sqrt{2}, -\frac{7\pi}{4}\right]$ (b) $\left[1, \frac{3\pi}{2}\right], \left[1, -\frac{\pi}{2}\right], \left[-1, \frac{\pi}{2}\right]$ (c) $\left[\sqrt{2}, \frac{5\pi}{4}\right], \left[\sqrt{2}, -\frac{3\pi}{4}\right], \left[-\sqrt{2}, \frac{\pi}{4}\right]$ (d) $\left[5, \frac{\pi}{3}\right], \left[-5, -\frac{2\pi}{3}\right], \left[5, -\frac{5\pi}{3}\right]$ (e) $\left[a\sqrt{2}, \frac{\pi}{4}\right], \left[-a\sqrt{2}, -\frac{3\pi}{4}\right]$,

$\left[a\sqrt{2}, -\frac{7\pi}{4}\right]$ (f) $\left[2, \frac{\pi}{6}\right], \left[2, -\frac{11\pi}{6}\right], \left[-2, -\frac{5\pi}{6}\right]$

3. $\left[-r, -\frac{\pi}{8}\right], \left[-r, \frac{\pi}{8}\right], \left[r, -\frac{\pi}{8}\right]$

5. $[-5, -0.93], \left[4, \frac{\pi}{2}\right], [5, 0.93]$

7.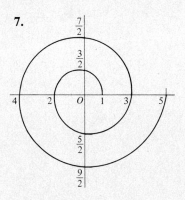

9.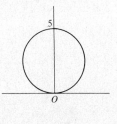

11.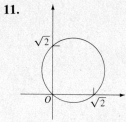

13.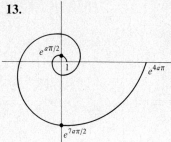

15. $\theta = 3\pi/4 = -\pi/4$

17. $r = \dfrac{7}{\cos \theta}$ for $-\pi < \theta < \pi$ and $\theta \neq -\dfrac{\pi}{2}, \dfrac{\pi}{2}$

SECTION 12.3 1. 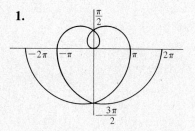 3.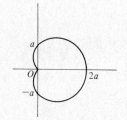

5.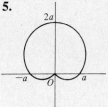

7.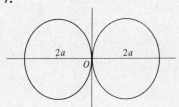

9.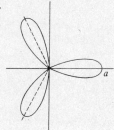

11.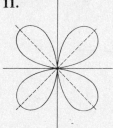

SECTION 12.4
1. πa^2
3. $\frac{3}{2}a^2\pi$
5. a^2
7. $\frac{1}{4}a^2\pi$
9. $\frac{3}{4}(e^{4\pi/3} - e^{2\pi/3})$
11. $8a$
13. $\frac{8}{3}[(1+\pi^2)^{3/2} - 1]$

SECTION 12.5
1. parabola: $y - \frac{1}{6} = -1(x + \frac{1}{2})^2$
   axis: $x = -\frac{1}{2}$
   vertex: $(-\frac{1}{2}, \frac{1}{6})$
   focus: $(-\frac{1}{2}, -\frac{1}{12})$
   directrix: $y = \frac{5}{12}$
3. parabola: $y - \frac{255}{16} = 4(x + \frac{1}{8})^2$
   axis: $x = -\frac{1}{8}$
   vertex: $(-\frac{1}{8}, \frac{255}{16})$
   focus: $(-\frac{1}{8}, 16)$
   directrix: $y = \frac{254}{16}$
5. parabola: $x + 1 = -y^2$
   axis: $y = 0$
   vertex: $(-1, 0)$
   focus: $(-\frac{5}{4}, 0)$
   directrix: $x = -\frac{3}{4}$
7. parabola: $x + 2 = 2(y + 1)^2$
   axis: $y = -1$
   vertex: $(-2, -1)$
   focus: $(-\frac{15}{8}, -1)$
   directrix: $x = -\frac{17}{8}$
9. You have the equations $(a, b + 1/4m) = (0, -3)$ and $y = 1 = b - 1/4m$, which give

$$a = 0$$
$$b + \frac{1}{4m} = -3$$
$$b - \frac{1}{4m} = 1$$

The solution is $b = -1$ and $m = -\frac{1}{8}$, and hence you get
parabola: $y + 1 = -\frac{1}{8}x^2$
axis: $x = 0$
vertex: $(0, -1)$
focus: $(0, -3)$
directrix: $y = 1$

11. From the equations $(a, b + 1/4m) = (2, 2)$
and $y = 0 = b - 1/4m$, you get
parabola: $y - 1 = \frac{1}{4}(x - 2)^2$
axis: $x = 2$
vertex: $(2, 1)$
focus: $(2, 2)$
directrix: $y = 0$

13. parabola: $x - \frac{7}{2} = \frac{1}{2}y^2$
axis: $y = 0$
vertex: $(\frac{7}{2}, 0)$
focus: $(4, 0)$
directrix: $x = 3$

15. The parabolas are $y + \frac{1}{2} = \frac{1}{2}x^2$ and $y - \frac{1}{2} = -\frac{1}{2}x^2$. They are found from the general equation $y - b = m(x - a)^2$, where the constants are found from the equations
$$0 - b = m(-1 - a)^2$$
$$0 - b = m(1 - a)^2$$
$$\left(a, b + \frac{1}{4m}\right) = (0, 0)$$

17. $y = \frac{1}{2}x^2 + \frac{1}{2}$ and $x = \frac{1}{2}y^2 + \frac{1}{2}$

SECTION 12.6

1. center: $(0, 0)$
major axis: $(0, 2)$ to $(0, -2)$
minor axis: $(-1, 0)$ to $(1, 0)$
$e = \frac{\sqrt{3}}{2}$
$F^+ = (0, \sqrt{3})$
$F^- = (0, -\sqrt{3})$
$l^+: y = \frac{4\sqrt{3}}{3}$
$l^-: y = -\frac{4\sqrt{3}}{3}$

3. center: $(1, -1)$
major axis: $(1, 1)$ to $(1, -3)$

minor axis: $(1-\sqrt{3}, -1)$ to $(1+\sqrt{3}, -1)$
$e = \frac{1}{2}$
$F^+ = (1, 0)$
$F^- = (1, -2)$
$l^+: y = 3$
$l^-: y = -5$

5. center: $(0, 0)$
major axis: $\left(-\frac{\sqrt{3}}{3}, 0\right)$ to $\left(\frac{\sqrt{3}}{3}, 0\right)$
minor axis: $(\frac{1}{2}, 0)$ to $(-\frac{1}{2}, 0)$
$e = \frac{1}{2}$
$F^+ = \left(\frac{\sqrt{3}}{6}, 0\right)$
$F^- = \left(-\frac{\sqrt{3}}{6}, 0\right)$
$l^+: x = \frac{2\sqrt{3}}{3}$
$l^-: x = -\frac{2\sqrt{3}}{6}$

7. center: $(0, 1)$
major axis: $(-1, 1)$ to $(1, 1)$
minor axis: $\left(0, 1+\frac{\sqrt{2}}{2}\right)$ to $\left(0, 1-\frac{\sqrt{2}}{2}\right)$
$e = \frac{\sqrt{2}}{2}$
$F^+ = \left(\frac{\sqrt{2}}{2}, 1\right)$
$F^- = \left(-\frac{\sqrt{2}}{2}, 1\right)$
$l^+: x = \sqrt{2}$
$l^-: x = -\sqrt{2}$

9. center: $(1, -1)$
major axis: $(1, 3)$ to $(1, -5)$
minor axis: $(-2, -1)$ to $(4, -1)$
$e = \frac{\sqrt{7}}{4}$
$F^+ = (1, -1+\sqrt{7})$
$F^- = (1, -1-\sqrt{7})$
$l^+: y = -1 + \frac{16\sqrt{7}}{7}$
$l^-: y = -1 - \frac{16\sqrt{7}}{7}$

11. $y - \sqrt{3} = -\frac{2\sqrt{3}}{3}\left(x - \frac{1}{2}\right)$

13. $\frac{(x+2)^2}{8} + \frac{(y+6)^2}{24} = 1$

15. $\dfrac{x^2}{16} + \dfrac{y^2}{7} = 1$

17. $\dfrac{x^2}{\left(\dfrac{10\sqrt{15}}{17}\right)^2} + \dfrac{(y - 10/17)^2}{\left(\dfrac{40}{17}\right)^2} = 1,\ \dfrac{x^2}{\left(\sqrt{\dfrac{20}{3}}\right)^2} + \dfrac{(y + 2/3)^2}{\left(\dfrac{8}{3}\right)^2} = 1$

SECTION 12.7

1. center: $(0, 0)$
axis: $y = 0$
vertices: $(1, 0)$ and $(-1, 0)$
asymptotes: $y = \pm\sqrt{2}\,x$
$e = \sqrt{3}$
$F^+ = (\sqrt{3}, 0)$
$F^- = (-\sqrt{3}, 0)$
$l^+\colon x = 1/\sqrt{3}$
$l^-\colon x = -1/\sqrt{3}$

3. center: $(1, -1)$
axis: $y = -1$
vertices: $(2, -1)$ and $(0, -1)$
asymptotes: $y + 1 = \pm 2(x - 1)$
$e = \sqrt{5}$
$F^+ = (\sqrt{5} + 1, -1)$
$F^- = (-\sqrt{5} + 1, -1)$
$l^+\colon x = \dfrac{1}{\sqrt{5}} + 1$
$l^-\colon x = -\dfrac{1}{\sqrt{5}} + 1$

5. center: $(-1, 1)$
axis: $y = 1$
vertices: $\left(\dfrac{1}{\sqrt{2}} - 1, 1\right)$ and $\left(-\dfrac{1}{\sqrt{2}} - 1, 1\right)$
asymptotes: $y - 1 = \pm\dfrac{\sqrt{2}}{2}(x + 1)$
$e = \dfrac{\sqrt{6}}{2}$
$F^+ = \left(\dfrac{\sqrt{3}}{2} - 1, 1\right)$
$F^- = \left(-\dfrac{\sqrt{3}}{2} - 1, 1\right)$
$l^+\colon x = \dfrac{1}{\sqrt{3}} - 1$
$l^-\colon x = -\dfrac{1}{\sqrt{3}} - 1$

7. center: $(\tfrac{1}{2}, \tfrac{1}{4})$
axis: $y = \tfrac{1}{4}$
vertices: $(1, \tfrac{1}{4})$ and $(0, \tfrac{1}{4})$

asymptotes: $y - \frac{1}{4} = \pm\frac{1}{2}(x - \frac{1}{2})$

$e = \frac{\sqrt{5}}{2}$

$F^+ = \left(\frac{\sqrt{5}}{4} + \frac{1}{2}, \frac{1}{4}\right)$

$F^- = \left(-\frac{\sqrt{5}}{4} + \frac{1}{2}, \frac{1}{4}\right)$

$l^+: x = \frac{1}{\sqrt{5}} + \frac{1}{2}$

$l^-: x = -\frac{1}{\sqrt{5}} + \frac{1}{2}$

9. This is a degenerate conic consisting of the two parallel lines $y = 1 - \sqrt{2}$ and $y = 1 + \sqrt{2}$.
11. This equation can be written as $(x+1)^2 + (y-1)^2 = 0$ and we see that $(x, y) = (-1, 1)$ is the only point satisfying it. Hence, this is a one-point degenerate conic.
13. This is a degenerate conic with no points, since the equation has no real solutions.
15. Same as Exercise 13 above.
17. This equation gives the hyperbola
$$-\frac{(x+\frac{1}{2})^2}{\frac{7}{4}} + \frac{y^2}{\frac{7}{4}} = 1$$
19. This equation gives the parabola $y - \frac{31}{16} = 4(x - \frac{1}{8})^2$.
21. $y = \pm\frac{\sqrt{17}}{3}x + 1$
23. hyperbola: $\frac{x^2}{10} - \frac{y^2}{90} = 1$

center: $(0, 0)$
axis: $y = 0$
vertices: $(\sqrt{10}, 0)$ and $(-\sqrt{10}, 0)$
asymptotes: $y = \pm 3x$
$e = \sqrt{10}$
$F^+ = (10, 0)$
$F^- = (-10, 0)$
$l^+: x = 1$
$l^-: x = -1$

25. The equation gives the right-hand branch of
$$\frac{x^2}{a^2} - \frac{y^2}{b^2} = 1$$

SECTION 12.8

1. parabola: $2u^2 - 4v + 1 = 0$

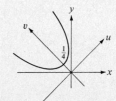

3. parabola: $2v^2 - u + 1 = 0$

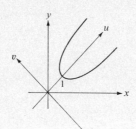

5. hyperbola: $\dfrac{v^2}{3^2} - \dfrac{u^2}{4^2} = 1$

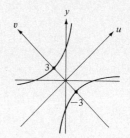

7. ellipse: $\dfrac{u^2}{\frac{161}{32}} + \dfrac{(v - \frac{1}{4})^2}{\frac{161}{16}} = 1$

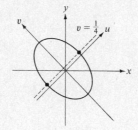

9. parabola: $\sqrt{2}\,v^2 - 4v - u + 5 = 0$

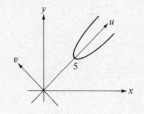

11. circle: $(u - 1)^2 + (v + 1)^2 = 1$

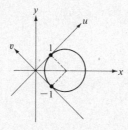

13. degenerate conic: $4v^2 - u^2 = 0$

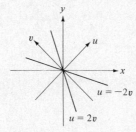

15. degenerate conic: $u(u - v) = 0$

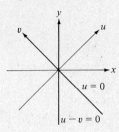

QUIZ 12.1
1. $y = x^{3/2}$
2. $x = \frac{3}{4} + \frac{1}{4} \cos 2t$
   $y = \sin t$

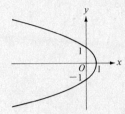

3. $\int_0^{\pi/2} (1 + \frac{9}{4} \sin t)^{1/2} \cos t \, dt = \frac{8}{27}\left[\left(\frac{13}{4}\right)^{3/2} - 1\right]$
4. $[\sqrt{2}, -\pi/4]$, $[\sqrt{2}, 7\pi/4]$, $[-\sqrt{2}, 3\pi/4]$
   $[2, \pi/4]$, $[2, -7\pi/4]$, $[-2, 5\pi/4]$
5. $r = 4 \sin 6\theta$

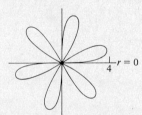

6. $6 \cdot \frac{1}{2} \int_0^{\pi/6} 16 \sin^2 6\theta \, d\theta = 4\pi$

7. parabola: $x = (y - 2)^2$
   axis: $y = 2$
   vertex: $(0, 2)$
   focus: $(\frac{1}{4}, 2)$
   directrix: $x = -\frac{1}{4}$

8. The parabola we seek is of the form $y - b = m(x - a)^2$. The constants are determined by $(a, b + 1/4m) = (-2, -4)$ and $b - 1/4m = 1$. The parabola is $y + \frac{3}{2} = -\frac{1}{10}(x + 2)^2$.

9. center: $(-1, 2)$
   major axis: $(-1, 0)$ to $(-1, 4)$
   minor axis: $(-1 - \sqrt{2}, 2)$ to $(-1 + \sqrt{2}, 2)$
   $e = \sqrt{2}/2$
   $F^+ = (-1, 2 + \sqrt{2})$

$F^- = (-1, 2 - \sqrt{2})$
$l^+: y = 2\sqrt{2} + 2$
$l^-: y = -2\sqrt{2} + 2$

10. $\dfrac{x^2}{20} + \dfrac{(y+4)^2}{4} = 1$

11. center: $(1, -1)$
    axis: $y = -1$
    vertices: $(1 + \sqrt{2}, -1)$ and $(1 - \sqrt{2}, -1)$
    asymptotes: $y + 1 = \dfrac{\sqrt{6}}{2}(x - 1)$ and $y + 1 = -\dfrac{\sqrt{6}}{2}(x - 1)$
    $e = \sqrt{10}/2$
    $F^+ = (\sqrt{5} + 1, -1)$
    $F^- = (-\sqrt{5} + 1, -1)$
    $l^+: x = \dfrac{2\sqrt{5}}{5} + 1$
    $l^-: x = -\dfrac{2\sqrt{5}}{5} + 1$

12. **(a)** ellipse: $\dfrac{u^2}{3} + \dfrac{v^2}{4} = 1$ **(b)** circle: $(x - 1)^2 + (y + 3)^2 = 3^2$

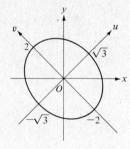
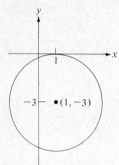

SECTION 13.1
1. 1
3. $2\sqrt{3}$
5. $(x + 1)^2 + (y - 2)^2 + (z + 3)^2 = 1$
7. Yes; they have the one point $(1, 0, 0)$ in common.
9.

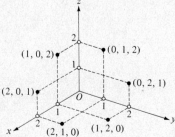

SECTION 13.2   **1.** The domain = all points (x, y). The range = all real numbers.

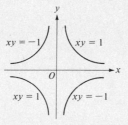

3. The domain = all points (x, y) with $y \neq 0$. The range = all real numbers.

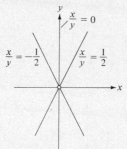

5. The domain = all points (x, y). The range = all nonnegative real numbers.

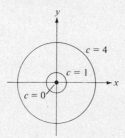

7. The domain = all points (x, y). The range = all real numbers.

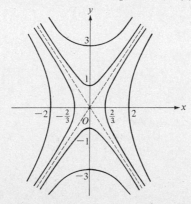

9. The domain = all points (x, y). The range = all real numbers.

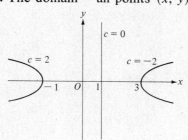

11. The domain = all points (x, y). The range = the interval $(-\infty, 1]$.

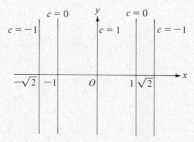

SECTION 13.3

1. xy plane: the point $(0, 0)$; xz plane: the lines $z = \pm 2x$; yz plane: the lines $z = \pm 2y$.
3. xy plane: the point $(0, 0)$; xz plane: the parabola $z = \frac{1}{3}x^2$; yz plane: the parabola $z = \frac{1}{4}y^2$.
5. xy plane: the parabola $y = x^2$; xz plane: the line $x = 0$; yz plane: the line $y = 0$.
7. xy plane: the lines $y = \pm 1$; xz plane: the lines $z = \pm \frac{1}{2}$; yz plane: the ellipse $y^2 + 4z^2 = 1$.

SECTION 13.4

1. $u_x = y$, $u_y = x$
3. $u_x = \dfrac{2y}{(x+y)^2}$, $u_y = -\dfrac{2x}{(x+y)^2}$
5. $u_x = 3x^2y^2$, $u_y = 2x^3y$
7. $u_x = 1$, $u_y = 0$
9. See text.
11. $u_x = f'\left(\dfrac{x}{y}\right)\dfrac{1}{y}$, $u_y = -f'\left(\dfrac{x}{y}\right)\dfrac{x}{y^2}$
13. $u_x = u_y = f'(x+y)$
15. (a) and (b) follow from the following computations.
$u_x = 3x^2 - 3y^2$, $u_{xx} = 6x$, $u_y = -6xy$, $u_{yy} = -6x$.
$v_x = 6xy$, $v_{xx} = 6y$, $v_y = 3x^2 - 3y^2$, $v_{yy} = -6y$.
17. $u_x = \exp(x + cy) + \exp(x - cy) = u$, and thus $u_{xx} = u_x = u$.

$u_y = c \exp(x+cy) - c \exp(x-cy)$, $u_{yy} = c^2 \exp(x+cy) + c^2 \exp(x-cy) = c^2 u = c^2 u_{xx}$.

SECTION 13.5

1. 1
3. 0
5. $y \neq 0$ and $\dfrac{x}{y} \neq \dfrac{(2n+1)\pi}{2}$ $n = 0, \pm 1, \pm 2, \ldots$
7. All points (x, y)
9. $x \neq 0$
11. It is continuous at $(0, 0)$. It is defined for all (x, y), $f(0, 0) = 0$ and $\lim\limits_{\substack{x \to 0 \\ y \to 0}} f(x, y) = 0$.
13. It is not continuous at $(0, 0)$, as can be seen from the fact that $f(x, 0) = 1/x$ does not approach a limit as $x \to 0$.

SECTION 13.6

1. All points (x, y, z).
3. All points (x, y, z) *not* in the plane $z = 0$.
5. All points (x, y, z) *not* lying on the cylinders $x^2 + y^2 = (2n+1)\pi/2$ for $n = 0, \pm 1, \pm 2, \ldots$.
7. All points *not* in the planes $u = \pm x$ and $z = \pm y$.
9. $u_x = \dfrac{2x}{(x^2+y^2+z^2)^2} u$, $u_{xy} = \dfrac{4xy[1 - 2(x^2+y^2+z^2)]}{(x^2+y^2+z^2)^4} u$

SECTION 13.7

1. $\ln(2x+y) \approx \ln(0.3) + \tfrac{20}{3}(x - 0.1) + \tfrac{10}{3}(y - 0.1)$
3. $\dfrac{1}{\sqrt{x^2+y}} \approx \dfrac{1}{\sqrt{5}} - \dfrac{2}{\sqrt{125}}(x-2) - \dfrac{1}{2\sqrt{125}}(y-1)$
5. $\left(\dfrac{1+x^2}{1+y^2}\right)^{1/2} \approx \dfrac{1}{\sqrt{2}} + \dfrac{\sqrt{2}}{5}(x-2) - \dfrac{3\sqrt{2}}{20}(y-3)$
7. $(z^2 + xy)^{1/2} \approx \dfrac{\sqrt{17}}{4} + \dfrac{1}{\sqrt{17}}(x-2) + \dfrac{4}{\sqrt{17}}\left(y - \tfrac{1}{2}\right) + \dfrac{1}{\sqrt{17}}\left(z - \tfrac{1}{4}\right)$
9. $(1 - x^2 - y^2 - z^2)^{1/2} \approx \tfrac{1}{2} - (x - \tfrac{1}{2}) - (y - \tfrac{1}{2}) - (z - \tfrac{1}{2})$
11. Take $x_0 = 2.3$, $y_0 = 0$, and $z_0 = 0$.   1.97.

SECTION 13.8

1. $\dfrac{2(1 - \ln t)}{(\ln t + t)^2}$
3. $2e^{2t}(\cos 2t - \sin 2t)$
5. $\left(\dfrac{\sec t}{1 + \tan t}\right)^2 \sec^2\left(\dfrac{\tan t}{1 + \tan t}\right)$
7. $w_r = \dfrac{\sin \theta - \cos \theta}{r^2}$, $w_\theta = -\dfrac{\sin \theta + \cos \theta}{r}$
9. $w_r = \dfrac{2w}{r^3(e^{2\theta} + e^{-2\theta})}$, $w_\theta = \dfrac{2w}{r^2(e^{2\theta} + e^{-2\theta})^2}(e^{2\theta} - e^{-2\theta})$

11. $w_r = 2r \sec^2 \theta - 20r \tan^2 \theta$, $w_\theta = -18r^2 \tan \theta \sec^2 \theta$

13. $\dfrac{y}{x}$

15. $\dfrac{5y - 3x^2}{3y^2 - 5x}$

17. $\dfrac{\sin(x+y) - y \cos xy}{x \cos xy - \sin(x+y)}$

SECTION 13.9

1. The basic computations are $u_x = e^y \cos x$, $u_{xx} = e^y(-\sin x)$, $v_x = -e^y \sin x$, $v_{xx} = -e^y \cos x$, $u_y = e^y \sin x$, $u_{yy} = e^y \sin x$, $v_y = e^y \cos x$, $v_{yy} = e^y \cos x$.

3. Basic equations:
$$F_{rr} = F_{xx} \cos^2 \theta + 2F_{xy} \sin \theta \cos \theta + F_{yy} \sin^2 \theta$$
$$\frac{1}{r^2}F_{\theta\theta} = F_{xx} \sin^2 \theta - 2F_{xy} \sin \theta \cos \theta + F_{yy} \cos^2 \theta - \frac{1}{r}F_r$$

5. $F_{xx} = \dfrac{3x^2}{(x^2 + y^2 + z^2)^{5/2}} - \dfrac{1}{(x^2 + y^2 + z^2)^{3/2}}$ and similarly for F_{yy} and F_{zz}.

7. $e^{-2r}F_{rr} = F_{xx} \cos^2 \theta + 2F_{xy} \cos \theta \sin \theta + F_{yy} \sin^2 \theta$
$\qquad + e^{-r}(F_x \cos \theta + F_y \sin \theta)$
$e^{-2r}F_{\theta\theta} = F_{xx} \sin^2 \theta - 2F_{xy} \cos \theta \sin \theta + F_{yy} \cos^2 \theta$
$\qquad - e^{-r}(F_x \cos \theta + F_y \sin \theta)$

SECTION 13.10

1. Abs. min. at $(2, -2)$; no local max.
3. Abs. min. at $(0, 0)$; no local max.
5. Local min. at $(\sqrt{\frac{2}{3}}, 0)$; no local max. and no abs. min.
7. Local min. at $(-\frac{1}{12}, \frac{1}{6})$; no local max. and no abs. min.
9. No local max. or min.
11. No local max. or min.
13. $(\frac{1}{2}, \frac{1}{8})$
15. $x = y = 22$, $z = 33$
17. $x = 50$ tons, $y = 1$ ton/week.

QUIZ 13.1

1. Center $(1, -1, -2)$, radius 1.
2.

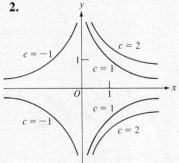

3. (a) $u_x = \dfrac{2y}{(x+y)^2}$, $u_y = -\dfrac{2x}{(x+y)^2}$ (b) $u_x = \dfrac{1}{\sqrt{y^2-x^2}}$,

$u_y = -\dfrac{x}{y\sqrt{y^2-x^2}}$ (c) $u_x = \dfrac{x}{\sqrt{x^2+y^2}}$, $u_y = \dfrac{y}{\sqrt{x^2+y^2}}$

4. 0
5. $x^2 + y^2 > z$
6. $f(x, y) \approx -\tfrac{5}{4}\ln 2 + (\tfrac{1}{2} - \ln 4)(x - \tfrac{1}{2}) + (2 - \ln 2)(y - \tfrac{1}{4})$
7. $\dfrac{2}{(\ln t)^2 + t^2}\left(\dfrac{\ln t}{t} + t\right)$
8. $w_r = \cos x \sin^{-1} y \cos\theta + \sin x \dfrac{1}{\sqrt{1-y^2}} \sin\theta$

$w_\theta = -r\cos x \sin^{-1} y \sin\theta + r\sin x \dfrac{1}{\sqrt{1-y^2}} \cos\theta$

9. y/x
10. We find that $F_{yy} = -4[\sin(x+2y) + \cos(x-2y)] = 4F_{xx}$.
11. (a) No maxima or minima. (b) Local minimum at $(0, 1)$.
    (c) Local minimum at $(\tfrac{4}{3}, \tfrac{1}{3})$.

SECTION 14.1

1. $\tfrac{15}{2}$
3. $5(e-1)/4$
5. $\tfrac{8}{3}$
7. $27 \ln 3 - 12$
9. 0
11. 0
13. 0
15. $3/2(\ln 2)^2$

SECTION 14.2

1. $\tfrac{1}{4}$
3. $\tfrac{5}{8}$
5. $4(e - e^{-1})$
7. 0
9. $2\iint_A \sqrt{r^2 - (x^2 + y^2)}\ dx\ dy$, where A is the disk $x^2 + y^2 \le r^2$
11. $\displaystyle\int_a^c f(x)\ dx = \int_a^b f(x)\ dx + \int_b^c f(x)\ dx$

SECTION 14.3

1. $\tfrac{4}{3}$
3. $\tfrac{14}{15}$
5. $\tfrac{5}{42}$
7. $\tfrac{5}{108}$
9. $\alpha = y^2$, $\beta = y^{1/3}$

11. It reduces to $\int_0^1 xe^{x^2}\,dx = \frac{1}{2}(e-1)$.

SECTION 14.4
1. 4; $(0, 0)$
3. $\frac{8}{3}$; $(0, 0)$
5. $\frac{1}{12}$; $(\frac{3}{5}, \frac{1}{2})$
7. $\frac{1}{6}$; $(\frac{2}{5}, \frac{2}{5})$
9. 5; $(1.1, 1.1)$

SECTION 14.5
1. $\pi/12$
3. π
5. 0
7. $a^3\pi/6$
9. $\pi/2$; $(0, 4/3\pi)$
11. $\pi/8$; $(0, 32/15\pi)$
13. $\pi/9$
15. 26

SECTION 14.6
1. 2
3. $\frac{1}{16}$
5. $\frac{7}{2}$
7. $\frac{9}{2}(e-1)(e^2-1)$

SECTION 14.7
1. Some crucial calculations:
$$\iiint_T y\,dx\,dy\,dz = \int_0^1 \left[\int_0^{1-x} y(1-x-y)\,dy\right]dx$$
$$= \int_0^1 \frac{(1-x)^3}{6}\,dx = \frac{1}{24}$$
$$\iiint_T z\,dx\,dy\,dz = \int_0^1 \left[\int_0^{1-x} \frac{(1-x-y)^2}{2}\,dy\right]dx$$
$$= \int_0^1 \frac{(1-x)^3}{6}\,dx = \frac{1}{24}$$
3. $abc/6$
5. $\pi/10$
7. $\frac{1}{10}$

SECTION 14.8
1. 2π; $(0, 0, \frac{2}{3})$
3. 16π; $(\frac{1}{4}, 0, \frac{1}{2})$
5. $\frac{3\pi}{20}$; $\left(-\frac{10}{9\pi}\left(\frac{\pi}{6}-\frac{\sqrt{3}}{8}\right), \frac{10}{9\pi}\left(\frac{\pi}{6}-\frac{\sqrt{3}}{8}\right), \frac{5}{8}\right)$
7. $\frac{9\pi}{64}$; $\left(-\frac{8\sqrt{3}}{45\pi}, \frac{8\sqrt{3}}{45\pi}, \frac{28}{45}\right)$
9. $4\pi(9 - \frac{5}{3}\sqrt{5})$
11. 8π

QUIZ 14.1
1. $\frac{1}{420}$
2. $\pi a^4/8$
3. $-\frac{1}{2}$
4. $\pi/10$, $(0, 0, \frac{1}{6})$

SECTION 15.1

1. $(a_1, a_2, 0)$, $(a_1, 0, a_3)$, $(0, a_2, a_3)$
3. $\mathbf{a} + \mathbf{b} = (-1, 0)$, $\mathbf{a} - \mathbf{b} = (-1, -2)$

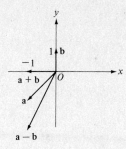

5. $\mathbf{a} + \mathbf{b} = (2, 0)$, $\mathbf{a} - \mathbf{b} = (0, 4)$

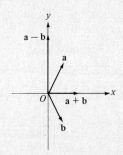

7. $\mathbf{a} + \mathbf{b} = (-3, -3)$, $\mathbf{a} - \mathbf{b} = (1, -1)$

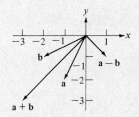

9. $\mathbf{a} + \mathbf{b} = (2, 1, 1)$, $\mathbf{a} - \mathbf{b} = (0, 1, -1)$
11. $\mathbf{a} + \mathbf{b} = (\frac{1}{6}, \frac{5}{6}, \frac{5}{6})$, $\mathbf{a} - \mathbf{b} = (\frac{5}{6}, -\frac{1}{6}, \frac{1}{6})$
13. $\mathbf{a} + \mathbf{b} = (0, 0)$, $\mathbf{a} - \mathbf{b} = (-4, 8)$

SECTION 15.2

1. $|\mathbf{a}| = \frac{5}{12}$, $(-\frac{4}{5}, -\frac{3}{5})$
3. $|\mathbf{c}| = 2$, $(-1/\sqrt{2}, -1/\sqrt{2})$
5. $|\mathbf{e}| = \sqrt{2}$, $(-1/\sqrt{2}, 0, 1/\sqrt{2})$
7. $|\mathbf{b}| = 2|k|$, $\left(\dfrac{1-k}{2|k|}, \dfrac{1+k}{2|k|}, \dfrac{\sqrt{2k^2-2}}{2|k|}\right)$
9. $|\mathbf{d}| = x\sqrt{21}$, $(1/\sqrt{21}, 2/\sqrt{21}, 4/\sqrt{21})$
11. $\mathbf{a} = -2\mathbf{v}$, $\mathbf{b} = 3\mathbf{v} - \frac{4}{3}\mathbf{w}$, $\mathbf{c} = 2\mathbf{v} - \frac{2}{3}\mathbf{w}$

SECTION 15.3
1. -15
3. 2
5. -13
7. $-1/\sqrt{42}, 4/\sqrt{42}, 5/\sqrt{42}$
9. $1/\sqrt{2}, 0, -1/\sqrt{2}$
11. $1/\sqrt{3}$
13. $-12/\sqrt{820}$
15. -1
17. $5/\sqrt{34}, -1/\sqrt{10}, 7/\sqrt{85}$

19. $\mathbf{a} \cdot (\mathbf{b} + \mathbf{c}) = (a_1, a_2) \cdot (b_1 + c_1, b_2 + c_2)$
$= a_1(b_1 + c_1) + a_2(b_2 + c_2)$
$(a_1 b_1 + a_2 b_2) + (a_1 c_1 + a_2 c_2) = \mathbf{a} \cdot \mathbf{b} + \mathbf{a} \cdot \mathbf{c}$

21. It equals $\mathbf{a} \cdot \mathbf{b}/|\mathbf{a}||\mathbf{b}|$ = cosine of the angle between $\mathbf{a}$ and $\mathbf{b}$.

SECTION 15.4
1. $\mathbf{f}'(t) = \left(e^t(1 + t), -e^{-t}\left(\frac{1}{t} + \frac{1}{t^2}\right)\right)$;
$\mathbf{f}''(t) = \left(e^t(2 + t), e^{-t}\left(\frac{1}{t} + \frac{2}{t^2} + \frac{2}{t^3}\right)\right)$

3. $\mathbf{f}'(t) = (-2 \sin t, -2 \sin 2t)$; $\mathbf{f}''(t) = (-2 \cos t, -4 \cos 2t)$

5. $\mathbf{f}'(t) = \dfrac{1}{2\sqrt{t}(\sqrt{t} + 1)^2} (\mathbf{i} + \mathbf{j})$; $\mathbf{f}''(t) = -\dfrac{(3\sqrt{t} + 1)}{4t^{3/2}(\sqrt{t} + 1)^3} (\mathbf{i} + \mathbf{j})$

7. $\mathbf{f}'(t) = (\cos t - t \sin t, \sin t + t \cos t, \cos t + \sin t + t(\cos t - \sin t))$; $\mathbf{f}''(t) = (-(2 \sin t + t \cos t), 2 \cos t - t \sin t, 2(\cos t - \sin t) - t(\cos t + \sin t))$

9. $\mathbf{f}'(t) = \mathbf{0}$; $\mathbf{f}''(t) = \mathbf{0}$

11. The result follows from the fact that
$$|\mathbf{f}(t)|^2 = \frac{1}{1 + t^2} + \frac{t^2}{1 + t^2} = 1$$

13. $h(t)\mathbf{g}''(t) + 2h'(t)\mathbf{g}'(t) + h''(t)\mathbf{g}(t)$

15. $\dfrac{2}{(t + 1)^2} \mathbf{f}'\left(\dfrac{t - 1}{t + 1}\right)$

17. $2\mathbf{f}(t) \cdot \mathbf{f}'(t)$

19. $\sqrt{2}(e^{3\pi} - 1)$

21. $\sqrt{2}$

SECTION 15.5
1. $\mathbf{T}(t) = \left(\dfrac{\cos t - \sin t}{\sqrt{2}}, \dfrac{\cos t + \sin t}{\sqrt{2}}\right)$;
$\mathbf{N}(t) = \left(-\dfrac{\cos t + \sin t}{\sqrt{2}}, \dfrac{\cos t - \sin t}{\sqrt{2}}\right)$

3. $\mathbf{T}(t) = \left(-\dfrac{2t}{t^2 + 1}, \dfrac{t^2 - 1}{t^2 + 1}\right)$; $\mathbf{N}(t) = \left(\dfrac{t^2 - 1}{t^2 + 1}, \dfrac{2t}{t^2 + 1}\right)$

5. $\mathbf{T}(t) = \left(\dfrac{\sqrt{1 - \cos t}}{\sqrt{2}}, \dfrac{\sin t}{\sqrt{2}\sqrt{1 - \cos t}}\right)$;
$\mathbf{N}(t) = \left(\dfrac{\sin t}{\sqrt{2}\sqrt{1 - \cos t}}, -\dfrac{\sqrt{1 - \cos t}}{\sqrt{2}}\right)$

7. $\mathbf{T}(t) = (1/\sqrt{11}, 1/\sqrt{11}, 3/\sqrt{11})$. Since $\mathbf{T}'(t) = \mathbf{0}$, $\mathbf{N}(t)$ is undefined.

9. For $t < 0$, $\mathbf{T}(t) = (-1/\sqrt{2}, 1/\sqrt{2}, 0)$. $\mathbf{N}(t)$ is undefined.

SECTION 15.6

1. $\mathbf{T}(t) = \dfrac{1}{\sqrt{1+9t^4}}(1, 3t^2)$; $\mathbf{N}(t) = \dfrac{1}{\sqrt{1+9t^4}}(-3t^2, 1)$;
$\kappa(t) = \dfrac{6t}{(1+9t^4)^{3/2}}$; $\mathbf{a} = \dfrac{18t^3}{\sqrt{1+9t^4}}\mathbf{T}(t) + \dfrac{6t}{\sqrt{1+9t^4}}\mathbf{N}(t)$

3. $\mathbf{T}(t) = \dfrac{1}{\sqrt{9t^2+4}}(3t, 2)$; $\mathbf{N}(t) = \dfrac{1}{\sqrt{9t^2+4}}(2, -3t)$;
$\kappa(t) = \dfrac{6}{t(9t^2+4)^{3/2}}$; $\mathbf{a} = \dfrac{18t^2+4}{\sqrt{9t^2+4}}\mathbf{T}(t) + \dfrac{6t}{\sqrt{9t^2+4}}\mathbf{N}(t)$

5. $\mathbf{T}(t) = \dfrac{1}{\sqrt{e^{2t}+e^{-2t}}}(e^t, -e^{-t})$; $\mathbf{N}(t) = \dfrac{1}{\sqrt{e^{2t}+e^{-2t}}}(e^{-t}, e^t)$;
$\kappa(t) = \dfrac{2}{(e^{2t}+e^{-2t})^{3/2}}$; $\mathbf{a} = \dfrac{e^{2t}-e^{-2t}}{\sqrt{e^{2t}+e^{-2t}}}\mathbf{T}(t) + \dfrac{2}{\sqrt{e^{2t}+e^{-2t}}}\mathbf{N}(t)$

7. $\mathbf{T}(t) = \left(-\dfrac{\sin t}{\sqrt{2}}, \dfrac{\cos t}{\sqrt{2}}, \dfrac{1}{\sqrt{2}}\right)$; $\mathbf{N}(t) = (-\cos t, -\sin t, 0)$;
$\kappa(t) = \tfrac{1}{2}$; $\mathbf{a} = 0 \cdot \mathbf{T}(t) + 1 \cdot \mathbf{N}(t) = \mathbf{N}(t)$

9. $\mathbf{T}(t) = \left(\dfrac{\cos t + \sin t}{\sqrt{11}}, \dfrac{\cos t - \sin t}{\sqrt{11}}, \dfrac{3}{\sqrt{11}}\right)$;
$\mathbf{N}(t) = \left(\dfrac{\cos t - \sin t}{\sqrt{2}}, -\dfrac{\cos t + \sin t}{\sqrt{2}}, 0\right)$; $\kappa(t) = \dfrac{\sqrt{2}}{11}e^{-t}$;
$\mathbf{a} = \sqrt{11}e^t\mathbf{T}(t) - \sqrt{2}e^t\mathbf{N}(t)$

11. $\mathbf{T}(t) = \left(\dfrac{1}{\sqrt{2}}, \dfrac{1}{\sqrt{2}}, 0\right)$; $\mathbf{N}(t)$ is undefined. $\mathbf{a} = 2\sqrt{2}\mathbf{T}(t)$

SECTION 15.7

1. $x + 2y = 0$
3. $y = 0$
5. $-x + y + z = 0$
7. $x + y + z = -3$
9. $x = 0$
11. $x + y + z = -1$
13. $x + 2y + 3z = 6$
15. $x - 4y + 5z = 1$

SECTION 15.8

1. Normal $(108, -51, -36)$; tangent plane $36(x + 2) - 17(y - 3) - 12(z - 1) = 0$
3. Normal $(-1, 3, 1)$; tangent plane $-x + 3y + z = -6$
5. Normal $(2\sqrt{3}, 2, 0)$; tangent plane $\sqrt{3}(x - \sqrt{3}) + (y - 1) = 0$
7. Normal $(-\tfrac{1}{2}, -\tfrac{3}{4})$; tangent line $2x + 3y = -2$
9. Normal $(-1/\sqrt{2}, -1/\sqrt{2})$; tangent line $x + y = \pi/4$
11. $(3x^2y^2, 2x^3y - 3z^4, -12yz^3)$
13. $(e^y, xe^y + e^z, ye^z)$
15. $(-e^{-x}y, e^{-x})$
17. $(2, -4, -5)$
19. $(2, 0, 0)$

SECTION 15.9
1. $|(\partial F/\partial x, \partial F/\partial y, \partial F/\partial z)| = \sqrt{(\partial F/\partial x)^2 + (\partial F/\partial y)^2 + (\partial F/\partial z)^2}$
3. 0
5. $(2, -4) \cdot (\frac{3}{5}, \frac{4}{5}) = -2$
7. $(3, 1, 1) \cdot (-1/2, -1/2, \sqrt{2}/2) = -2 + \sqrt{2}/2$
9. $(-2, 3, -1) \cdot (\cos\theta, \sin\theta, 0) = -2\cos\theta + 3\sin\theta$

SECTION 15.10
1. $(0, 0, 1)$
3. $(-8, -8, 0)$
5. $(-1, -1, 1)$
7. -5
9. 0
11. $-44\mathbf{i} + 14\mathbf{j} + 2\mathbf{k}$
13. $\mathbf{a} \times \mathbf{a} = (a_1^2 - a_1^2, a_2^2 - a_2^2, a_3^2 - a_3^2) = \mathbf{0}$
15. $\mathbf{a} \times (\mathbf{b} + \mathbf{c}) = (a_2(b_3 + c_3) - a_3(b_2 + c_2), a_3(b_1 + c_1)$
    $- a_1(b_3 + c_3), a_1(b_2 + c_2) - a_2(b_1 + c_1))$
    $= (a_2 b_3 - a_3 b_2, a_3 b_1 - a_1 b_3, a_1 b_2 - a_2 b_1)$
    $+ (a_2 c_3 - a_3 c_2, a_3 c_1 - a_1 c_3, a_1 c_2 - a_2 c_1)$
    $= \mathbf{a} \times \mathbf{b} + \mathbf{a} \times \mathbf{c}$
17. The first coordinate of $(\mathbf{a} \times \mathbf{b}) \times \mathbf{c}$ is $(a_3 b_1 - a_1 b_3)c_3$
    $- (a_1 b_2 - a_2 b_1)c_2$, which upon simplification gives $(\mathbf{a} \cdot \mathbf{c})b_1$
    $- (\mathbf{b} \cdot \mathbf{c})a_1$; this is the first coordinate of $(\mathbf{a} \cdot \mathbf{c})\mathbf{b} - (\mathbf{b} \cdot \mathbf{c})\mathbf{a}$.
    The other two coordinates of the equation are checked similarly.

SECTION 15.11
1. $(1 + 4\pi^2 + \pi^4)^{1/2}(1 + \pi^2)^{-3/2}$
3. $1/\sqrt{2}$
5. $2/5^{3/2}$
7. $2^{-3/2}$
9. This follows from equation (2) in the text, using $\mathbf{B} = -(\mathbf{N} \times \mathbf{T})$, $\mathbf{v} \times \mathbf{a} = -(\mathbf{a} \times \mathbf{v})$ and dividing by $v^3(t)\kappa(t)$.

QUIZ 15.1
1. $\left(\dfrac{1}{\sqrt{30}}, \dfrac{5}{\sqrt{30}}, \dfrac{4}{\sqrt{30}}\right)$
2. $\int_0^2 [1 + e^{2t}(2 - \sin 2t + 4 \sin 2t \cos 2t + 3 \cos^2 2t)]^{1/2} \, dt$
3. $6 \times 5^{-3/2}$
4. $(3 + 2\pi, \pi^2, 4\pi)$. $(3 + 2\pi)(x - \pi) + \pi^2(y - 1) + 4\pi(z - 2) = 0$
5. $\mathbf{A} = (8, -5, -2)$; two points on the plane are $(1, 2, -1)$ and $(2, 2, 3)$.

INDEX

Absolute maxima and minima, 95, 547
Absolute value, 14
Acceleration, 78
  centripetal, 635
  vector, 634
Alternating series, 411
Angle, 289
  radian measure, 289
Antiderivative, 173
Arc length, 446, 624
Area, 154, 574
  signed, 159
  between two curves, 167
Asymptotes, 106, 481
Axis (axes), 17, 466, 472, 481, 498
  rotation of, 489

Bhaskara, 23
Binormal, 656
Bounded
  function, 116
  sequence, 375

Cardioid, 459
Cartesian coordinates, 18

Celsius, 25
Center
  of ellipse, 472
  of gravity, 575, 590
  of hyperbola, 481
Chain rule, 81, 533, 538
Circle, 21
Closed interval, 10
Comparison test, 415
Composition of functions, 80
Concave up and down, 100
Conditional convergence, 411
Cone, 511, 513
Conic sections, 466
  degenerate, 487
Constant of proportionality, 246
Continuity, 96, 113, 115, 520, 521
Contour curve, 505
Convergence of sequences, 368
Convergence of series, 401
  absolute, 411
  comparison test, 415
  conditional, 411
  integral test, 406
  radius of, 424
  ratio test, 419

Coordinate
  axes, 17, 498
  cylindrical, 525, 592
  plane, 498
  polar, 450, 577
  spherical, 526, 595
  system, 18, 497
Cosecant (csc), 303
  inverse, 316
Cosine (cos), 287
  addition formulas, 291
  direction, 612
  inverse, 314
  law of, 614
  between vectors, 615
Cost, 127, 245
Cotangent (cot), 303
  inverse, 316
Critical point, 96
Cross section, 508
Curvature, 632, 655
Curve
  contour, 505
  curvature of, 632, 655
  length of, 446 624
  level, 504, 649

Curve (continued)
  normal vector to, 629, 644
  tangent vector to, 627–628, 644
  vector form, 619
Cycloid, 442
Cylinder, 509–512
Cylindrical coordinates, 525, 592

Decreasing
  function, 90, 91, 99
  sequence, 374
Definite integral. See Integral
Degenerate conic, 487
Density, 573, 589
Derivative, 41, 43, 56, 76
  chain rule, 81, 533
  directional, 647–648
  implicit, 84, 535
  Leibnitz formula, 79
  notation, 59, 514
  partial, 513–516
  power rule, 70
  product rule, 65
  quotient rule, 68
  reciprocal rule, 66
  of vector function, 621
Determinant, 651
Differential, 150, 188, 339
Differentiation
  formulas, 74
  implicit, 84, 535
  logarithmic, 226
  of power series, 428
Directional derivative, 647
  and gradients, 648
Direction cosine, 612
Directrix, 467, 473, 482
Disk, 517
  truncated, 519
Distance formula, 21, 499
Divergence
  of sequences, 368
  of series, 401
Domain, 28, 502
Dot (inner) product, 613
Double integral, 555, 563
Dummy variable, 166

Eccentricity, 472, 482
Ellipse, 472, 476
Ellipsoid, 511

Elliptic,
  cylinder, 513
  paraboloid, 513
Exponential, 214–220
  function, 224
  growth and decay, 246

Factorial ($n!$), 372
Fahrenheit, 25
Fermat's principle, 321
Focus (foci), 467, 473, 482
Function, 26
  bounded, 116
  composition, 80
  concave, 100
  continuous, 96, 113, 115, 520
  decreasing, 90, 91, 99
  derivative, 41, 43, 56, 76
  differentiable, 119
  domain, 28, 502
  even, 31, 288
  exponential, 213, 224
  graph, 29, 89, 99, 103, 108, 456–460, 503
  hyperbolic, 489, 676–677
  implicit, 83–84, 535
  increasing, 90, 91, 99
  inverse, 306
  logarithm, 204, 221
  odd, 32, 288
  periodic, 289
  piecewise continuous, 194
  range, 28, 504
  rational, 105
  three or more variables, 522
  trigonometric, 286
  two variables, 497
  vector, 619
Fundamental theorem of calculus, 154, 179

General cone, 263
General quadratic equation, 478, 487
Gradient, 644, 648
Graph, 29, 89, 99, 108, 456, 503

Harmonic series, 405
Helix, 620
Hyperbola, 481, 484
Hyperbolic function, 489, 676–677
Hyperbolic paraboloid, 507, 513
Hyperboloid, 512

Implicit differentiation, 84, 535
Implicit function, 83–84
Increasing function, 90, 91, 99
Increasing sequence, 374
Increment, 34
Indefinite integral, 183
Indeterminate form, 431
Inequalities, 7
Infinite series, 399
Infinity, 11
Inflection point, 101
Initial value problem, 239
Inner product, 613
Integral
  approximation with rectangles, 275
  definite, 157
  double, 555, 563
  improper, 267, 272
  indefinite, 183
  iterated, 557, 568, 584
  triple, 583
Integral test for series, 406
Integration
  by parts, 230
  by substitution, 187
  of partial fractions, 356
  of power series, 426
  of trigonometric functions, 326
  numerical, 274
Interval, 10
Inverse function, 306
Iterated integral, 557, 568, 584
Iteration, 143

Kepler, 203

Law of cosines, 614
Left and right (one-sided) limits, 111
Leibnitz, 79, 203
  formula for derivatives, 79
Length of a curve (arc length), 446, 624
Level curve, 504, 649
L'Hospital's rule, 431, 433
Limit, 41, 48, 53, 517
  at infinity, 51
  left and right (one-sided), 111
Linear approximation, 146, 528
  error, 543
Linear function, 33
Lines
  parallel, 37

perpendicular, 39
point-slope formula, 35
slope-intercept formula, 35
two-point formula, 36
Logarithmic differentiation, 226
Logarithmic spiral, 465
Logarithms
  common, 221
  natural, 204
  to other bases, 221

Maclaurin
  approximations, 388
  series, 396
Magnitude of a vector, 606
Major axis, 472
Marginal
  analysis, 127, 245
  cost, 127, 245
  revenue, 129, 245
Mass, 573, 589
Maxima and minima
  absolute, 95, 547
  local, 92, 99, 102, 547–548
  tests, 96, 99, 102
Mean value theorem, 121
Minimum. See Maxima and minima
Minor axis, 472
Monotonic sequence, 374

$n!$ (n factorial), 372
Napier, 203
Natural logarithms, 204
Newton, 203
Newton's method, 143
Normal, 629, 640
Normalized vector, 610
Number line, 5
Numbers
  irrational, 215
  transcendental, 215
Numerical integration, 274

One-sided limits, 110
Open interval, 10
Open region, 521
Ordered pair, 18
Orthogonal, 616, 624, 640, 653

Parabola, 466
Parabolic cylinder, 513
Parallelogram, 607–608

Parameter, 441
Parametric equation, 442
Partial fractions, 352
Partial sum, 400
Plane, 498, 636
  coordinate, 498
  normal to, 637
  tangent, 641–642
Planetary orbits, 480
Polar coordinates, 450, 577, 579
  arc length, 463
  area, 461
  double integral, 576, 579
  graphing, 456–460
Power series, 423
Price, 130
Principal normal, 629
Production cost, 245
Production rate, 243
Profit, 132, 552
Pythagoras
  theorem of, 20, 23, 499, 614

Quadrant, 19

Radian measure, 289, 290
Radius of convergence, 424
Range, 28, 504
Rate of change, 138
Rate of proportional change, 247–248
Real line, 5
Real numbers, 2
Rectangular coordinates, 18
Reduction formula, 233
Region of integration, 556, 566
Related rates, 139, 318
Revenue, 129–130, 552
Rolle's theorem, 119
Root, 4

Saddle point, 547
Scalar product, 613, 617
Scientific notation, 3
Secant (sec), 303
  inverse, 316
Sequence, 367
Series, 399–400
  absolute convergence, 411
  alternating, 411
  boundedness theorem, 403
  comparison test, 415
  conditional convergence, 411

convergence and divergence, 401
  harmonic, 405
  integral test, 406
  Maclaurin, 396
  partial sum, 400
  power, 423
  radius of convergence, 424
  ratio test, 419
  Taylor, 392
Sigma notation, 162
Simpson's rule, 278
Sine (sin), 287
  addition formulas, 291
  inverse, 312
Slope, 33, 41, 513
Snell's law of refraction, 322
Solid of revolution, 251
Speed, 634
Sphere, 500
Spherical coordinates, 526, 595
Summation index, 162
Surface, 503
  cross section, 508
  of revolution, 511
  saddle point, 547
Symmetry, 30, 31, 33

Tangent (tan), 39, 42, 43, 303
  inverse, 314
  plane, 641–642
  vector, 627–628
Taylor
  approximations, 366, 387
  series, 392
Tetrahedron, 590
Transcendental number, 215
Translations, 468, 477, 489
Trapezoidal rule, 277

Unit circle, 22
Unit normal, 629
Unit vector, 609

Variable
  dependent, 27
  dummy, 166
  independent, 27
  separated, 559
Vector, 600, 605

Vector (*continued*)
- acceleration, 634
- binormal, 656
- components, 601
- derivative, 621
- difference, 604, 608
- equivalence, 600
- function, 619
- initial point, 600
- magnitude, 600, 606
- normal, 629, 642, 644
- normalized, 610
- orthogonal, 616
- product, 650
- sum, 602–607
- tangent, 627–628
- terminal point, 600
- unit, 609
- velocity, 634
- zero, 602

Velocity, 77
- vector, 634

Vertex (vertices), 466, 481

Volume
- by slicing, 260
- solid of revolution, 253
- under a surface, 555, 558

75 76 77 9 8 7 6 5 4 3 2 1

PROPERTIES OF NATURAL LOGARITHMS

1. $\frac{d}{dx}(\ln x) = \frac{1}{x}$
2. $\ln 1 = 0$
3. $\ln(ab) = \ln a + \ln b$
4. $\ln \frac{1}{a} = -\ln a$
5. $\ln \frac{a}{b} = \ln a - \ln b$
6. $\ln a^r = r \ln a$
7. $\ln a < \ln b$ when $a < b$

PROPERTIES OF EXPONENTIALS

1. $\ln(\exp a) = a$
   $\exp(\ln b) = b$
2. $\exp 0 = 1$
3. $\exp a = e^a$
4. $\exp(a + b) = (\exp a)(\exp b)$
5. $\exp(-a) = \frac{1}{\exp a}$
6. $\exp a > 0$
7. $\exp a < \exp b$ whenever $a < b$

PROPERTIES OF LIMITS

1. If c is any constant, then $\lim_{x \to x_0} c = c$.
2. $\lim_{x \to x_0} x^r = x_0^r$ for all numbers $r \neq 0$.

    If $\lim_{x \to x_0} f(x) = A$, $\lim_{x \to x_0} g(x) = B$, then

3. $\lim_{x \to x_0} d \cdot f(x) = d \cdot A$ for any constant d.
4. $\lim_{x \to x_0} [f(x) + g(x)] = A + B$.
5. $\lim_{x \to x_0} [f(x) - g(x)] = A - B$.
6. $\lim_{x \to x_0} [f(x) \cdot g(x)] = A \cdot B$.
7. $\lim_{x \to x_0} \frac{f(x)}{g(x)} = \frac{A}{B}$, provided that $B \neq 0$.